I0882356

Theory of CMOS Digital Circuits and Circuit Failures

Masakazu Shoji

Princeton University Press
Princeton, New Jersey

Theory of CMOS Digital Circuits
and Circuit Failures

Copyright © 1992 by Princeton University Press
Published by Princeton University Press, 41 William Street,
Princeton, New Jersey 08540

All Rights Reserved

Library of Congress Cataloging-in-Publication Data

Shoji, Masakazu, 1936–
Theory of CMOS digital circuits and circuit failures / Masakazu Shoji.
p. cm.
Includes bibliographical references and index.
ISBN 0–691–08763–6 (cl)
1. Metal oxide semiconductors, Complementary. 2. Semiconductors—Failures. 3. Digital integrated circuits—Design and constgruction—Data processing. I. Title.
TK7871.99.M44S525 1992 92–12354
621.39'5—dc20

Princeton University Press books are printed on acid-free paper, and meet the guidelines for permanence and durability of the Committee on Production Guidelines for Book Longevity of the Council on Library Resources

Printed in the United States of America

10 9 8 7 6 5 4 3 2 1

Contents

Preface and Acknowledgments

A CMOS circuit designer conceived a new circuit. He thought over a number of possible conditions to which the new circuit would be subjected and concluded that the new circuit would work under all realistic conditions. He designed a CMOS chip and requested production of a silicon wafer. However, when he tested the processed silicon chips, he found that the circuit did not work properly. The new circuit failed in a way he never dreamed of. Any experienced designer has had a few experiences like the one briefly sketched above. It is a designer's nightmare, but it is not a new problem. In the days of vacuum tube electronics, the problem was how to control spontaneous oscillation of radio frequency tuned amplifiers. Just as it was difficult to identify which minute coupling caused the oscillation, it is difficult to identify which small design imperfection causes CMOS logic circuit failure.

The period from the 1930s to mid-1960s was the time of vacuum tube electronics. During that period, circuit design know-how relevant to preventing circuit failure was disseminated through informal verbal communications, trade journals, and, quite significantly, in amateur journals (like "Radio Amateur's Handbook"). A problem such as high-frequency amplifier oscillation was dealt with by experienced wiremen, who used techniques such as one-point grounding and shielding. If the amplifier still oscillated, an ad hoc remedy, such as power supply bypassing or neutralization, was used. Digital electronics were born during that period, but their use was very limited.

The period from the 1960s to mid 1970s was the time of discrete semiconductor device electronics. During this period, digital electronics grew so rapidly that it overtook traditional analog electronics. In the 1970s, the beginnings of modern large-scale digital integrated electronics were seen. At the beginning of this period, digital LSI circuits were designed using the traditional approach. However, several new problems that did not exist in discrete semiconductor device electronics soon emerged. The new problems were difficulties in diagnosing VLSI circuit failures, and the high cost of design iteration. The two new problems demanded that circuit failures be studied systematically as a scientific system of knowledge to guide designers. Significant improvement in circuit diagnostic capability, especially techniques using electron beams, did not change the situation. We have no alternative to insuring reliable circuit operation on the first mask.

The lesson learned from vacuum tube electronics — that circuit design is never a straightforward mechanical process — has been forgotten *many* times. Bitter experience learned from failure has not always been transferred from senior designers to junior designers effectively enough to prevent repeating the same mistakes. The basic reason for poor communication has been that knowledge gained by precious design failure experiences has never been systematized. What is worse, as circuit simulation capability became widely available in the 1980s, the entire problem area was perceived as resolved and is now left unattended. For this false optimism, one sobering thought is enough: one circuit failure in a VLSI chip is too many. The objective of this book is to ask why circuit design errors occur. The answer to this question is simple: because we do not understand how CMOS circuits work well enough. Thus we must ask, How do we reach a better understanding of CMOS circuit operation?

The mainstream effort of the present ASIC (application specific integrated circuit) industry is to automate the entire chip design process and to eliminate design errors, and ultimately chip designers. This objective has been partially attained. If a circuit is designed following the vendor's guidelines, the customer is likely to obtain working chips the first time. This design style, however, does not produce high-performance chips. The present design automation means *low performance* chip design. From the circuit designer's viewpoint, however, this is a sorry situation, because carefully optimized circuits perform many times better than automatically designed circuits. Modern scaled-down CMOS technology has tremendous signal processing capability, only a small part of which is used in automatically designed chips. The general public perception that CMOS circuits are slower than bipolar circuits is a bias originating from comparing automatically designed chips. If the performance objective is met by the automatic design method, the need to understand individual circuits ceases to exist. In reality, customers are forced into this less than desirable situation. This dominant trend created stagnation in the science of circuits. What I wish to achieve is the creation of a small sidestream in order to break the stagnation, to reactivate enthusiasm in circuit research, and to exploit the capabilities of high performance CMOS circuits. Readers should understand that my motivation in writing this book is to design the highest performance CMOS circuits possible. Newly conceived, high-performance circuits almost always have hidden problems — problems that make the circuit fail. The first requirement, then, is to understand circuit failure causes. Such is the objective of this book. Research into CMOS circuit failure is indeed the first step towards simultaneously achieving two desirable but contradictory objectives, namely, error-free design and high-performance design. An example of this fundamental theoretical work that has practical significance is the high-precision waveform control in the digital crosspoint switch, which was recently achieved in AT&T's Bell Laboratories. A deep understanding of the switching mechanism of gate circuits (Sections 6.13 and 6.14) and the technique for solving the problem subject to certain assumptions (Sections 5.14 and 5.15) leads to successful design.

A peculiar aspect of research into circuit failure is that basic electrical phenomena that cause one circuit to fail can be used advantageously to obtain desired operation for other circuits. From this viewpoint, the objective of research into circuit failure is not to identify *undesirable* or *nuisance* phenomena, but rather to assure that basic principles of physics and circuits are used properly to attain desired objectives. Therefore we need a thorough understanding of the operating mechanisms of circuits. Study of CMOS circuit failure provides the opportunity to investigate operational modes of circuits that are not exercised under ordinary circumstances. When a CMOS circuit fails, the circuit operates in an *abnormal mode*. The *abnormal modes* of CMOS circuit operation magnify problems hidden in normal operation. Studies of CMOS circuit failure, therefore, provide an opportunity to study CMOS circuits in great depth.

To change the predominant culture, several new approaches had to be introduced into CMOS digital circuit theory. Closed-form analysis of many basic logic circuits are presented in this book. Some readers may wonder why lengthy closed-form circuit analysis is necessary if the entire circuit can be simulated. There is an important difference between the two approaches. Closed-form analysis forces the designer to think systematically about the working mechanisms of CMOS circuits. Circuit simulation is essentially a mechanical procedure that can be carried out automatically, without thinking. The verification of integrated circuits by simulation alone requires an infinite number of conditions. Examples of specific simulation exist that would have indicated circuit problems if the designer had looked at the outputs. Because the designer did not expect any trouble, he did not look. Fabricated circuits then failed. Closed-form circuit analysis had been used regularly until about the 1960s. Vacuum tube circuit papers published before that time contain elaborate mathematics and complete analysis. The complete analysis and deep understanding obviously helped circuit designers to prevent circuit failures. In CMOS digital circuit theory, however, this fundamental step has never been worked out. The present complex digital CMOS circuits are built on rather weak theory — theory that does not provide a systematic method of circuit analysis. Another reason to support closed-form analysis is more subtle: the problem of circuit failure is ultimately a matter concerning the designer's confidence. Unless his confidence is augmented by back-of-envelope computation, no useful decision can be reached on any design issue. As a practicing designer, I often felt that conclusions about circuits that cannot be derived using simple physical models are unusable, as are any simulation results that cannot be confirmed by back-of-envelope calculations. A CMOS digital circuit theory that is useful for such purposes has never existed. This book intends to fill that gap. To achieve this objective, I needed to present closed-form analysis of practically all the elementary CMOS digital circuits. Such analysis is the main body of this book. I have been a consultant to CMOS IC designers of AT&T. If a circuit fails, I must be able to explain systematically why it failed. The theory of this book was created to make a foundation for that work. In the current, system-oriented culture of VLSI technology, circuit issues are often treated as secondary: in-depth understanding of a circuit's operational mechanisms is required only when the circuits fail, at which time the system's project comes to a screeching halt. For this reason, I organized a new CMOS digital circuit theory using circuit failure as the central issue.

Since even simple CMOS digital circuits are significantly more complex than the vacuum tube circuits of the 1950s, simplification in active device (FET) characteristics is required to make closed-form analyses feasible. We use a model called the collapsible current generator model of FETs in this book. If this model is used, many complex digital circuit problems can be solved in closed form. The collapsible current generator model gives a clear identification of the FET operating point in any of the saturation, triode, or nonconducting region. This brings digital circuit operation into perspective in the following sense. Each FET, and the *digital* circuit assembled from them, may be considered to be in one of many possible *states* at any time. Each *state* of a digital circuit is an *analog* and *linear* circuit that evolves with time following Kirchhoff's laws, without changing the structure (configuration, or circuit connectivity). As the evolution reaches a critical point, a change in the circuit configuration takes place, and the digital circuit changes state. The digital circuit may then be modeled as a sequence of analog circuits that follow one after another as time proceeds. Using this approach, CMOS digital circuit models were established, and the characteristics of the models were analyzed to reveal every detail of the circuit, including failure problems. In this book I study properties of circuits using the collapsible current generator model to its limits. I believe that this method, which may look artificial to some, is the only way to reach a deep understanding of the properties of CMOS circuits that emerge from the connectivity of components. I am aware of the criticism that the models of CMOS digital circuits used in this monograph are too simple: indeed, these simple models were found to be the only way to reach the real depth of the problem. This simplification may reflect my background as a physicist who is trained to see Nature in the simplest possible way. Although device characteristics are simplified, the other, more essential, complexity of a circuit's connectivity is always reflected accurately in the circuit analysis.

Development of CMOS digital circuit theory that is based on the new model makes this monograph quite different from my previous book entitled *CMOS Digital Circuit Technology*, published by Prentice-Hall, Inc., in 1987. That book and this monograph have practically no overlapping material. In its basic philosophy, however, this monograph is an intensive continuation of the previous book. From the nature of the subject of the monograph, whose stress is on many details, this monograph is intended for reference by practicing CMOS circuit designers, and not for a course textbook. Because this monograph addresses specific subjects of CMOS VLSI technology in depth and is significantly different from the more general textbooks on VLSI technology, it can be useful to graduate-level research.

The science of testing digital circuits and techniques of evaluating completeness of test coverage have grown into a successful area of investigation, but the study of circuit failure itself has never followed such a prosperous route. There is a fundamental difference between these two areas: they appear similar only because they both address the same aspect of VLSI technology—chip failures—but otherwise they share practically nothing in common. The two areas of investigation are really complementary, and they are both important. This book concentrates on circuit failures originating from design and does not include material related to testability.

Professor Fred Rosenberger pointed out an idiosyncrasy of this book that missed my attention. I avoided using the parameters that have clear meanings in logic analysis but are hard to define in electrical circuit theory. Setup/hold times of latch, and noise margin of gate are the examples. Following his suggestions, significant rewriting was made, but the trace of the original bias is still there. I consider that proper theoretical treatment of these difficult parameters is a problem of future.

With sorrow, my first acknowledgment is to Professor Aldert van der Ziel and Dr. William Shockley, who both passed away during the period when this book was being written. There is no way to measure the influence that these two giants of modern electronics imprinted in the mind of the young graduate student in Minnesota, and the starting member of the technical staff at Murray Hill, by actually showing him how to solve technical problems.

Professor F. Rosenberger of Washington University, St. Louis, Missouri, went through the entire manuscript, edited English, corrected errors and gave a number of technical suggestions. Since this book presents many unorthodox viewpoints, his comments were important in recognizing many biases that were not obvious to me. The areas of the text where his comments made significant improvements are indicated there. I am truly grateful for his contributions: without him this book would never be completed. Professor S. Kang of the University of Illinois, Professor O. Wing and Professor C. Zhukowski of Columbia University and Professor E. Chenette of the University of Florida read the first manuscript of this book and gave valuable comments and encouragement. Dr. R. R. Troutman of the IBM T. J. Watson Research Center reviewed Section 6.8 (which is his original work) and gave valuable comments. A stimulating discussion with Dr. B. Dalal of Amdahl Corporation created the material of Section 7.9.

This work became possible with the support and encouragement from the staff and management of the Computing Science Research Center, AT&T Bell Laboratories, Murray Hill, N.J. The Center recognized and supported basic theoretical research in integrated circuit electrical phenomena. This is quite an exceptional foresight in the current technical climate of the United States. I am grateful especially to the Center management, T. G. Szymanski, A. Aho, R. Sethi, and A. G. Fraser, and the staff, J. H. Condon, and M. F. Jukl.

Help from my close friends can not be overemphasized. H. Khorramabadi reviewed the manuscript, gave constant encouragements, and lead me to study the material presented in the last two sections of Chapter 6. A. Walcheski of the AT&T Solid-State Technology Center read the manuscript and gave valuable comments. R. H. Krambeck and G. P. Sampson III arranged for me to work as a technical consultant in the Allentown-Cedarcrest Facility of AT&T's Bell Laboratories. This assignment gave me an opportunity to test many ideas and methods of CMOS circuit analysis. Joint work with the staff there, and the excitement of design project success, are the most pleasing experiences of a VLSI designer's comradeship. Once I was a designer in Krambeck's organization. J. Fields, R. Scavuzzo, T. Poon and M. S. Tsai allowed me to work with their staffs. Design work, especially with Jyoti Sabnis, M. Saniski, P. Argade, and T. Gabara brought many successes. There were many fundamental theoretical problems of high-speed VLSI circuits whose answer appeared only after many hours of intense research. This experience showed me the technical prowess of the high-performance CMOS VLSI design organization most clearly, as well as the good cooperation between research and development which we established. T. Wik and M. DePaolis helped me join in many of their design projects in the Allentown-Union Blvd. facility of AT&T Bell Laboratories. Many successes in design projects with them, especially with R. J. Wozniak and F. Barber in the digital crosspoint switch, can never be forgotten. There the design team achieved success that might be called a miracle. G. Daryanani and P. Wilford, in the former transmission systems design area, provided many stimulating problems that fueled these successes.

Lilya Lorrin of Princeton University Press gave me warm support and every kind of help. Being a rather unique and unorthodox technical work, it was difficult and painful to push this work forward. This book came into existence because of her constant encouragement, which I will remember forever.

When I look back to my distant past, my first interest in electronics was stimulated by my late uncle Tsurukichi Shoji. He, a co-founder of the Datsun (the present Nissan) Automobile Company, may be considered as one of the earliest electronics pioneers in Japan. One of the subjects he helped me to learn in 1940s was a method of preventing oscillation of 1-V-2 shortwave radios (this notation used at that time means RF amplifier, regenerative detector, AF preamplifier, AF power amplifier and rectifier, 5-tube radio). This was my first encounter with problems of electronic circuit failure. I never forgot his lesson; how difficult it is to make an electronic circuit work, and how joyful it is to see a circuit working! My parents warmly supported my enthusiasm for electronics. With their support I gained practical experience from audio to microwave electronics in my early years from 1940 to the mid-1950s. In my college days, Professor K. Shimoda, the pioneer of quantum electronics in Japan, taught me the most advanced electronic circuit techniques. My encounters with Professor van der Ziel and Dr. Shockley in the 1960s followed that background. I am grateful to all my teachers and friends, who have always guided me as well as possible. For me, electronic circuits were not lifeless objects — they have been my close friends also. Since they have rudimentary intelligence, they are often quite unpredictable. A systematic method of understanding them, a *psychology* of electronic circuits, never existed. The problem addressed in this book has been my half-century-old problem, the solution to which only the first step has now been taken.

List of Mathematical Symbols

Symbol	Definition	Value and unit
α	Rate of voltage increase	***
B_N	NFET transconductance	***
B_P	PFET transconductance	***
β	Ratio of transconductances of NFET and PFET	***
C_0	Output node capacitance	***
C_1	Input node capacitance	***
C_A	Capacitance per unit area of PN junction	***
C_m	Miller effect capacitance	***
C_P	Capacitance per unit periphery pf PN junction	***
C_L	Load capacitance	***
$C_{X,A,B}$	Gate's internal node capacitance	***
ε_0	Dielectric constant of vacuum	8.854×10^{-14} *Farad/cm*
ε_{OX}	Dielectric constant of SiO_2	0.345×10^{-12} *Farad/cm*
ε_S	Dielectric constant of silicon	1.04×10^{-12} *Farad/cm*
G	Gain, or conductance	***
L	Length of FET gate	***
μ_0	Magnetic permeability of vacuum	1.25×10^{-8} *Henry/cm*
μ_N	Mobility of electrons in inversion channel	***
μ_P	Mobility of holes in inversion channel	***
N_A	Acceptor density	***
N_D	Donor density	***
n_i	Intrinsic carrier density	***
ω	Angular frequency	***
Q	Charge	***

q	Electronic charge	1.602×10^{-19} *Coulomb*
S	Inverse of complex time constant	***
T_D	Delay time of gate	***
T_{OX}	Thickness of gate oxide of FET	***
T_{FOX}	Thickness of field oxide	***
T_{COX}	Thickness of composite oxide	***
V_0	Output node voltage	***
V_1	Input node voltage	***
V_{BE}	Threshold voltage of PN junction	***
V_D	Drain voltage	***
V_{DD}	Power supply voltage	***
V_{dif}	Diffusion potential	***
V_G	Gate voltage	***
V_{GSW}	Switching threshold voltage of static gate	***
V_H	Logic high level voltage	***
V_L	Logic low level voltage	***
V_{THN}	NFET threshold voltage	***
V_{THP}	PFET threshold voltage	***
v_∞	Saturation drift velocity of carriers	***
W	Width (size) of FET	***
X	Dimension of source/drain island of FET	***
Y	Admittance	***
Z	Impedance	***

Theory of CMOS Digital Circuits
and Circuit Failures

1 Physics of CMOS Integrated Circuits

1.1 Introduction

In this chapter we prepare the background materials for this book — concepts and methods necessary for detailed studies of CMOS digital circuit failures that will be carried out in the later chapters. Electrical properties of basic components such as field-effect transistors (FETs), resistors, capacitors, inductors, and interconnects in CMOS integrated circuits are reviewed, and characteristics of elementary circuits built by interconnecting these components, such as CMOS static and dynamic gates and linear and nonlinear amplifiers, are summarized. The study of circuit failure requires closed-form theoretical analyses of rather complex circuits. Therefore, two methods of closed-form analysis of CMOS circuits containing FETs and other passive components are discussed. These techniques of closed-form analysis become the basic tools in later chapters. The first method is to represent all active devices, components and interconnects by equivalent *linear* resistors and capacitors, whose values are determined from each component's physical model. Electrical properties of the equivalent resistance-capacitance circuit are analyzed using many useful analytical methods of conventional linear circuit theory (summarized in the later part of this chapter), and the results are translated back into the properties of digital CMOS circuits. In the second method, FETs are represented by gate voltage controlled *collapsible* current generators (devices that generate constant current if terminal voltages are not zero). As we study the second method of analysis in detail, we find that a digital circuit may be represented by an analog circuit, whose circuit configuration (connectivity of devices and components) changes with time. In this representation we recognize that information about how this change of circuit configuration takes place is more useful in understanding the circuit than the detailed numerical information or an algebraic formula of time-dependent voltages of the nodes. To make the best use of the information about the changes of the circuit configuration, we introduce the concept of circuit *states*, called *micro-states*. Both of the methods (linear resistor model and the collapsible current generator model) are mathematically simple: Using one method or the other, most CMOS digital circuits can be analyzed in closed form.

To study CMOS digital circuits we require a good FET model, as well as a reasonable method of representing an integrated circuit by a lumped constant equivalent circuit. This chapter examines how to determine the best possible equivalent circuit and the approximations imposed by the equivalent circuits.

1.2 Field-Effect Transistors

MOSFET characteristics have been studied in great detail during thirty years of research [01] [02] [03] [04] [05]. Sophisticated FET models are used in numerical simulation of MOS circuits. For application to studies of circuit failure, however, very high numerical precision is usually unnecessary. For research into circuit failure, simple device models that allow closed-form analysis are more desirable than better numerical accuracy attained by precise but complex models. This is because fundamental circuit characteristics are determined more from interconnection of devices than from the details of device characteristics. When the conclusions of theoretical circuit research are applied to practical IC design, the results obtained using the simple device model are calibrated using more accurate simulation results (this can be done by reinterpreting device model parameters). This two-step approach is more straightforward, more useful and often more accurate than *brute-force* circuit simulation.

Figure 1(a) shows a cross-section of N-channel MOSFET (NFET) in CMOS integrated circuit. The processing technology required to fabricate this structure is not discussed in detail. Interested readers are referred to standard references on IC processing [06]. The substrate of a conventional CMOS IC is N-type silicon, on the surface of which a P-type diffused area called a P-tub is formed. Thin oxide (gate oxide) of thickness T_{OX} is grown on the P-tub surface, and polysilicon gate material is deposited on the top of the thin oxide layer. Polysilicon gate area (cross-hatched) is patterned using photolithography. Using the polysilicon gate

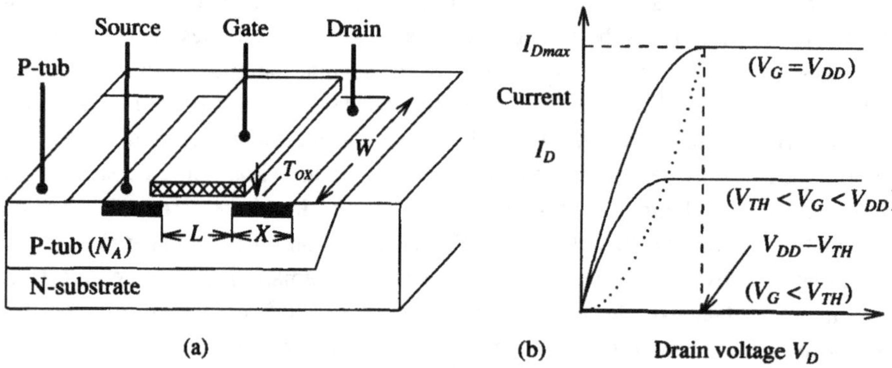

Figure 1 MOS FET characteristics

features as mask, N+ drain and source impurities are implanted and diffused. Channel length L (often called *electrical* channel length) is approximately the same, but it is less than the gate polysilicon width (called *designed* channel length). Following further deposition of oxide insulation (intermediate oxide) source, gate, drain, and substrate are contacted by metallic (usually aluminum) conductors by cutting holes (windows) through the intermediate oxide. The finished NFETs have channel length L, width (often called FET size) W, and source-drain island width X, as defined in Fig. 1(a). The process of fabricating PFETs is interwoven with the NFET process. The only essential difference is that a PFET is fabricated on N-substrate (doped to make N-tub). An FET is a four-terminal device (Source, Drain, Gate, and Substrate). A PFET substrate terminal is common for all the PFETs on the same chip. We described the traditional N-substrate CMOS technology. Recently, P-type substrates are often used as starting material. In this case NFETs are fabricated on a globally connected P-tub, and PFETs are fabricated on isolated N-tubs. If the substrate is P-type, the NFET is a device having three independent terminals and a common substrate terminal, and a PFET has four independent terminals.

Suppose that the substrate and the source terminals are grounded, the gate is biased to positive voltage V_G, and the drain to positive voltage V_D, both relative to the grounded source. According to the theory by Weimer, the drain current I_D of the NFET is given by [07]

$$I_D = B_G(V_G - V_{TH} - \frac{1}{2}V_D)V_D \quad \text{if} \quad V_D < V_G - V_{TH} \tag{1a}$$

$$= \frac{B_G}{2}(V_G - V_{TH})^2 = I_{Dmax} \quad \text{if} \quad V_D \geq V_G - V_{TH} \quad \text{where} \quad B_G = \frac{\varepsilon_{OX}\mu_N W}{T_{OX}L} \tag{1b}$$

where ε_{OX} is the dielectric constant of silicon dioxide ($0.345 \times 10^{-12} F/cm$) and μ_N is the mobility of electrons in the surface channel. V_{TH} is called the threshold voltage. If $V_{TH} > 0$ as in conventional *enhancement mode* CMOS FETs, a FET carries no drain current if the gate voltage is zero. Equations (1a) and (1b) are plotted in Fig. 1(b). Equations (1a) and (1b) are valid, subject to the following conditions:

1) The electric field parallel to the length of FET channel is low, so that the surface electron drift velocity is proportional to the local field within the channel, directed from drain to source (low-field condition).
2) The source/drain diffused island is shallow. The gate field (that induces channel charge) is approximately perpendicular to the channel, and the drain-source field (within the channel) is approximately parallel to the silicon surface. The channel is quite shallow (long-channel condition).
3) The substrate is lightly doped, so that the charge that exists between the conducting surface channel and the neutral bulk substrate is negligible.

In Eq. (1), the B_G parameter that specifies FET current has special suffix G, which stands for the gradual channel, low-field FET model. For NFETs we use B_{GN}, and for PFETs we use B_{GP}.

Characteristics of PFETs have the same form as Eqs. (1a) and (1b) except that V_D is replaced by $V_S - V_D$ and V_G by $V_S - V_G$, where V_S is the source voltage (the highest voltage relative to ground). In Eq. (1b), μ_P becomes the mobility of the surface holes. Enhancement PFET threshold voltage V_{THN} is defined as a positive number. This choice of sign is convenient in writing the circuit equations. Equations (1a) and (1b) give *idealized* FET characteristics. To explain characteristics of *real* MOSFETs in scaled-down CMOS ICs the assumptions used to derive Eqs. (1a) and (1b), and the parameters contained in them, must be qualified and reinterpreted. We must consider the effects of substrate doping, shape of the channel, identification of source and drain, and the high-field carrier transport effects.

1.3 FET Threshold Voltage

The simple FET characteristics of Eqs. (1a) and (1b) are derived assuming a lightly doped substrate. If the substrate is *doped* P-type [diffused P-tub shown in Fig. 1(a)], FET current-voltage characteristics deviate from Eqs. (1a) and (1b) because of the space-charge that exists between the channel and the doped substrate (or tub). The mechanism is as follows: Since (1) the surface channel is N-type (because it consists of thin surface layer heavily populated with electrons); (2) the substrate is P-type (that contains free holes); and (3) the electrons and the holes do not mix, there must be a layer depleted of holes immediately below the surface channel. Since the layer depleted of majority (positive) holes has negative *ionic* charge, the *depletion* layer carries negative charge. The surface conducting channel also carries negative electronic charge. The negative charge of the depletion layer must be induced in addition to the electron charge in the surface channel by the applied gate voltage. The most remarkable effects are positive shift (*increase*) in the threshold voltage V_{TH} of a FET (increases if substrate doping is increased), and dependence of V_{TH} on the substrate bias voltage (source-substrate potential difference). This second effect is called the *back-bias effect.*

With reference to Fig. 2(a), NFET MN1 is fabricated on a grounded P-tub, and the source is biased to V_S volts relative to the grounded P-tub (where $V_S > 0$). A small drain voltage relative to the source, V_{DS} (where $V_{DS} > 0$) is applied, and channel current I_D is measured. As gate to source voltage, V_{GS}, is increased from zero, I_D begins to flow when V_{GS} equals threshold voltage V_{TH}. According to the theory of the back-bias effect, V_{TH} is an increasing function of source voltage V_S, and the dependence is written as

$$V_{TH} = V_{TH}(V_S) \tag{2}$$

Since the back-bias effect has significant impact on circuit design, we determine the function of Eq. (2) using simple physical reasoning. Let the threshold voltage of a *reference* NFET fabricated on a lightly doped substrate, but otherwise same as the first NFET, be V_{TH0}. Suppose that V_{TH0} is given. Suppose that the substrate of the NFET under consideration is doped to N_A $(acceptors/cm^3)$. Then the Fermi level of the P-type substrate is shifted down by [05]

$$\phi_F = \frac{kT}{q} \log (N_A/n_i)$$

from the center of the forbidden band, where k is Boltzmann's constant, T is the absolute temperature, q is the charge of an electron and n_i is the intrinsic carrier density of silicon. We note that the Fermi level of the reference NFET substrate is at the center of the forbidden

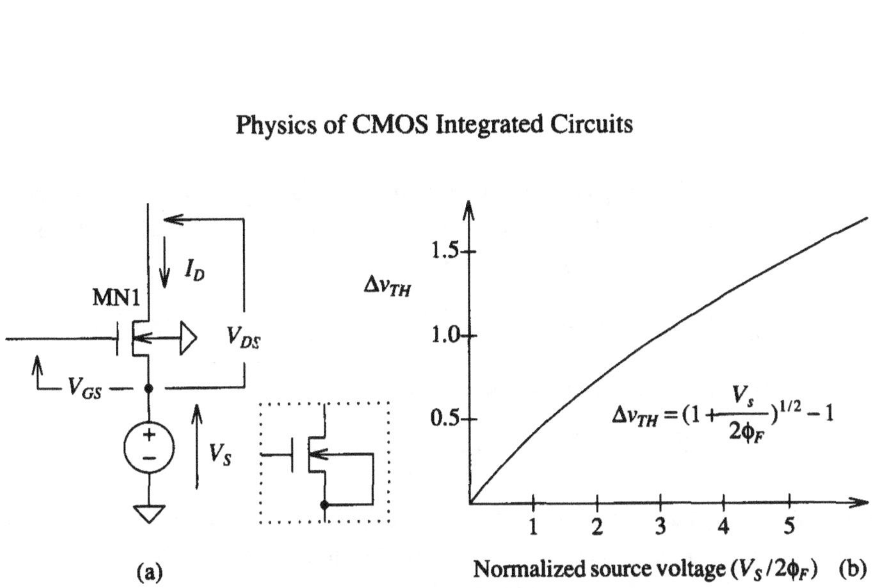

Figure 2 Back-bias effect

band. To create a conducting surface channel, the P-tub silicon surface must be converted to N-type *inverted* surface having comparable surface electron concentration ($\approx N_A$). The potential of the silicon surface must be pulled up by $2\phi_F$, and this potential must be supplied from the gate. Therefore a term ($2\phi_F$) must be added to V_{TH0}. As we see from this term, doping P-tub *increases* the NFET threshold voltage, or the NFET becomes *less* conductive. Intuitively an NFET fabricated on *more heavily doped* P-type substrate is *harder* to turn on. This explains the mechanism of positive shift of the threshold voltage, V_{TH}, if substrate doping is increased.

Let us consider the effects of source-substrate voltage. When the substrate and the source are held at the same potential [inset in Fig. 2(a)], the negative space-charge layer beneath the conducting channel sustains potential $2\phi_F$. Charge density per unit area of the space-charge layer is given by electrostatics as (P-tub is assumed uniformly doped)

$$Q_0 = [2\varepsilon_S \cdot qN_A \cdot (2\phi_F)]^{1/2} . \qquad \text{(3a)}$$

If source to substrate voltage V_S ($V_S > 0$) is applied, Q_0 increases to Q. Q is found by replacing $2\phi_F$ in Eq. (3a) by $2\phi_F + V_S$ as

$$Q = [2\varepsilon_S \cdot qN_A \cdot (2\phi_F + V_S)]^{1/2} \qquad \text{(3b)}$$

and therefore the incremental charge in the space-charge layer, ΔQ, is given by

$$\Delta Q = Q - Q_0 = (2\varepsilon_S \cdot qN_A)^{1/2}[(2\phi_F + V_S)^{1/2} - (2\phi_F)^{1/2}] .$$

Since this extra charge ΔQ must be induced by the positively biased gate, threshold voltage increases by

$$\Delta V_{TH} = \frac{\Delta Q}{C_{OX}}$$

where $C_{OX} = \varepsilon_{OX}/T_{OX}$ is the gate oxide capacitance per unit gate area. Therefore, the real threshold voltage V_{TH} is given by adding the two corrections to V_{TH0} (threshold voltage of the reference FET fabricated on lightly doped substrate) as

$$V_{TH}(V_S) = V_{TH0} + 2\phi_F + \frac{(2\varepsilon_S \cdot qN_A)^{1/2}}{C_{OX}} [(2\phi_F + V_S)^{1/2} - (2\phi_F)^{1/2}]. \tag{4a}$$

If V_S dependence is separated out for convenience

$$\frac{V_{TH}(V_S) - V_{TH}(0)}{V_0} = (1 + \frac{V_S}{2\phi_F})^{1/2} - 1 = \Delta v_{TH}(V_S/2\phi_F) \tag{4b}$$

where $V_{TH}(0) = V_{TH0} + 2\phi_F$ and $V_0 = (2\varepsilon_S qN_A \cdot 2\phi_F)^{1/2}/C_{OX}$. Figure 2(b) shows function $\Delta v_{TH}(V_S/2\phi_F)$ defined in Eq. (4b) versus normalized source voltage $V_S/2\phi_F$.

If we assume $N_A = 2 \times 10^{16}$ *acceptors/cm*^{3}, the Fermi level at room temperature is

$$\phi_F = 25mV \times \log \frac{2 \times 10^{16}}{2 \times 10^{10}} = 345mV \approx 0.35V$$

below the center of the forbidden band, where thermal voltage $kT/q = 25\,mV$ and $n_i = 2 \times 10^{10}\,cm^{-3}$ at room temperature. We have

$$V_0 = \frac{T_{OX}(2\varepsilon_S qN_A \cdot 2\phi_F)^{1/2}}{\varepsilon_{OX}}$$

$$= \frac{2.5 \times 10^{-6} \times (2 \times 10^{-12} \times 1.6 \times 10^{-19} \times 2 \times 10^{16} \times 0.7)^{1/2}}{0.345 \times 10^{-12}} = 485mV.$$

Therefore if $V_S/2\phi_F = 5$, threshold voltage V_{TH} increases by about 0.75 volt.

Substrate doping is the standard technique for FET threshold voltage control. Enhancement mode NFETs in digital CMOS ICs are now fabricated exclusively by doping P-tubs. If a lightly doped substrate is used, the threshold voltage of NFETs becomes negative, or NFETs conduct even at zero gate voltage (such NFETs are called *depletion* mode NFETs). There is immobile positive charge in the gate oxide immediately above the channel (conventionally called Q_{SS}) that induces a conducting surface channel in NFETs, even at zero gate voltage. We understand that enhancement mode NFETs in CMOS are inevitably associated with the back-bias effect.

The back-bias effect creates significant difficulties in setting the DC bias point of CMOS circuits that have series-connected FETs and that carry DC currents. Such a circuit structure is often seen if analog circuits are integrated into digital CMOS circuits. The differential amplifier shown in Fig. 3(a) is used to convert a small-amplitude differential signal into a single-ended CMOS level signal. Since this circuit is complex, full analysis will be given later (Section 2.11). Here we consider the most relevant points only. If both input FETs MN+ and MN− are to split current generated by MN0 equally at the *balanced* operating point and are to operate as a linear differential amplifier at the bias point, the range of input voltage allowed is limited by the increased threshold voltage of NFET MN+ and MN− due to the back-bias effect. The range of input voltages allowed for linear operation is schematically shown in Fig. 3(b). If V_{IN+} and V_{IN-} are reduced below the limit, common source voltage V_C is reduced too

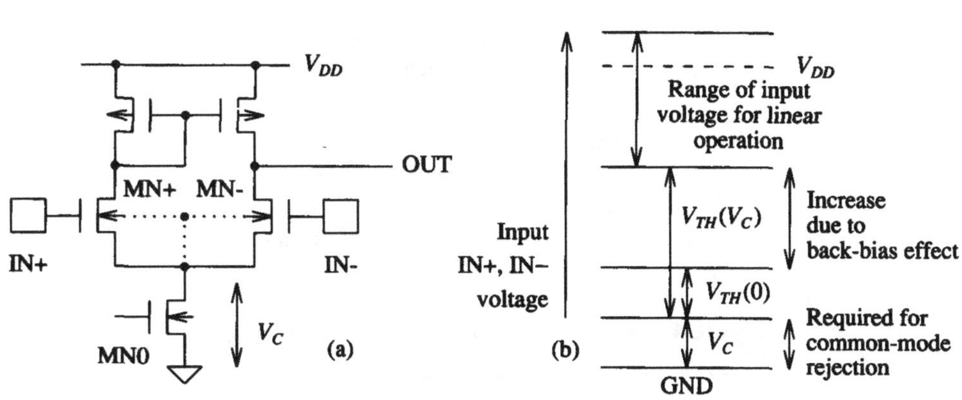

Figure 3 CMOS differential amplifier

much. A minimum V_C voltage must be maintained at the common source node of the differential NFET pair to guarantee that the current generator, MN0, is well into the saturation region and the current is independent of V_C. If this condition is not satisfied, the differential amplifier has poor CMRR (common mode rejection ratio-change of the two input voltages in the same polarity) . The problem is aggravated if the power supply voltage is low.

One way of avoiding the back-bias effect is to use a separate P-tub for NFETs MN+ and MN− and connect it to the respective sources, as shown by the dotted line in Fig. 3(a). This technique works only if identification of source and drain of the FETs is maintained under all the possible operating conditions of the circuit. Therefore the technique cannot be used everywhere. For instance, it cannot be used in transmission gates (Section 1.10). In many digital CMOS circuits, source-drain identity cannot be maintained. Most of the MOSFETs in analog circuits, however, maintain source-drain identity. The DC operating point of linear CMOS circuits must be verified using DC simulation that accurately reflects the back-bias effect. If circuits are designed using only back-of-envelope calculations, and if the circuit operation is not verified by high-precision DC simulation, complex DC circuits may not work at all because of wrong DC operating points. Bipolar differential amplifiers do not suffer from back-bias effects, and therefore the performance is generally better than CMOS differential amplifiers. Emitter-collector *identity crisis* similar to that in MOS digital circuits takes place in bipolar digital circuits as well.

1.4 FETs Having Irregular Shapes

Conventional MOSFET theory assumes rectangular FET channels, having *constant* channel width, W, and channel length, L. In this section we consider FETs that have arbitrary, or irregular channel shapes. Such FETs are used in high performance (fast) circuits, in linear circuits,

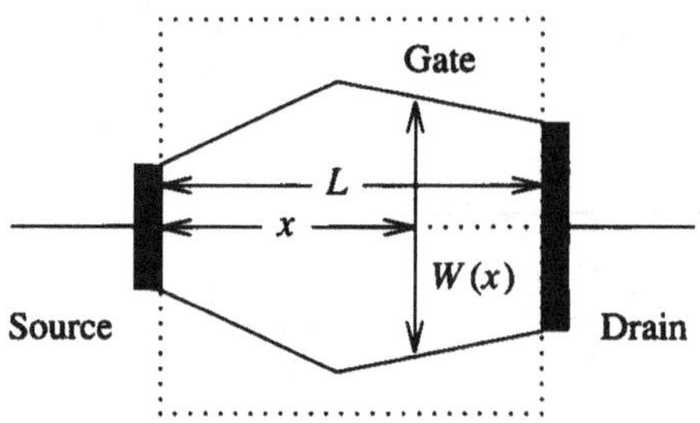

Figure 4 FET having *approximately* regular shape

or in densely packed digital circuits. If FET channel shape is as shown in Fig. 4, a source-drain axis of the FET and width of channel $W(x)$ measured perpendicular to the axis at distance x from the source can be defined [08]. Let the source of the FET be grounded, the gate be biased to positive voltage, V_G, and the drain to small positive voltage, V_D. Current at any cross section equals channel (or drain) current I_D.

$$I_D = W(x)\mu_N \left[\frac{\varepsilon_{OX}[V_G - V(x) - V_{TH}]}{T_{OX}} \right] \frac{dV(x)}{dx} \tag{5}$$

where $V(x)$ is the potential of the channel at distance x from the source. The expression within the large square bracket is mobile charge density induced on silicon surface, and $dV(x)/dx$ is the electric field in the channel. Equation (5) can be integrated as

$$I_D \int_0^L \frac{dx}{W(x)} = \left(\frac{\varepsilon_{OX}\mu}{T_{OX}} \right) \int_0^L [V_G - V(x) - V_{TH}] \frac{dV(x)}{dx} dx = \left(\frac{\varepsilon_{OX}\mu}{T_{OX}} \right) \int_0^{V_D} (V_G - V - V_{TH}) dV$$

and we obtain

$$I_D = \left(\frac{\varepsilon_{OX}\mu}{T_{OX}} \right) \frac{1}{\int_0^L \frac{dx}{W(x)}} (V_G - V_{TH} - \frac{1}{2} V_D) V_D . \tag{6}$$

This formula is of the same form as the conventional I-V characteristics of a rectangular channel FET derived using the same low-field, gradual channel approximation

$$I_D = \left(\frac{\varepsilon_{OX}\mu}{T_{OX}} \right) \left(\frac{W}{L} \right) (V_G - V_{TH} - \frac{1}{2} V_D) V_D$$

if the following substitution is made.

$$\frac{L}{W} \rightarrow \int_0^L \frac{dx}{W(x)} = N_{sq} .$$

N_{sq} is a parameter called *number of squares*. If the channel of the FET were a uniform resistance sheet having sheet resistance $1/\sigma$ *ohms/square*, the resistance of the two-terminal *resistor* shown in Fig. 4 is given by $(1/\sigma)N_{sq}$. N_{sq} is determined from the standard field analysis, if the shape of the resistor body and of the contacts are specified. The number of squares is conventionally used to compute resistance of diffused or deposited resistors on monolithic and hybrid integrated circuits.

If Eq. (6) is written as

$$I_D = \left(\frac{\varepsilon_{OX}\mu}{T_{OX}}\right)\left(\frac{1}{N_{sq}}\right)(V_G - V_{TH} - \frac{1}{2}V_D)V_D \tag{7}$$

it is valid for *any* channel shape [09][10]. Equation (7) is accurate: it does not include any approximation other than the three assumptions stated in section 1.2. We prove this statement.

Suppose that an irregularly shaped FET is biased in the same way as before, as shown in Fig. 5(a). Within the boundary of the FET channel, the potential is function of position *vector* $\vec{r}$, which is written as $V(\vec{r})$. At location $\vec{r}$ the density of mobile electron charge induced on the channel is given by $(\varepsilon_{OX}/T_{OX})[V_G - V(\vec{r}) - V_{TH}]$. The current across curve γ (which separates the FET channel into two parts) equals drain current I_D and is given by

$$I_D = \int \vec{J}(\vec{r}) \cdot \vec{n}\, ds \tag{8}$$

where $\vec{J}(\vec{r})$ is the surface channel current density vector defined by

$$\vec{J}(\vec{r}) = \sigma(\vec{r}) \cdot \nabla V(\vec{r})$$

and where $\vec{n}$ is the unit vector within the channel, which is perpendicular to curve γ at location $\vec{r}$. In this equation

$$\sigma(\vec{r}) = \left(\frac{\varepsilon_{OX}\mu}{T_{OX}}\right)[V_G - V(\vec{r}) - V_{TH}]$$

is the conductivity of the channel at location $\vec{r}$. Line element ds in Eq. (8) is taken along

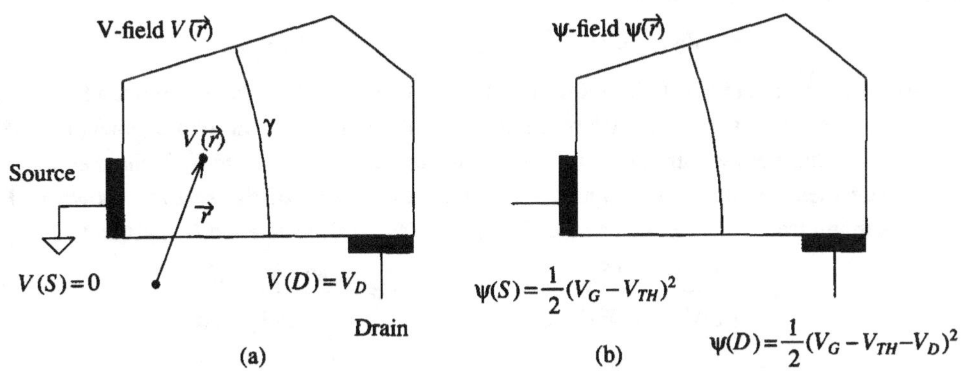

Figure 5 General theory of FET having arbitrary shape

curve γ, and integration is carried out along the entire curve γ. We note that Eq. (8) has the same physical meanings as Eq. (5). Since we consider DC only, $\vec{J}(\vec{r})$ satisfies the current conservation condition

$$\nabla \cdot \vec{J}(\vec{r})=0.$$

Since

$$\nabla \cdot [\sigma(\vec{r})\nabla V(\vec{r})]=\left(\frac{\varepsilon_{OX}\mu}{T_{OX}}\right)\nabla \cdot [[V_G - V(\vec{r}) - V_{TH}] \cdot \nabla V(\vec{r})]=0$$

we define

$$\phi(\vec{r})=V_G - V(\vec{r}) - V_{TH}.$$

Then we have

$$\nabla \cdot [\phi(\vec{r})\nabla\phi(\vec{r})]=\nabla \cdot [\frac{1}{2}\nabla\phi(\vec{r})^2]=0.$$

Therefore we further define

$$\psi(\vec{r})=\frac{1}{2}\phi(\vec{r})^2=\frac{1}{2}[V_G - V(\vec{r}) - V_{TH}]^2$$

and $\psi(\vec{r})$ satisfies the Laplace equation

$$\nabla^2\psi(\vec{r})=0.$$

On the *boundary* of the source and the drain contacts, the potential field $V(\vec{r})$ takes constant values [source voltage (= 0) and drain voltage (= V_D), respectively].

If $V(\vec{r})$ = *constant*, $\psi(\vec{r})$ is also *constant*. Then on the two contacts,

$$\psi(S)=\frac{1}{2}(V_G - V_{TH})^2 \quad \text{on the source} \qquad \psi(D)=\frac{1}{2}(V_G - V_{TH} - V_D)^2 \quad \text{on the drain}. \tag{9}$$

On the free (side) edges of the channel, current flows parallel to the edge. If the unit vector normal to the edge is $\vec{N}$, we must have

$$\vec{N} \cdot \nabla V(\vec{r})=0.$$

From this we are able to show that

$$\vec{N} \cdot \nabla\psi(\vec{r})=\phi(\vec{r})\vec{N} \cdot \nabla\phi(\vec{r})=-\phi(\vec{r})\vec{N} \cdot \nabla V(\vec{r})=0.$$

Therefore $\psi(\vec{r})$ is a potential field within the device, which satisfies the same boundary condition on the free edges and takes different but constant potential on the two contacts [Eq. (9)]. $\psi(\vec{r})$ is a potential field within the *resistor* that has the same shape as the FET channel.

We use these results and compute the number of squares of the *resistor* shown in Fig. 5(b). Assuming uniform sheet resistance $(1/\sigma)$(*ohms*/*square*), resistance R is given by

$$R=\frac{\psi(D)-\psi(S)}{\int_\gamma \sigma\nabla\psi(\vec{r}) \cdot \vec{n}\, ds} \quad \text{or} \quad N_{sq}=R \cdot \sigma=\frac{\psi(D)-\psi(S)}{\int_\gamma \nabla\psi(\vec{r}) \cdot \vec{n}\, ds}. \tag{10}$$

If we consider field $V(\vec{r})$ of Fig. 5(a), we have

$$I_D = \int_\gamma \vec{J}(\vec{r}) \cdot \vec{n}\, ds = -\left(\frac{\varepsilon_{OX}\mu}{T_{OX}}\right) \int_\gamma \nabla\psi(\vec{r}) \cdot \vec{n}\, ds\,. \tag{11}$$

From Eqs. (9), (10) and (11) we have

$$N_{sq} = \frac{1}{I_D}\left(\frac{\varepsilon_{OX}\mu}{T_{OX}}\right)(V_G - V_{TH} - \frac{1}{2}V_D)V_D$$

and this is equivalent to Eq. (7). Therefore Eq. (7) has been proven.

Equation (7) shows that if two FETs having number of squares of channel $N_{sq}(1)$ and $N_{sq}(2)$ are connected in series, the *composite* FET has effective number of squares $N_{sq}(1S2)$ (S stands for *series*) given by

$$N_{sq}(1S2) = N_{sq}(1) + N_{sq}(2)\,, \tag{12a}$$

and if they are connected in parallel

$$[N_{sq}(1P2)]^{-1} = [N_{sq}(1)]^{-1} + [N_{sq}(2)]^{-1} \tag{12b}$$

where P stands for *parallel*. Therefore, FET channels satisfy the same rules for series and parallel connections as *linear* resistors. This relationship is the basis of linear resistor models of FETs discussed in Section 1.11.

Equation (7) is valid when substrate doping is small (as in the characteristics of Section 1.2) and if drift velocity of channel carriers is proportional to the local electric field. If this latter relationship (low-field condition) is not satisfied as in submicron MOSFETs, Eq. (7) is only an approximation. If many FETs are connected in series, however, the voltage that each FET sustains decreases, and therefore the channel field decreases. Then the assumption of low field in channel is satisfied well.

An application of the theory of this section is to the closed CMOS (C^2MOS) FET shown in Fig. 6. This FET may be characterized by the size of the drain island X and the gate length L. The number of squares N_{sq} is determined by the ratio L/X using a function f as

$$N_{sq} = f(L/X)$$

where the function $f(L/X)$ determined from numerical analysis is plotted in Fig. 7.

Closed CMOS FETs are used to design high performance databus drivers. If N is the

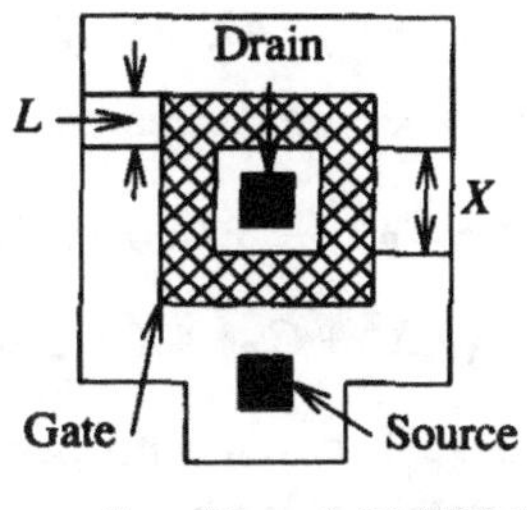

Figure 6 Closed CMOS FET

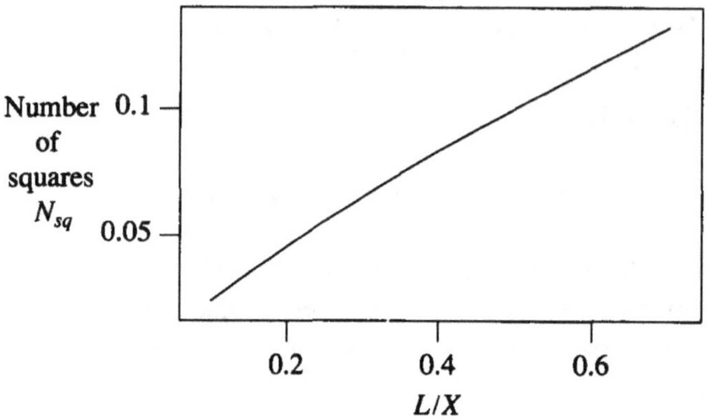

Figure 7 N_{sq} versus L/X

number of input-output (I/O) interface ports communicating with the bus, the delay time of the databus, T_D, is given by

$$T_D = g \cdot NV_{DD} \cdot \frac{C_D}{I_D} \tag{13}$$

where C_D is the capacitance of a single output driver that is disabled (*not* conducting), I_D is the current of a single output driver that is conducting, and g is a numerical factor of the order of unity. The quality of a FET as an output driver is determined by how small the ratio C_D/I_D can be. A closed CMOS FET has significantly smaller C_D/I_D ratio than a regular FET if design and process technology are optimized. A regular FET, shown in Fig. 1, and a closed CMOS FET, shown in Fig. 6, can be compared. Drain capacitance of a regular FET is given by

$$C_D = C_A \cdot WX + C_P \cdot 2(W+X) + (1/2)C_{OX} \cdot (L' - L)W$$

where C_A is the drain diffused island capacitance per unit area, C_P is the drain peripheral capacitance per unit length, C_{OX} $(=\varepsilon_{OX}/T_{OX})$ is the gate oxide capacitance per unit area, and $L' - L$ is the length of overlap of gate electrode on drain and source [note factor (1/2) to account for symmetrical overlaps to the source and to the drain]. The drain capacitance of a closed FET is

$$C_D(CCMOS) = C_A \cdot X^2 + C_P \cdot 4X + C_{OX} \cdot 4 \cdot (1/2)(L' - L)X \, .$$

If the minimum size FETs of the both types are compared we may set $W = X$. Then

$$C_D = C_A \cdot X^2 + C_P \cdot 4X + C_{OX} \cdot (1/2)(L' - L)X$$

$$C_D(CCMOS) = C_A \cdot X^2 + C_P \cdot 4X + C_{OX} \cdot 2(L' - L)X \, .$$

Let the current of a regular CMOS FET be i_D amp per unit width of channel. Then the currents of the two FETs are approximately given by

$$I_D = X \cdot i_D \qquad I_D(CCMOS) = 4X \cdot i_D$$

and therefore

$$\frac{C_D}{I_D} = \frac{C_A X + 4C_P}{i_D} + \frac{C_{OX}(L' - L)}{2i_D} \tag{14a}$$

$$\frac{C_D(CCMOS)}{I_D(CCMOS)} = \frac{C_A X + 4C_P}{4i_D} + \frac{C_{OX}(L' - L)}{2i_D} .$$

This result shows that if drain-gate overlap capacitance effect is neglected $(L' - L \approx 0)$, the closed CMOS FET is 4 times faster than the regular FET. In reality, the ratio of improvement is significantly less than that due to gate-drain overlap capacitance, but this type FET deserves much attention in high-performance CMOS circuits in the future. Applications to CMOS dynamic PLA, Domino CMOS, and output drivers may be considered.

1.5 Gate Capacitance

The gate electrode of a MOSFET has capacitance to the other electrodes (source, drain and substrate). The total capacitance looking into the gate terminal is about $C_G = C_{OX}WL = \varepsilon_{OX}WL/T_{OX}$, where the parameters have their conventional meanings. How the total capacitance is divided among two-terminal capacitances to each of the other electrodes is not obvious. How capacitance is divided can be resolved only by studying the electrical transients within the FET channel.

For gate voltages less than the threshold voltage, no channel exists. The parts of the gate that are directly above the source and the drain diffused islands contribute *overlap* capacitances C_{OS} and C_{OD}, respectively, to the source and to the drain. The rest of the gate capacitance is to the substrate. The equivalent circuit of a nonconducting FET is shown in Fig. 8(a). Once the channel forms, the *overlap* capacitances remain the same, but the gate to channel capacitance should be assigned to the source and drain as well. This is the logical consequence of allowing *no* internal structure of a FET, which is the fundamental assumption in our present FET modeling methods. Capacitance from the conducting channel to the substrate is also assigned to the source and to the drain. The equivalent circuit of a conducting FET is shown in Fig. 8(b). Capacitances C_S and C_D include the capacitance of the conducting channel to substrate. Since this back gate capacitance is qualitatively the same, we consider gate to channel capacitance only. In order to determine the division of the channel capacitance to the source and drain, it is necessary to model the channel of the FET as an RC chain circuit as shown in Fig. 9. We need to consider the electrical transient within the equivalent RC chain [11].

The electrical transient on the RC chain circuit is determined from a *wave equation*. We derive the fundamental equation as follows. We define the voltage on the channel at distance x away from the source at time t, $V(x,t)$, and current in the drain-source direction, $I(x,t)$, as shown in Fig. 9. The instantaneous voltage and the current satisfy

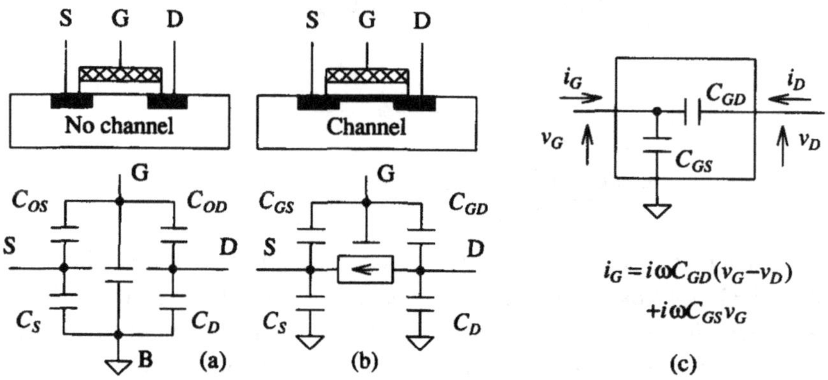

Figure 8 Gate capacitance of FETs

$$I(x,t) = \frac{\varepsilon_{OX}\mu W}{T_{OX}} [V_G - V_{THN} - V(x,t)] \frac{dV(x,t)}{dx}. \tag{15}$$

$V(x,t)$ and $I(x,t)$ are written as the sum of the time-independent DC parts $V_0(x)$ and I_0, and the time-dependent small signal AC parts $\Delta V(x,t)$ and $\Delta I(x,t)$, as

$$V(x,t) = V_0(x) + \Delta V(x,t) \qquad I(x,t) = I_0 + \Delta I(x,t).$$

When we consider a grounded-gate configuration, V_G does not depend on time. The DC parts $V_0(x)$ and I_0 satisfy

$$I_0 = \frac{\varepsilon_{OX}\mu W}{T_{OX}} [V_G - V_{THN} - V_0(t)] \frac{dV_0(x)}{dx}.$$

Since the gate is insulated, the only current that flows between the channel and the gate is the displacement current of the gate capacitance. Current conservation law for the two nearby locations on the channel at x and $x + \Delta x$ is written as

$$\Delta I(x + \Delta x, t) - \Delta I(x,t) = W \cdot C_{OX} \cdot \Delta x \cdot \frac{d\Delta V(x,t)}{dt}$$

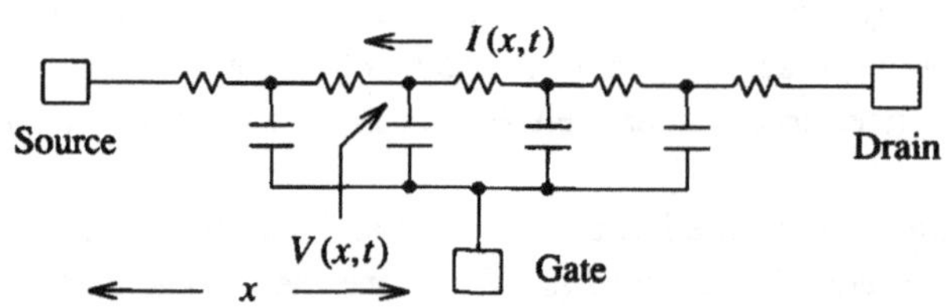

Figure 9 Transmission line model of FET channel

where $C_{OX} = \varepsilon_{OX}/T_{OX}$ is gate to channel capacitance per unit channel area. By combining the three equations, and after some algebraic manipulation,

$$\mu \frac{d^2}{dx^2}[[V_G - V_{THN} - V_0(x)]\Delta V(x,t)] = \frac{d}{dt}\Delta V(x,t) \tag{16a}$$

and

$$\Delta I(x,t) = \frac{\varepsilon_{OX}\mu W}{T_{OX}}\frac{d}{dx}[[V_G - V_{THN} - V_0(x)]\Delta V(x,t)]\,. \tag{16b}$$

We replace (d/dt) by $i\omega$ and $V(x,t)$ by the Fourier transform $V(x,\omega)$. We use

$$V_0(x) = V_G - V_{THN} - [(V_G - V_{THN})^2 - (2I_0T_{OX}/\varepsilon_{OX}\mu W)x]^{1/2}$$
$$\psi(x,\omega) = [V_G - V_{THN} - V_0(x)]\Delta V(x,\omega)\,.$$

The wave equation is written as

$$\frac{d^2\psi(\xi,\omega)}{d\xi^2} = \frac{i\omega\tau}{\sqrt{\xi_0 - \xi}}\psi(\xi,\omega) \quad \text{where} \quad x = L\xi \tag{17}$$

where ξ is the normalized coordinate taken along the channel, $\xi_0 = I_{Dmax}/I_0$, I_{Dmax} = $[(\varepsilon_{OX}\mu W)/(2T_{OX}L)](V_G - V_{THN})^2$ and $\tau = [\varepsilon_{OX}WL^3/2I_0T_{OX}\mu]^{1/2}$. We solve the wave equation in the limit of $\omega \to 0$ (since digital circuits work from DC to the maximum frequency). If $\omega = 0$, the solution is given by $\psi = \alpha + \beta(\xi_0 - \xi)$, where α and β are the constants of integration. If ω is small this first-order approximation is substituted to the right-hand side of the wave equation and integrated twice to obtain the second-order approximation. We then have

$$\psi(\xi,\omega) = \alpha[1 + (4/3)i\omega\tau(\xi_0 - \xi)^{3/2}] + \beta[(\xi_0 - \xi) + (4/15)i\omega\tau(\xi_0 - \xi)^{5/2}]\,.$$

We use the grounded-source parameters,

$$\Delta V(0,\omega) = -v_G \quad \Delta V(L,\omega) = v_D - v_G \quad \text{where} \quad \Delta V(x,\omega) = \left(\frac{\varepsilon_{OX}\mu W}{2I_0T_{OX}L}\right)^{1/2}\frac{\psi(\xi,\omega)}{\sqrt{\xi_0 - \xi}}$$

and where v_G and v_D are the small signal AC variations of the gate and the drain voltages relative to the grounded source. Integration constant α and β are given in terms of v_G and v_D. Since the algebra is straightforward but is very tedious, only the results are given here:

$$\Delta I(x,\omega) = -\left(\frac{2I_0\varepsilon_{OX}\mu W}{T_{OX}L}\right)^{1/2}\frac{1}{\Delta}[f_G(\xi_0,\xi)v_G + f_D(\xi_0,\xi)v_D] \tag{18}$$

where

$$f_G(\xi_0,\xi) = \frac{1}{\sqrt{\xi_0 - 1}} - \frac{1}{\sqrt{\xi_0}} + i\omega\tau[-\frac{4}{3} + 2(\sqrt{\xi_0} - \sqrt{\xi_0 - 1})(\xi_0 - \xi)^{1/2}$$
$$+\frac{2}{3}(\frac{1}{\sqrt{\xi_0 - 1}} - \frac{1}{\sqrt{\xi_0}})(\xi_0 - \xi)^{3/2}]$$

$$f_D(\xi_0,\xi) = \frac{1}{\sqrt{\xi_0}} + i\omega\tau[\frac{4}{3}\xi_0 - 2\sqrt{\xi_0}(\xi_0 - \xi)^{1/2} + \frac{2}{3\sqrt{\xi_0}}(\xi_0 - \xi)^{3/2}]$$

$$\Delta = -\frac{1}{\sqrt{\xi_0}\sqrt{\xi_0 - 1}} + o(\omega)\,.$$

We note that

$$\Delta I(0,\omega) = -i_S \quad \Delta I(L,\omega) = i_D \quad \text{and} \quad i_G = -i_S - i_D$$

where i_S, i_C, and i_D are the small signal AC currents that flow into the respective electrodes. We have

$$i_G = \frac{1}{\Delta}\left(\frac{2I_0\varepsilon_{OX}\mu W}{T_{OX}L}\right)^{1/2} i\omega\tau[g_G(\xi_0)v_G + g_D(\xi_0)v_D] \tag{19}$$

where

$$g_G(\xi_0) = -\frac{8}{3}\xi_0 + \frac{4}{3} - \frac{2}{3}\frac{(\xi_0-1)^{3/2}}{\sqrt{\xi_0}} - \frac{2}{3}\frac{\xi_0^{\ 3/2}}{\sqrt{\xi_0-1}} + 4\sqrt{\xi_0}\sqrt{\xi_0-1}$$

$$g_D(\xi_0) = \frac{2}{3}\frac{(\xi_0-1)^{3/2}}{\sqrt{\xi_0}} - 2\sqrt{\xi_0}\sqrt{\xi_0-1} + \frac{4}{3}\xi_0$$

We have, with reference to Fig. 8(c),

$$i_G = i\omega\left(\frac{\varepsilon_{OX}WL}{T_{OX}}\right)\cdot\left[\frac{g_G(\xi_0)}{\Delta}v_G + \frac{g_D(\xi_0)}{\Delta}v_D\right] = i\omega(C_{GD}+C_{GS})v_G - i\omega C_{GD}v_D$$

where

$$C_{GD} = -\left(\frac{\varepsilon_{OX}WL}{T_{OX}}\right)\frac{g_D(\xi_0)}{\Delta} \qquad C_{GS} = \left(\frac{\varepsilon_{OX}WL}{T_{OX}}\right)\frac{g_G(\xi_0)+g_D(\xi_0)}{\Delta}. \tag{20}$$

C_{GD} and C_{GS} versus normalized drain voltage $V_D/(V_G - V_{THN})$ are plotted in Fig. 10. If drain to source voltage is zero, source and drain are symmetrical with respect to the gate. Then $C_{GD} = C_{GS} = 0.5(\varepsilon_{OX}WL/T_{OX})$. If $V_D = V_G - V_{THN}$ FET is in saturation, and in the idealized model neglecting overlap capacitance, C_{GD}, becomes zero. C_{GS} increases with increasing V_D because source and drain become asymmetric with respect to the gate, and capacitance

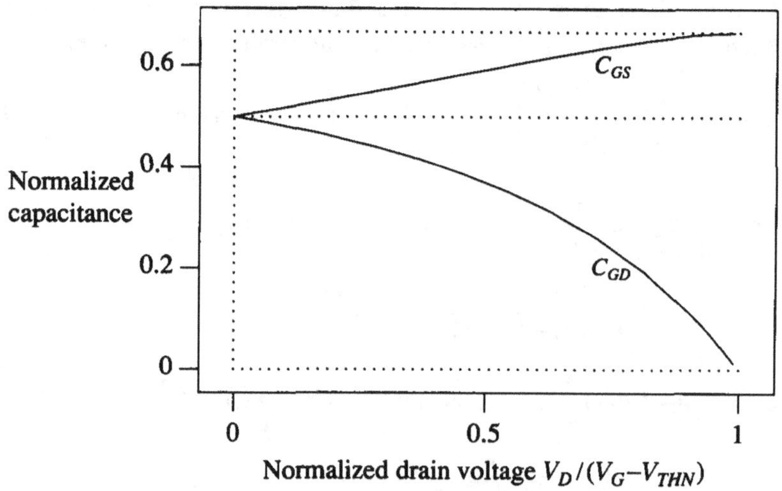

Figure 10 Gate capacitance versus drain voltage of FET

between the channel and the drain becomes less. Then more of the capacitance between the gate and the channel is credited to the source; the maximum is $(2/3)(\varepsilon_{OX} WL/T_{OX})$ at saturation. Since the capacitance derived here is gate to channel capacitance, overlap capacitances must be added to obtain total capacitance to the source and to the drain.

1.6 Channel Carrier Drift Velocity Saturation Effect

Now we consider the effect that makes FET characteristics deviate from the simple characteristic of Eq. (1). Channel carrier drift velocity does not increase proportionally to electric field for high fields. The drift velocity of carriers in the channel of FET can be approximated by a function of the field E

$$v(E) = \frac{\mu E}{1 + (\mu / v_\infty) E} \qquad v(E) \to v_\infty \;\; (E \to \infty) \tag{21}$$

where μ is the low-field mobility and v_∞ is the saturation drift velocity of the carriers at the high field limit. Estimates of the parameter values are as follows: $\mu_N = 400\, cm^2/V\, sec$, $\mu_P = 100\, cm^2/V\, sec$, and $v_\infty(electrons) \approx v_\infty(holes) \approx 8 \times 10^6\, cm/sec$. At fields higher than v_∞/μ, drift velocity saturates gradually to v_∞. This has the following interpretation: If the drain voltage is higher than about $Lv_\infty/\mu = V_0$, an extra mechanism of current saturation is operative: channel current saturates because carriers are unable to move faster than v_∞. If Eq. (21) is used instead of the simple relationship $v(E) = \mu E$, we obtain the following results. Channel current I_D is a function of gate voltage V_G and drain voltage V_D given by [12]

$$I_D = \left(\frac{\mu C_{OX} W}{L}\right) \frac{(V_G - V_{TH}) - (1/2) V_D}{1 + (\mu / L v_\infty) V_D} V_D \quad \text{if} \quad V_D < V_{Dmax} \tag{22a}$$

and

$$I_D = I_{Dmax} = \frac{1}{2}\left(\frac{\mu W C_{OX}}{L}\right) V_{Dmax}{}^2 \;\; \text{if} \;\; V_D > V_{Dmax} \tag{22b}$$

where

$$V_{Dmax} = (Lv_\infty/\mu)[[1 + 2\frac{\mu}{Lv_\infty}(V_G - V_{TH})]^{1/2} - 1] \tag{22c}$$

and where $C_{OX} = \varepsilon_{OX}/T_{OX}$ is defined prior, W is the width and L is the length of the channel of FET. Equations (22a)–(22c) show that FET current is proportional to mobility μ only if ratio μ/v_∞ is constant. Since the electrons and holes have about the same v_∞, the ratio of NFET current and PFET current is *not* simply equal to the mobility ratio.

Next we introduce parameter $V_0 = Lv_\infty/\mu$. In the limit of large V_0, $V_{Dmax} \to V_{DMAX} = V_G - V_{TH}$, and $I_{Dmax} \to I_{DMAX} = (\mu W C_{OX}/2L)(V_G - V_{TH})^2$. This is the case of low channel field. In the limit of small V_0, however,

$$V_{Dmax} \rightarrow \sqrt{2(Lv_\infty/\mu)(V_G - V_{TH})} \text{ and } I_{Dmax} \rightarrow WC_{OX}v_\infty(V_G - V_{TH}).$$

The high field FET saturation current is proportional to $(V_G - V_{TH})$ rather than $(V_G - V_{TH})^2$, as in Eq. (1b). This dependence is one justification of the collapsible current generator model, which will be introduced in the next section and which plays the key role in this book. Equations (22a) and (22b) are plotted in Fig. 11(a). Current I_D is normalized to I_{DMAX} and drain voltage V_D to V_{DD} (the power supply voltage). Gate voltage V_G is set at V_{DD}. Several curves are shown for values of $v_0 = V_0/V_{DD}$ (normalized V_0) ranging from 0.1 to infinity. If v_0 is large (low-field condition), the FET saturates at $V_D = V_{DMAX} = V_G - V_{TH}$. If v_0 is small, however, FET saturates at drain voltage significantly lower than V_{DMAX}.

Figure 11(b) shows plots of Eqs. (22a) and (22b) for $v_0 = V_0/V_{DD} = 0.1$, for several values of normalized gate voltages $v_G = V_G/V_{DD}$ ranging from 0.2 to 1.0. Saturation current varies almost linearly with v_G. Figure 11(c) shows I_{Dmax}/I_{DMAX} versus v_G for several values of v_0. The relationship is parabolic for large v_0, but it is approximately linear for small v_0.

For $L = 1.0 \times 10^{-4}\, cm$, $v_\infty = 8 \times 10^6\, cm/sec$ and $\mu\,(electrons) = 400\, cm^2V/sec$, we have $V_0 = 2\,V$, which is significantly less than $V_{DD}(5\,V)$ (we have $v_0 = V_0/V_{DD} = 0.4$). Thus current saturation of FETs is controlled quite significantly by saturation of drift velocity at high fields. Linear dependence of saturation current on gate voltage is more realistic than the quadratic dependence of the low-field model to represent scaled-down CMOS FETs of feature size less than about 1.75 μm.

The theory of Section 1.4 showed that FETs with arbitrary shaped channels have I-V characteristics determined only by the number of squares of the channel. A consequence of the theory is that FETs have identical current-voltage characteristics if source and the drain are exchanged. This conclusion does not hold if the fundamental assumptions of the theory

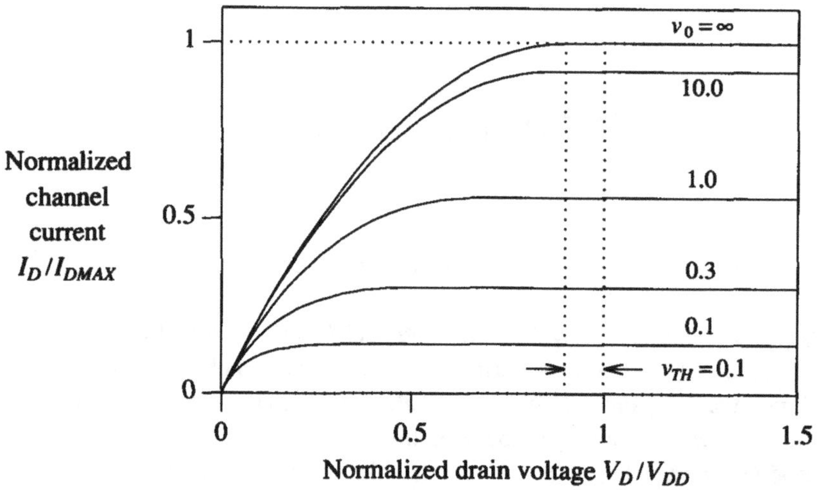

Figure 11(a): Channel carrier velocity saturation and I-V characteristics of FET $V_G = V_{DD}$ and current I_D is normalized to $I_{DMAX} = (B/2)(V_{DD} - V_{TH})^2$

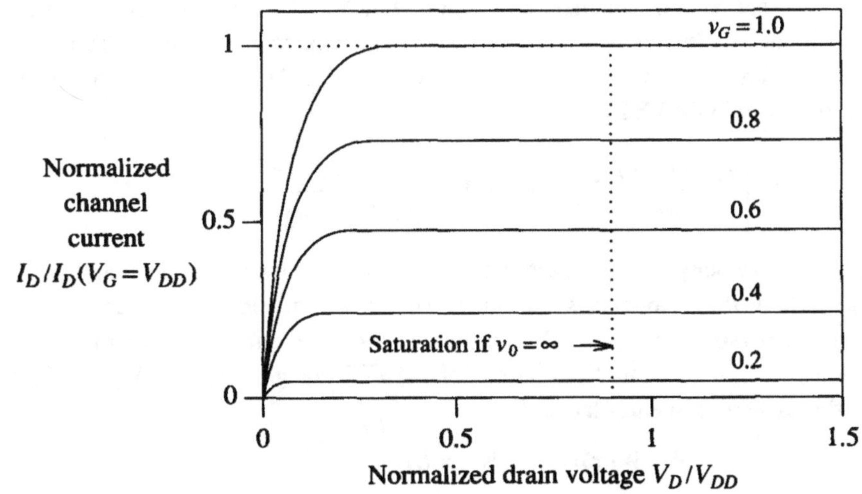

Figure 11(b) Linear dependence of drain current on gate voltage in scaled-down CMOS FET
$v_{TH} = 0.1$, and $v_0 = 0.1$

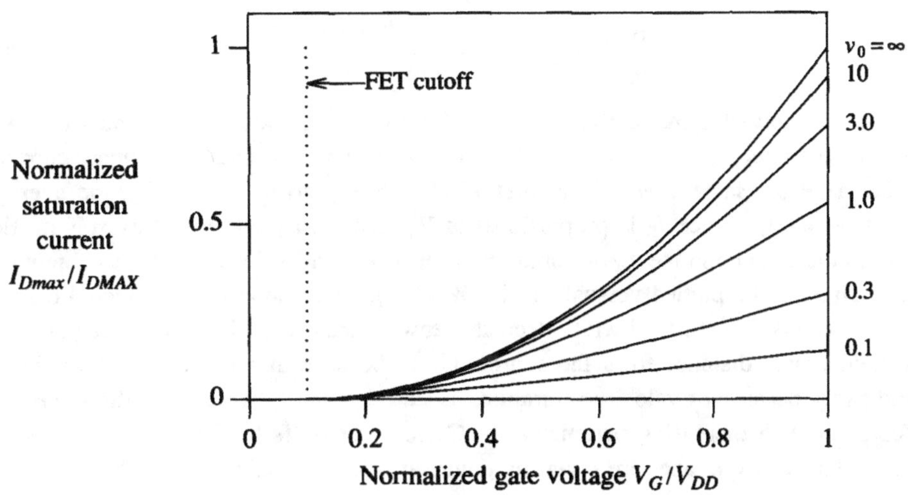

Figure 11(c): FET saturation current versus gate voltage
$v_{TH} = 0.1$

of Section 1.4 are not satisfied. There are two mechanisms that create asymmetry. The first mechanism is velocity saturation of channel carriers at high fields (discussed in this section).

If velocity saturation is included in the analysis of I-V characteristics of FETs having nonuniform channel width $W(x)$ (width is defined as shown in Fig. 4), the equation that determines potential profile within the FET channel is

$$I=\left[\frac{\varepsilon_{OX}\mu W(x)}{T_{OX}}[V_G-V_{TH}-V(x)]-\frac{I\mu}{v_\infty}\right]\frac{dV(x)}{dx}, \tag{23}$$

where we assumed the velocity-field characteristic of Eq. (21). Since channel length is L, coordinate x taken along the channel has the range $0<x<L$. Equation (23) is used in the following numerical analysis. For a given $W(x)$, the current I that satisfies $dV(x)/dx \rightarrow +\infty$ when $x \rightarrow L$ is sought. I is the saturation current of the FET for gate voltage V_G and high drain voltage. If the channel is linearly tapered

$$W(x)=W_0-W_1(x/L),$$

we use normalization

$$\xi=x/L \quad i=I/I_0 \quad v(\xi)=V(x)/(V_G-V_{TH})$$

and define parameters

$$I_0=\frac{\varepsilon_{OX}\mu(V_G-V_{TH})^2W_0}{2T_{OX}L} \qquad w=\frac{W_1}{W_0} \qquad \sigma=\frac{\mu(V_G-V_{TH})}{2Lv_\infty}.$$

Then we obtain this normalized equation

$$\frac{dv}{d\xi}=\frac{i}{2[(1-w\xi)(1-v)-\sigma i]}. \tag{24}$$

This equation was solved numerically, subject to the initial condition $v(0)=0$ and $v'(\xi)\rightarrow\infty$ for $\xi\rightarrow 1$. To obtain the effects of channel shape on the channel current, we compare currents of FETs that have the same *average* channel widths. For a given $w=W_1/W_0$, the average width is $W_0[1-(w/2)]$. Since I_0 is proportional to W_0, current i [determined by solving Eq. (24) subject to the initial and the *final* conditions] must be divided by $1-(w/2)$ to obtain the normalized current to be plotted versus $w\,(=W_1/W_0)$. Figure 12 shows the corrected current versus w. If w is positive, channel width decreases toward the drain. The field in the channel increases rapidly with distance from the source, partly because of reduction in the induced charge, and partly because of velocity saturation. Because of these two effects, the saturation current decreases with increasing parameter w. Currents of FETs having the same average channel width but opposite taper are given by points at w and $-w$ of Fig. 12. The current of FETs with a narrow drain is less than that with a narrow source. By exchanging the source and the drain, saturation current is different. This analysis is intended to show the source-drain asymmetry in simplest terms. The analysis does not rigorously include the space-charge dynamics in the channel of the FET, and therefore the conclusion is only qualitative: In accurate analysis, diffusion current must be included, and channel thickness should not be assumed to be negligible.

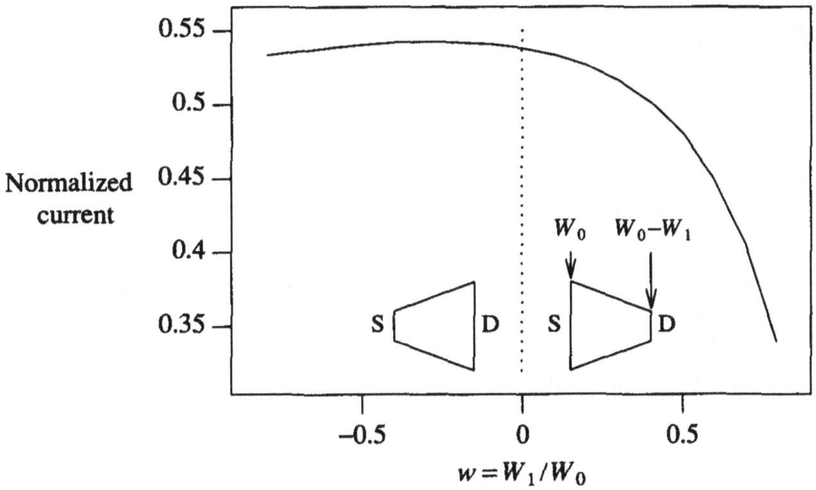

Figure 12 FET saturation current versus taper of channel

The second mechanism of FET source-drain asymmetry is that substrate potential is *pulled* by capacitive coupling from the source, gate, or drain. Substrate potential pulling effect exists even in regular shaped FETs, as shown in Section 4.20. Asymmetry is created by the difference in the size and the shape of the source and the drain diffused islands. If the structure of FET is symmetrical, however, when source and drain islands are included, both polarities of FET operation are identical (by generating the same substrate effects). If a FET has essentially asymmetrical structure, however, one bias polarity has more effect than the other and asymmetry in the current emerges. The effect is enhanced in a structure that isolates the part of the substrate immediately beneath the channel from the rest of the substrate. This effect has been observed in vertical FET structures.

1.7 Collapsible Current Generator Model of FET

FET current-voltage (I-V) characteristics given by Eq. (1a) and (1b) are simple enough to solve some digital circuit analysis problems in closed form. However, in problems where the input voltage, V_G, is function of time t, the FET characteristic is still too complex to carry out closed-form analysis. Mathematically the circuit equations cannot be solved by separation of variables, because the structure of the terms in the circuit equation in which V_G, which depends on time t, is multiplied by dependent variable V_D. Further simplification of FET characteristics to obtain closed-form solutions for practically any problem can be made as follows. Figure 13(a) shows approximation of Eqs. (1a) and (1b) by using two straight line segments. Although the piecewise linear FET characteristic gives closed-form solutions to many waveform-dependent problems, the analyses are still too complex to be generally useful. One

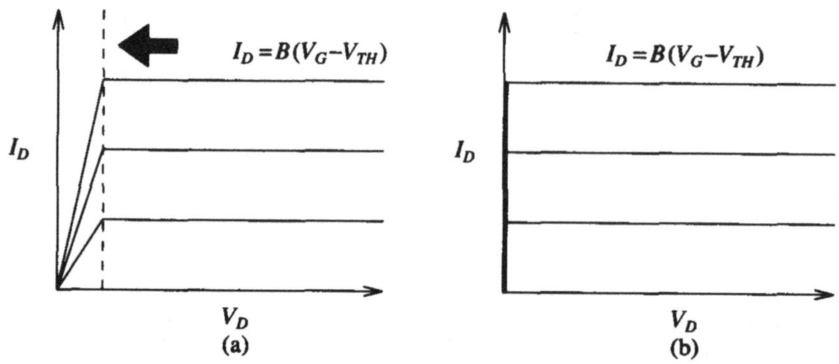

Figure 13 Simplified I-V characteristic of NFETs

more step of simplification is possible by moving the breakpoint of the two line segments to the left, as shown in Fig. 13(b). With this FET characteristic, NFET has drain current $I_D = B(V_G - V_{TH})$ if the drain voltage V_D is positive. NFETs saturate at negligibly small positive drain voltage: $V_D = 0$ if current less than $B(V_G - V_{TH})$ is forced through the FET channel (note that the characteristic has a restrictive condition at $V_D = 0$). We assumed that I_D is a linear function of V_G. Parameter B is the FET transconductance. This simplification is justified for scaled-down CMOS FETs, as shown in Section 1.6. Generally I_D in the saturation region can be any function of V_G, depending on the characteristics of the device. Thus even bipolar transistors (BJT) can be modeled using the *collapsible* current generator model, as we will see in Section 6.8.

The current generator vanishes, or *collapses*, if the drain and the source have same voltage. This additional feature makes the *Collapsible Current Generator model* of FETs quite different from a simple controlled current generator, as is frequently used in analog circuit analysis. If a simple controlled current generator is used to represent a FET, circuit node voltages may exceed the range between ground and V_{DD}. In analog circuit small-signal analysis, node voltages do not increase or decrease indefinitely. In digital circuits, node voltages deviate from the averaged bias voltage, and device nonlinearity is required to limit the deviation. If the collapsible current generator model is used, circuit node voltages are clamped automatically at ground and at V_{DD}, correctly reflecting nonlinear circuit operation. This is a significant advantage in digital CMOS circuit analysis.

The collapsible current generator model has the following interesting characteristics: The FET characteristic splits clearly into the three distinct regions defined by gate, source, and drain voltages V_G, V_S, and V_D, respectively, as

Triode region: $V_D = V_S$ and $V_G > V_S + V_{TH}$
Saturation region: $V_D > V_S$ and $V_G > V_S + V_{TH}$
Nonconducting region: $V_G < V_S + V_{TH}$ $(V_D \geq V_S)$.

If this model is used, a FET has three distinct states that can be identified without ambiguity [in the FET model of Eq. (1) whether or not a FET is in *triode* region is not clearly defined]. Any MOSFET digital circuit has its own *state*: the digital circuit *state* is characterized by the *states* of all its FETs. Later the traditional concept of *state* in logic circuits (such as that represented by the state diagram of sequential logic circuits) can be extended down one level to the MOSFET circuit level in order to describe the working mechanisms of CMOS digital circuits.

In CMOS circuits analyzed using gradual channel, low-field FET models, the concept of *state* of the circuit is not crisply defined. Two circuit nodes that are connected by a FET in the triode region (the channel is equivalent to a small resistance) behave approximately as a single *fused* node. Two nodes that communicate through a FET in the saturation region are separate nodes, actively interacting. Two nodes that are connected by a nonconducting FET are separate nodes. Grouping the circuit nodes into *same*, *different*, and *disconnected* nodes based on FET states establishes a new, high-level order into CMOS digital circuits. The grouping changes with time. This method of describing CMOS digital circuits is very convenient in studies of circuit failure. A full development of this concept is carried out in Section 1.15 and subsequent sections.

The conventional model of FETs in gate-level timing simulators is to replace the channel of any conducting FET by a linear resistance. The significance of the equivalent linear resistance model of FETs in circuit theory is clear if the linear resistance model is compared with the collapsible current generator model. The linear resistor model is the most linearized model of FETs, and the collapsible current generator model is the most non-linearized. The linear model hides the existence of different regions, triode, saturation, and nonconducting, while the collapsible current generator model highlights the differences. In this sense the two models represent the two opposite extrema of linearity and nonlinearity. The two models are, however, the opposite extrema in many other criteria, as we will see later in this book, especially in Section 1.10. The two models are therefore truly complementary in digital circuit theory.

Later in this book it becomes necessary to study the effects of FET current-voltage characteristics on DC CMOS logic gate characteristics. Collapsible current generator models or linear resistor models are too simplified and specialized for this kind of DC parametric study. The high-field model of Section 1.6 is, however, too complex. The following model gives the entire spectrum between the collapsible current generator model and the opposite extremum, the linear resistor model in a simple interpolating formula. Using a parameter V_A that has dimension of voltage, the FET channel current I_D is given by

$$I_D = B_H \frac{(V_G - V_{THN})V_D}{V_D + V_A} \quad (V_G > V_{THN},\ V_S = 0) \quad = 0 \ (V_G < V_{THN}). \tag{25}$$

Equation (25) is a simplified version of Eq. (22a), derived by neglecting the second term in the numerator. That term represents FET current saturation originating from pinch-off of channel due to increasing drain voltage. In Eq. (25), device nonlinearity is due to the velocity-field characteristics [Eq. (21)] of surface carriers alone. Therefore Eq. (25) is the FET characteristic close to, but not at the limit of, the high channel field. If $V_A = 0$, the model becomes the collapsible current generator model. This is the *ultimate* high-field FET characteristic. We call this FET model the simplified high-field model. When $V_A \rightarrow \infty$, however,

$$I_D = \frac{B_H}{V_A}(V_G - V_{THN})V_D$$

and this is equivalent to substituting a linear resistor whose value is $V_A/B_H(V_G - V_{THN})$ for a conducting FET. The model covers the entire range. Using normalization

$$v_D = \frac{V_D}{V_{DD}} \quad v_G = \frac{V_G}{V_{DD}} \quad v_A = \frac{V_A}{V_{DD}} \quad v_{THN} = \frac{V_{THN}}{V_{DD}}$$

we have

$$i_D = \frac{I_D}{I_0} = \frac{1 + v_A}{v_D + v_A} v_D \text{ where } I_0 = \frac{B_H(V_{DD} - V_{THN})V_{DD}}{V_{DD} + V_A}.$$

Figure 14 shows FET characteristics using this model for several values of nonlinearity parameter v_A. There are several circuit analyses in this book for which the simple collapsible current generator does not provide the necessary accuracy. Examples are in Sections 2.4, 3.9, 3.10, 4.9 and 4.11. In such special cases, we use either the low-field gradual channel model or the simplified high-field model. In Eq. (25) the B_H parameter that specifies FET current has special suffix H, that stands for the high-field FET model. For NFETs we use B_{HN}, and for PFETs we use B_{HP}. B_H is the FET transconductance in the limit as $V_D \to \infty$. To use this characteristic at low V_D, it is convenient to normalize I_D using I_{Dmax}, which is the FET current for $V_D = V_G = V_{DD}$, as is done in Section 4.11. Then we have

$$I_D = I_{Dmax} \frac{(V_{DD} + V_A)(V_G - V_{THN})V_D}{V_{DD}(V_{DD} - V_{THN})(V_D + V_A)} \quad \text{where} \quad I_{Dmax} = B_H \frac{(V_{DD} - V_{THN})V_{DD}}{V_{DD} + V_A}.$$

As for the collapsible current generator model, we use parameter B, whose suffix begins with a character other than G or H. The suffix specifies a particular FET, or FETs, in the circuit. B is the FET transconductance.

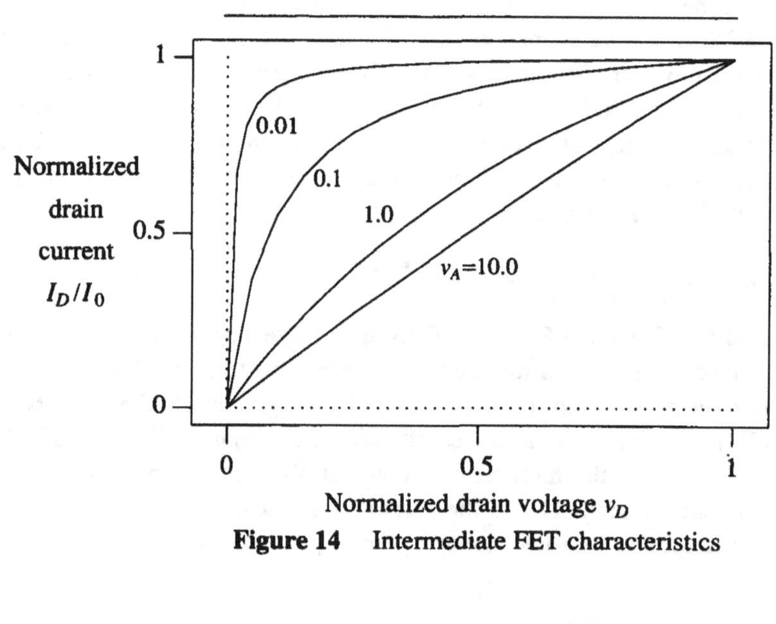

Figure 14 Intermediate FET characteristics

The collapsible current generator model eliminates the gradual change of channel current in the triode region of the gradual channel, low-field model. In Section 1.5 we showed that gate to drain (C_{GD}) and gate to source (C_{GS}) capacitances depend on the drain voltage in the triode region. If the range of drain voltage of the triode region is collapsed, there should be a discontinuous change in the gate to source and gate to drain capacitances. It is clear that the discontinuity in capacitance originates from the simplification that a FET has no *internal* structure. Inclusion of the discontinuity creates many difficulties in circuit analysis. In the collapsible current generator model used in this book, we neglect this discontinuity and use averaged capacitances. The collapsible current generator model is *not* a simple limit of a FET where the internal structure is *continuously* reduced to nothing. Although the use of averaged capacitance is a significant additional simplification, it does not lead to unrealistic consequences in digital circuits. In this book, we study the properties of circuits originating from device connectivity by using a simplified device model that has *no* internal structure.

1.8 Capacitors

Now we change our subject to passive components on CMOS chips. In CMOS digital circuits, capacitors are seldom used to achieve intended signal processing objectives: they are rather unwanted *parasitics*. Exceptions are load capacitance of dynamic CMOS gate (Section 1.10) and capacitance needed to trim signal delay accurately. In analog CMOS circuits, however, capacitors are the principal passive components used to synthesize circuit functions. Capacitors that are fabricated on semiconductor silicon are in one of the two possible states, depending on the condition of the silicon surface: (1) Accumulation mode, or (2) Depletion mode.

Figure 15(a) shows a capacitor fabricated on N-type silicon surface doped to N_D (*donors/cm*^{3}). The upper capacitor plate is metal, heavily doped polysilicon or polysilicide. Oxide is any of gate, field, or composite oxide. Field oxide or gate oxide isolates the silicon substrate from polysilicon or polysilicide layer, and the composite oxide isolates the metal layer from the substrate. If positive voltage is applied to the upper capacitor plate, the population of electrons at the surface of the N-type semiconductor silicon, where the potential energy of electrons within the semiconductor bulk is minimum, increases. The electrons are

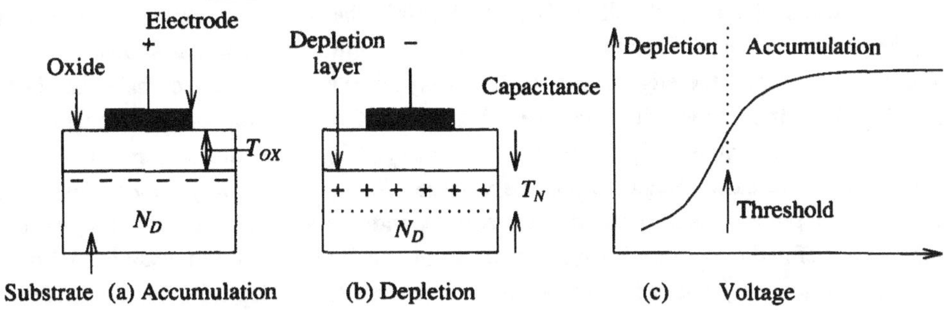

Figure 15 MOS capacitor, accumulation, and depletion modes

confined within a narrow surface layer having a width approximately equal to the Debye length, and the electron density in the thin layer is quite high.

The Debye length can be understood as follows. Since electrons in a semiconductor have thermal energy, they need not be confined to the surface where the electrostatic potential is the minimum. The thickness of the layer within which electrons exist is called the Debye length. Debye length is estimated as follows. In Fig. 15(a), the silicon substrate is doped to N_D (*electrons/cm*^{3}), and therefore mobile charge density is qN_D (*coulomb/cm*^{3}). Suppose that the negative mobile charge has moved a small distance l upward relative to the positive charge of the donor ions fixed to the silicon lattice. This *charge separation* creates surface charge density $\sigma = qN_D l(Coulomb/cm^2)$. An electric field $E = \sigma/\varepsilon_S\,(V/cm)$ is created within the *dipole* layer. Electrostatic energy per unit volume stored in the layer is given by $\varepsilon_S E^2/2\,(Joule/cm^3)$. Since the cause of the separation of charge is the thermal energy of electrons, we have

$$\varepsilon_S \frac{E^2}{2} = \frac{1}{2} N_D kT$$

using the equipartition law (each electron carries average energy $kT/2$ per degree of freedom in thermal equilibrium, where k is Boltzmann's constant and T is the absolute temperature). From this equation we find Debye length l as

$$l = (\varepsilon_S kT/q^2 N_D)^{1/2}\,. \tag{26}$$

If $N_D = 2\times 10^{18}$ (*electrons/cm*^{3}), $l = 0.3\times 10^{-6}\,cm$. Since the Debye length is several orders of magnitude smaller than typical oxide thickness (10^{-4} - 10^{-5} cm), capacitance is determined by the oxide thickness alone and is given by $C_A = \varepsilon_{OX} S/T_{OX}$, where S is the area of the upper plate and T_{OX} is the thickness of the oxide. This type of capacitor is called an *accumulation mode* capacitor.

If a negative voltage is applied to the upper plate, however, electrons are repelled from the semiconductor surface, thereby leaving ionized and positively charged donors on the substrate surface. The maximum volume density of the positive charge equals the doping density of silicon times the electron charge, qN_D. The charge density can be significantly smaller than that attained by the accumulation mode surface charge of electrons. If doping N_D is high, depleting a thin surface layer creates enough positive charge, and an electric field that is high enough to sustain the applied voltage is created within the oxide. If doping N_D is low, however, the surface of the semiconductor is depleted to a significant depth, as shown in Fig. 15(b) (the dotted line shows the edge of the depletion layer). Part of the silicon substrate becomes *insulating* dielectric. Since the effective thickness of the *insulator* of the capacitor is then more than T_{OX}, capacitance C_D is less than $C_A = \varepsilon_{OX} S/T_{OX}$, and furthermore the capacitance decreases with increasing negative applied voltage. Figure 15(c) shows *small signal* capacitance versus applied voltage (since capacitance depends on voltage, only small-signal capacitance can be defined clearly). This type of capacitor is called a *depletion mode* capacitor.

If the surface of the semiconductor substrate is depleted to depth T_N [see Fig. 15(b)], the charge stored in this capacitor, Q (*Coulomb/cm*^{3}) is given by $Q = qN_D T_N$ or $T_N = Q/qN_D$. The field in the oxide is Q/ε_{OX} and the averaged field in the *depletion* layer is $Q/2\varepsilon_S$. Since the capacitor sustains external voltage V, we have

$$V + V_{dif} = \left(\frac{T_{OX}}{\varepsilon_{OX}}\right)Q + \frac{Q}{2\varepsilon_S}\frac{Q}{qN_D} \tag{27}$$

where V is the externally applied voltage and V_{dif} is called diffusion potential. V_{dif} is the potential established by exchanging electrons between semiconductor and metal (of the upper capacitor plate) when they are contacted. The potential difference is created because the materials of the two capacitor plates have different *affinity* to electrons. V_{dif} is invisible if the two electrodes of the capacitor are made of the same material. In our present problem they are not. Solving Eq. (27) for Q we obtain [13]

$$Q = qN_D T_{OX}(\varepsilon_S/\varepsilon_{OX})[\sqrt{1 + (\varepsilon_{OX}/\varepsilon_S)^2[(V + V_{dif})/V_0]} - 1]$$

where $V_0 = qN_D T_{OX}{}^2/2\varepsilon_S$ is the voltage that would be developed across the depletion layer if $T_N = T_{OX}$. We then have

$$C = \frac{\partial Q}{\partial V} = \frac{(\varepsilon_{OX}/T_{OX})}{\sqrt{1 + (\varepsilon_{OX}/\varepsilon_S)^2[(V + V_{dif})/V_0]}}. \tag{28}$$

If $(V + V_{dif})/V_0 < (\varepsilon_S/\varepsilon_{OX})^2$ the capacitance decreases significantly, starting from the constant value at $C_{OX} = \varepsilon_{OX}/T_{OX}$ with increasing *negative* applied voltage to the capacitor. We explained the voltage dependence of a MOS capacitor using metal on N-type silicon. The same mechanism works in PN junction capacitors as well. In a reverse-biased PN junction capacitor, both capacitor plates work as depletion mode capacitors.

Voltage dependence of capacitance is often undesirable. A standard technique of reducing voltage sensitivity of capacitors is to dope the silicon surface heavily, as shown in Fig. 16(a). Standard digital CMOS processes allow fabrication of an N+ layer (same as the source-drain of NFETs) underneath *intermediate* oxide. The upper capacitor plate is metal. A problem of this capacitor is that intermediate oxide of a conventional CMOS process is thick (>1 μm), and therefore a large area is required to obtain only a few pF of capacitance. CMOS processes adapted for analog ICs have two layers of heavily doped polysilicon, and capacitance between the two polysilicon layers is used, as shown in Fig. 16(b). The oxide layer

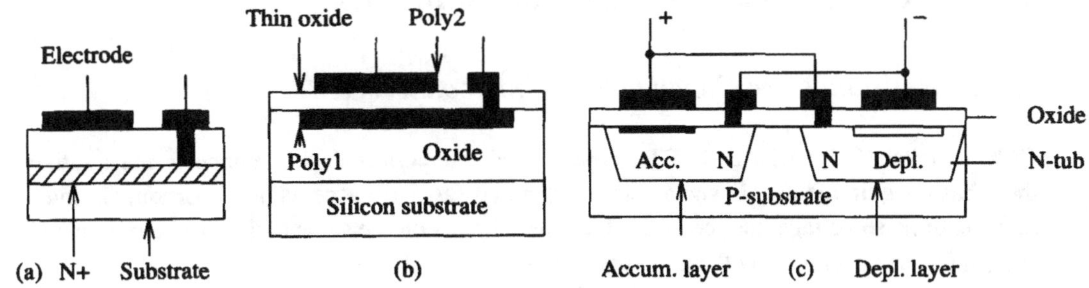

Figure 16 Bias independent capacitor

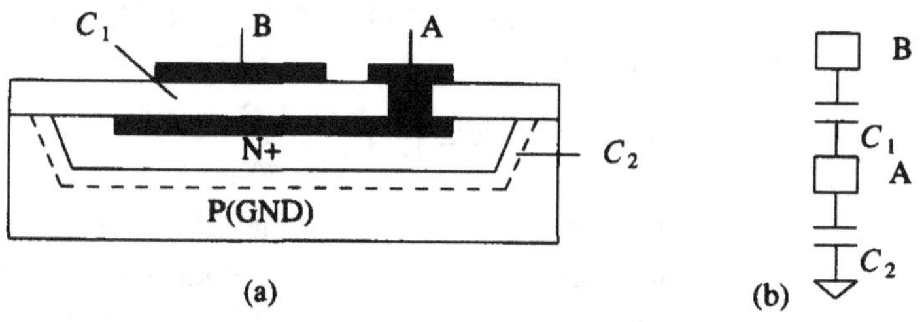

Figure 17 Parasitic of CMOS capacitor

between polysilicon layers is thin (typically about 1000 angstroms). A crude compensation of voltage dependence can be achieved by connecting two capacitors in parallel, as shown in Fig. 16(c). If one capacitor is in depletion mode, the other is in accumulation mode, and the sum of the two capacitors remains approximately voltage independent.

Voltage dependence of capacitance is not always a disadvantage. Voltage dependence of the PFET's drain island capacitance helps to reduce the noise sensitivity of dynamic CMOS logic circuits in the precharged state. The beneficial use of voltage dependence of CMOS capacitance is a challenge for the future. A capacitor that is fabricated on a conductive substrate always has capacitance to the substrate. Figure 17(a) shows a metal to N+ capacitor fabricated on an isolated N-tub. The capacitance of the reverse-biased PN junction between the N-tub and the grounded P-type substrate is an unwanted parasitic capacitance. The equivalent circuit of the capacitor is shown in Fig. 17(b). Capacitance C_1 and C_2 can be, in the worst case, comparable. Then a circuit that designed assuming C_1 alone suffers from the parasitic capacitance C_2. Parasitic capacitance C_2 is unimportant in some circuits. If one side of the capacitor should be grounded, that side is made the lower plate. If one terminal is a time-dependent voltage source (driven by low internal impedance driver), that node should be the lower plate (terminal A) of Fig. 17(b). Then capacitance C_2 adds only to the loading of the low impedance driver, but the circuit performance is otherwise not compromised. If both terminals of a capacitor must have a high impedance level to ground, the active capacitance compensation technique that will be discussed in Section 5.18 can be used.

1.9 Interconnects, Resistors, and Inductors

1.9.1 Resistance and Capacitance of Interconnects

CMOS digital ICs consist of PFETs, NFETs, and interconnects. Interconnects must satisfy three basic requirements: (1) signals are transmitted fast with acceptable distortion, (2) interference, or noise voltage, induced from the other circuits on the same chip must be minimized, and (3) the area occupied by the interconnects is minimized.

Interconnects are designed using conductive wires. There are several different *grades* of wires according to their conductivities. Metal wires, having series resistance 0.03–0.06 ohms/square and capacitance of about 2 pF/cm are best as interconnects. Modern CMOS technology provides more than one level of metal interconnect, some up to four levels. Multilevel metal CMOS technology is, however, expensive, and the conductivity of multilevel metal is less than that of single-level metal (because thickness is reduced to planarize the surface). Therefore the number of metal levels should be minimized if performance, interconnect capability, or any other essential design requirements are not to be compromised.

Figure 18 shows a schematic of metal or polysilicon wires in a CMOS IC. We consider the capacitance of the center conductor, C. The center conductor of the three-conductor system, with the conducting silicon substrate grounded, is characterized by three capacitance coefficients between the conductors and the substrate, C_{CA}, C_{CB}, and C_C. Capacitive coupling among multiple conductors is characterized by using these capacitance coefficients. C_{CA}, C_{CB}, and C_C are the only relevant capacitance coefficients of the four-conductor system. If conductor A and B stay at ground potential and conductor C makes a ground to V_{DD} transition, the driver of conductor C feels capacitance $C_C + C_{CA} + C_{CB}$. This is the *typical* capacitance of wire C in a digital integrated circuit. If conductors A and B make a ground to V_{DD} transition simultaneously with conductor C, capacitance components C_{CA} and C_{CB} have no effect: therefore the driver only feels capacitance C_C. This is the optimistic estimate. If conductors A and B are originally at V_{DD}, and if they make a high-to-low transition simultaneously with conductor C making a low-to-high transition, capacitance coefficients C_{CA} and C_{CB} are doubly effective: the driver feels capacitance $C_C + 2(C_{CA} + C_{CB})$. This is the *pessimistic* estimate of the interconnect capacitance. Capacitances C_{CA} and C_{CB} characterize mutual coupling of the signals as well. The effect of mutual coupling is that voltage is induced on conductor C from A and B. If center wire C is tristated (the node is disconnected from the rest of the circuit), and if the potential of wire A changes from 0 to V_{DD}, the potential of wire C changes by

$$\Delta V_C = \frac{C_{CA}}{C_{CA} + C_{CB} + C_C} V_{DD} \,. \tag{29}$$

This is the estimate of noise induction from one node to the other, which will be studied in chapter 4.

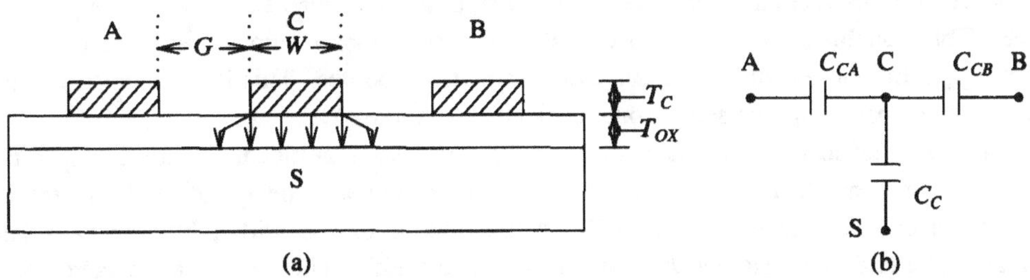

Figure 18 Wiring capacitance

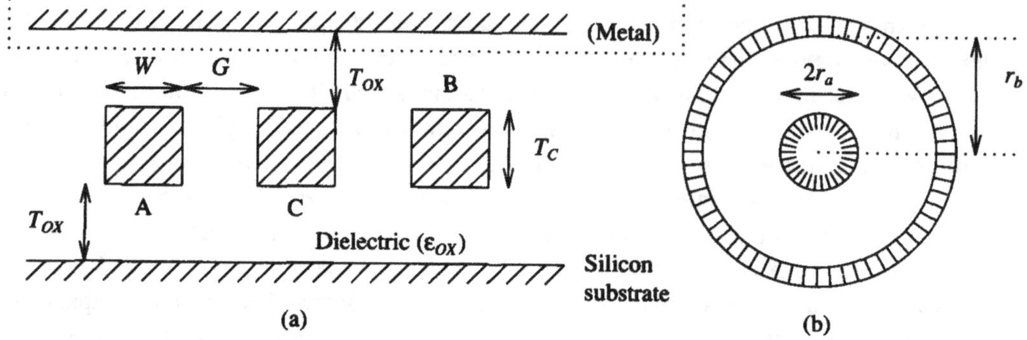

Figure 19 Limit of scaling interconnects

Next we consider *typical* wiring capacitance. Suppose conductors A and B in Fig. 19(a) are grounded. Since conductor C is surrounded on three sides—by the substrate, conductor A, and conductor B—capacitance of conductor C is approximated by that of a coaxial cable, whose center conductor has radius $r_a = (W + T_C)/4$ and whose cavity has inner diameter $r_b = r_a + (G + T_{OX})/2$, as shown in Fig. 19(b). This approximation is good in scaled-down CMOS, where thickness (T_C)/width (W) ratio and gap (G)/width (W) ratio of the conductor are approximately 1. Capacitance of the *coaxial cable* shown in Fig. 19(b) approximates the typical capacitance of conductor C of the three-conductor system, $C_C + C_{CA} + C_{CB}$. We have,

$$C_C + C_{CA} + C_{CB} = \frac{2\pi\varepsilon_{OX}}{\log(r_b/r_a)} \tag{30}$$

where $r_b/r_a \approx 3$ if $T_{OX} = W = G = T_C$. Then we may conclude $\log(r_b/r_a) \approx 1$ in scale-down CMOS. Since $\varepsilon_{OX} \approx 0.345 pF/cm$, the minimum wiring capacitance attained by scaledown is limited to about 2pF/cm. Equation (30) gives insight into the capacitance of the first-level metal wire, if the second-level metal exists above [the second metal level enclosed within the dotted box of Fig. 19(a)]. According to Eq. (30), capacitance of the minimum width, minimum spacing wire of the first-level metal does not increase significantly. A narrow, first-level metal wire has so much fringing capacitance to silicon substrate and to the nearby conductors, that addition of the second-level metal does not significantly increase the already high capacitance. This conclusion is in contrast to the capacitance of a wide, first-level metal wire ($W \gg T_{OX}$). In this case the capacitance approximately doubles. This is an important point to note if a wide metal wire is used to route critical signals like clock.

For delay-critical applications, even metal wire has significant resistance-capacitance delay. The standard technique used in this case is to increase the width of the metal wire. Then resistance decreases proportionally to the inverse of the width, but the capacitance increases at a rate much slower than the proportional width. Then we obtain a factor of 2–4 improvement in delay. This technique is fundamental in clock distribution system design; it will be discussed in Section 5.5.

Now we consider wires less conductive than metal wires. They can be used either as interconnects or as resistors. Polysilicon wires have resistance 1–5 ohms/square (polysilicide) and 15–30 ohms/square (doped polysilicon). Capacitance of polysilicon wires is comparable to metal wires (2pF/cm) in any small feature size CMOS because of the *coaxial cable* mechanism discussed before. In 1–2 μm CMOS, diffused silicon (N+, P+) wires have high series resistance (about 30 ohms/square for N+ diffusion and 100–150 ohms/square for P+ diffusion). In submicron CMOS the resistance is less, often a few ohms/square (silicided). Isolated tubs (P-tub or N-tub, depending on the type of the substrate) has 1–2 kohms/square resistance, and therefore high resistor values (up to a few hundred kohms) can be designed using long tubs. Diffused resistors have significant dependence on voltage between the resistor body and the surrounding silicon substrate. The voltage coefficient requires experimental determination or numerical analysis: simple estimates assuming doping-independent mobility gives inaccurate underestimates of the coefficient.

The capacitance of N+ or P+ diffused silicon wires is PN junction capacitance between the resistor body and the substrate. The capacitance is significantly higher than metal or polysilicon wire capacitance (10–15pF/cm). This high capacitance originates from the following causes. Three sides of a diffused wire are isolated from the substrate by PN junction depletion layers. The dielectric constant of silicon is approximately three times higher than that of silicon dioxide. The insulating layer, be it dielectric or depletion layer, has about the same thickness to maintain the structural integrity of integrated circuit, or to allow rational layout rules. In the present integrated circuit structure, the parameter values can be chosen to maintain the structural integrity (why this is possible is not obvious—this would be an interesting subject of research).

If resistance or capacitance is not parasitic but is used for some signal processing purposes, we must determine its *quality*. On-chip resistors have parasitic capacitance to ground (or V_{DD}). The structure, the distributed and the lumped equivalent circuits of an on-chip resistor are shown in Figs. 20(a), (b), and (c), respectively. Quality of resistor is measured by delay time T_D defined by $T_D=(1/2)RC$ where R and C are given by $R=(L/W)\rho$ and $C = C_A \cdot WL + C_P \cdot 2(W+L)$ where L is the length and W is the width of the resistor, and the effect of the contact islands are neglected. C_A is the capacitance per unit area of the resistor body, and C_P is the peripheral *fringing* effect capacitance per unit peripheral length of the resistor body. For a switching transient that lasts time T_D, the resistance and the parasitic

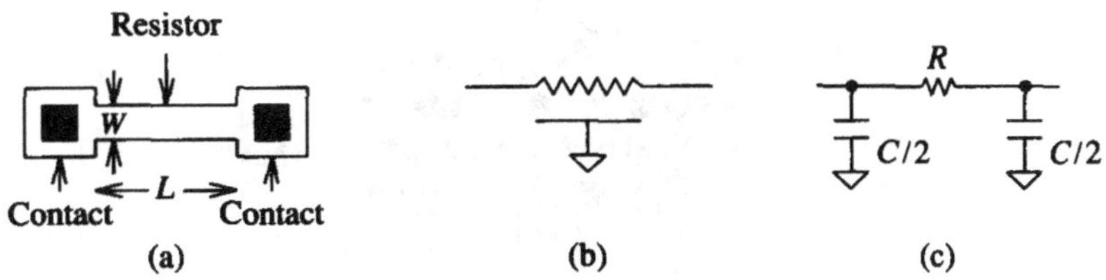

Figure 20 Equivalent circuit of on-chip resistor

capacitance present comparable impedance. If T_D is smaller, the resistor is better in quality. With R given,

$$T_D = \frac{R}{2}[C_A \frac{R}{\rho} W + 2C_P(1 + \frac{R}{\rho})]W . \tag{31}$$

This equation shows that T_D is less if width W is less. High performance resistors should be designed as small as the layout rule, matching to the other resistors, and the power rating permit. If W is set at the minimum, T_D is reduced by increasing ρ. Then a better quality resistor can be designed using P+ diffusion (130 ohms/squar) than N+ diffusion (25 ohms/square) for 1.75 μm CMOS. For typical values of $C_A = 3 \times 10^{-4} pF/\mu m^2$, $C_P = 3 \times 10^{-4} pF/\mu m$, $R = 1000\, ohms$, $W = 10\,\mu$m, and $\rho = 130 ohms/square$, $T_D = 140\, psec$. If $\rho = 25\, ohms/square$, $T_D = 720\, psec$. This result shows that the limit of diffused resistor will be reached soon.

Capacitance of diffused silicon wires can be defined clearly, because diffused wires are buried in the fixed-potential substrate or tub. Signals carried by diffused wires are effectively *shielded* on three sides by the constant potential substrate. Although this is a significant merit, voltage-dependent and large capacitance, and high series-resistance make diffused wires the most unattractive interconnect. Diffused wires can be used only to transmit slow signals. Diffused wires *cannot* be overlaid by polysilicon wires, since in the modern self-aligned MOS technology the intersection of the two wires automatically becomes a FET. Diffused wires and polysilicon wires can be laid out parallel and interleaved, however, and this arrangement allows very high wiring density in one-level metal CMOS technology, as shown in Fig. 21. For slow CMOS devices in a critically crowded area, this is a useful layout option.

In Fig. 21 polysilicon and diffused wires are perpendicular to metal wires. The perpendicular wires shown in Fig. 22(a) appear to have major capacitance contributed from the vicinity of intersection, P. The capacitance of the intersection, however, cannot be defined clearly, because the parts of the a wires distance away from the intersection contribute slowly decreasing capacitance with the increase of distance. How the two intersecting wires end must be reflected in the capacitance computation. This leads to an inevitable conclusion that three-dimensional field analysis is necessary to determine accurate capacitance values. If simplified one- or two-dimensional field analysis is used, significant error in capacitance remains.

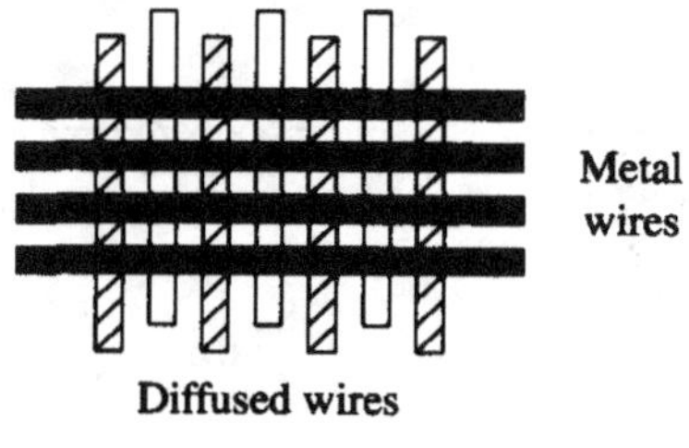

Figure 21 Dense interconnect scheme

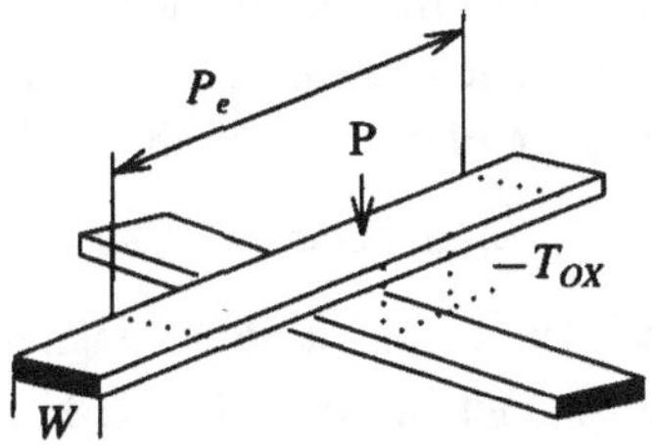

Figure 22 Capacitance of wires intersecting at right angle

1.9.2 Inductance of Interconnects

Next we study the general properties of on-chip inductance. Inductance is estimated as follows [14]. With reference to Fig. 23(a), self inductance of the circular loop equals the magnetic flux created *within* the loop by unit current around the loop. If a conductor has significant cross-sectional area, magnetic flux exists inside the conductor where the current flows. The self inductance must be separated into two parts, the *external* and the *internal* inductances. External inductance is computed by concentrating all the current at the center line of the conductor [in Fig. 23(a) the circle with tangential arrow indicating current] and by computing magnetic flux that exists within the inner circle B (radius b). External inductance can be computed by elementary field analysis, but the algebra is usually tedious. Internal inductance is approximated by breaking the loop open, by assuming that the conductor is infinitely long, and by computing the magnetic field within the wire. In this field analysis we compute magnetic field using Ampere's law

$$\int_C \vec{H} d\vec{s} = I$$

where C is the closed circuit that surrounds current I. In this analysis, magnetic field $\vec{H}$

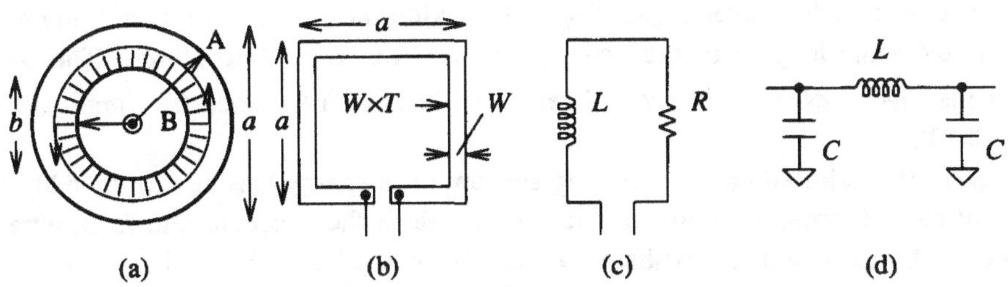

Figure 23 On-chip inductance and parasitics

created by current I can be identified. Internal inductance is determined by integrating flux linkage to the fraction of the current over the entire volume of the conductor. Internal inductance of a long wire having a circular cross section is given by

$$L_i(circular) = \mu_0/8\pi \tag{32}$$

per unit length of the wire, where μ_0 is the magnetic permeability of free space ($\mu_0 = 4\pi/10^9$ *henry/cm*). The self inductance of a circular loop having inner and outer radius b and a, respectively, where $|a-b| \ll a$, as shown in Fig. 23(a), is

$$L = L_i + L_e = \mu_0 a[\log(16a/(a-b)) - 1.75]. \tag{33a}$$

If we consider a square loop having size $a \times a$ shown in Fig. 23(b), inductance of the loop is approximately

$$L = \frac{\mu_0 a}{\pi}\left[2\log\frac{2a}{(1+\sqrt{2})r} + (2\sqrt{2} - \frac{7}{2})\right] \tag{33b}$$

where r is the radius of the conductor (having a circular cross section) of the loop. If the conductor has an approximately rectangular cross section, $r \approx \sqrt{WT/\pi}$. If $a = 0.1\, cm$ and $r = 1.8\, \mu m$, $L = 4.6\, nH$. These are the estimates of inductance for typical integrated circuits.

On chip inductance has significant parasitics. If the inductance is fabricated on an insulating substrate like that of SOS CMOS devices, resistance of the metal conductor is the only significant parasitic. The resistance of the square loop is estimated from $R = (4a/W)\rho$ where ρ is the sheet resistance of the metal. If $a = 0.1\, cm$, $W = 10\, \mu m$, and $\rho = 0.04\, ohms/square$, $R = 16\, ohms$. Impedance of the loop is the series connection of the inductance L and the parasitic series resistance R, as shown in Fig. 23(c). Quality of the inductance as a circuit component can be evaluated from the inductance/resistance time constant $T_{LR} = L/R$. If an electrical transient lasts longer than time T_{LR}, the effects of inductance are negligible. The longer the time constant, the better the inductance. If we use the parameter values obtained before, $T_{LR} = 4.6\,(nH)/16\,(ohms) = 290\, psec$. In fast CMOS SOS circuits with large conductor loops, the effect of inductance can be more than the effect of series resistance.

On-chip inductance has other parasitics. If the conductor loop is fabricated on bulk CMOS, the effects of parasitic capacitance to the substrate are characterized from the period of resonance of the loop, T_{LC}. The equivalent circuit of the inductance is shown in Fig. 23(d). Capacitance C is estimated from $C = (1/2) \cdot 4aW \cdot (\varepsilon_{OX}/T_{OX})$ where ε_{OX} is the dielectric constant of silicon dioxide insulator and T_{OX} is the oxide thickness. This is a crude estimate. If we use the same loop as before and $T_{OX} = 1\, \mu m$, we have $C = 0.69\, pF$. The period of resonance is $T_{LC} = 2\pi\sqrt{LC} = 390\, psec$. Again the effects of inductance are significant in submicron CMOS.

Inductance fabricated on a conducting substrate has energy loss due to the eddy current induced in the substrate, as shown in Fig. 24(a). Since the substrate allows penetration of magnetic field, the current is distributed within the bulk silicon. We estimate the effects of eddy current crudely, using the equivalent circuit shown in Fig. 24(b). The circuit equations of the equivalent circuit of Fig. 24(b) are

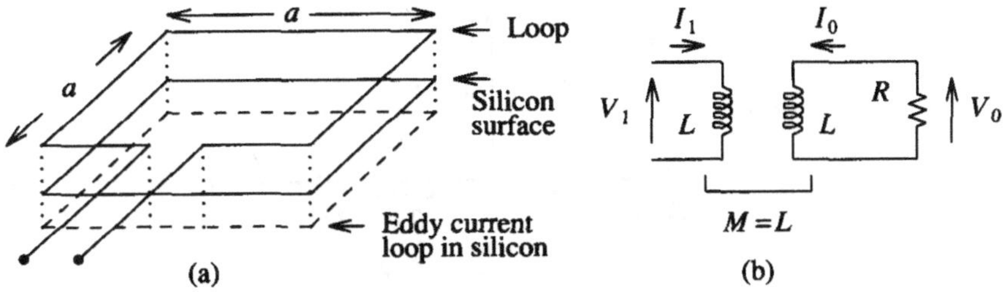

Figure 24 Eddy current loss of on-chip inductance

$$V_1 = L_1 \frac{dI_1}{dt} + M \frac{dI_0}{dt} \qquad V_0 = M \frac{dI_1}{dt} + L_2 \frac{dI_0}{dt} \qquad V_0 = -RI_0$$

We assume that the semiconductor substrate is modeled using a single loop that has the same dimension. The two loops are assumed to be perfectly coupled magnetically (no leakage flux). Then we have $L_1 = L_2 = M = L$. Assuming sinusoidal variation we eliminate V_0 and I_0 and obtain

$$V \frac{1}{I_1} = i\omega L + \frac{\omega^2 L^2}{i\omega L + R} \rightarrow i\omega L[1 - (\omega L/R)^2] + (\omega^2 L^2/R) \quad (\omega \rightarrow 0). \tag{34}$$

Equation (34) shows that the inductance has frequency-dependent series resistance $\omega^2 L^2/R$. Resistance R can be estimated as follows. The semiconductor area underneath the loop is equivalent to a one-turn loop having circumference $4a$ and width a. The resistance of a column that has length $4a$ and cross-sectional area $a \times l$, where l is the thickness of the substrate, is given by $R = (4/l) \cdot \rho$ where ρ is the resistivity of the substrate. We assume that the magnetic field penetrates the entire thickness of the substrate. If $\rho = 0.01$ *ohm cm* and $l = 500\ \mu$m, $R = 0.8$ *ohms*. If we use $L = 4.6$ *nH*, the time constant is 6 nsec. We note that the three time constants characterizing the on-chip inductance are all within one order of magnitude. The relative importance of parasitics depends on the shape and vertical structure of the chip. On-chip inductance is a complex and unclean circuit element. We showed that the inductive impedance is comparable to resistive impedance of the interconnect. The level of inductive impedance is, however, still significantly lower than the impedance level of the FETs. Indeed, 4.6 nH of inductance presents 28 ohms at 1GHz, which is more than the parasitic resistance. Typical device impedance is, however, several hundred ohms, or more. The two impedance levels become closer as CMOS device speed increases.

1.9.3 Analysis of Interconnects and Parasitics

It is sometimes convenient to analyze digital circuits in the frequency domain. A simple guideline for this type of analysis is as follows. We determine the frequency spectrum contained in the waveform shown in Fig. 25 [15]. The waveform is defined by

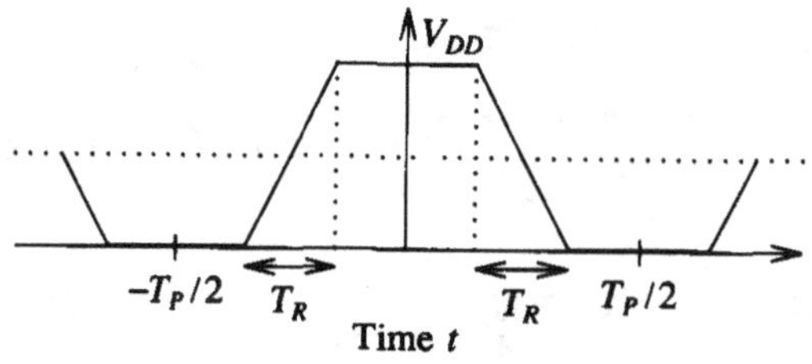

Figure 25 Fourier analysis of digital signal waveform

$$V(t) = (V_{DD}/T_R)[t + (T_P/4) + (T_R/2)] \quad -(T_P/4) - (T_R/2) < t < -(T_P/4) + (T_R/2)$$

$$V(t) = V_{DD} \quad -(T_P/4) + (T_R/2) < t < (T_P/4) - (T_R/2)$$

$$V(t) = V_{DD} - (V_{DD}/T_R)[t - (T_P/4) + (T_R/2)] \quad (T_P/4) - (T_R/2) < t < (T_P/4) + (T_R/2)$$

$$V(t) = 0 \quad (T_P/4) + (T_R/2) < t < (T_P/2).$$

By using the standard Fourier analysis technique, we may write

$$V(t) = (V_{DD}/2) + \sum_{k=1}^{\infty} a_k \cos[(2\pi k/T_P)t] \tag{35}$$

where a_n are determined as

$$\frac{a_{4n+1}}{(V_{DD}/2)(T_R/T_P)} = F[(4n+1)\pi(T_R/T_P)] \qquad \frac{a_{4n+3}}{(V_{DD}/2)(T_R/T_P)} = -F[(4n+3)\pi(T_R/T_P)]$$

$$a_{4n} = a_{4n+2} = 0$$

where n is integer index. Function F is given by

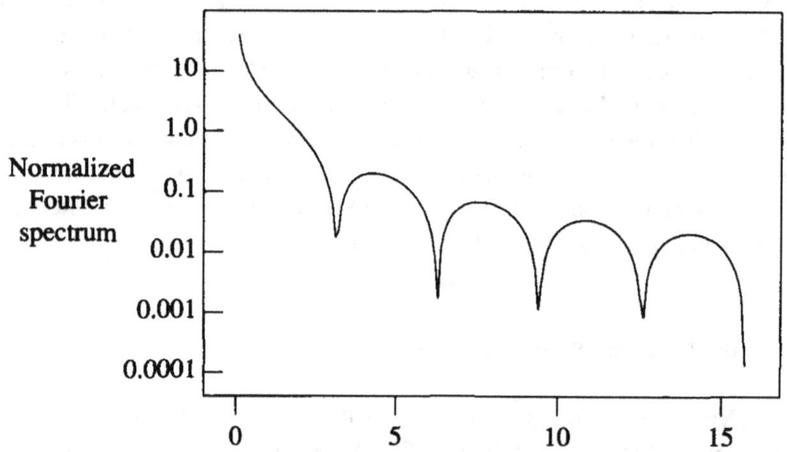

Figure 26 Function $F(x)$

$$F(x) = \frac{4}{x}\left(\cos x + \frac{\sin x - x\cos x}{x}\right).$$

Figure 26 shows $F(x)$ versus x.

The roots of $F(x)$, x_m, that are solutions of $F(x_m) = 0$, are found by inspection as

$$x_m = m\pi \quad m = 1, 2, \ldots.$$

Then the first *zero* of the odd-order Fourier coefficient is at $m = 1$. The index of Fourier coefficient k corresponding to the roots is given by

$$k = T_P/T_R$$

Figure 27 shows the case $T_R/T_P = 0.1$. The waveforms shown in the figure are the sum of the Fourier components up to $k = 3, 5$ and $10 (= T_P/T_R)$. If the first 10 harmonics are summed, the waveform approximation is quite good. Up to the first five harmonics are acceptable for rough estimates. Summing only the first three terms is too crude for any analysis. Generally if Fourier components up to frequency $(1/T_R)$ (Hz) are included, the analysis is accurate. This conclusion is independent of the fundamental period T_P and provides ideas about how to correlate frequency domain and time domain analyses.

Before concluding this section we note the following. In integrated circuits it is sometimes difficult to distinguish active components from passive interconnects from the circuit layout. For instance, is a long polysilicon gate of a FET part of the FET or a part of the wiring? From electrical characterization the difference between wiring and active devices like FETs is clear, due to the following essential differences. The effects of an interconnect on electrical response can be reduced to nothing if the length of the interconnect is reduced. If an interconnect has series resistance $R_I\,(ohms/cm)$ and parallel capacitance $C_I\,(pF/cm)$, the *unloaded* delay of a signal through the interconnect is approximately $R_I C_I l^2$, where l is the length of the wire. If the desired maximum delay is $T_{\max}$, l should be reduced below

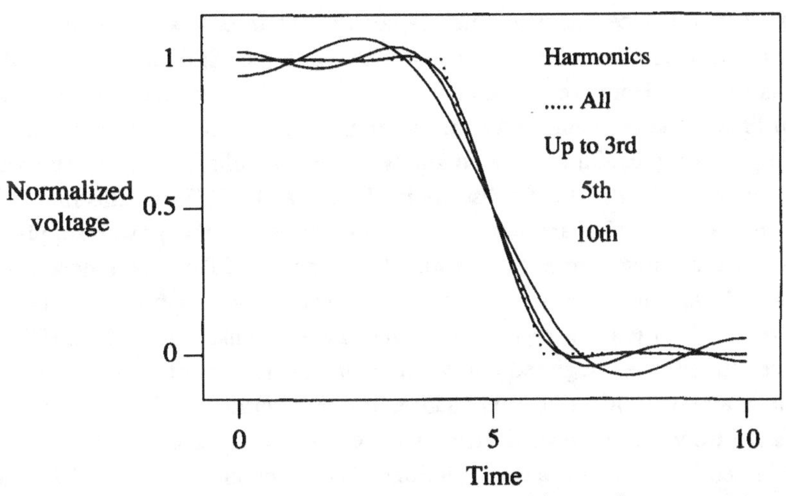

Figure 27 Fourier component and waveform approximation

$\sqrt{T_{max}/(R_l C_l)}$. Theoretically it is always possible to reduce l below the limit by using the proper circuit layout. The delay of a three-terminal active device is different. A certain minimum delay must be tolerated every time an active device exists in the signal path. We make the following general observation. In an integrated circuit technology, the minimum dimension of an active device in the direction parallel to the current is determined by the processing technology. The only adjustable device dimension is perpendicular to the direction of current (usually called device size). If a device has transconductance $g_m(mho/cm)$ and parasitic capacitance $C_D(pF/cm)$ and size $W\,(cm)$, the intrinsic device delay is given by $C_D \cdot W/g_m \cdot W = C_D/g_m$, which is independent of W. Device delay has an absolute minimum because scaling in the direction of current flow is not allowed by reasons of technology integrity.

High-temperature superconductors are now considered the future material for interconnect. This new development poses a question: which of capacitance, resistance, or inductance are the *essential* parasitics of interconnect? If a signal path consists of conductors having nonzero size, they add inductance and capacitance that create the signal delay. This signal delay is required from the propagation velocity of light. There is a fundamental reason that inductance and capacitance parasitics exist. Resistance, however, can be removed if superconductors are used. From this viewpoint, inductance and capacitance are the *essential*, or more fundamental, parasitics. Essential parasitics cannot be reduced to zero. As an example, if a FET has nonzero size, it has nonzero parasitic capacitance, that limits its minimum delay.

1.10 CMOS Circuits and Logic Gates

1.10.1 CMOS Gate Circuits

Integrated circuits are active device intensive: they typically contain only a small number of significant parasitic components. We summarize the structure of the most fundamental CMOS digital circuits in this section. Here we consider the most general interconnection of active devices. NFETs and PFETs can be connected entirely arbitrarily, but such is not done in real circuits, especially in practically useful circuits on silicon chips. In bulk CMOS, the substrate connections of one type of FET (PFETs if substrate is N-type, and NFETs if substrate is P-type) are common, and all the substrate terminals are connected to the power supply or ground, respectively. The substrate connection of the other type of FET can be independent, but they usually share tubs and are usually connected to the other power supply level in order to save silicon area required for extra wiring. Source and drain terminals of a MOSFET can be identified only when the FET is integrated into a circuit and operating voltages are applied. In some circuits, source and drain identification changes during operation. MOSFETs having rectangular channels are truly symmetrical devices with respect to the source and the drain, and the FET can be biased in either of the two polarities. Characteristics of regular FETs for the two polarities are identical.

In digital CMOS circuits, FETs are used as *switches* that connect or disconnect two circuit nodes, and the state of a switch is controlled by the FET gate to source voltage. The absence of an *absolute* identification of the source and drain is compatible with application of the FET as switch. There is a significant difference between a FET or bipolar transistor and a toggle switch or electromagnetic relay. This difference originates from the gate voltage of the FET being referenced to its source voltage. Electromagnetic relays do not have this kind of reference: the input current to the relay winding can be at any voltage level. FETs and bipolar transistors are three-terminal switching devices, and relays and optoelectronic switches (combination of a phototransistor and a light-emitting diode) are four-terminal switches. In semiconductor logic circuits, a voltage that controls one switching device is generated by other switching devices (exactly speaking, from the output terminal of *other* devices relative to a common reference voltage level like ground). Then the maximum control voltage available is V_{DD}, the power supply voltage. Since a FET is a three-terminal device, the conductance of a FET depends on the voltage difference between the control input of the switch (gate) and the reference voltage within the switch (source). The reference within the switch, and the reference of the control voltage are therefore not common. This results in the following. If single FET *switch* is used, as shown in Fig. 28(a), PFET MP1 is able to charge capacitor C_1 up to V_{DD}, but NFET MN2 is able to charge capacitor C_2 up to only $V_{DD} - V_{THN}$, where V_{THN} is the threshold voltage of NFET including the back-bias effect. Suppose that capacitance C_2 was originally discharged. NFET MN2 (whose gate voltage is V_{DD}) charges capacitor C_2 initially, but as voltage V_2 approaches $V_{DD} - V_{THN}$ the NFET current decreases to zero. Further charging is impossible. If V_2 is originally higher than $V_{DD} - V_{THN}$, NFET MN2 is turned off, and V_2 remains unchanged (we neglect leakage currents). Similarly node voltage V_3 can be driven to ground by NFET MN3, but node V_4 can be driven down only to V_{THP}, where V_{THP} is the PFET threshold voltage including the back-bias effect. Because of this problem, a CMOS circuit that is equivalent to an on-off switch over the range from ground to V_{DD} is not a single FET but is a parallel combination of a PFET and an NFET, as shown in Fig. 28(b). The gates of the NFET and PFET are driven by complementary digital control signals. The combination is called a transmission gate. This circuit is able to connect with low impedance any two nodes whose voltages V_A, V_B are within in the range $0 \leq V_A, V_B \leq V_{DD}$.

A transmission gate switch is integrated into various CMOS gates as shown by the following examples. In the following we consider CMOS gates only; any gate that carries

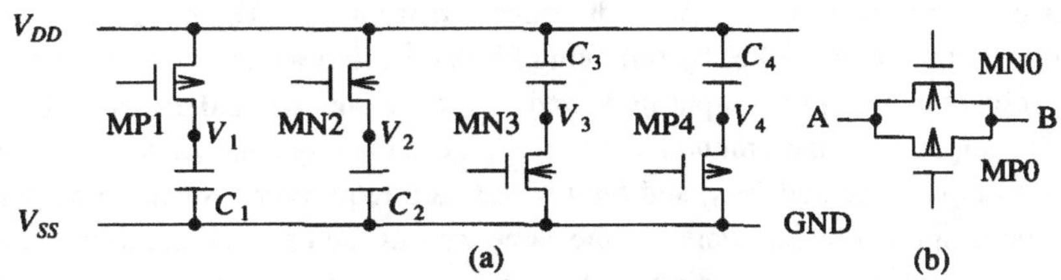

Figure 28 FETs as switches

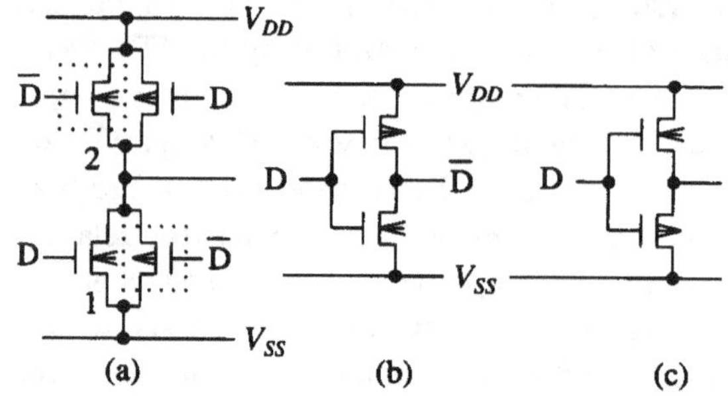

Figure 29 CMOS inverter

standby (DC) current like pseudo NMOS gate is not included. A CMOS inverter can be constructed by connecting two switches as shown in Fig. 29(a). If input signal D is high and its complement D̄ is low, switch 1 connects the output node to ground, and the second switch 2 turns off. If input signal D is low and D̄ is high, however, switch 2 connects the output node to V_{DD} and switch 1 turns off. In this circuit, signal D and its logical complement signal D̄ are both required, and there is a question of how the complement is generated. However, this question is not essential to the study of the most fundamental circuit connectivity. The inverter can be simplified by removing the two FETs enclosed in the dotted boxes. The simplified circuit shown in Fig. 29(b) operates exactly in the same way as the original transmission gate circuit. This circuit generates signal $\bar{D}$ from D. If FETs that are not in the dotted boxes are removed, as shown in Fig. 29(c), we get noninverting symmetrical source-follower circuit. This is not a desirable digital circuit. As for the rules of simplification, NFETs between the output node and V_{DD}, and PFETs between the output node and V_{SS} are removed, but there are cases where the rules lead to confusion (Schmitt trigger circuit, Section 4.17).

A CMOS NAND3 gate is constructed as shown in Fig. 30(a). The output node of the gate is connected to ground only if all the three input signals A, B, and C are high (and their complements are low). Then all three series-connected switches are on, and all three switches that connect the output node to V_{DD} are off. The output node is connected to V_{SS} and is disconnected from V_{DD}. If any one or more of the three input signals A, B, or C are low (and their respective complements are high), however, at least one of the series-connected switches is off, and at least one of the three switches that connect the output node to V_{DD} is on. The output node is connected to V_{DD} and is disconnected from V_{SS}. The circuit of Fig. 30(a) can be simplified to that of Fig. 30(b) by removing FETs enclosed within the dotted boxes. All the NFETs that exist between the output node and V_{DD} are removed, and all the PFETs that exist between the output node and ground are removed. As a consequence, only PFETs are required between the output node and V_{DD}, and only NFETs are required between the output node and ground. The simplified circuit works in the same way as the original circuit as far as the output node is concerned. The two circuits are different, however, if the two internal nodes, X and Y, are taken into consideration. What sort of logic variables are generated at nodes X and

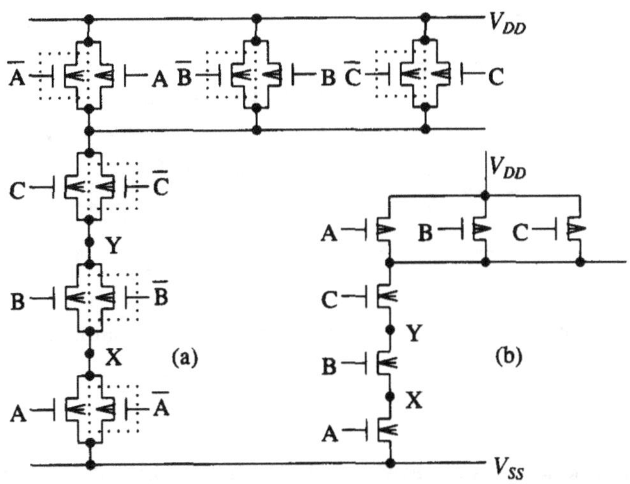

Figure 30 CMOS static gate

Y is an interesting question. Node X is high if and only if signals B and C are both high at least sometime after signal A made the last high-to-low transition. Node X is low if and only if signal A is high at least sometime after either signal B or C made the last high-to-low transition. These nodes can be tristated, and therefore the variable represented by the node retains memory of the past. We note that nodes X and Y of the simplified circuit can be pulled up only to $V_{DD} - V_{THN}$. Nodes in the *simplified* gate that cannot be pulled up to V_{DD} or that cannot be pulled down to V_{SS} do not represent a Boolean combination of the *present* values of the input logic variables. This is consistent in that the node voltages do not settle at the CMOS logic levels in steady states. In the simplified CMOS gate, the reverse also holds: if a gate is designed such that an internal node takes only 0 or V_{DD} in steady state, the node represents a Boolean combination of the present input variables. This is a multiple output CMOS gate. A CMOS gate that has multiple outputs must satisfy a few extra requirements. An example of such a circuit is given in Section 3.3.

A CMOS dynamic inverter circuit is shown in Fig. 31(a). The dynamic inverter circuit of Fig. 31(b) is simplified from the complete circuit shown in Fig. 31(a) by removing FETs in the dotted boxes. The circuit of Fig. 31(b) works as follows. First, logic signal D is disabled to default low logic level. Clock CK pulls down and charges up the load capacitance of the output node. After the capacitance is *precharged*, CK goes down. Signal D is then enabled. If signal D stays low, the output node stays high, and it is high impedance. If signal D makes a low-to-high transition, the output node is pulled down. The complement of the input signal is created at the output.

In CMOS technology of about 1 μm feature size, an NFET is about twice as conductive as a PFET of the same dimensions. Suppose that PFETs and NFETs in a CMOS gate have the same dimensions. Then a chain of series-connected PFETs of a CMOS multi-input static NOR

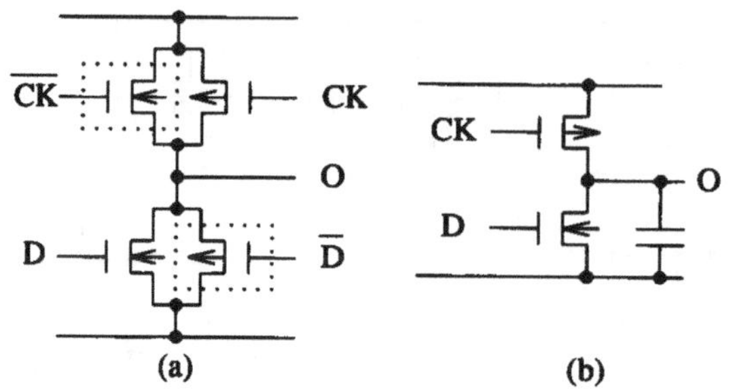

Figure 31 CMOS dynamic inverters and dynamic gates

gate pulls the output node up very slowly, and therefore NOR gates having many inputs are slow in pull-up transition. Dynamic NOR gates require clocking, but they are significantly faster than static NOR gates.

We need to clarify how CMOS static circuits and CMOS dynamic circuits are structurally related. Previously we introduced conventional CMOS static circuits as simplified versions of the more *complete* static circuits built from transmission gates rather than from discrete FETs as the switching elements. Following the same line of thought, we show that CMOS dynamic circuits are simplified versions of conventional CMOS static circuits. Figure 32(a) shows cascaded conventional CMOS static NAND2, INV, NAND3, NOR2, and INV gates. Cascaded gates consist of FET blocks A, B, C, ..., I, and J. The circuit is quite complex. The circuit can be simplified while maintaining the same functionality. By removing some FET blocks and replacing them by precharge FETs. In Fig. 32(b), PFET blocks of every other gate (blocks B, F and J) are removed and replaced by single-precharge PFETs. In this circuit, dynamic gates and static gates are cascaded alternately. This circuit is called Domino CMOS. The conventional Domino CMOS uses only inverters for static gates [16]. This restricted version of the Domino CMOS logic circuit was established as a reliable CMOS logic circuit family, and it is widely used to build microprocessor datapath logic. Even greater simplification of the circuit of Fig. 32(a) can be made by removing PFET blocks D and H, and NFET blocks A, E, and I. PFET blocks are replaced by single-precharge PFETs and NFET blocks are replaced by single-predischarge NFETs, as shown in Fig. 32(c). This logic circuit is called NORA (No-Race) dynamic logic circuit [17].

This simplification scheme of the CMOS static logic circuit by substitution of precharge and predischarge FETs can be generalized. There are many varieties of CMOS dynamic logic circuits depending on how FET block substitutions are made. If a single pair of clock signals (CK and $\overline{CK}$) is to be used everywhere, the rules that must be followed are that between two NFET dynamic gates an odd number of inverting static gates must exist, and between NFET dynamic gate and PFET dynamic gate (and vice versa) an even number of inverting static gates (including zero, or no gate) must exist. Domino CMOS and NORA logic circuits are only two special examples of the large number of varieties of CMOS static-dynamic logic circuits.

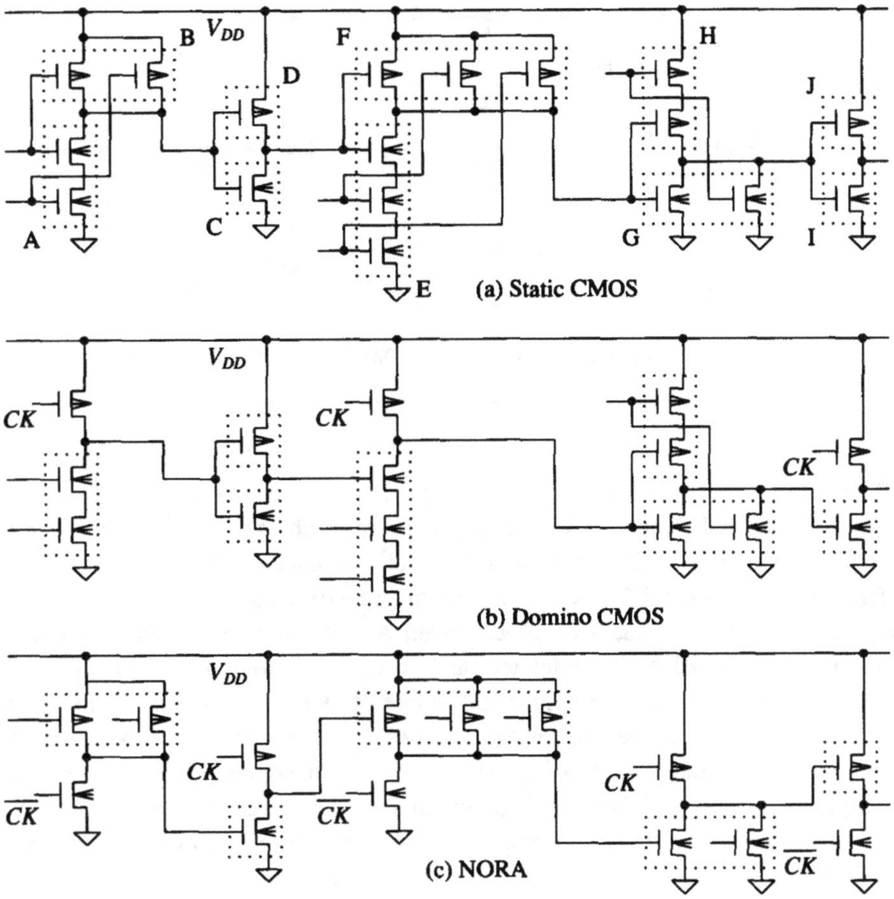

Figure 32 CMOS logic gates

Figures 32(b) and (c) do not include logic gates that are at the beginning of the gate chain. In order to precharge or predischarge the dynamic nodes completely, input signals to the NFET dynamic gates must be low during precharge, and input signals to the PFET dynamic gates must be high during predischarge. The *default* initial logic levels can be set all through the gate chain, if all the first-stage dynamic gates of the gate chain have an extra clocked FET in series with the FETs that accept the input data. In the case of an NFET logic gate, the special NFET is called a ground switch. The gate circuit shown in Fig. 33(a) has a ground switch and precharge PFET, both driven by clock CK. If the first stage is a PFET gate, the complement of the clock, $\overline{\text{CK}}$, drives a predischarge NFET and the PFET that is in series with the PFETs that accept the input signals. The special NFET and PFET are shown in the dotted boxes of Fig. 33(a) and (b).

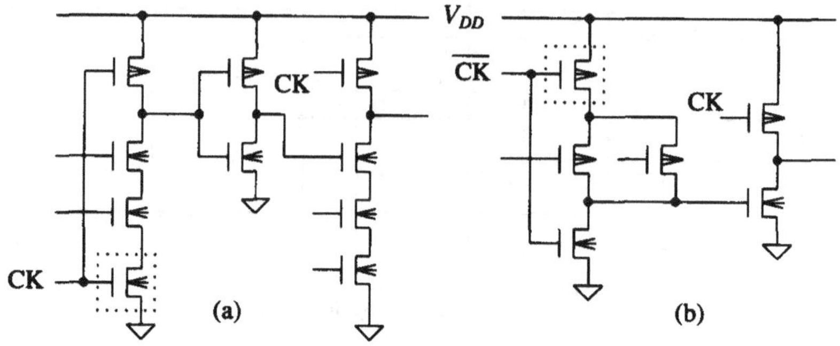

Figure 33 First stage of Domino CMOS and NORA gate chain

The CMOS static logic circuit shown in Fig. 32(a) can be simplified by removing all the PFET blocks B, D, F, H, and J, and replacing them by precharge PFETs as shown in Fig. 34. This circuit is a variation of CMOS dynamic logic circuit, but cascaded stages must be clocked by different clocks, CK1, CK2, CK3, ..., . CK2 must be delayed relative to CK1 long enough to complete pull-down of node 1 by NFET chain A. CK3 must be delayed to CK2 so that pullbown by NFET block C is completed, and so on. This logic circuit requires a ground switch for every stage to guarantee secure precharge of dynamic nodes before logic operation begins. The circuit is equivalent to multiple cascaded CMOS dynamic PLAs. The logic circuit schemes of Figs. 32(b), (c) and Fig. 34 can be intermixed, and varieties of logic circuits that are hybrids of static and dynamic logic circuits can be designed. Static and dynamic circuits are not independent circuit concepts: they are the two extrema of a wide spectrum of CMOS logic circuits.

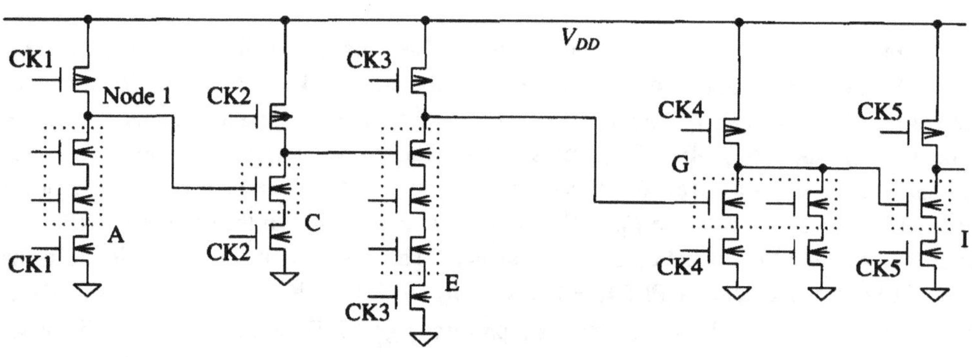

Figure 34 Generalized cascaded CMOS PLA

1.10.2 Characteristics of CMOS Gates and Equivalent Circuits

A logic gate circuit derives two basic characteristics from semiconductor devices used to build the gate. They are (1) gain in the *active* range of the input voltage and (2) *nonlinearity* in the extrema range. The gain of a CMOS inverter shown in Fig. 35(a) at the active input voltage range has the effect of pushing the low logic-level away from the high logic-level. Nonlinearity has the effect of limiting the difference between the two logic levels. Nonlinearity of gate circuit originates from the mechanism that the output voltage of a CMOS gate is unable to exceed the range set by the power supply voltage. This is because a FET (or its paired combination) is essentially an on-off switch. In order to reflect this fundamental requirement in the circuit theory, a FET cannot be modeled by a simple current generator as in small-signal circuits. An extra feature must be added: the current generator must collapse if the terminal voltage is zero.

The gain of a CMOS gate, G, is defined at the highest-gain bias point (usually close to $V_0 \approx V_1 \approx V_{DD}/2$) of the inverter as

$$G = \left(\frac{\partial V_0}{\partial V_1}\right)_{MAX} \tag{36}$$

where V_0 is the output voltage, V_1 is the input voltage, and the partial derivative is taken at the highest gain point. If the transconductance of the FETs at the bias point is g_{mP} (PFET) and g_{mN} (NFET) and if the drain conductances are r_{DP} (PFET) and r_{DN} (NFET), respectively,

$$G = -r_D g_m \quad \text{where} \quad g_m = g_{mP} + g_{mN} \quad \text{and} \quad \frac{1}{r_D} = \frac{1}{r_{DP}} + \frac{1}{r_{DN}}.$$

Gain (negative) remains constant approximately up to frequency f_B and then decreases with increasing frequency. Gain reaches unity at frequency f_U. Frequencies f_B and f_U are given by

$$f_B = 1/2\pi C_L r_D \qquad f_U = g_m/2\pi C_L \qquad f_U = |G| \cdot f_B \tag{37}$$

where C_L is the load capacitance of the inverter, which includes the parasitic capacitances of the FETs themselves. Equation (37) means that the amplifier becomes insensitive to signal

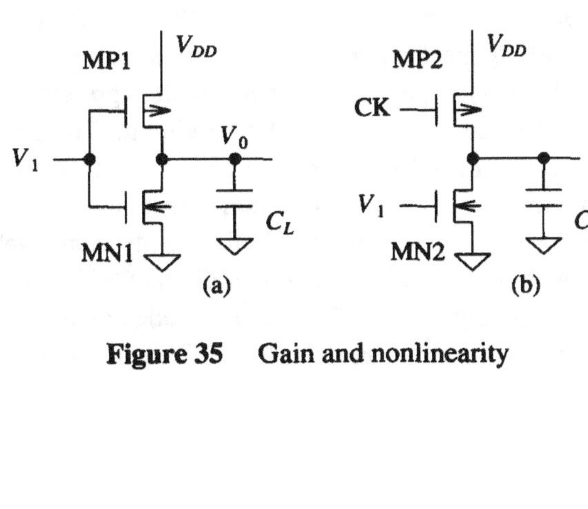

Figure 35 Gain and nonlinearity

frequencies above f_B, which depends on the gain. If the gain is higher, f_B is lower. In digital circuits, we consider nonlinear, large amplitude operation. The gain is interpreted as the average gain over the range of signal swing. Digital gate generates output signal that has the same amplitude as the input signal, and in this case $|G| = 1$. This is the fastest operation as an amplifier. Then we have $f_B = f_U$. The nonlinear amplifier does not respond to an input signal change until at least delay time t_D, which is of the order of one period of sinusoid at frequency f_U. Suppose for the moment that the load capacitance of an amplifier consists entirely of the capacitance of the FET itself (amplifier is not loaded). Then C_L and g_m are both proportional to the width of the device used to build the amplifier. Gate delay time t_D is then independent of the device size, and this is the lowest possible delay. This is the most significant difference between passive interconnects and active devices, as we discussed in the previous section.

The dynamic inverter shown in Fig. 35(b) is first precharged by pulling clock CK down. During precharge, the gate of MN2 is held low. After precharge is over, clock CK is pulled up, and the input signal is turned on. If the input signal goes high, the gate of MN2 makes a low-to-high transition. The output node is pulled down, thereby executing logic inversion. *Gain G* of the dynamic inverter shown in Fig. 35(b) is determined from the following consideration. If the gate of NFET MN2 is biased to voltage V_1 above ground, MN2 carries current $I_N = B(V_1 - V_{TH})$. We use the collapsible current generator FET model. The charge in capacitor C_L is drained by this current, and after time t the output node voltage drops by $B(V_1 - V_{TH})t/C_L$. Then gain G is given by

$$G = \left(\frac{\partial V_0}{\partial V_1}\right) = -\frac{Bt}{C_L}. \tag{38}$$

Gain G increases with elapsed time. This definition of dynamic gate gain may seem strange, but it is useful in many theoretical analysis, including noise analysis. The time-dependent gain of Eq. (38) shows the mechanism of amplification very clearly: gain is created by accumulating effects caused by the input voltage for a long time. Furthermore, to get *unity* gain (required for regeneration of digital signal at the output node), time $t_D = C_L/B$ is necessary. A static inverter requires this much delay also. The difference in delay time of dynamic and static gates is not due to any fundamental difference in the gate delay mechanism: rather, dynamic gates have significantly less load capacitance than static gates, because the circuit is simpler.

Analyses of circuits on silicon chips are carried out using equivalent circuits. Equivalent circuits are physical models used in circuit research. As such, the equivalent circuit should represent the essential working mechanisms of the circuit accurately, and it should be simple enough to be manageable analytically. Equivalent circuit representation is never complete since various approximations are made to simplify the circuit. The simplicity requirement is fundamental since we need to understand the circuit operation in terms of conventional circuit theory. If we investigate the nature of approximations made, we will arrive at the conclusion that several different equivalent circuits are required to study completely the properties of a single circuit on a silicon chip. Equivalent circuits of digital CMOS circuits are subject to the following two restrictions: (1) digital CMOS circuits include nonlinear active devices (MOSFETs) whose models are never simple, and (2) digital CMOS circuits include essentially no components other than MOSFETs. If other components do exist, either they are required to improve circuit performance or they cannot be reduced beyond a certain minimum by circuit layout constraints (especially in interconnect parasitics).

The complexity of FET models has the following effects. The simple FET models that have been presented in the previous sections give MOSFET DC characteristics only. The models are obviously incomplete. AC characteristics of MOSFETs can be derived, in principle, by including the internal structure of the FET channel, as was done in Section 1.5. Unfortunately the model developed in this way is so complex that it is unmanageable. If we try to circumvent the physically sound approach and still wish to arrive at a consistent dynamic model of a FET, we need the following. If we require a device model that is complete for both DC and AC operations of FET, the model must satisfy the consistency requirement imposed from the fundamental law of physics. This is the requirement of over-all device charge neutrality [18]. Figure 36(a) shows a cross-sectional view of an N-channel MOSFET. The ± signs indicate charges that exist at different parts of the device. The charges are assigned to the four terminals, drain (Q_D), gate (Q_G), source (Q_S), and substrate (Q_B). The sum of all the charges must be zero, or

$$Q_D + Q_G + Q_S + Q_B = 0. \tag{39}$$

The charge neutrality condition (or charge conservation) means physically the following. If the condition is violated, the device has net positive or negative charge. Then there must be charge of the opposite polarity somewhere else. The positive and negative charges create electric field outside the device that carries electrostatic energy. Then the operation of the circuit that includes the device would be controlled by the energy that exists *outside* the circuit. This is a violation of an assumption always adhered to, namely, *lumped* circuit representation. If such an external electric field should exist, the field sets in motion the current carriers that exist between the locations of the positive and the negative charges, and the resulting currents modify the charge distribution such that over-all charge neutrality is restored to within about the dielectric relaxation time of the substrate semiconductor material. Since the dielectric relaxation time is 1 picosecond or less, the charge neutrality of the device, Eq. (39), is restored

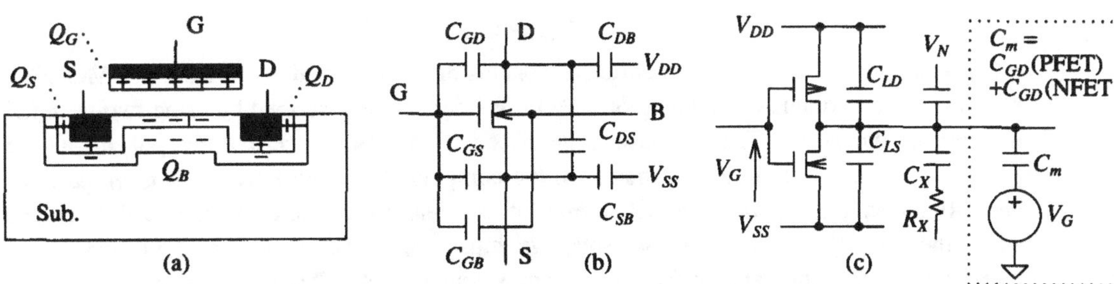

Figure 36 Equivalent circuit

to well within the switching delay time of the CMOS circuit, or less than 10 picoseconds. This is certainly the case at present.

Since the charge neutrality condition is fundamental, the theory of MOSFETs should be built using the four charges rather than the four terminal voltages V_D, V_G, V_S, and V_B as the variables. By using the four charges the terminal voltages are written as

$$V_D = V_D(Q_D, Q_G, Q_S, Q_B) \quad V_G = V_G(Q_D, Q_G, Q_S, Q_B)$$

$$V_S = V_S(Q_D, Q_G, Q_S, Q_B) \quad V_B = V_B(Q_D, Q_G, Q_S, Q_B).$$

If charges fluctuate centered around their respective DC values

$$Q_D \to Q_D(0) + \Delta Q_D \quad Q_G \to Q_G(0) + \Delta Q_G \quad Q_S \to Q_S(0) + \Delta Q_S \quad Q_B \to Q_B(0) + \Delta Q_B$$

the current that flows into the MOSFET is given by

$$I_D = \frac{\Delta Q_D}{\Delta t} \quad I_G = \frac{\Delta Q_G}{\Delta t} \quad I_S = \frac{\Delta Q_S}{\Delta t} \quad I_B = \frac{\Delta Q_B}{\Delta t} \quad \text{where } I_D + I_G + I_S + I_B = 0$$

and where Δt is the time required for the changes to take place. We have

$$\Delta V_D = (\partial V_D/\partial Q_D)\Delta Q_D + (\partial V_D/\partial Q_G)\Delta Q_G + (\partial V_D/\partial Q_S)\Delta Q_S + (\partial V_D/\partial Q_B)\Delta Q_B \tag{40}$$

$$\cdots\cdots\cdots\cdot$$

$$\Delta V_B = (\partial V_B/\partial Q_D)\Delta Q_D + (\partial V_B/\partial Q_G)\Delta Q_G + (\partial V_B/\partial Q_S)\Delta Q_S + (\partial V_B/\partial Q_B)\Delta Q_B.$$

Let us consider interaction between the gate and the drain terminals. If all but the drain terminal are AC-grounded ($\Delta V_D \neq 0, \Delta V_G = \Delta V_S = \Delta V_B = 0$), ΔV_D versus ΔQ_G is found from Eq. (40) as

$$\Delta Q_G = C_{GD}\Delta V_D \quad \text{or} \quad I_G = C_{GD}\frac{\Delta V_D}{\Delta t}$$

where C_{GD} is a coefficient having dimension of capacitance. If all but the gate terminal are AC-grounded ($\Delta V_G \neq 0, \Delta V_D = \Delta V_S = \Delta V_B = 0$),

$$\Delta Q_D = C_{DG}\Delta V_G \quad \text{or} \quad I_D = C_{DG}\frac{\Delta V_G}{\Delta t}$$

where C_{DG} is a similar coefficient. If in the FET model the two terminals (gate G and drain D) were equivalently connected by conventional *two-terminal* capacitance

$$C_{DG} = C_{GD}$$

must be satisfied, but this quite restrictive conclusion cannot be derived from the MOSFET physics. Gate to drain AC coupling of a MOSFET cannot be described by using two-terminal capacitance parasitics. The response is described by a capacitance-like component, whose properties are quite unfamiliar. Above all, the component has more than two terminals. To simplify the equivalent circuit model, to make circuit analysis feasible, and above all to understand the circuit's working mechanisms, we must *approximate* the real MOSFET using two-terminal parasitic capacitances in the equivalent circuit. This is the most fundamental approximation involved in the equivalent circuit. The consequence is that a circuit on a silicon chip must be represented by different equivalent circuits for different problems.

If the AC response of a MOSFET device is represented using two-terminal components only, how the components are connected and what the component values should be, must be determined to approximate the device operation in the particular circuit. Capacitance assignment must be carried out relying on judgment based on the MOSFET physics and circuit operational mechanisms. As to the connection of the capacitance parasitics, the requirement that the DC characteristics of the device should be represented by a single controlled current generator that does not have any internal structure is crucial. Two-terminal capacitance parasitics can be connected only between the four terminals, so that the maximum number of independent capacitance parasitics is $4!/(2!\times 2!)=6$. The following list contains all of them.

1) Gate-source capacitance C_{GS}

Gate-source capacitance consists of bias-independent *overlap capacitance* between the gate and the source diffused island, and the capacitance between the gate and the channel, which depends on the biasing conditions. The theory of Section 1.5 (assuming the gradual-channel, low-field model) shows that the gate-source capacitance of conducting FET at zero drain-source voltage is 1/2 the gate oxide capacitance ($= \varepsilon_{OX} LW/T_{OX}$ in the conventional notation). The capacitance increases with increasing drain-source voltage. The maximum capacitance at current saturation is 2/3 of the gate oxide capacitance.

2) Gate-drain capacitance C_{GD}

Gate-drain capacitance of a conducting FET consists of bias-independent overlap capacitance between the gate and the drain diffused island, and a bias-dependent component originating from the gate-channel capacitance. The latter is 1/2 the gate oxide capacitance if the drain-source voltage is zero. The capacitance decreases with increasing drain-source voltage, and vanishes at saturation.

3) Drain-source capacitance C_{DS}

Since the semiconductor substrate is a conductor that effectively shields the two electrodes, capacitive coupling between the drain and the source is primarily due to coupling through the dielectric that exists above the active part of the FET. If interconnect wires exist, C_{DS} can be primarily due to the capacitance between the wiring. The capacitance is voltage-independent.

4), (5) Drain-substrate, and source-substrate capacitances C_{DB} and C_{SB}

These are primarily reverse-biased PN-junction capacitances of the source/drain diffused islands. The capacitance decreases with increasing reverse-bias voltage. In a conducting FET, channel-substrate capacitance is split to source and drain. Contribution from the interconnect capacitance is also added to the junction capacitance.

6) Gate-substrate capacitance C_{GB}

Direct coupling capacitance between gate and substrate exists only if the substrate is exposed to the gate, or if the conducting channel does not exist. The capacitance C_{GB} is a series connection of the gate oxide capacitance and the depletion layer capacitance underneath the silicon surface. The capacitance decreases with increasing gate voltage until the channel is formed.

An Individual FET has the equivalent circuit shown in Fig. 36(b). If a CMOS digital circuit is assembled from the FETs as shown in Fig. 36(c), the node capacitances consist of the capacitances associated with the FETs and with wiring. For instance, capacitance C_{LS} of Fig. 36(c) includes C_{DS}(NFET) and C_{DB}(NFET) as the most obvious component. C_{LS} includes wiring capacitance, which is laid out on the grounded P-tub, and C_{LD} includes capacitance of wiring laid out on the N-tub. These capacitances may have significant series resistance as shown by capacitance C_X and resistance R_X of Fig. 36(c). The *other* terminal of the two-terminal component is not always at constant ground or V_{DD} potential. If input voltage to the inverter of Fig. 36(c) makes an *instantaneous* low-to-high transition and stabilizes at V_{DD}, and the switching transient at the output node that follows the input transition is studied, C_{LD} includes $C_{GD}(PFET)$ and $C_{GD}(NFET)$ also. If voltage V_G makes a gradual high-to-low transition, the same device capacitances $C_{GD}(PFET)$ and $C_{GD}(NFET)$ are components of the Miller capacitance C_m. The other terminal of the Miller capacitance is connected to input voltage V_G, as shown equivalently in the dotted box of Fig. 36(c). As we observe, component connections and their values in the simplified equivalent circuit depend on how the circuit is excited.

Where the parasitic capacitances should be connected depends on the circuit operation. Figure 37(a) shows that currents that flow in ground-bus section GA–GB, and power-bus section PA–PB depend on how the load capacitances C_{LD} and C_{LS} are assigned. Current during an output pull-down transition is indicated by the arrow marks in Fig. 37(a). Capacitance C_{LD} determines current derived from the power supply, and C_{LS} determines current that closes the local loop. For switching delay analysis, however, load capacitance C_{LD} and C_{LS} can be connected between the output node and any constant voltage, as long as the sum $C_{LD}+C_{LS}$ is held constant. Where to credit Miller capacitance of Fig. 37(b), and how much, are tricky

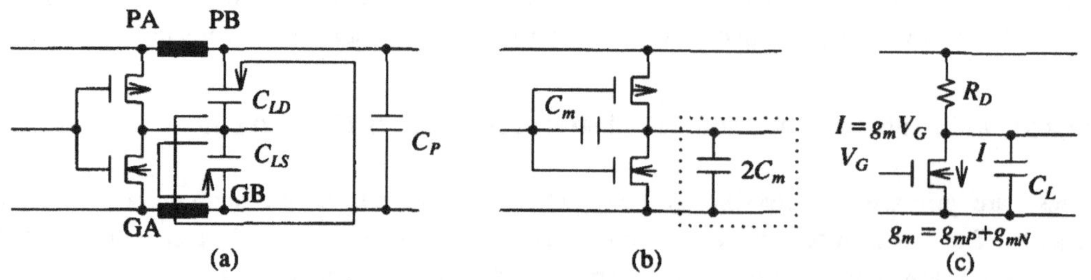

Figure 37 Equivalent circuit depends on the problem studied

problems. In delay analysis, Miller capacitance is approximately twice as effective as load capacitance connected between the output node and any constant voltage. Therefore $2C_m$ added between the output node and ground (dotted box) substitutes for the Miller capacitance. If the reverse signal flow from the output to the input is to be studied, the Miller capacitance must be included at the location, shown in Fig. 37(b).

In certain problems of digital circuits, the equivalent circuit must reflect small-signal performance accurately. For this purpose inclusion of load resistance as shown in Fig. 37(c) is required. The effective load resistance is determined from the Early effect of the FETs. The length of the FET channel decreases with increasing drain voltage, and accordingly the FET saturation current increases with increasing drain voltage. This mechanism creates a nonzero drain conductance in the FET saturation region. The nonzero drain conductance determines the DC gain, which is infinity in the model without it [Eq. (38)]. The Early effect theory will be discussed in Section 4.2. The *linear* equivalent circuit is drawn in simplified form: PFETs and NFETs driving the output node can be represented by a single FET with transconductance equal to the sum of the transconductances of the two FETs at the small-signal operating point, as shown in Fig. 37(c).

1.11 Electrical Phenomena in the Equivalent Linear RC Chain Circuits

Figures 32–34 show that even a simple CMOS logic circuit contains many FETs connected in a complex web, and analysis of the circuit is likewise complex. This complexity due to connectivity must be retained in the theory to understand the *circuit* phenomena that originate from the interconnection of devices. Circuit analysis that is based on realistic FET models like Eqs. (22a) and (22c) becomes so complex that the only practical way to obtain useful results is to simulate the circuit using software like SPICE. Simulation, however, is like experiments: Without a guiding theory simulation results reveal few details about normal circuit operation or circuit failures.

Useful closed-form theories that serve not only as guiding theories but also provide semi-quantitative results can be built using drastically simplified FET models. The FET model of Eqs. (22a) to (22c) is either made more nonlinear like that of Fig. 13 (the collapsible current generator model) or is made more linear, to the equivalent linear resistor representing the conducting channel. The first model is the major subject of this book. Therefore, we give a short review of the second model. In the second model, a conducting FET is represented by a linear equivalent resistor and capacitor, and a nonconducting FET is represented by a linear capacitor only. This linear resistor model allows us to correlate gate circuit structure directly with gate delay time by giving closed-form answers. The linear resistor model requires, however, many fundamental clarifications. Especially we need to prescribe how to determine the equivalent FET resistance. Detailed discussion of this problem, and the assumptions made to introduce equivalent resistance are not required in this book, and the interested reader is referred to the author's book *CMOS digital circuit technology* (chapter 3). Here we mention a basic restriction of the linear equivalent resistor model. The model gives the delay time of a digital signal to the gate's *output node* only. It does this by simply replacing each FET by an *equivalent* resistor that can be determined directly from the number of squares of the FET channel. This model is simple and useful, but it does not give any information about the gate's *internal* nodes [for example node X and Y in Fig. 30]. The theory is not convenient for circuit failure study. The fundamental reason is that the FET conductance is determined by gate-to-source voltage, and not by gate-to-ground voltage. Therefore all the FETs of a single pull-down NFET chain must be considered as a single, long-channel equivalent FET.

We give a short review of the linear resistor model of FETs to show how the model works. Using this model, a CMOS digital circuit can be represented by many segments of simple RC chains (or meshs, in general) circuits. FETs are three-terminal devices. The resistance of the channel depends on the gate-to-source voltage. Since a FET is represented by a *single* linear resistor, the effects of the rate of variation of gate voltage (input voltage slew rate) cannot be included cleanly in the theory in such a way that the physical meanings will be clear. Therefore the model is able to give delay time only when the input node switches rapidly. Implicit in this assumption is that interaction between two cascaded gates cannot be analyzed by using this model. Therefore with this model, CMOS VLSI circuits are split into segments that contain only one driving gate, the capacitive load and the associated interconnects, and signal propagation within each segment are analyzed *separately*. We point out that the simplicity of the linear resistor model is partially a mirage: the model does not have the capability of analyzing strongly interacting multiple gates. Figure 38(a) is an example of a

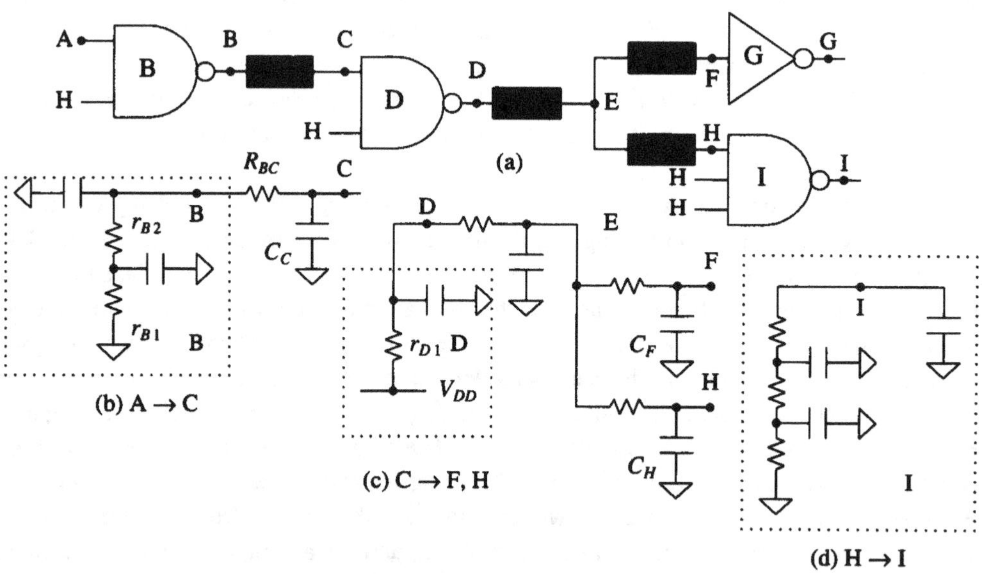

Figure 38 Equivalent circuit of CMOS VLSI chip

chain of logic gates and interconnects, and 38(b)–(d) are the equivalent circuits at different times. Shadowed rectangles in Fig. 38(a) show resistive interconnects and capacitive parasitics. Immediately after signal A makes an *instantaneous* low-to-high transition, the segment of the circuit between nodes A and C is equivalently described by the circuit of Fig. 38(b). The section of the circuit to the right of node C has no representation at this time. The two resistors within the dotted box represent two series connected pull-down NFETs in the NAND2 gate B. The resistor outside the box, R_{BC}, is interconnect resistance that has a different physical nature than r_{B1} and r_{B2}. R_{BC} is a real, linear resistance of the interconnect, and r_{B1} and r_{B2} are *equivalent* resistances of the NFET channels. C_C is the capacitance of the interconnect. In this model, voltage dependence of node capacitance is neglected: capacitances of nodes are average capacitance values over the range of voltages. The model allows inclusion of the effects of complex resistive interconnects that exist, for example, between nodes B and C: This is a convenient feature of the equivalent resistor model. Load capacitance C_C was originally charged to V_{DD}. When the voltage of node C falls to $V_{DD}/2$ by discharging capacitor C_C, the segment of circuit between nodes C to F and H is equivalently described by the circuit shown in Fig. 38(c). Here we have the implicit assumption that the voltage of node C had been V_{DD} until that time, and then it dropped instantly to zero. The equivalent resistance of PFET, r_{D1}, is defined based on the same assumption. Capacitors C_F and C_H are completely discharged at that time. When capacitor C_H is charged to $V_{DD}/2$, gate I is equivalently described by the circuit shown in Fig. 38(d). Again we make an implicit assumption that the node voltage changed instantly from 0 to V_{DD}. Discharge characteristics of each circuit section can be analyzed using the standard linear circuit theory. This method is convenient in gate-level timing simulators.

We showed how a CMOS digital integrated circuit is represented equivalently by a complex web of resistances and capacitances. In the following we summarize electrical phenomena in linear RC circuits and discuss the analysis methods. We present a simple interpretation of complex electrical phenomena in RC networks in the *time* domain, intending to show what happens in RC networks in real time [19]. In order to carry out closed-form analysis, it is convenient not to distinguish distributed and lumped RC networks too rigorously. Many lumped RC networks can be described equivalently by a nonuniform distributed parameter RC network as shown in Fig. 39. In this figure resistance $R(x)$ and capacitance $C(x)$ depend on

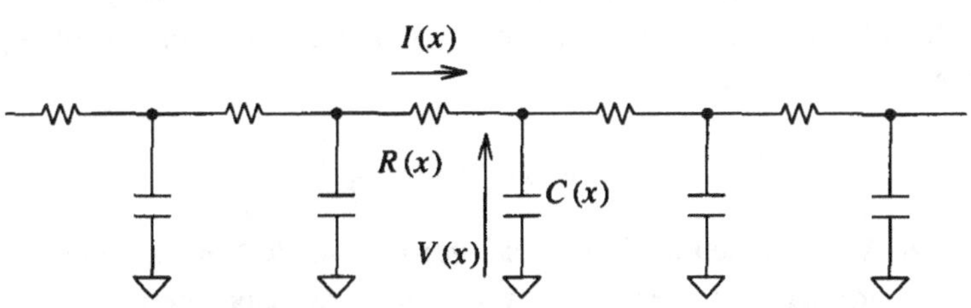

Figure 39 Distributed RC chain circuit

location x. In this continuous RC network model, it is not possible to study the effects of a very abrupt change of resistance and capacitance in a lumped RC network. We study the effects of *gradual* change with x. Location x is a variable introduced for indexing the RC chain, and therefore x need not be a parameter having dimension of length. Voltage developed on a *distributed* node at x, $V(x,t)$, and the current at location x, $I(x,t)$ satisfy (for convenience, parameter dependence of V and I is not shown)

$$\frac{\partial V}{\partial x} = -R(x)I \qquad \frac{\partial I}{\partial x} = -C(x)\frac{\partial V}{\partial t}. \tag{41}$$

The two circuit equations are combined to give

$$\frac{\partial V}{\partial t} = D(x)\frac{\partial^2 V}{\partial x^2} - D(x)\lambda(x)\frac{\partial V}{\partial x} \tag{42}$$

where

$$D(x) = \frac{1}{R(x)C(x)} \quad \text{and} \quad \lambda(x) = \frac{1}{R(x)}\frac{dR(X)}{dx}.$$

Suppose, for the moment, that $D(x)$ and $\lambda(x)$ are constants, D and λ, respectively. We use the following transformation,

$$V(x,t) \to \Psi(\xi,t) \text{ where } x \to \xi = x - vt \text{ and } t \to t.$$

This is a Galileian transformation to coordinate ξ that moves toward the right with velocity v relative to the original coordinate x. Since

$$\frac{\partial \Psi}{\partial \xi} = \frac{\partial V}{\partial x} \qquad \frac{\partial \Psi}{\partial t} = -v\frac{\partial V}{\partial x} + \frac{\partial V}{\partial t}$$

we have

$$\frac{\partial \Psi}{\partial t} = D\frac{\partial^2 \Psi}{\partial \xi^2} - (D\lambda - v)\frac{\partial \Psi}{\partial \xi}.$$

This equation is the simplest if

$$v = D\lambda. \tag{43}$$

Then

$$\frac{\partial \Psi}{\partial t} = D\frac{\partial^2 \Psi}{\partial \xi^2} \tag{44}$$

and this is an ordinary diffusion equation. First we consider this ordinary diffusion equation. We study how the diffusion effect smoothes out the initial, nonuniform potential profile of the RC chain. From Eq. (44) we have

$$\Psi(\xi, t+\Delta t) = \Psi(\xi,t) + D\frac{\partial^2 \Psi(\xi,t)}{\partial \xi^2}\Delta t. \tag{45}$$

In Fig. 40(a), curve A shows the initial potential profile of the RC chain. Since the time rate of change of Ψ is proportional to the second derivative, and the second derivative is the curvature of curve A, the potential profile after time Δt, $\Psi(\xi, t+\Delta t)$, can be drawn as shown by curve B.

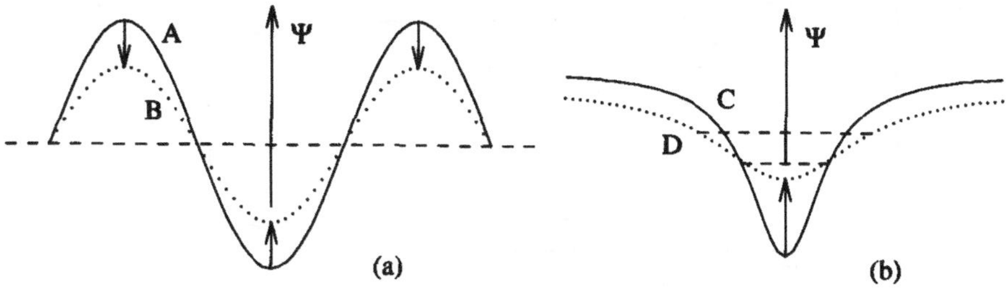

Figure 40 Effects of diffusion mechanism

The amplitude of a periodic nonuniformity decreases, while the shape (or period) remains unchanged. In Fig. 40(b), curve C shows the initial potential profile. Using Eq. (45), we find that curve C evolves to curve D. In this case the diffusion effect broadens narrow, isolated features of potential profile, thereby making the potential profile smooth (in addition to reducing the amplitude). The diffusion effect smoothes out the potential profile by the two mechanisms.

Equations (43) and (44) show that the potential profile in an RC chain circuit is smoothed out by the diffusion effect, to which drift motion of the potential profile is superposed if resistance of the chain is not uniform. The voltages of all the nodes that are connected through resistors tend to the same final voltage. A significant, but less obvious, alternative mechanism of smoothout is that the potential profile *moves* in the direction of increasing resistance of the RC chain. Let us consider a simple example. In Eq. (42), $D(x)$ and $\lambda(x)$ are both constant if $R(x) = R_0 exp(\lambda x)$ and $C(x) = C_0 exp(-\lambda x)$. This is the case of an exponentially scaled chain. Figures 41(a) and (b) show evolution of the potential profile of a discharging RC chain circuit. Figure 41(a) shows potential profiles in an unscaled RC chain ($\lambda = 0$). The chain was originally charged to unit voltage, and the left end of the chain was grounded. The chain discharges following the ordinary *diffusion* process. Figure 41(b) shows the potential profile within an exponentially scaled chain ($\lambda > 0$): the profile moves at about constant velocity to the

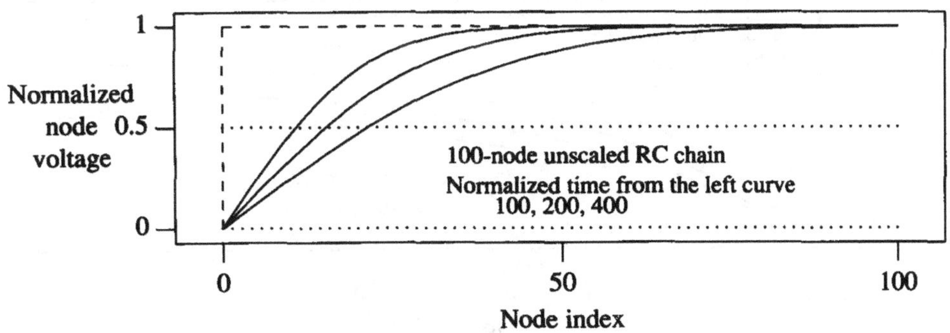

Figure 41(a) Time evolution of potential profile within exponentially scaled chain

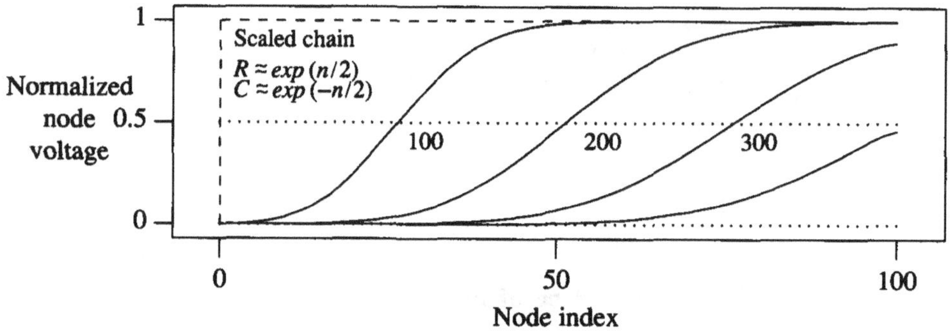

Figure 41(b) Time evolution of potential profile within exponentially scaled chain

---

right. The constant velocity is given by Eq. (43). Both mechanisms achieve smoothout of the potential profile.

Let us consider the second special case, namely, that $R(x) = R(constant)$ and $C(x)$ is x-dependent. We have then $\lambda(x) = 0$, and Eq. (42) becomes

$$\frac{\partial V}{\partial t} = D(x)\frac{\partial^2 V}{\partial x^2}. \tag{46}$$

In this case the diffusion coefficient varies spatially. Let us consider a 25-member RC chain that has $R(x) = 1$ and $C(x) = exp(-0.1x)$. The chain is charged to unit potential, and the left end of the chain is grounded at time $t = 0$. Figure 42 shows numerical solution of Eq. (46). If $C(x)$ were constant, the length of the part of the RC chain discharged after time t would be proportional to $\sqrt{t}$, as we observe from Fig. 41(a). In Fig. 42 the chain discharges faster than that: the half-voltage point on the potential profile moves approximately with constant velocity to the right. Since $D(x)$ is a rapidly increasing function of x, the voltages of the nodes that are close to the right end of the RC chain equalize rapidly. The spatially uniform node voltage of the right half of the chain decreases with time, as the left side of the chain draws current to ground. The discharge mechanism due to a nonuniform diffusion coefficient can be inter-

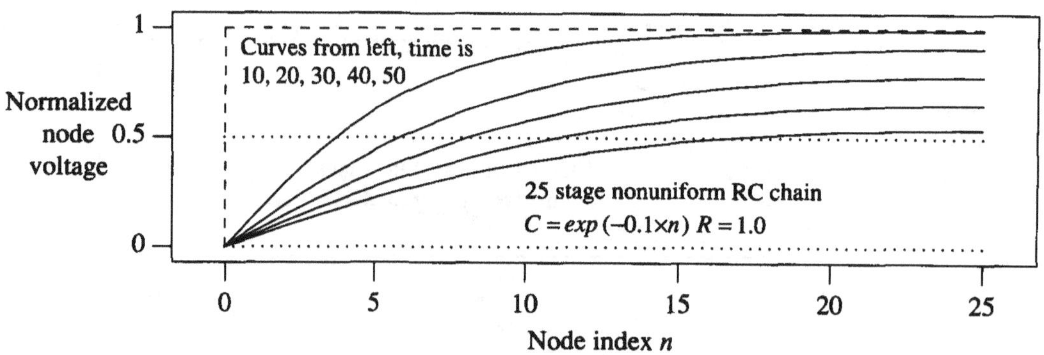

Figure 42 Effects of nonuniform capacitance

preted by assuming that the length of the RC chain were effectively reduced, because the diffusion coefficient $D(x)$ increases rapidly with increasing x. This mode of discharge is less elegant than the drift mechanism, but this mechanism exists more widely in nature (the thermoelectric effect is a direct analog of this effect).

Discharge mechanisms, potential profile drift and nonuniform diffusion coefficient, can be explained further as follows. Let us consider three consecutive circuit nodes at location $x - \Delta x$, x and $x + \Delta x$. Node voltages at the beginning are linearly graded as

$$V(x - \Delta x) = V(x) - \Delta V \quad V(x + \Delta x) = V(x) + \Delta V .$$

If the resistance gradient is positive, current into node x, $I(x)$ is given by

$$I(x) = \frac{V(x+\Delta x) - V(x)}{R'} + \frac{V(x-\Delta x) - V(x)}{R} .$$

Since $R' > R$ we have $I(x) < 0$. Then the voltage of node x decreases relative to the voltages of nodes $x \pm \Delta x$. A positive resistance gradient has the effect of increasing the curvature of the potential profile, as shown in Fig. 41(b). We note that curvature at the left end of the RC chain becomes concave upwards. This is a requirement for a potential profile to detach from the left-side end by creating *virtual* ground *inside* the RC chain circuit, which *moves* to the right. If $R' = R$ (no resistance gradient), this effect is absent, as shown in Fig. 41(a) and 42.

1.12 Elmore's Delay Formula

There exists a simple formula that gives the discharge delay time of a uniformly charged RC chain circuit [20]. Since this formula will be used frequently in CMOS delay analysis, two derivations, one simple and the other complete, will be given here. A third derivation is given in chapter 6 of the author's book, *CMOS digital circuit technology*. Consider a two-stage cascaded RC chain circuit shown in Fig. 43(a). Capacitors C_1 and C_2 are originally charged to V_{DD}. If the left end of the chain is grounded, how long a time (T_D) does it take for the voltage of node 2 to drop to V_{DD}/e? This discharge delay time, T_D, is a function of four parameters R_1, R_2, C_1, and C_2. If R_1 and R_2 are both multiplied by factor k, T_D is multiplied by the same factor. If C_1 and C_2 are both multiplied by factor k, T_D is multiplied by the same factor. Therefore T_D is a bilinear expression of $R's$ and $C's$ as follows:

$$T_D = a_{11} R_1 C_1 + a_{12} R_1 C_2 + a_{21} R_2 C_1 + a_{22} R_2 C_2 .$$

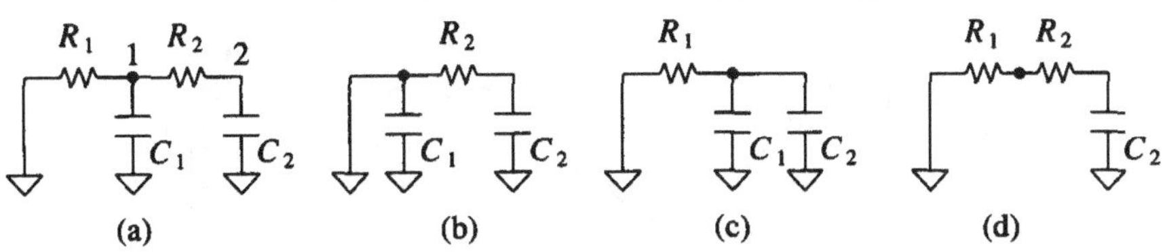

Figure 43 Derivation of Elmore's formula

We consider the limit of $R_1 \to 0$ and $C_1 \to \infty$ [Fig. 43(b)]. Since T_D must remain finite, $a_{21} = 0$. If $R_2 \to 0$, Fig. 43(c) must give $T_D = R_1(C_1 + C_2)$ and therefore $a_{11} = a_{12} = 1$. Similarly if $C_1 \to 0$, Fig. 43(d) gives $T_D = (R_1 + R_2)C_2$ and we have $a_{22} = 1$. We then have

$$T_D = R_1(C_1 + C_2) + R_2C_2 = (R_1 + R_2)C_2 + R_1C_1 . \tag{47}$$

The two special cases $R_2 = 0$ and $C_1 = 0$ indicate that Eq. (47) is a generalization of the ordinary RC delay formula. We cascade an extra RC stage consisting of R_3 and C_3 to the two-stage RC chain of Fig. 43(a) as shown in Fig. 44(a). The first two stages consisting of R_1, R_2, C_1, and C_2 have delay time T_D given by Eq. (47). If these two stages are approximated by a one-stage RC chain circuit consisting of R_X and C_X as shown in Fig. 44(b), we select

$$C_X = T_D/R_X$$

to make the delay of the two circuits equal. Furthermore we require that *DC responses* of the two circuits are the same. Then we must have

$$R_X = R_1 + R_2$$

Then the delay time of the circuit shown in Fig. 44(b) is approximated by repeated use of the formula as

$$T_D = R_X(C_X + C_3) + R_3C_3 = R_1C_1 + (R_1 + R_2)C_2 + (R_1 + R_2 + R_3)C_3$$
$$= R_1(C_1 + C_2 + C_3) + R_2(C_2 + C_3) + R_3C_3$$

The formula can be generalized to an N-stage cascaded RC chain as

$$T_D = \sum_{i=1}^{N} R_i \sum_{j=i}^{N} C_j . \tag{48}$$

We derived Elmore's delay formula by using only the fundamental properties of the chain delay time. We need to understand more about this formula and to generalize the formula by following the original method of derivation used by Elmore. Consider the N-link RC chain circuit shown in Fig. 45. The RC chain has arbitrary R_i and C_i for $1 \le i \le N$. Originally all the capacitors are completely discharged. At time $t = 0$ the left end of the RC chain is pulled up to unit potential (1 volt) within a negligibly short time, and the potential is maintained at that level. The voltage of node N_k (for $1 \le k \le N$), $V_k(t)$, changes as shown by the curve of Fig. 46(a), and its derivative $V'_k(t)$, as shown in Fig. 46(b). We have

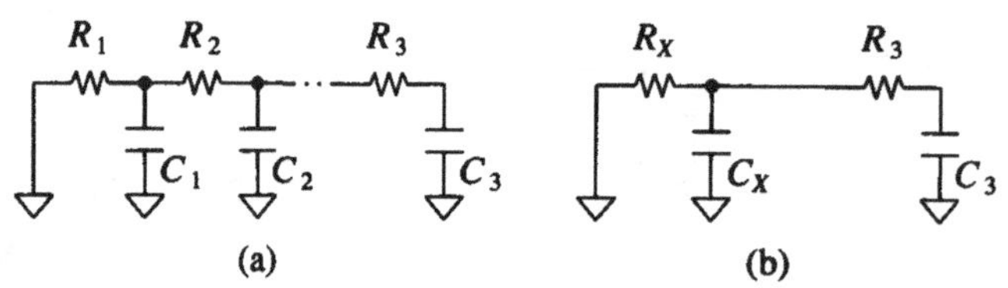

Figure 44 Derivation of Elmore's formula (continued)

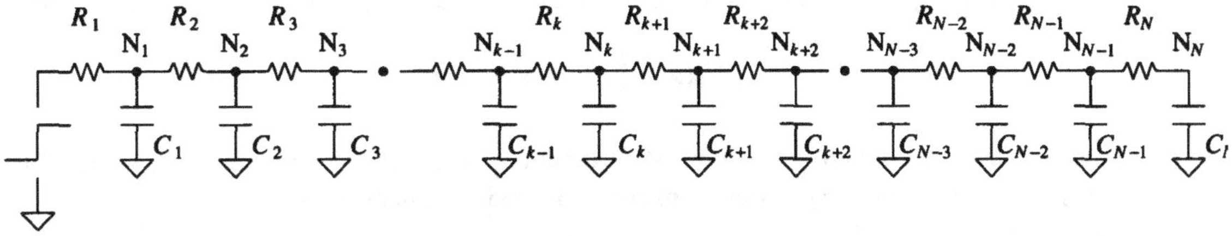

Figure 45 Analysis of arbitrary RC chain circuit

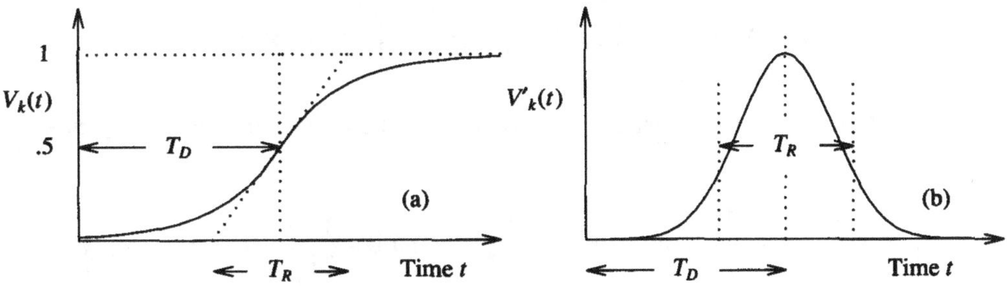

Figure 46 Elmore's definition of delay time

$$V_k(0)=0.0 \quad \lim_{t\to\infty} V_k(t)=1.0.$$

Elmore defined delay time T_D and rise time T_R of node k voltage $V_k(t)$ by the following *average* formula, which uses $V_k'(t)$ as the normalized weight function.

$$T_D=\int_0^\infty tV_k'(t)dt \quad T_R=[const.\int_0^\infty (t-T_D)^2V_k'(t)dt]^{1/2} \tag{49}$$

$$=[const.[\int_0^\infty t^2V_k'(t)dt-T_D{}^2]]^{1/2}$$

where *const.* is a numerical factor that will be determined later, and where

$$\int_0^\infty V_k'(t)dt=V_k(\infty)-V_k(0)=1.0.$$

The meaning of Eqs. (49) is obvious from Fig. 46(b), but there is one important point that calls for our attention. $V_k'(t)$ should tend to zero sufficiently *fast* in the limit of $t\to\infty$ so that major contribution to the integrals of Eqs. (49) originate from the vicinity of the maximum of $V_k'(t)$. If this condition is not satisfied, Eqs. (49) gives delay and rise times that are useless overestimates for digital circuit analysis. Here we assume that $V_k'(t)\to 0$ when $t\to\infty$ fast enough to guarantee rapid convergence of the integrals.

We write Laplace transforms of time-dependent variable $V_k(t)$ $(0 \le k \le N)$ as $V_k(s)$, where

$$V_k(s) = \int_0^\infty exp(-st) V_k(t) dt .$$

Since node 0 is driven by a unit step function $[V_0(t) = 0\ (t<0)$, and $V_0(t) = 1\ (t>0)]$, $V_0(s) = 1/s$. The *response function* from node 0 to node k is defined by

$$G_k(s) = \frac{V_k(s)}{V_0(s)} = sV_k(s) \tag{50}$$

or

$$G_k(s) = sV_k(s) = \int_0^\infty V_k(t) s\, exp(-st) dt = \int_0^\infty V_k'(t) exp(-st) dt . \tag{51}$$

If $V_k'(t)$ tends to zero rapidly when $t \to \infty$ as we assumed, $exp(-st)$ can be expanded into power series and

$$G_k(s) = \sum_{n=0}^{\infty} \frac{(-1)^n}{n!} \int_0^\infty V_k'(t) t^n dt \cdot s^n = 1 - \left[\int_0^\infty V_k'(t) t dt \right] s + \frac{1}{2} \left[\int_0^\infty V_k'(t) t^2 dt \right] s^2 + o(s^3) .$$

The response function $G_k(s)$ is derived using elementary linear circuit theory. $G_k(s)$ is a positive real function of s that may be written as

$$G_k(s) = \frac{N(s)}{\Delta(s)} = 1 - (b_1 - a_1)s + [a_2 - b_2 - b_1(a_1 - b_1)]s^2 + o(s^3)$$

where

$$N(s) = 1 + a_1 s + a_2 s^2 + \dots \quad \Delta(s) = 1 + b_1 s + b_2 s^2 + \dots$$

and where the maximum power of polynomial $N(s)$ is lower than that of polynomial $\Delta(s)$. Once $G_k(s)$ is determined by linear circuit analysis, power series expansion of $G_k(s)$ is elementary.

$$T_D = \int_0^\infty t V_k'(t) dt = b_1 - a_1 \tag{52a}$$

$$T_R = \left[const. (b_1{}^2 - a_1{}^2 + 2a_2 - 2b_2) \right]^{1/2} . \tag{52b}$$

The two formulas are conventionally called the Elmore's formula. Elmore has chosen the constant in the second equation 2π, but for convenience of discussion we leave the value undetermined.

It is necessary to determine the response function $G_k(s)$ of an N-member RC chain using the linear circuit theory. In the case of a four-member chain $(N=4)$, the Laplace transform of node voltages, $V_0(s)$ ..., $V_4(s)$ satisfies the following recurrence formula.

$$V_3(s)=V_4(s)+R_4C_4sV_4(s) \tag{53}$$

$$V_2(s)=V_3(s)+R_3[C_3sV_3(s)+C_4sV_4(s)]$$

$$V_1(s)=V_2(s)+R_2[C_2sV_2(s)+C_3sV_3(s)+C_4sV_4(s)]$$

$$V_0(s)=V_1(s)+R_1[C_1sV_1(s)+C_2sV_2(s)+C_3sV_3(s)+C_4sV_4(s)]\,.$$

From the recurrence formula we have

$$V_1(s)=\frac{1+a_{11}s+a_{12}s^2+\cdots}{1+b_1s+b_2s^2+\cdots}V_0(s)\ \ldots,\quad V_4(s)=\frac{1+a_{41}s+a_{42}s^2+\cdots}{1+b_1s+b_2s^2+\cdots}V_0(s)\,.$$

Coefficients b_1, a_{11}, a_{21}, a_{31}, and a_{41} are the sum of terms of the form (R_iC_j). It is convenient to show all the possible terms of the form (R_iC_j) on a checkerboard, as in Fig. 47(a). Coefficient a_{41} is zero, or the sum of *no* term on the checkerboard. Coefficient a_{31} is the sum of the term indicated by sign A, or $a_{31}=R_4C_4$. Coefficient a_{21} is the sum of the terms indicated by signs A and B, or

$$a_{21}=R_4C_4+R_3C_3+R_3C_4$$

and similarly coefficient a_{11} is sum of the terms A, B, and C or

$$a_{11}=R_4C_4+R_3C_3+R_3C_4+R_2C_2+R_2C_3+R_2C_4$$

Finally b_1 is the sum of all the the terms A, B, C, and D, or

$$b_1=R_4C_4+R_3(C_3+C_4)+R_2(C_2+C_3+C_4)+R_1(C_1+C_2+C_3+C_4)\,.$$

Similarly b_2, a_{12}, a_{22}, a_{32}, and a_{42} are sums of the terms of the form $(R_iC_j)(R_kC_l)$ indicated on the checkerboard of Fig. 48(a). On the checkerboard the sum rule works as follows: $a_{42}=0$, and $a_{32}=0$ and a_{22} is the sum of term A or

$$a_{22}=(R_3C_3)(R_4C_4)$$

a_{12} is the sum over terms A and B as

(a)

	C_1	C_2	C_3	C_4
R_1	D	D	D	D
R_2		C	C	C
R_3			B	B
R_4				A

(b)

	C_1	C_2	C_3	C_4	C_5
R_1	E	E	E	E	E
R_2		D	D	D	D
R_3			C	C	C
R_4				B	B
R_5					A

Figure 47 Linear terms of four- and five-member RC chain

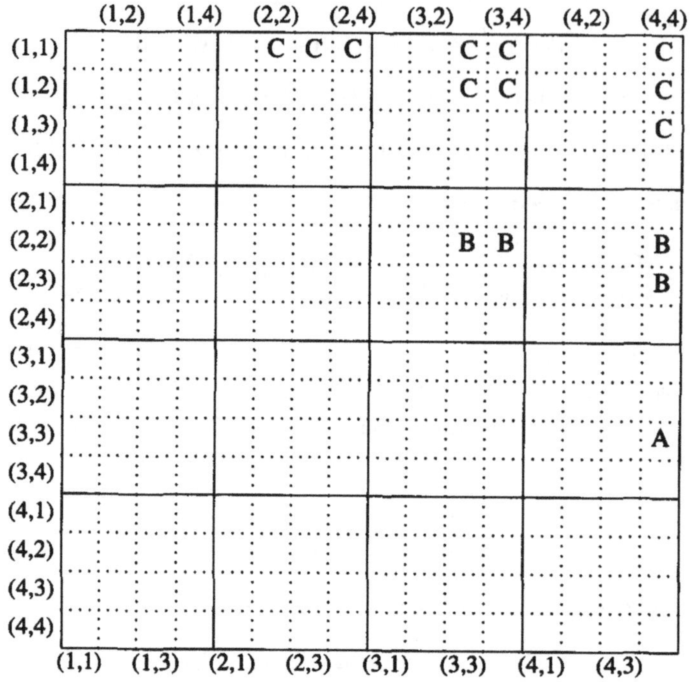

Figure 48(a) Product terms of four-member RC chain response function

$$a_{12} = (R_3C_3)(R_4C_4)$$

$$+(R_2C_3)(R_4C_4) + (R_2C_2)(R_3C_3) + (R_2C_2)(R_3C_4) + (R_2C_2)(R_4C_4)$$

and b_2 is the sum over terms A, B, and C. This rule can be extended to a more complex RC chain. An example of five-member RC chain is shown in Fig. 47(b) and Fig. 48(b).

It is interesting to compare Elmore's formula with numerical results to establish the limit of accuracy. Before doing so, we briefly review the method of simple numerical analysis of RC chain circuits.

1.13 Numerical Analysis of RC Chain Circuit

Simulation of RC chain circuits shown in Fig. 45 can be carried out as follows. The following is an elementary method that is easy to implement in research programs. Nodes (1, 2, ..., N), series resistances R_i, parallel capacitances C_i, node voltages V_i, and the total number of RC chain links N are defined in Fig. 45. The current conservation law at node i is written as

$$\frac{dV_i(t)}{dt} = \frac{1}{C_iR_i} V_{i-1} - \frac{1}{C_i}\left(\frac{1}{R_i} + \frac{1}{R_{i+1}}\right)V_i + \frac{1}{C_iR_{i+1}} V_{i+1} \quad (1 \le i < N). \tag{54}$$

If this equation is differentiated once with respect to time t

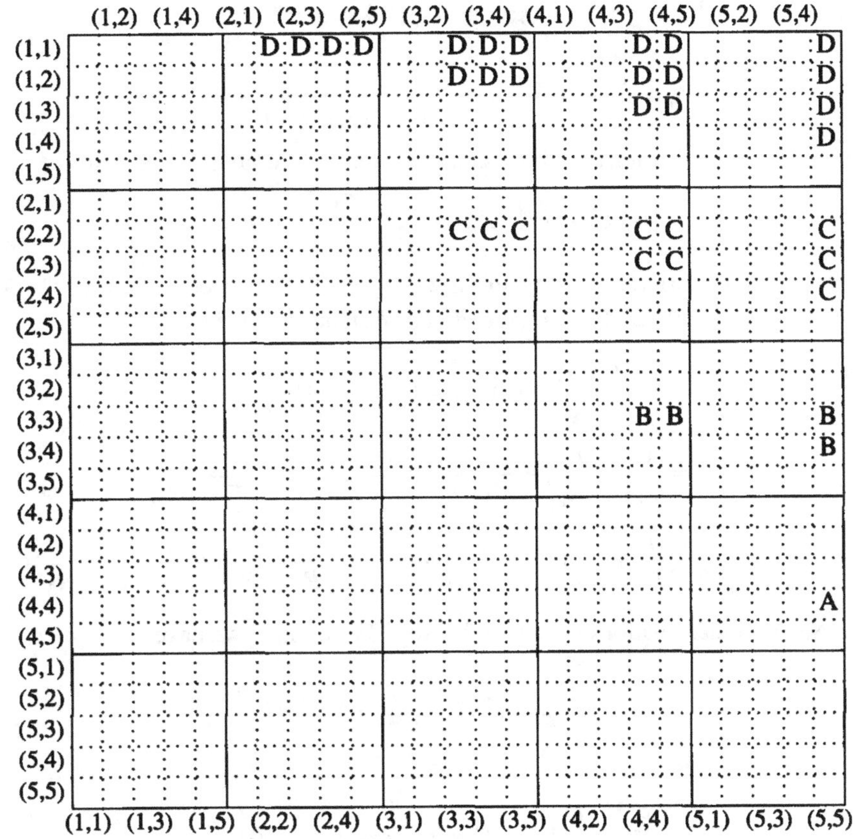

Figure 48(b) Product terms of five-member RC chain response function

---

$$\frac{d^2V_i(t)}{dt^2} = \frac{1}{C_iR_i}\frac{dV_{i-1}}{dt} - \frac{1}{C_i}\left(\frac{1}{R_i} + \frac{1}{R_{i+1}}\right)\frac{dV_i}{dt} + \frac{1}{C_iR_{i+1}}\frac{dV_{i+1}}{dt}.$$

If (dV_{i-1}/dt), (dV_i/dt) and (dV_{i+1}/dt) are substituted from Eq. (54) (with increased or decreased index i) we obtain

$$\frac{d^2V_i(t)}{dt^2} = A_{i-2}V_{i-2} + A_{i-1}V_{i-1} + A_iV_i + A_{i+1}V_{i+1} + A_{i+2}V_{i+2}$$

where

$$A_{i-2} = \frac{1}{C_iC_{i-1}R_iR_{i-1}} \qquad A_{i+2} = \frac{1}{C_iC_{i+1}R_{i+1}R_{i+2}} \tag{55a}$$

$$A_{i-1} = -\frac{1}{C_iR_i}\left[\frac{1}{C_{i-1}}\left(\frac{1}{R_{i-1}} + \frac{1}{R_i}\right) + \frac{1}{C_i}\left(\frac{1}{R_i} + \frac{1}{R_{i+1}}\right)\right] \tag{55b}$$

$$A_{i+1} = -\frac{1}{C_i R_{i+1}} [\frac{1}{C_i} (\frac{1}{R_i} + \frac{1}{R_{i+1}}) + \frac{1}{C_{i+1}} (\frac{1}{R_{i+1}} + \frac{1}{R_{i+2}})] \tag{55c}$$

$$A_i = \frac{1}{C_i} [\frac{1}{C_{i-1} R_i^2} + \frac{1}{C_i} (\frac{1}{R_i} + \frac{1}{R_{i+1}})^2 + \frac{1}{C_{i+1} R_{i+1}^2}] . \tag{55d}$$

Node voltages can be expanded into power series of a small time increment Δt as

$$V_i(t + \Delta t) = V_i(t) + \frac{dV_i(t)}{dt} \Delta t + \frac{1}{2!} \frac{d^2 V_i(t)}{dr^2} (\Delta t)^2 + o[(\Delta t)^3] . \tag{56}$$

Equations (54) and (55a)–(55d) are valid for $2 \le i \le N-2$. For nodes 1 and $N-1$, Eq. (54) is still valid, but Eqs. (55a)–(55d) are modified. For instance for node 1

$$\frac{d^2 V_1(t)}{dt^2} = A_0 V_{DD} + A_1 V_1 + A_2 V_2 + A_3 V_3$$

$$A_1 = \frac{1}{C_1} [\frac{1}{C_2 R_2^2} + \frac{1}{C_1} (\frac{1}{R_1} + \frac{1}{R_2})^2] \qquad A_3 = \frac{1}{C_1 C_2 R_2 R_3}$$

$$A_0 = -\frac{1}{C_1^2 R_1} (\frac{1}{R_1} + \frac{1}{R_2}) \qquad A_2 = -\frac{1}{C_1 R_2} [\frac{1}{C_1} (\frac{1}{R_1} + \frac{1}{R_2}) + \frac{1}{C_2} (\frac{1}{R_2} + \frac{1}{R_3})] .$$

For node $N-1$ similar relations are derived easily. For node N we have

$$\frac{dV_N(t)}{dt} = \frac{1}{C_N R_N} (V_{N-1} - V_N)$$

and

$$\frac{d^2 V_N(t)}{dt^2} = A_{N-2} V_{N-2} + A_{N-1} V_{N-1} + A_N V_N$$

where

$$A_{N-2} = \frac{1}{C_N C_{N-1} R_N R_{N-1}} \qquad A_N = \frac{1}{C_N R_N^2} (\frac{1}{C_{N-1}} + \frac{1}{C_N})$$

$$A_{N-1} = -\frac{1}{C_N R_N} [\frac{1}{C_{N-1}} (\frac{1}{R_{N-1}} + \frac{1}{R_N}) + \frac{1}{C_N R_N}] .$$

Using these derivatives, Eq. (54) is integrated using the series expansion of Eq. (56). If the initial voltage profile of the chain is given, the time evolution of the voltage profile can be determined by numerical integration. This simple method is easiest for those who are not familiar with methods of elaborate numerical analysis.

Numerical confirmation of Elmore's formula is as follows. Since we are interested in the delay time of reaching voltage $V_{DD}/2$ in digital CMOS circuits, Eq. (48) is modified to

$$T_D(Elmore) = \log_e(2) \sum_{i=1}^{N} R_i \sum_{j=i}^{N} C_i . \tag{57}$$

The delay time determined from numerical analysis is written as $T_D(Sim)$, and we define the ratio $T_D(Sim)/T_D(Elmore) = r(S/E)$. Ratios for various chain configurations are compared.

1) Uniform chain: For $N=1$ the formula is accurate $[r(S/E)=1.0]$. For $N=2$ $r(S/E)=1.065$, for $N=3$ $r(S/E)=1.077$ and for $10 \geq N \geq 4$ $r(S/E)=1.082-1.086$. Here we compared the last node of each chain.

2) Graded chain: Long, linearly graded chain ($N=10$). Let $R_i=1.0$. If $C_i = 1.0-0.05(i-5)$, $r(S/E)=1.094$. If $C_i=1.0+0.05(i-5)$, $r(S/E)=1.078$. Let $C_i=1.0$. If $R_i = 1.0-0.05(i-5)$, $r(S/E)=1.077$ and if $R_i=1.0+0.05(i-5)$, $r(S/E)=1.094$. Grading resistance and capacitance have different effects on $r(S/E)$.

3) Bump inside the chain: If $R_i=C_1=1.0$ except for $C_5=10.0$, $r(S/E)=1.071$. If $R_i=C_1=1.0$ except for $R_5=10.0$, $r(S/E)=1.054$.

4) Delays to the internal nodes of the chain and rise/fall times of the nodes: Let us choose a four-member RC chain ($N=4$ in Fig. 45) as an example. We compare delay times and rise times of all the nodes computed from Elmore's formula and those determined by solving the circuit equations numerically. The second and third columns of Table 1 show delay and rise times of nodes 1, 2, 3, and 4 determined from Eqs. (52a) and (52b), respectively [with constant in Eq. (52b) set at 1.0]. The fourth and fifth columns show delay and rise times, respectively, determined by numerical analysis. The sixth column shows ratios of the simulated and the computed delay times. According to Eq. (57) the ratio should be $\log 2=0.693$. This is indeed the case for nodes close to the end of the RC chain (nodes 3,4). For nodes close to the driving end, however, the ratio is much less than 0.693 (nodes 1,2). Elmore's formula gives a significant overestimate of the delay times. This is the problem mentioned at the beginning of Section 1.12 that originates from Elmore's definition of the delay time. Figure 49 shows voltage waveforms of the node 1 and node 4 of the four-member chain. The voltage of node 1 increases rapidly until about 0.6. After that time the rate of increase slows down quite significantly. Then the time derivative of node voltage, $V_1'(t)$, does not tend to zero fast enough, and therefore the contribution to the integral of Eq. (52a) at large t gives an

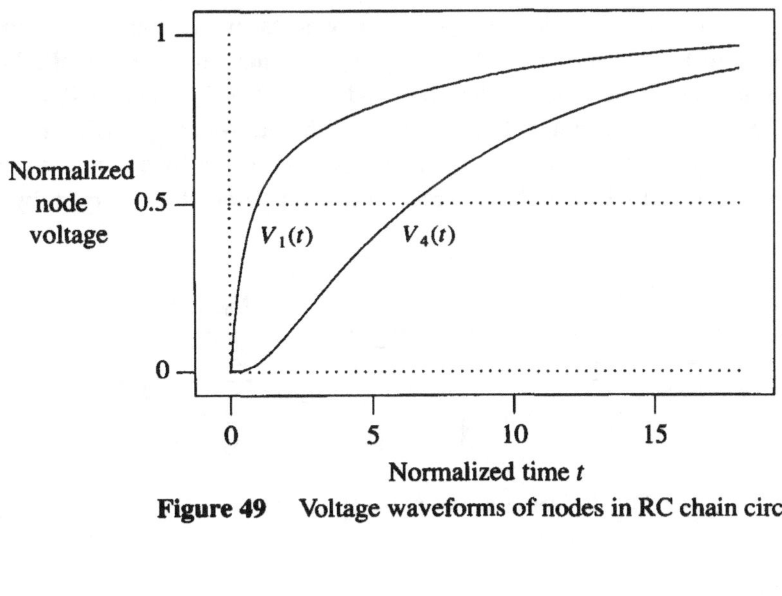

Figure 49 Voltage waveforms of nodes in RC chain circuit

Table 1: Delay time/rise time comparison						
—	Elmore formula		Simulation		Ratio (S./E.)	
Node	Delay time	Rise time	Delay time	Rise time	Ratio (delay)	Ratio (rise)
1	4.000	6.633	1.107	5.223	0.277	0.787
2	7.000	7.937	4.050	15.19	0.578	1.915
3	9.000	8.303	6.407	16.44	0.712	1.981
4	10.00	8.367	7.459	16.48	0.746	1.971

unnaturally large delay time. The definition of delay time using Eq. (52a) becomes improper for digital circuit analysis in such cases. As for the nodes close to the end of the RC chain, the time derivative of node voltage tends to zero rapidly. Definition of the delay time using Eq. (52a) is *proper*, and therefore Elmore's formula gives proper delay time.

The last column of Table 1 shows ratios of the simulated and computed rise times. The factor to be multiplied to get delay time is, according to Elmore, $\sqrt{2\pi}=2.51$. The ratios for nodes close to the end of the RC chain is about 2.0. The two numbers are close enough, considering crudeness of the approximation. For nodes close to the beginning of the RC chain, however, the ratio is significantly less. For rise times of nodes, we see the similar problems as we saw in delay times that originate from Elmore's definition.

1.14 Two Effects of Series Resistance in CMOS Circuits

In the equivalent linear resistor FET model, CMOS digital circuits are modeled using a resistance-capacitance network. Because CMOS circuits carry no standby (DC) current, the network should have no conductive connection to V_{DD} or to ground except through a single FET: connection to ground or to V_{DD} power-bus is always through the internal resistance of a FET. The assumption of *rapid switching* requires that the conducting path is either to V_{DD} or to ground, but not to both at any one time. The structure of the equivalent CMOS circuit is shown schematically in Fig. 50(a), where the driver FET is represented by equivalent resistance R_D. The entire circuit is discharged through resistance R_D to ground, or charged (by the

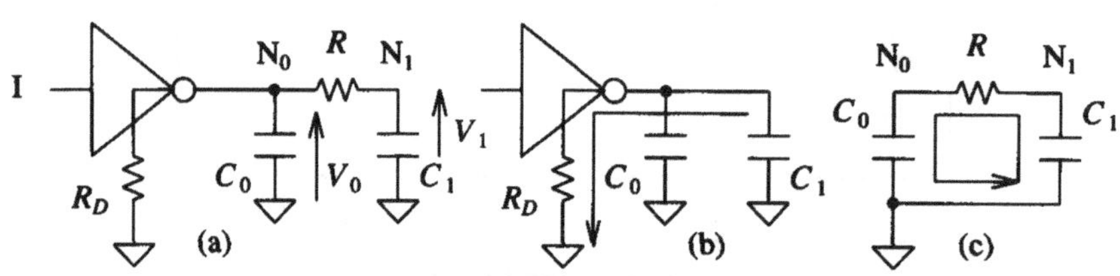

Figure 50 Driver and series resistance effects

complementary equivalent resistor) to V_{DD}. Resistance R of Fig. 50(a) represents wiring resistance of polysilicon crossovers and diffused or metal interconnects. R is a conventional resistor. The circuit of Fig. 50(a) shows the general equivalent circuit of the CMOS driver and resistive and capacitive interconnects that distribute the signal. We study the effects of the series resistance R in this circuit.

The series resistance R has two effects. If $R=0$, node N_0 and node N_1 charge and discharge simultaneously. If $R>0$, however, node N_1 charges and discharges later than node N_0. Resistance R *delays* the response of node N_1 relative to node N_0. This effect is certainly well known. The resistance R has another effect. The driver has nonzero *equivalent* resistance R_D, and therefore the response of node N_0 does not take place immediately after the input node (I) switching. Delay time from signal input I to node N_0 is *reduced* if resistance R is increased. This is because capacitance C_1 becomes *less* effective in delaying the response of node N_0. This effect, that series resistance sometimes speeds up the response of certain nodes, is often overlooked. As we see later, this effect is quite important in understanding CMOS circuits. In this section we analyze two effects of a resistor using two different FET models. This section is intended as the interface to the next section, where we introduce the methods of circuit analysis that are based on the other device model, the collapsible current generator model.

We analyze this problem first using the equivalent linear resistor model of a FET. Two basic mechanisms are operative in the process of a discharge of the circuit shown in Fig. 50(a). The one is discharge of *joint* capacitances C_0 and C_1 through the driver's equivalent resistance R_D as shown in Fig. 50(b), and the other is equalization of node voltages N_0 and N_1 through resistance R as shown in Fig. 50(c). The arrows in Figs. 50(b) and (c) show directions of currents by the respective transient processes. The first and the second processes are characterized by time constants T_D and T_P, respectively, that are given by

$$T_D = R_D(C_0 + C_1) \qquad T_P = \frac{RC_0C_1}{C_0 + C_1}. \tag{58}$$

The two time constants T_D and T_P play the central role in the following analysis. To specify the problem of Fig. 50(a) completely, one more parameter is required, namely, the ratio of capacitance C_1 to C_0, $\gamma = C_1/C_0$. Note that T_D and T_P are symmetrical with respect to C_0 and C_1. Parameter γ specifies the asymmetry. The three parameters specify the problem completely.

The circuit equations satisfied by node voltages V_0 and V_1 of Fig. 50(a) are

$$C_0\frac{dV_0}{dt} + (\frac{1}{R_D} + \frac{1}{R})V_0 - \frac{1}{R}V_1 = 0 \qquad -\frac{1}{R}V_0 + C_1\frac{dV_1}{dt} + \frac{1}{R}V_1 = 0. \tag{59}$$

The initial conditions are $V_0(0) = V_1(0) = V_{DD}$. Assuming time variation of the form $V_0, V_1 \approx exp(St)$, parameter S is determined from Eq. (59) as $S = \alpha$ and β (where we choose $\beta < \alpha$ and they are both negative) given by

$$\alpha = -\frac{1}{2}(\Gamma - \Delta) \quad \beta = -\frac{1}{2}(\Gamma + \Delta) \text{ where } \Gamma = \frac{1}{T_P} + \frac{(1+\gamma)}{T_D} \quad \Delta = (\Gamma^2 - \frac{4}{T_D T_P})^{1/2}$$

and the solution of Eq. (59) is given by

$$\frac{V_0(t)}{V_{DD}}=\frac{\beta[1+(1+\gamma)T_P\alpha]}{\beta-\alpha}\,exp(\alpha t)-\frac{\alpha[1+(1+\gamma)T_P\beta]}{\beta-\alpha}\,exp(\beta t) \tag{60a}$$

$$[V_1(t)/V_{DD}]=[\beta/(\beta-\alpha)]exp(\alpha t)-[\alpha/(\beta-\alpha)]exp(\beta t)\,. \tag{60b}$$

In the limit of $T_p \rightarrow \infty$ $\alpha \rightarrow -1/[(1+\gamma)T_P]$ and $\beta \rightarrow -(1+\gamma)/T_D$.

$$[V_0(t)/V_{DD}] \rightarrow exp[-(1+\gamma)t/T_D] \tag{61a}$$

$$\frac{V_1(t)}{V_{DD}} \rightarrow \frac{(1+\gamma)^2T_P\exp[-t/[(1+\gamma)T_P]]-T_D\exp[-(1+\gamma)t/T_D]}{(1+\gamma)^2T_P-T_D}\,. \tag{61b}$$

Delay times (to voltage $V_{DD}/2$) of nodes N_0 and N_1 are T_0 and T_1, respectively, and they are given by

$$T_0=\frac{\log 2}{(1+\gamma)}T_D=\log 2\cdot(C_0R_D)$$

$$T_1=\frac{T_D}{(1+\gamma)}+\log 2\cdot(1+\gamma)T_P=(C_0R_D)+\log 2(C_1R)$$

where delay time T_1 was derived by neglecting the term proportional to $exp(\beta t)$. This is justifiable, since $\alpha \rightarrow 0$ when $T_P \rightarrow \infty$, and therefore $|\beta| \gg |\alpha|$.

In the limit of $T_P \rightarrow 0$ $(R \rightarrow 0)$

$$\frac{V_0(t)}{V_{DD}}=\frac{V_1(t)}{V_{DD}} \rightarrow \exp[-t/T_D] \text{ and}$$

$$T_0=T_1=\log 2\cdot T_D=\log 2\cdot R_D(C_0+C_1)\,. \tag{61c}$$

The dotted curves in Fig. 52 plot delay times of nodes N_0 and N_1 versus T_P/T_D, assuming $C_1/C_0=1.0$. Node N_1 delay time increases with increasing T_P/T_D, but node N_0 delay time decreases slightly with increasing T_P/T_D and soon becomes independent of T_P/T_D as the lower limit is reached. The delay times plotted in Fig. 52 were determined by solving Eqs. (60a) and (60b) numerically. This numerical analysis is simple: closed-form approximations can be made using Eqs. (61a) to (61c). We summarize the analysis based on the equivalent resistor model: using closed-form analysis and approximation methods, delay times can be computed easily. The model is *delay time oriented.*

The equivalent resistor FET model allows closed-form circuit analysis yielding many useful numerical results. The other model, the collapsible current generator model introduced in Section 1.7 is just as convenient as the equivalent resistor model, and the model gives quite detailed information of circuit operation and semiquantitative node voltage waveforms. Figure 51(a) shows the same equivalent circuit as Fig. 50(a), except that NFET in the CMOS driver is modeled using the collapsible current generator model. Current I_D of the NFET is given by $I_D=B(V_{DD}-V_{TH})$ if $V_0>0$, where B is the transconductance of the NFET and the input node of the driver is assumed to have been pulled up from ground to V_{DD} instantaneously at time $t=0$. Time constants T_D and T_P are now redefined as

$$T_D=\frac{C_0+C_1}{I_D}V_{DD} \quad T_P=\frac{RC_0C_1}{C_0+C_1}\,. \tag{62}$$

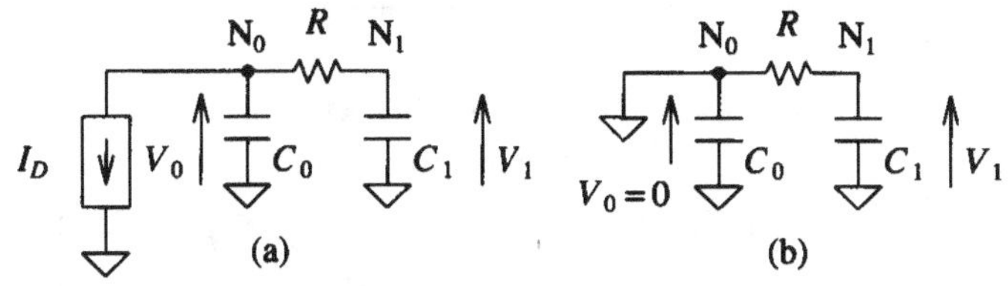

Figure 51 Analysis using collapsible current generator model

Nodes N_0 and N_1 are charged to V_{DD} at time $t = 0$. Until node N_0 voltage reaches zero, the circuit equations satisfied by node voltages V_0 and V_1 are

$$C_0 \frac{dV_0}{dt} + I_D = \frac{V_1 - V_0}{R} \qquad C_1 \frac{dV_1}{dt} + \frac{V_1 - V_0}{R} = 0 \tag{63}$$

where the initial conditions are $V_0(0) = V_1(0) = V_{DD}$. The set of equations have an integral satisfying the initial condition

$$C_0 V_0(t) + C_1 V_1(t) = (C_0 + C_1) V_{DD} - I_D t\,.$$

Using Eq. (63) we find the following solution:

$$\frac{V_0(t)}{V_{DD}} = 1 - \frac{t}{T_D} - \gamma \frac{T_P}{T_D}[1 - \exp(-t/T_P)]$$

$$\frac{V_1(t)}{V_{DD}} = 1 - \frac{t}{T_D} + \frac{T_P}{T_D}[1 - \exp(-t/T_P)]\,. \tag{64a}$$

The set of equations Eq. (63) is valid until time t_C. At that time $V_0(t_C) = 0$ is reached. For $t > t_C$ we have the equivalent circuit shown in Fig. 51(b) and the node voltages are found by solving

$$V_0(t) = 0 \quad C_1 \frac{dV_1}{dt} = -\frac{V_1}{R} \quad \text{or} \quad V_1(t) = V_1(t_C) \exp\left[-\frac{t - t_C}{(1+\gamma) T_P}\right] \tag{64b}$$

where $(1+\gamma)T_P = RC_1$ is the time constant consisting of resistor R and capacitor C_1. Solid curves of Fig. 52 show plots of delay times determined from waveforms Eqs. (64a) and (64b) numerically. The linear resistor model and the collapsible current generator model give results that are very much alike. The two results are different only by a constant factor, which originates from the difference in the definitions of T_D used for normalization of time [see Eq. (58) and Eq. (62)].

Equations (64a) and (64b) are plotted by solid curves in Fig. 53. Node voltage $V_0(t)$ decreases almost linearly with time and reaches $V_0(t) = 0$ at $t = t_C$ $(= 0.6T_D)$. After that time $V_0(t) = 0$. The change of *state* that takes place at $t = t_C$ is obvious from the $V_0(t)$ waveform. Although invisible, different functions represent $V_1(t)$ for $t < t_C$ and $t > t_C$. What happens in the circuit at time $t = t_C$ can be understood by rewriting Eq. (63) as

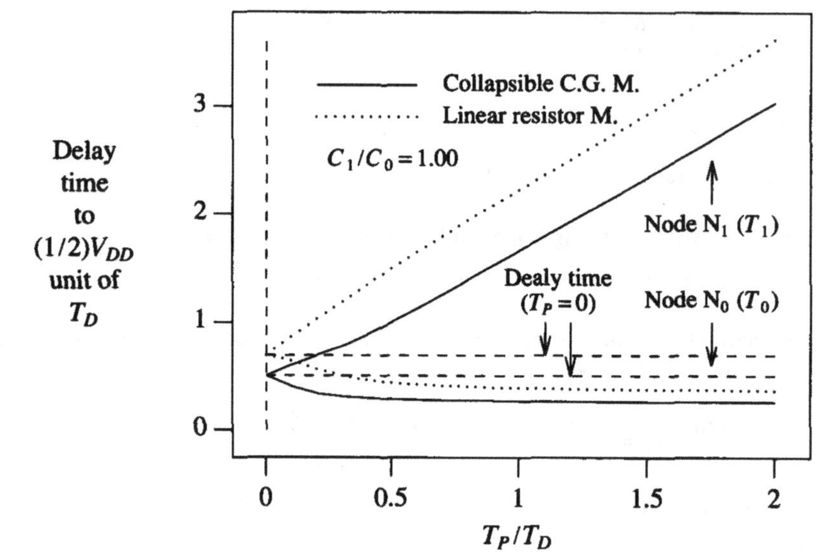

Figure 52 Delay times of nodes N_0 and N_1

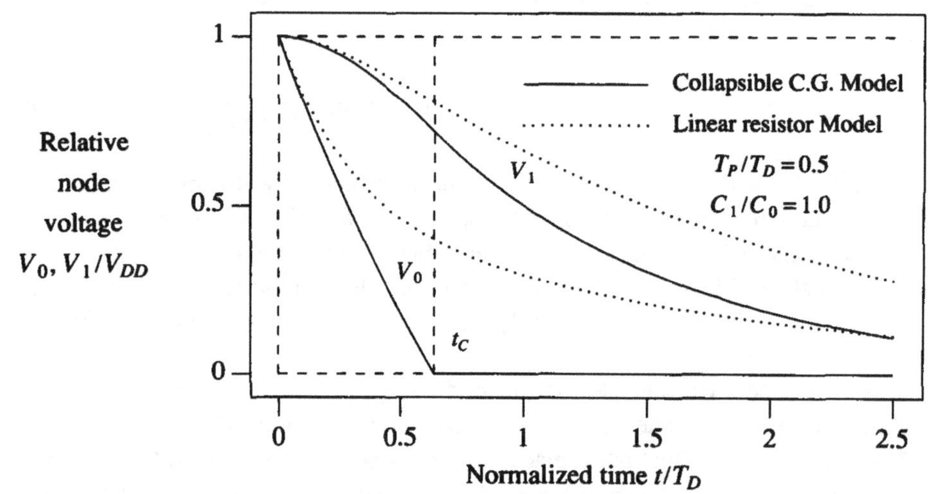

Figure 53 Node voltage waveforms

$$I_D = \frac{V_1 - V_0}{R} - C_0 \frac{dV_0}{dt}$$

The first term on the right-hand side is the current that flows from node 1 to node 0, and the second term is the discharge current of capacitor C_0. Since $(dV_0/dt) < 0$ for $t < t_C$, the second term (with negative sign) adds to the first term, which is the positive current from node 1, to make up NFET current, I_D. At $t = t_C$ the NFET goes into the triode region. Node 0 voltage change halts [(dV_0/dt) vanishes], the NFET current decreases from I_D to $I_D - |C_0(dV_0/dt)|_{t=t_C}$, and the reduced current holds the NFET in the triode region. The two events at $t = t_C$, that is the NFET moving into the triode region and the halt of node voltage change, hold each other after time $t = t_C$, and the circuit enters into a new state. We may say that the circuit is in different *states* before and after time $t = t_C$, since the circuit has *effectively* different configurations as is evident from Figs. 51(a) and 51(b). The results of the equivalent resistor model, Eqs. (60a) and (60b), are plotted in Fig. 53 by dotted curves. We see no unusual feature in the curves. Thus we may conclude as follows. If a MOSFET circuit is modeled using the linear resistor model of FETs, the circuit is described by a single set of circuit equations [Eq. (59)], and the plots of node voltages versus time over the entire range show no unusual feature. This is the linear circuit representation of the digital circuit. If the same circuit is modeled using the collapsible current generator model of FETs, the circuit is described by more than one set of circuit equations, and one set of equations succeeds the other at the critical time (like t_C). The second model contains *more details* of the circuit characteristics than the first model. Therefore the collapsible current generator model is convenient for more detailed analysis of a circuit's operational mechanisms.

1.15 Circuit Analysis Based on the FET Collapsible Current Generator Model

Up to the preceding section, we discussed methods of CMOS digital circuit analysis using the FET *linear equivalent* resistor model. In this section we consider circuit analysis using the model that is the opposite extremum of device representation. The simplified model of FET introduced in Section 1.7 was referred to as the collapsible current generator model. In this model, a FET is modeled as a current generator having infinite internal resistance if the drain-source voltage is positive (the FET *saturation* region). If the drain-source voltage is zero, however, the FET channel presents no resistance for channel currents less than the maximum current in the saturation region for the same gate-to-source voltage (the FET *triode* region). If the gate-to-source voltage is less than the threshold voltage, the FET is turned off (*nonconducting* region). The FET is clearly in one of the three distinct regions. Clear identification of the FET's operational *mode* is convenient, above all, in understanding CMOS digital circuit operation.

If this model is used with one additional simplification, that the FET threshold voltage is independent of the source voltage (no back-bias effect), practically any CMOS digital circuit problem can be solved in closed form using the well-established theory of the linear simultaneous differential equations. Although this assumption may not be accurate enough for numerical data required for circuit design, the advantage of theoretical analysis of a circuit is so great that the simplified model warrants in-depth investigation. The mathematical simplicity of the model can be understood from the following example. With reference to the circuit diagram shown in Fig. 54, NFETs MNA and MNB are in saturation and NFET MNC is in the nonconducting region. The voltage of node A, V_A, satisfies equation

$$C_A \frac{dV_A}{dt} = B_A(V_{IA} - V_A - V_{THN}) - B_B(V_{IB} - V_B - V_{THN}) \tag{65a}$$

where we assume that $V_B < V_A < V_Z$. Suppose that NFET MNC turns on (or goes into the saturation region) because voltage V_{IC} goes up. Then node A voltage satisfies equation

$$C_A \frac{dV_A}{dt} = B_A(V_{IA} - V_A - V_{THN}) - B_B(V_{IB} - V_B - V_{THN}) - B_C(V_{IC} - V_C - V_{THN}) \tag{65b}$$

where we assume that $V_C < V_A$. If NFET MNC moves into the triode region, NFET MNC presents zero resistance. Node A and C *fuse* together, and the combined node has capacitance $(C_A + C_C)$. The fused node (A, C) voltage is determined from the equation [21]

$$(C_A + C_C) \frac{dV_{AC}}{dt}$$
$$= B_A(V_{IA} - V_{AC} - V_{THN}) - B_B(V_{IB} - V_B - V_{THN}) - B_D(V_{ID} - V_D - V_{THN}) \tag{65c}$$

Since there is one node equation for each node of the circuit, the problem of circuit analysis amounts to solving a set of simultaneous linear differential equations having *constant* coefficients, subject to the specified initial conditions and input voltage waveforms. The simultaneous equations can be solved in closed form using the standard techniques of linear, simultaneous differential equations. Many examples of solutions are given later in this book.

The problems of circuit analysis using this FET model can be reinterpreted from Eqs. (65a)–(65c) as follows. For example, NFET MNC must have gone into the present state (triode region), because NFET MND was unable to discharge node C rapidly, and NFET MNB discharged node A rapidly (node A originally had a higher voltage than node C). Therefore

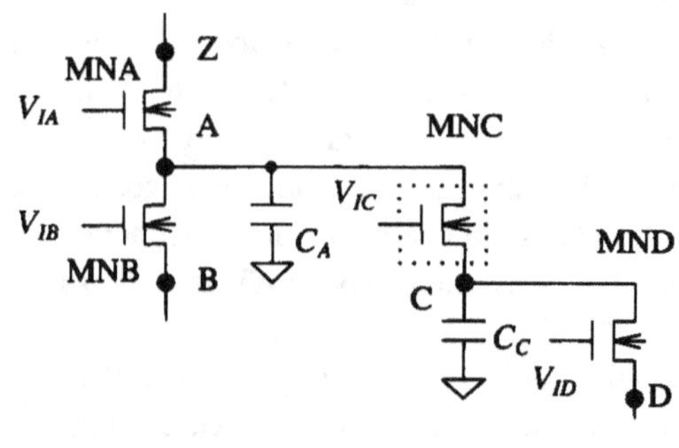

Figure 54 Application of collapsible current generator model of FETs

node A and node C approach the same voltage. NFET MNC moves into the triode region at the moment when the two voltages V_A and V_C become equal. Let that time be t_C. Then for $t < t_C$ we have a set of simultaneous equations that distinguish nodes A and C, and for $t > t_C$ we have circuit equations that do not. Two different sets of simultaneous circuit equations govern two different regimes of circuit operations. Both of the two sets of simultaneous equations are simple enough that closed-form solutions are available. The two sets of solutions are connected at time $t = t_C$, such that the node voltages of both match at time t_C. This mathematical procedure means the following: There is a *macroscopic* structure in circuit operation that is reflected in successive changes of circuit configuration governed by different sets of circuit equations. For understanding circuit operation, qualitative information on how the changes of circuit configuration take place is more important than the actual solution of the circuit equations.

Using the collapsible current generator model of FETs and the model of digital circuits discussed above, *analog* (*linear*) circuit and *digital* (*nonlinear*) circuits are clearly distinguished as follows. An analog circuit in its entire range of operation can be represented by only one set of simultaneous equations. An analog circuit does not change circuit configuration during its operation. The main reason why analog circuits can be analyzed easily is that circuit connectivity does not change during operation. A digital circuit is represented by more than one set of simultaneous equations, and their solutions are connected together at the *critical* time, when the circuit configuration changes. From this viewpoint a digital circuit can be represented by several different analog circuits in succession. This change of circuit configuration depends, among other parameters, on the initial condition of the circuit, and this makes analysis very complex.

Mathematical difficulties in solving this problem in closed form are limited only in the following areas. The set of simultaneous equations satisfied by voltages of circuit nodes N_0, N_1, ..., N_N; V_0, V_1, ..., V_N, respectively, are of the form

$$C_0 \frac{dV_0}{dt} = G_{00}V_0 + G_{01}V_1 + \ldots + G_{0N}V_N + X_0 \tag{66}$$

$$\ldots\ldots\ldots\ldots$$

$$C_N \frac{dV_N}{dt} = G_{N0}V_0 + G_{N1}V_1 + \ldots + G_{NN}V_N + X_N$$

where $C_0, \ldots, C_N$, $G_{00}, \ldots, G_{NN}$ and $X_0, \ldots, X_N$ are constants. If the original equations are not of this form, the equations can be solved for dV_i/dt $(0 \le i \le N)$. So we may assume the form of Eq. (66). The differential equations can be solved as follows. The general solution is of the form

$$V_i(t) = H_i(t) + P_i \quad (0 \le i \le N)$$

where $H_i(t)$ is the solution of the *homogeneous* part of the simultaneous equations obtained by setting $X_i = 0$ $(0 \le i \le N)$. $H_i(t)$ contains $N+1$ *independent* constants of integration that must be determined from the initial conditions. The homogeneous part of the equation can be solved in the from

$$H_i(t) = \sum_{j=0}^{N} A_{ij} exp(S_i t) . \tag{67}$$

In Eq. (67) not all the coefficients A_{ij} are independent. There are only $N+1$ independent coefficients, and the others are represented by the independent ones. In Eq. (67) the time dependence of the voltages is the sum of terms of the form $exp(St)$. S_i is determined from the *secular* equation,

$$\begin{bmatrix} G_{00} - C_0 S & \ldots\ldots & \ldots\ldots \\ \ldots\ldots & \ldots\ldots & \ldots\ldots \\ \ldots\ldots & \ldots\ldots & G_{NN} - C_N S \end{bmatrix} = 0 \tag{68}$$

If $N > 3$ this polynomial equation cannot be solved in general in closed form for root S (Abel's theorem). This difficulty does not limit the usefulness of this approach. There are many simple problems that do not suffer from this difficulty. Furthermore some circuits can be simplified so that N can be reduced below 4. If necessary, approximate solutions can be found by using techniques like infinite series expansion, based on the physical insight into the problem. If Eq. (68) has multiple roots, or if the input signal waveforms are specified, the solution is more complex, but the mathematical procedure is known [22]. We leave these details to standard textbooks on differential equations. The constants P_i are a special solution of the original equation. They are a set of constants determined from the simultaneous equation

$$0 = G_{00} P_0 + G_{01} P_1 + \ldots. + G_{0N} P_N + X_0 \tag{69}$$

$$\ldots\ldots\ldots.$$

$$0 = G_{N0} P_0 + G_{N1} P_1 + \ldots. + G_{NN} P_N + X_N .$$

In order to determine the critical time t_C when the circuit changes configuration, generally a transcendental equation of the form

$$a + \sum_{i=0}^{N} (b_i + c_i t + \ldots) exp(S_i t_c) = 0 \tag{70}$$

must be solved. Again the solution can be derived using various approximation methods. Even if a numerical solution has to be used, computation is not extensive, and the nature of the solution is known. We may conclude that practically all the problems of CMOS digital circuits can be solved in closed form using this model, as is shown in this book. The convenience of this opens new possibilities for circuit research, since closed-form analysis of CMOS logic gates has not yet been performed.

1.16 Microstate Sequence of CMOS Digital Circuits

In the last section we learned that the switching process of CMOS digital circuits can be understood as a sequence of changes in the circuit structure, or the component's connectivity. Understanding the sequence of changes, bringing the switching process into perspective, and defining the problem of circuit analysis precisely are difficult because the structural changes of the circuit are complex. In this section we give a method to help understand this process, by using a diagram to represent the circuit's structural change: the method of *microstate sequence* of CMOS circuits.

In order to introduce the basic concept of *microstate* in digital CMOS circuits, we consider first the operation of a CMOS microprocessor-based computer. The highest level of operation, like solving a differential equation, is carried out by using the operators of a higher level language, like $+, -, \times$, and $/$. When the higher level language is compiled, an operator function is carried out by a group of microprocessor machine language instructions. When a machine language code is given to the microprocessor, the control section of the microprocessor *compiles* the machine language code into a set of *microprogram instructions*, which controls the individual computational resource of the microprocessor, like registers and adders. One instruction of the microprogram specifies a *state* of the microprocessor: the microprocessor goes through a sequence of *states* to execute one machine language instruction. For every microprogram instruction, the logic circuit executes the specified logic operation. As we saw in the previous section, the logic circuit goes through a sequence of *microstates*, each of which corresponds to a specific logic circuit connectivity. Therefore a circuit configuration may be considered as a *microstate* of the circuit, and a sequence of *microstates* of the circuit executes a single logic operation. By this observation, the microstate of a CMOS circuit may be understood as a level one step lower than the *state* of the logic circuit. Microstate belongs to the domain of electrical circuit, in contrast to conventional *state*, that belongs to the domain of logic. There is no level lower than microstates of logic circuits, since below that the electrical phenomena are on the level of analog circuits where the concept of discrete, or *digital* parameters that are essential to identify any *state*, loses its meanings.

As we discussed in the preceding section, there are three essential states, or regions, of current-voltage characteristics of a FET. They are: the nonconducting region (N), the conducting in saturation region (S), and the conducting in triode region (T). In states N and S, the source and the drain of the FET are separate nodes. In state T the source and the drain are fused into a single electrical node. There are six ways ($6 = 3 \times 2$ is the number of ways to select two items in specific order from a set of 3) a state changes, as shown in Fig. 55 (a)–(f). Figure 55(a) shows that NFET MN1 changed from nonconducting region N to saturation region S. An interesting change of FET state is the exchange of source and drain identification, shown in Fig. 55(k). We observe that the output node of a logic gate may settle at logic-high or logic-low level, thereby splitting the logic circuit into two parts. Therefore the four symbols shown in Figs. 55(g)–(j) will be used. Figure 55(g) shows that the output node of inverter INVA settled at V_{SS}. Figure 55(h) shows that the output node began to make low-to-high transition from the logic-low state. A logic circuit that was split at the node is now integrated (we note that if a node in a chain of logic gates is settled at a logic level voltage, the chain is effectively broken off at the node).

The symbols have the following meanings:

a) NFET MN1 changes from nonconducting to saturation region. Current is not zero.

b) MN1 changes from saturation to nonconducting region. Current becomes zero.

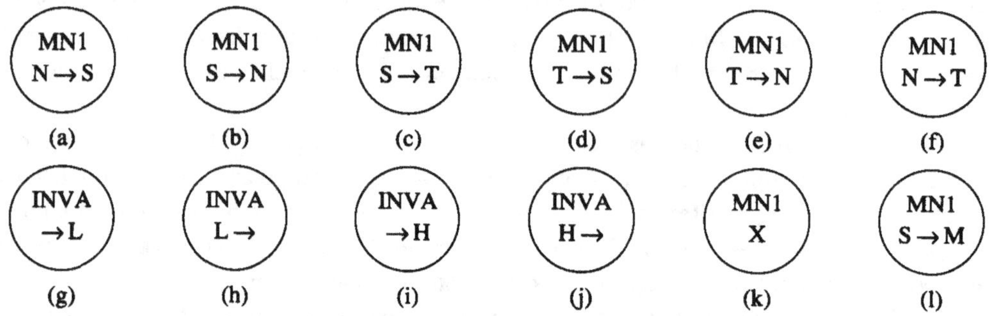

Figure 55 Symbols used in microstate diagram of CMOS circuits

---

c) MN1 changes from saturation to triode region. The source and the drain of the FET become electrically the same node.
d) MN1 changes from triode to saturation region. The source and the drain separate to different nodes.
e) MN1 changes from triode to nonconducting region. Circuit separates.
f) MN1 changes from nonconducting to triode region. Drain and source merge into single node.
g) Output node of inverter INVA settles to ground potential.
h) Voltage of output node of INVA begins to increase from zero.
i) Output node of INVA settles at V_{DD} potential.
j) Voltage of output node of INVA begins to decrease from V_{DD}.
k) Source and drain identification of NFET MN1 exchanges.
l) NFET changes from saturation to marginally conducting region.

The last symbol, included for convenience, requires explanation. If capacitance is charged through a source-follower circuit (Section 2.8), the FET current decreases as the capacitor voltage approaches $V_G - V_{TH}$. The state the FET arrives at after a long time is the exact boundary between the saturation region and nonconducting region (or in saturation region with zero saturation current). Since this state occurs frequently, we make a special identifier M (M stands for marginal conduction). Figure 55(l) shows the final phase of the chargeup process of a capacitor by a source-follower circuit. If state M is used, we assume no leakage current in the FET.

The symbols shown in Fig. 55(a)–(k) specify the FET and the type of state change. We need to specify how successive changes take place. The arrows between the circles in Fig. 56(b) indicate that the two successive changes take nonzero time between the events: the change indicated by the circle to the left is followed by the change indicated by that to the right, and the arrow indicates the logic circuit is in the microstate at least for some time. If two circles are directly connected, or connected by a thick line, the two changes occur simultaneously. This situation takes place if two signals are simultaneously changing, or if one

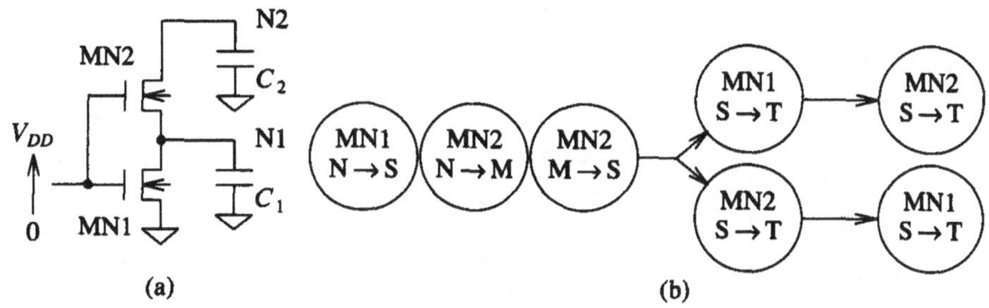

Figure 56 Microstate sequence of NFET chain discharge—I

change induces the other change immediately. This may happen either by rapid change of input signal, or by a peculiarity of the state involved (especially marginally conducting, M-state). We show several examples of the microstate sequence of simple circuits in the following discussion.

Figure 56 shows the microstate sequence of NFET chain discharge. Nodes N1 and N2 of the NFET chain of Fig. 56(a) were originally charged to $V_{DD} - V_{THN}$ and V_{DD}, respectively. Gate voltages of the two NFETs MN1 and MN2 were at ground potential. The initial condition is specified by the states of the two NFETs as

MN1(N) MN2(N)

In this initial microstate, nodes N1, N2, and ground are all distinct circuit nodes. The initial condition may be indicated in the diagram if so desired, but often that is not really necessary. We did not do so here for simplicity. At time $t = 0$ the gate voltage of the two NFETs makes an instantaneous transition from 0 to V_{DD}. Then NFETs MN1 and MN2 are both driven into the saturation region. This is indicated by the first three contacted circles. The three events take place in rapid succession: First, NFET MN1 turns on. Second, NFET MN2 goes into the M region, which is the exact boundary between the nonconducting and the saturation regions. Third, even an infinitesimal reduction of node N1 voltage by the current of NFET MN1 causes NFET MN2 to go into the saturation region. The first three events are arranged in order, to show the cause-effect relationships. The circuit of Fig. 56(a) ends up with a final state in which both nodes N1 and N2 are grounded. In this state both FETs are in the triode (T) region. Therefore two transitions, MN1 (S → T) and MN2 (S → T), must be completed before the final state is reached. It is possible, however, for either of the two transitions to be first. Therefore the microstate sequence of this circuit is as shown in Fig. 56(b). Which of the two branches of the sequence is followed is determined by many factors, including FET scaling and capacitance values. The microstate sequence of many circuits can be drawn as shown by this example: If the change of the node voltages is monotonic, microstate transitions must follow the order N → S → T or T → S → N. The entries of a microstate sequence need to include all the possible state changes.

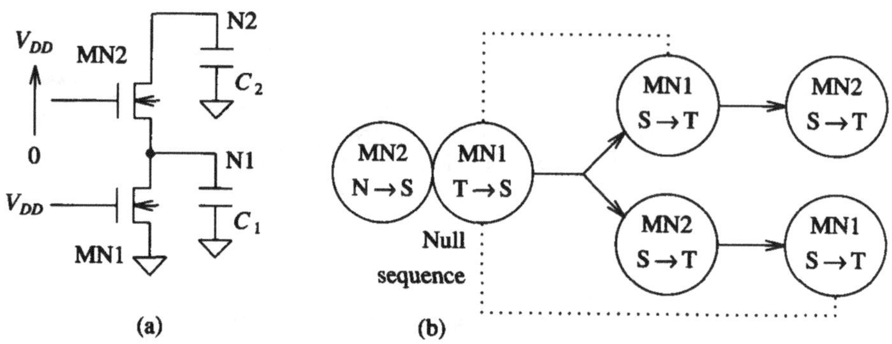

Figure 57 Microstate sequence of NFET chain discharge — II

An example of a case where the change of node voltage is not unidirectional is shown in Fig. 57. The gate of NFET MN1 is already at V_{DD}, and node N1 and ground are fused together. Node N2 was at a positive voltage, and the gate of MN2 was at ground. At time $t=0$ the gate of MN2 makes an instantaneous low-to-high transition. The final state of this circuit is that both nodes N1 and N2 are fused to ground. Both NFETs must then be in the triode (T) region. Immediately after $t=0$, NFET MN2 changes to the saturation region (S). NFET MN1 may have either of two possibilities: (1) if NFET MN1 is larger than NFET MN2, MN1 remains in the triode region (T), since MN1 is able to conduct the current from MN2 to ground without developing drain to source voltage. (2) If MN1 is smaller than MN2, however, MN1 changes to the saturation region: Since MN1 is unable to conduct current from MN2, MN1 goes into the saturation region and node N1 voltage increases. The rate of increase of the node voltage is (*Current of* MN2 – *Max current of* MN1)/C_1. In this case the microstate sequence is shown in Fig. 57(b). Since MN1 goes into the saturation region immediately after $t=0$, the first two circles are contacted in the microstate sequence. After that time either MN1 goes back to the triode region first, or MN2 goes into the triode region first. If MN2 is much larger than MN1, the first case is unlikely. We note that the microstate change MN1 (T → S) followed by MN1 (S → T) is a sequence that *does* and *undoes* a change. Such a *null* sequence creates a non-monotonic node N1 voltage waveform. Null sequences complicate microstate sequences quite significantly, since each null sequence increases the number of events in the microstate sequence by 2. If a microstate sequence has N events, and if any permutation of the events represents possible circuit operation, the number of possible microstate sequence increases by a factor of $(N+2)!/N! = (N+2)(N+1)$. Since a null sequence is ordered, $(N+2)(N+1)/2$ times more possible sequences are created. We recognize that null sequences are the major source of complexity in circuit operation. If microstate sequences that include an arbitrary number of null sequences must be considered, the problem becomes too complex to be manageable. Practically, the best approach is to consider microstate sequences that contain no null sequence (we call such a microstate sequence a *skeleton* sequence) and then add a small number of null sequences and investigate their effects.

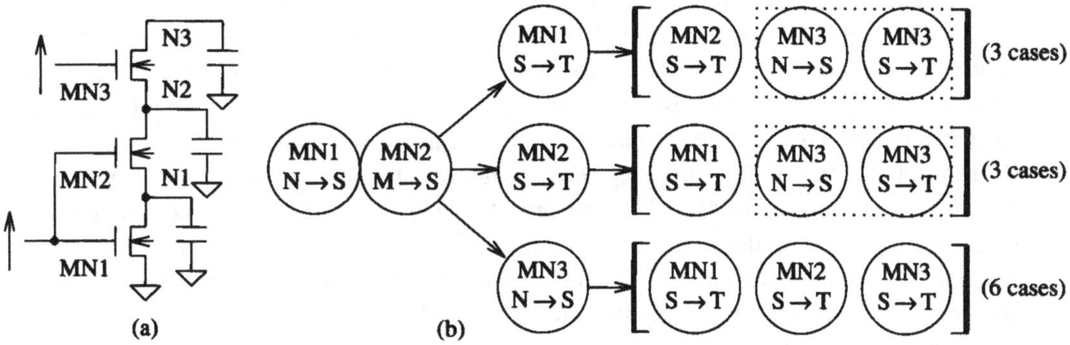

Figure 58 Microstate diagram of three-member NFET chain

Figure 58 shows the discharge process of a three-member NFET chain. Nodes N1, N2 and N3 are charged originally to $V_{DD} - V_{THN}$, $V_{DD} - V_{THN}$ and V_{DD}, respectively. At time $t = 0$ the gates of NFETs MN1 and MN2 make an instantaneous low-to-high transition, and after some time, the gate of NFET MN3 makes the same transition. NFET MN2 originally sustains no voltage. Therefore immediately after the gate voltage change, the FET is in the marginal conduction (M) region. This state, however, changes immediately into the saturation region since NFET MN1 pulls node N1 down, and the reduced node N1 voltage removes NFET MN2 from marginal conduction. Since the two changes take place in rapid succession, the two circles MN1 (N → S) and MN2 (M → S) are contacting in the microstate sequence. From this initial state, three possibilities follow. (1) If NFET MN1 is large, MN1 goes into triode region first. (2) If MN2 is large, MN2 may go into triode region first. (3) If the delay of the signal driving gate of MN3 is small, MN3 turns on and goes into saturation region before the other two possibilities occur. In this last case, all NFETs MN1, MN2 and MN3 go subsequently into the triode region before the discharge process completes. Since any FET may go into the triode region first, the three changes — MN1 (S → T), MN2 (S → T), and MN3 (S → T) — may take place in any order. Therefore we must draw all six (= 3!) possibilities. Since this is too cumbersome, we use square brackets [***] to indicate all permutations of the entries in the brackets. Similarly, the first case (NFET MN1 goes into the triode region first) has three subcases, because NFET MN3 goes from the nonconducting region to the saturation region and then to the triode region in that order. In order to maintain the order of two events but otherwise allow free permutation, dotted squares shown in the top and the middle sequence are used. Within the bracket of the top sequence of Fig. 58(b), event MN2: S → T can be anywhere while the order of the other two events (MN3: N → S and MN3: S → T) are maintained. As can be seen, there are many possible microstate sequences.

Why must so many microstate sequences be included to describe the operation of a simple circuit? Unfortunately circuit operation has that much complexity, and the complexity cannot be avoided. Event MN2: S → T in the first sequence may be in three possible locations, depending on the circuit parameters. If parameters are given, one particular order of events is determined. Let us make the problem more realistic and suppose that the node capacitance has voltage dependence. Voltage dependent node capacitance has the additional effect of shifting *local* time (to which the microstate events involving the node refer) to global time defined using the average capacitance for all the nodes. This problem is discussed later in Section 2.8. Since the microstate sequence is determined assuming that all node capacitances are average, interpretation of real circuit operation requires inclusion of sequences that are close enough to cover the realistic variations. We note that Fig. 58(b) shows the *skeleton* sequence of the three-member NFET chain. Possibilities exist that a small number of null sequences could be added (for instance, if NFET MN2 was in the triode region, MN2 may return to the saturation region when NFET MN3 turns on). Whether or not this actually takes place must be determined by circuit analysis: The microstate sequence shows clearly where to investigate such possibilities.

The concept of microstates in CMOS digital circuits and their representation, microstate sequences, are new concepts that have not been seen in earlier references. In order to familiarize readers with the concept, the internal structure of microstate sequences are studied as a *mathematical* problem of combinations. This is useful, since what appears to be complex is in reality very simple from a different viewpoint, and we are able to understand why the circuit problem looks so very complex and difficult to understand. Microstate sequences consist of indications of changes of the state of *all* FETs in the circuit. We represent changes of FET state by roman characters A, B, C, and so on. A microstate sequence is an *ordered* array of the characters or *objects*. We consider the example shown in Fig. 59(a). Some objects must be arranged in order, like sequence $A \rightarrow B \rightarrow C$ as shown in line 1 of Fig. 59(a). An example of such an ordered sequence is an NFET that is turned on ($N \rightarrow S$) and then goes into the triode region ($S \rightarrow T$), in a circuit that makes monotonic node voltage changes. There are objects that can be arranged in any order, as shown by D, E, F, G of line 2. Objects that can be arranged in an arbitrary order are, for example, change of states of FETs in different branches of the circuit. Depending on how FETs are scaled, and bias or signal voltages are applied,

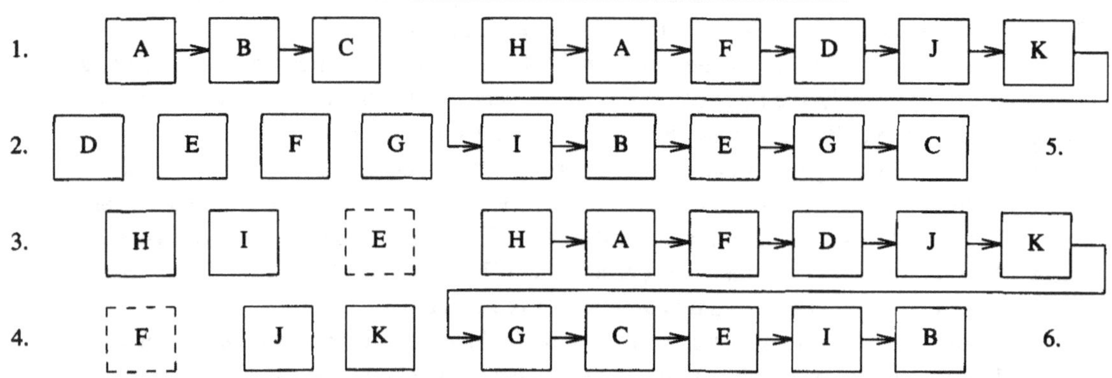

(a) (b)

Figure 59 Merging several microstate sequences together

state changes may take place in any order. If the set of directed objects A, B, and C and the set of objects that may have arbitrary order, D, E, F, and G are *merged*, they are arranged in an arbitrary order, except that order $A \rightarrow B \rightarrow C$ is maintained. Let us consider the ordered sequence 1. There are four gaps between the objects (to the left of A, between A and B, between B and C, and to the right of C). Four objects D, E, F, and G are placed in the slots in arbitrary order, and one slot may have any number of elements (including *no* element). There are objects, H and I, of line 3 of Fig. 59 that must always be placed to the left of an object, say E, but the order of objects H and I is not specified. The objects H and I are the *causes* of the event represented by object E. There are objects J and K that are the *effect* of the other object, say F that is their cause, and therefore J and K must be placed to the right side of F.

To assemble a complete microstate sequence from the subsequences of Fig. 59(a), subsequences 1–4 are placed in order, following the restrictions mentioned above. Sequence 5 of Fig. 59(b) shows an acceptable microstate sequence satisfying the requirements. Sequence 6 is unacceptable, since the order of C and B and the order of E and I are not correct. Our problem is to find all, or as many possible, sequences like 5, and to study the corresponding electrical transient phenomena in the circuit. Each sequence of events 1–4 of Fig. 59(a) is simple. Complexity originates from the huge number of possible microstate sequences generated by combinations. Complexity in digital circuit phenomena is that of *multiple exposure* of many simple phenomena, but which may happen in any order. This point, which is simple and obvious now, has not been universally recognized.

Microstate sequences of a circuit may have many possibilities, but not all of them take place in reality. A microstate sequence that includes many null sequences is unlikely. It is the objective of circuit analysis to reduce the number of sequences using specifics of the problem (for instance, circuit parameter values). Even if the number of microstate sequences is reduced, there are still many sequences that must be considered. Many microstate sequences end up with the final state that gives a correct logic answer. The difference between two such microstate sequences may reflect circuit performance like delay. Some microstate sequences may lead to undesirable circuit operation. Since all the possible modes of evolution of circuits can be presented in perspective, the microstate sequence is a powerful tool for describing and classifying circuit operating mechanisms.

Microstate sequences represent the causes of an effect, and the effects of a cause. When we draw microstate sequences, we identify which of two events is the *stronger* cause of the effect. Grading is given to the cause-effect relationships. Weak causes and irrelevant effects are deemphasized. The microstate sequence is an ideal symbolic tool to carry out such evaluation. Furthermore, microstate sequences can be used to define a *circuit problem* under consideration in precise terms. Microstate sequences may be considered as the *language* to describe operation of CMOS digital circuits.

1.17 Circuit Failure Analysis Using Microstate Sequence

In the last section we developed the concept of microstate sequence of CMOS digital circuits. The concept is useful in understanding the working mechanisms of CMOS digital circuits. In this section we show that the concept of microstates gives a clear definition of *digital* circuit failures. As we saw, the microstate of a circuit is a representation of the circuit configuration. Starting from the specified initial microstate, the circuit goes through a number of microstates and arrives at the final microstate. If the final microstate is not the correct one, obviously the circuit must have failed. Even if the final microstate is the correct one, the order that circuit configurational changes took place may matter, especially to performance. If n circuit configuration changes are required, there can be a maximum of $n!$ possibilities. Even if n is small, the number of possible microstate sequences can be huge, and because of this multiplicity, failure analysis is quite complicated.

Node voltages of the final microstate represent the answer to the logic operation executed. If the final microstate is not reached subject to the specified condition (like allocated timing budget), circuit failure took place. The microstate sequence of the inverter in Fig. 60(a) [whose node waveforms are shown in Fig. 60(b)] is given in Fig. 61, with corresponding notations (A, B, I, and II). Various possible node waveforms are shown in Fig. 60(b). If the delay

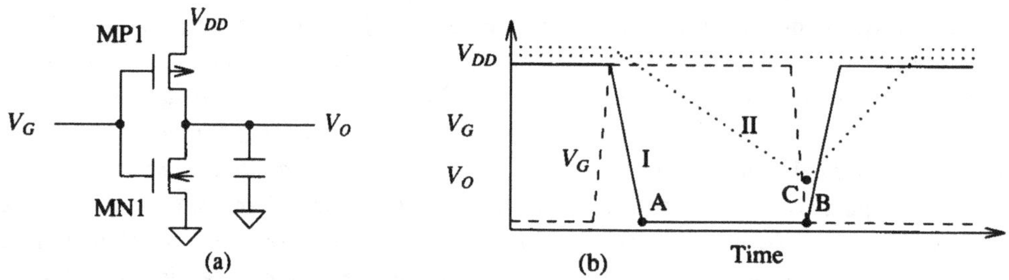

Figure 60 Inverter driven by up-going pulse

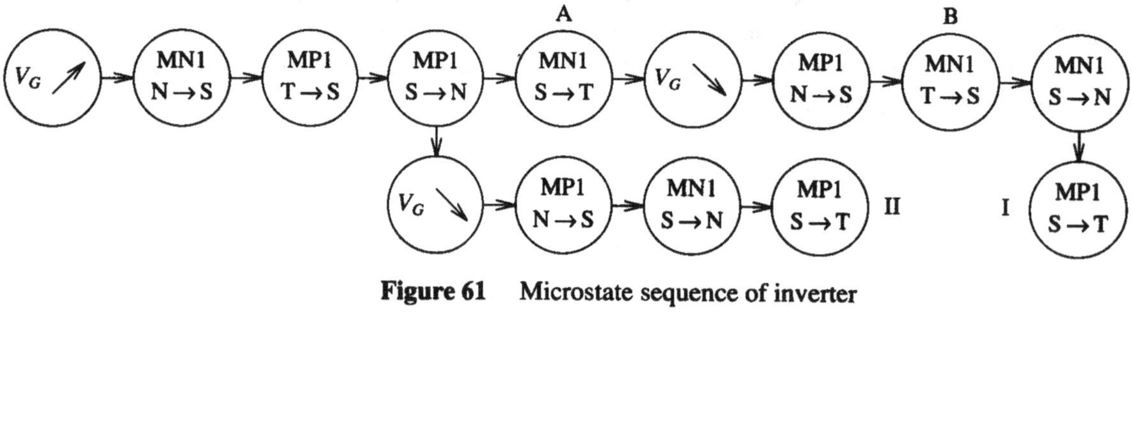

Figure 61 Microstate sequence of inverter

time of the inverter is short, the output node waveform is shown by curve I of Fig. 60(b), and the microstate sequence is shown by sequence I of Fig. 61. NFET MN1 is in the triode region from microstate change A to B of Fig. 61, and the downgoing output pulse has a flat bottom region [A–B of Fig. 60(b)]. The input pulse [V_G of Fig. 60(b)] was correctly inverted. If the delay of the inverter is long, the waveform is shown by curve II of Fig. 60(b), and since NFET MN1 never reaches the triode region, the microstate sequence is as shown by sequence II of Fig. 61. The output pulse does not have a flat bottom region, and this case can be identified as *fail*: pass/fail criteria defined in this way are crisp.

Switching waveforms derived using the collapsible current generator model of FETs are uniquely convenient in determining *pass* and *fail* of CMOS circuit tests. In order to define pass/fail of a circuit using other FET models, node voltage ranges that are interpreted as logic 0 and 1 must be determined. There is arbitrariness in this definition. Node waveforms have a gradual *exponential* tail, and it takes an infinitely long time to settle at the final voltage level. If the collapsible current generator model is used in the theory, however, CMOS circuit nodes arrive at the final voltage level within a finite time. There is no ambiguity originating from the *exponential* approach of node voltage to the final, steady-state node voltage. Only those circuit nodes that settle to the final voltage within a finite time represent *valid* CMOS logic variables. Internal nodes such as the output node of a source-follower (see Section 1.10) take infinite time to settle to the final voltage level. Such nodes do not represent a valid combination of input Boolean variables to the circuit, but they affect switching delay times of the (output) nodes that represent Boolean variables. Ambiguities in pass-fail criteria originating from voltage and time *uncleanliness* can be removed from CMOS digital circuit theory by using the collapsible current generator model. The example of Fig. 60 shows that the inverter fails if there is not enough time to complete the pull-down operations. The circuit model shows *existence* of failure mode if the correct final microstate is not reached. The reader will be persuaded that this is the only way to build unambiguous circuit failure theory. The node voltage waveforms used in the analysis that is based on the collapsible current generator model are somewhat between digital and analog variables: The waveforms are voltagewise digital, and timewise analog. Therefore it is possible to examine the state of circuit at a specified time, and to determine whether or not the circuit failed. Furthermore, failure analysis of circuits using the concept of microstates reflects closely the circuit properties derived from *connectivity* of the devices in the circuit. Properties of FETs are reflected through only a small number of parameters that have clear physical meanings (like saturation current and threshold voltage). Then it is possible to identify the cause of circuit failure cleanly, whether device or circuit is responsible. It might be thought that the collapsible current generator model of a FET is too idealized to represent FETs in real circuits. This is not the case. The idealized model and real FET characteristic can be correlated in a way that the former represents the latter very well. When the theoretical results derived from analysis that is based on microstate sequences is used to interprete the characteristic of real CMOS circuits, this point is taken into consideration. Circuit failure modes identified using microstate sequence has clear correspondence to real circuit issues.

In this book analysis of the CMOS digital circuit will be carried out in most cases in closed form, and the results will be plotted in normalized form to show the physical parameter dependence clearly. Complication of this analysis is limited to the determination of the roots of algebraic or transcendental equations as mentioned in Section 1.15. It is possible to simplify this procedure by using computers. In numerical analysis the complexity is reduced to triviality. To do so efficiently, however, the remaining analysis must also be carried out numerically. In order to study microstate sequences, it is convenient to write an independent program for each microstate. The program accepts a set of initial conditions and circuit parameters (like FET transconductance), and then it prints out node voltages. If a microstate transition is detected, the program outputs node voltages at the time of the configuration change. The output data are used as the initial condition of the program for the next microstate. Since programs are run interactively, what happens in the circuit can be understood very clearly.

References

[01] Sze, S. M. *Physics of semiconductor devices.* New York: John Wiley and Sons, 1981.

[02] Brews, J. R. *Physics of the MOS transistor.* Applied solid state science, supplement 2A. New York: Academic Press, 1981.

[03] Nicollian, E. H., and Brews, J. R. *MOS (Metal-Oxide-Semiconductor) physics and technology.* New York: John Wiley and Sons, 1982.

[04] Grove, A. S. *Physics and technology of semiconductor devices.* New York: John Wiley and Sons, 1967.

[05] Richman, P. *MOS field-effect transistors and integrated circuits.* New York: John Wiley and Sons, 1973.

[06] Wolf, S., and Tauber, R. N. *Silicon processing for the VLSI era.* Sunset Beach, CA: Lattice Press, Sunset Beach, 1986.

[07] Borkan, H., and Weimer, P. K. "An analysis of the characteristic of insulated-gate thin-film transistors." *RCA Review* 24 (June, 1963): 153–65.

[08] Shoji, M. *CMOS digital circuit technology.* Englewood Cliffs, NJ: Prentice Hall, 1988.

[09] Grignoux, P., and Geiger, R. L. "Modeling of MOS transistors with nonrectangular-gate geometries." *IEEE Trans. on Electron Devices* vol. ED-29 (August, 1982): 1261–69.

[10] de May, G. "A comment on 'Modeling of MOS transistors with nonrectangular gate geometries.'" *IEEE Trans. on Electron Devices* vol. ED-30 (July, 1983): 862–63.

[11] Shoji, M. "Analysis of high-frequency thermal noise of enhancement mode MOS field-effect transistors." *IEEE Trans. on Electron Devices* vol. ED-13 (June, 1966): 520–24.

[12] Hoeneisen, B., and Mead, C. A. "Current-voltage characteristics of small-size MOS transistors." *IEEE Trans. on Electron Devices* vol. ED-19 (March, 1972): 382–83.

[13] Carr, W. N., and Mize, J. P. *MOS/LSI design and application.* New York: McGraw Hill, 1972.

[14] Cullwick, E. G. *The fundamentals of electromagnetism.* Cambridge: Cambridge University Press, 1966.

[15] Houpis, C., and Lubelfeld, J. *Outlines of pulse circuits.* New York: Regents Publishing Co., 1966.

[16] Krambeck, R. H., Lee, C. M., and Law, H. S. "High-speed compact circuits with CMOS." *IEEE J. Solid-State Electronics* vol. SC-17 (June, 1982): 614–19.

[17] Goncalves, N. P., and de Man, H. J. "NORA: a racefree dynamic CMOS technique for pipelined logic structures." *IEEE J. Solid-State Circuits* vol. SC-18 (June, 1983): 261–68.

[18] Divekar, D. A. *FET modeling for circuit simulation.* Boston: Kluwer Academic Publishers, 1988.

[19] Shoji, M. "FET scaling in Domino CMOS gates." *IEEE J. of Solid-State Circuits* vol. SC-20 (October, 1985): 1067–71.

[20] Elmore, W. C. "The transient response of damped linear networks with particular regard to wideband amplifiers." *J. Appl. Phys.* 19 (January, 1948): 55–63.

[21] Professor F. Rosenberger suggested to set up a special symbol for fused node voltages, like V_A && V_C. This notation will have many advantages. The notation was not used in this book because the suggestion came too late to change the text.

[22] Ross, S. L. *Introduction to ordinary differential equation.* New York: John Wiley & Sons, 1966.

2 Method of Analysis of CMOS Circuit Failures

2.1 Introduction

An attempt to build a practical theory of CMOS digital circuit failures is frustrated by the fact that circuit failures occur rarely, more or less like spontaneous phenomena, and the varieties of known failures are limited. Evidence upon which scientific systematization should be carried out is scarce and fragmentary. This situation is made worse, since not many published records of circuit failures exist, and the records are widely scattered. A CMOS VLSI designer cannot feel a false sense of security from this situation, since any new circuit could fail in a way never expected. What is worse, the designer may be repeating the same failure someone else committed before. The basic aspect of circuit failure is that the issue includes psychological problems of the process of circuit invention, especially what is *not* obvious to and what is neglected by the mind of a designer-inventor. An observation made in chapter 1 that the complexity of circuit operation is basically that of *multiple exposure*, or of many events happening simultaneously, is revealing. If a systematic method of understanding complex electrical phenomena in a digital circuit is established, the problems can be reduced significantly. For of this reason, a theory of circuit failures must provide the methods of understanding CMOS circuits in great detail.

The first approach to the problems that are at the boundary between physical and human sciences is to compile records of past failure and to systematize them, as is practiced in biological and sociological sciences. This procedure establishes directions for research. In this book, such records are used as the starting point. This material, however, is not enough. After we find general patterns of circuit failure, we must generalize them to predict yet-to-be experienced failures. In order to carry out this step, we must generalize meager experience with the help of *gedanken experiments*. In a *gedanken* experiment, the behavior of the object, that is, a CMOS circuit, is determined not by a real *test pattern* experiment, but by analysis using the basic principles of the circuit theory. Efforts are made to determine and analyze conditions that cause the circuit to fail. If a set of conditions that causes circuit failure is found, its ramifications, especially how wide and how general is the failure mechanism, are studied. Important failure mechanisms have been generalized to cover a wide range of cases by this method. We classify circuit failures identified in this way into a group: CMOS circuits that are relevant to each major failure category are studied in detail in the following chapters.

A powerful method of gaining detailed insight into circuit failures is to characterize a combination of several FETs that comprise a fundamental circuit building block (typically logic gates, simple amplifiers, etc.) using the closed-form analysis methods, to the point that many circuit failure problems are revealed in the process of analysis. This is the traditional approach used until the 1960s. I wish to revive this method. There are many examples of this approach in this book. In some cases this approach need not go all the way to the final, closed-form results; understanding the microstate sequence of the circuit is often sufficient. This macroscopic characterization that utilizes the concept of a microstate sequence is perhaps the most productive method for future research into circuit failure. Whatever the approach we take, a thorough understanding of CMOS circuits is fundamental.

As the first step toward analysis of circuit failure, we require precise definition of the concepts involved, especially what is meant by *digital circuit failure*. Furthermore, we need to study various details of MOSFET circuits used to execute logic functions, thereby familiarizing ourselves with circuit operation and problems. Therefore, we begin this chapter with sections dealing with a quite general consideration of the definitions of digital circuit failure and their classification, and in following sections we study methods of analysis.

2.2 Definition of Logic Circuit Failures

It is necessary to define clearly what is meant by the term *digital circuit failures*. In a logic circuit built using semiconductor devices, Boolean variables are represented by node voltages. In an idealized model of a gate used in logic theory, the gate is able to take only two output voltage levels, V_L or V_H, where V_H is higher than V_L, or $V_H > V_L$. V_H is called the high (or 1) logic level, and V_L is called the low (or 0) logic level. This definition of logic voltage levels contains the following subtle details. In the inverting logic gate circuit shown in Fig. 1(a), the output node, O, is at voltage V_L when the input node, I, is at voltage V_H, and the output node is at voltage V_H when the input node is at voltage V_L. The two voltage levels V_L and V_H are defined *self-consistently* using the inverter by the theoretical model. How this definition is made and what its practical significance is in real logic circuits are the problems to be studied.

There are two clean ways to define the pair of logic voltage levels. The first is to use a cross-coupled latch circuit, as shown in Fig. 1(b). The circuit has two stable states: in the one, node Q is high (V_H) and node $\bar{Q}$ is low (V_L), and in the other, node Q is low (V_L) and node $\bar{Q}$ is high (V_H). Using this circuit, the pair of logic levels V_L and V_H can be defined clearly. The second way is to use cascaded multistage inverters, as shown in Fig. 1(c). In this inverter chain (having a large number of cascaded stages N), the N-th node voltage V_N can be either V_L or V_H, depending on input voltage V_{IN}.

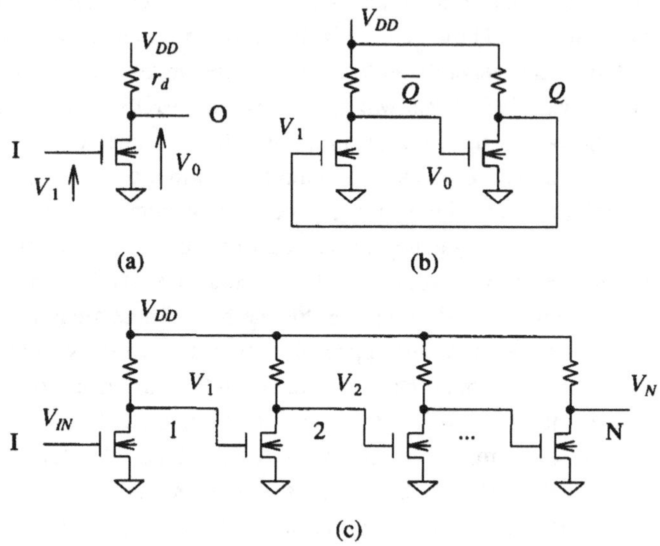

Figure 1 Definition of logic level

The two definitions give the same V_H and V_L. The two examples show that the logic level voltages can be defined clearly only for rather restricted conditions, when the logic circuit contains a data storage device like a latch, or when the circuit consists of long chain of the same type gates. If neither a storage device nor a long chain of cascaded gates exists, the logic voltage levels V_L and V_H cannot be defined clearly. In order to define the logic levels clearly, the signal must go through a large number of *amplifying* stages of gates, so that the final signal levels are determined by the balance between amplification and the *essential nonlinearity* of the *active circuit*. In Fig. 1(b) the signal circulates many times around the feedback loop, and in Fig. 1(c) the signal goes through a large number of stages, thereby accumulating voltage gain along the open, or closed paths, ultimately bound by the circuit's nonlinearity. We need to investigate the gain and nonlinearity of *active circuits*.

We investigate how node voltages of the cross-coupled latch circuit of Fig. 1(b) settle. Figure 2(a) shows input voltage V_1 versus output voltage V_0 of an inverter having a resistive load as shown in Fig. 1(a). The same inverter is used as both stages of the cross-coupled latch shown in Fig. 1(b). The NFETs are modeled using the collapsible current generator model, and the threshold voltage V_{THN} is positive. We then have

$$V_0 = V_{DD} \quad (0 < V_1 < V_{THN}) \quad = V_{DD} - r_d B_N (V_1 - V_{THN}) \quad (V_{THN} < V_1 < V_{1max}) \tag{1}$$

where B_N is transconductance of the NFET and where $V_{1max} = (V_{DD}/r_d B_N) + V_{THN}$. If $V_{1max} < V_{DD}$,

$$V_0 = 0 \quad (V_{1max} \le V_1 \le V_{DD}).$$

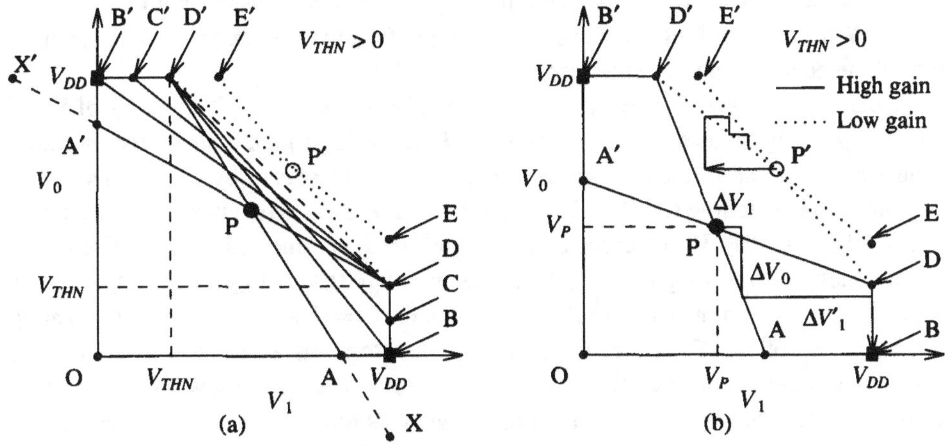

Figure 2 Enhancement mode MOSFET logic and stability of logic levels

---

Figure 2(a) shows V_0 versus V_1 of the first-stage inverter of the latch circuit, for different values of r_d and B_N. Parameter $G = r_d B_N$ is the *gain* of the FET *amplifier* with the resistive load.

Some typical cases shown in Fig. 2(a) are as follows.

> Curve (B′ D′ A B) is for $V_{THN} > 0$, and $V_{1max} < V_{DD}$ (large gain G, $V_{1max} = \overline{OA}$)
> Curves (B′ D′ B), (B′ D′ C), (B′ D′ D) and (B′ D′ E) are for $V_{THN} > 0$ and successively smaller gain G.

We note that the slope of the curve is the amplifier gain ($G = r_d B_N$, unsigned).

> Curve (B′ C′ D) for reduced V_{THN}.
> Curve (B′ D) for $V_{THN} = 0$ and for small gain G.

In the symmetrical cross-coupled circuit of Fig. 1(b), the second stage has the same input-output characteristics, which are plotted on the same V_1 and V_0 coordinate system of Fig. 2(a). Thus curve (B′ A′ D B) is the characteristic of the second-stage inverter having the same parameter values as curve (B′ D′ A B) of the first stage. The two characteristics are *mirror images* with respect to the diagonal line $V_0 = V_1$ of the coordinate system: one is derived from the other by exchanging V_0 and V_1. If the characteristics of the two stages are superposed as shown in Fig. 2(a) the voltages at which the cross-coupled circuit settles are determined from the intersections of the two characteristics.

There are two different types of intersections. Intersections like P and P′ are on the diagonal line of the coordinate plane ($V_1 = V_0$), and intersections like B and B′ are on the V_1 and V_0 coordinate axis, respectively. The two different types of intersections reflect different characteristics of the circuit. Two pairs of curves (B′ D′ A B), (B D A′ B′) for large G and curves (B′ D′ E), (B D E′) for small G are copied to Fig. 2(b). Intersections of type P (closed circle) and

P′ (open circle) always exist, but intersections of type B and B′ (closed square) do not exist if G is small. Whether or not intersections of type B and B′ (closed squares) exist determines the stability of the operating point of the type P and P', as follows.

Suppose that voltages V_0 and V_1 of the circuit of Fig. 1(b) are both set at V_P of Fig. 2(b). This is the operating point given by intersection P of the high gain characteristics shown by the solid curves (closed circle). Suppose that voltage V_1 increased by ΔV_1. Then using the $V_0 - V_1$ characteristic [curve (B′ D′ A B)], V_0 decreases by ΔV_0. Then using $V_1 - V_0$ characteristic [curve (B D A′ B′)], V_1 increases by $\Delta V_1'$. The steps are repeated, and the circuit operating point finally moves to point B (closed square, $V_0 = 0, V_1 = V_{DD}$) and settles there. If $\Delta V_1 < 0$, the operating point settles at B′ (closed square, $V_0 = V_{DD}$, and $V_1 = 0$) by repeating the similar *stairstep* motion. The operating point P is therefore unstable if $G > 1$. If $G < 1$, however, operating point P′ (open circle), which is the intersection of the low gain characteristics shown by the dotted curves, is stable, and this is shown as follows. If V_1 decreases, voltage V_0 and V_1 follow the stairway curve around point P′ (open circle) of Fig. 2(b), and the operating point returns to point P′. If $G < 1$, the intersection P′ is a stable operating point. A circuit is *active* if an intersection like P is an unstable bias point, and it is *inactive* (or passive) if P′ is a stable bias point. We conclude that if $G > 1$ the cross-coupled circuit is active, and it works as a bistable latch. If G is smaller, the circuit is inactive, and it settles at a single, stable operating point. The boundary between the two different circuit characteristics occurs when gain G is unity, which is the critical gain. The critical $V_1 - V_0$ characteristic is given by curve (B′ D′ D) of Fig. 2(a). This critical operating condition means that if the input voltage of the inverter of Fig. 1(a) is V_{DD}, the output voltage is V_{THN}, just low enough to turn off the next stage inverter. The latch circuit gives a stable logic level

$$V_H = V_{DD} \quad 0 < V_L < V_{THN} \text{ if } 1 < G(= r_d B_N) < V_{DD}/(V_{DD} - V_{THN})$$

$$V_L = 0 \text{ if } G > V_{DD}/(V_{DD} - V_{THN}).$$

Figures 3(a) and 3(b) show similar curves for $V_{THN} = 0$ and for $V_{THN} < 0$, respectively. We examine Fig. 3(a) first. If gain G is high [curves (B′ A B) and (B A′ B′)], intersection P (closed circle) gives an unstable operating point. In this case, the cross-coupled circuit settles at either of the two operating points B or B′ (closed squares). If gain G is low, however, curves (B′ C) and (B C′) have only one stable operating point P′ (open circle), where the circuit settles. This case is similar to the case $V_{THN} > 0$. Logic level voltages are

$$V_H = V_{DD} \quad V_L = 0 \text{ if } G > 1.$$

If $V_{THN} < 0$, Fig. 3(b) shows the similar plots. We have

$$V_0 = (V_{DD} - G|V_{THN}|) - GV_1 \quad (0 < V_1 < V_{1max}) \quad = 0 \quad (V_{1max} < V_1) \tag{2}$$

where $V_{1max} = (V_{DD}/G) - |V_{THN}|$. Curves (D′ E) and (D E′) are for gain $G < 1$ and curves (B′ A B) and (B A′ B′) for gain $G > 1$. If $G < 1$ there is one stable bias point, P′ (open circle). If $G > 1$ there is one unstable bias point, P (closed circle), and there are two stable bias points, B and B′ (closed squares). We have

$$V_H = V_{DD} - G|V_{THN}| \quad V_L = 0 \text{ if } G > 1.$$

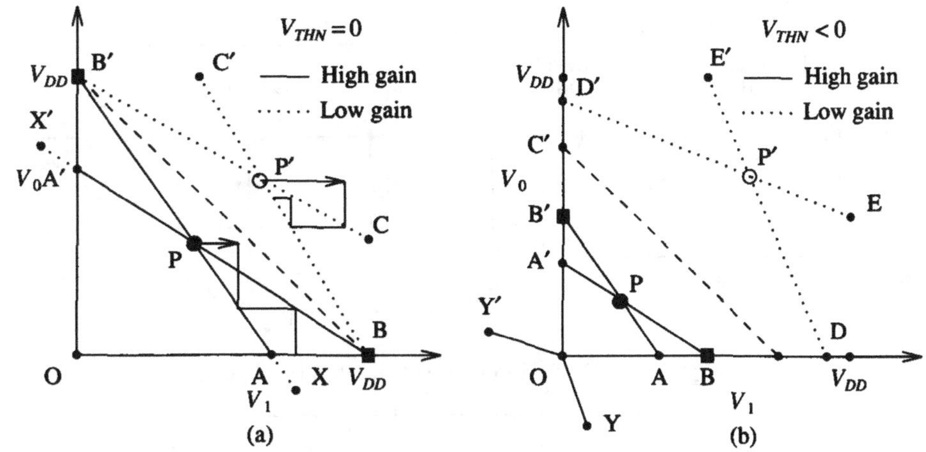

Figure 3 Depletion mode MOSFET logic and logic level problem

Figures 2 and 3 show how logic levels are determined in logic circuits built using enhancement (Fig. 2) and depletion (Fig. 3) mode FETs. In the enhancement mode logic circuit, logic levels are set at 0 and V_{DD} if gain G is larger than unity. The circuit design is simple and straightforward. In depletion mode logic circuits, the high logic level is $V_{DD} - G|V_{THN}|$. Since gain $G > 1$ is required, increasing gain means rapidly decreasing logic high level. Stable logic circuit design is still possible but is very difficult, and furthermore logic-high and logic-low levels cannot be maintained if device parameters (G and V_{THN}) fluctuate. The mechanism that determines the logic level is a balance between gain and nonlinearity, and the mechanism is complex. The complexity reflects directly in stability (reliability) of a logic circuit. If different types of logic gates are cascaded freely, it is practically impossible to pay attention to the logic levels of each gate. Therefore the simplest case is the only practical case. That is the enhancement mode circuit, with high gain. The CMOS logic circuit uses a pair of P-channel and N-channel enhancement mode FETs for pull-up and pull-down, respectively. Since both FETs are active, transconductances of the two FETs add. There is no resistance intentionally connected. In reality, internal resistance (drain resistance) of the FETs is not infinite, and therefore finite resistance loading exists, but the gain G of CMOS inverters is still high. Since CMOS logic circuits have enhancement mode FETs and the gain is high, the circuit is nearly perfect from the stability criteria. The logic levels are

$$V_H = V_{DD} \quad V_L = 0\ (V_{SS}).$$

If only depletion mode FETs are available, the gate circuit design becomes difficult. A well-established technique of avoiding the difficulties of depletion mode FET logic circuits is to use a floating voltage source for logic level shifting as shown in Fig. 4. If DC voltage

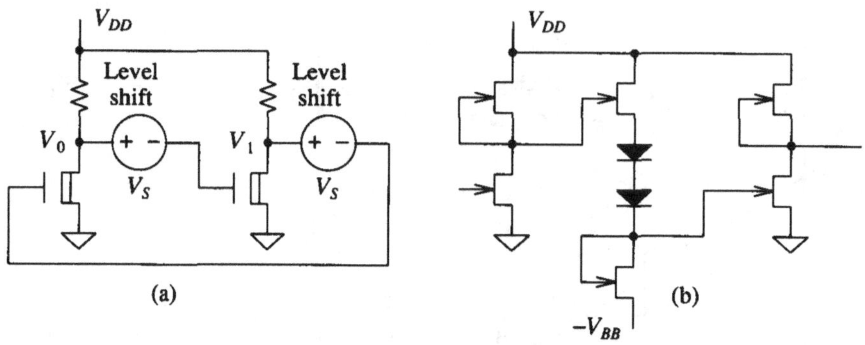

Figure 4 Depletion mode logic with level shift

source V_S (note polarity) is inserted at the output node of the depletion mode FET as shown in Fig. 4(a), V_0 and V_1 are related by

$$V_1 = V_{DD} - G(V_0 + |V_{THN}| - V_S)\,. \tag{3}$$

In effect the voltage level shift by V_S increases the threshold voltage of the second FET by V_S thereby converting a depletion mode FET into an enhancement mode FET. A GaAs MESFET logic circuit called BFL (buffered FET logic) shown in Fig. 4(b) uses this technique [01]. Power supply voltage $-V_{BB}$ is less than ground. The series-connected diodes provide the voltage level shift.

Let us consider CMOS gates. In the realistic device models, the devices are always represented by a resistance-like components. Settling of a logic node is effected by the resistance (the model of the device) charging the capacitance of the node (capacitance is an essential parasitic, Section 1.9). The node voltage approaches the logic level exponentially, and it takes infinite time to reach the logic level. This is called an *exponential tail.* Node voltage never reaches V_H or V_L due to the exponential tail. Then if node voltage is only slightly higher than V_L because of the exponential tail, should we say that the circuit never reached the low level and that it failed? This is not realistic. The collapsible current generator model removes the exponential tail in the theory. This is a significant advantage in theory, but because this is an advantage in theory only, we need to investigate many other aspects of the logic-level problem. Especially, are logic levels as a pair of *single* values acceptable?

To begin with, let us consider whether or not the voltages between V_L and V_H have Boolean values [02]. This is the issue of quantization. Output of a digital gate takes any voltage in the range between the extrema. The *analog* output voltage must be *quantized* to derive the digital, or the Boolean value represented by that. One way to do this is as follows. If the output node voltage is higher than a digital threshold voltage V_{LOGIC} the Boolean value is 1, and if lower, 0. This *quantization* process has several fundamental problems. The first problem is how V_{LOGIC} is chosen. A digital inverter has static input voltage (V_1)-output voltage [$V_0(V_1)$ is a function of V_1] characteristic as shown in Fig. 5(a) (see also Figs. 9 and 10 of Sec-

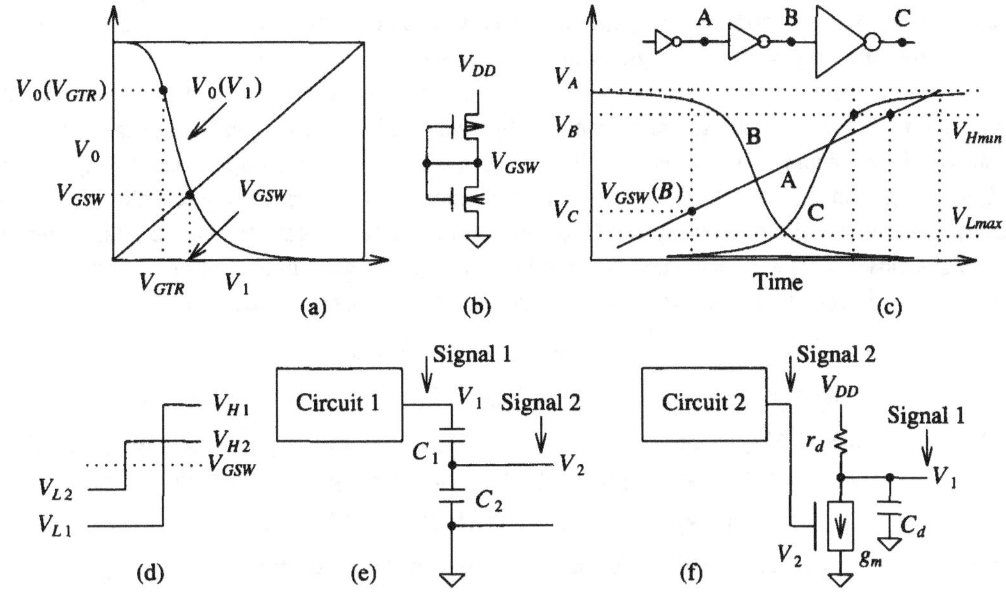

Figure 5 Amplitude and delay of digital signal

tion 2.4 as the real examples). Small-signal gain $G=|dV_0(V_1)/dV_1|$ is the slope of $V_0(V_1)$, which is low at small V_1, increases with increasing V_1, attains a maximum at V_{GTR}, and decreases and becomes low again at the maximum input voltage. The most obvious selection will be to set V_{LOGIC} at V_{GTR}. This is certainly rational, since the highest gain bias point is the boundary between the two low-gain regions that represent the high and the low levels. We need to note, however, that $V_1=V_{GTR}$ never means $V_0(V_1)=V_{GTR}$. Since generally $V_0(V_{GTR})\neq V_{GTR}$, both the input and the output of an inverter could have the same Boolean value. This is quite inconvenient. Input and output of an inverter must take complementary logic values in a steady state.

To satisfy this requiremernt, we look for the bias point where the input and the output of the inverter have the same voltage. Mathematically this is to seek for a solution of the equation, V_{GSW}

$$V_0(V_{GSW})=V_{GSW}\,.$$

The solution is shown by the intersection of the static characteristic $V_0=V_0(V_1)$ and the diagonal $V_0=V_1$ of Fig. 5(a). In MOS gates, V_{GSW} is the common voltage reached if the input and output are connected, as shown in Fig. 5(b). Generally $V_{GSW}\neq V_{GTR}$, but V_{GSW} is in the range where the small-signal gain is high. According to this choice, any voltage less than $V_{LOGIC}=V_{GSW}$ has Boolean value 0, and any voltage higher, 1. This simple definition, that we call *unguardbanded* quantization, still has the following problems.

1) V_{GSW} depends on the type of the gate, the device sizes and the many other circuit parameters. This uncertainty of V_{GSW} creates complicated problems: For instance, a voltage that is low for one gate may be high by the other gates.

2) If the gate is biased close to V_{GSW}, the gate becomes noise sensitive. The output voltage is controlled by input signal as well as by noise. If the gate input has noise $\pm V_N$, it is impossible to decide the Boolean value of the signal in the range from $V_{GSW} - V_N$ to $V_{GSW} + V_N$. Determination of the Boolean value can then be carried out excluding the region dominated by noise, and also by the V_{GSW} variation due to the gate types, etc. The range of voltages close to V_{GSW} of the gates is excluded from the Boolean quantization, and the *acceptable* range of high (V_H) and low (V_L) level voltages are defined, respectively, as

$$V_H > V_{Hmin} \text{ and } V_L < V_{Lmax}$$

$$\text{where } V_{Hmin} > V_{GSW} + V_N + \Delta V_{GSWH} \quad V_{Lmax} < V_{GSW} - V_N - \Delta V_{GSWL}$$

where ΔV_{GSWH} and ΔV_{GSWL} are the variations of V_{GSW} due to the gate types, etc., and V_{Hmin} and V_{Lmax} must still satisfy the compatibility requirement using the inverter characteristics,

$$V_0(V_{Hmin}) \leq V_{Lmax} \text{ and } V_0(V_{Lmax}) \geq V_{Hmin} .$$

In this *guardbanded* quantization, the input voltages in the range from V_{Lmax} to V_{Hmin} do not have Boolean values. V_{Hmin} and V_{Lmax} may be interpreted as *noise margin*. The range must be taken wide enough to cover V_{GSW} dependence on gate types, etc., as well as the noise effects. Where to set V_{Hmin} and V_{Lmax} is a problem. If they are set too close to V_{GSW}, the gate delay times determined by setting all the enable signals at the voltage levels become useless overestimates. If they are set too far away from V_{GSW}, the allowed ranges of V_H and V_L (that are limited by the power supply voltages as well) become too narrow. To strike a compromise between the two extrema, V_{Hmin} and V_{Lmax} can be chosen about

$$V_{Hmin} \approx V_{DD} - V_{THP} \quad V_{Lmax} \approx V_{THN} .$$

This definition is practical in CMOS chip testing. The automatic testing machines have the capability to force input signals thus specified, and to determine whether or not the output high and low level voltages are within the specified ranges. We rephrase this as follows: The automatic test machines have *intelligence* to identify the high and low levels defined by the guardband. The high and low voltage *bands* are convenient in logic analysis in gate level, which uses the simplest gate representation. This definition is, however, awkward in circuit level study. Definition of V_{Hmin} and V_{Lmax} is quite complicated, since it includes the maximum and minimum over the gate types, etc., which has no relevance to the operating mechanism of individual circuits. If we use this definition, many subtle electrical phenomena are buried under the ambiguity of the definition, and this we must avoid.

As we saw, the guardbanded logic level definition is adequate in the chip's input-output functional integrity test. The digital circuit failures are, however, the chip's internal circuit problems: we must determine whether or not a specific circuit in a logic gate chain (and not the entire chain) failed. Exclusion of voltages close to V_{GSW} from the Boolean quantization, although convenient for I/O characterization, puts an unnatural constraint to the interpretation of the circuit operation. A gate circuit responds to continuous input voltage waveform, and it has only one well-defined threshold voltage, V_{GSW}, of its own. Figure 5(c) shows the node A, B, and C waveforms of an inverter chain, with V_{Hmin} and V_{Lmax}. Each gate does not *know* the V_{GSW} of the other gates: The circuit inside a chip has no *capability*, or *intelligence* to identify and react to the *guardbanded* high and low voltage levels. Furthermore, the transients of the gate chain overlap, as are shown by waveforms A, B, and C of Fig. 5(c) (corresponding to the nodes A, B, and C of the inverter chain). This may be rephrased as follows. Voltages in the excluded range (from V_{Lmax} to V_{Hmin}) do not have Boolean values during the early phase of the transient, but they *acquire* their Boolean values after the signal propagated through the gate stages was *interpreted* by the gates, and they even may have produced the outputs at the chip's terminal. A signal at a node that has not yet reached the final logic level can produce a *digital* signal at a *downstream* node that reached the final logic level (and acquired the Boolean value). An upstream signal that is *not* an acceptable digital signal may produce a digital signal at the logic chain downstream. This is a violation of causality, which is unacceptable. We need to examine *real* information flow in digital circuits. Information moves from gate A to gate B when the output of gate A reached V_{GSW} of gate B. This is the time when the state of gate B is significantly perturbed by gate A output. At that time the node voltage has not yet reached the threshold of the guardbanded quantization to produce the Boolean value. Gate A, however, *intends* to change the node voltage in the direction, and that can be judged from the node A waveform as a whole. The destination of this intended change is the Boolean value that is *projected* to the future. Whether or not gate A worked properly is determined by the *projected* Boolean value. This revives the *unguardbanded* quantization to a limited extent. In this book we use the pass-fail criteria of a gate defined in this way. The Boolean value generated by the *unguardbanded* quantization is not the *real* but rather the *projected* or *interpreted* value. Whether or not the Boolean value projection is correct can be determined from the same circuit analysis that gives, above all, detailed node waveforms.

To avoid misunderstanding, we rephrase the conclusion of the last paragraph as follows. Suppose that a logic gate pulls down. Using the unguardbanded quantization, the operation result is generated when the voltage is reduced from V_H to V_{GSW} of the gate receiving the voltage. This is *not* sufficient evidence to assure the low Boolean value. Input voltage at V_{GSW} has no Boolean value, whatever the quantization scheme may be. If the voltage is held at V_{GSW}, there is no way to determine the *intention* of the node driver. As we will see in the following examination, a voltage close to V_{GSW} takes a long time to determine its Boolean value, and as such, the voltage has no digital information. In this case, the information is carried by the time derivative of the voltage, and not by the voltage itself. If the node voltage has been decreasing for some time, and if there is no mechanism operating to revert the change, the *intention* of the circuit driving the node is certain. Then it makes sense to reinstate the *unguardbanded* quantization, with assurance from the node waveform. The electrical signal that passed the node produces an output digital signal at the chip's terminal some time later. We consider that the chip's digital output is produced from the projected Boolean value of the internal node signal: the projected Boolean value becomes real some time later. At the time of projection, the Boolean value is not *real*: It is an *interpretation* of the changing node state. We note that the chip's internal node Boolean values are all of this *interpretation*, which can be inferred from the details of the internal node transients, and which produce the processed digital signal at the output terminal, whose integrity is determined using the guardbanded node voltage quantization. In this way a violation of the causality of the digital signal propagation is avoided.

The problems of quantization may sound as if it were useful only as an intellectual curiosity. This is not the case. Since logic gates are complex analog circuits used in a very simple digital operation, the electrical complexity *creeps* into the *logic* simplicity, and confusing problems appear. Depending on the setting of a gate's V_{GSW} and the rise/fall times of the signal, the final result of processing can be generated earlier than the intermediate results, as shown by signals A and C of Fig. 5(c) (see more in Section 5.16). In logic circuit analysis that does not include time, this is harmless, but it is inconvenient, at the least, in the circuit analysis that always includes time.

An input voltage, V_{IN}, that is higher than V_L but lower than V_{GSW} has in many cases the same effect as the true logic low voltage V_L in a long logic chain. In the same manner, an input voltage, V_{IN}, in the range $V_H > V_{IN} > V_{GSW}$ may produce the same logic answer at the output terminal as input voltage V_H. In CMOS logic circuits the range of *allowed* input voltages is very wide. CMOS gates have very crisp switching characteristics. Figure 9 of Section 2.4 shows the static switching characteristic of a CMOS inverter. Ten percent of possible input voltage variation centered at the switching threshold voltage swings the output node voltage more than 80% of the difference between logic 1 and logic 0 voltage levels. The input voltages close to V_{GSW} have the following fundamental problem.

If the amplitude of the input signal is reduced to small value centered at V_{GSW}, the time required to generate digital signal at the output that has the required amplitude increases. It takes successively longer time to acquire Boolean value 0 or 1 as the input amplitude decreases. If the input amplitude is reduced more, the swing of the output signal voltage becomes less than the digital amplitude, and the guardbanded quantization becomes impossible. Thus it is quite natural to consider that the signal that passes V_{GSW} *acquires* its Boolean value not at the time, but *later*. Furthermore, we see that the amplitude of the input signal has a measure of *quality*, and the larger amplitude signal has better quality.

Consider two logic circuits. The first circuit generates logic voltage levels (V_{H1}, V_{L1}) where $\Delta V_1 = V_{H1} - V_{L1}$ is the logic amplitude, and the second circuit generates (V_{H2}, V_{L2}) where $\Delta V_2 = V_{H2} - V_{L2}$ is the logic amplitude, as shown in Fig. 5(d). We assume $\Delta V_1 > \Delta V_2$. We assume that both digital signals are available from low internal impedance sources. We ask a question: Which one is the higher *quality* digital information? This problem can be answered as follows. We attempt to generate the second signal from the first, and the first signal from the second. If one conversion requires essentially *something* extra, the difference determines the *quality* of the signals. The conversion can be carried out using the circuit shown in Fig. 5(e) and (f), respectively. To generate the second (small amplitude) signal from the first (large amplitude) signal, a simple capacitive voltage divider circuit of Fig. 5(e) is sufficient. The voltage division ratio is

$$\Delta V_2 / \Delta V_1 = C_1 / (C_1 + C_2).$$

The divider circuit delay can be made as short as we wish. This means that the circuit consists of passive components alone, and the delay can be made arbitrarily small by reducing the dimension of the components. To generate the first signal from the second, we need an amplifier which could consist of a controlled current generator having transconductance g_m and a load resistance r_d. As we discussed in Section 1.9, any active device has parasitic capacitance C_d such that the device time constant $T_d = C_d/g_m$ is independent of the size of the device (since capacitance is an *essential* parasitic, this conclusion cannot be avoided). T_d is essentially a lower bound of the delay time of a signal through the amplifier, when input and output signals have the same amplitude. If gain $G = \Delta V_1/\Delta V_2$ is larger than 1, the delay time is increased to $G \cdot T_d$. Signal 1 appears at the output of the amplifier delay time $T_D = (\Delta V_1/\Delta V_2)T_d$ later. We may conclude as follows: Since there is no way of *advancing* the signal in time (otherwise causality is violated!), the signal that arrives earlier is better, and that criteria can be traded with signal amplitude. Signal 1 is *better* than signal 2, by time T_D. Signal 1 becomes closer to signal 2 if a better amplifying device is available, but the cost difference will never be zero. The assumption of low internal impedance source is justifiable since the quality is measured by time, and not by electrical energy.

Some readers may wonder whether the argument holds if an ideal transformer is available. This is certainly the case, but the really crucial point of the argument is that an amplifier is always non-ideal, since the *essential* parasitic capacitance [C_d of Fig. 5(f)] is associated with the transconductance. By the same token, transformer windings are always associated with the *essential* parasitic capacitance, and if that is included, the delay is inevitable (the *cost* factor may not be the same, however). The crucial point is that the parasitics of the capacitance of the voltage divider can be made small, but those of the amplifier and the transformer cannot.

Pass-fail of digital circuits can be determined only after defining an exchange ratio between the time and the amplitude. Input voltage in the range $V_0 < V_{IN} < V_{GSW}$ or $V_{GSW} < V_{IN} < V_1$ gives the same logic output if delay time is neglected. If the *quality* of the information measured in the delay time is included, logic low level input that is higher than V_L, and logic high level input that is lower than V_H have poorer quality, because such input signals generate output signals later than full amplitude logic signals. To carry out valid and consistent theoretical study of circuit failures, we must provide an input signal level (V_H, V_L) to the circuit, and we need to examine whether or not an output signal level (V_H, V_L) is reached within the allowed time.

Before concluding this section, we define the difference between *digital signal* and *signal*, which are frequently used in this book. In CMOS digital circuits, a signal is carried by a charge stored in a circuit node, which is measured as the node voltage and the node capacitance serves as the conversion factor. The flow of signal is therefore the flow of charge, or electrical current. CMOS circuits that generate large amplitude digital signal voltages are able to respond to continuous input signals. Consider two independent circuit nodes. The first node and the second node interact through circuits that are provided by the designer, or through spurious circuits that are not desired. If a logic design error does not exist, interaction of the two nodes by sending and receiving *signals* having large amplitude is impossible. This point may be explained as follows. In order to send a large amplitude digital signal, the node must receive current that is comparable to the maximum current a typical size FET is able to conduct. Such a large current can be injected only if there is misconnection of the device (logic error), or if the FETs that are connected correctly work differently from the designer's intention (an example is timing failure, as it will be discussed in the next section). If the current is smaller than that, it *appears* impossible to create a voltage change that is comparable to the logic voltage swing, at least within the period of time of one clock cycle. If the first node voltage has one digital signal level (say V_H), and if the second node has the other logic signal level (V_L), interaction through passive circuits tends to reduce the difference between the two voltage levels. This type interaction has been considered unimportant in CMOS digital circuits, since logic levels are widely separated. Such is the assumption made by circuit designers when new circuits are conceived, but it is not a realistic assumption in scaled-down CMOS circuits. In many circuit failures, one node affects another node through *small signal* interaction. Digital circuit failures happen when one or more digital nodes send small *current signals* to another node, and the small effect is *accumulated* by the receiving node. Because of the multiplicity of such interactions, or because of the hypersensitivity of the receiving node that allows accumulation of the small effects and amplification, an error at the digital signal level may be created. *Signal* in circuit failure study means *small* amplitude current from one digital node to another. Indeed, the unguardbanded quantization requires interpretation of signal transfer from one gate to the other as many small amplitude signal transfers.

2.3 CMOS Digital Circuit Failure Mechanisms

The first step in studying CMOS circuit failure will be to classify circuit failures into categories. In the following we introduce four categories of circuit failure. I believe that the method by which the classification was established is less important than classification itself. Therefore the classification is presented as the starting point for this section. During the years when I worked as a CMOS digital circuit designer, I made several circuit design errors, and I heard the case histories of many other design projects that had circuit failures. When I asked myself whether or not there exists some general pattern of circuit failures, the answer appeared in the positive. An example is the charge-sharing problem of Domino CMOS circuits, which is aggravated in NORA dynamic logic circuits. This failure mechanism can be generalized to include many other circuit failures. The first failure category that is discussed in this section—anomaly in signal flow—was established by generalizing Domino circuit charge-sharing.

[A] Anomaly in signal flow

The basic mechanism of this circuit failure is as follows. A CMOS digital circuit is a web of connected active devices and passive components. Many currents flow between the circuit nodes through the active devices and the passive components. The logic operation that we wish to execute by the circuit, however, utilizes only certain current flows. Most of the current

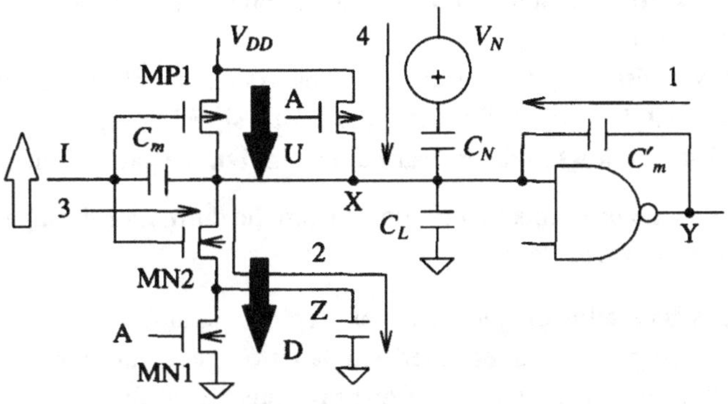

Figure 6 Signal flow in CMOS circuits

flows that take place in the circuit are either irrelevant to the logic operation, or they are positively harmful. This point is clear from the following observation. Since capacitance is an *essential* parasitic (Section 1.9), any circuit node having nonzero size has parasitic capacitance to all the other nodes. Although the capacitance is small, it is never zero. If the capacitance is considered, any two circuit nodes are capacitively coupled. In a CMOS gate, some of the internal nodes of the gate are connected by conducting FETs. Since any pair of circuit nodes is connected through finite impedance, the flow of currents among the circuit nodes is quite complex. Figure 6 shows two cascaded CMOS static NAND2 gates (one is shown by FET symbols and the other by logic symbol). Suppose that the first stage NAND gate is enabled (signal A is a high level). If input signal I switches, only two currents U (for pull-up, PFETs) and D (for pull-down, NFET chain current) are needed for the desired gate operation (they are shown by thick arrows). The problem is that in this circuit many currents other than currents U and D flow, as shown in Fig. 6.

A short explanation of the extra currents is required. If the NAND2 gate shown by the logic symbol switches, the change of the output node Y voltage is transmitted back to the input node X through Miller capacitance C'_m. This current is shown by arrow 1. If the first stage NAND2 gate is disabled (signal A is a low level), and if input signal I makes a low-to-high transition, current 2 flows from the output node X to the gate's internal node Z, if the initial voltage level of node Z is low. The current is shown by arrow 2. If input node I makes a transition, the Miller capacitance C_m transmits the change of input voltage directly to the output, as current, indicated by arrow 3. Currents 1 and 3 flow through similar input-output coupling capacitance (Miller capacitance), but for convenience of presentation they are explained in different gates. Voltage changes of the circuit nodes that are outside the cascaded NAND2 gate circuit induce digital circuit *noise* at node X through coupling capacitance C_N, as shown by arrow 4. This signal is qualitatively different from signals 1–3 since the signal is effectively random in timing, in amplitude, and in waveform.

To summarize the observations from Fig. 6, circuit failure may be caused by the extra currents that flow through the gate circuit. The extra currents are classified as follows:

1) Current flows from the output of a gate to the input (signal flow backward to the digital signal flow direction).
2) Current flows within a gate in such a way as to degrade digital information at the output node (signal flow perpendicular to the digital signal flow).
3) The digital signal flows in the forward direction, but the gate's control function is lost.

If we generalize the problems that originate from the current or the signal flow anomalies, we must include the following items:

4) spontaneous generation of signal (self-excited oscillation);
5) signal that is *stagnant*, or interrupted at a data storage circuit (metastable state);
6) mechanical or electrical breakdown that prevents signal flow due to open or short circuits. Both open and short circuits may prevent signal flow.

Mechanism (6) involves *mechanical* breakdown of connections caused by process imperfections or severe external influences like electrostatic discharge (ESD). A similar effect results from process variations that shift normal DC operating point of the circuit so much that the gate fails to respond to the input voltage change.

[B] Digital noise

Scaled CMOS technology has circuit nodes so lightly loaded by capacitance that even digital circuits are vulnerable to noise. Digital circuit noise is, however, not due to thermodynamic fluctuation or shot effect, as in linear circuits, but rather to spurious voltage induction from the other nodes of the circuits. Noise phenomena belong to anomalous signal flow, but there is one significant difference. The source of noise does not belong to the circuit under consideration, and, as such, noise waveforms are unknown. In spite of that, digital noise is not, in its strict sense, a random phenomenon. If a very powerful simulator becomes available, digital noise effects can be analyzed. From a practical viewpoint, however, digital noise has every attribute of a random and unpredictable noise phenomenon that originates from the system's *complexity*. Digital noise can be classified into three types, depending on the mechanisms.

1) Induced noise: noise voltage induction to one circuit node of a chip from the other circuit nodes on the same chip;
2) Power-bus noise: discharge current spike of on-chip or off-chip circuit creates power-bus voltage fluctuation, which affects all, or a significant part, of the circuits on the chip;
3) Communication noise: noise voltage induced on the transmission lines and wires that connect the chips together. Communication noise is different from induced noise in the node's impedance level, logic amplitude, mechanisms of noise induction, and the effects of noise.

In scaled-down CMOS VLSI chips, noisy circuits on a chip cannot be effectively isolated from the rest of the circuits, and sensitive circuits cannot be protected effectively by shielding.

[C] Recombination of split digital signals

Circuit failure mechanisms that are due to signal flow anomaly are essentially independent of time. Circuit failures that involve time are our next category. At time $t = 0$ the signal source provides data to a logic circuit. At a later time, $t = T_D$, voltage levels at the outputs of the logic circuit are examined and the levels determined. The signal sources are CMOS D-latches (Section 5.6). When the clock of the D-latch makes a low-to-high transition, the latch delivers the data to the logic circuit. The electrical signals representing the source data are copies of the source latch clock, whose levels reflect the input data. The destinations that examine the output data are CMOS D-latches also. D-latches are really testing devices of circuit. The clock that drives the destination D-latches is generated by delaying the clock signal that drives the signal source latches. The clock is delayed either by on-chip or off-chip circuits. In the case where the clock is delayed by the off-chip circuit (by the clock generator circuit), it is not obvious whether the destination clock is a delayed version of the source clock, but it really is. The logic circuit itself is a cascaded logic gate chain. The structure of a logic circuit together with the clocking arrangement is shown in Fig. 7. Chain A is the logic circuit, and chain B is the clock delay circuit. The source signal (in this case, clock) is split into two, and the split clocks travel through the two different logic gate chains. They are recombined by the CMOS D-latch that works as a digital waveform *mixer* circuit. If the chain A signal arrives earlier than the chain B signal, the circuit produces the correct answer. If the chain A signal arrives later than the chain B signal, the logic operation has error. We note that the circuit fails if the delay relationship between the two logic chains is not right. By generalizing this observation, the two logic chains A and B need not be the logic circuit and the clock delay circuit, and the circuit that combines the two signals needs not be the D-latch: Any logic circuit that splits the signal, processes the two signals separately, and later combines the signals, has similar problems.

Circuit failures originating from recombination of split signals include the following cases:

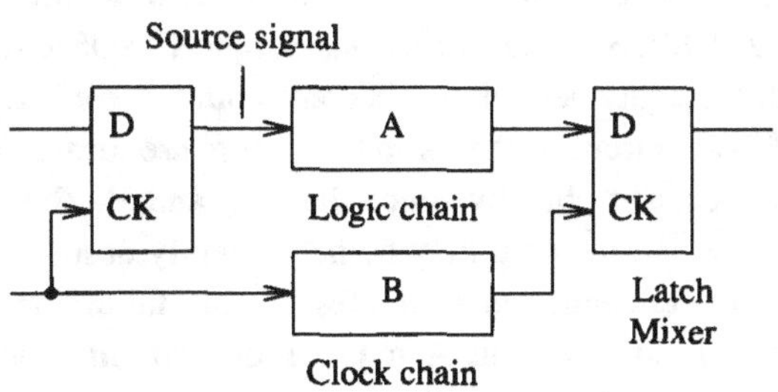

Figure 7 Recombination of split signals

1) failure to store processed data into the destination storage circuit. This is the mechanism that sets the maximum clock frequency of the circuit.

At the ends of the gate chains there is a waveform mixer that is different from the frequency mixer in a superheterodyne receiver (which is sensitive to the envelope of waveform, but not to the waveform itself); a *digital* waveform mixer is input waveform and phase (delay) sensitive. From this extra sensitivity a number of circuit failures originate, including the following:

2) generation of extra pulse(s) or glitch(es);
3) signal waveform distortion, which results in signal delay uncertainty;
4) increased sensitivity of circuit delay to process, temperature, and power supply voltage variations.

Forward transmission of signal through a gate that has lost control because of input signal timing could be classified to the mechanism of anomalous signal flow, or to the mechanism of recombination of split signals. We classify this failure mechanism to the latter for the following reasons. First, two input signals are involved. Both signals originate from latches that are controlled by the system clock. Therefore the failure mechanism involves recombination of split signals. Control failure of the gate transmits a large signal from input to output. The concept of anomalous signal flow, however, assumes a small signal. Therefore the failure mechanism is better classified to the mechanism of recombination of split signals.

[D] Circuit connectivity uncertainty

Real circuits always contain capacitive, inductive or resistive parasitics. Circuits that are built on a conductive silicon substrate may even have parasitic *active* devices like bipolar transistors and thick-oxide MOSFETs. Under ordinary conditions, the parasitic components may degrade circuit performance, but they do not cause catastrophic circuit failures. There are circuits, however, that are sensitive to such parasitics. A completely different and, in its own right, stable and reliable circuit emerges if the parasitics are *activated.* Subject to a small disturbing influence, the circuit may switch from one configuration to another, thereby causing catastrophic circuit failure. A CMOS output driver is a simple CMOS circuit. If parasitic bipolar transistors are included, the equivalent circuit becomes quite complex. This circuit works as a bipolar transistor (BJT) latch circuit if the extra transistors are turned on. If the latch is turned on, the MOS driver fails catastrophically, since the V_{DD} and V_{SS} buses are effectively short-circuited through the BJTs, and the chip may be permanently destroyed. This phenomenon is called latchup. Frequently encountered examples of this failure mode have the following mechanism. Normal operation of circuit A requires that circuit B is working normally, but circuit B works only if circuit A works. This type of circuit requires a power-up reset, or initialization mechanism, that is often overlooked. Another example is a circuit that has gain, and parasitics add a path for positive feedback. Spontaneous oscillation, or latching action, results.

[E] Parametric uncertainty

It is often believed that simulation of circuit level delay is accurate enough for verification of any circuit design. In reality it is not. Unexpected problems may arise from uncertainty or unreliability of delay simulation, which may lead to wrong decisions by the designer. Circuit delay depends on the conditions of so many digital signals that not all the combinations of their states can be covered by simulation that deals with one case at a time. The *worst-case* condition is often unidentifiable or quite misleading. Some of the information that is necessary for delay simulation may not be available when the simulation is carried out.

Delay and parameter uncertainty can be classified as follows:

1) essential delay uncertainty of gate (dependence of delay on digital data and on digital signal dependent loading conditions);
2) uncertain design data (such as process and temperature conditions);
3) uncertain operational specifications (input signal timing, simultaneous switching signal, slew rate, Miller effect);
4) certain parasitics that cannot be determined precisely;
5) certain circuit parameters cannot be determined by simulation under certain conditions.

Item (5) originates from many circuit parameters (such as relative delay of two logic chains) that are not *positive definite*, and therefore the limit of simulation error cannot be established. This is predominantly a mathematical issue, but it has a significant impact on circuit design, as it will be shown in Section 6.16.

Based on the observations of this section, CMOS digital circuit failures can be classified as follows:

I. Circuit failures
- A) Failures due to abnormal signal flow
  - a) Backward signal flow — Miller effect
  - b) Unprocessed signal bypassing a gate — Feedforward
  - c) Signal deterioration within a gate — Charge-sharing
  - d) Spontaneous signal generation — Self-excited oscillation
  - e) Interruption of signal propagation
    - α) Classical logic gate failure, or mechanical breakdown
    - β) Metastable state
    - γ) Wrong DC bias point setting
- B) Noise in digital circuits
  - a) Local induced noise
  - b) Global circuit failure due to power-bus noise
  - c) System and communication noise
- C) Failures due to recombination of split signals
  - a) Circuit failures at the frequency limit
  - b) Hazard or generation of glitch
  - c) Waveform distortion due to digital signal recombination

- D) Unstable circuit connectivity
  - a) Inclusion of parasitics changes equivalent circuit structure drastically
- E) Circuit design failures due to insufficient or inaccurate design information
  - a) Essential uncertainty of gates—digital data pattern dependence
  - b) Essential limit of numerical simulation
  - c) Improper assessment of the effects of process, voltage and temperature variations
    - α) Bias point setting problems
  - d) Environmental problems
    - α) Ordinary hostile environment—high temperatures, ESDs, etc.
    - β) Ionizing radiation—neutrons and γ rays, dose effect and single event upset

Items (1) and (3) correspond to the two quality criteria of logic signal, "How large a signal is available and how soon?" that was discussed in Section 2.2. Item (3), recombination of split signal is, in essence, a procedure of measuring *time*, which is a relative parameter: When a signal is split and then recombined later, we are effectively comparing delay times of the two signal paths, or measuring delay time of one path using that of the other as the standard.

In order to cover CMOS circuit failures completely, we need to include circuit failures originating from the system level design and human aspects such as project planning and organization failures. Based on the author's past experience, the following items can be listed:

II. System design and project failures
- A) System assembly failures
  - a) System's conceptual failures
  - b) System integration failures
    - α) Chip-to-chip communication failures due to noise and metastability
- B) Wrong technology selection
  - a) Technology other than CMOS should have been used
  - b) Merits of CMOS not fully exploited
- C) Development project execution failures

Although the subjects summarized above will be discussed using CMOS as the example, many failure problems are common to other IC technologies.

A difficulty of classifying circuit failures into groups is that some failures can be classified into more than one group. Digital circuit noise and abnormal signal flow have many aspects in common. In this book, ambiguous subjects will be discussed only in the section where the strongest connection exists, and the related classification will be mentioned there.

The basic strategy adopted in this book is to investigate in depth those electrical phenomena that belong to categories (1)–(5), and to understand the phenomena well enough to be able to discover problems of circuit operation during design. I believe that this approach will be more efficient and more systematic than describing the known cases of circuit failures one by one, and thereby avoiding the confusion of trivial and repetitive discussions. The cost of this approach is a heavy dependence on mathematical analysis.

2.4 Static Characteristics of CMOS Gates

In this chapter we intend to make a gradual transition from the conventional gradual channel, low-field FET model and its close associate, the equivalent linear resistor model, to the new collapsible current generator FET model. Therefore we discuss some results from the conventional model and compare them with results derived using the collapsible current generator model. We begin with *static* characteristics of CMOS inverters that are frequently used in this book.

The static relationship between the input (V_G) and the output (V_D) voltages of CMOS inverter can be determined using the gradual channel, low-field FET model of chapter 1 as follows [03]. With reference to the CMOS inverter circuit shown in Fig. 8, the current of NFET MN1, I_N, is given by

$$I_N = B_{GN}(V_G - V_{THN} - \frac{1}{2} V_D)V_D \tag{4a}$$

if $V_D < V_G - V_{THN}$. Or if the NFET is in saturation,

$$I_N = \frac{B_{GN}}{2}(V_G - V_{THN})^2 \quad (V_D > V_G - V_{THN}). \tag{4b}$$

For the PFET,

$$I_P = B_{GP}[V_{DD} - V_G - V_{THP} - \frac{1}{2}(V_{DD} - V_D)](V_{DD} - V_D)$$

if $V_D > V_G + V_{THP}$. Otherwise the PFET is in saturation,

$$I_P = \frac{B_{GP}}{2}(V_{DD} - V_G - V_{THP})^2 \quad (V_D < V_G + V_{THP}).$$

The definition of parameters B_{GN} and B_{GP} are given in Section 1.2.

If $V_D < V_G - V_{THN}$, the law of current conservation requires that

$$B_{GN}(V_G - V_{THN} - \frac{1}{2} V_D)V_D = \frac{B_{GP}}{2}(V_{DD} - V_G - V_{THP})^2$$

and if $V_D > V_G + V_{THP}$

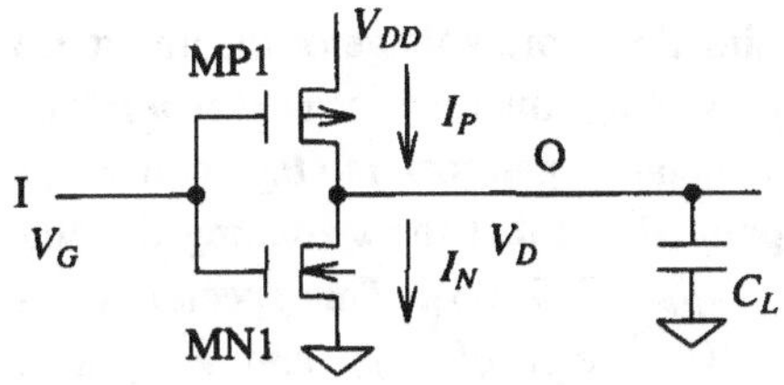

Figure 8 Analysis of CMOS inverter static characteristics

$$B_{GP}[V_{DD} - V_G - V_{THP} - \frac{1}{2}(V_{DD} - V_D)](V_{DD} - V_D) = \frac{B_{GN}}{2}(V_G - V_{THN})^2 .$$

Solving these equations for V_D we obtain

$$V_D = (V_G - V_{THN}) - [(V_G - V_{THN})^2 - \frac{B_{GP}}{B_{GN}}(V_{DD} - V_G - V_{THP})^2]^{1/2} \text{ if } V_D < V_G - V_{THN} \tag{5}$$

$$V_{DD} - V_D = (V_{DD} - V_G - V_{THP}) - [(V_{DD} - V_G - V_{THP})^2 - \frac{B_{GN}}{B_{GP}}(V_G - V_{THN})^2]^{1/2} \text{ if } V_D > V_G + V_{THP} .$$

Since the two equations contain V_G within the square root sign, $\partial V_D/\partial V_G$ diverges if the content of the square root tends to zero. We have $\partial V_D/\partial V_G \rightarrow \infty$ if $V_G = V_{GSW}$ where

$$V_{GSW} = \frac{\sqrt{B_{GP}}V_{DD} + \sqrt{B_{GN}}V_{THN} - \sqrt{B_{GP}}V_{THP}}{\sqrt{B_{GN}} + \sqrt{B_P}} \tag{6}$$

$$\text{or } v_{GSW} = \frac{V_{GSW}}{V_{DD}} = \frac{1 + \sqrt{\beta}v_{THN} - v_{THP}}{1 + \sqrt{\beta}}$$

where v_{THN} and v_{THP} are the normalized FET threshold voltages defined by $v_{THN} = V_{THN}/V_{DD}$ and $v_{THP} = V_{THP}/V_{DD}$, respectively, and where $\beta = B_{GN}/B_{GP}$ is the ratio of the FET's beta values. Normalizing the voltages using $v_D = V_D/V_{DD}$ and $v_G = V_G/V_{DD}$,

$$v_D = (v_G - v_{THN}) - [(v_G - v_{THN})^2 - \frac{1}{\beta}(1 - v_G - v_{THN})^2]^{1/2} \quad (v_G - v_{THN} > v_D) \tag{7}$$

$$1 - v_D = (1 - v_G - v_{THP}) - [(1 - v_G - v_{THP})^2 - \beta(v_G - v_{THN})^2]^{1/2} \quad (v_D > v_G + v_{THP}) .$$

Solid curves in Fig. 9(a) show v_D versus v_G, for $\beta = 1/3$, 1 and 3, assuming typical threshold voltages of $v_{THN} = 0.12$, and $v_{THP} = 0.2$. In the triangular area above the dotted line P of Fig. 9(a), PFET MP1 of the inverter is in the triode region. Below line P the PFET is in the saturation region. Similarly, below the dotted line N, the NFET MN1 is in the triode region and is in the saturation region above. In the region between lines N and P both PFET and NFET are in the saturation region. A FET in the saturation region acts as a current generator. Therefore in this region equilibrium can be reached only for a single gate voltage where the current of the NFET and of the PFET are equal.

This voltage, the switching threshold voltage of the inverter, V_{GSW}, is given by Eq. (6). In the case $B_{GP} = B_{GN}$ and $V_{THP} = V_{THN}$, the inverter is *symmetrical.* Unequal threshold voltages of FETs or unequal transconductance parameters (B_{GP} and B_{GN}) cause asymmetrical switching characteristics. Burns pointed out that the switching threshold voltage is independent of beta and is $V_{DD}/2$ if $v_{THN} = v_{THP} = 0.5$ [03]. This special case is interesting but has not yet found practical application. The range of v_{THP} and v_{THN} in conventional bulk CMOS is 0.1–0.2. If $v_{THN} + v_{THP}$ is larger than 1.0, and if the inverter is loaded by capacitance only as shown in Fig. 8, the inverter has hysteresis as shown in Fig. 9(b) [04] where we assumed

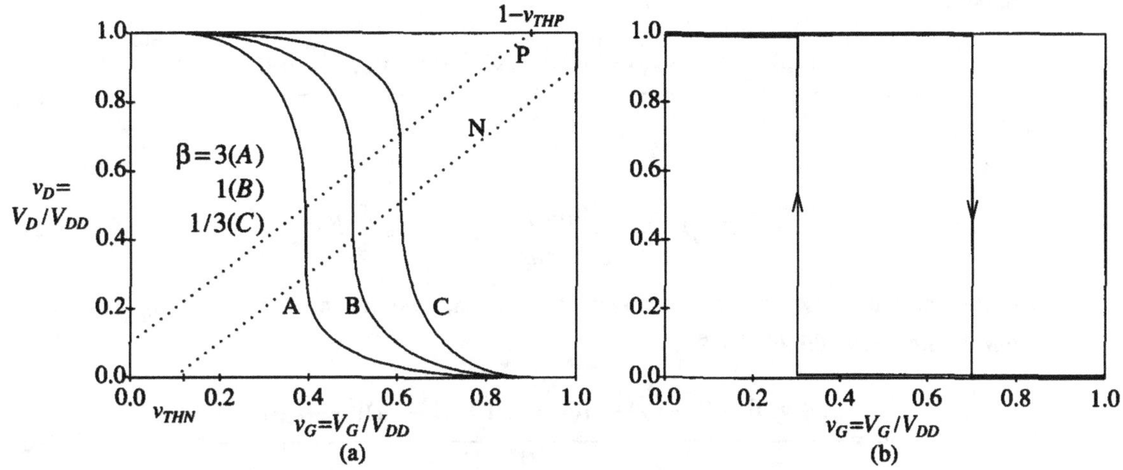

Figure 9 Static characteristic of CMOS inverter

$v_{THP} = v_{THN} = 0.7$. Pull-up and pull-down transitions occur quite slowly because one of the FETs is turned off and the other is only weakly conducting to change the state of the output node.

The steepness of the static switching characteristic of Fig. 9(a) depends on the non-linearity of the FET current-voltage characteristic. In order to study the effects of the FET characteristics on the gate's static characteristic, we use the simplified high-field FET model introduced in Section 1.7 [Eq. (25) of chapter 1].

$$I_N = B_{HN}\frac{(V_G - V_{THN})V_D}{V_D + V_{AN}} \qquad I_P = B_{HP}\frac{(V_{DD} - V_G - V_{THP})(V_{DD} - V_D)}{(V_{DD} - V_D) + V_{AP}} \tag{8}$$

where V_{AN} and V_{AP} are the parameters having the dimension of voltage. If $V_{AP} = V_{AN} = 0$, we have the collapsible current generator model of FET. If $V_{AP} = V_{AN} \to \infty$, the characteristic approximates that of a gate voltage controlled *linear* resistor. $V_{AN} = V_{AP} = 0$ represents the maximum FET nonlinearity, and $V_{AN} = V_{AP} = \infty$ the minimum FET nonlinearity. Our problem is to investigate the dependence of the gate's static characteristic on V_{AN} and V_{AP}. The static switching characteristic is derived by requiring that $I_N = I_P$ and finding V_D versus V_G as the solution of quadratic equation,

$$(\Gamma - 1)V_D{}^2 + [V_{DD} - V_{AN} - \Gamma(V_{DD} + V_{AP})]V_D + V_{DD}V_{AN} = 0$$

where

$$\Gamma = \frac{B_{HN}(V_G - V_{THN})}{B_{HP}(V_{DD} - V_G - V_{THP})}.$$

By using the following normalization

$$v_D = V_D/V_{DD} \;\; v_G = V_G/V_{DD} \;\; v_{THN} = V_{THN}/V_{DD}$$

$$v_{THP} = V_{THP}/V_{DD} \;\; v_{AN} = V_{AN}/V_{DD} \;\; v_{AP} = V_{AP}/V_{DD}$$

We have

$$v_D = \frac{(1+v_{AP})\Gamma-(1-v_{AN})-\left[[(1+v_{AP})\Gamma-(1-v_{AN})]^2-4v_{AN}(\Gamma-1)\right]^{1/2}}{2(\Gamma-1)} \tag{9a}$$

where

$$\Gamma=\beta\frac{(v_G-v_{THN})}{(1-v_G-v_{THP})} \quad \text{and} \quad \beta=\frac{B_{HN}}{B_{HP}}. \tag{9b}$$

In the special case of a *symmetrical* inverter, we have $\beta=1.0$, $v_{AN}=v_{AP}=v_A$, $v_{THN}=v_{THP}=v_{TH}$ and therefore

$$v_D = \frac{(1+v_A)\Gamma-(1-v_A)-\left[[(1+v_A)\Gamma-(1-v_A)]^2-4v_A(\Gamma-1)\right]^{1/2}}{2(\Gamma-1)}. \tag{9c}$$

Figure 10 shows v_D versus v_G determined from Eq. (9c) for v_A from 0.01 to 10. If $v_A \to 0$ the static switching characteristic can be approximated by a step function at the inverter's switching threshold voltage ($v_G=0.5$). In the limit as $v_A \to \infty$, Eq. (9c) approaches

$$v_D=1 \quad (v_G<v_{TH})$$

$$=(1-v_G-v_{TH})/(1-2v_{TH}) \quad (v_{TH}<v_G<1-v_{TH}) \quad =0 \quad (v_G>1-v_{TH}).$$

This is the case of the minimum nonlinearity, or the case of the equivalent linear resistor model for the FETs. Figure 10 shows that nonlinearity of the FET I_D-V_D characteristics (that creates high transconductance and high load resistance only within narrow input voltage region) causes steep transitions, and a well-defined threshold voltage for CMOS gates.

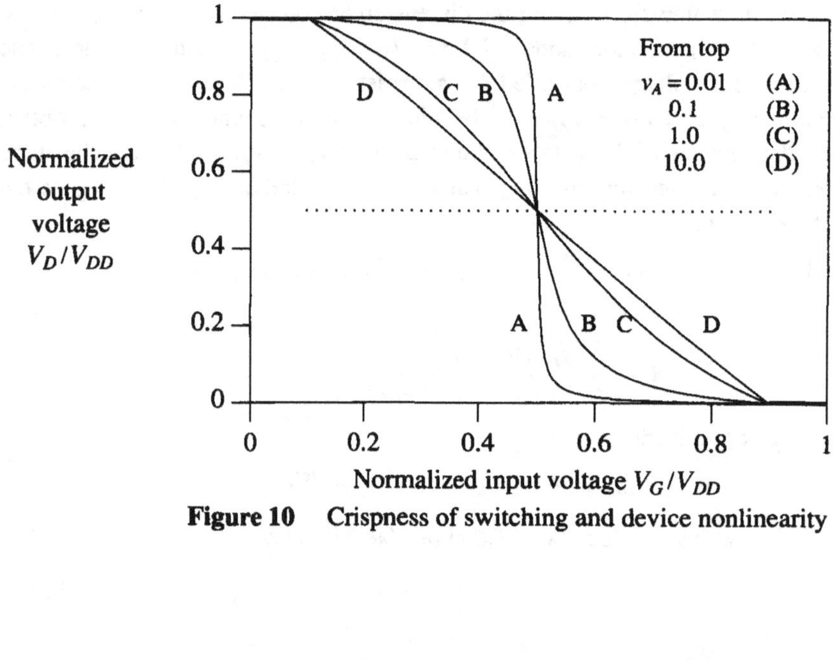

Figure 10 Crispness of switching and device nonlinearity

2.5 DC Characteristics of FET Chain

A CMOS logic gate consists of series-connected and parallel-connected FETs. Terminal characteristics of series-connected FETs are complex. The simplest and the most useful case is shown in Fig. 11 and analyzed as an example. To simplify the analysis we assume that the back-bias effect is small (threshold voltage of the NFETs does not depend on the source voltage). NFET MN1 is driven by gate voltage source V_G, and the gate of MN2 is connected to the power supply voltage (V_{DD}). The NFET chain is *enabled* by this biasing arrangement. We seek the relationship between current I and voltage V_D. This relationship is useful in analyzing the characteristics of NAND gates in Section 2.9.

2.5.1 Collapsible Current Generator Model

We assume that NFETs MN1 and MN2 have the same size. We use the collapsible current generator model first. We have

$$I = 0 \quad \text{if} \quad V_G \le V_{TH} . \tag{10a}$$

If $V_G > V_{TH}$, NFETs MN1 and MN2 are both in the saturation region and carry the same current, if V_X satisfies $V_D > V_X > 0$, and $V_X = V_{DD} - V_G$. If $V_G < V_{DD}$, $V_X > 0$. This bias point is stable (since if $V_X < V_{DD} - V_G$ MN2 carries more current than MN1 and therefore V_X increases, and vice versa). Then

$$I = B(V_G - V_{TH}) \quad \text{if} \quad V_D > V_X (= V_{DD} - V_G) > 0 \tag{10b}$$

$$\text{and} \quad V_{DD} \ge V_G \ge V_{TH} .$$

If $V_D < V_{DD} - V_G$, MN2 is in the triode region. Then $V_X = V_D$, and the current is given by Eq. (10b). If $V_{DD} - V_G < 0$, MN1 is in the triode region and the current is determined by NFET MN2 as

$$I = B(V_{DD} - V_{TH}) \quad V_G > V_{DD} > V_{TH} \quad V_D > 0 . \tag{10c}$$

This result is applied to understand the switching process of a tristatable inverter, shown in Fig. 12(a). The tristatable inverter is assumed to be symmetrical: All FETs have the same

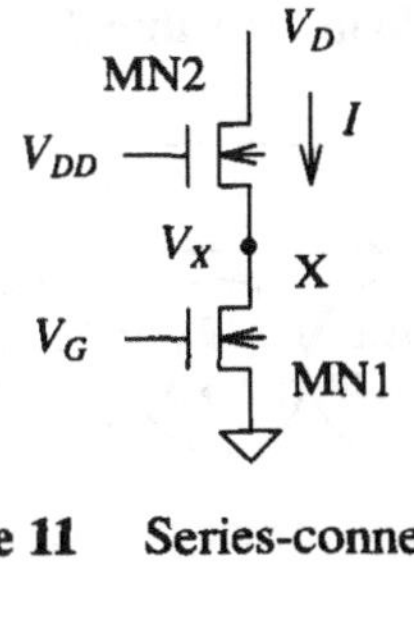

Figure 11 Series-connected FETs

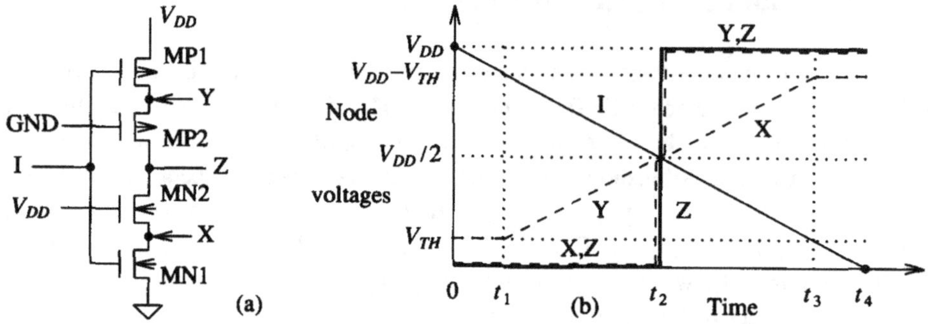

Figure 12 Switching of tristatable inverter

beta and the same threshold voltage. Figure 12(b) shows node voltage waveforms, when the voltage of input node I decreases gradually and uniformly starting from V_{DD}, as shown by curve I. We consider static switching characteristics. We assume input voltage V_I changes with time as

$$V_I(t)=V_{DD}-V_{DD}(t/t_4)\,.$$

At time $t=t_1$ the input node voltage is $V_I(t_1)=V_{DD}-V_{TH}$, where V_{TH} is the threshold voltage of PFET MP1. PFET MP1 turns on. Node Y, which was originally voltage V_{TH} above ground, is pulled up by the current from MP1, and PFET MP2 turns on (MP2 was originally in the M-state, the marginally conducting state). These two events occur simultaneously at time t_1, as shown in the microstate sequence of Fig. 13. Note that we assume slow switching [or negligible capacitance at the nodes X, Y, and Z of Fig. 12(a)]. After time t_1, node Y voltage increases with time t as

$$V_Y(t)=V_{DD}-V_I(t)$$

because MP1 and MP2 have the same beta and the same threshold voltage, they carry equal current. Current from the pull-up PFET chain (MP1, MP2) is still less than the maximum current of the NFET chain consisting of MN1 and MN2. Therefore nodes X and Z remain at ground potential. The PFET chain current increases while the maximum current of the NFET chain decreases with decreasing $V_I(t)$. At time $t=t_2$, $V_I(t)=V_{DD}/2$, at which time the two currents are equal. If this *switching threshold voltage* is exceeded, PFET MP1 moves into the triode region and node Y is pulled up to V_{DD}. Then the maximum current of MP2 becomes higher than the current of the NFET chain, and therefore MP2 moves into the triode region

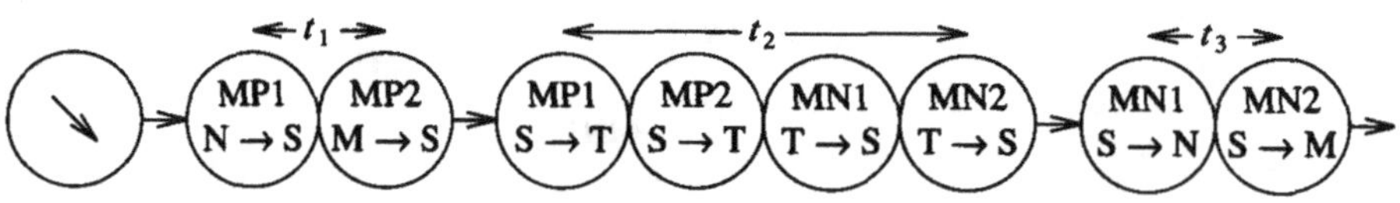

Figure 13 Microstate sequence of switching tristate inverter

also and node Z is pulled up. Consequently NFETs MN1 and MN2 move out of the triode region and node Z makes the transition from 0 to V_{DD}. From this moment on, the NFET chain determines the current. V_X increases from 0 to $V_{DD}/2$ at the moment of switching, to maintain current equilibrium within the NFET chain. These four events happen in rapid succession at time t_2, as shown in the microstate sequence of Fig. 13. After t_2, both NFETs MN1 and MN2 are in saturation. V_X increases with time as

$$V_X(t) = V_{DD} - V_I(t)$$

and both MN1 and MN2 turn off at time t_3 when $V_I(t_3) = V_{TH}$. MN2 goes into the marginally conducting M-state.

2.5.2 *Gradual Channel, Low-field Model*

If the gradual channel, low-field FET model is used in the analysis, the analysis of series-connected FETs becomes more complex. We consider two different cases, $V_D > V_{DD} - V_{TH}$ and $V_D < V_{DD} - V_{TH}$ with reference to Fig. 11. In the first case, NFET MN2 is in saturation [saturation and triode regions of FET are defined in accordance with the gradual channel, low-field model (Section 1.2): If the gate voltage is higher than the threshold voltage and if the drain voltage is higher than the gate voltage minus threshold voltage (both referred to the source), the NFET is in the saturation region]. In the second case, it is in the triode region. For either case, MN1 can be either in saturation or in triode region.

A) $V_D > V_{DD} - V_{TH}$ (MN2 is in saturation)

If MN1 is also in saturation, V_X is determined from

$$I = \frac{B_{G2}}{2}(V_{DD} - V_X - V_{TH})^2 = \frac{B_{G1}}{2}(V_G - V_{TH})^2 \tag{11}$$

where B_{G1} and B_{G2} are the betas of NFET MN1 and MN2, respectively. Here MN1 and MN2 may have different sizes: The following analysis holds even if there are N series-connected NFETs instead of MN2, whose gates are all connected to V_{DD}. If there are N series-connected MN2s, B_{G2} should be divided by N. This substitution makes the present results applicable to NAND gates having multiple inputs. V_{TH} is assumed to be independent of the source voltage. We have, from Eq. (11),

$$V_X = V_{DD} - V_{TH} - \sqrt{B_{G1}/B_{G2}}(V_G - V_{TH})$$

and MN1 is in saturation if $V_X > V_G - V_{TH}$. This condition is satisfied if $V_G < V_{Gm}$, where V_{Gm} is given by

$$V_{Gm} = \frac{V_{DD} + (B_{G1}/B_{G2})^{1/2} V_{TH}}{1 + (B_{G1}/B_{G2})^{1/2}}.$$

Current I is determined by V_G alone, from Eq. (11). If $V_G > V_{Gm}$ MN1 is in the triode region, and Eq. (11) is modified to be

$$I = \frac{B_{G2}}{2}(V_{DD} - V_X - V_{TH})^2 = B_{G1}[V_G - V_{TH} - (1/2)V_X]V_X \tag{12}$$

and by solving Eq. (12) for V_X we obtain

$$V_X = [B_{G1}/(B_{G1} + B_{G2})][V_G + V_S - [(V_G + V_S)^2 \tag{13}$$

$$-(B_{G2}/B_{G1})[1 + (B_{G2}/B_{G1})](V_{DD} - V_{TH})^2]^{1/2}]$$

where

$$V_S = (B_{G2}/B_{G1})V_{DD} - [1 + (B_{G2}/B_{G1})]V_{TH}\,.$$

By substituting Eq. (13) into Eq. (12), I is determined.

B) $V_D < V_{DD} - V_{TH}$ (MN2 in triode region)

Similarly MN1 is in saturation if

$$V_G < V_{Gm} =$$

$$\frac{-V_Y + [V_Y^2 - [(B_{G1}/B_{G2}) - 1][(V_{DD} - V_D - V_{TH})^2 - V_{DD}^2 + (B_{G1}/B_{G2})V_{TH}^2]]^{1/2}}{(B_{G1}/B_{G2}) - 1}$$

where $V_Y = V_{DD} - (B_{G1}/B_{G2})V_{TH}$ and I is given by $I = (B_{G1}/2)(V_G - V_{TH})^2$. If $V_G > V_{Gm}$, however, MN1 is in the triode region, and V_X is given by

$$V_X = \frac{B_{G1}}{B_{G1} + B_{G2}}[V_G + V_S - [(V_G + V_S)^2$$

$$-2\frac{(B_{G1} + B_{G2})B_{G2}}{B_{G1}^2}(V_{DD} - V_{TH} - (1/2)V_D)V_D]^{1/2}]$$

and the current I is given by

$$I = B_{G1}[V_G - V_{TH} - (1/2)V_X]V_X\,. \tag{14}$$

These results are used to derive static characteristics of NAND gates in later sections. One interesting point we need to make is that if the NFET chain is used as the pull-down circuit of a multi-input CMOS NAND gate, and if the FETs are modeled using the low-field gradual channel model, the current of the NFET chain and of the PFETs may be equal, even if none of the NFETs are in the saturation region.

2.6 Essential Uncertainty in the Switching Threshold Voltage of CMOS Gates

CMOS gates, and in general any conventional logic gates, are *multi-input* amplifier circuits. Since there is more than one input to the amplifier, signals from different inputs interact, creating effects that do not exist in *single-input* amplifier circuits. The simplest CMOS gate, a CMOS inverter [shown in Fig. 14(c)], has PFET MP1 and NFET MN1, both driven by V_G.

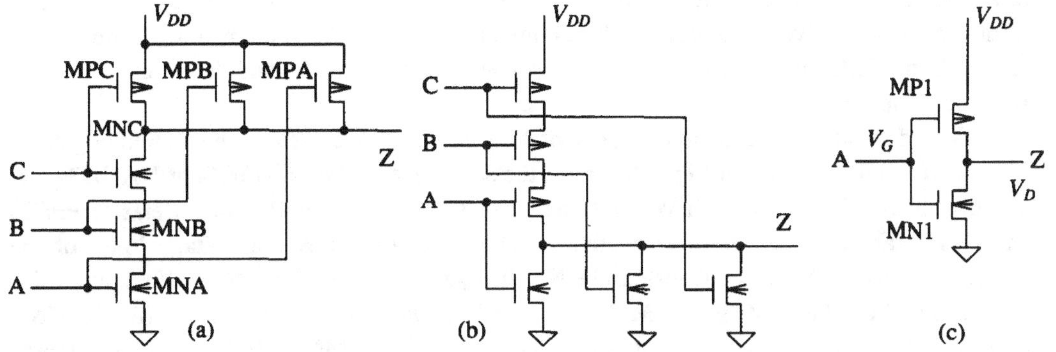

Figure 14 Typical CMOS static gates

The relationship between input voltage, V_G, and output voltage, V_D, of a CMOS inverter was determined in Section 2.4. With reference to Fig. 14(c) and using the gradual channel, low-field FET model, the switching threshold voltage of the inverter is given by

$$V_{GSW} = \frac{V_{DD} - V_{THP} + \sqrt{B_{GN}/B_{GP}}V_{THN}}{1 + \sqrt{B_{GN}/B_{GP}}}. \tag{15}$$

The relationship between V_G and V_D was shown in Fig. 9(a), for several values of B_{GN}/B_{GP}. The switching threshold voltage depends on the beta ratio of the NFET and PFET (B_{GN}/B_{GP}) and the FET threshold voltages, V_{THN} and V_{THP}. If $V_{THP}, V_{THN} \ll V_{DD}$, $V_{GSW} = V_{DD}/(1 + \sqrt{B_{GN}/B_{GP}})$. If $B_{GN} = B_{GP}$ and $V_{THP} = V_{THN}$, $V_{GSW} = V_{DD}/2$. If $W_{N,P}$ is the width (or size) of NFET and PFET, and $\mu_{N,P}$ are the electron and the hole mobilities, respectively (suffix N for NFET, and P for PFET),

$$\frac{B_{GN}}{B_{GP}} = \frac{\mu_N}{\mu_P}\frac{W_N}{W_P}. \tag{16}$$

$B_{GN} = B_{GP}$ if $W_N/W_P = \mu_P/\mu_N \approx 0.4$ if mobility of holes, μ_P, is 2.5 times smaller than that of electrons, μ_N. If $B_{GN} = B_{GP}$ and $V_{THP} = V_{THN}$, the PFET and NFET have identical characteristics, and the inverter is *symmetrical.* Output voltage V_D changes abruptly from $V_{GSW} + V_{THP}$ to $V_{GSW} - V_{THN}$, if input voltage V_G increases from slightly less than V_{GSW} to slightly more than V_{GSW}.

A discontinuous change of V_D at $V_G = V_{GSW}$ in the CMOS inverter exists because the NFET and PFET are both in the saturation region, and because saturated FETs are equivalent to current generators. This discontinuous voltage change in static characteristics is convenient for a definition of the switching threshold voltage, but a discontinuity may not exist in NAND or NOR gates. If series-connected FETs are modeled using the gradual channel FET model with arbitrary input voltages changes, NAND and NOR gates may not have a discontinuous static switching characteristic. Using the collapsible current generator model, however, at least one of the series-connected FETs is in the saturation region if the voltage that is applied to the chain is positive. Therefore, a discontinuous change in the static characteristic always exists. In the following we use the gradual channel, low-field model and derive the most useful conclusions from the model.

V_{GSW} depends on the details of the switching process. We assume that NAND3 gate of Fig. 14(a) and the inverter of Fig. 14(c) have the same size FETs (all NFETs of NAND3 gate, and the NFET of the inverter have the same dimensions, as do the PFETs). Betas of PFETs and NFETs are B_{GP} and B_{GN}, respectively. The switching threshold voltage V_{GSW} of the inverter is given by Eq. (15). Consider the NAND3 gate shown in Fig. 14(a). When input signals B and C are high, MPB and MPC are turned off, and the two NFETs MNB and MNC (that are fully turned on) act as resistance in series with NFET MNA. If the chain of the three NFETs is replaced by an equivalent single, *long-channel* NFET, the NFET has nonuniform gate voltage (2/3 of the drain-side channel has a higher gate voltage than the rest). Since λ NFETs are connected in series (in our example $\lambda = 3$) and since only a single PFET is active, B_{GP} in Eq. (15) stays the same, but B_{GN} can be approximated by dividing B_{GN} of the inverter by λ. The switching threshold voltage V'_{GSW} after this substitution into Eq. (15) is

$$V'_{GSW} = \frac{V_{DD} - V_{THP} + \sqrt{B_{GN}/\lambda B_{GP}}V_{THN}}{1 + \sqrt{B_{GN}/\lambda B_{GP}}} \,.$$

This gate switching voltage is, however, an overestimate of the true switching threshold voltage because the effective gate voltage of the *long-channel* NFET that is set equivalent to the three-member NFET chain is higher than the voltage to input terminal A. Therefore

$$V_{GSW}(B = H, C = H) \le V'_{GSW} \,. \tag{17}$$

Let us consider the second case that three signals A, B, and C all change simultaneously. Then the switching threshold voltage is given by Eq. (15), with substitution

$$B_{GN} \to B_{GN}/\lambda \quad \text{and} \quad B_{GP} \to \lambda B_{GP}$$

as

$$V_{GSW}(A = B = C) = \frac{V_{DD} - V_{THP} + (1/\lambda)\sqrt{B_{GN}/B_{GP}}V_{THN}}{1 + (1/\lambda)\sqrt{B_{GN}/B_{GP}}} \,. \tag{18}$$

We have

$$V_{GSW}(A = B = C) > V_{GSW}(B = H, C = H) \,.$$

We showed that the switching threshold voltage of a multi-input gate depends on how each of the input signals changes with time. The same conclusion is derived for the NOR3 gate shown in Fig. 14(b). This is an example of uncertainty in the switching threshold voltage of digital gates, and such uncertainty is a fundamental property of any logic gate circuit that has switching devices connected in series or in parallel. As a second example, a similar uncertainty of the switching threshold voltage exists in a bipolar ECL gate (shown in Fig. 15 of chapter 7). In the ECL gate, if only one input out of N inputs switches and the rest stay at the logic-low level, the gate switches at $V_{GSW} = V_{REF}$. If all N inputs switch simultaneously, the gate switches at $V_{GSW} = V_{REF} - (kT/q)\log N$, where (kT/q) is the *thermal* voltage (25mV at the room temperature). One immediate consequence is that the swing of logic voltage must be large enough to cover the possible range of uncertainty in the logic threshold voltage. It is impossible to reduce the swing of the logic voltage below the uncertainty limit.

2.7 Static Characteristics of 2-Input NAND Gates

In the previous section we showed that V_{GSW} of multi-input gates depends on all inputs. We study this problem completely here as a simple example. We study the simplest case, a CMOS NAND2 gate. The problem can be stated as follows. Since there are two independent input voltages, V_A (V_A is the input voltage to the NFET closest to ground) and V_B, and one dependent output voltage, V_0, which varies from 0 to V_{DD}, the multivariable functional relationship that gives V_0 as a function of V_A and V_B can be plotted as a *surface* within the cubic region $0 < V_A, V_B, V_0 < V_{DD}$ of the $V_A - V_B - V_0$ coordinate system, $V_0 = V_0(V_A, V_B)$. Our problem is to determine the surface.

2.7.1 Gradual-channel, Low-field Model

Figure 15 shows the results of analysis using the gradual-channel, low-field FET model, and the theory of series-connected FET chains developed in Section 2.5. We assume that all the

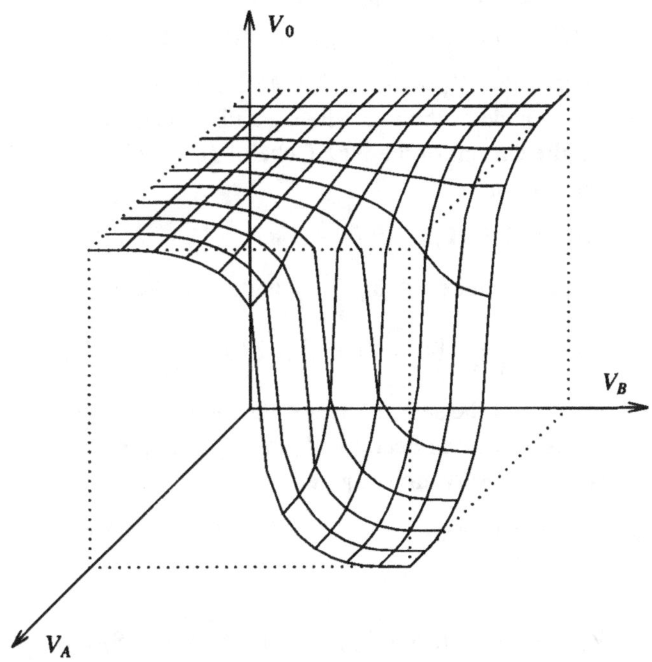

Figure 15 Static switching characteristics of NAND2 gate

PFETs and NFETs have the same beta and that the FET threshold voltage is independent of the source voltage. The discontinuous static characteristic transition of inverters observed in Fig. 9(a) has generally disappeared, and the transition becomes continuous and gradual. We mentioned before that the discontinuous transition of inverters is due to FET saturation. As we see in Fig. 15, a gradual transition between logic levels takes place, even if the FETs saturate perfectly. The following explanation clarifies this point. In NAND2 gate switching, suppose $V_B = V_{DD}$ (input for NFET MNB and PFET MPB), and V_A (input for NFET MNA and PFET MPA) increases from 0 to V_{DD}. PFET MPB is turned off. NFET MNA and PFET MPA have the same current. The other NFET MNB sustains positive voltage also. That voltage reduces the drain voltages of NFET MNA and PFET MPA, thereby forcing them into the triode region. Therefore the steep transition is either reduced, or eliminated. The steep transition from logic level 0 to logic level 1 as observed in inverters is not observed in multi-input gates, except when the switching process and the betas of the FETs are special. If the two NFETs have the same size, and if V_A and V_B switch simultaneously ($V_A = V_B$), a steep transition at the switching threshold remains.

2.7.2 Collapsible Current Generator Model

Since the problem of interacting input signals is complex, we show a second, simpler, analysis. We use the collapsible current generator model and neglect the back-bias effect. Let the output node voltage V_0 be neither ground nor V_{DD}, and the two input voltages be in the range $V_{TH} < V_A, V_B < V_{DD} - V_{TH}$. Two PFETs supply current

$$I_P = B_P(V_{DD} - V_A - V_{TH}) + B_P(V_{DD} - V_B - V_{TH})$$

to output node O. Suppose that $V_A > V_B$. NFET MNA carries more current to ground than NFET MNB to the internal node of the two NFETs. In DC operation the internal node is then at ground potential, and the current of the NFET chain, I_N equals that of NFET MNB. Then the NAND2 gate switches when

$$I_P = I_N \quad \text{or} \quad 2B_P(V_{DD} - V_{TH}) - B_P(V_A + V_B) = B_N(V_B - V_{TH})$$

is satisfied. V_B versus V_A is given by solving this equation for V_B as

$$V_B = \frac{1}{B_P + B_N}\,[2B_P(V_{DD} - V_{TH}) + B_N V_{TH} - B_P V_A]\,. \tag{19}$$

If $V_B > V_A$, the two NFETs carry the same current if the internal node voltage is $V_B - V_A$, which is positive in this case. As we saw in Section 2.5, this operating point is stable. Therefore I_N is $B_N(V_A - V_{TH})$ and current conservation requires

$$I_P = I_N \quad \text{or} \quad 2B_P(V_{DD} - V_{TH}) - B_P(V_A + V_B) = B_N(V_A - V_{TH})\,.$$

This equation gives

$$V_A = \frac{1}{B_P + B_N}\,[2B_P(V_{DD} - V_{TH}) + B_N V_{TH} - B_P V_B]\,. \tag{20}$$

If we set $V_A = V_B$ we get

$$V_A = V_B (= V_C) = \frac{2B_P(V_{DD} - V_{TH}) + B_N V_{TH}}{2B_P + B_N}.$$

If $V_A > V_{DD} - V_{TH}$ and $V_{DD} - V_{TH} > V_B$,

$$V_B = V'_B = \frac{B_P(V_{DD} - V_{TH}) + B_N V_{TH}}{B_P + B_N} \tag{21a}$$

and this result is the same as Eq. (19) with $V_A = V_{DD} - V_{TH}$. If $V_B > V_{DD} - V_{TH}$ and $V_{DD} - V_{TH} > V_A$,

$$V_A = V'_A = \frac{B_P(V_{DD} - V_{TH}) + B_N V_{TH}}{B_P + B_N} \tag{21b}$$

and this result is the same as Eq. (20) with $V_B = V_{DD} - V_{TH}$. Figure 16 shows the static switching characteristic of the NAND2 gate, plotted in the $V_A - V_B$ coordinate system. Figures 15 and 16 show that the discontinuous transition of the output node voltage of NAND2 gates takes place if the FETs are represented by the collapsible current generator model, the most nonlinear representation of FETs. Figure 16 shows clearly that the switching threshold voltage for V_A depends on the value of V_B, and vice versa. This mutual dependence of the switching threshold voltage results in an essential uncertainty in the switching delay time, as will be discussed in detail in chapter 6.

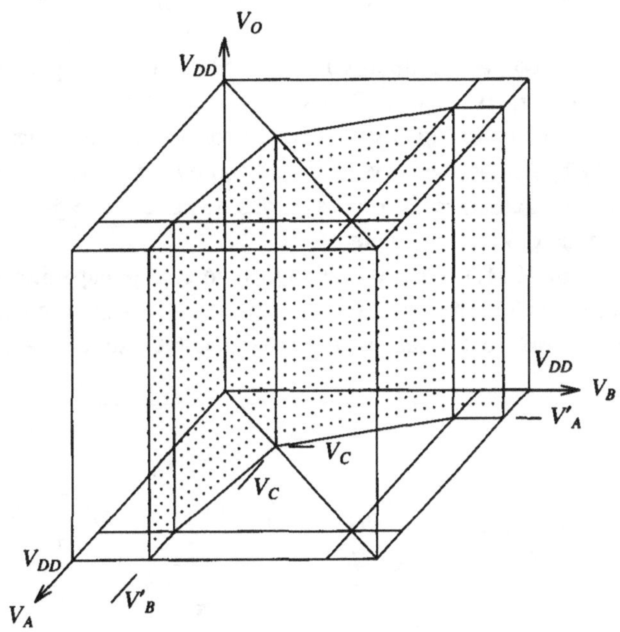

Figure 16 Static switching characteristic of NAND2 gate

2.8 Transient Operation of Simple CMOS Circuits

In this section we study transient, or dynamic, characteristics of CMOS circuits. In chapter 1 the concept of a microstate sequence of CMOS digital circuit operation was introduced using the collapsible current generator FET model. This model is powerful for analyzing transient characteristics of CMOS circuits. In this section, we begin with the simplest CMOS circuits and proceed systematically to ever more complex circuits.

2.8.1 Capacitor Charge/Discharge by FET

Figure 17(a) shows a charged capacitor C_L that is discharged through NFET MN1. The gate of MN1 was initially at 0, and capacitor C_L was initially charged to V_{DD}. The gate voltage of NFET MN1 makes a rapid low-to-high transition (from 0 to V_{DD}) at time $t=0$. At the beginning, MN1 is in the nonconducting (N) region. After discharge is completed, MN1 is in the triode region (T). Therefore two changes of states, N $\rightarrow$ S, and S $\rightarrow T$, must follow in that order, where S stands for the saturation region of an NFET. This change is monotonic: There is no mechanism in this circuit that allows the FET to go back and forth between two different regions. The microstate sequence is shown in Fig. 17(b). The discharge process may be described mathematically as follows. After time $t=0$, $V_1(t)$ satisfies the circuit equation

$$C_L \frac{dV_1(t)}{dt} = -B_1(V_{DD} - V_{THN}) \quad \text{or} \quad V_1(t) = V_{DD} - \frac{B_1(V_{DD} - V_{THN})}{C_L} t \tag{22}$$

This solution is valid as long as the drain voltage of MN1, $V_1(t)$, is positive. This condition requires that $t < t_1$, where $t_1 = (C_L V_{DD})/[B_1(V_{DD} - V_{THN})]$. If $t > t_1$ $V_1(t) = 0$, and MN1 is in the triode region. V_1 is shown in Fig. 18(a). Changes of the circuit's microstate are indicated by A and B in Fig. 17(b) and on the curve of Fig. 18(a) by the same symbols. The discharge of capacitor C_L may be regarded as the process of *fusing* node N1 to ground: one node of the circuit vanishes when the process is completed.

Another fundamental CMOS circuit operation is to charge capacitor C_L through NFET MN1, as shown in Fig. 17(c). We assume that capacitor C_L was originally discharged. At time $t=0$ the gate of NFET MN1 makes a low-to-high transition ($0 \rightarrow V_{DD}$). Immediately

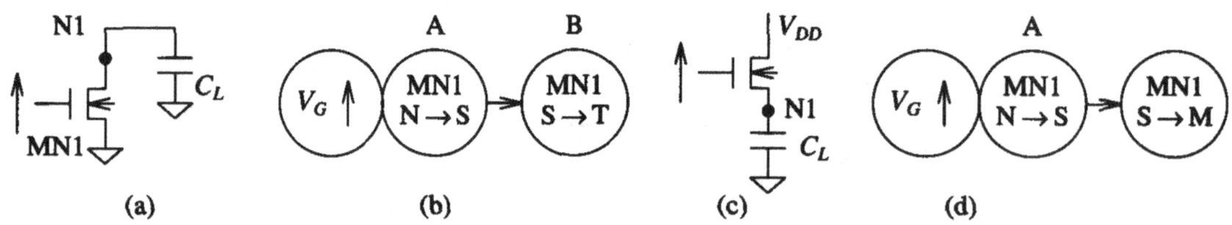

Figure 17 Elements of MOSFET circuits

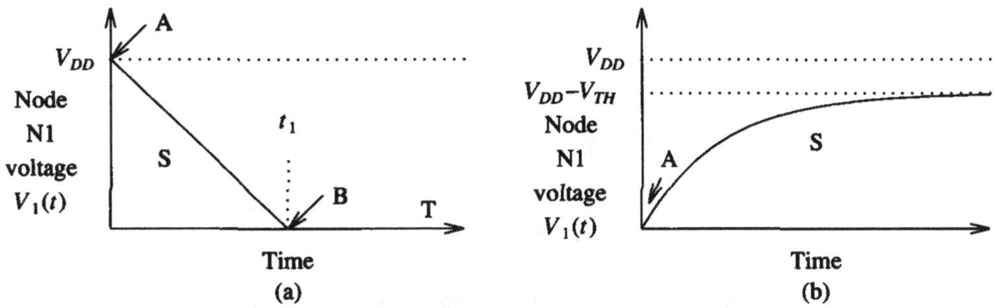

Figure 18 Node waveforms

after the transition, MN1 is in the saturation region. NFET MN1 stays in the saturation region all the way through the chargeup process: no change of state of NFET MN1 takes place. After a long time, V_1 approaches $V_{DD} - V_{THN}$, where V_{THN} is NFET threshold voltage including the back-bias effect. This is state M, the marginally conducting state of the FET discussed in Section 1.16. In state M, NFET MN1 is still in the saturation region, but the current is zero. This process can be described mathematically as follows. Immediately after $t=0$, node N1 voltage $V_1(t)$ satisfies

$$C_L \frac{dV_1(t)}{dt} = B_1[V_{DD} - V_1(t) - V_{THN}]\,. \tag{23a}$$

This equation can be integrated subject to initial condition $V_1(0)=0$ as

$$V_1(t) = (V_{DD} - V_{THN})[1 - exp[-(B_1/C_L)t]]\,. \tag{23b}$$

The V_1 waveform is shown in Fig. 18(b). Since node N1 is connected to V_{DD} through the marginally conducting NFET, MN1, the node has high impedance: the node is more susceptible to external noise induction than if connected to V_{DD} or ground by a FET in the triode region. For instance, if by any reason V_1 is pulled above $V_{DD} - V_{THN}$, the node stays at this voltage, and the NFET goes into the nonconducting (N) region. In this charge-up process, the number of nodes in the circuit does not change. The microstate sequence is shown in Fig. 17(d). The NFET works as if it were a linear resistor that has no *state*. This fact is reflected to the microstate sequence: The FET stays in the saturation region. The marginally conducting state M is not a distinct state, but one extreme of the S state. The microstate sequence of the two single-FET circuits were quite simple. The complexity of microstate sequences increases rapidly as the number of FETs increases. In the following we give several examples of two-FET circuits.

2.8.2 Inverter Response

We analyze the CMOS inverter circuit shown in Fig. 19 including input voltage waveform. The input signal changes with time as

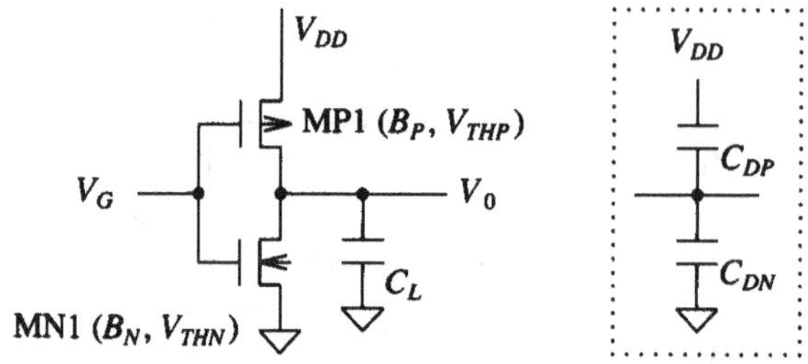

Figure 19 Analysis of inverter switching process

$$V_G(t)=0\ (t<0)\ =\alpha t\ (0<t<t_4)\ =V_{DD}\ (t>t_4)\,.$$

Times t_0, t_2, and t_4 are defined in Fig. 21 (dashed waveform) as

$$t_0=V_{THN}/\alpha \quad t_2=(V_{DD}-V_{THP})/\alpha \quad t_4=V_{DD}/\alpha\,.$$

The switching threshold voltage of the inverter of Fig. 19 is given by V_G where MP1 and MN1 currents equal.

$$V_{GSW}=\frac{B_P(V_{DD}-V_{THP})+B_N V_{THN}}{B_N+B_P}\,. \tag{24}$$

Until V_G reaches the switching threshold voltage V_{GSW}, V_0 is held at V_{DD} by PFET MP1 which is in the triode region. The time when pull-down of the output node begins, t_1 of Fig. 21, is given by

$$t_1=V_{GSW}/\alpha\,.$$

At time t_1 PFET MP1 moves from the triode region to the saturation region, and both MP1 and MN1 are in the saturation region and the pull-down transient begins. The circuit equation describing the pull-down process is

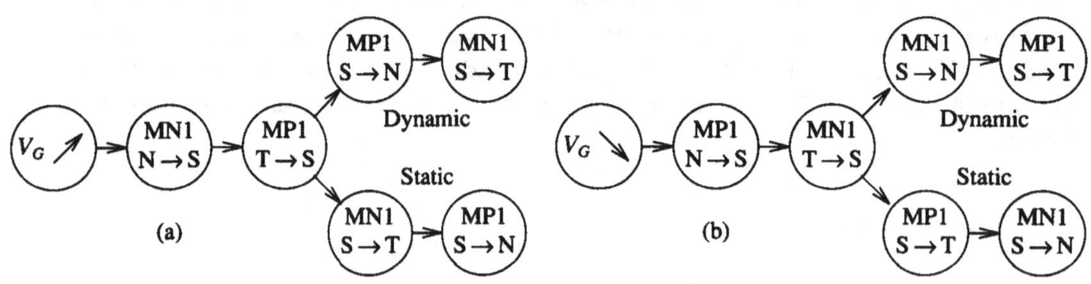

Figure 20 Microstate sequence of inverter

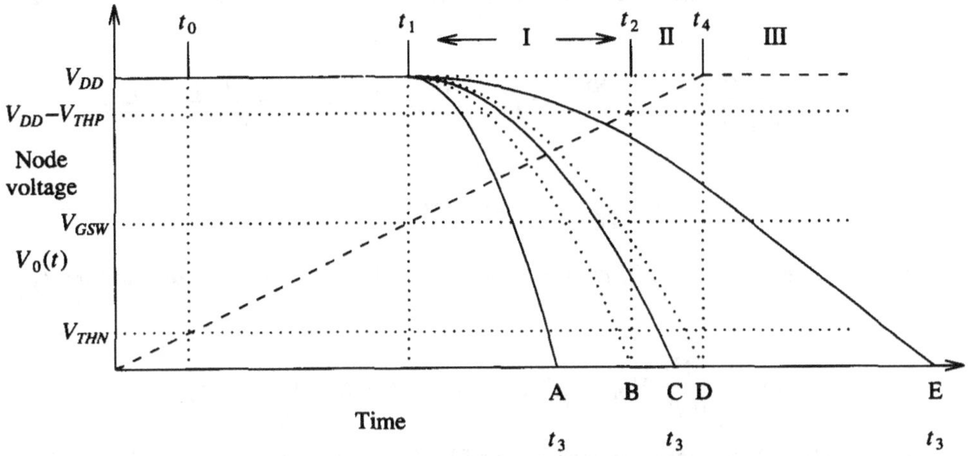

Figure 21 Inverter switching waveform

$$C_L \frac{dV_0}{dt} = (B_N + B_P)[V_{GSW} - V_G(t)] \tag{25a}$$

which is solved as

$$V_0(t) = V_{DD} - \frac{(B_N + B_P)\alpha}{2C_L}(t - t_1)^2. \tag{25b}$$

We set $V_0(t_3) = 0$, and determine t_3, which is the time when the output node voltage reaches zero. We have, from Eq. (25b),

$$t_3 = t_1 + \left[\frac{2C_L V_{DD}}{(B_N + B_P)\alpha}\right]^{1/2}.$$

Time t_3 must be less than t_2 for the present analysis to be valid, which requires that α must be less than a critical limit α_{C1} given by

$$\alpha_{C1} = \frac{{B_N}^2}{2C_L V_{DD}} \frac{(V_{DD} - V_{THN} - V_{THP})^2}{B_N + B_P}.$$

If α is less than this critical limit, MN1 moves into the triode region at time t_3. After that time the V_0 remains at zero, as shown by curve A of Fig. 21. If α is more than α_{C1}, $V_0(t_2)$ will be positive. Since PFET MP1 turns off at t_2, the circuit equation for time $t_2 < t < t_4$ is given by

$$C_L \frac{dV_0}{dt} = -B_N\alpha(t - t_0) \tag{26a}$$

which is solved as

$$V_0(t) = V_0(t_2) + [(B_N\alpha)/C_L]\left(\frac{{t_2}^2}{2} - t_0 t_2 - \frac{t^2}{2} + t_0 t\right). \tag{26b}$$

In this solution, $V_0(t)$ reaches zero before time t_4 if α is less than the second critical limit α_{C2} given by

$$\alpha_{C2}=\frac{1}{V_{DD}}\left[\frac{B_N}{C_L}\left[V_{DD}\left(\frac{V_{DD}}{2}-V_{THN}\right)-(V_{DD}-V_{THP})\left(\frac{V_{DD}-V_{THP}}{2}-V_{THN}\right)\right]\right.$$

$$\left.+\frac{B_N+B_P}{2C_L}(V_{DD}-V_{THP}-V_{GSW})^2\right]$$

It can be shown that $\alpha_{C2}>\alpha_{C1}$. If α is between α_{C2} and α_{C1}, V_0 reaches zero at time t_3 given by

$$t_3=t_0+\left[\frac{2C_L}{B_N\alpha}V_{DD}-\frac{B_P}{B_N}(t_2-t_1)^2+(t_1-t_0)(2t_2-t_1-t_0)\right]^{1/2}$$

and if $t>t_3$, $V_0(t)=0$, as shown by curve C of Fig. 21. After time t_4, V_G settles to V_{DD}. We next consider the case when $\alpha>\alpha_{C2}$. The circuit equation is given by

$$C_L\frac{dV_0}{dt}=-B_N(V_{DD}-V_{THN})$$

which is solved as

$$V_0(t)=V_0(t_4)-\frac{B_N}{C_L}(V_{DD}-V_{THN})(t-t_4) \tag{27a}$$

where

$$V_0(t_4)=V_{DD}-\frac{B_N+B_P}{2C_L}\alpha(t_2-t_1)^2+\frac{B_N}{C_L}\alpha\left(\frac{t_2^2}{2}-t_0t_2-\frac{t_4^2}{2}+t_0t_4\right) \tag{27b}$$

and the inverter reaches ground potential at time t_3 given by

$$t_3-t_4=\frac{C_L}{B_N(V_{DD}-V_{THN})}V(t_4).$$

Curve E of Fig. 21 shows the waveform.

Microstate sequences of the CMOS inverter are shown in Fig. 20(a) and (b). For an output pull-down transition [Fig. 20(a)], NFET MN1 turns on first, and then PFET MP1 makes a state change from the triode region to the saturation region. Since then all the FETs that are connected to the output node are in the saturation region, the output node voltage goes down (note that if the output node is fused to V_{DD} or ground by a FET in the triode region, the voltage does not change). There are two possibilities that may follow. If input voltage V_G makes a rapid low-to-high transition, PFET MP1 turns off first, and NFET MN1 continues to pull the output node down. This is the case of *dynamic* switching. If input voltage V_G makes slow transition, NFET MN1 completes the pull-down before PFET MP1 turns off. This is the case of *static* switching. In this case, significant overlap current flows through the two FETs (Section 5.12). These are two distinct modes of switching of CMOS inverter. In static switching, the input node voltage continues to change after the output node voltage settles. The cause-effect relationship appears to be reversed from the observation of casual node voltage alone.

The input and output transients are essentially simultaneous, or they are two different aspects of the *same* electrical transient. Dynamic switching involves delay. The output voltage changes by a delay time later than the input voltage. By observation of voltage, the cause-effect relationship can be established. Input and output transients are essentially different, and they are causally related. The difference in the switching mechanism shows up dramatically in a ringoscillator discussed in Section 6.4. Figure 20(b) shows the microstate sequence of a CMOS inverter for an output pull-up transition. This sequence can be interpreted in a similar manner.

2.8.3 Voltage Dependence of Node Capacitance

The voltage dependence of load capacitance C_L was not included in the closed-form analysis. Here we consider voltage dependence of the drain island capacitance C_{DP} and C_{DN} shown in the dotted box of Fig. 19. The capacitance of a reverse-biased abrupt PN junction, one side of which is doped to N (*carriers*/*cm*^{3}), and the other side is doped much higher than that, is given by $\sqrt{\varepsilon_S qN/2(V+V_{dif})}$ per unit area, where ε_S is dielectric constant of silicon (10^{-12} F/cm), V is the voltage applied to the PN junction, and V_{dif} is the diffusion potential (Section 1.8). If the *total* area of the drain island of the PFET and NFET are S_{Peff} and S_{Neff}, respectively, the output node capacitance of the inverter that is due to the drain islands $C_{DP}+C_{DN}$ is given by

$$C_L = C_{DP} + C_{DN} = C\left[\frac{S_{PN}}{\sqrt{V_{DD} - V_D + V_{dif}}} + \frac{1}{\sqrt{V_D + V_{dif}}}\right] \tag{28}$$

where $C=\sqrt{\varepsilon_S qN/2}\cdot S_{Neff}$ and $S_{PN}=S_{Peff}/S_{Neff}$. This assumes the P-tub and N-tub are doped approximately the same. If the drain island is a rectangle $W \times X$ in size, and if the PN junction depth is l, $S_{Peff}=W\cdot X+2l\cdot(W+X)$ and S_{Neff} is similar. The second term approximates crudely the *side wall* capacitance of the drain island. Higher capacitance that is due to the higher tub surface doping is neglected. Normalized total capacitance C_L/C_N (where $C_N=C/\sqrt{V_{DD}}$) versus normalized voltage v_D for several values of S_{PN} is shown in Fig. 22. The capacitance has a minimum at a voltage between 0 and V_{DD}. The voltage for minimum capacitance is $V_{DD}/2$ if $S_{PN}=1.0$.

The switching transient of an inverter whose load capacitance is given by Eq. (28) can be analyzed as follows. With reference to Fig. 19, we consider the special case where the input voltage of the inverter makes a rapid low-to-high transition at time $t=0$. The circuit equation of the transient, written by using the normalized voltages $v_D=V_D/V_{DD}$, $v_{dif}=V_{dif}/V_{DD}$ and $v_{THN}=V_{THN}/V_{DD}$, is

$$C_N\left(\frac{S_{PN}}{\sqrt{1-v_D+v_{dif}}} + \frac{1}{\sqrt{v_D+v_{dif}}}\right)\frac{dv_D}{dt} = -B_N(1-v_{THN}) \qquad C_N = C/\sqrt{V_{DD}}$$

which is solved subject to the initial condition, $v_D(0)=1$, as

$$S_{PN}(\sqrt{1-v_D+v_{dif}} - \sqrt{v_{dif}}) + (\sqrt{1+v_{dif}} - \sqrt{v_D+v_{dif}}) = (B_N/2C_N)(1-v_{THN})t\,. \tag{29}$$

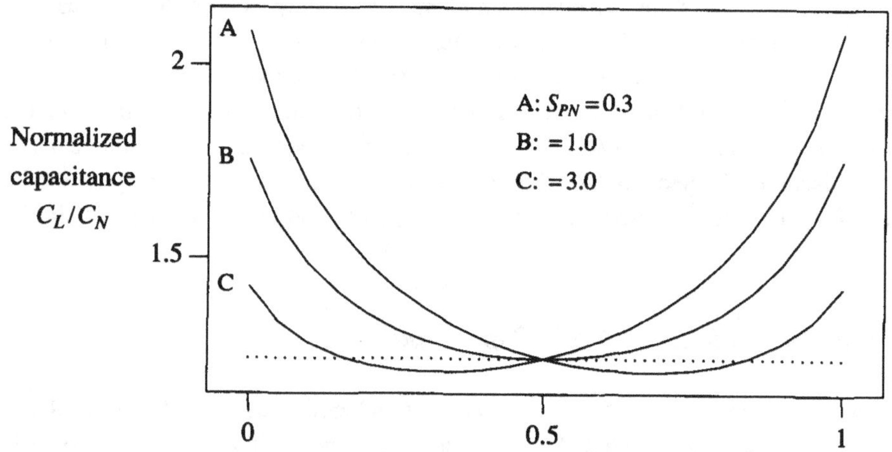

Figure 22 Drain island capacitance at the output node of inverter

Equation (29) is plotted in Fig. 23. If v_D is found as a function of t, $v_D(t)=0$ at $t=t_{END}$. At that time the NFET goes into the triode region. By substituting $v_D=0$ into Eq. (29) we have

$$t_{END}=2\frac{C_N}{B_N}\frac{(S_{PN}+1)(\sqrt{1+v_{dif}}-\sqrt{v_{dif}})}{1-v_{THN}}$$

and $v_D(t)=0$ if $t>t_{END}$. This result can be interpreted as follows. If the *effective* voltage-independent capacitance, C_{eff}, were defined such that the time at which the NFET goes into the triode region is the same

$$C_{eff}=2C_N(S_{PN}+1)(\sqrt{1+v_{dif}}-\sqrt{v_{dif}}).$$

Then using the effective capacitance for $0<t<t_{END}$, the relationship between t and v_D would be

$$1-v_D=(B_N/C_{eff})(1-v_{THN})t. \tag{30}$$

Equations (29) and (30) are plotted in Fig. 23, for $v_{dif}=v_{THN}=0.1$ and 0.0, assuming $S_{PN}=1$. The two equations, Eqs. (29) and (30), give different switching waveforms. The difference may be interpreted by assuming that the two equations refer to a different *time t* contained in their expressions. If time t contained in Eq. (30) is the *global* time, time in Eq. (29) is a *local* time that is a stretched and shrunk version of the global time, due to the voltage-dependence of the node capacitance. The difference is given by

$$\Delta t=\frac{2C_N}{B_N(1-v_{THN})}[-S_{PN}(\sqrt{1+v_{dif}}-\sqrt{1-v_D+v_{dif}})-(\sqrt{v_D+v_{dif}}-\sqrt{v_{dif}})$$
$$+(S_{PN}+1)(\sqrt{1+v_{dif}}-\sqrt{v_{dif}})v_D]$$

and Δt versus v_D is plotted in Fig. 24 for $S_{PN}=1$. Figure 24 shows that the time that the output node of the real inverter measures is delayed with respect to the global time determined by

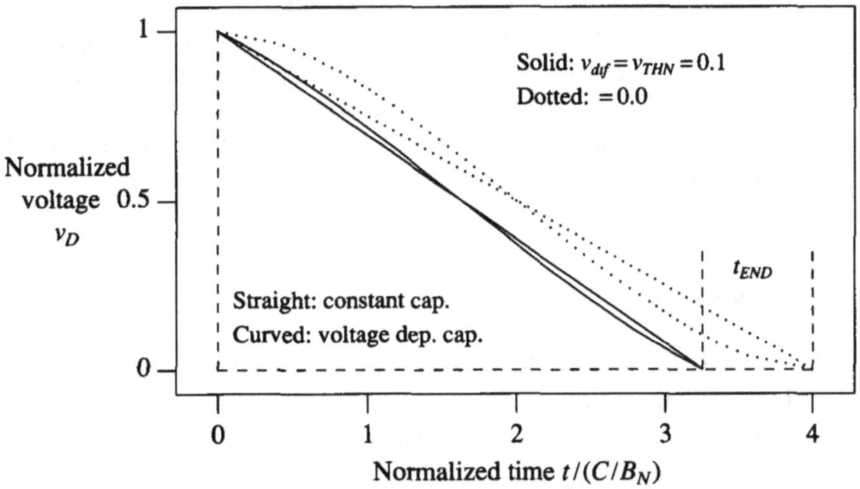

Figure 23 Output node voltage of inverter versus time

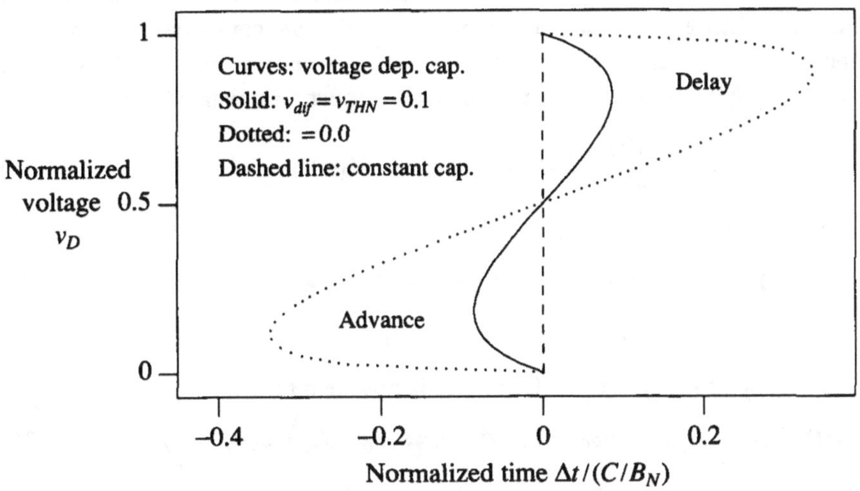

Figure 24 Output node voltage of inverter versus difference between local and global times

the averaged effective capacitance at the beginning of the switching, and then the local time catches up at $v_D = 0.5$. After that the local time is ahead of the global time, and at the end of the switching the two times align again (this is because C_{eff} is so defined). We note that the voltage dependence of capacitance advances and delays *the local* time with respect to *the global* reference time. This observation brings with it an interesting point in the interpretation of the microstate sequence. In the representation of a microstate sequence, there are two different types of relationships. Two events may be ordered (causal) or unordered. For instance, in the inverter pull-down transient, NFET: N → S always precedes PFET: T → S. There are cases where the two events are not related at all. FET events on two different parallel logic chains is an example. Delay/advance of local time due to secondary effect like voltage dependence of capacitance affects the order of the second type microstate events, but never changes the order of the first type of events. In this sense, causal order in the microstate sequence is *invariant* to voltage dependence of node capacitance.

2.8.4 *Decay of Virtual Ground*

As the last example of simple circuit analysis, we study the decay process of *virtual* ground. In Fig. 25(a), NFET MNA has beta B_A, and its gate is biased at V_G. NFET MNB is in series with NFET MNA, and its drain is biased to V_{DD}. V_I is an increasing function of time given by

$$V_I(t)=0 \ \ (t<0) \quad V_I(t)=\alpha t \ \ (t>0)$$

where α is the rate of increase in the input voltage. NFETs MNA and MNB are modeled using the collapsible current generator model with *zero* threshold voltage, to simplify the analysis. Immediately after $t=0$, MNB current is less than MNA current, therefore, node A voltage is 0. This state of node A is called *virtual* ground. This state continues until time t_1 is reached, when increasing MNB current equals the constant maximum current of MNA. Time t_1 is given by $t_1=(B_A V_G)/(B_B \alpha)$. After that time the circuit equation satisfied by $V_A(t)$ is

$$C_A \frac{dV_A(t)}{dt} + B_A V_G = B_B[V_I(t) - V_A(t)] \tag{34}$$

and the initial condition is $V_A(t_1)=0$. The solution is

$$V_A(t)=[(C_A\alpha)/B_B][exp[(B_B/C_A)(t_1-t)]-1]+\alpha(t-t_1) \tag{35}$$

$$\text{where } V_A'(t)=0 \text{ if } t=t_1 .$$

If normalized time $(\alpha/V_{DD})(t-t_1)=\theta$ is used, Eq. (35) is rewritten as

$$[V_A(t)/V_{DD}]=\theta_0[exp(-\theta/\theta_0)-1]+\theta \ \text{ where } \theta_0=(C_A\alpha)/(B_B V_{DD}). \tag{36}$$

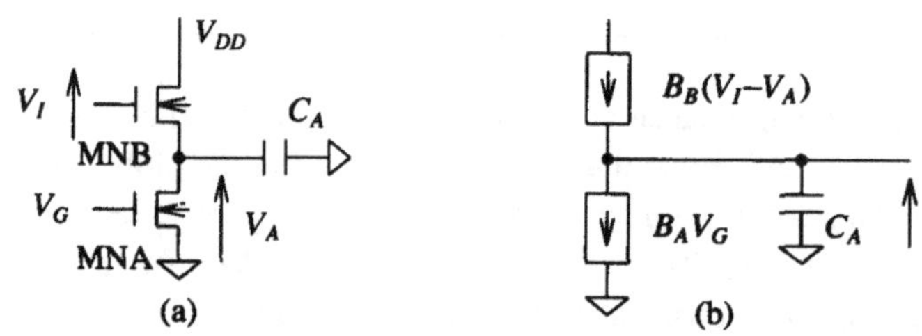

Figure 25 Decay of virtual ground

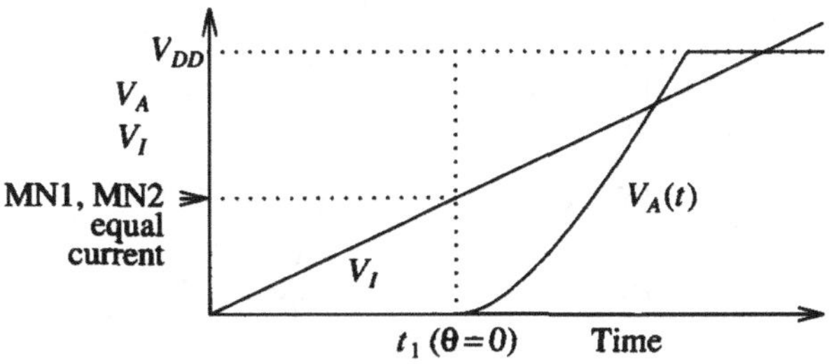

Figure 26 Virtual ground problem, node waveforms

Figure 26 shows node voltages versus time. After $t = t_1$ the MNB current becomes higher than the MNA current and the virtual ground breaks down. The breakdown process does not proceed rapidly, because the current available to charge C_A at time $t = t_1$ is zero. The node voltage increases as $V_A(t) \approx (t - t_1)^2$. As is shown in these examples, the mechanism of FET chain discharge is quite complex, even using the simple FET model. For analysis of circuit failure, a detailed understanding of this complex process is required. A collapsible current generator model of a FET, and a linear resistance model of a FET are both simple enough to clarify the detailed mechanisms of discharge in closed-form analysis, but the former emphasizes many interesting details (like formation of virtual ground), while the latter tends to deemphasize such crucial peculiarities.

2.9 Circuits of Medium Complexity

2.9.1 CMOS Static NAND2 Date

As the first example of an analysis of medium complexity CMOS circuits, we analyze the switching process of a 2-input NAND gate. We study the pull-down transition first, where the input signal makes a rapid low-to-high transition. Since the PFET(s) are turned off instantly, we need to consider the NFET chain only, which is shown in Fig. 27(b). NFETs MN1 and MN2 generally have different transconductances, B_1 and B_2, respectively. Figure 27(a) shows two modes of switching. Input 1 goes to MN1, and input 2 is to MN2. In the first case, input 2 is already at a logic-high level (V_{DD}), and signal 1 makes a rapid low-to-high transition. Since NFET MN1 was not conducting before, $V_X(0) = V_{DD} - V_{THN}$. Since the leakage current of MN2 can usually be neglected, this is a realistic initial condition for the switching process. We use the collapsible current generator model of a FET with constant threshold voltage.

The microstate sequence of the pull-down switching process is shown in Fig. 28(a). Immediately following the turn on of NFET MN1, V_X decreases, NFET MN2 moves from the marginally conducting region to the saturation region. Both NFETs are then in the saturation

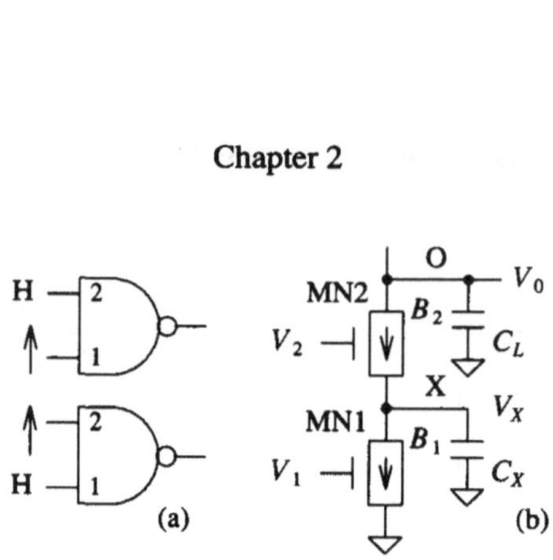

Figure 27 Analysis of NAND2 pull-down delay

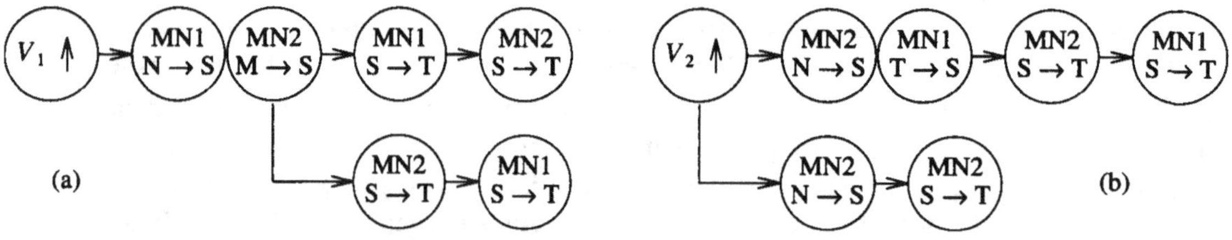

Figure 28 Microstate sequence of NFET pull-down chain

region. Either of them can go into the triode region first, and accordingly there are two possible cases shown in the microstate sequence. If both NFETs MN1 and MN2 are in the saturation region, the circuit equations are

$$C_L \frac{dV_O}{dt} = -B_2(V_{DD} - V_X - V_{THN}) \tag{37}$$

$$C_X \frac{dV_X}{dt} = (B_2 - B_1)(V_{DD} - V_{THN}) - B_2 V_X .$$

The solution of $V_X(t)$ satisfying $V_X(0) = V_{DD} - V_{THN}$ is

$$V_X(t) = [1 - (B_1/B_2)](V_{DD} - V_{THN}) + (B_1/B_2)(V_{DD} - V_{THN}) exp[-(B_2/C_X)t] . \tag{38}$$

Node O is originally at V_{DD}. The solution satisfying $V_O(0) = V_{DD}$ is given by integrating once again as

$$V_O(t) = V_{DD} - \frac{B_1}{C_L}(V_{DD} - V_{THN})t + \frac{B_1 C_X}{B_2 C_L}(V_{DD} - V_{THN})[1 - exp[-(B_2/C_X)t]] . \tag{39}$$

Depending on the ratio (B_1/B_2) there are two different possibilities. If $(B_1/B_2) > 1$, $V_X(t)$ becomes negative if $t > t_1$, where $t_1 = (C_X/B_2)\log[B_1/(B_1 - B_2)]$. After $t > t_1$ node X is fused to ground or becomes virtual ground, and $V_X(t) = 0$. Then node voltage V_O satisfies equation

$$C_L \frac{dV_O}{dt} = -B_2(V_{DD} - V_{THN}) \text{ or } V_O(t) = V_O(t_1) - \frac{B_2}{C_L}(V_{DD} - V_{THN})(t - t_1) . \tag{40}$$

Solutions of $V_0(t)$ for $t<t_1$ and for $t>t_1$ are connected at $t=t_1$. Since this case is simple, node waveforms are not shown. If $(B_1/B_2)<1$, $V_X(t)>0$ when $t\to\infty$. This limit is unrealistic: since $V_0(t)$ decreases monotonically, NFET MN2 moves into the triode region, and nodes O and X fuse at time $t=t_2$. Time t_2 is found by solving equation $V_X(t_2)=V_0(t_2)$ [Eqs. (38) and (39)] for t_2. After that time the voltage of the fused node decreases as

$$V_0(t)=V_X(t)=V_0(t_2)-\frac{B_1(V_{DD}-V_{THN})}{C_L+C_X}(t-t_2). \tag{41}$$

We use the following normalization

$$\theta=(B_2/C_X)t \quad \theta_2=(B_2/C_X)t_2 \quad \theta_3=(B_2/C_X)t_3$$

$$v_0(\theta)=V_0(t)/V_{DD} \quad v_X(\theta)=V_X(t)/V_{DD}$$

and at $t=t_3$ all the nodes fuse to ground, or $V_0(t_3)=V_X(t_3)=0$. We define normalized parameters

$$v_{THN}=V_{THN}/V_{DD} \quad b_{12}=B_1/B_2 \quad c_{XL}=C_X/C_L$$

and plot $v_0(\theta)$ and $v_X(\theta)$ versus θ. Figure 29 shows typical node waveforms analyzed using this method. Delay time t_D can be determined by solving $V_0(t)=V_{DD}/2$. Delay time is defined here as the interval between the time when the input signal reaches $V_{DD}/2$ and the time when the output node voltage reaches $V_{DD}/2$: $V_{DD}/2$ is *defined* as the switching threshold voltage of the CMOS gate [05].

If NFET MN1 had input voltage V_{DD}, and if the input of NFET MN2 switches, either of the following two possibilities exist. If node X of Fig. 27(b) remains at ground potential (*virtual* ground),

$$V_0(t)=V_{DD}-(B_2/C_L)(V_{DD}-V_{THN})t. \tag{42}$$

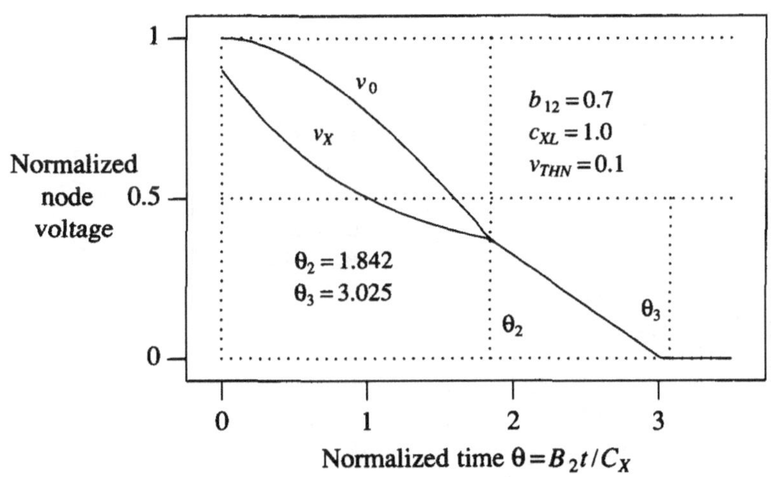

Figure 29 Node waveforms

$(B_1/B_2)>1$ is required to maintain the virtual ground. Delay time t_D is given in closed-form as

$$t_D=(1/2)(C_L/C_X)(C_X/B_2)[1/(1-v_{THN})].$$

If the *virtual* ground is destroyed by the current from MN2, node X voltage is

$$V_X(t)=[1-(B_1/B_2)](V_{DD}-V_{THN})[1-exp[-(B_2/C_X)t]]. \quad (43)$$

This takes place only if $(B_1/B_2)<1$. In this case, node voltage V_0 is given by

$$V_0(t)=V_{DD}-\frac{B_1}{C_L}(V_{DD}-V_{THN})t \quad (44)$$

$$-\frac{C_X}{C_L}\frac{B_2-B_1}{B_2}(V_{DD}-V_{THN})[1-exp[-(B_2/C_X)t]]$$

At time $t=t_2$, $V_X(t_2)=V_0(t_2)$, and nodes O and X fuse. After that time the fused node voltage is given by

$$V_0(t)=V_X(t)=V_0(t_2)-\frac{B_1(V_{DD}-V_{THN})}{C_X+C_L}(t-t_2). \quad (45)$$

Figure 30 shows $V_0(t)$ and $V_X(t)$ in the same normalized form as before, versus normalized time. From the node waveforms of Figs. 29 and 30, the gate delay can be determined as shown in Fig. 31. Normalized pull-down delay time $t_D/(C_X/B_2)$ is plotted versus the NFET beta ratio B_1/B_2. Solid curves show high internal node capacitance (C_X), and dotted curves show medium internal node capacitance, where $c_{XL}=C_X/C_L$ is the normalized internal node capacitance. Curves A and B show NFET MN1 on, when MN2 switches. Curves C and D show NFET MN2 on, when MN1 switches. Curves C and D show that the delay increases with decreasing transconductance of MN1. This dependence is well known. Curve A shows that the delay time is practically independent of the beta ratio. In this case, switching of the

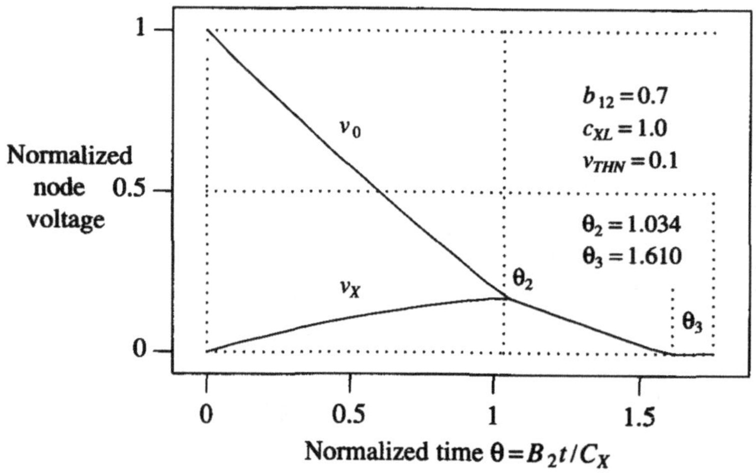

Figure 30 Node voltage waveforms

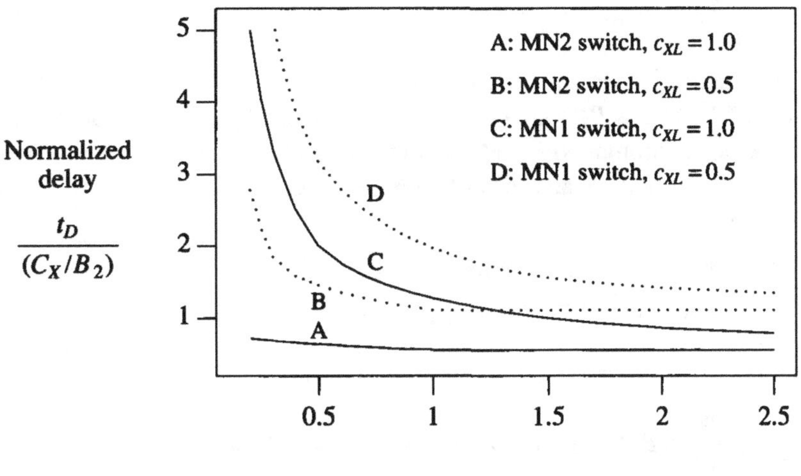

Figure 31 Pull-down delay versus the ratio of the NFET currents

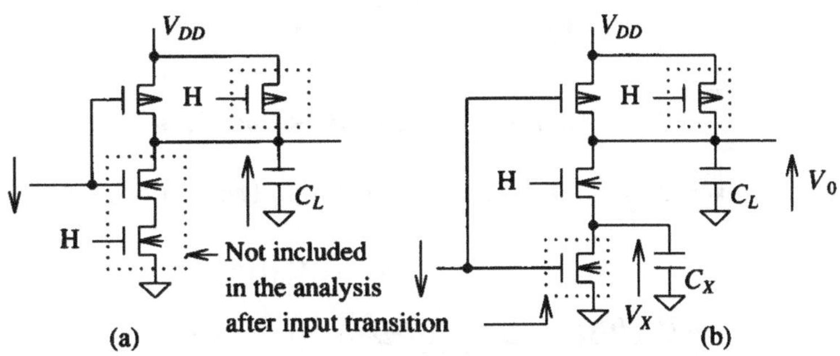

Figure 32 Pull-up transition of NAND2 gate

gate proceeds almost entirely by charge-sharing between nodes O and X. This is not a real switching of the gate. This pseudo switching takes place if C_X is large. A consequence of large C_X, a related phenomenon of internal hazard, will be discussed in Section 5.11.

The pull-up transition of a NAND2 gate has two different cases. Figure 32(a) shows the simplified equivalent circuit for the case when input 1 is kept at a high level and input 2 makes an instantaneous high-to-low transition. This case is trivial since the NFET chain is disconnected. In Fig. 32(b), input 2 is kept at a high level and input 1 makes high-to-low transition. In this case the gate's internal node X is charged by the MN2 source-follower circuit. The *effective* capacitance that must be charged up is more than that of Fig. 32(a), C_L. It is interesting to study how the switching proceeds in this case.

Immediately after time $t = 0$, capacitance C_X is still completely discharged. Node X and the output node are fused. Then the circuit equation and the solution are

$$(C_X + C_L)\frac{dV_0}{dt} = B_P(V_{DD} - V_{THP}) \qquad V_0(t) = V_X(t) = \frac{B_P(V_{DD} - V_{THP})}{C_X + C_L}t \tag{46}$$

respectively. The pull-up delay time, t_D, is given by $t_D = (C_X + C_L)/[2B_P(1 - v_{THP})]$. The maximum current of NFET MN2 is $B_2[V_{DD} - V_X(t) - V_{THN}]$, where B_2 is transconductance of MN2. The current that flows through NFET MN2 is given by $C_X \cdot dV_X(t)/dt$. If the current exceeds the maximum current, node X and the output node break apart at time t_1. By substituting $V_X(t)$, t_1 is found as

$$t_1 = [(C_X + C_L)/B_P][(V_{DD} - V_{THN})/(V_{DD} - V_{THP})] - (C_X/B_2).$$

After t_1 V_0 and V_X are determined from the circuit equations

$$C_X\frac{dV_X}{dt} = B_2[V_{DD} - V_X(t) - V_{THN}]$$

$$C_L\frac{dV_0}{dt} = B_P(V_{DD} - V_{THP}) - B_2(V_{DD} - V_X - V_{THN})$$

which are solved subject to the initial condition that the node voltages match at $t = t_1$. The solutions are

$$V_X(t) = V_{DD} - V_{THN} - [V_{DD} - V_{THN} - V_0(t_1)]exp[-(B_2/C_X)(t - t_1)] \tag{47}$$

$$V_0(t) = V_0(t_1) + \frac{B_P(V_{DD} - V_{THP})}{C_L}(t - t_1)$$

$$-\frac{C_X}{C_L}[V_{DD} - V_{THN} - V_0(t_1)][1 - exp[-(B_2/C_X)(t - t_1)]].$$

If $V_0(t)$ of Eq. (47) exceeds V_{DD}, $V_0(t)$ is equal to V_{DD}. The two solutions are normalized using

$$v_X(\theta) = V_X(t)/V_{DD} \qquad v_0(\theta) = V_0(t)/V_{DD}$$

$$t = (C_X/B_2)\theta \qquad t_D = (C_X/B_2)\theta_D \qquad t_1 = (C_X/B_2)\theta_1$$

and the normalized solutions are plotted in Fig. 33. The parameter values used are shown in the figure. Node X charges only to $V_{DD} - V_{THN}$.

2.9.2 *Cascode Voltage Switch*

We study a second example of medium complexity circuits. A cascode voltage switch is a symmetrical CMOS logic circuit, shown in Fig. 34(a), that generates both the output signal and its complement [06]. The pull-down chains (NFETs MN1 and MN2) can be replaced by NFET circuits that accept complementary pairs of input signals, and that generate complementary output signals. We assume that signals D and $\bar{D}$ make up- and down-going transitions, respectively, at time $t = 0$, within negligibly short rise and fall times. We assume zero threshold voltages of FETs for simplicity, and we assume that the two pull-down branches make a symmetrical cross-coupled circuit.

The circuit equations immediately after the D and $\bar{D}$ transitions are

$$C_L\frac{dV_1(t)}{dt} = B_P[V_{DD} - V_2(t)] - B_N V_{DD} \qquad C_L\frac{dV_2(t)}{dt} = B_P[V_{DD} - V_1(t)] \tag{48}$$

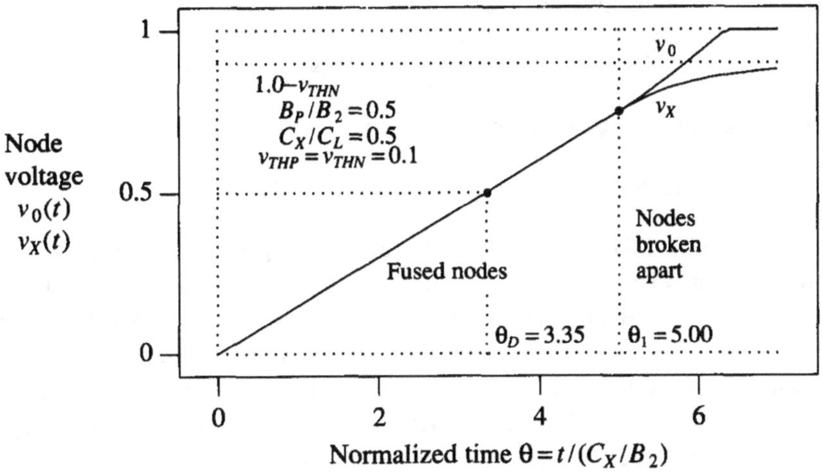

Figure 33 Node waveforms of NAND2 gate pull-up

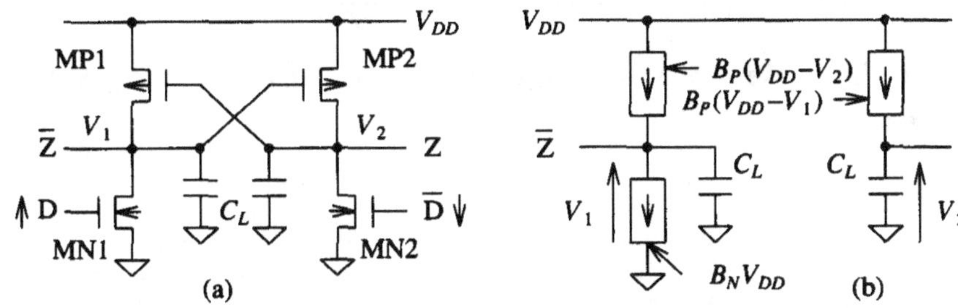

Figure 34 Cascode voltage switch inverter delay analysis

which are solved subject to the initial conditions

$$V_1(0)=V_{DD} \qquad V_2(0)=0$$

as follows

$$V_1(t)=V_{DD}[1-(b/2)[exp(B_Pt/C_L)-exp(-B_Pt/C_L)]] \tag{49a}$$

$$V_2(t)=[(bV_{DD})/2][exp(B_Pt/C_L)+\exp(-B_Pt/C_L)-2] \tag{49b}$$

where $b=(B_N-B_P)/B_P$. We note that if $b \le 0$, the cascode voltage switch circuit is unable to switch. If $b<0$ or $B_N<B_P$, the maximum pull-down current of node $\overline{Z}$ is less than the current PFET MP1 is able to supply while maintaining itself in the triode region, and therefore the node is stuck at V_{DD}. The cascode voltage switch is a *ratioed* logic circuit.

Node voltages $V_1(t)$ reaches 0 at time $t=t_{GND}$ given by

$$t_{GND}=\frac{C_L}{B_P}\log\frac{1+\sqrt{1+b^2}}{b}. \tag{50}$$

For time $t > t_{GND}$ node $\bar{Z}$ is virtual ground. Then we have

$$V_1(t)=0 \quad V_2(t)=V_2(t_{GND})+\frac{B_p V_{DD}}{C_L}(t-t_{GND}) \quad (t>t_{GND}) \tag{51}$$

and $V_2(t)$ reaches V_{DD} at time t_{END} that is determined from

$$V_{DD}=V_2(t_{GND})+\frac{B_P V_{DD}}{C_L}(t_{END}-t_{GND}).$$

For $t_{GND}<t<t_{END}$, NFET MN1 must conduct current $B_P[V_{DD}-V_2(t)]$, and since $0<V_2(t)<V_{DD}$, this current is less than $B_P V_{DD}$. Since $B_N > B_P$, the maximum current NFET MN1 is able to conduct is greater than that. Therefore MN1 stays in the triode region. For $t>t_{END}$, $V_1(t)=0$ and $V_2(t)=V_{DD}$. At $t=t_{END}$ the switching is completed. The microstate sequence of the switching process is shown in Fig. 35. Since we assumed zero FET threshold voltage, the microstate changes MP1: T → S and MP2: N → S take place simultaneously. If the threshold voltage is small but positive MP1: T → S takes place first. In Fig. 35 we assumed small, but nonzero threshold voltage of FET, so that the *causal* relationships of the events are shown clearly. Equation (49) may be written as

$$v_1(\theta)=\frac{V_1(t)}{V_{DD}}=1-b\sinh(\theta) \quad v_2(\theta)=\frac{V_2(t)}{V_{DD}}=b[\cosh(\theta)-1] \tag{52}$$

where $\theta=t/(C_L/B_P)$. We define $\theta_{GND}=t_{GND}/(C_L/B_P)$ and $\theta_{END}=t_{END}/(C_L/B_P)$. Then for $\theta_{GND}<\theta<\theta_{END}$

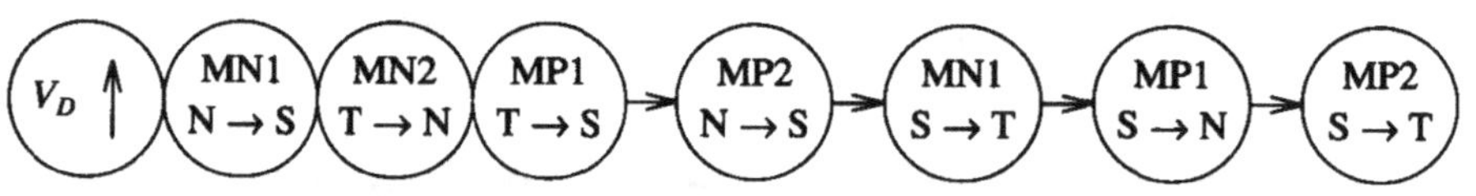

Figure 35 Microstate sequence of cascode voltage switch inverter

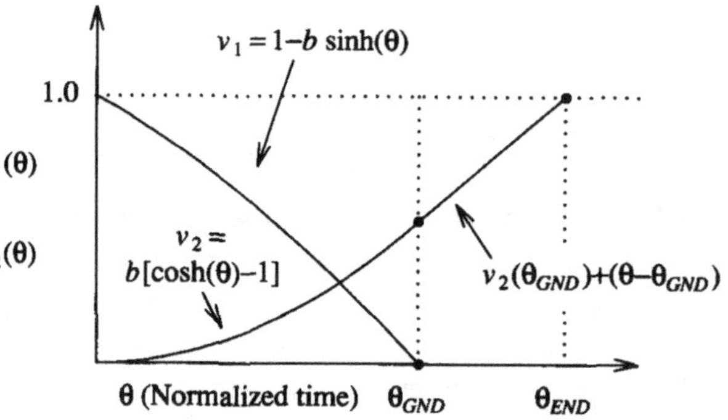

Figure 36 Switching waveform of cascode voltage switch

$$v_2(\theta) = v_2(\theta_{GND}) + (\theta - \theta_{GND})$$

Figure 36 shows $v_1(\theta)$ and $v_2(\theta)$ versus θ. At the beginning, positive feedback tries to maintain the present state of the circuit: v_2 begins to change quite slowly [$v'_2(0)=0$ in Fig. 36]. The rate of change of v_2 later increases by the same positive feedback mechanism. CMOS gates that have positive feedback have this type of response, which is not advantageous for circuit speedup, except for unusually heavily loaded nodes (Section 6.12).

2.10 Complex Circuits

Discharge of series-connected NFET chains is an important fundamental process that is relevant to many CMOS circuits. This problem has a solution that uses the equivalent linear resistor model, presented in Section 1.11. Here we study the same problem using the collapsible current generator model and compare the results. We show the two different analyses because the linear resistor model and the collapsible current generator model give two extreme discharge results for a FET chain. In Section 1.11 we saw that the potential profile within a scaled RC chain has drift motion. The drift motion was found using rather circuitous mathematics that made the conclusion quite obscure. Indeed the model tends to hide the drift motion of the potential profile in marginal cases. The analysis of a linear resistor model consists of integrating the FETs of a FET chain one by one from the grounded end into a single, long-channel FET, as shown in Figs. 37(a) and (b). Details of this process are described in the author's book *CMOS digital circuit technology* (chapter 3). That the procedure ends up with a single, long channel FET means that all the FETs are simultaneously involved in the discharge process. In the *low-field* limit where the linear resistor model is valid, FETs at different locations of the FET chain do not have significantly different channel *sheet* resistance. Therefore, in the process of integrating FETs, no *peculiarity* emerges. FET chain current is determined by all the FETs, from the beginning of the discharge to the end.

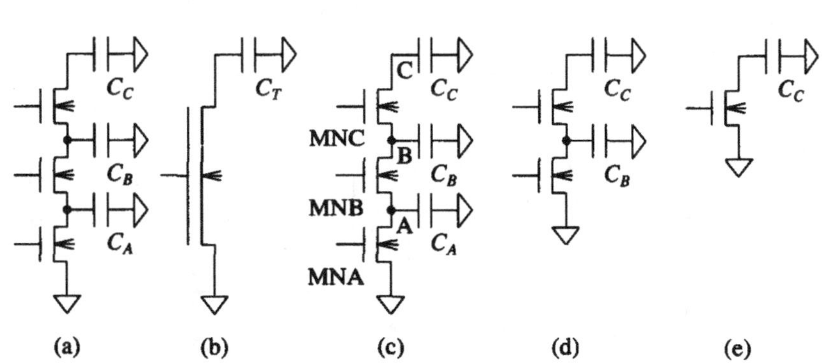

Figure 37 Linear resistor model and collapsible current generator model

Analysis of scaled FET chain discharge is simple if the collapsible current generator model is used. Suppose, in Fig. 37(c), that MNA is larger than MNB, and MNB is larger than MNC. The internal nodes are initially charged to V_{DD}. Since NFET MNA is larger than MNB, node A of Fig. 37(c) is pulled down to virtual ground, and the circuit changes from Fig. 37(c) to 37(d). Since NFET MNB is larger than MNC, node B is pulled down to virtual ground and the circuit changes to Fig. 37(e). In this model, FETs in the scaled FET chain are *eliminated* one by one from the grounded end, rather than integrated into a single, long-channel FET. The simplicity of this model suggests that this is the alternative, and the essential mechanism of FET chain discharge. In the high-field limit where the collapsible current generator model is valid, the FET chain current is determined only by the FET supporting the maximum voltage, and which is deep in the saturation region. Other FETs in the low-field region have high channel conductance, and they develop no significant voltage drop across the source and drain terminals. They are all in the triode region.

Let us consider this problem analytically. In the long NFET chain of Fig. 38, we assume that the gates of all NFETs except that of MN1 are held at V_{DD}. The FET threshold voltages are assumed to be zero for simplicity. Gate of MN1 was originally at ground potential. Internal nodes, N1, N2, ..., are charged to V_{DD}. The gate of MN1 makes a low-to-high transition (from 0 to V_{DD}) at time $t = 0$, thereby initiating discharge of the NFET chain. Analysis of this process without any simplification is too complex to be manageable. There are so many possible microstate sequences of the FET chain discharge process that covering all cases is quite impossible. In this section, one special microstate sequence is studied in detail to gain insight into this problem. In Fig. 39, at the moment of switching, all the FETs change from the marginally conducting state to the saturation region. This is of course due to the peculiar definition of the marginally conducting state. Then discharge starts from the left (grounded) end and moves to the right. NFET MN1 goes into the triode region first, followed by MN2, MN3, ..., in that order. This is the mechanism of the drift motion of the potential profile, introduced in Figs. 37(c)–(e).

In order to make this problem tractable, we assume the following. Node voltages V_n are functions of node index n. V_n is a *gradual* function of node index n. Then instead of index n we may use *coordinate* x, and we may write the node voltages as a function of location x and

---

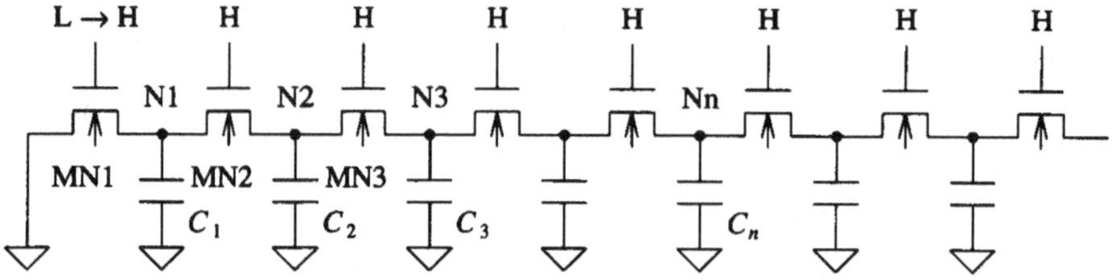

Figure 38 NFET chain discharge process

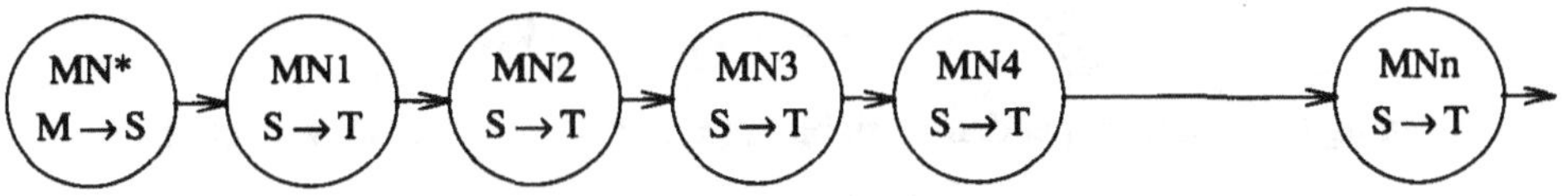

Figure 39 FET chain discharge microstate diagram

time t, as $V(x,t)$. We consider transconductance of FETs and node capacitances to be functions of *coordinate* x, as $B(x)$ and $C(x)$, respectively (*continuous* approximation). The equivalent circuit of the FET chain near location x is shown in Fig. 40, where $V(x,t)$ is monotonically increasing function of x. A node equation can be written as

$$C(x)\frac{\partial V(x,t)}{\partial t} = B(x+1)[V_{DD} - V(x,t)] - B(x)[V_{DD} - V(x-1,t)]. \tag{53}$$

We assumed that V_{THN} is zero. If we use approximations

$$V(x-1,t) = V(x,t) - V'(x,t) \qquad B(x+1) = B(x) + B'(x) \tag{54}$$

where primes indicate derivatives by x. We have

$$\frac{\partial V(x,t)}{\partial t} = -\frac{B(x)}{C(x)}\frac{\partial V(x,t)}{\partial x} + \frac{B'(x)}{C(x)}[V_{DD} - V(x,t)]. \tag{55}$$

We consider the capacitance of NFET MNx as all credited to node x. Since the transconductance of the FET and capacitance of the FET are both proportional to the FET size (width), we have

$$\frac{B(x)}{C(x)} = constant = \zeta. \tag{56}$$

We then have $B'(x)/C(x) = \zeta C'(x)/C(x)$. We consider the simplest case of an *exponentially* scaled chain. Then we have

$$\frac{C'(x)}{C(x)} = const = \gamma \text{ or } C(x) = const \times exp(\gamma x). \tag{57}$$

Equation (55) becomes

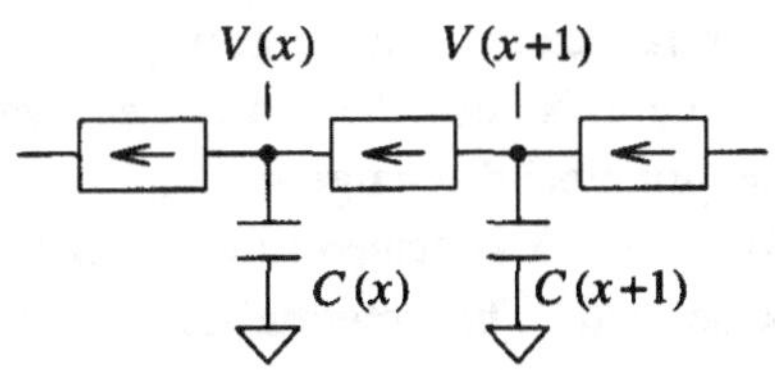

Figure 40 Derivation of circuit equation

$$\frac{\partial V(x,t)}{\partial t} + \zeta \frac{\partial V(x,t)}{\partial x} = \zeta\gamma[V_{DD} - V(x,t)]. \tag{58}$$

By transforming to the new dependent variable, $\phi(x,t) = V_{DD} - V(x,t)$, we obtain

$$P(x,t)\frac{\partial \phi(x,t)}{\partial t} + Q(x,t)\frac{\partial \phi(x,t)}{\partial x} = -\zeta\gamma\,\phi(x,t) \quad P(x,t)=1 \;\; Q(x,t)=\zeta .$$

This equation can be solved by standard techniques [07]. The auxiliary differential equation of this partial differential equation is

$$\frac{dt}{1} = \frac{dx}{\zeta} = \frac{d\phi}{-\zeta\gamma\phi} .$$

The two independent solutions of this auxiliary equation are

$$x - \zeta t = const(1) \text{ and } x + \frac{1}{\gamma}\log\phi = const(2)$$

and the general solution of the equation is

$$G(x-\zeta t, x + \frac{1}{\gamma}\log\phi) = 0 \text{ or } V(x,t) = V_{DD} - exp(-\gamma x)g(x-\zeta t) \tag{59}$$

where G and g are arbitrary functions. Let us explain what Eq. (59) means physically. If the FET chain is uniform, $\gamma = 0$, and therefore $V(x,t) = V_{DD} - g(x-\zeta t)$. The potential profile moves to the right with constant velocity ζ. Drift motion exists even in an unscaled chain. Since this solution contains an *arbitrary* function g, it may be thought that even a discontinuous step function may propagate with constant velocity. Since derivation of the partial differential equation, Eq. (57), assumes that node voltages vary gradually with node index, or coordinate x, a discontinuous solution is not consistent with the assumption. Therefore g must be a gradual function of the argument. To obtain a simple, graphical interpretation of the solution, we write the general solution by using different g-functions as

$$V(x,t) = V_{DD} - exp(-\zeta\gamma t)g(x-\zeta t). \tag{60}$$

Curve A of Fig. 41 shows the initial potential profile. After time t, the profile moves to the right, to location Q. If γ is negative (FET size decreases toward the right), the exponential factor in Eq. (60) is larger than unity. Then curve B is scaled downward to curve C. Curve C intersects the horizontal axis at point T. To the left of point T, the voltage given by Eq. (60) becomes negative. A negative node voltage is not acceptable. Therefore to the left side of point T, curve C is replaced by a straight line $\overline{OT}$. We note that rightward motion of potential profile $\overline{OT}$ is the sum of $\overline{OQ}$ and $\overline{QT}$. $\overline{OQ}$ is contributed by the collapsible current generator model itself, and $\overline{QT}$ by exponential scaling. In the collapsible current generator model, drift motion of potential profile is more enhanced than in the equivalent linear resistance model, and the mechanism becomes the principal discharge mechanism of the FET chain.

If γ is positive in Fig. 41, curve A is scaled upward to curve D. The potential profile between point O and S is not defined. The potential profile in this region can be found by requiring

$$V(0,t) = 0$$

as

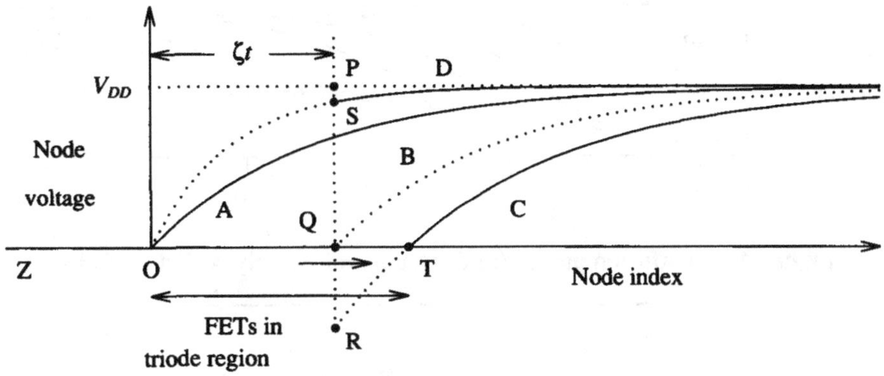

Figure 41 Potential profile of FEET chain

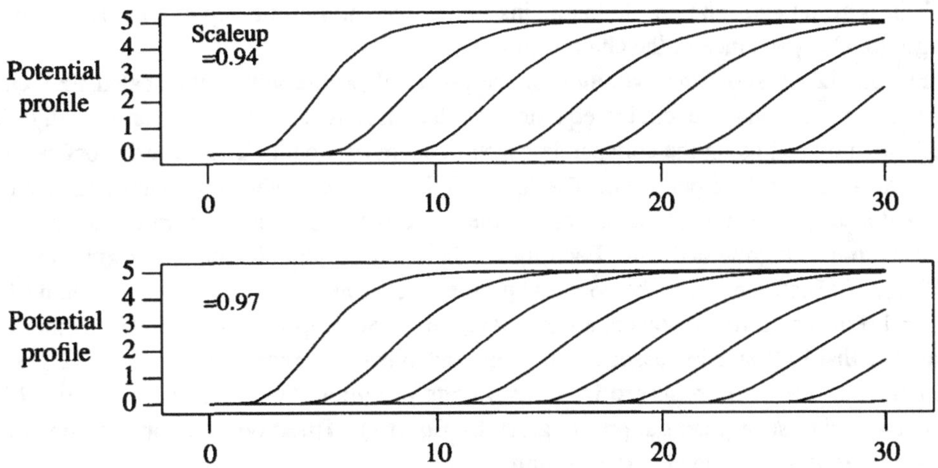

Figure 42 Drift motion of potential profile in scaled FET chain

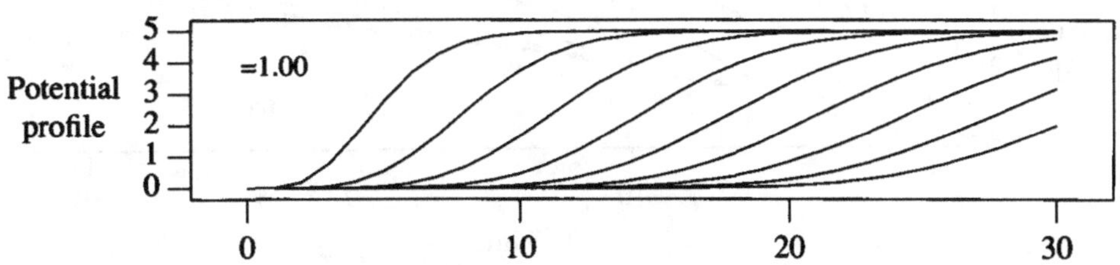

Figure 43 Mixed drift-diffusion discharge process in unscaled FET chain

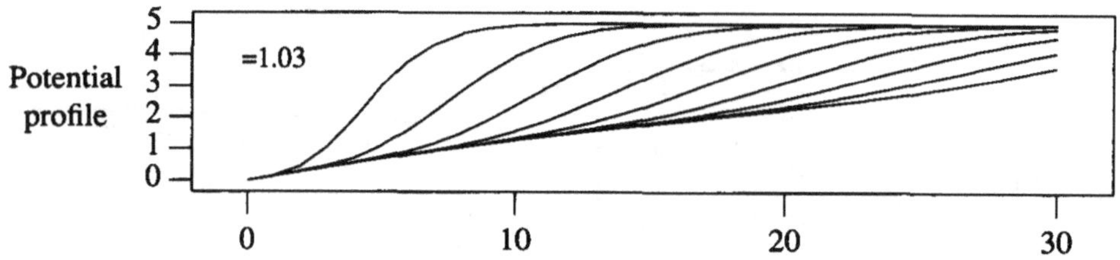

Figure 44 Diffusion process of discharge in inversely scaled FET chain

$$V(x,t) = V_{DD}[1 - exp(-\gamma x)]\,. \tag{61}$$

This is the steady-state potential profile established within the FET chain if the FET size increases exponentially to the right. If the chain is infinitely long and is charged to V_{DD}, then the steady-state potential profile is established within the chain sometime after the left end of the chain is grounded. The profile maintains itself, since an infinite supply of charge exists to the right (total capacitance of the chain is infinite).

Figures 42–45 show the evolution of the potential profile within the scaled FET chain, determined by solving the circuit equation of the discrete NFET chain numerically. The NFETs are modeled using the collapsible current generator model. The scale factors are 0.94, 0.97, 1.00, 1.03, and 1.06 per stage. The first NFET from the grounded end has unit size (unit transconductance and unit capacitance). If the scale factor is less than unity, the FET size decreases from the grounded end. Each figure shows the potential profile at normalized time 3, 6, 9, ..., 27. Drift motion of the potential profile exists even in an unscaled FET chain (Scale factor = 1.00). The drift motion enhances with scaling (Scaleup = 0.94 and 0.97). If the chain is scaled so that FET size increases from the grounded end, the potential profile converges to a steady-state profile, as is observed in the two bottom figures (Scale factor = 1.03 and 1.06). This is the steady-state potential profile given by Eq. (61). These observations agree with the closed-form theory developed in this section.

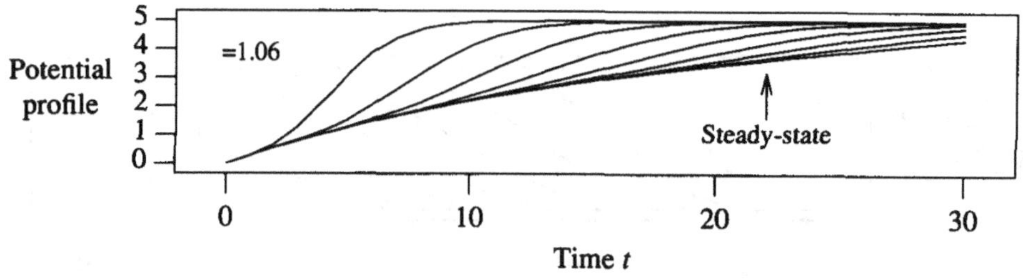

Figure 45 Approach to steady state potential profile by diffusion process

2.11 Differential Amplifier

The differential amplifier shown in Fig. 46(a) is often used to convert a pair of bipolar, reduced-amplitude digital input signals to a single-ended, large-amplitude CMOS signal [08]. As an example of complex CMOS circuits, we study this circuit. We use the collapsible current generator model, and we neglect the back-bias effect.

2.11.1 *Static Characteristics*

NFET MN0 works as a current generator driving the common source tree, whose current, I_0, is given by

$$I_0 = B_{N0}(V_G - V_{THN}) \tag{62}$$

if $V_0 > 0$. The transconductances of NFETs MN1 and MN2 are B_N, and those of PFETs MP1 and MP2 are B_P. We consider a *symmetrical* (MN1 and MN2 are identical, as are MP1 and MP2) differential amplifier and study the static switching characteristics first. The part of the circuit that consists of MP1 and MP2 is called a current *mirror*. MP2 reproduces the current of MP1, if the drain-source voltage of MP2 is not zero (in the collapsible current generator model). Suppose that the two input voltages V_{I+} and V_{I-} are equal to V_I (*balanced* operation). NFETs MN1 and MN2, and PFETs MP1 and MP2 carry equal current. Then we have $I_0/2 = B_P(V_{DD} - V_{0-} - V_{THP})$ by considering the current flowing through PFET MP1. Solving this equation for V_{0-} we find

$$V_{0-}(Balance) = V_{DD} - V_{THP} - (I_0/2B_P)\,. \tag{63}$$

If we consider the limit as V_I increases, NFETs MN1 and MN2 are both forced into the triode region. Then $V_{0-}(Balance) > V_0$ must hold. V_0 must be positive (above ground). Then $V_{0-}(Balance)$ must be positive also. We require

$$B_P(V_{DD} - V_{THP}) > (I_0/2) \tag{64}$$

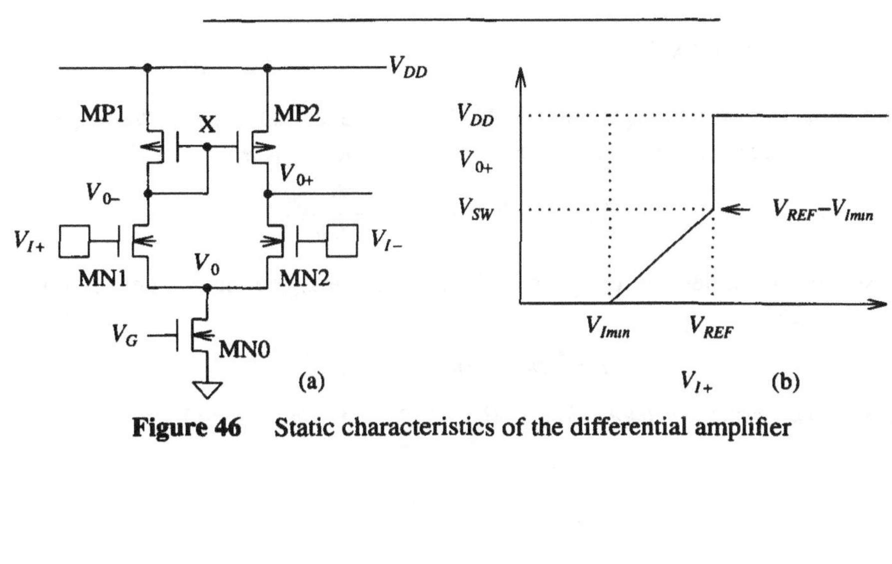

Figure 46 Static characteristics of the differential amplifier

or the *maximum current* of the PFET current mirror circuit must be higher than $I_0/2$. The common source node voltage, V_0, has a maximum at

$$V_0(Balance) \le V_{0-}(Balance)$$

and the equal sign holds when V_I is large. In order to reach this limit it is unnecessary that V_I actually increases to infinity. If the saturation current of NFETs MN1 and MN2, whose source voltage is $V_0(Balance)$, is higher than $I_0/2$, they enter the triode region. This takes place if V_I is higher than V_{Imax} given by

$$B_N[V_{Imax} - V_0(Balance) - V_{THN}] = \frac{I_0}{2}.$$

V_{Imax} is found as

$$V_{Imax} = V_{DD} - V_{THP} + V_{THN} + \frac{I_0}{2}\left(\frac{1}{B_N} - \frac{1}{B_P}\right). \tag{65a}$$

If $V_I > V_{Imax}$, NFETs MN1 and MN2 lose control and the differential amplifier has no gain. Similarly there is a lower limit to the common input voltage V_I. Since NFETs MN1 and MN2 must be able to carry currents higher than $I_0/2$ while maintaining $V_0 > 0$, $(I_0/2) \le B_N(V_I - V_{THN})$. Therefore

$$V_{Imin} = V_{THN} + (I_0/2B_N) < V_I. \tag{65b}$$

The differential amplifier is balanced and has small-signal gain if the common input voltage, V_I, is in the range from V_{Imin} to V_{Imax}.

Suppose that V_{I-} is set at the reference voltage V_{REF}, where V_{REF} satisfies

$$V_{Imin} < V_{REF} < V_{Imax}$$

and V_{I+} is higher than V_{I-}, or $V_{I+} > V_{REF}$. Then NFET MN1 carries more current than MN2, and therefore MP1 carries more current than MP2. Then MP2 is in the triode region, sustaining no voltage. V_{0+} is at V_{DD}. In this case MP1 and MP2 do not work as a current *mirror*. If V_{I+} is lower than V_{REF}, however, MP1 and MP2 work as a current mirror. In this case V_0 adjusts so that MN1 carries $I_0/2$. We have

$$\frac{I_0}{2} = B_N(V_{I+} - V_0 - V_{THN}) \quad \text{or} \quad V_0 = V_{I+} - V_{THN} - \frac{I_0}{2B_N}$$

and since the maximum current of NFET MN2 is higher than that of MN1, MN2 goes into the triode region, sustaining no voltage. V_{0+} equals the common source voltage V_0, as

$$V_{0+} = V_0 = V_{I+} - V_{THN} - \frac{I_0}{2B_N}.$$

In this case the amplifier has unity gain. This operational regime reaches its limit when $V_0 = 0$. We then have

$$V_{I+} = V_{THN} + (I_0/2B_N) = V_{Imin}.$$

This static characteristic is shown in Fig. 46(b). High gain occurs when $V_{I+} = V_{REF}$.

2.11.2 Dynamic Characteristics

The dynamic switching characteristics of a CMOS differential amplifier can be analyzed using the circuit that includes the node capacitances shown in Fig. 47. Common source node S, the current mirror's internal node X, and output node O are loaded by capacitances C_S, C_C, and C_L, respectively. The collapsible current generator I_0 represents the current generator NFET (MN0 of Fig. 46). The gate of MN2 is held at V_{REF}, and the gate of MN1 switches from V_L ($V_L < V_{REF}$) to V_H ($V_H > V_{REF}$) (low-to-high transition). We analyze the transient waveforms of nodes S, X, and O.

Prior to switching, the voltages of nodes S, X, and O are given by

$$V_0(INIT) = V_L - V_{THN} - (I_0/2B_N) \tag{66}$$

$$V_{0-}(INIT) = V_{DD} - V_{THP} - (I_0/2B_P)$$

$$V_{0+}(INIT) = V_0(INIT) = V_L - V_{THN} - (I_0/2B_N).$$

Before switching, $V_{I+} = V_L\,(low)$, and $V_{I+} > V_{Imin}$ [Eq. (65b)]. The differential amplifier is in the range where the input-output relationship is linear and has unity gain [Fig. 46(b)]. The initial conditions of Eqs. (66) are interpreted as follows. Since $V_{REF} > V_{I+} > V_{Imin}$ the PFET current mirror forces the same current for MN1 and MN2 ($I_0/2$). $V_{0-}(INIT)$ and $V_0(INIT)$ are derived from that requirement. Since $V_{I+} < V_{REF}$, MN2 is in the triode region, sustaining no voltage. As for the final state after switching, node voltages will be as follows:

$$V_0(FINAL) = (1/2)(V_H + V_{REF} - 2V_{THN}) - (I_0/2B_N) \tag{67a}$$

$$V_{0-}(FINAL) = V_{DD} - V_{THP} - (I_0/2B_P) - [B_N/(2B_P)](V_H - V_{REF}) \quad V_{0+}(FINAL) = V_{DD}$$

if MN1 is not in the triode region and

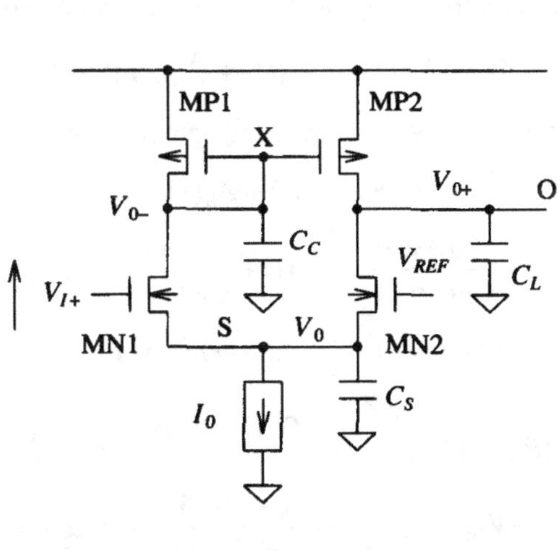

Figure 47 Switching of differential amplifier

$$V_0(FINAL) = V_{0-}(FINAL) \tag{67b}$$

$$= \frac{B_P(V_{DD} - V_{THP}) + B_N(V_{REF} - V_{THN}) - I_0}{B_N + B_P} \qquad V_{0+}(FINAL) = V_{DD}$$

if MN1 is in the triode region. MN1 is in the triode region if

$$V_H > \frac{2B_P(V_{DD} - V_{THP} + V_{THN})}{B_N + B_P} + \frac{B_N - B_P}{B_N(B_N + B_P)}(B_N V_{REF} - I_0).$$

Both NFETs carry current, but MN1 carries more current than MN2, and MP2 goes into the triode region. Then PFET MP2 carries current, but it develops no voltage. We study how the dynamic transition from the state specified by Eq. (66) to that by Eq. (67) takes place.

Before switching, MN2 is in the triode region (nodes O and S are fused). Until MN2 moves out of the triode region, the state of the circuit is determined from a pair of equations that have two independent node voltages V_0 and V_{0-} as

$$(C_S + C_L)\frac{dV_0}{dt} = -I_0 + B_N(V_H - V_0 - V_{THN}) + B_P(V_{DD} - V_{0-} - V_{THP}) \tag{68}$$

$$C_C\frac{dV_{0-}}{dt} = B_P(V_{DD} - V_{0-} - V_{THP}) - B_N(V_H - V_0 - V_{THN})$$

The first equation governs the two nodes having the same potential: $V_0(t)$ and $V_{0+}(t)$. The set of equations is solved subject to the initial conditions, Eq. (66). The solution is the superposition of the special solution

$$V_0(SP) = V_H - V_{THN} - \frac{I_0}{2B_N} \qquad V_{0-}(SP) = V_{DD} - V_{THP} - \frac{I_0}{2B_P}$$

and of the general solution of the homogeneous parts of the circuit equations

$$(C_S + C_L)\frac{dV_0}{dt} = -B_N V_0 - B_P V_{0-} \qquad C_C\frac{dV_{0-}}{dt} = B_N V_0 - B_P V_{0-}.$$

If we assume solutions of the form V_0, $V_{0-} \approx exp(St)$, parameter S satisfies the secular equation

$$C_C(C_S + C_L)S^2 + [C_C B_N + (C_S + C_L)B_P]S + 2B_N B_P = 0$$

whose two solutions are α and β. We then have the homogeneous solution

$$V_0 = A_0(+)exp(\alpha t) + A_0(-)exp(\beta t)$$

$$V_{0-} = \frac{B_N A_0(+)}{C_C\alpha + B_P}exp(\alpha t) + \frac{B_N A_0(-)}{C_C\beta + B_P}exp(\beta t)$$

and the complete solution that satisfies the initial conditions is

$$V_0 = V_H - V_{THN} - \frac{I_0}{2B_N} + \frac{V_H - V_L}{C_C(\beta - \alpha)}\Big[(C_C\alpha + B_P)exp(\alpha t) - (C_C\beta + B_P)exp(\beta t)\Big] \tag{69}$$

$$V_{0-} = V_{DD} - V_{THP} - \frac{I_0}{2B_P} + \frac{(V_H - V_L)B_N}{C_C(\beta - \alpha)}[exp(\alpha t) - exp(\beta t)]$$

where

$$\alpha(+), \beta(-) = \frac{-[(C_S + C_L)B_P + C_C B_N] \pm \sqrt{[(C_S + C_L)B_P + C_C B_N]^2 - 8C_C(C_S + C_L)B_N B_P}}{2C_C(C_S + C_L)}$$

We note that α and β may be complex numbers. Then we take the real parts of Eq. (69). Appearance of a complex number in the solution does not mean oscillatory voltage change: Before completing one cycle of oscillation, the circuit will change state.

There are several possibilities that may terminate the initial microstate. If V_H is not excessively high, the solution is valid until NFET MN2 moves out of the triode region. This takes place when

$$\text{MN2 current } [= B_P(V_{DD} - V_{0-} - V_{THP}) - C_L(dV_0/dt)] < B_N(V_{REF} - V_0 - V_{THN})$$

is no longer maintained. The inequality determines the critical time t_0. The equation for t_0 is rewritten as

$$B_N(V_H - V_{REF}) = \frac{V_H - V_L}{C_C(\beta - \alpha)}\Big[[C_L C_C \alpha + (B_P C_L - B_N C_C)]\alpha exp(\alpha t_0)$$
$$+ [-C_L C_C \beta - (B_P C_L - B_N C_C)]\beta exp(\beta t_0)\Big].$$

Solving for t_0 of this equation gives the time when the set of equations Eq. (68) ceases to be valid. After t_0 the voltages of nodes S and O are no longer equal. All three nodes are now distinct and we have a new set of equations,

$$C_S \frac{dV_0}{dt} = B_N(V_H - V_0 - V_{THN}) + B_N(V_{REF} - V_0 - V_{THN}) - I_0 \quad (70)$$
$$C_C \frac{dV_{0-}}{dt} = B_P(V_{DD} - V_{0-} - V_{THP}) - B_N(V_H - V_0 - V_{THN})$$
$$C_L \frac{dV_{0+}}{dt} = B_P(V_{DD} - V_{0-} - V_{THP}) - B_N(V_{REF} - V_0 - V_{THN}).$$

Since the first equation contains V_0 only, it can be solved immediately, as

$$V_0(t) = X_0 + [V_0(t_0) - X_0]exp[(2B_N/C_S)(t_0 - t)] \quad (71a)$$

where

$$X_0 = [(V_H + V_{REF})/2] - V_{THN} - (I_0/2B_N) = V_0(FINAL)$$

and where $V_0(t_0)$ is the initial condition for Eq. (70) obtained by substituting $t = t_0$ in Eq. (69). This solution is substituted into the second equation of Eq. (70) and $V_{0-}(t)$ is found as

$$V_{0-}(t) = X_2 + [V_{0-}(t_0) - X_2 - \frac{V_0(t_0) - X_0}{(B_P/B_N) - 2(C_C/C_S)}]exp[(B_P/C_C)(t_0 - t)] \quad (71b)$$
$$+ \frac{V_0(t_0) - X_0}{(B_P/B_N) - 2(C_C/C_S)} exp[(2B_N/C_S)(t_0 - t)]$$

where

$$X_2 = V_{DD} - V_{THP} + (B_N/B_P)(X_0 - V_H + V_{THN})$$

and similarly output node voltage $V_{0+}(t)$, after considerable algebra, is found as

$$V_{0+}(t)=V_{0+}(t_0)+\frac{1}{C_L}[B_P(V_{DD}-V_{THP})-B_N(V_{REF}-V_{THN})+B_NX_0-B_PX_2](t-t_0) \quad (71c)$$
$$+\frac{1}{2}\frac{C_S}{C_L}[V_0(t_0)-X_0][1-exp[(2B_N/C_S)(t_0-t)]]$$
$$-\frac{C_C}{C_L}[V_{0-}(t_0)-X_2-\frac{V_0(t_0)-X_0}{(B_P/B_N)-2(C_C/C_S)}][1-exp[(B_P/C_C)(t_0-t)]]$$
$$-\frac{1}{2}\frac{B_P}{B_N}\frac{C_S}{C_L}\frac{V_0(t_0)-X_0}{(B_P/B_N)-2(C_C/C_S)}[1-exp[(2B_N/C_S)(t_0-t)]]\,.$$

The microstate (all the FETs in the saturation region) may end when PFET MP2 moves into the triode region. After MP2 moves into the triode region we have a closed-form solution for V_0 and V_{0-} as

$$V_0(t)=X_0+[V_0(t_1)-X_0]exp[2(B_N/C_S)(t_1-t)] \quad (71d)$$

where t_1 is the time when MP2 enters the triode region, and

$$V_{0-}(t)=X_3+[V_{0-}(t_1)-X_3]exp[(B_P/C_C)(t_1-t)] \quad (71e)$$
$$+[V_0(t_1)-X_0]\frac{exp[2(B_N/C_S)(t_1-t)]-exp[(B_P/C_C)(t_1-t)]}{(B_P/B_N)-2(C_C/C_S)}$$
$$+(B_N/B_P)X_0[1-exp[(B_P/C_C)(t_1-t)]]$$

where

$$X_3=(V_{DD}-V_{THP})-(B_N/B_P)(V_H-V_{THN})\,.$$

Figure 48 shows normalized node voltages versus normalized time. Circuit parameters used in this example are shown in the figure. V_0 increases with time, and it saturates at about

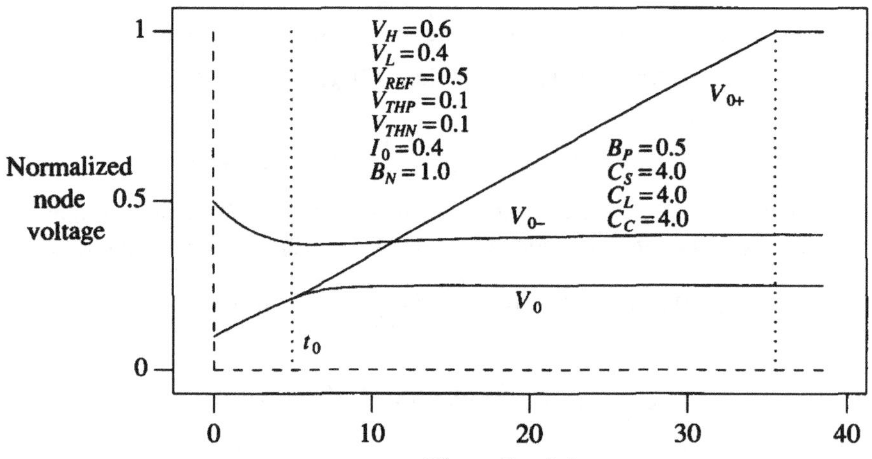

Figure 48 Node voltage waveforms of the differential amplifier

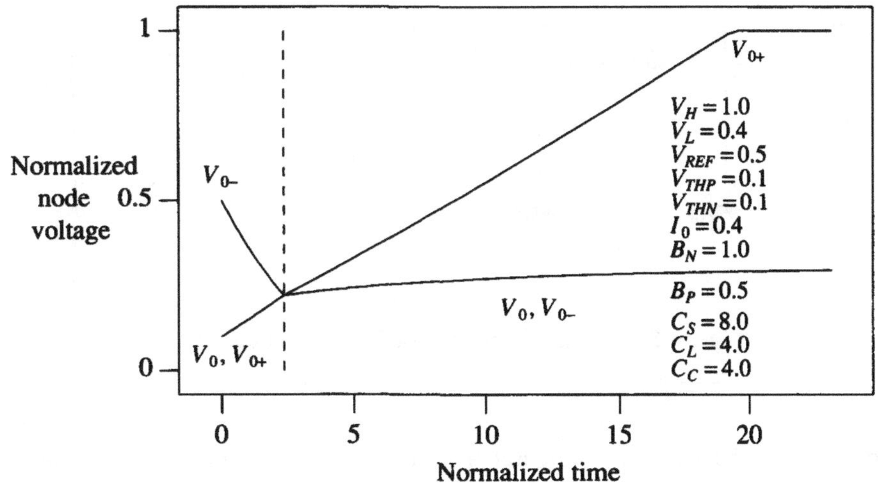

Figure 49 Node voltage waveforms of differential amplifier

$V_0(\infty) = 0.25$. The $V_{0-}(t)$ decreases with time, and it also saturates but at $V_{0-}(\infty) = 0.4$. Since $V_{REF} = 0.5$ and $V_{THN} = 0.1$, NFET MN2 starts from the triode region and ends in the saturation region. Microstate change MN2: T → S takes place at about normalized time $t = 4$. At the final state, MN1 current is 0.25 and MN2 current is 0.15. Both NFETs are in the saturation region. The microstate sequence of switching is shown in sequence A of Fig. 50. Output node voltage $V_{0+}(t)$ increases almost linearly with time, and it reaches V_{DD} at normalized time $t = 35$. Since MP1 current is higher than MP2 current, MP2 goes into the triode region, thereby completing the switching process.

Figure 49 shows a second example. The high level input voltage is higher than the first example, at 1.0. Because of the higher input voltage, NFET MN1 goes into the triode region first. Immediately before NFET MN1 goes into the triode region, NFET MN2 is in the triode region. The current that flows through MN2 in the triode region is less than the maximum current allowed for MN2 to develop a voltage between the drain and the source. Then the microstate sequence of this switching process should look like sequence B of Fig. 50. Figures

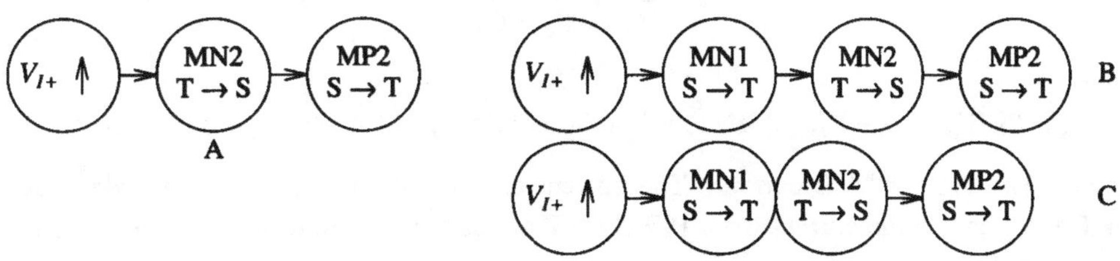

Figure 50 Microstate sequence of differential amplifier (pull-up)

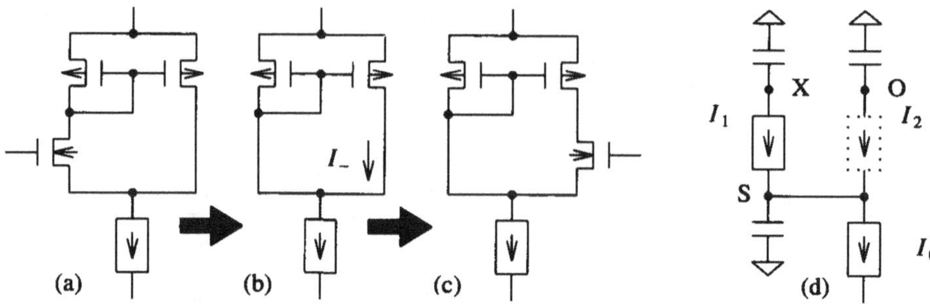

Figure 51 Multiple circuit configurational change

51(a) and (b) show the circuit's configurational change. In this figure, FETs in the triode region are not shown. If we attempt to connect the solutions of the circuits of Fig. 51(a) and (b), however, we find the following difficulty. The solution of the circuit of Fig. 51(b) gives current I_- (the current flowing through NFET MN2 that is in triode region) larger than the maximum current allowed for MN2 to stay in the triode region. Therefore the connection of solutions of these two circuit configurations is impossible.

This problem can be explained as follows. After the first microstate transition, the new microstate is inconsistent. The assumption that NFET MN2 is in the triode region is not correct. The transition to the third microstate occurs at the same time, which gives the circuit shown in Fig. 51(c). In this circuit MN2 is in the saturation region. Node waveforms in this example are shown in Fig. 49. The microstate boundary shown by the dashed line of Fig. 49 indicates two microstate transitions consisting of MN1: S $\to$ T and MN2: T $\to$ S. We note that the common source node and the internal node of the current mirror circuit are fused at normalized time $t = 2.4$, and they remain fused through all the later times. At the time when the two nodes fuse, the first derivative of the node voltage is not continuous, and that discontinuity creates the double microstate transition. With reference to Fig. 49 we consider current $I_{MN2}(Before)$, the current that flows through NFET MN2 immediately before the transition from the circuit configuration of Fig. 51(a) to that of Fig. 51(b). Before the transition, current I_X flows into the fused node that consists of the common source node and the output node, which has capacitance $C_S + C_L$. The rate of increase in the node voltage is given by

$$\frac{\Delta V_0}{\Delta t} = \frac{I_X}{C_S + C_L}$$

where

$$I_X = B_N(V_H - V_0 - V_{THN})(MN1) + B_P(V_{DD} - V_0 - V_{THP})(MP2) - I_0(Current\ Gen.).$$

We used $V_0 \approx V_{0-}$ for MP2 current. The current exchanged among the fused nodes through NFET MN2 equals the current from PFET MP2 minus the current drained to ground through capacitance C_L, as

$$I_{MN2}(Before) = B_P(V_{DD} - V_0 - V_{THP}) - C_L \frac{\Delta V_0}{\Delta t}$$

$$= \frac{C_S B_P(V_{DD} - V_0 - V_{THP}) - C_L[B_N(V_H - V_0 - V_{THN}) - I_0]}{C_S + C_L} \tag{72a}$$

Similarly current that flows immediately after the transition, $I_{MN2}(After)$, is given by

$$I_{MN2}(After) = \frac{(C_C + C_S)B_P(V_{DD} - V_0 - V_{THP}) - C_L[B_P(V_{DD} - V_0 - V_{THP}) - I_0]}{C_C + C_S + C_L} \tag{72b}$$

We see $I_{MN2}(Before) \neq I_{MN2}(After)$ in general. Then, depending on the magnitude of the two currents, the state represented by the circuit of Fig. 51(b) may become inconsistent. If this should take place, NFET MN2 goes into the saturation region, and the second microstate transition follows immediately. Such *multiple* microstate transitions take place because of the discrete FET current change when the drain-to-source voltage vanishes. Although artificial and somewhat unnatural, this is an inevitable consequence of the collapsible current generator model that makes a transition from current generator to 0 volt voltage source. The microstate sequence should have contacting circles for the events MN1: S → T and MN2: T → S, as shown by sequence C of Fig. 50. This double microstate transition can be understood better from the example shown in Fig. 51(d). In this equivalent circuit, current generators I_0 and I_1 are identical, and current generator I_2 is small and is collapsed. Nodes O and S are fused. Node S voltage holds at a constant level as long as current generators I_0 and I_1 work jointly as a voltage divider. If current generator I_1 collapses, however, node S voltage decreases, and the collapsed current generator I_2 uncollapses. The collapse of I_1 and reappearance of I_2 take place simultaneously, and the rate of change of node S voltage has discontinuity. Generally, if the rate of change of a node voltage has a discontinuity, either the gate voltage of a FET connected to the node had a step change, or a FET made a state change, to or from the triode region.

2.12 Memory Cell

We study a CMOS memory cell as the last example of complex circuit analysis. A memory cell undergoes read and write operations, and we consider only read operation in this section. The objective of this analysis is to show that in the *balanced* memory cell access scheme, there is a mechanism that protects the data existing in the memory cell. We study this mechanism using the equivalent circuit shown in Fig. 52(a). In a steady state, node voltages V_0 and V_1 represent the digital value stored in the cell. We use the collapsible current generator model with zero threshold voltage. Nodes N_0 and N_1 are loaded with equal capacitance, C.

Before the memory cell is *read*, the pair of bit lines, B and $\bar{B}$ of Fig. 52(a), are both precharged to V_{DD}, and then they are tristated. Node N_0 of the memory cell is initially high (V_{DD}) and node N_1 is initially low. The word line clock, CK, makes a low-to-high transition at time $t = 0$ thereby initiating the memory cell read process. After the read process is over, the state of the memory cell should remain the same as before. During the read process the

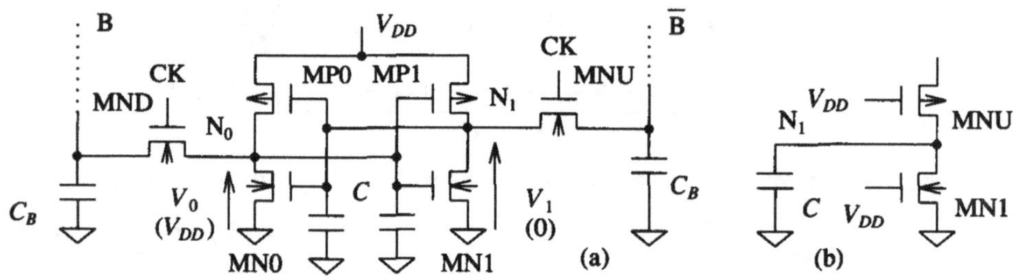

Figure 52 Memory cell read process analysis

memory cell is *perturbed.* Node N_1 may be pulled up by pass transistor MNU. Node N_0, however, is held at a high level by NFET MND as well as by PFET MP0. Therefore the *balanced* memory cell read access scheme has some *protection* against destruction of the original data.

CK makes a low-to-high transition and nodes N_0 and N_1 are connected to bit lines B and $\overline{\text{B}}$, respectively. Node N_0 is *not* perturbed since nodes B and N_0 are both at V_{DD}. Node N_1 is not influenced at all if current that is injected by NFET MNU is less than the current NFET MN1 is able to conduct to ground. If

$$B_N\,(\text{MN0},\text{MN1}) > B_{NT}\,(\text{MNU},\text{MND})$$

NFET MN1 remains in the triode region, and bit line $\overline{\text{B}}$ is pulled down to virtual ground, node N_1 as

$$V_{\overline{B}}(t) = V_{DD} - \frac{B_{NT}}{C_B}\,t \quad (0<t<t_a=\frac{C_B V_{DD}}{B_{NT}}) \qquad =0 \quad (t_a<t). \tag{73}$$

The microstate sequence of this process is shown as sequence A of Fig. 53. Since this is the trivial case of absolutely stable memory cell read process, we assume that $B_{NT} > B_N$ in the following non-trivial cases. If $B_{NT} > B_N$, node N_1 is pulled up. If the maximum voltage of node N_1 is less than the switching threshold voltage of the inverter made of MN0, MP0 is given by

$$V_{GSW} = [B_P/(B_P + B_N)]V_{DD}\,.$$

Node N_0 is held at V_{DD}. $V_1(t)$ is determined from

$$C\frac{dV_1(t)}{dt} = B_{NT}[V_{DD} - V_1(t)] - B_N V_{DD}$$

subject to the initial condition $V_1(0) = 0$ we find $V_1(t)$ as

$$V_1(t) = (1 - \frac{B_N}{B_{NT}})V_{DD}[1 - exp[-(B_{NT}/C)t]]\,. \tag{74}$$

If the maximum reachable voltage, $[1 - (B_N/B_{NT})]V_{DD}$, is less than V_{GSW}, the circuit operates as follows: V_1 increases, node N_1 and bitline node $\overline{\text{B}}$ fuse, and then the fused node voltage decreases. Node $\overline{\text{B}}$ voltage, before the two nodes fuse, $V_{\overline{B}}(t)$, is given by

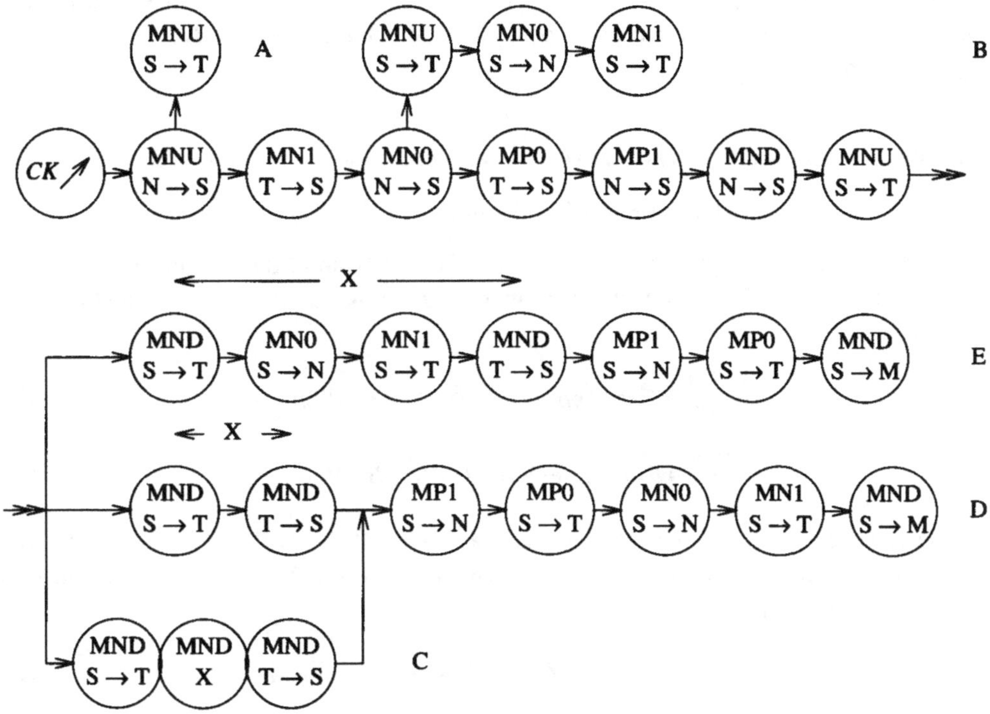

Figure 53 Microstate sequence of memory cell read process

$$V_{\overline{B}}(t) = V_{DD} - \frac{B_N}{C_B} V_{DD} t - \frac{C}{C_B}(1 - \frac{B_N}{B_{NT}}) V_{DD}[1 - exp[-(B_{NT}/C)t]] \tag{75}$$

and nodes N_1 and $\overline{B}$ fuse at time t_0 given by

$$t_0 = \frac{C}{B_{NT}} \theta \left[-\frac{(B_N/B_{NT})(C/C_B)}{(1+(C/C_B))(1-(B_N/B_{NT}))}, 1 - \frac{1}{(1+(C/C_B))(1-(B_N/B_{NT}))} \right]$$

where the θ-function has two arguments, κ and χ, as $\theta(\kappa, \chi)$. The value of the θ-function is determined by solving the transcendental equation

$$exp(-\theta) + \kappa\theta = \chi$$

for θ. The details of this function are given in the Appendix. After the two nodes fuse, the joint node voltage decreases with time as

$$V_1(t) = V_1(t_0) - \frac{B_N V_{DD}}{C + C_B}(t - t_0). \tag{76}$$

We note that once the two nodes fuse, the common node voltage decreases. The microstate sequence for this case is shown by sequence B of Fig. 53. This sequence contains two null sequences MN1: T $\to$ S, MN1: S $\to$ T and MN0: N $\to$ S, MN0: S $\to$ N. Microstate sequence B of Fig. 53 is followed if the maximum possible node voltage $[1-(B_N/B_{NT})]V_{DD}$ is less than V_{GSW} or if condition

$$B_P + B_N > B_{NT}$$

is satisfied. If this condition is *not* satisfied, V_1 exceeds V_{GSW} at time $t = t_1$ where

$$t_1 = (C/B_{NT}) \log[[(1-(B_N/B_{NT}))V_{DD}]/[(1-(B_N/B_{NT}))V_{DD} - V_{GSW}]]$$

and after that time, the inverter that consists of MP0, MN0, and MND begins to pull node N_0 down. After $t = t_1$ all six FETs of the memory cell are in the saturation region, and the circuit equations are

$$C\frac{dV_1}{dt} = B_{NT}(V_{DD} - V_1) - B_N V_0 + B_P(V_{DD} - V_0) \tag{77a}$$

$$C\frac{dV_0}{dt} = B_{NT}(V_{DD} - V_0) - B_N V_1 + B_P(V_{DD} - V_1)\,. \tag{77b}$$

This state, however, does not last. Either pass transistor, MND or MNU, moves into the triode region, and the circuit connectivity changes again. If NFET MNU moves into the triode region and nodes 1 and $\bar{\mathrm{B}}$ fuse,

$$(C + C_B)\frac{dV_1}{dt} = -B_N V_0 + B_P(V_{DD} - V_0) \qquad V_{\bar{B}}(t) = V_1(t) \tag{78a}$$

$$C\frac{dV_0}{dt} = B_{NT}(V_{DD} - V_0) - B_N V_1 + B_P(V_{DD} - V_1) \qquad C_B\frac{dV_B}{dt} = -B_{NT}(V_{DD} - V_0)\,. \tag{78b}$$

The microstate sequence can be studied using a simple computer program that solves Eqs. (77) and (78). Figure 54 shows the four node waveforms for the given set of cell parameter values and for several values of transconductance of the pass transistors, B_{NT}. In the figure, solid curves show $V_0(t)$ (above) and $V_1(t)$ (below) waveforms, and the associated dotted curves show $V_B(t)$ and $V_{\bar{B}}(t)$. From the node waveforms, the microstate sequences of the memory cell read process can be determined as follows. $V_1(t)$ increases due to the current from pass transistor MNU, and the threshold voltage of the inverter consisting of MN0 and MP0 is exceeded. PFET MP0 moves into the saturation region (MP0: T $\to$ S) and node 0 begins to pull down. Then pass transistor MND turns on (MND: N $\to$ S). After that MNU moves into the triode region (MNU: S $\to$ T). Up to this point the microstate sequences are all the same.

Then, node 1 begins to pull down. Node 0 continues to pull down, but as V_1 decreases far enough, node 0 changes to pull up. After some time, node 0 and node B voltages equalize. The equalization may be for just a moment, if the pass transistor is small (small B_{NT}). This case is shown by microstate sequence C of Fig. 53, and the curve for $B_{NT} = 1.5$ in Fig. 54. If pass transistor MND is large, MND stays in the triode region for a short time. As the voltage of node 0 approaches V_{DD}, MND returns to the saturation region in the reverse polarity. This case is shown by microstate sequence D of Fig. 51, and the curve for $B_{NT} = 1.8$ in Fig. 54. If the pass transistor is still larger, MND is in the triode region for a long time, as is observed from microstate sequence E of Fig. 53 and the curve for $B_{NT} = 2.5$ in Fig. 54.

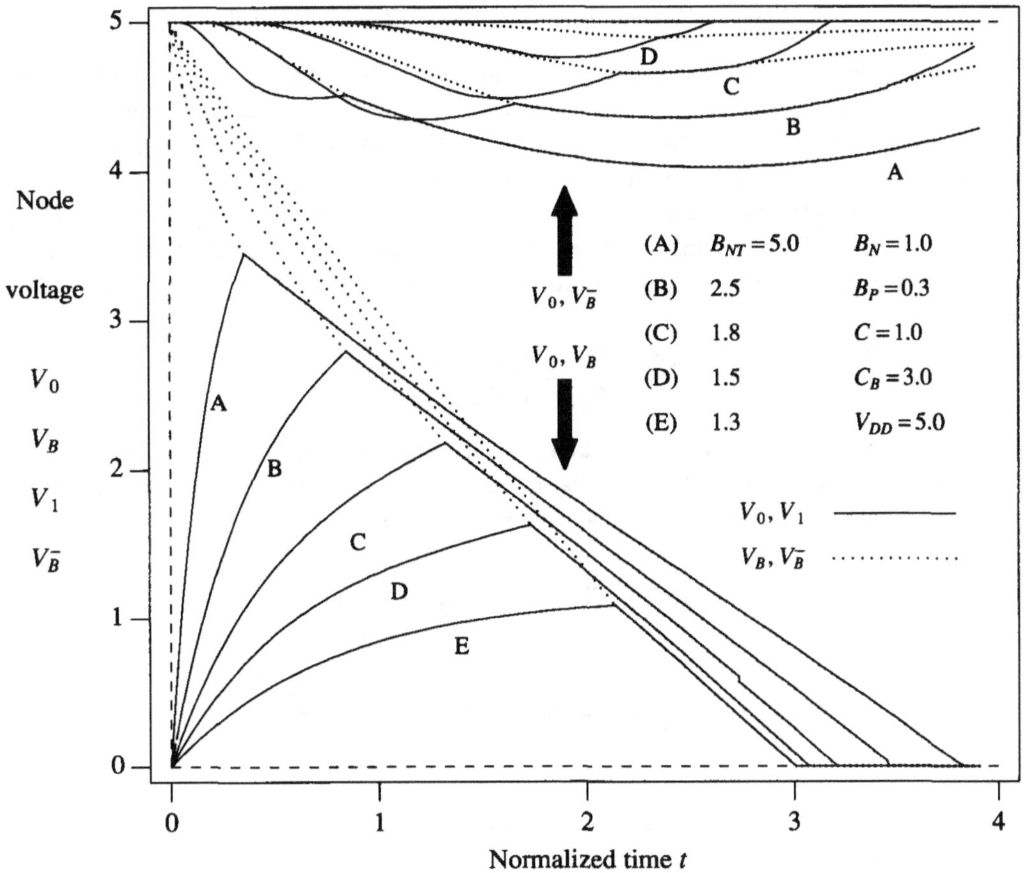

Figure 54 Memory cell read process, node waveforms

2.12.1 *Phase Plane Analysis*

Since Eqs. (77a) and (77b) do not contain time explicitly, the set of equations can be studied using the V_0-V_1 *phase* plane. Since the pair of equations are symmetrical with respect to V_1 and V_0, we use transformation

$$\phi_1 = V_1 - V_A \qquad \phi_0 = V_0 - V_A \quad \text{where} \quad V_A = \frac{(B_{NT} + B_P)V_{DD}}{B_{NT} + B_N + B_P}$$

in Eqs. (77a) and (77b) to obtain

$$C\frac{d\phi_1}{dt} = -(B_N + B_P)\phi_0 - B_{NT}\phi_1 \quad C\frac{d\phi_0}{dt} = -B_{NT}\phi_0 - (B_N + B_P)\phi_1 \,. \tag{79}$$

If ϕ_1 and ϕ_0 depend on time as $\phi_1, \phi_0 \approx exp(St)$, parameter S can be determined by solving the *secular* equation, as follows:

$$S=\alpha \text{ or } \beta \text{ where } \alpha=-\frac{1}{C}(B_{NT}-B_N-B_P) \quad \beta=-\frac{1}{C}(B_{NT}+B_N+B_P). \tag{80}$$

We note that both α and β are negative, since $B_{NT} > B_N + B_P$ is required to get into the state. The positive feedback mechanism of cross-coupled circuits is *not* operative during this phase of the memory read process: The pass transistors MNU and MND apply enough loading to override the positive feedback that originates from the cross-coupling.

It is interesting to bring varieties of memory cell circuit responses into perspective in the V_0-V_1 phase plane. We start with the transformed equations Eq. (79). We eliminate independent variable t and consider the *phase trajectories* on the ϕ_0-ϕ_1 phase plane. Equations (79) are combined to

$$\frac{d\phi_0}{d\phi_1}=\frac{(B_P+B_N)\phi_1+B_{NT}\phi_0}{B_{NT}\phi_1+(B_P+B_N)\phi_0}. \tag{81}$$

If we assume $\phi_0=a\phi_1$ where a is a constant, $a=\pm 1$. Therefore the two *diagonals* $a_\pm$ of Fig. 55(a) are the symmetry axis of the phase trajectories on the phase plane. If we rotate the ϕ_0-ϕ_1 coordinate system by 45 degrees using the diagonal transformation

$$\psi_1=\frac{1}{\sqrt{2}}(\phi_1+\phi_0) \quad \psi_0=\frac{1}{\sqrt{2}}(-\phi_1+\phi_0)$$

we have

$$\gamma\psi_0+\psi_1\frac{d\psi_0}{d\psi_1}=0 \text{ where } \gamma=-\frac{B_{NT}-(B_P+B_N)}{B_{NT}+(B_P+B_N)}.$$

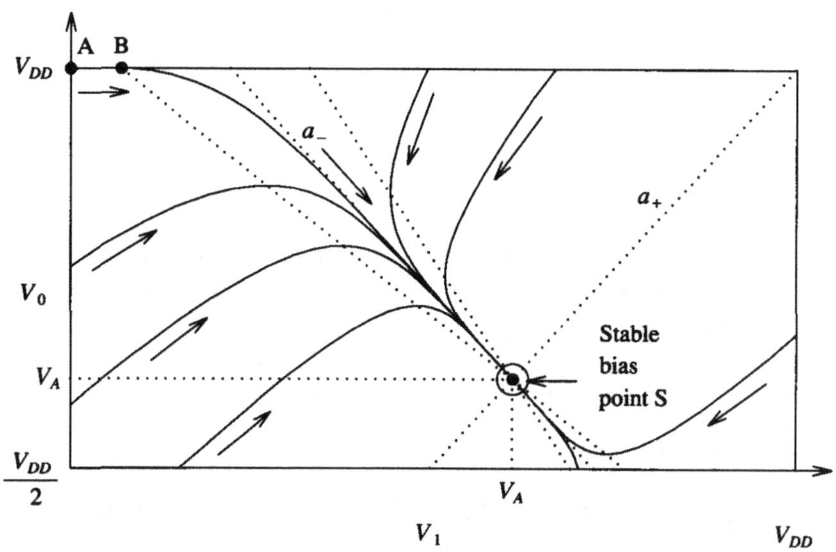

Figure 55(a) Phase trajectories of memory read process

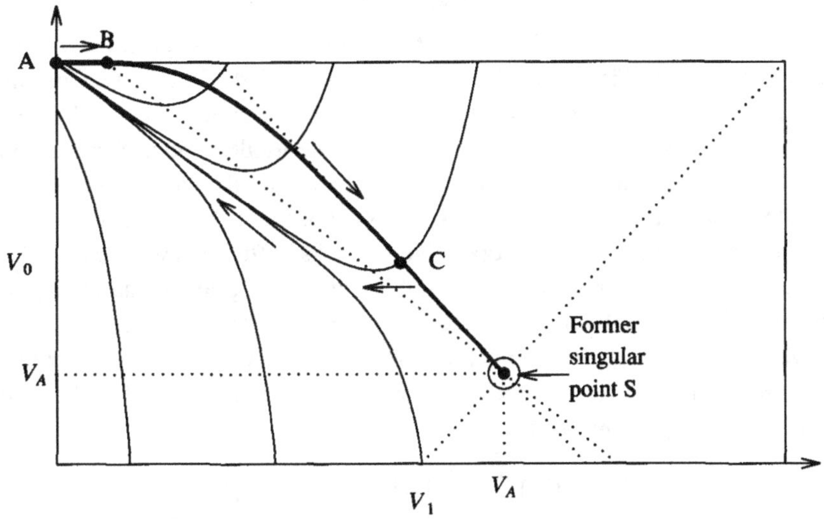

Figure 55(b) Phase trajectories of memory read process

Since $B_{NT} > B_N + B_P$, $\gamma < 0$. Then

$$\psi_0 = const. \psi_1^{|\gamma|} \quad 0 < |\gamma| < 1 .$$

Phase trajectories on the V_0-V_1 phase plane are found by transforming back variables by $\psi \rightarrow \phi \rightarrow V$. The trajectories are shown in Fig. 55(a). The phase point, whose coordinates are V_1 and V_0, moves on the trajectories. The motion of the phase point gives an idea of how the state of the latch evolves with time. Peculiar features on the phase plane help to understand the time evolution. The singular point represents a *stable* bias point where the trajectories converge. Time constants α and β that determine growth or decay rate of the perturbation around the stable bias point are negative. Loading by the bit lines suppresses the positive feedback mechanism of the cross-coupling. This regime continues as long as both pass transistors MNU and MND stay in saturation. Since MNU and MND will eventually go into the triode region, the stable bias point exists only temporarily. This is another interesting observation in the circuit analysis that is based on the collapsible current generator model, namely, formation of a temporary stable bias point. As the circuit configuration change takes place, the singular point may move or disappear. The phase trajectories of Fig. 55(a) have the following meanings. The operation that is followed between the phase points A and B is described by the closed-form theory at the beginning of this section [Eqs. (73)–(76)]. At the end of the process, all six FETs of the memory cell go into the saturation region. Then among the varieties of phase trajectories of Fig. 55(a), the trajectory that starts at point B is followed as shown by the arrow. The other trajectories are followed if different initial conditions are established. All the trajectories approach the singular point S after a long time. The singular point is a *stable* bias point. It takes time for the phase point to reach the stable bias point. If Eq. (80) is substituted into Eq. (79) we see that

$$\text{For } \alpha\text{: } \phi_1 = -\phi_0 \qquad \text{For } \beta\text{: } \phi_1 = \phi_0 .$$

For the *differential* variation of node voltage that is required to approach the stable bias point indicated by the arrow in Fig. 55(a), the time constant is $1/|\alpha|$. This time constant is long if $B_{NT} \approx B_N + B_P$. Then the phase point has not reached the vicinity of the stable bias point, at the time when pass transistors MNU or MND go into the triode region. Some time before the phase point reaches the stable bias point, pass transistor MNU goes into the triode region. Positive feedback sets in if the pass transistor MNU goes into the triode region. At the time when the positive feedback becomes operative, there is sufficient *information* of the latch's original state left in the latch circuit to reestablish the state prior to the read process. The original memory data are well *preserved* until the positive feedback reinforces it.

Positive feedback brings the memory cell back to the original state. A return to the original state begins as soon as NFET MNU goes into the triode region and nodes 1 and $\bar{B}$ fuse together. The circuit equations become

$$(C + C_B)\frac{dV_1}{dt} = B_P(V_{DD} - V_0) - B_N V_0 \tag{82}$$

$$C\frac{dV_0}{dt} = B_P(V_{DD} - V_1) + B_{NT}(V_{DD} - V_0) - B_N V_1 \tag{83}$$

and the phase trajectories are determined from

$$\frac{dV_0}{dV_1} = \frac{C + C_B}{C}\,\frac{B_P(V_{DD} - V_1) + B_{NT}(V_{DD} - V_0) - B_N V_1}{B_P(V_{DD} - V_0) - B_N V_0} . \tag{84}$$

The new singular point is located at (V_X, V_{GSW}) of the V_0-V_1 phase plane, where V_X and V_{GSW} are given by

$$V_{GSW} = \frac{B_P V_{DD}}{B_P + B_N} \qquad V_X = V_{DD} + \frac{B_N(B_{NT} - B_P - B_N)}{(B_P + B_N)^2} V_{DD} .$$

Since $B_{NT} > B_P + B_N$, $V_X > V_{DD}$. The phase trajectories are as shown in Fig. 55(b) (fine solid curves). The new singular point is outside the region shown in Fig. 55(b). The memory cell read process follows the arrows in Fig. 55(b) from A to B to C and heading back to A.

When both pass transistors MNU and MND go into the triode region, positive feedback becomes fully operative. The circuit equations become

$$(C_B + C)\frac{dV_1}{dt} = B_P(V_{DD} - V_0) - B_N V_0 \qquad (C_B + C)\frac{dV_0}{dt} = B_P(V_{DD} - V_1) - B_N V_1 . \tag{85}$$

If we use variables $\phi_1 = V_1 - V_{GSW}$ and $\phi_0 = V_0 - V_{GSW}$ where $V_{GSW} = B_P V_{DD}/(B_P + B_N)$, the phase trajectories are determined from

$$\frac{d\phi_1}{dt} = -\alpha\phi_0 \quad \frac{d\phi_0}{dt} = -\alpha\phi_1 \quad \text{where } \alpha = \frac{B_P + B_N}{C_B + C}$$

$$\frac{d\phi_0}{d\phi_1} = \frac{\phi_1}{\phi_0} \quad \text{or } \phi_0 = \pm\sqrt{\phi_1^2 - const.} . \tag{86}$$

The phase trajectories of Eq. (86) are shown by fine solid curves in Fig. 55(c). They are a set

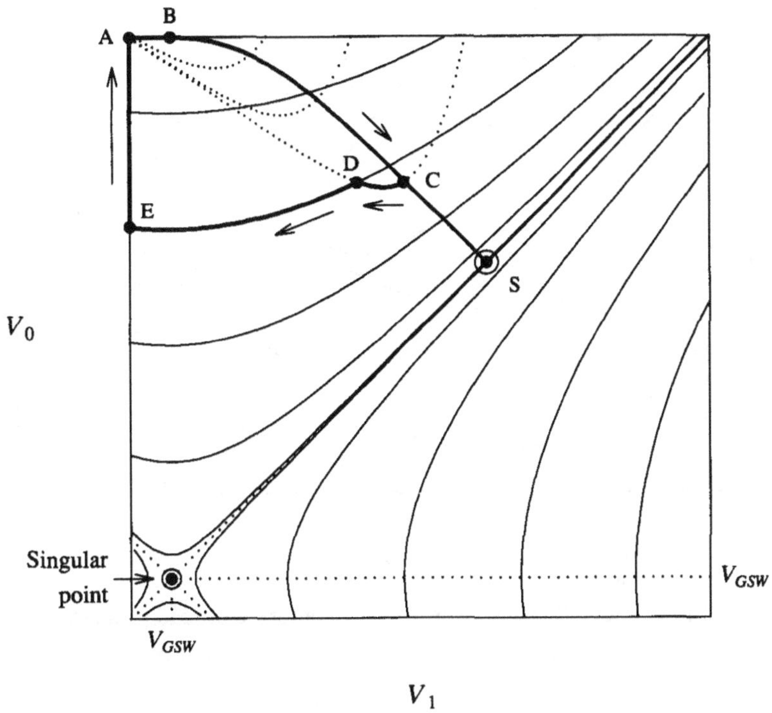

Figure 55(c) Phase trajectories of memory read process

---

of hyperbolas. The phase trajectories of the earlier phase [Figs. 55(a) and (b)] are superposed on the hyperbolas. The time rates of the voltage change in this phase are $\pm\alpha$, and one of the time constants becomes positive. As soon as both pass transistors MNU and MND go into the triode region, full positive feedback sets in as indicated by the emergence of the positive time constant α, and the singular point moves to (V_{GSW}, V_{GSW}) on the $V_0 - V_1$ phase plane as shown in Fig. 55(c). The new singular point is unstable, as is observed from the behavior of the phase trajectories.

The memory cell read process can be summarized from the phase trajectories of Fig. 55(c) as follows. The process up to phase point C was explained by referring to Fig. 55(b). At some time before the phase point reaches the first stable bias point, pass transistor MNU goes into the triode region, and the phase trajectories change to curve $\overline{CD}$. The new trajectories generally head back to the original state of the latch. After some time, pass transistor MND goes into the triode region, and phase trajectories change to curve $\overline{DE}$ of Fig. 55(c). Some time later, node 1 voltage goes down to zero (point E), and after that node 0 continues to pull up, returning the original state of the latch (point A). Changeover of phase trajectories [A, B, C, D, and E of Fig. 55(c)] indicates microstate changes. As seen from this example, phase trajectories show over-all operation of the circuit quite well. Figure 56 shows phase trajectories on the V_0-V_1 phase plane for several values of B_{NT}. The trajectories for $B_{NT} = 2.5$ and 1.8 are quite similar to that shown in Fig. 55(c).

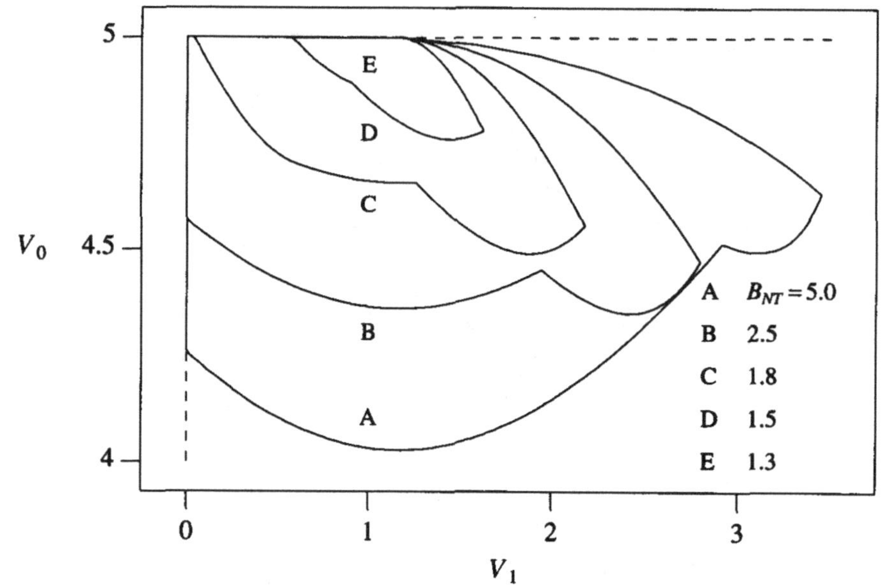

Figure 56 Phase trajectories of memory cell read process

2.13 Effects of Ionizing Radiation on Memory Cell

Energetic ionized particles can penetrate into the semiconductor and lose energy by creating electron-hole pairs in the semiconductor bulk. Electron-hole *plasma* is generated within a cylindrical region of about 100 angstrom radius, centered on the track of the ionized particle. When the track begins and ends within a uniform and featureless semiconductor, electrons and holes recombine within several ten's of microseconds. If the track goes across a reverse-biased PN junction as shown in Fig. 57(a), however, holes within the plasma cylinder in the N-type substrate are collected by the P+ region through the reverse-biased PN junction, by the same mechanism as the collector of a bipolar transistor. The track goes across the space-charge layer, and it stretches on both sides of the PN junction. Electrons generated on the track that goes through the P+ region are similarly collected by the N-type substrate. The number of such electrons is small, however, because the P+ region has many free holes, which recombine quickly with the electrons generated by the ionized particle. Since the N-substrate is lightly doped, electron density is low. Then many holes generated by the ionized particle are collected by the reverse-biased PN junction before being lost by recombination.

The number of holes collected by the P+ region depends on the vertical structure of the chip. Figure 57(b) shows the cross section of a chip that has a P+ substrate, lightly doped N-tub, and P+ diffused island. After the ionized particle strikes, holes are pulled into the P+ substrate, as well as into the P+ diffused island. The number of holes collected by the P+ diffused island (above) is significantly less. The substrate structure shown in Fig. 57(c) is also effective

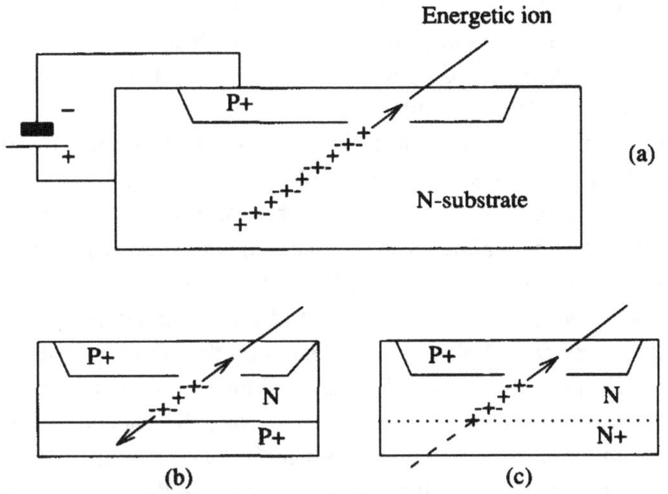

Figure 57 Generation of electron-hole plasma by ionizing radiation

in reducing hole collection. Holes generated in the portion of the particle track within the N+ substrate recombine rapidly with electrons. Only these holes that are generated within the lightly doped N-tub are swept into the P+ region.

We study the discharge process of a circuit node through a reverse-biased PN junction that is struck by ionizing radiation. Figure 58(a) shows a charged capacitor C and a reverse-biased PN junction connected in parallel. Capacitance C includes the capacitance of the PN junction itself. Capacitor C is charged to a positive voltage at the time the ionizing radiation strikes. Electron-hole pairs are generated within the space-charge layer of the reverse-biased PN junction. By the field within the space-charge layer, electrons move upward into the N+ diffused island, and holes move downwards to the P-type substrate. The electron and hole currents discharge the capacitor, and the capacitor voltage decreases. If only a small number of electron-hole pairs are created, the capacitor voltage change is small, and the PN junction is still reverse-biased at somewhat reduced voltage. If there are more electron-hole pairs than

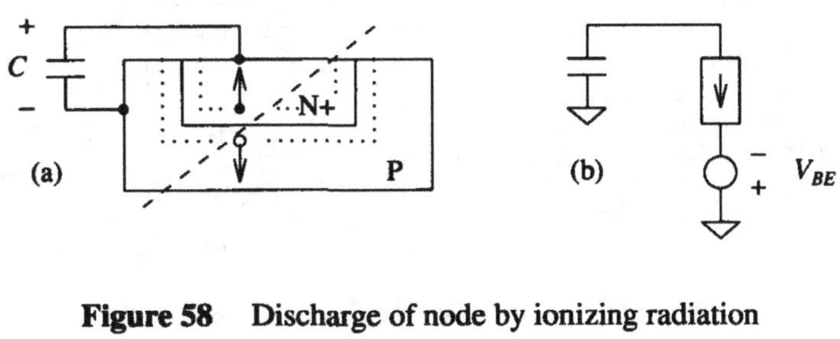

Figure 58 Discharge of node by ionizing radiation

required to discharge the capacitor completely, however, the PN junction works effectively as a *photovoltaic cell*: The PN junction generates *reverse* voltage [09].

Suppose that the capacitor is discharged completely, and the terminal voltage becomes zero. If excess electron-hole pairs still exist, electrons are captured by the N+ island, and holes are captured by the P-substrate, since N-type material has affinity for electrons (and vice versa, for holes). By the excess electrons and holes, the N+ island is now charged negatively and the P-substrate is charged positively. The lower plate of the capacitor is now charged positively with respect to the upper plate. The PN junction is now biased in the *forward* direction. If electron-hole pairs were generated at a constant rate, the PN junction would be forward-biased to the voltage at which the forward current of the PN junction equals the current of the generated and captured electron-hole pairs. At that point the capacitor voltage reaches equilibrium. This voltage is about the threshold voltage of forward conduction of the PN junction, V_{BE}. Immediately after strong ionizing radiation strikes, the capacitor voltage drops to the lowest voltage (the minimum is $-V_{BE}$), and then the voltage gradually returns to zero as the charge that is deposited on the capacitor is removed through the forward-biased PN junction, or by any other kind of leakage mechanism.

2.13.1 Dynamic Characteristics

Using this model, we consider how a static node of a CMOS digital circuit responds when ionizing radiation strikes. We consider the equivalent circuit shown in Fig. 59(a), and we model the ionization current using the collapsible current generator $I(t)$, whose terminal is connected to a voltage source $-V_{BE}$. We assume that the current generator has a flat-top current pulse with height I_0 and width t_2 as

$$I(t) = I_0 \quad (0 \le t \le t_2) \qquad = 0 \text{ otherwise}.$$

PFET MP1 current is $B_P(V_{DD} - V_{THP})$ if $V_0(t) < V_{DD}$. The microstate sequences of this process are shown in Fig. 60. If $B_P(V_{DD} - V_{THP}) > I_0$ the output node of the inverter is not affected at all [dotted curve A of Fig. 59(c), and microstate sequence HIT → END of Fig. 60]. If $B_P(V_{DD} - V_{THP}) < I_0$, node voltage $V_0(t)$ decreases with time as

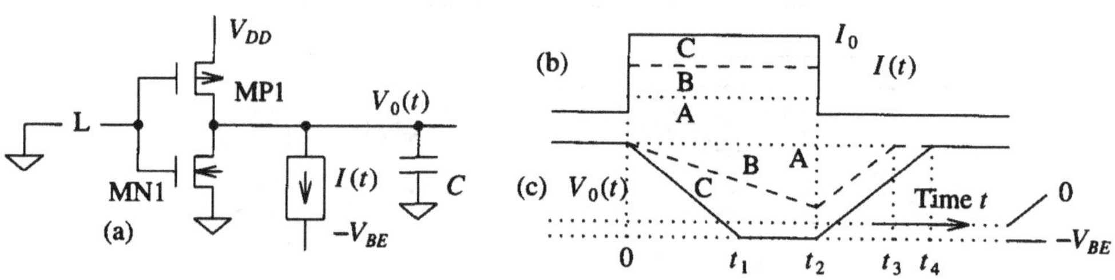

Figure 59 Temporary discharge of static node

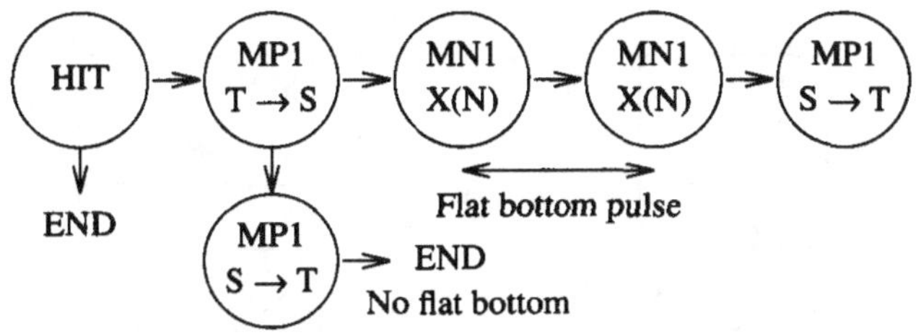

Figure 60 Microstate sequence of SEU of inverter

$$V_0(t)=V_{DD}-\frac{I_0-B_P(V_{DD}-V_{THP})}{C}\,t\,.$$

$V_0(t)=-V_{BE}$ when $t=t_1$, where t_1 is given by $t_1=C(V_{DD}+V_{BE})/[I_0-B_P(V_{DD}-V_{THP})]$. If $t_1>t_2$, $V_0(t)$ is a triangular pulse having no flat bottom. If $V_0(t)$ becomes negative, NFET MN1 changes its source-drain identification. Since MN1 is in the N-region, that does not matter. This is an example of an *irrelevant* null sequence involving a nonconducting FET. If $V_{BE}>V_{THN}$, however, MN1 may turn on in the reverse direction, but we do not consider this rather unlikely case. If $t_1<t_2$ the downgoing voltage pulse at the output node has a flat bottom, as shown by curve C of Fig. 59(c). Then for $t_1<t<t_2$, $V_0(t)=-V_{BE}$. For $t>t_2$, $V_0(t)$ is given, for both cases by

$$V_0(t)=V_0(t_2)+\frac{B_P(V_{DD}-V_{THP})}{C}\,(t-t_2)$$

until time t_3 or t_4 is reached, when $V_0(t_3)=V_0(t_4)=V_{DD}$. We have

$$t_3,t_4=t_2+\frac{C[V_{DD}-V_0(t_2)]}{B_P(V_{DD}-V_{THP})}\,.$$

Circuit failure that is caused by a single strike of ionizing radiation is called single event upset (SEU) [10]. SEUs of data storage circuits like a D-latch and memory cell are especially harmful. Figure 61(a) shows a conventional cross-coupled memory cell. If the node that is high is struck by ionizing radiation and discharges, the second node that was low begins to charge. The second node may charge up completely before exercising effects to restore the first node to the high level. Then the state of the latch is changed. The memory cell of Fig. 61(a) is vulnerable to SEU failures. The memory cell shown in Fig. 61(b) is, however, resistant to the SEU failure mechanism. Suppose that nodes P and Q were originally high (V_{DD}), and nodes $\bar{P}$ and $\bar{Q}$ were originally low (ground). The drain island of NFET MN1 (node Q) is struck by ionizing radiation and discharged. The potential of node Q decreases to $-V_{BE}$ in the worst case. Because the high resistance R prevents node P from discharging immediately, there is the possibility that V_Q can be restored. If node Q is driven up before V_P goes down significantly, SEU leaves no permanent effect on the state of the storage cell. To make sure

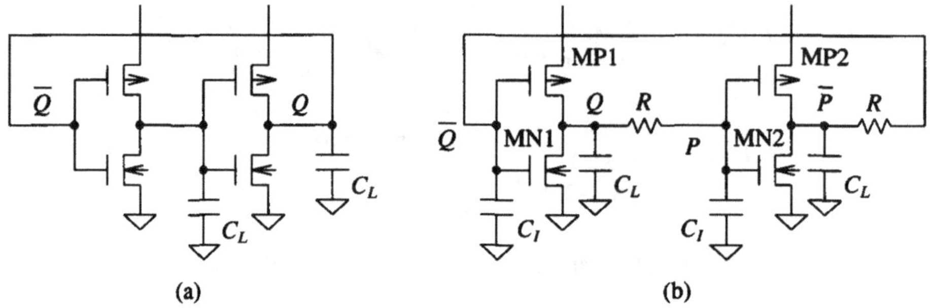

Figure 61 Memory cell resistant to SEU

that this protection mechanism works, the time constant of the node P discharge, $C_I R$, must be longer than a certain minimum. The microstate sequence of the SEU process is shown in Fig. 62. The microstate sequence is structurally similar to that of latchup (Fig. 27 of chapter 6): The SEU transient can be classified by noting how many FETs changed state before the transient reverses to the original state. As the SEU proceeds, the FETs change state in the order MP1 (T → S), MP2 (N → S), MN2 (T → S), and MN1 (N → S). The order of events MP1 (S → N) and MP2 (S → T), and of MN2 (S → N) and MN1 (S → T) are due to positive FET threshold voltages.

Immediately after the drain of NFET MN1 is struck by ionizing radiation, the node voltages V_Q and V_P of Fig. 61(b) satisfy the circuit equation

$$C_I \frac{dV_P}{dt} = \frac{V_Q - V_P}{R} \qquad C_L \frac{dV_Q}{dt} = B_P(V_{DD} - V_{THP}) - \frac{V_Q - V_P}{R}. \tag{87}$$

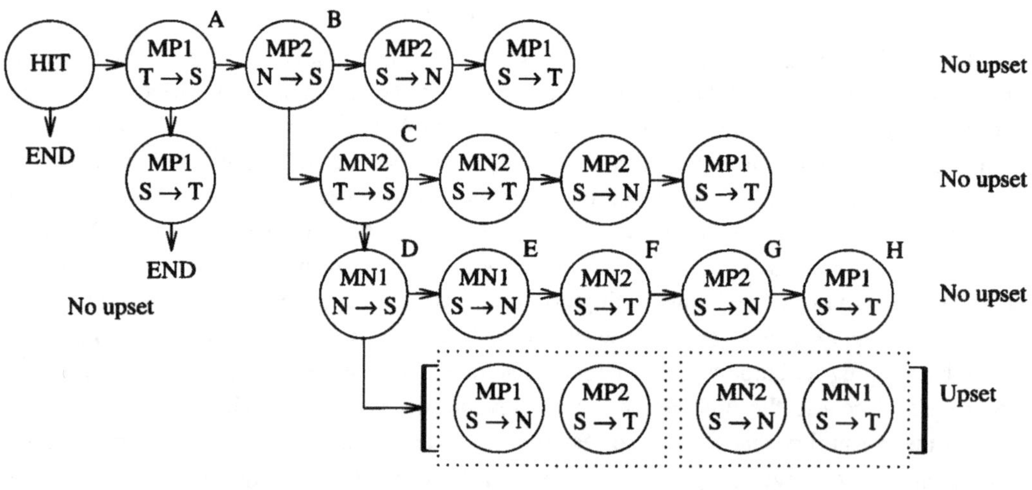

Figure 62 Microstate sequence of D-latch SEU

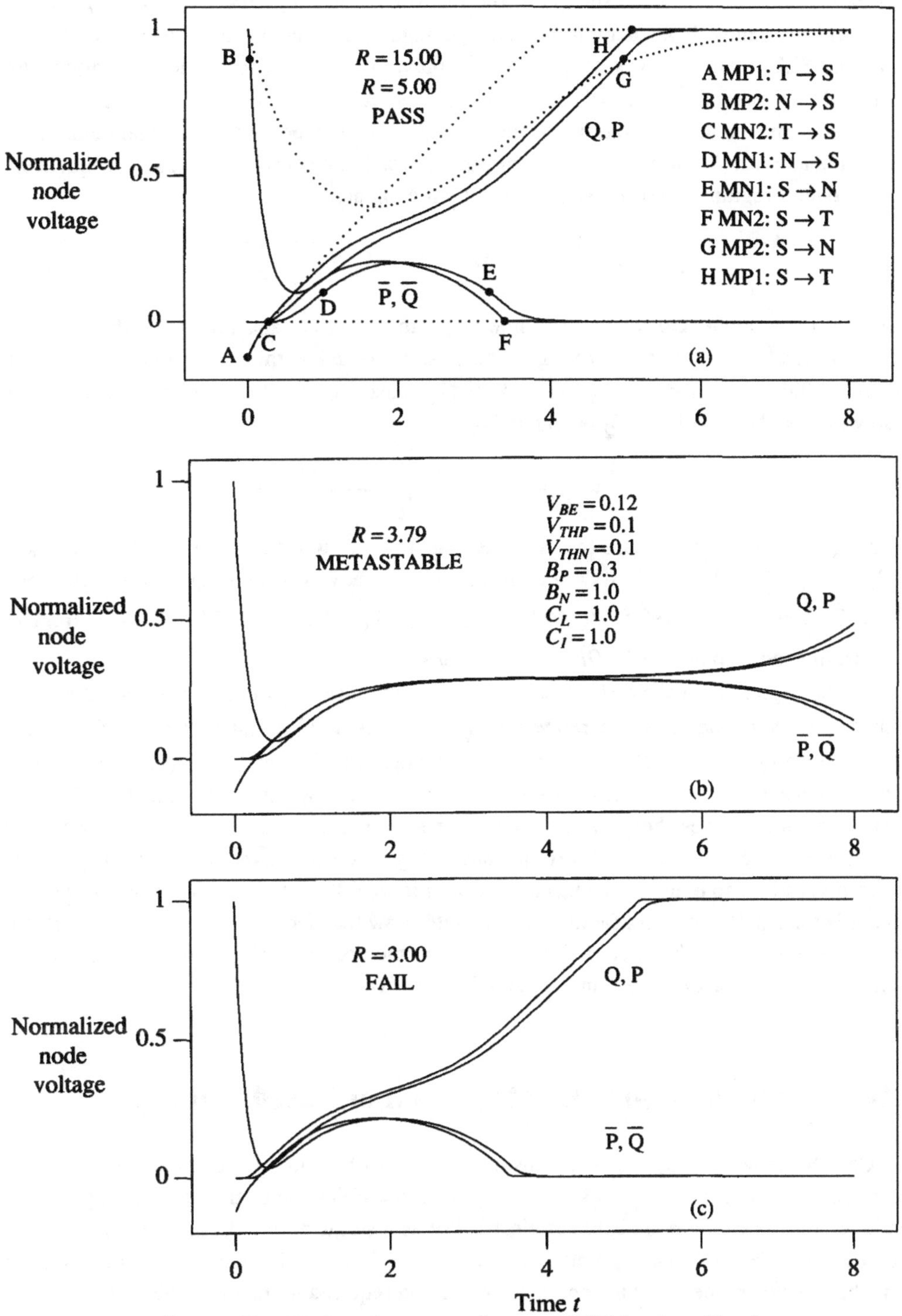

Figure 63 Node voltage waveforms of SEU-hardened latch

The set of equations is solved subject to the initial condition $V_Q(0) = -V_{BE}$ and $V_P = V_{DD}$. If V_P never falls below the switching threshold voltage of the inverter that consists of MP2 and MN2, the SEU event involves nodes P and Q only. The latch recovers to the original state, since nodes $\bar{P}$ and $\bar{Q}$ are undisturbed. If the SEU event is more intense, PFET MP2 turns on, NFET MN2 moves into the saturation region, etc. To write down the complete closed-form solution is tedious. We use a computer program to track the complex microstate sequence. In the computer program, node voltages V_P and $V_{\bar{Q}}$ are updated by

$$V_P \to V_P + \frac{V_Q - V_P}{C_I R}\Delta t \qquad V_{\bar{Q}} \to V_{\bar{Q}} + \frac{V_{\bar{P}} - V_{\bar{Q}}}{C_I R}\Delta t \tag{88}$$

where Δt is the numerical integration time step. In numerical analysis using the collapsible current generator model, node voltages must be restricted within the range for which the model is valid. Since nodes P and $\bar{Q}$ are loaded by capacitance only, no restriction is set on the node voltages. Node voltage V_Q is updated by

$$V_Q \to V_Q + \frac{I(MP1) - I(MN1) - I(\vec{QP})}{C_L}\Delta t \tag{89}$$

where $I(MP1)$ and $I(MN1)$ are the currents that flow through FETs MP1 and MN1, respectively. If $V_Q > 0$, $I(MN1) = B_N(V_{\bar{Q}} - V_{THN})$. If $V_Q < 0$, however, node Q becomes the source of NFET MN1, and therefore $I(MN1) = -B_N(V_{\bar{Q}} - V_Q - V_{THN})$. $I(\vec{QP})$ is the current that flows from node Q to node P $[I(\vec{QP}) = (V_Q - V_P)/R]$.

At the beginning of the SEU transient, $V_Q < 0$, but once V_Q becomes positive, the voltage is subsequently restricted to the range $0 \le V_Q \le V_{DD}$. A similar equation and restriction are used for node $\bar{Q}$. Figure 63(a)–(c) show the numerical results. In Fig. 63(a) the node waveforms and the microstate sequence of Fig. 62 are labeled by the same alphabetic symbols, A–H. The dotted curves show the waveforms for $R = 15$, and the solid curves for $R = 5$. In the two cases, the latch recovers to its original state. Figure 63(b) shows the marginal case where the latch goes into an extended metastable state. Figure 63(c) shows the case where the series resistor is too small to prevent failure. The results show that series resistance is quite effective in preventing circuit failures caused by SEU. The memory cell of Fig. 61(b) requires, however, much more time to write than the cell of Fig. 61(a).

2.14 Integration and Disintegration of Fused Nodes

If a CMOS digital circuit is modeled using the collapsible current generator model, several nodes may be fused together in a single node when the FETs connecting the nodes go into the triode region. Then the node behaves electrically as a single node. As the state of the circuit evolves, the node may break up into its component nodes. In this section we study the condition where breakup does not happen, or the condition required to maintain the node as a single node.

2.14.1 Fused Nodes

Figure 64 shows that two nodes A and B are connected by N NFETs, whose gate voltages are V_{G1}, V_{G2}, ..., V_{GN}, respectively. The two nodes have the same potential V_C. Since the collapsible current generator model is used, i-th NFET is able to carry the maximum current

$$J_i = B_i(V_{Gi} - V_C - V_{TH}) \quad \text{if } V_{Gi} > V_C + V_{TH}$$

in the triode region. Nodes A and B have capacitances to ground, C_A and C_B, respectively, and currents I_A and I_B are injected into nodes A and B, respectively. We assumed that nodes A and B were at common potential V_C at time $t=0$. After a short time, Δt, node A and node B voltages, V_A and V_B, respectively, are given by

$$V_A = V_C + \frac{(I_A + I_X)\Delta t}{C_A} \qquad V_B = V_C + \frac{(I_B - I_X)\Delta t}{C_B}$$

where I_X is the sum of the currents that flow from node A to node B through the N parallel-connected NFETs. If nodes A and B have the same voltage at $t=\Delta t$, we must have

$$\frac{I_A + I_X}{C_A} = \frac{I_B - I_X}{C_B} \quad \text{or} \quad I_X = \frac{C_A I_B - C_B I_A}{C_A + C_B}. \tag{90}$$

Current I_X must be less than the maximum current allowed (irrespective of polarity)

$$-I_{\max} < I_X < I_{\max} \quad \text{where } I_{\max} = \sum_{i=1}^{N} J_i. \tag{91}$$

This is the condition where nodes A and B remain *fused* after Δt. We note that the sum of the maximum currents of the N parallel connected FETs must be higher than I_X (the current required to maintain the two node voltages equal). How I_X splits between FETs is not specified at present, and this point raises an interesting issue later. Equations (72a) and (72b) derived in Section 2.12 are applications of Eq. (91).

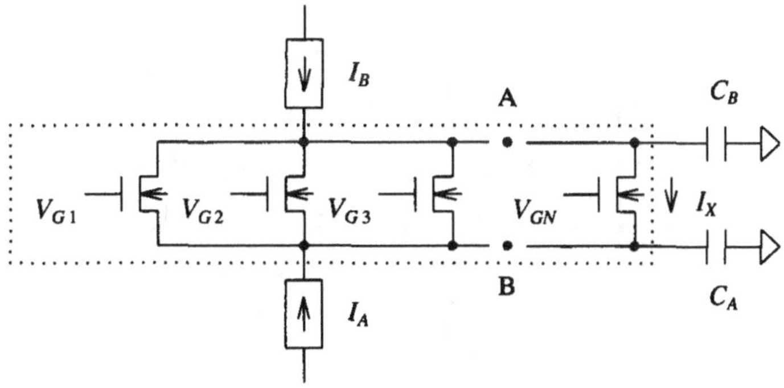

Figure 64 Stability analysis of two nodes connected by N NFETs

We attempt a similar analysis for several other configurations of *fused* nodes to gain more familiarity with the problem. Figure 65 shows N nodes 1, 2, 3, ..., N connected to node 0 in a *star* connection of N NFETs, which are all in the triode region. Suppose the i-th FET is able to carry a maximum current J_i without developing source-drain voltage, where J_i is given by

$$J_i = B_i(V_{Gi} - V_C - V_{TH}) \text{ if } V_{Gi} > V_C + V_{TH}$$

and where V_C is the voltage common to all the nodes at time $t=0$. At time Δt later, V_0 is given by

$$V_0 = V_C + \frac{\Delta t}{C_0}\sum_{i=1}^{N} I_{Xi}$$

where I_{Xi} is the current in the i-th FET (current from node i to node 0). V_i $(N \geq i \geq 1)$ is given by

$$V_i = V_C + \frac{1}{C_i}(I_i - I_{Xi})\Delta t\,.$$

If all node voltages remain the same

$$\frac{\Delta t}{C_0}\sum_{i=1}^{N} I_{Xi} = \Delta V_C \text{ (Node 0)} \qquad \frac{\Delta t}{C_i}(I_i - I_{Xi}) = \Delta V_C \quad (1 \leq i \leq N)$$

where ΔV_C is increase in the *common* node voltage within period Δt. There are $N+1$ simultaneous equations, which are the *node* equations of the circuit, and the number of equations is the same as the number of the nodes. We have a closed-form solution of the simultaneous equations:

$$\frac{\Delta V_C}{\Delta t} = \frac{\sum_{i=1}^{N} I_i}{\sum_{i=0}^{N} C_i} \quad \text{where} \quad I_{Xi} = I_i - \frac{\Delta V_C}{\Delta t} C_i \tag{92}$$

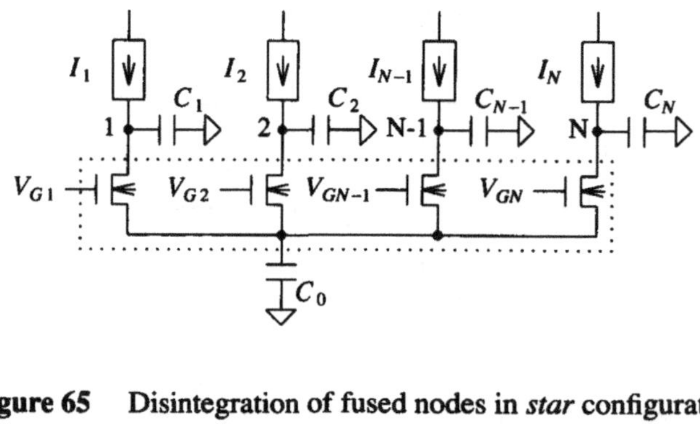

Figure 65 Disintegration of fused nodes in *star* configuration

If I_{Xi} is less than the maximum current of i-th NFET, the nodes remain fused, or if

$$-J_i < I_{Xi} < J_i \quad (1 \le i \le N)$$

the $N+1$ nodes remain fused at time Δt.

Next we consider a chain of $N+1$ nodes that are fused into a single node by N series connected FETs in the triode region, as shown in Fig. 66. Current from node i to node $i-1$ (through i-th NFET) is $I(i, i-1)$. At time $t=0$ the entire $N+1$ nodes are assumed to be fused at common node voltage V_C. We seek the condition that keeps the nodes fused at time $t=\Delta t$. We have a set of equations satisfied by the node voltages as follows:

$$V_0 = V_C + \frac{I_0 + I(1,0)}{C_0}\Delta t$$

$$V_1 = V_C + \frac{I_1 + I(2,1) - I(1,0)}{C_1}\Delta t$$

$$\cdots\cdots\cdots\cdots\cdots\cdots\cdots\cdots\cdots\cdots$$

$$V_{N-1} = V_C + \frac{I_{N-1} + I(N, N-1) - I(N-1, N-2)}{C_{N-1}}\Delta t$$

$$V_N = V_C + \frac{I_N - I(N, N-1)}{C_N}\Delta t\,.$$

Using ΔV_C as before, we have

$$\Delta V_C = \Delta t\,\frac{I_0 + I(1,0)}{C_0} = \Delta t\,\frac{I_1 + I(2,1) - I(1,0)}{C_1} = \ldots, = \Delta t\,\frac{I_N - I_N(N, N-1)}{C_N}$$

which may be written in matrix form using $q = \Delta V_C / \Delta t$ as

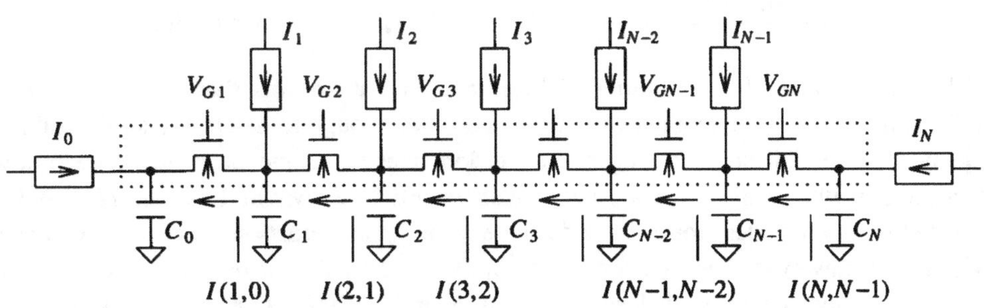

Figure 66 Stability of N + 1 nodes connected by N + 1 NFETs

$$\begin{bmatrix} 1 & 0 & 0 \cdots\cdots & -C_0 \\ -1 & 1 & 0 \cdots\cdots & -C_1 \\ 0 & -1 & 1 \cdots\cdots & -C_2 \\ 0 & 0 & -1 \cdots\cdots & \cdots \\ \cdots\cdots & & \cdots\cdots & -C_i \\ \cdots\cdots & & \cdots 1 & -C_{N-1} \\ \cdots\cdots & & \cdots -1 & -C_N \end{bmatrix} \begin{bmatrix} I(1,0) \\ I(2,1) \\ I(3,2) \\ \cdots \\ \cdots \\ I(N,N-1) \\ q \end{bmatrix} = - \begin{bmatrix} I_0 \\ I_1 \\ I_2 \\ \cdots \\ \cdots \\ I_{N-1} \\ I_N \end{bmatrix}$$

This equation can be solved in the following form. For $\Delta V_C/\Delta t$

$$q = \frac{\Delta V_C}{\Delta t} = \frac{S}{\Delta} \quad \text{where} \quad S = \sum_{i=0}^{N} I_i \quad \Delta = \sum_{i=0}^{N} C_i .$$

We define

$$S_k = \sum_{i=0}^{k} I_i \quad \Delta_k = \sum_{i=0}^{k} C_k .$$

Then in general

$$I(k,k-1) = \frac{\Delta_{k-1} S - S_{k-1} \Delta}{\Delta} . \tag{93}$$

All $N+1$ nodes are fused if

$$-J_k < I(k,k-1) < J_k \qquad (1 \le k \le N)$$

where J_k is the maximum current that can flow through k-th NFET

$$J_k = B_k(V_{Gk} - V_C - V_{TH}) \quad \text{if} \quad V_{Gk} > V_C + V_{TH}$$

2.14.2 Loop of Fused Nodes

From the three examples it might be thought that whether or not the nodes are kept fused can always be determined. There are exceptions to this conclusion. We consider fused nodes that make an N-member loop, as shown in Fig. 67(a) and (b). The circuit equations that determine current of NFETs $I(i,i-1)$ [$i = 1,2,\ldots,N$ (N equations)] are

$$\frac{I_i - I(i,i-1) + I(i+1,i)}{C_i} \Delta t = \Delta V_C \quad i = 1,2,\ldots,N \ (N \text{ equations})$$

and there are a total of N equations. There are $N+1$ unknowns that must be determined [$I(i,i-1)$ and ΔV_C]. The number of equations are not enough to determine all the unknowns uniquely. Where did the extra degree of freedom creep into the problem? The following observation is illuminating. In a linear chain of FETs shown in Fig. 66, N NFETs are connected to $N+1$ nodes; for each node there is one node equation. In an N-member loop configuration, however, there are the same number of NFETs, but only N nodes. Obviously the number of node equations is less by one. In a circuit that contains two loops as shown in Fig. 67(b), there are 7 FETs and 8 unknowns. The number of nodes is only 6. The number of equations is less by 2.

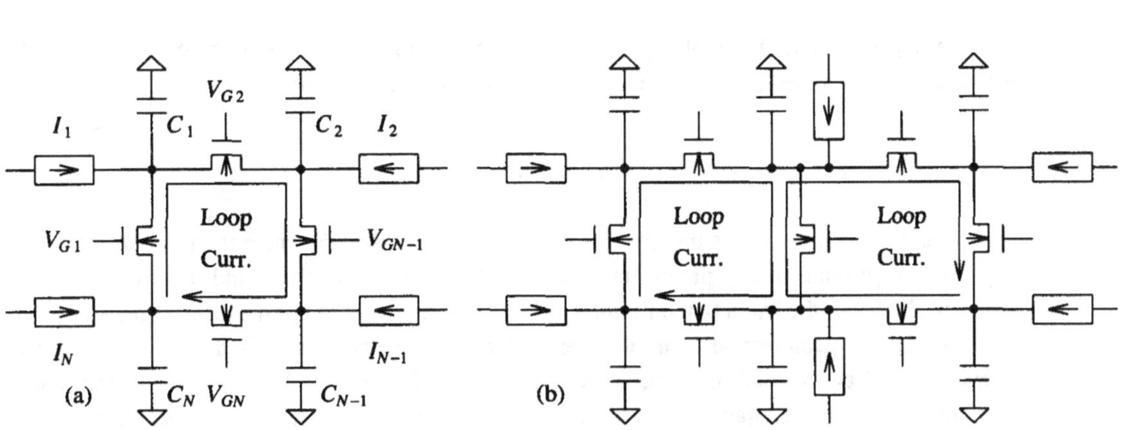

Figure 67 Closed loops of FETs and loop currents

Why the FET currents are not determined uniquely can be understood from the following observation. In the circuit of Fig. 67(a), suppose that one set of currents that satisfies all the available node equations was found. Then another set of currents satisfying the same set of equations can be found by adding equal current to all. This is physically equivalent to superposing the loop current shown in Fig. 67(a) to the first solution. If several closed loops of FETs are fused together to form a fused node as shown in Fig. 67(b), loop current can flow around each of the closed loops. The loop current is controlled by the currents injected from the other, independent circuit nodes that are communicating with any one of the component nodes, and also dependent on the past history of the loop. If the loop current exceeds the maximum, the loop breaks open, and the circuit configuration changes. The *extra degrees of freedom* that appeared from the reduced number of node equations can be explained by one or more loop currents shown in Fig. 67(a) and (b). The problem that originates from loop current has already been seen in the example of Fig. 64. In the solution, the only requirement for nodes A and B to remain fused is that the *total* current flowing through N parallel connected NFETs is less than a certain limit. How much current is carried by each NFET cannot be determined. This is because the circuit can be reinterpreted as having $N-1$ loops, each of which consists of two NFETs. Loop current is a *required* consequence of the collapsible current generator model of FET. If two parallel current paths have zero resistance, how the current is split into the two paths can be determined only by including the inductances of the paths that are the *essential* parasitics (Section 1.9.3).

If a circuit includes fused nodes, the circuit connectivity changes if the fused node disintegrates. In order to decide whether or not disintegration takes place, current distribution within the fused node must be followed simultaneously with the over-all evolution of the state of the entire circuit. Our circuit model separates the problem of circuit analysis clearly into global and local (within fused node) problems. This is a nice feature of the model based on the collapsible current generator. The model must be used carefully. It is tempting to make the simplifying assumption that a group of nodes remain fused once they fuse. This is a risky assumption: it is necessary to make sure that the assumption is valid.

2.14.3 Loop Resistance and Inductance

The concept of loop current has many similarities to *permanent* current in superconducting rings, which provides many theoretical ideas. A loop has inductance, and if current is established in the loop, the inductance maintains it. Since inductance is an essential parasitic, any FET loop has inductance that maintains the current. If resistance is zero, the loop current lasts forever (called permanent current). A real loop of FETs has small, but not zero, resistance. The small residual resistance is crucial in determining how fast the loop current decays. Therefore in the collapsible current generator model, permanent loop current is an idealization, but the concept of *semi*-permanent current (that lasts long enough to create observable effects) allows useful interpretation of transient phenomena in high-speed ICs. The effects of inductance, that is usually small in an ordinary CMOS circuit, shows up in special, highly conductive FET loop problems because of the peculiarity of the device model. This peculiarity is essential to the model, and even if the effect is not significant at present, it will be so in future higher-speed CMOS ICs. Therefore we need to understand this problem.

Inductance must be included in the analysis if loop resistance is small. Loop current I in the resistance (R)-inductance (L) loop is determined from the loop equation

$$L\frac{dI(t)}{dt} + RI(t) = 0 \quad \text{as} \quad I(t) = I(0)exp(-t/t_{LR}) \tag{94}$$

where $t_{LR} = L/R$ is the inductance-resistance time constant of current decay and $I(0)$ is the current at time $t = 0$. Time constant t_{LR} is the time required for the loop current to attenuate to $1/e$ times the original current. The inductance of a circular current loop having radius r and made of wire having diameter $2a$ is approximately

$$L \approx \mu_0 r[\log(8r/a) - (7/4)] .$$

Let us consider a small loop. If $r = 10\,\mu$m and $a = 1\,\mu$m, $L = 33$ pH. If the series resistance is 1.26 ohms, $t_{LR} = 26.3$psec. This is the typical response time of a small local loop that consists of local wiring and small FETs. The loop current decays within a few times of t_{LR}. If the MOS circuit switching delay is 100 psec or more the loop current decays within a negligibly short time. As soon as loop current decays, the *steady state* current is determined by the loop resistance alone. As for a large loop, if $r = 1000\,\mu$m, $a = 5\,\mu$m, $L = 7.07\,nH$. If the series resistance is 2.51 ohms $t_{LR} = 2.82\,nsec$. The loop current lasts longer than typical delay time of submicron CMOS gates. As shown from the numerical estimates, the effects of inductance and permanent current is still small, but may not be negligible. A current loop that includes the circuit that is external to a chip is an example. Problems of power-bus noise are therefore the first problems in CMOS chips that involve inductance.

We study how current distribution in the FET loop is determined, and how the *missing* node equations can be supplemented. Figure 68(b) shows a four FET loop, all FETs being in the triode region. If the FETs are modeled using the collapsible current generator model, all

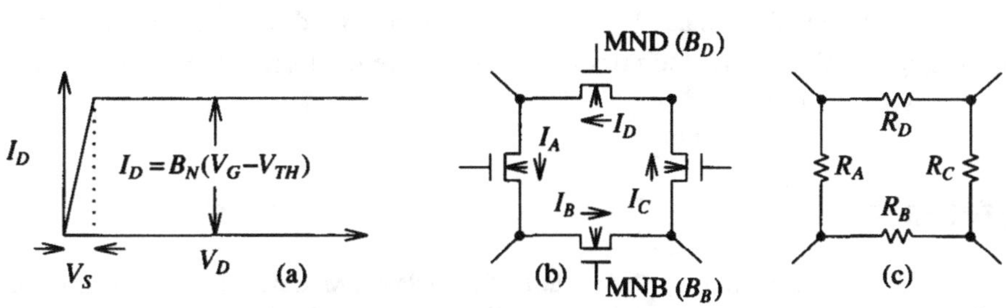

Figure 68 Resistance dominated FET loop

FET resistances are zero, and consequently I_A, I_B, I_C, and I_D are not unique. This conclusion is unrealistic in that channel resistance of a FET in the triode region is small, but not zero. If the FET loop has series resistance and the semipermanent loop current decays rapidly, the current distribution within the loop is determined by the FET channel resistance as follows. Fortunately, the channel resistance can be determined to be consistent with the collapsible current generator model if all FETs have same channel length, and the same other device parameters. In the current-voltage characteristic shown in Fig. 68(a), we assume that current saturation of FETs takes place at a small, but nonzero *saturation* voltage V_S. Then resistance of the channel of FET in triode region is

$$R_A = \frac{V_S}{B_A(V_{GA} - V_C - V_{TH})} \quad R_B = \frac{V_S}{B_B(V_{GB} - V_C - V_{TH})} \quad \cdots\cdots$$

Then we have

$$R_A I_A + R_B I_B + R_C I_C + R_D I_D = 0 \tag{95a}$$

or

$$\frac{I_A}{B_A(V_{GA} - V_C - V_{TH})} + \cdots + \frac{I_D}{B_D(V_{GD} - V_C - V_{TH})} = 0 \,. \tag{95b}$$

This equation supplements the missing node equation, and it uniquely determines all FET currents in the loop. The equation is convenient since it is independent of the choice of V_S. We recognize that if the loop inductance is included, the loop nodes are not *fused* in the exact sense of the word. Two fused nodes are connected by a collapsed current generator. The current generator, that has nonzero size, has nonzero inductance in series with the (collapsed) current generator. The inductance is an essential parasitics (Section 1.9.3). There are small voltage differences between the two nodes, which are due to current flowing through the essential inductance.

Generally the problem of loop current, or the problem of *missing* equations to determine the circuit variables uniquely, can be resolved as follows. The parasitics that were missing in the original circuit model (like inductance and FET channel resistance that are not included in the collapsible current generator model) are added to the equivalent circuit and the *normal modes* of the circuit are determined. Some of the normal mode frequencies are higher than the limit of the circuit model. If these high frequency normal modes are assumed to have settled within a negligibly short time, the missing equations are obtained from the remaining inductance and resistance terms.

References

[01] Soares, R., ed. *Application of GaAs MESFETs*. Dedham, MA: Artech House Inc., 1983.

[02] The text of the rest of this section was modified following Professor Rosenberger's comments. The issue was that V_{GSW} is the voltage where the gate is critically influenced by the input voltage, but if it is used as the boundary between Boolean values 0 and 1, many problems happen. He pointed out that the original writing was incomplete, suggested major addition, and helped me by providing his views. I added my view that the internal node Boolean values are not *real*, but are *interpretation*, and added more discussions on this issue. I think that this addition was a significant improvement of this section. I am grateful for Professor Rosenberger's help.

[03] Burns, J. R. "Switching response of complementary-symmetry MOS transistor logic circuits." *RCA Review* 25 (December, 1964): 627–61.

[04] Elmasry, M. I. *Digital MOS integrated circuits: a tutorial*. Digital MOS integrated circuits. New York: IEEE Press, 1981.

[05] Shoji, M. *CMOS digital circuit technology*. Englewood Cliffs, NJ: Prentice Hall, 1988.

[06] Heller, L. G., Griffin, W. R., Davis, J. W., and Thoma, N. G. "Cascode voltage switch logic: a differential CMOS logic family." *ISSCC84 Digest* (February, 1984): 16–17.

[07] Miller, F. H. *Partial differential equations*. New York: John Wiley & Sons, 1941.

[08] Giacoletto, L. J. *Differential amplifiers*. New York: Wiley-Interscience, 1970.

[09] Johnston, W. D., Jr. *Solar voltaic cells*. New York: Marcel Dekker, Inc., 1980.

[10] Buchanan, B. L. "Radiation effects and radiation hardening of VLSI circuits" in *VLSI handbook* ed. N. G. Einspruch. Orlando, FL: Academic Press, 1985.

Appendix

The $\theta(\kappa, \chi)$ function introduced in Section 2.12 is useful in many problems in this book, and therefore its properties are summarized here. By definition we write

$$\theta = \theta(\kappa, \chi) \quad \text{if} \quad exp(-\theta) + \kappa\theta = \chi$$

$\theta(\kappa, \chi)$ is a multivalue function as shown in the graphs of Figs. 69 and 70.

The following approximate formula can be derived:

If $\theta < 0$ and $|\theta|$ is large,

$$\theta(\kappa, \chi) = -\log(\chi + \kappa \log(\chi + \kappa \log(\chi + \ldots)))$$

If $\theta > 0$ and $|\theta|$ is large,

$$\theta(\kappa, \chi) = \frac{1}{\kappa}(\chi - \exp(-\frac{1}{\kappa}(\chi - \exp(-\frac{1}{\kappa}(\chi + \ldots\ldots))))$$

If $|\theta|$ is small,

$$\theta(\kappa, \chi) = \frac{\chi - 1}{\kappa - 1}$$

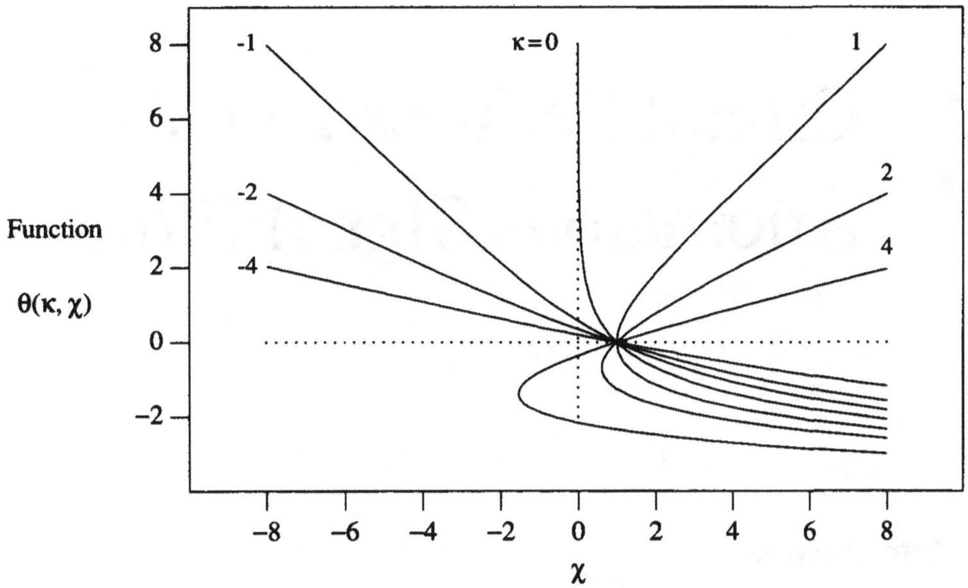

Figure 69 Graph of θ function

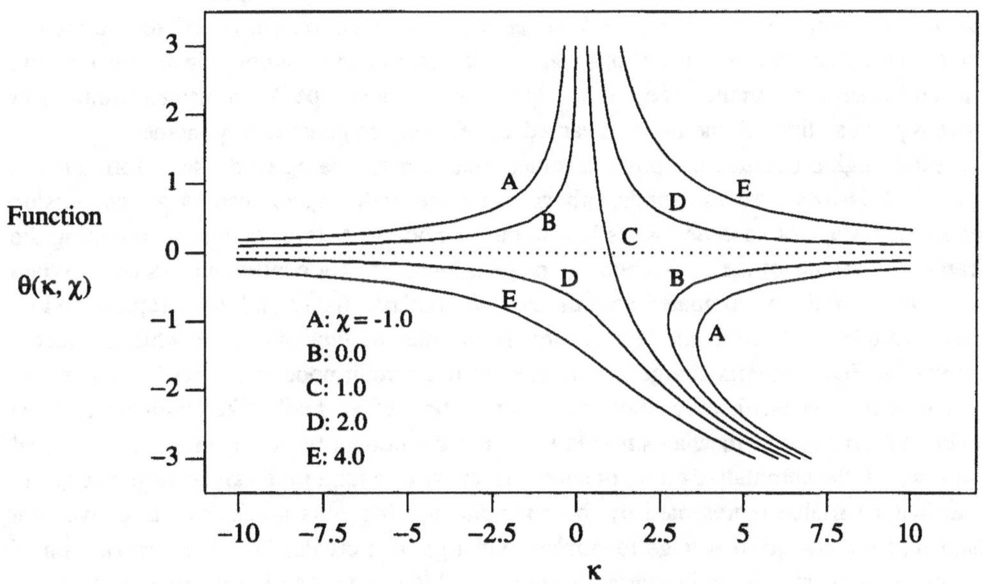

Figure 70 Graph of θ function

3 Circuit Failures Due to Anomalous Signal Flow

3.1 Introduction

In the previous chapter we showed that there are circuit failures that originate from anomaly in small-amplitude *electrical* signal flows, or currents. What this short definition implies is very broad. If there is a misdirection of a large-amplitude signal, there should certainly be a digital circuit failure, but this is better called logic design error. Although this is obviously one kind of circuit failure, we do not discuss it further. We restrict our discussions to only a few limited cases that I believe are not trivial, and not obvious. We concentrate on less dramatic, small-signal effects. If the normal direction of *logic* signal flow (large amplitude) is defined as the *forward* direction, the *small-amplitude* signal may flow forward, backward, to other nodes within the same gate, and to nodes of other gates. Sometimes small-amplitude signals may not flow at all because of circuit's mechanical breakdown (short, open), unfortunate timing, or wrong bias point setting. Sometimes unwanted signals may be generated spontaneously.

In MOS digital circuits, the physical object that carries analog or digital information is charge, which is measured as a node voltage using the node capacitance as the conversion medium. Each stage of cascaded gates has an output node that stores charge representing the Boolean logic variable of the output node. Charge is an analog (or continuous) variable: When the continuous variable is compared with the *digital threshold*, the digital value represented by the analog variable is determined. Node charge is modified by *flow of charge*, which is electrical current. A FET converts charge (or voltage) of one circuit node into electrical current in and out of other nodes, thereby allowing accumulation of a small effect with time. This unidirectional conversion mechanism creates signal directionality and gain. Even if signal current is small, the cumulative effect of small current over a long time can be large enough to upset a digital variable represented by the node charge. It is this *unidirectional* conversion mechanism, from charge to voltage to current to charge, that creates both the *desired* gain of the circuit and undesirable digital circuit failures. A logic gate may be regarded as a device that controls the supply of charge to the output node by the charge at each input node. The charges of input nodes A, B, C, ..., of Fig. 1 control the charge of output node O through the

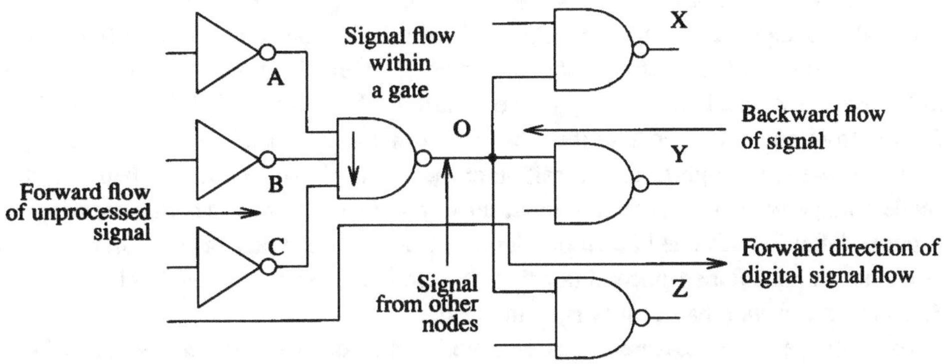

Figure 1 Schematic of digital signal flow

circuit of the NAND3 gate, and this control should be unidirectional, or A, B, C → O. The charge of node O should affect the charge of nodes X, Y, and Z, but the charge of nodes X, Y, and Z should not affect the charge of node O. In a real circuit, this unidirectionality is never complete. The nodes X, Y, Z, ..., influence node O by reverse signal flow. The flow mechanism, intensity of reverse signal, and details of circuit failures are subjects of this study.

There is more to the problem of signal flow direction. Circuit nodes within a gate are able to influence the charge at the output node of that gate. The internal nodes of a logic gate store charge that is determined by present and past input variable values. In effect, current within a logic gate executes effectively a *logic operation* between the present and the past values of the input variables. The internal nodes exist, in the circuit diagram, in the transverse direction to, adjacent to the main logic signal flow. Such internal logic operation creates circuit failures, as well as delay uncertainty. The *reverse* and the *transverse* signal flow is further generalized as *unwanted* signal flow. Forward signal flow of incorrectly processed signals is another case included in this type of failure. Furthermore, we must include spontaneous signal generation, or self-excited oscillation, and interruption of signal by improperly biased or mechanically broken gates. In this chapter, circuit failures originating from this wide range of mechanisms are discussed.

Before starting we need to clarify what is meant by *directionality*. Directionality of signals is often a confusing concept. We saw that directionality involves unidirectional conversion of one type of parameter to another. Is this the only mechanism operating in nature? Before going into the discussions, I wish to clarify the real meanings of the word, thereby preparing for the later theoretical developments.

3.2 Directionality of Signals

Directionality means that an *identified object* moves from the first location to the second location, but never from the second location to the first location. In electronic circuits, *location* means a circuit node. The *object* can be *tagged* charged carrier(s) like electrons, but more frequently it is a *feature* in the waveform of a *node* variable like voltage, which may be called *excitation*. Low-to-high or high-to-low transition and an isolated pulse are examples of excitation in node voltage waveforms. Directionality of charged carriers is seen infrequently [for example, electrons in CCD (charge-coupled device)]. Directionality of electronic circuits is almost always associated with *motion* of the *feature* in the waveform of the node variable, or *excitation*, from one circuit node to the other. If a low-to-high transition of the input signal to a gate is followed by a high-to-low transition of the output signal, and if the change in the output node voltage never influences the input node, this *inverting* gate has directionality. Even this simple definition has one hidden problem: We note that identity of the *feature* in a waveform cannot be maintained unconditionally. A discussion on this point is given in Section 6.13. Here we consider that identity is maintained.

Node voltage is the measure of charge stored in the node capacitance. Motion of the *feature* in the node voltage waveform from the first to the second circuit node could be interpreted as the corresponding motion of the electrons from one node to the other. This idea is incorrect. In connected signal paths, all electrons that exist in the signal path between the two circuit nodes shift from their original locations to varying distances, such that charge is accumulated at the second circuit node. In other circuits there can be a barrier for electron motion. In this case, the electrons of one side of the barrier may control the electron motion on the other side. Then, whether or not directionality exists from the first circuit node to the second circuit node is decided only after the structure of the signal path between the two nodes is examined.

3.2.1 Directionality of Circuits

Circuits built from linear two-terminal components like resistance, inductance, and capacitance are *bidirectional*. This means the following. There is the well-known *reciprocity* theorem of linear passive circuits [01]. Suppose that an N-terminal *black box* assembled from the linear components has terminal voltages V_1, V_2, ..., V_N and currents at the terminals (measured positive if current flows into the box) I_1, I_2, ..., I_N. The currents and voltages are related by a set of linear equations

$$I_1 = G_{11}V_1 + G_{12}V_2 + \dots + G_{1N}V_N \tag{1}$$

$$I_2 = G_{21}V_1 + G_{22}V_2 + \dots + G_{2N}V_N$$

$$\cdots\cdots\cdots\cdots$$

$$I_N = G_{N1}V_1 + G_{N2}V_2 + \dots + G_{NN}V_N$$

where the reciprocity theorem sets restrictions in coefficients G_{ij}. We must have

$$G_{mn} = G_{nm} \quad 1 \le m, n \le N\,. \tag{2}$$

This is the condition of *reciprocity* or of *complete* bidirectionality. Currents $I_1 - I_N$ and voltages $V_1 - V_N$ are generally the Laplace transforms of the currents and the voltages. In circuits that consist of resistance alone, currents and voltages may be considered as real time-dependent variables. Let us consider this case. Suppose that inputs 1, 3, 4, ..., of the black box shown in Fig. 2(a) are held at ground potential and $V_2 = \Delta V$. According to Eq. (1), I_1 is given by $G_{12}\Delta V$. If inputs 2, 3, ..., N are held at ground potential and if $V_1 = \Delta V$, I_2 is given by $G_{21}\Delta V$. Since $G_{12} = G_{21}$, the two currents are the same. With reference to Fig. 2(a), the reciprocity theorem means that the two terminals 1 and 2 are connected equivalently by a linear, passive two-terminal resistor R_{12} ($= -1/G_{12}$) (– sign is due to the direction of current: R_{12} is positive). From this structure of the equivalent circuit, we are sure that the circuit is indeed structurally bidirectional. The rest of the circuit in the black box (in dotted box) does not matter, since, in the assumption we just made, all but one terminal voltage is held at ground. There is no *directionality*, or the circuit is *completely* bidirectional if the circuit consists of two-terminal linear components. In order to create directionality the admittance matrix $[G_{mn}]$ must be asymmetric, or $G_{mn} \neq G_{nm}$ at least for some suffix (m, n).

We note that asymmetry of the admittance matrix indicates that the circuit is *not* completely bidirectional. Another point to note: The matrix asymmetry never means that the circuit is *unidirectional*. The circuit is unidirectional from terminal 1 to terminal 2 only if $G_{21} \neq 0$ and $G_{12} = 0$. Then input V_1 is able to control *output* I_2, but output V_2 is unable to control input I_1. The requirement $G_{12} = 0$ indicates that this is a special case. Generally, the relative magnitude $\Gamma = |G_{12}/G_{21}|$ may take any value between 0 and ∞. If $\Gamma > 1$ there is directionality from terminal 2 to terminal 1, and if $\Gamma < 1$ from terminal 1 to terminal 2. Therefore there are different *degrees* of directionality: bidirectionality, limited directionality, and unidirectionality. Complete bidirectionality and limited directionality exist in nature, but *complete* unidirectionality exists only for ideal circuits at DC, since coupling capacitance between the input and the output nodes can never be reduced to zero [note that capacitance is an *essential* circuit parasitic (Section 1.9)]. Unidirectionality exists only as an idealization.

In the bidirectional circuit of Fig. 2(a), nodes 1 and 2 are connected by linear resistor R_{12}. The linear resistor has the identical characteristic when the two terminals are interchanged. What happens if a two-terminal device that has different characteristics for the two different polarities is substituted for R_{12}? For the simplest example we replace the linear resistor by an *ideal* diode that is a short-circuit if the anode is positive and an open-circuit if the anode is negative, with reference to the cathode. The apparent directionality of the circuit that includes the diode is a mirage: If the diode conducts, voltage V_1 affects current I_2, and V_2 affects current I_1. The circuit is bidirectional. If the diode turns off, the two nodes are isolated. Later we will show that two isolated circuit nodes should be regarded as bidirectional, since capacitance is an *essential* parasitic.

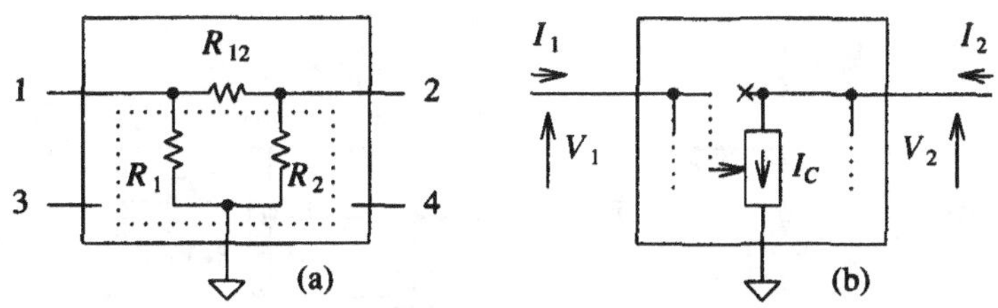

Figure 2 Reciprocal and non-reciprocal circuits

Attempts to create directionality by replacing the linear resistor R_{12} of Fig. 2(a) by a general two-terminal component are unsuccessful. In order to prevent V_2 from controlling I_1, the connection between terminals 1 and 2 must be *broken* somewhere, as shown in Fig. 2(b). If the connection is broken, however, voltage V_1 is unable to influence current I_2. Therefore, the current I_2 must be *affected* by some current control mechanism. The connection between terminal 1 and 2 must have the structure schematically shown in Fig. 2(b). Voltage V_1 is able to control the current generator I_C, but voltage V_2 is unable to control current I_1. Current generator I_C must be inserted between terminal 2 and a circuit node other than node 1, most likely ground (the voltage reference). This is required because only voltage V_1 determines current I_2: no return current from anywhere else is allowed to modify the voltage that determines current I_2. The circuit between terminals 1 and 2 *cannot be* a two-terminal circuit. The circuit structure of Fig. 2(b) is fundamentally different from that of Fig. 2(a): In order to accumulate a substantial charge at node 2, electrons to the left of the broken point × need not move significantly. If we wait long enough, the device pumps any amount of charge into node 2.

The three-terminal controlled current generator is a generic model of an active device like a FET. In a CMOS FET, electrons (or holes) that carry the channel current are *induced* and *controlled* by the gate voltage through electrostatic attraction. In Fig. 3(a) we show the most fundamental unidirectional circuit that includes a FET modeled using the controlled current generator. FET current, I, is given by $I = B(V_1 - V_{TH})$ if $V_1 > V_{TH}$, where V_{TH} is the threshold voltage and B is the transconductance. Suppose that the input voltage V_1 and output voltage V_0 are written as

$$V_1 = V_{1S} + v_1 \quad V_0 = V_{0S} + v_0$$

where V_{1S} and V_{0S} are the DC operating points and v_1 and v_0 are small variations. Similarly, input and output currents I_1 and I_0 are written as [the positive directions are defined in Fig. 3(a)]

$$I_1 = I_{1S} + i_1 \quad I_0 = I_{0S} + i_0 .$$

We write the relationship among small variations as

$$i_0 = G_{00}v_0 + G_{01}v_1 \quad i_1 = G_{10}v_0 + G_{11}v_1 . \tag{3}$$

The coefficients G_{ij} $(0 \le i, j \le 1)$ at the low frequencies are determined as

$$G_{00} = 1/R_0 \quad G_{01} = g_m \quad G_{10} = G_{11} = 0$$

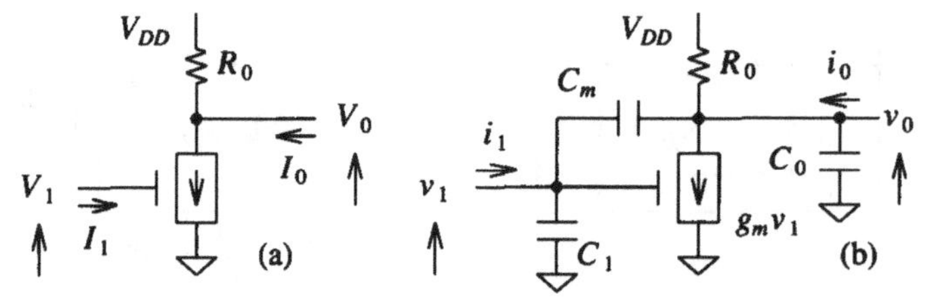

Figure 3 Nonreciprocal MOS amplifier circuit

where g_m is the transconductance of the device. We find $G_{10} \neq G_{01}$. Since $G_{10} = 0$ this circuit has *unidirectionality* at DC. From Eq. (3), a change of input voltage v_1 generates a change in output current i_0, but a change in output voltage v_0 does not influence the input current i_1 (a change in v_0 creates a change in i_0: this is the response of the *internal impedance* of the amplifier circuit). Since the reciprocity theorem is violated, and since the device can never be described by an equivalent circuit like Fig. 2(a), directionality of signal from input to output exists in this circuit.

Next we study directionality of the MOS amplifier circuit at high frequencies. The simple equivalent circuit of Fig. 3(a) must be supplemented by parasitic capacitances C_0, C_1, and C_m. C_0 is the load capacitance to the output, C_1 is the input capacitance of the device, and C_m is the coupling capacitance between the input and the output. The capacitances are the *essential* parasitics. They cannot be reduced below certain limits determined by the device size. If the lower limit is reached using device design ingenuity, C_0, C_1 and C_m are of the same order of magnitude. The equivalent circuit of Fig. 3(b) is described by equations

$$i_0 = [i\omega(C_0 + C_m) + (1/R_0)]v_0 + (g_m - i\omega C_m)v_1$$

$$i_1 = -i\omega C_m v_0 + i\omega(C_1 + C_m)v_1 .$$

Directionality is measured by

$$\Gamma = |G_{01}/G_{10}| = \sqrt{1 + (g_m/\omega C_m)^2} \tag{4}$$

and the no-load voltage gain (setting $i_0 = 0$) is given by

$$G = |v_0/v_1| = G_0 \left[\frac{1 + (\omega C_m/g_m)^2}{1 + [\omega(C_0 + C_m)R_0]^2} \right]^{1/2} \tag{5}$$

where $G_0 = g_m R_0$ is the DC gain. G decreases with increasing frequency and drops by 3 db at angular frequency $\omega_B = g_m/[(C_0 + C_m)G_0]$, which is the *bandwidth* of the amplifier. G is unity at angular frequency $\omega_U = (g_m/C_m)\ [1 - (1/G_0)^2]^{1/2}\ [(1 + (C_0/C_m))^2 - 1]^{-1/2}$. The product of DC gain, G_0, and bandwidth, ω_B, is a constant at $g_m/(C_0 + C_m)$. If higher gain is required, the bandwidth is less. As for directionality, $\Gamma = \sqrt{2}$ at $\omega = \omega_D = g_m/C_m$. If $G_0 \gg 1$ and if C_0 and C_m are of the same order of magnitude, $\omega_U \approx \omega_D$. The amplifier loses directionality at the frequency where voltage gain is lost. By generalizing this observation, the mechanism of amplification is conversion of input voltage to device current, and accumulation of the device current into the output capacitance to generate the output voltage. The conversion-generation mechanism creates voltage gain and directionality. Since gain and time of accumulation of device current are proportional, the gain-bandwidth product is a constant of the device. At high frequencies, voltage gain is lost, and so is directionality. Two isolated circuit nodes are always connected bidirectionally by the parasitic capacitance. At high frequencies, the bidirectional parasitic capacitance determines the response.

We recognize that the directionality of conversion, such as voltage to current, that takes place within the electron device creates the directionality of signal. This is the mechanism of directionality of three-terminal, amplifying devices like FETs, BJTs, and vacuum tubes. In this mechanism there is a common reference of voltage, which is ground. Since conversion from input voltage to device current is carried out using electrostatic attraction that works between the control electrode and the charged carriers, an *increase* in input voltage generates more current that is directed to the voltage reference (source of NFET, or emitter of NPN BJT). Since the output voltage is generated at the drain or at the collector by accumulating the current in a capacitance, more positive input voltage means more negative output voltage. This type of directionality is always associated with *inversion* of signal. Gain of the circuit, directionality, and inversion originate from the single conversion-generation mechanism of amplification. Signal directionality of CMOS linear amplifiers and of CMOS static gates is based on this mechanism alone.

We need to ask whether or not directionality of conversion among variables is the only mechanism to create signal directionality. If we study the assumptions of the present theory, we find the following. The definition of directionality of signal, or nonreciprocity, of the circuit of Fig. 3 is established only by linearizing the circuit, assuming small signal amplitude. If we look for a different mechanism of directionality, we need to search for *essentially* nonlinear unidirectional electrical phenomena. Study of the mechanisms of various digital circuits reveals that the second type of directionality of *digital* signal originates from a quite specific mechanism, namely, directionality due to *burnout*.

This is the mechanism of directionality widely observed in nature. Propagation of fire along a fuse, falling dominoes, and excitation of a nerve axon are a few examples. The media of signal propagation has two states. In the case of a fuse, the first state is a new fuse, and the second state is a string of ash left after the fuse has burned. Change from the first state to the second state is possible, but the reverse is impossible. If the state change begins at one end of the long structure, the state change propagates along the structure. The best example is a flame propagating along a fuse. Therefore, we call this mechanism a *burnout* directionality. In many examples of this type directionality, the initial state holds some form of energy, and the final state does not (for example, fuse). Therefore once the state transition takes place, no reverse change is possible. We note that this type of directionality exists only in large amplitude signals that have fixed amplitude. Small-signal, variable amplitude directionality cannot be created by this mechanism. In a fuse, small-amplitude temperature fluctuation along the fuse propagates bidirectionally. Since the changes of state at all locations have the same direction, logic is not inverting. Because of this peculiarity, logic circuits built using this type directionality are not as versatile as logic circuits built using inverting logic devices available from the conversion directionality mechanism [02]. Figure 4(a) shows such a long structure. Circuits within boxes A, B, ..., can be unidirectional circuits as shown in Fig. 4(b) (Domino CMOS) or bidirectional circuit as shown in Fig. 4(c) (tunnel diode logic circuit). As is seen from the example, in this directionality mechanism, interaction between the first node (where signal starts) and the second node (where signal arrives) *can be* completely bidirectional. The initial state is established by precharge of Domino CMOS circuit of Fig. 4(b), and by temporarily reducing the power supply voltage of the tunnel diode circuit of Fig. 4(c). Conventional Domino CMOS dynamic circuits operate essentially by the mechanism of burnout. Since a CMOS gate is inverting, two stages of gates (in Domino CMOS, one is dynamic and the other is static) must be cascaded to construct a noninverting gate that works as a unit cell.

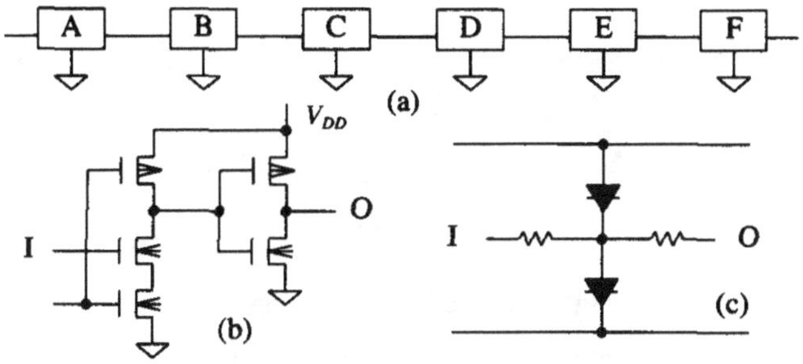

Figure 4 Directionality by burnout

An interesting feature of this type of directionality is that speed of signal propagation can be increased, subject to certain restrictions, to a very high speed. Suppose that the bidirectional circuit of Fig. 4(c) was built using fast negative resistance diodes. If the circuit is biased at the critical point of switching, only a small trigger voltage is sufficient to start switching all the diodes in the chain. If triggered, a small-amplitude voltage wave propagates at high speed, and all the diodes start switching almost simultaneously. An example of the related phenomena is a fuse that is held at a temperature close to the spontaneous ignition point: If one end is ignited the flame propagates very rapidly. Such a phenomenon was actually observed in negative resistance bulk material like GaAs [03]. A fast carry generator circuit was built using the mechanism [04]. In Domino CMOS, such high-speed propagation does not take place.

If we include directionality associated with the motion of identified charged carriers, the following directionality mechanism is sometimes used. This is the mechanism often called three-phase clocking. Three storage locations A, B, and C are set up, and information carrying object like charge is moved from A to B, B to C, and C to A. While the object is moved from A to B, connections between B and C, and C and A are disabled. Similarly while the object is moved from B to C, connections between A and B, and C and A are broken. This scheme is used in charge-coupled devices and in magnetic logic *parametrons*. Since the object is carried from the first location to the second unaltered, the logic is *basically* noninverting. The parametron is an exception. Because it is a magnetic logic, transformer inversion is available, and therefore the parametron logic allows inversion. This three-phase transport scheme has not yet been applied to CMOS logic circuits.

3.2.2 Diode's Directionality

Diodes do not create directionality, but they provide useful functions to improve circuit operations. Let us consider the circuit shown in Fig. 5. Suppose that nodes A and B are driven by CMOS drivers. In order to OR signals A and B, the two signals are tied at node C through two

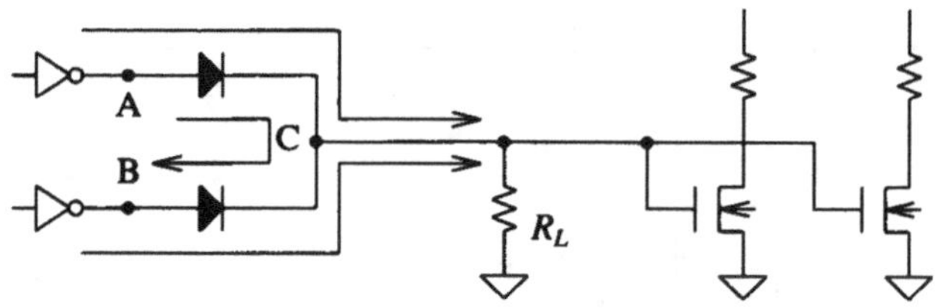

Figure 5 Diode directionality

diodes, and the node is grounded through resistor R_L [05]. The circuit works if $A=L$ and $B=L$, or $A=H$ and $B=H$. When both A and B are low, both diodes are turned off, and node C is pulled down by resistor R_L alone. If both nodes A and B are high, both diodes are on, and the circuit works. If $A=L$ and $B=H$ or $A=H$ and $B=L$, however, the two drivers have conflicting output voltage levels. If the diodes are short-circuited, node C voltage will be some value between V_{DD} and ground. Whether or not the voltage level is accepted as the high level depends on the switching threshold voltage of the signal destination, as well as the internal impedance of the CMOS drivers. If diodes are recovered, the one that is connected to the inverter that drives the output node to a logic-low level turns off automatically, and conflict of the drivers is avoided. When the output voltages of the drivers are different, the diodes prevent current flow from one driver to the other. In a few *troublesome* combinations of the signal levels, the circuit reconfigures itself automatically with the help of properly inserted diodes. Nonlinear devices like diodes are able to *change* digital circuit *connectivity* such that conflicting drives of the circuit are avoided. This operation is confusing with signal directionality, but it is fundamentally different. The FET amplifier of Fig. 5 establishes the direction of signal, and the role of the diodes is to prevent *minor* circuit problems.

3.3 Direction of Digital Signal Flow Established by Digital Control

There are certain digital CMOS circuits in which the direction of digital signal flow is established by control exercised by the other digital signals through special circuit structures. In this section we study several examples of this type.

3.3.1 Carry Generator Trees

Figure 6 shows a FET chain circuit which forms a multi-input, multi-output logic gate. Each stage consists of a PFET MP(i) that precharges node N_i to V_{DD}, an NFET MN(i) that discharges the node to ground potential, and an NFET MNX(i) that connects node N_i to node N_{i-1}. The signal controlling the gate of NFET MN(i) is G_i, and the signal controlling MNX(i)

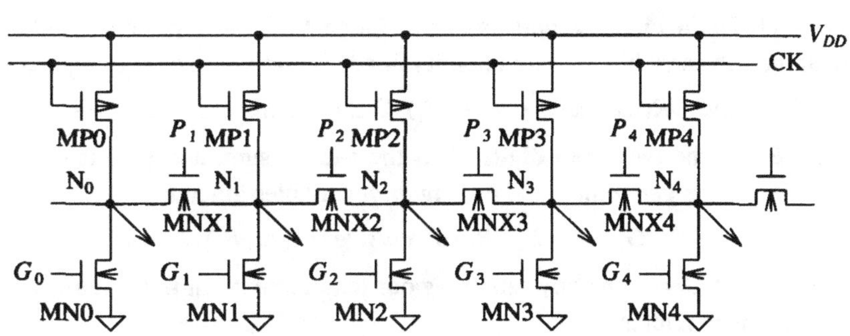

Figure 6 FET chain multi-input/multi-output gate structure

is P_i. Clock CK drives the gates of precharge PFETs MP(i). This circuit operates as follows. Before logic operation, all signals P_i and G_i $(0 \le i \le M)$ are forced to a logic-low level (ground). Therefore all nodes are disconnected from each other. CK makes a high-to-low transition, the PFETs conduct, and all nodes N_i $(0 \le i \le M)$ are precharged to V_{DD}. CK returns to logic high (V_{DD}) and PFETs MP(i) turn off. Then all signals P_i and G_i are turned on to drive the gates of their respective FETs. Then some nodes discharge and others remain precharged. After some time nodes $N_0 - N_M$ settle at their respective final logic levels, which are the outputs of the logic gate. The outputs have the following Boolean expressions.

$$N_0 = \overline{G_0 + P_1G_1 + P_1P_2G_2 + \cdots + (P_1P_2 \cdots P_MG_M)} \tag{6}$$

$$N_1 = \overline{G_0P_1 + G_1 + P_2G_2 + P_2P_3G_3 + \cdots + (P_2P_3 \cdots P_MG_M)}$$

$$\cdots\cdots\cdots\cdots$$

$$N_M = \overline{(G_0P_1P_2 \cdots P_M) + (G_1P_2 \cdots P_M) + \cdots + G_{M-1}P_M + G_M}\,.$$

If node N_i is pulled down by NFET MN(i) (input signal G_i is high), node N_{i-1} and node N_{i+1} are pulled down if signals P_i and P_{i+1}, respectively, are high. The digital signal propagates to the right as well as to the left sides. It is sometimes convenient to restrict the direction of signal from node N_i to node N_{i+1}, but not from N_i to node N_{i-1}. This is achieved by restricting logic variables, such that $P_i = Low$ if $G_i = High$ or $P_iG_i = 0$ all the time. If this condition is imposed,

$$N_0 = \overline{G_0} \tag{7}$$

$$N_1 = \overline{G_0P_1 + G_1}$$

$$N_2 = \overline{G_0P_1P_2 + G_1P_2 + G_2}$$

$$\cdots\cdots\cdots\cdots\cdots$$

$$N_M = \overline{(G_0P_1 \cdots P_M) + (G_1P_2 \cdots P_M) + \cdots + G_{M-1}P_M + G_M}\,.$$

The circuit looks as if the signals propagate from left to right. This circuit has an important application [06] [07]: the circuit generates carries for an M-bit binary adder. Node N_i signal is the inverse of the i-th carry bit, C_i, of an M-bit full adder that executes logic operation

$$S_i = Exclusive\ OR(Exclusive\ OR(A_i, B_i), C_{i-1}) \quad \text{where } C_{i-1} = G_{i-1} + C_{i-2}P_{i-1} \tag{8}$$

where A_i and B_i are the two input digits, S_i is the i-th bit sum, and C_{i-1} is the $(i-1)$-th bit carry. G_i and P_i are the *generation* and *propagation* variables given by

$$G_i = A_i \cdot B_i \quad P_i = Exclusive\ OR(A_i, B_i)\,. \tag{9}$$

We note that $P_i = 0$ if $G_i = 1$, thereby satisfying our requirement. In this circuit, the often used simpler propagation variables

$$P_i = A_i + B_i \tag{10}$$

cannot be used, because then the signal could propagate both forward and *backward* along the chain.

3.3.2 Multi-output Gates

Similar problems exist in multi-output CMOS gates. In multi-output gates, failure is caused by signal flow from one output node to the other output node. Because of the direct interaction of the logic output voltages, either or both output node voltages are severely affected if the logic levels of the two outputs are different. This failure mechanism is simple to understand: Since two nodes within a single gate are connected together by a conducting path, the two interacting output node voltages tend to equalize. Let us consider the multi-output Domino CMOS circuit shown in Fig. 7. In this circuit, signals A_1, A_2, B_1, and B_2 are the first and the second digits of two binary numbers that are added. The circuit generates carries for the first and the second bit positions, C_1 and C_2 [08]. Boolean expressions for the variables at nodes C_1 and C_2 are given by

$$C_1 = A_1 \cdot B_1 \qquad C_2 = A_2 \cdot B_2 + (A_1 \cdot B_1) \cdot Exclusive\ OR(A_2, B_2)\,.$$

The Exclusive OR term of the second expression is the carry *propagation* variable generated within the gate. This circuit works as it should if all the input signals $A_1 - \overline{B_2}$ stabilize to their final voltage levels before the clock CK makes a low-to-high transition. If the Exclusive OR term is replaced by the OR term as shown by the circuit in the inset of Fig. 7, however, the circuit fails for certain input signal level combinations. This is the case if either or both A_1 or B_1 is low, and both A_2 and B_2 are high. Use of the Exclusive OR instead of OR is to single out this special case and prevent failure by a *patch*, which was fortunately possible in this case. The situation is analogous to the use of diodes in logic circuits (last section). How the patch is designed can be understood from a simplified version of this gate. Since the problem is the same for dynamic and static gates, we consider a static gate shown in Fig. 8(a).

Figure 8(a) shows a CMOS static AOI (AND-OR-Invert) gate having three inputs A, B, and C. If we make the following substitution in Fig. 7, we see that the gates of Fig. 7 and of Fig. 8(a) are logically equivalent.

$$A_1 \cdot B_1 \rightarrow A \quad Exclusive\ OR(A_2, B_2) \rightarrow B \quad A_2 \cdot B_2 \rightarrow C\,.$$

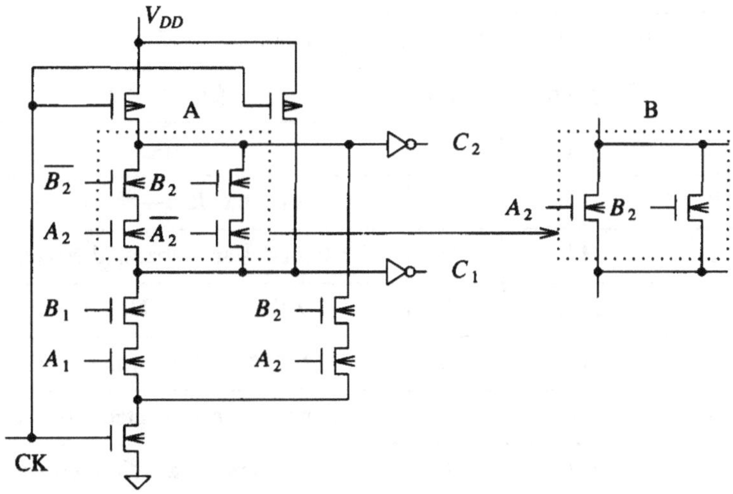

Figure 7 Multiple output Domino CMOS gate

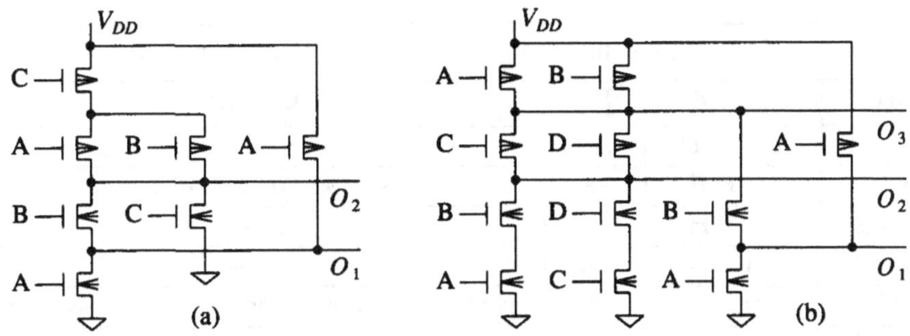

Figure 8 Static multi-output gates

The static gate of Fig. 8(a) has an output conflict. How the conflict occurs can be studied systematically by compiling a table of logic levels for all the cases. In Table 1, all combinations of input signals and output signals are shown. For combinations of input levels that produce the same output levels, the two output nodes may interact through conducting FETs. This is not a problem, so far as the integrity of the logic voltage levels is concerned (there may be timing problems, however). These cases are shown by type A in the third column of Table 1. When the two output nodes have different logic levels, the FET circuit that connects the two outputs may or may not be conducting. If the MOSFET circuit that connects the two output nodes is not conducting, there is no conflict. This case is identified by type B in Table 1. When the two output nodes have different logic levels and the two output nodes are connected by a conducting MOSFET circuit, conflict takes place. Type C in Table 1 shows that current

Table 1: Conflict of multiple gate outputs of Fig. 8(a)			
Input data	Output data	Type	Comments
$A\ B\ C$	$O_1\ O_2$	—	—
0 0 0	H H	A	All same data—No conflict
0 0 1	H L	B	$B=0$, nodes are separated-No conflict
0 1 0	H H	A	All same data-No conflict
0 1 1	H L	C	*****$B=1$, Conflict*****
1 0 0	L H	B	$B=0$, nodes are separated—No conflict
1 0 1	L L	A	All same data—No conflict
1 1 0	L L	A	All same data—No conflict
1 1 1	L L	A	All same data—No conflict

Table 2: No conflict of multiple gate outputs of Fig. 8(b)			
Input data	Output data	Type	Comments
$A\ B\ C\ D$	$O_1\ O_2\ O_3$	—	—
0 0 0 0	H H H	A	All same data—No conflict
0 0 0 1	H H H	A	—
0 0 1 0	H H H	A	—
0 0 1 1	H L H	B	Node O_2 separated from the others—No conflict
0 1 0 0	H H H	A	All same data
0 1 0 1	H H H	A	—
0 1 1 0	H H H	A	—
0 1 1 1	H L H	B	Node O_2 separated
1 0 0 0	L H H	C	$O_1 \neq O_3$ but $B=0$ nodes are separated
1 0 0 1	L H H	C	No conflict
1 0 1 0	L H H	C	—
1 0 1 1	L L H	B,C	All output nodes are separated
1 1 0 0	L L L	A	—
1 1 0 1	L L L	A	—
1 1 1 0	L L L	A	—
1 1 1 1	L L L	A	—

flows from output node O_1 to O_2 through a conducting NFET. The voltages of the two nodes cannot be determined from the circuit diagram alone: the sizes and conductivity ratios of the FETs are required to determine the voltage levels. If the NFET that connects O_1 and O_2 is small, the circuit may work, but integrity, reliability, and speed of the circuit are subject to serious question.

There are other multi-output static gates that do not suffer from the conflict problem. Figure 8(b) shows an example that was given in the author's previous book. Since there are four input variables and three output nodes, Table 2 has sixteen rows. Exhaustive study of the table reveals the following. Type A entries in the third column show cases in which all the nodes have the same voltages. In this case there is no conflict. In type B, node O_2 is separated from the other outputs, and therefore O_2 is able to take a different voltage from the other output nodes. In case C, nodes O_1 and O_3 are disconnected, and the two nodes may take different voltages. Table 2 shows there is no conflict between the output nodes. Examination of gate circuits using the complete logic table is a simple and systematic way to discover conflicts in multi-output gates. If the number of inputs of a multi-output gate is M, 2^M entries of the logic table must be examined. If M is large, the examination is time consuming or even impossible. This will not happen practically, since large multi-output gates are slow, and they will never be used.

3.4 Forward Transmission of Unprocessed or Improperly Processed Signals in CMOS Gate

We use the word *signal* to mean small amplitude current that carries information. Circuit failure can be caused by even small signal flow if small effects can be accumulated to create a large effect. A small signal may flow through any path in a CMOS digital circuit. There are so many paths where small currents may flow, that understanding circuit failures due to this mechanism is quite complicated. It is therefore convenient to classify small signal flow with reference to the direction of the *intended* digital signal flow in the circuit. The simplest case is that the signal flows in the same direction as the digital signal. At first sight it seems that such a signal flow could not create circuit failures, but it does. A CMOS gate should generate *processed* signals at the output node. If the CMOS gate transmits *unprocessed* signals directly from the input to the output, the circuit may fail. This problem can be understood from the following example.

A CMOS tristatable inverter circuit has two variations, shown in Figs. 9(a) and (b). If CK is low and $\overline{CK}$ is high, both circuits disconnect output node Z from input node D, but the degree of isolation is different. In the circuit of Fig. 9(a), a change of input signal D that takes place when CK is low and $\overline{CK}$ is high does not alter the potential of output node Z, because the gates of NFET MNA and PFET MPA, which are held at a fixed-gate potential, effectively isolate the output node from the input node D. In the circuit of Fig. 9(b) there is direct coupling through capacitance C_m, and the output node potential changes if input signal D makes a transition. The variation of V_Z, ΔV_Z for a 0 to V_{DD} input transition, is given by

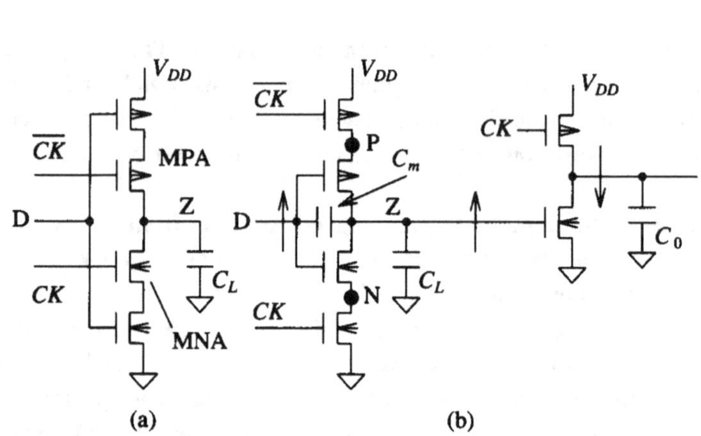

Figure 9 Two variations of CMOS tristatable inverters

$$\Delta V_Z = \frac{C_m}{C_m + C_L} V_{DD} \tag{11}$$

where C_m is the *Miller* effect capacitance between the input and the output nodes, and C_L is the load capacitance shown in Fig. 9(b). Practically, this voltage change can be as much as 20% of V_{DD} if the output node of the gate is lightly loaded. There is another effect associated with the circuit of Fig. 9(b). The charge of output capacitor C_L may be *shared* with the parasitic capacitances of internal nodes P or N of the tristatable inverter. Suppose node N was originally at ground potential, input D was low, node Z was high, and the gate was tristated. If signal D makes a low-to-high transition, part of the charge stored in capacitance C_L is transferred to node N and the potential of node Z falls. The transfer of charge among the nodes within a gate is called charge-sharing, and it will be discussed in Section 3.8. The circuit of Fig. 9(b) shows a dynamic inverter that follows the dynamic *latch* consisting of the tristatable inverter and capacitive load C_L. If V_Z is driven above the NFET threshold voltage by capacitive coupling or by charge sharing or any other effects, the dynamic inverter pulls down, and circuit failure occurs. Note that this dynamic inverter has a much lower switching threshold voltage than a static inverter.

3.4.1 Transmission Gates

Tristatable inverters and transmission gates are the most fundamental dynamic gate circuits. In transmission gates there is a similar problem of clock signal leakage to the floating output node, which results in a *residual voltage* originating from unbalance in the capacitive induction and clock skew. Output waveform of a transmission gate is not *clean*. We need to understand the basic operation of a transmission gate, and we can do so with reference to the circuit diagram of Fig. 10 and the microstate sequences of Fig. 11. Depending on the timing relationship and waveforms of the clocks CKP, CKN, and input signal D, there are a number of differ-

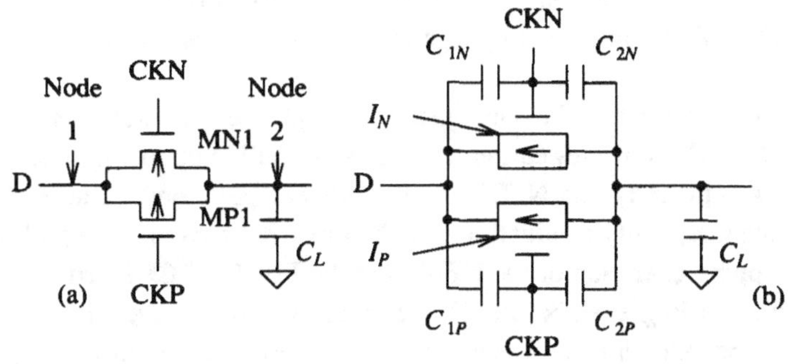

Figure 10 Transmission gate analysis

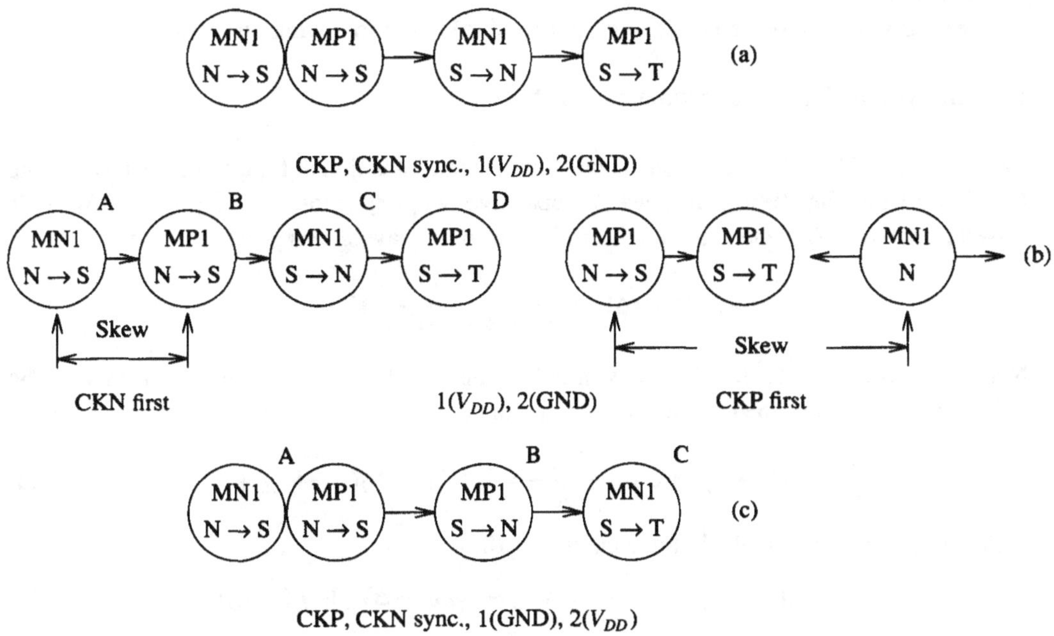

Figure 11 Transmission gate microstate sequence

ent ways a transmission gate works. Figure 11 shows microstate sequences of some of the typical transmission gate operations. Sequence (a) takes place if clocks CKP and CKN are perfectly synchronized, and they make rapid transitions to turn on the transmission gate. Input node 1 is held at V_{DD} by a CMOS driver that may be regarded as voltage source, and load capacitance C_L was completely discharged. When the clocks make transitions, both NFET MN1 and PFET MP1 of Fig. 10(a) move into the saturation region, and node 2 starts to pull up.

NFET MN1 works as a source-follower. NFET MN1 turns off as node 2 voltage approaches $V_{DD} - V_{THN}$. PFET MP1 continues to pull up node 2 to V_{DD}, and then MP1 goes into the triode region.

Sequence (b) is subject to the same setup as sequence (a), except that clocks CKP and CKN are skewed. If CKN arrives earlier, node 2 is not pulled up to V_{DD} until CKP arrives. If CKP is delayed significantly, the NFET source-follower pulls node 2 up to $V_{DD} - V_{THN}$ and MN1 goes into the marginally conducting state before CKP arrives. The pull-up is carried out by the nonoverlapping operation of the NFET and the PFET. If CKP arrives earlier, however, node 2 is pulled up to V_{DD} even before CKN arrives. If CKN arrives much later NFET MN1 remains in the nonconducting state, and MN1 plays no role. Sequence (c) is the same as sequence (a), except that node 1 is held at ground potential and capacitor C_L was originally charged to V_{DD}. The roles of NFET and PFET are exchanged. The microstate sequences of Figs. 11(a) and (c) differ only in the FET identification. This is an example of *symmetry* in microstate sequence.

We carry out detailed analysis of case (b) and (c) in the following discussion.

I) Analysis of Fig. 11(b) sequence (CKN first)

Suppose that CKN makes an instantaneous low-to-high transition at time $t = 0$, thereby turning NFET MN1 on [Fig. 10(a)]. Because of capacitive coupling through C_{2N} [Fig. 10(b)], V_2 is driven from 0 to ΔV_N, where ΔV_N is determined from the law of charge conservation as

$$V_2(0+) = \Delta V_N = \frac{C_{2N}}{C_L + C_{2P} + C_{2N}} V_{DD}.$$

Node 2 is pulled up, starting from this initial voltage, by the current from NFET MN1. The pull-up process is described by the circuit equation

$$(C_L + C_{2P} + C_{2N}) \frac{dV_2(t)}{dt} = B_N [V_{DD} - V_2(t) - V_{THN}] \tag{12a}$$

which is solved subject to the initial condition $V_2(0) = \Delta V_N$ as

$$V_2(t) = (V_{DD} - V_{THN}) - [(V_{DD} - V_{THN}) - \Delta V_N] exp(-t/t_D) \tag{12b}$$

where $t_D = (C_L + C_{2P} + C_{2N})/B_N$. If $t \to \infty$, $V_2(t) \to V_{DD} - V_{THN}$ (MN1 to the marginally conducting state).

At time $t = t_A$ CKP makes a high-to-low transition, and PFET MP1 turns on. We have $V_2(t_A) < V_{DD} - V_{THN}$: the NFET has not yet reached the marginally conducting state. At $t = t_A$ a negative voltage of magnitude ΔV_P

$$\Delta V_P = \frac{C_{2P}}{C_L + C_{2P} + C_{2N}} V_{DD}$$

is induced on node 2. Node 2 voltage immediately after the MP1 turn-on is computed using Eq. (12b) as $V_2(t_A+) = V_2(t_A-) - \Delta V_P$. For $T > t_A$ both the PFET and NFET conduct, and the circuit equation is

$$(C_L + C_{2P} + C_{2N}) \frac{dV_2(t)}{dt} = B_N[V_X - V_2(t)] \tag{13}$$

where $V_X = V_{DD} - V_{THN} + (B_P/B_N)(V_{DD} - V_{THP})$. We note that the time constant t_D is determined by the NFET and the load capacitance. The effect of the PFET is to add a constant, positive rate of voltage increase. The right-hand side of Eq. (13) is the sum of both NFET and PFET currents. This equation is solved subject to the initial condition $V_2(t_A) = V_2(t_A+)$ as

$$V_2(t) = V_X - [V_X - V_2(t_A-) + \Delta V_P]exp[-(t - t_A)/t_D]\,. \tag{14}$$

At time $t = t_B$, $V_2(t_B) = V_{DD} - V_{THN}$. At that time, NFET MN1 turns off. The circuit equation changes to

$$(C_L + C_{2P} + C_{2N}) \frac{dV_2(t)}{dt} = B_P(V_{DD} - V_{THP})\,. \tag{15}$$

Equation (15) is valid until $V_2(t) = V_{DD}$ is reached at time t_1, and the solution is

$$V_2(t) = V_2(t_B) + \frac{B_P(V_{DD} - V_{THP})}{C_L + C_{2P} + C_{2N}}(t - t_B)\,. \tag{16}$$

Finally V_2 reaches V_{DD} at time $t = t_1$ and $V_2(t) = V_{DD}$ for $t > t_1$. The V_2 waveform is shown in Fig. 12. The microstate sequence of Fig. 11(b) and node waveform of Fig. 12 are correlated by the same alphabetic symbols A, B, C, and D.

II) Analysis of Fig. 11(c) sequence

CKP and CKN make simultaneous transitions, thereby turning the transmission gate on. Node 1 is held at ground potential, and node 2 was originally charged to V_{DD}. The circuit equation is

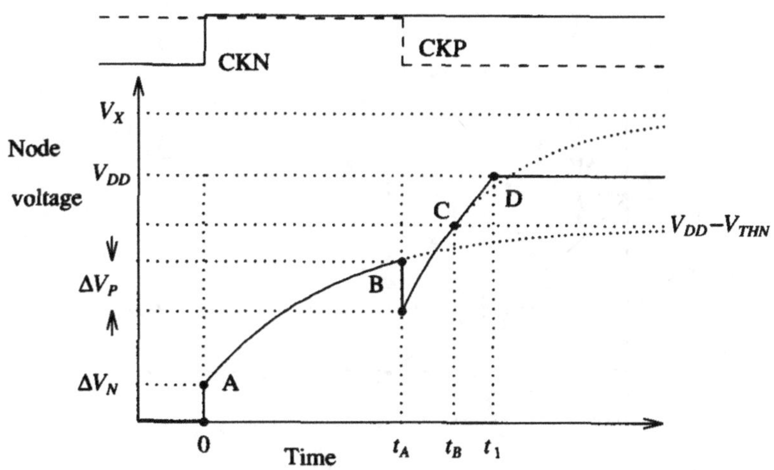

Figure 12 Analysis of transmission gate circuit waveform, CKN first

$$(C_L + C_{2P} + C_{2N})\frac{dV_2(t)}{dt} = -B_N(V_{DD} - V_{THN}) - B_P[V_2(t) - V_{THP}] \quad V_2(t) > V_{THP} \tag{17}$$

$$= -B_N(V_{DD} - V_{THN}) \quad 0 < V_2(t) < V_{THP}$$

Immediately following the clock transition, node 2 is discharged from V_{DD} to the initial voltage $V_2(0+)$ given by

$$V_2(0+) = V_{DD} - \Delta V_N \text{ where } \Delta V_N = \frac{C_{2P} - C_{2N}}{C_L + C_{2P} + C_{2N}} V_{DD}. \tag{18}$$

The solution that satisfies the initial condition is given by

$$V_2(t) = [V_{DD} - V_{THP} + (B_N/B_P)(V_{DD} - V_{THN}) - \Delta V_N]\exp(-t/t_{DP})$$
$$+V_{THP} - (B_N/B_P)(V_{DD} - V_{THN})$$

where $t_{DP} = (C_L + C_{2P} + C_{2N})/B_P$. $V_2(t)$ reaches V_{THP} at $t = t_1$, where

$$t_1 = t_{DP} \cdot \log\left[\frac{(B_P + B_N)V_{DD} - B_P[(C_{2P} - C_{2N})/(C_L + C_{2P} + C_{2N})]V_{DD} - B_P V_{THP} - B_N V_{THN}}{B_N(V_{DD} - V_{THN})}\right] \tag{19}$$

For $t > t_1$

$$V_2(t) = V_{THP} - \frac{1}{t_{DN}}(V_{DD} - V_{THN})(t - t_1) \text{ where } t_{DN} = \frac{C_L + C_{2P} + C_{2N}}{B_N} \tag{20}$$

and Eq. (18) is valid until $t = t_2$ where t_2 is given by $t_2 = t_1 + t_{DN}[V_{THP}/(V_{DD} - V_{THN})]$. $V_2(t) = 0$ if $t > t_2$. The node waveform is plotted in Fig. 12. The microstate sequence of Fig. 11(c) and Fig. 13 are correlated by the same alphabetic symbols A, B, and C.

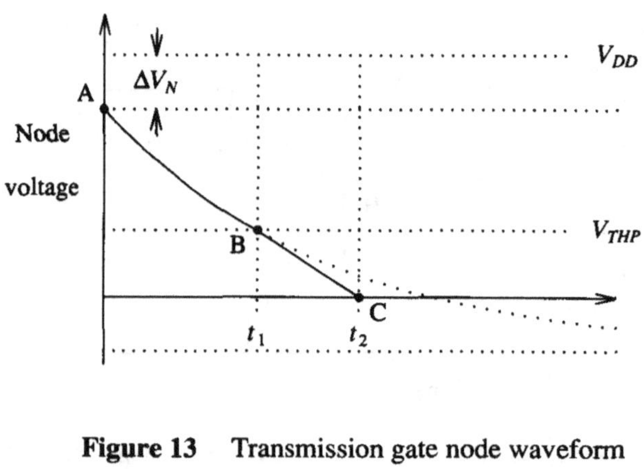

Figure 13 Transmission gate node waveform

3.4.2 Feedforward

Capacitance that couples the input and the output nodes of an inverter transmits the input signal directly to the output. This *feedforward* effect is always associated with the Miller effect, which will be discussed in the following sections. Here we consider the feedforward effect alone. Suppose that the input voltage of a CMOS static inverter, V_1, is zero, the output voltage, V_0, is V_{DD}, and the input makes a transition from 0 to V_{DD}. Before the output node begins to pull down, the output node is temporarily pulled above V_{DD} by the capacitive coupling from the input. The feedforward effect ends when the output node voltage returns to V_{DD}. *Since the output node voltage is higher than V_{DD}, it is confusing to use a conventional* circuit diagram: we redraw the equivalent circuit as shown in Fig. 14. Since the drain diffused island of the PFET and the N-type substrate forms a PN junction, the output node voltage is clamped at V_{DD} plus the threshold voltage of forward conduction of the PN junction (V_{BE}). If the voltage is higher than the conduction threshold voltage, many undesirable effects associated with minority carrier injection happen (Sections 6.6–6.8). If feedforward is small, and if this clamping action does not take place the circuit of Fig. 14 with non-zero transition time can be analyzed using the collapsible current generator model. After $t=0$ we have $V_{DD} \leq V_0(t) \leq V_{DD} + V_{BE}$ where V_{BE} is the conduction threshold voltage of forward biased PN junction. The initial condition is

$$V_1(0)=0 \quad V_0(0)=V_{DD}$$

Node voltage $V_0(t)$ satisfies the circuit equation

$$C_L \frac{dV_0(t)}{dt} + B_N V_1(t) + B_P[V_0(t) - V_1(t)] = C_m \frac{d}{dt}[V_1(t) - V_0(t)] \tag{21}$$

where B_N and B_P are the transconductances of MN1 and MP1, respectively, and the FET threshold voltages are zero. At the beginning of the input voltage transition, PFET current $B_P[V_0(t) - V_1(t)]$ is more significant than NFET current $B_N V_1(t)$ since $V_1(t)$ is still small. For feedforward effect to pull V_0 above V_{DD}, the sum of the NFET and PFET currents should be less than the capacitively induced current. Equation (21) is solved as

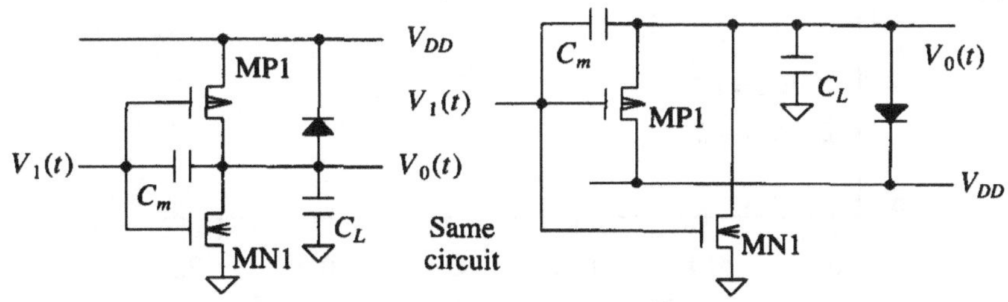

Figure 14 Feedforward of signal

$$V_0(t) = V_{DD} exp[-[B_P/(C_L + C_m)]t] + \frac{C_m}{C_L + C_m} V_1(t)$$

$$+ \frac{C_L(B_P - B_N) - C_m B_N}{(C_L + C_m)^2} exp[-[B_P/(C_L + C_m)]t] \int_0^t V_1(\xi) exp[[B_P/(C_L + C_m)]\xi] d\xi$$

If we use normalization

$$v_0(\theta) = V_0(t)/V_{DD} \quad v_1(\theta) = V_1(t)/V_{DD} \quad \theta = t/t_0 \tag{22}$$

where $t_0 = (C_L + C_m)/B_P$, we have

$$v_0(\theta) = exp(-\theta) + \frac{1}{(C_L/C_m) + 1} v_1(\theta) \tag{23}$$

$$+ \frac{(C_L/C_m)[1 - (B_N/B_P)] - (B_N/B_P)}{(C_L/C_m) + 1} exp(-\theta) \int_0^\theta v_1(\eta) \exp(\eta) d\eta .$$

If we use simple and manageable input waveform $V_1(t) = V_{DD}[1 - exp(-t/t_1)]$ as the input node waveform

$$\frac{v_0(\theta)}{[C_m/(C_L + C_m)]} \tag{24}$$

$$= (1 - \beta)(1 + \gamma) + [(1 + \gamma) + [(1 + \gamma)(1 - \beta) - 1][t_0/(t_1 - t_0)]] exp(-\theta)$$

$$-[(1 + \gamma)(1 - \beta) + [(1 + \gamma)(1 - \beta) - 1][t_0/(t_1 - t_0)]] exp[-(t_0/t_1)\theta]$$

where $\beta = B_N/B_P$ and $\gamma = C_L/C_m$. Figure 15 shows a plot of normalized voltage $v_0(\theta)$ versus θ. The feedforward does not pull V_0 above V_{DD} if the input signal slew rate is less than a certain limit. Then the PFET is able to absorb the capacitively injected current to the output node. The limit is given by $(B_P/C_m) > V_1'(t)|_{t=0}$.

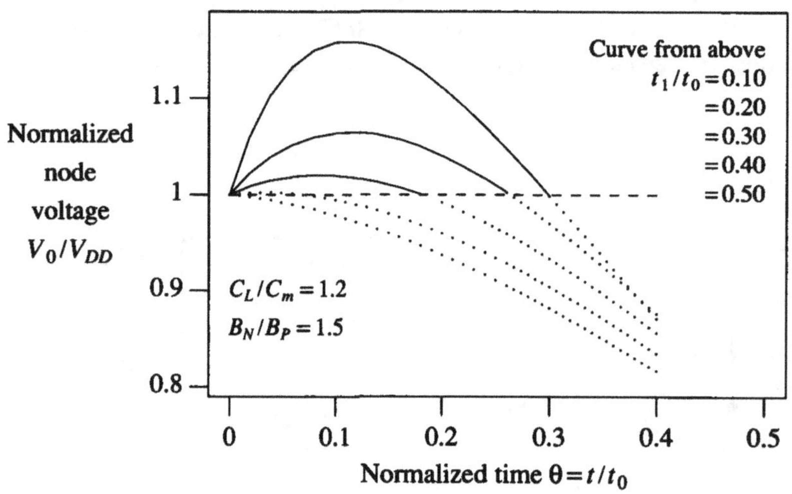

Figure 15 Feedforward node waveforms

3.5 Miller Effect and Reverse Signal Flow

Miller effect that is caused by capacitive coupling between the input and the output nodes of a linear amplifier is a negative feedback effect that increases the effective input capacitance of the *inverting* linear amplifier [09]. Digital CMOS gates are essentially nonlinear amplifiers. Nonlinearity affects many details of the Miller effect. Figure 16(a) shows a nonlinear amplifier with Miller effect feedback capacitance, C_m. The solid curve of Fig. 16(b) shows the static switching characteristic of the inverter. For simplicity we assume that the characteristic is symmetrical with respect to the maximum gain point, P, ($V_{GTR} = V_{GSW} = V_1 = V_0 = V_{DD}/2$). If the amplifier is biased at point P, the amplitude of the input signal, A, is small, and the frequency is low, we may write

$$V_1 = \mathrm{Re}[(V_{DD}/2) + A exp(i\omega t)] \qquad V_0 = \mathrm{Re}[(V_{DD}/2) - G_{\max} A exp(i\omega t)] \tag{25}$$

where $G_{\max}$ is the maximum gain of the amplifier, the slope of the static switching characteristic at point P [Fig. 16(b), dashed line]. If A increases, the output signal is distorted, since odd order harmonics are generated. If we neglect the higher harmonics and approximate the output waveform by the Fourier component at the fundamental frequency alone, we need to replace gain $G_{\max}$ of Eq. (25) by the *effective* gain, G_{eff}. The fundamental frequency component of the input current of the amplifier of Fig. 16(a) is

$$i_1 = C_m \frac{d}{dt}[V_1(t) - V_0(t)] = \mathrm{Re}[C_m(1 + G_{eff}) A exp(i\omega t)]\,. \tag{26}$$

Then the input characteristic of the amplifier is as if the capacitance increased from C_m to $C_m(1 + G_{eff})$ where $G_{eff} \rightarrow G_{\max}$ if the signal amplitude A is reduced, but $G_{eff} \rightarrow 1$ if $A \rightarrow V_{DD}/2$. In the limit of large A, the Miller capacitance is twice as effective as capacitance from the node to ground. The effective input capacitance of the amplifier depends on signal amplitude as shown in Fig. 16(c).

We showed how the Miller effect in a small-signal linear amplifier and in a large-signal nonlinear digital gate should be related. *Ideal* operation of a linear amplifier is that the input and the output transients take place simultaneously (this operation is the *static* switching of a digital inverter discussed in Section 2.8). To force the transient of the output node to be simul-

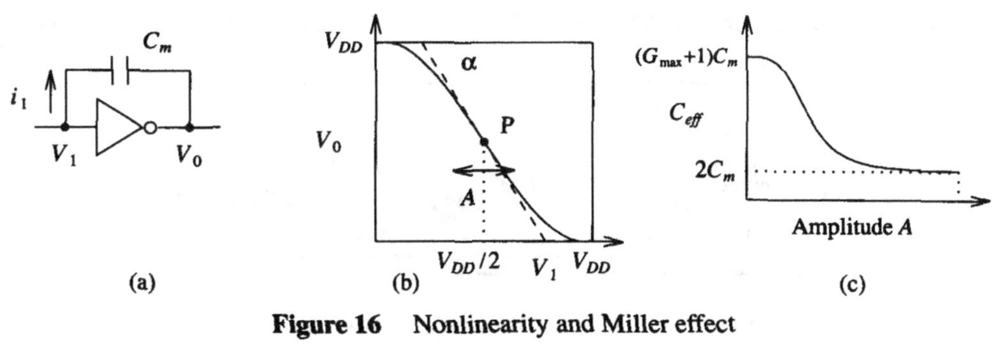

Figure 16 Nonlinearity and Miller effect

taneous with the transient of the input node, extra energy must be pumped into the amplifier from the signal source, if input-output coupling exists. The increased capacitive loading of the input node of the amplifier is interpreted as the energy *cost* of reducing the amplifier delay and thereby of forcing the input and output transients to be closely simultaneous. If the coupling capacitance exists ($C_m \neq 0$), the input node transition is resisted by the coupling (Section 2.2). Since the input and output transients are simultaneous, there is no time to generate the controlled energy to assist the output transient: then only the input signal source is able to *pay* the energy cost. In dynamic switching the transient of the output node is delayed with respect to the transient of the input node, and *ideally* the two transients do not overlap. In the ideal dynamic switching there is no Miller effect in its conventional sense. Miller effect exists if the switching is not ideally dynamic, or if the transient of the input and the output of the inverter are simultaneous.

Impedance of the circuit that includes Miller effect is quite complex. Miller effect is an application of the theory of generalized two-terminal impedance including active devices. This point of view will be developed fully in Section 5.18. Here we stress that Miller effect covers a very wide range of electrical phenomena. We show briefly the extent of coverage of Miller effect phenomena in digital circuits by using the most general equivalent circuit model. The input impedance that is observed from the left to the right side of the arrow of Fig. 17 is the node Miller effect impedance. The far end of capacitance C is connected to essentially any active or passive circuit that accepts any number of signals originating from the *same* signal source. Since all the master latch (or slave latch) signals originate from the master (or slave) clock and, as such, they may be considered as the *same* signal, there are quite a wide range of possibilities. The signal may be advanced or delayed relative to the driving signal. The circuit is generally nonlinear. The Miller effect in a digital circuit has two extrema cases: in the first case, the circuit that includes capacitor C may be considered as modified *impedance* to the driver, and in the second case, the active circuit is essentially an independent voltage source (Section 5.18).

Miller effect in a one-stage inverting gate is significant, but it does not always create catastrophic circuit failures, especially if the circuit is robust, or if there is operational margin. Often it is only a minor perturbation to normal circuit operation, because the mechanism is essentially negative feedback. Miller effect that takes place across a two-stage cascaded, non-inverting amplifier has a positive feedback mechanism, and that may cause disastrous circuit failures. Figure 18(a) shows cascaded inverters with feedback capacitor C_m. Signal $V_S(t)$ and noise $V_N(t)$ sources are shown. The input of the cascaded gate is driven by a fast-rising step

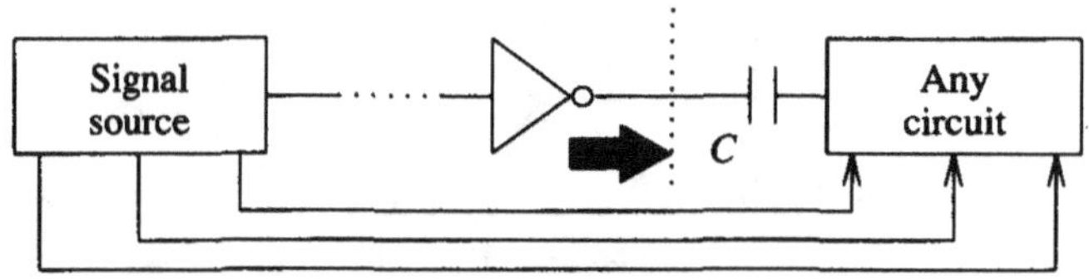

Figure 17 Generalized Miller effect

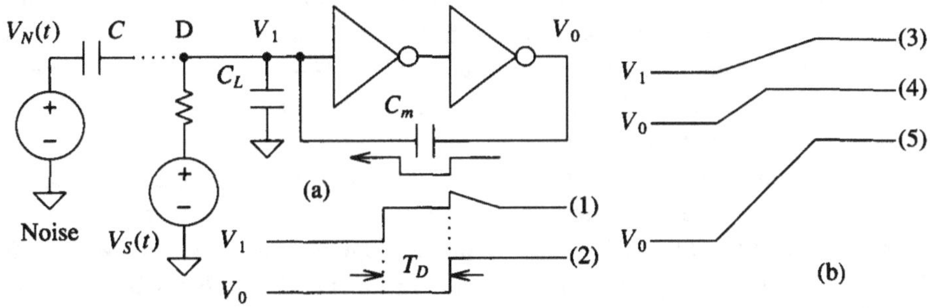

Figure 18 Positive Miller effect

voltage source $V_S(t)$. Suppose that input voltage V_1 makes a step increase as shown by waveform (1) of Fig. 18(b). The input node driver of the cascaded inverters sees total capacitive load $(C_L + C_m)$. After delay time T_D, V_0 makes a step increase as shown by waveform (2). At that moment the voltage spike as shown in waveform (1) develops at the input node. The current that flows through capacitor C_m is shown by the arrow in Fig. 18(a). This current could have helped to pull up the input node, but it does not, since it arrived too late. If input voltage V_1 makes a slow transition as shown by waveform (3), however, node voltages V_1 and V_0 increase together. If the output waveform is as shown by waveform (4), current that flows from output node to input node helps to pull up the input node at the beginning of the input transition. The cascaded inverters and Miller capacitor C_m work jointly as negative capacitance presented to the driving voltage source, D. The negative capacitance cancels a fraction of the *positive* capacitive load C_L of the input node. The effective capacitance is reduced from $(C_L + C_m)$ to

$$C_{eff} = C_L - (G-1)C_m \tag{27}$$

where G is the gain of the two-stage amplifier. The effective loading capacitance of node 1 becomes less than C_L if $G > 1$. This effect exists only until the output node voltage V_0 saturates at V_{DD}. After that time the driver of the input node D sees load capacitance $(C_L + C_m)$. In order to obtain capacitance compensation over the entire range of the operational voltage, the swing of the output node voltage of the cascaded inverters must be wider than the swing of the input voltage, as shown by waveform (5). Furthermore, signal delay through the cascaded inverters must be small. If the amplifier delay is longer than the transition time (time required for signal to make low-to-high transition), capacitance cancellation does not take place [waveforms (1) and (2)]. Numerical analysis shows that the amplifier delay must be 10% or less of the transition time to achieve significant capacitance cancellation [10].

Suppose that the delay of the cascaded inverters is small. Let us include the noise source $V_N(t)$ and consider noise voltage induced on the input node of the two-stage amplifier as shown in Fig. 18(a). If the amplifier is inactive, the induced noise voltage is

$$V_1(t) = \frac{C}{C_m + C_L + C} V_N(t) \tag{28}$$

where $V_N(t)$ is the noise source voltage, and C is the coupling capacitance. If the amplifier is active (DC-biased to the point that has significant small-signal gain), $C_L + C_m$ is partially compensated, so that the effective capacitance is $C_L - (G-1)C_m$. Then the induced noise voltage becomes higher than estimated from Eq. (28). Gates that receive the node voltage may fail because of excessive induced noise.

Miller effect in digital circuits is more complicated than that of linear circuits, because a digital gate is a multi-input amplifier. We consider how the Miller effect in a digital gate is *structurally* different from that of a linear amplifier. Figure 19(a) shows the Miller effect of a NAND2 gate. Input A is logic-high level, and the gate is enabled. Input B is originally low, and input B makes low-to-high transition (arrow 1). Output node Z is pulled down by the gate (arrow 2), and the decreasing voltage of the output node couples capacitively back to the input node (arrow 3). This coupling counteracts the pull-up of input B, thereby increasing the load capacitance of the driver of node B. This Miller effect (effect that involves input B alone) has a direct analog to the Miller effect in linear amplifiers.

A multi-input gate like a NAND2 can have Miller effect that is different from this. In Fig. 19(b), input B has already started a transition to the high-logic voltage level, and input A was originally low. When input A makes a low-to-high transition (arrow 1), output node Z pulls down (arrow 2), and the decreasing output node voltage is coupled to input node B through capacitance C_m (arrow 3). Similarly, decreasing node X voltage couples to input node B through capacitance C'_m. If input node B is still making the low-to-high transition, the Miller effect counteracts the pull-up. In this example, *cause* and *effect* of the Miller effect refer to different input terminals of a gate. By generalizing this observation, any input is able to cause Miller effect at any input(s). Whether or not such coupling really exists depends on the logic levels of the other signals. In Fig. 19(c), switching input C creates Miller effect to itself, but not to input A, since input B is at logic low level and the gate is disabled. Miller effect in this general definition is very complicated. Miller effect in digital circuits is a phenomenon that carries two node subscripts, one refers to the cause, and the other to the effect.

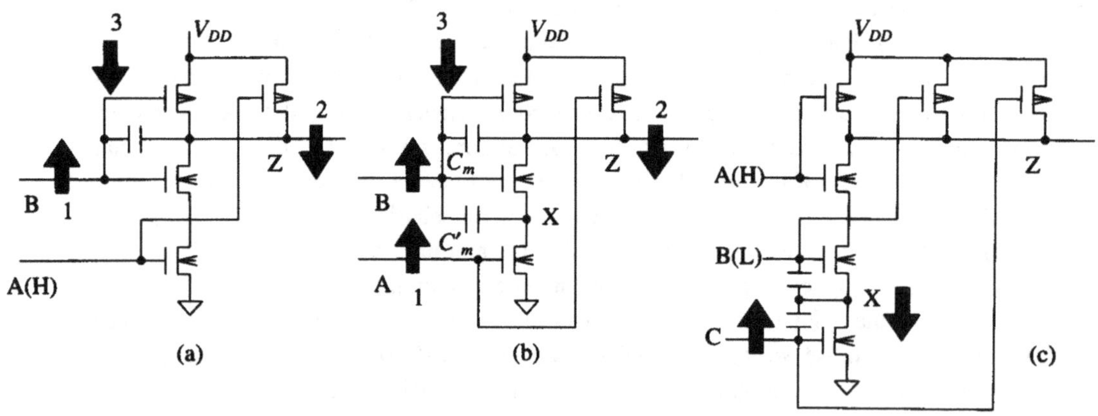

Figure 19 Miller effect in logic gates

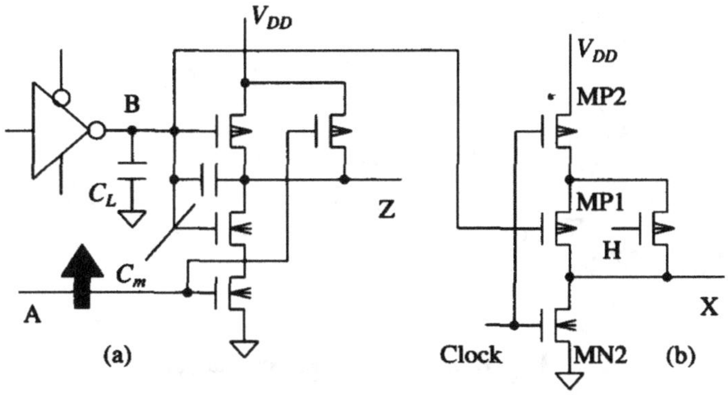

Figure 20 Miller effect failure mechanism

Since backward signal propagation by Miller effect is the most drastic misdirection of signal flow, many circuit failures originate from the mechanism. In the circuit shown in Fig. 20, a tristatable inverter drives node B to the logic-high level, and then the clock turns off the tristatable inverter. Capacitance C_L holds charge, and therefore node B stays at V_{DD}. Input node A of the static NAND2 gate was originally at ground potential. Therefore its output node was originally at V_{DD}. The output node of dynamic NOR2 gate X was predischarged to ground potential. After predischarge, the dynamic gate clock made a high-to-low transition: predischarge NFET MN2 is turned off, and PFET MP2 is turned on. Since node B voltage is high, PFET MP1 is turned off, and therefore output node X is held at ground potential. Since gate Z is a static gate, A may change at any time. Suppose that A makes a low-to-high transition. Then output node Z of the NAND2 gate pulls down. The voltage change is transmitted back to node B through C_m, and the voltage of node B falls from V_{DD} to $C_L V_{DD}/(C_m + C_L)$. This voltage drop is not compensated, since the tristatable inverter is already turned off. If node B drops more than the threshold voltage of PFET MP1, node X is pulled up by PFET MP1, thereby creating a logic error at node X. The logic error was caused because a signal was coupled from output node Z to input node B of the static NAND2 gate. This failure mechanism was first examined by M.S.C. Chung during the BellMac microprocessor project. This example shows that cascading CMOS dynamic gates directly is a risky design. Dynamic circuits called NORA CMOS dynamic logic suffer severely from this problem (Section 4.14).

3.6 Closed-form Theory of Miller Effect

3.6.1 Miller Effect in Static Gates

As we saw in the last section, Miller effect in digital circuits is very complex. It is necessary to provide a simple, closed-form circuit theory. We use the collapsible current generator model, with zero threshold voltage, $V_{TH} = 0$ (Section 1.7). We analyze the inverter of Fig.

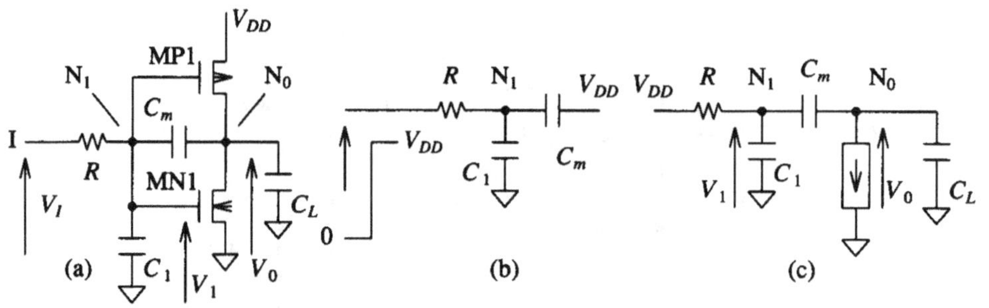

Figure 21 Miller effect analysis

21(a) with $B_N = B_P$. Node I is driven instantaneously from ground to V_{DD} by a voltage source, and resistance R connects node I and node N_1. We study the V_1 waveform. C_1 is the input capacitance, C_m the Miller effect capacitance, and C_L is the load capacitance, respectively, of the inverter. To make the closed-form analysis tractable, we assume the following. The Miller effect is always associated with the feed-forward effect (Section 3.4). If all details of the feed-forward are included, analysis becomes quite complex. Since the Miller effect is significant only when the input signal slew rate is low, we neglect those aspects of the feed-forward effect that are insignificant for low input slew rate. We assume that V_0 is held at V_{DD} until the moment the inverter begins to pull down at its switching threshold voltage. If the feedforward effect is accurately included, the capacitive current from N_1 to N_0, at the moment when the NFET and PFET currents are equal, shifts the switching threshold voltage of the inverter. We neglect this shift, which creates a slight, but tolerable inconsistency in the circuit equations. The microstate sequence that results from this assumption is shown in Fig. 22. In this microstate sequence the Miller effect exists if both the NFET and PFET are in the saturation region. If both MN1 and MP1 are in the saturation region, V_0 changes with time. Event C (MN1: S $\rightarrow$ T) must precede event D (MP1: S $\rightarrow$ N) (slow or static switching). If event D takes place earlier than event C, we will have Miller *aftereffect*. We concentrate on the electrical transient between events B and C first.

Until V_1 reaches the switching threshold voltage of the symmetrical CMOS inverter,

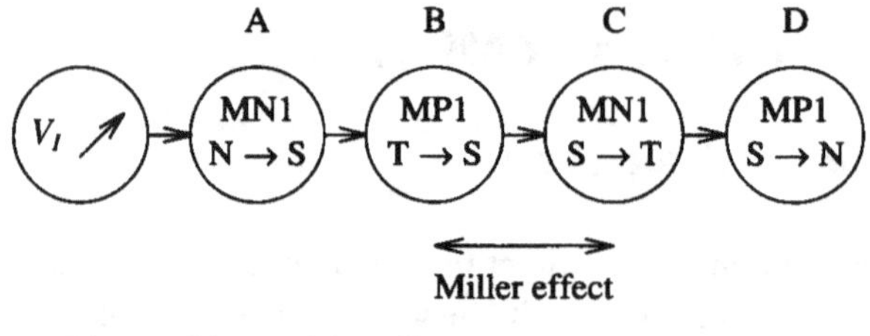

Figure 22 Miller effect microstate sequence

$V_{DD}/2$, V_0 stays at V_{DD}. The circuit of Fig. 21(a) may be simplified to that of Fig. 21(b). V_1 is given by

$$V_1(t) = V_{DD}[1 - exp[-t/R(C_1 + C_m)]] . \tag{29}$$

At time $t_{SW} = \log 2 \cdot R(C_1 + C_m)$, $V_1(t_{SW}) = V_{DD}/2$. At this time the inverter output begins to pull down. V_0 is driven by collapsible current generators I_N and I_P that model the NFET and PFET, respectively, where $I_N - I_P = 2B[V_1 - (V_{DD}/2)]$. The equivalent circuit at this point is shown by Fig. 21(c). The circuit equations that describe the switching transient are

$$C_1 \frac{dV_1}{dt} = \frac{V_{DD} - V_1}{R} - C_m \frac{d}{dt}(V_1 - V_0) \tag{30}$$

$$C_L \frac{dV_0}{dt} + 2B(V_1 - V\frac{DD}{2}) + C_m \frac{d}{dt}(V_0 - V_1) = 0 .$$

V_0 can be eliminated as

$$\frac{dV_1}{dt} = \frac{1}{T_M}(V_X - V_1) \tag{31}$$

where

$$T_M = \frac{R(C_L C_1 + C_L C_m + C_1 C_m)}{C_L + C_m + 2BRC_m} \quad \text{and} \quad V_X = \frac{C_L + C_m + BRC_m}{C_L + C_m + 2BRC_m} V_{DD} . \tag{32}$$

Equation (32) shows that input node voltage tends to V_X when $t \to \infty$. In Eq. (32), B is the transconductance of each FET, and R is the internal resistance of the signal source driving node N_1. If the BR product is small, node N_1 is driven by a low impedance source, and the inverter output response is slow compared to the input: this is the condition for dynamic switching, and for Miller aftereffect. If BR is large, however, node N_1 is driven by a high impedance source, and the inverter responds quickly. The transients at the input and the output of the inverter become, more or less, simultaneous. This is the condition for significant Miller effect. Equation (32) shows that

$$V_X \to V_{DD} \text{ when } BR \to 0 \quad V_X \to V_{DD}/2 \text{ when } BR \to \infty \tag{33}$$

Equation (31) shows that node voltage $V_1(t)$ is temporarily clamped at V_X, which is close to the switching threshold voltage, $V_{DD}/2$, if $BR \to \infty$. Equation (30) ceases to be valid some time after t_{SW}. At time $t_{MF} = t_{SW} + T_X$, the output node completes the high-to-low transition. After time t_{MF}, the V_1 is described again by the simple time constant process of Eq. (29). Thus

$$V_1(t) = V_{DD} - [V_{DD} - V(t_{MF})]exp[(t_{MF} - t)/R(C_1 + C_m)] . \tag{34}$$

The Miller effect is understood by studying Eqs. (31) and (32) for time $t_{SW} < t < t_{MF}$.

Equation (31) is solved as

$$V_1(t) = V_X + (V\frac{DD}{2} - V_X)exp[(t_{SW} - t)/T_M] \quad (t_{SW} < t < t_{MF}) \tag{35}$$

where t_{MF} is determined from the following analysis. V_0 satisfies equation

$$C_m \frac{dV_0}{dt} = (C_1 + C_m)\frac{dV_1}{dt} - \frac{V_{DD} - V_1}{R} .$$

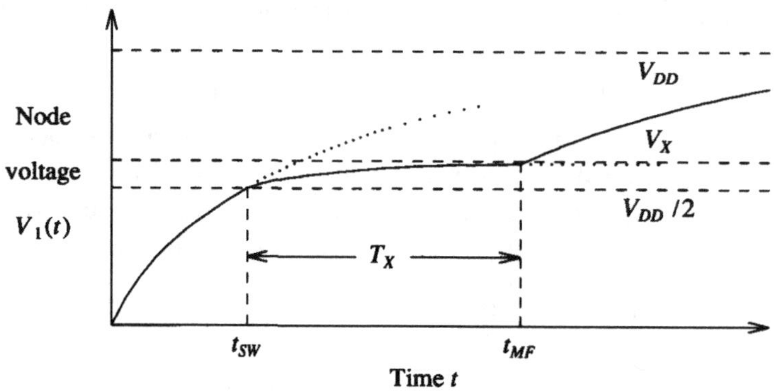

Figure 23 Miller effect waveform

$V_1(t)$ from Eq. (35) is substituted into this equation. After some algebra we find

$$\frac{dV_0}{dt} = -\chi[-\lambda exp[(t_{SW} - t)/T_M] + 1] \quad \text{where} \tag{36}$$

$$\chi = \frac{BV_{DD}}{C_L + C_m + 2BRC_m} \qquad \lambda = \frac{(C_L + C_m)[C_m + 2BR(C_m + C_1)]}{2BR(C_L C_1 + C_L C_m + C_1 C_m)}$$

This equation is solved subject to the initial condition $V_0(t_{SW}) = V_{DD}$ as

$$V_0(t) = -\chi \int_{t_{SW}}^{t} [-\lambda exp[(t_{SW} - \theta)/T_M] + 1] d\theta + V_{DD} \tag{37}$$

$$= \chi\lambda T_M[1 - exp[(t_{SW} - t)/T_M]] - \chi(t - t_{SW}) + V_{DD}$$

Since the feed-forward effect was not accurately included, $V_0'(t_{SW}) \neq 0$. Neglecting the small discrepancy, we set $V_0(t_{MF}) = 0$ and solve for T_X as

$$\frac{T_X}{T_M} = \theta(-\frac{1}{\lambda}, 1 - V\frac{DD}{\chi\lambda T_M}).$$

From this analysis, the Miller effect waveform is determined as shown in Fig. 23. A significant plateau is created at about the switching threshold voltage, $V_{DD}/2$. The Miller effect creates significant node waveform distortion, which affects delay characteristics of the circuit (Section 6.11).

3.6.2 Miller Effect in Dynamic Gates

The Miller effect takes place in CMOS dynamic logic circuits as well. Figure 24(a) shows the equivalent circuit of a CMOS dynamic inverter immediately after precharge. The circuit equations for node voltages V_0 and V_1 are

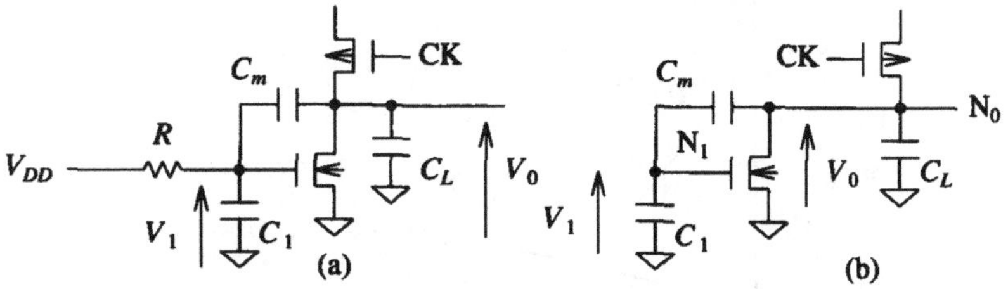

Figure 24 Miller effect of dynamic gate (a), and of a floating node (b)

$$C_1 \frac{dV_1(t)}{dt} = \frac{V_{DD} - V_1(t)}{R} + C_m \frac{d}{dt}[V_0(t) - V_1(t)] \tag{38a}$$

$$C_m \frac{d}{dt}[V_0(t) - V_1(t)] + B_N V_1(t) + C_L \frac{dV_0(t)}{dt} = 0 \tag{38b}$$

and the initial conditions are

$$V_1(0) = 0 \quad V_0(0) = V_{DD}.$$

Since Eq. (38b) contains $dV_0(t)/dt$ but no $V_0(t)$, the derivative can be substituted into the first equation, and thus $V_0(t)$ can be eliminated. $V_1(t)$ satisfies

$$\frac{C_m C_1 + C_1 C_L + C_L C_m}{C_m} \frac{dV_1(t)}{dt} \tag{39}$$

$$+(B_N + \frac{C_L + C_m}{RC_m})[V_1(t) - \frac{C_L + C_m}{(B_N R + 1)C_m + C_L} V_{DD}] = 0$$

Using transformation

$$\phi_1(t) = V_1(t) - V_X$$

we have

$$\frac{d\phi_1(t)}{dt} + \frac{1}{T_M}\phi_1(t) = 0 \tag{40}$$

where

$$T_M = \frac{R(C_m C_1 + C_1 C_L + C_L C_m)}{(B_N R + 1)C_m + C_L} \quad \text{and} \quad V_X = \frac{C_m + C_L}{(B_N R + 1)C_m + C_L} V_{DD} \quad (V_X < V_{DD})$$

The solution that satisfies the initial conditions is

$$V_1(t) = V_X[1 - exp(-t/T_M)] \tag{41a}$$

$$V_0(t) = V_{DD} - \frac{1}{RC_m}\int_0^t [V_{DD} - V_1(t)]dt + \frac{C_1 + C_m}{C_m} V_1(t) \tag{41b}$$

$$= V_{DD} - \left(\frac{V_{DD} - V_X}{RC_m}\right)t + V\frac{X}{RC_m}[(C_1 + C_m)R - T_M][1 - exp(-t/T_M)].$$

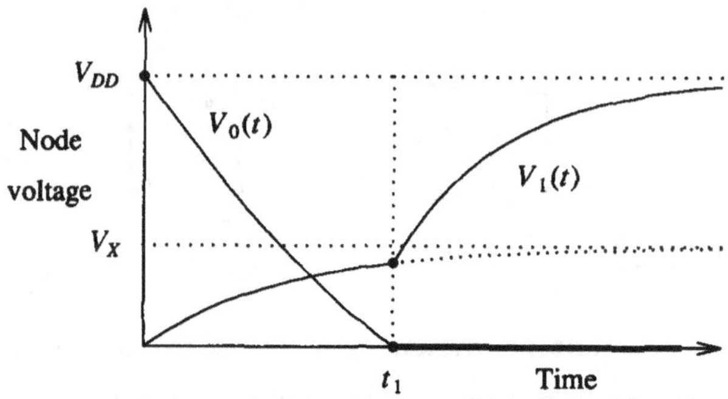

Figure 25 Dynamic gate Miller effect node waveforms

$V_0(t)$ reaches zero when $t=t_1$, where t_1 is given by

$$t_1 = T_M\theta\left\{\frac{(V_{DD}-V_X)T_M}{V_X[(C_1+C_m)R-T_M]}, 1+\frac{RC_mV_{DD}}{V_X[(C_1+C_m)R-T_M]}\right\}$$

where we use the θ-function defined in the Appendix of chapter 2. Node waveforms given by Eqs. (41a) and (41b) are plotted in Fig. 25. At time $t=t_1$ the NFET goes into the triode region. Before time t_1, $V_1(t)$ increases slowly, since the Miller effect capacitance loads the input node. After time t_1 the Miller effect loading vanishes, and the input node voltage increases rapidly. $V_1(t)$ after t_1 is given by

$$V_1(t)=V_{DD}-[V_{DD}-V_1(t_1)]exp[-(t-t_1)/R(C_1+C_m)] \tag{42}$$

3.6.3 *Floating Nodes*

If resistor R of Fig. 24(a) is very large, the circuit of Fig. 24(a) changes to the circuit of Fig. 24(b). This circuit shows *floating* circuit nodes N_0 and N_1. A floating node is a tristated node or a node that is disconnected from any driving source by electrical or mechanical failure. Suppose that at time $t=0$, node voltages $V_1(0)$ and $V_0(0)$ are both positive. If the NFET draws current, $V_0(t)$ decreases with time. Since nodes 0 and 1 are coupled by capacitance C_m, V_1 decreases as well. If $V_1(0)$ was low, V_1 reaches V_{THN} and the NFET turns off. After that, node voltages V_1 and V_0 no longer change. If $V_1(0)$ is high, however, node 0 discharges completely. Depending on the initial condition, the circuit ends up with a different final state. Conservation of charge at the floating node 1 requires

$$(C_1+C_m)V_1(t)-C_mV_0(t)=const.=(C_1+C_m)V_1(0)-C_mV_0(0).$$

The circuit equation is

$$(C_L+C_m)\frac{dV_0}{dt}-C_m\frac{dV_1}{dt}+B(V_1-V_{THN})=0. \tag{43}$$

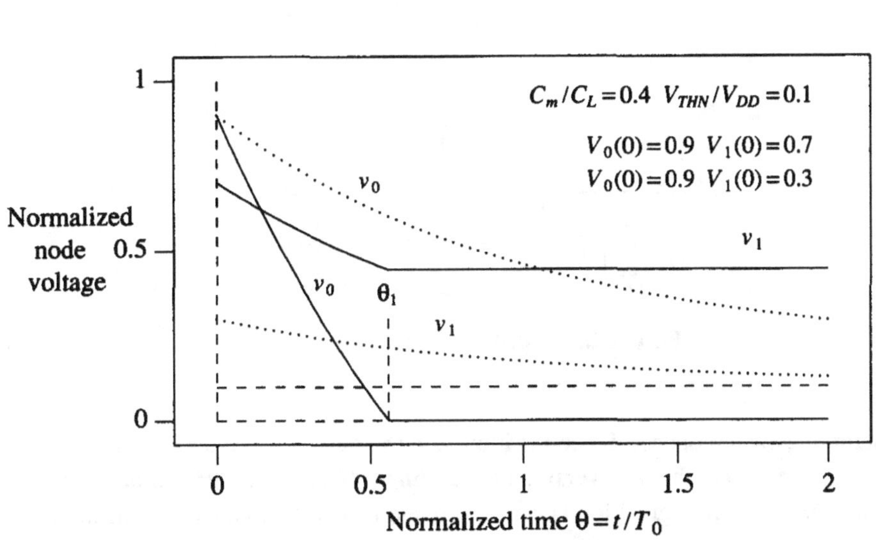

Figure 26 Miller effect of floating node

If we define $T_0=(C_LC_1+C_LC_m+C_1C_m)/C_mB$, we obtain

$$V_0(t)=V_0(\infty)+\frac{C_1+C_m}{C_m}[V_1(0)-V_{THN}]exp(-t/T_0) \tag{44a}$$

$$V_1(t)=V_{THN}+[V_1(0)-V_{THN}]exp(-t/T_0) \tag{44b}$$

where $V_0(\infty)=V_0(0)-[(C_1+C_m)/C_m][V_1(0)-V_{THN}]$. If $V_0(\infty)<0$, V_0 drops to ground potential and the NFET goes into the triode region. In Fig. 26, normalized node voltages $v_0=V_0(t)/V_{DD}$ and $v_1=V_1(t)/V_{DD}$ are plotted versus normalized time $\theta=t/T_0$. At $\theta_1=t_1/T_0$ where t_1 is given by

$$t_1=-T_0\log\left[1-\frac{C_m}{C_1+C_m}\frac{V_0(0)}{V_1(0)-V_{THN}}\right]$$

v_0 vanishes and the FET goes into the triode region. If $V_0(\infty)>0$ the NFET turns off before the completing discharge of node 0, as shown by the dotted curves of Fig. 26.

3.6.4 Miller Aftereffect

In this analysis of Miller effect, transitions of the input and output nodes of the gate take place simultaneously. If the output node transition is significantly delayed from the input node transition, we have Miller *after*-effect. The equivalent circuit of Fig. 27 is used to define and to study this effect. Figure 27 shows a CMOS inverter with an equivalent sampling switch at the input. Originally, the V_1 was zero, and V_0 was V_{DD}. The switch closes momentarily at time $t=0$ and V_1 changes from 0 to V_{DD} within a negligibly short time. Before the V_0 begins to

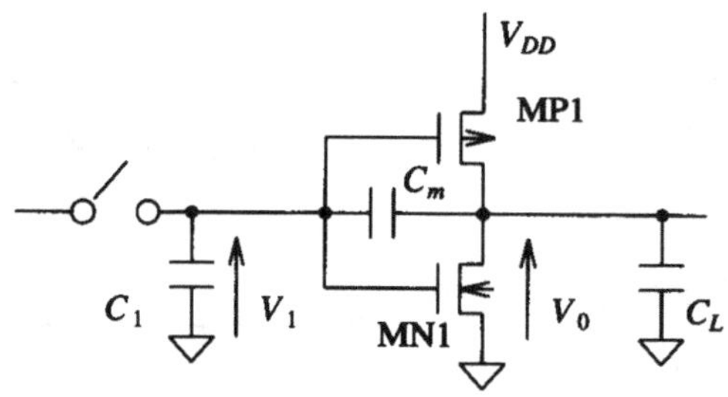

Figure 27 Miller aftereffect

decrease, the sampling switch opens. Since the input and the output nodes are coupled by C_m, V_1 decreases as V_0 decreases. This is a version of the Miller effect: This version of the Miller effect takes place *after* the input transition is completed and is therefore called the Miller aftereffect.

Immediately after the sampling, output node voltage V_0 becomes higher than V_{DD} because of the capacitive coupling from input to output. We have initial conditions

$$V_0(+0) = V_{DD} + \frac{C_m}{C_L + C_m} V_{DD} \quad \text{and} \quad V_1(+0) = V_{DD} . \tag{45}$$

We assume a realistic case where the excess voltage over V_{DD} of the output node is less than the threshold voltage of conduction of the drain island PN junction of PFET. From conservation of charge of N_1, node voltages $V_1(t)$ and $V_0(t)$ are related by

$$V_1(t) = \frac{C_1(C_L + C_m) - C_m^2}{(C_1 + C_m)(C_L + C_m)} V_{DD} + \frac{C_m}{C_1 + C_m} V_0(t) . \tag{46}$$

[Note that $C_m/(C_1 + C_m)$ is the output-input voltage division ratio, and the relationship is verified by substituting Eq. (45) at time $t = 0$]. Since $V_0(+0) > V_{DD}$, the output node is the *source*, and V_{DD} is the *drain* of MP1 at this time. As the output node is pulled down by MN1, the voltage reaches V_{DD} at time $t = t_1$. At that moment the source and the drain of MP1 reverse. We assume, for simplicity, that the inverter is symmetrical, and the FET threshold voltages are zero. The microstate sequence of this switching process is shown in Fig. 28. As V_1 makes step change from 0 to V_{DD}, three state changes take place in rapid succession. MP1 reverses the source/drain identification and then goes into the saturation region in the reverse polarity, and MN1 turns on. As the discharge of the output node proceeds, V_0 drops to V_{DD} at time t_1. At $t = t_1$ V_1 is less than V_{DD}. At this moment MP1 goes into the triode region and the source/drain identification of MP1 reverses again. MP1 returns to the saturation region in the normal polarity and stays in the saturation region (we assume that the drop of input voltage below V_{DD} is small). By the reversal of the source/drain identification, current flowing out from the output node of the inverter (by MP1) makes a stepwise decrease, ΔI, at the moment of the reversal

$$\Delta I = 2B[V_{DD} - V_1(t)] .$$

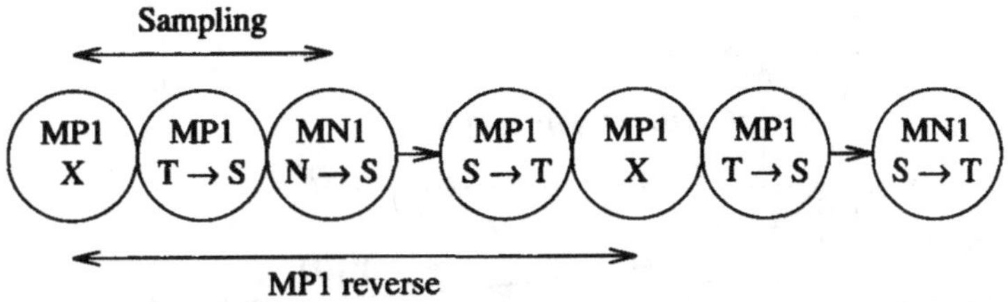

Figure 28 Microstate sequence of Miller aftereffect

The stepwise change in current occurs in the collapsible current generator model if the source/drain identity switches (discussion in Section 3.7). Pull-down of V_0 continues. At time t_2, NFET MN1 goes into the triode region and the transient is terminated.

Immediately after sampling of the input voltage, MP1 has reversed polarity, and the circuit equation is

$$(C_L + C_m)\frac{dV_0(t)}{dt} - C_m\frac{dV_1(t)}{dt} = -BV_1(t)\,(\text{NFET}) - B[V_0(t) - V_1(t)]\,(\text{PFET}) \qquad (47)$$

or using Eq. (46) we have

$$\frac{dV_0(t)}{dt} = -\frac{1}{T_{M1}}V_0(t) \quad \text{where} \quad T_{M1} = \frac{C_LC_1 + C_1C_m + C_mC_L}{B(C_1 + C_m)}. \qquad (48)$$

The solution satisfying the initial condition [Eq. (45)] is given by

$$V_0(t) = V_{DD}\left(1 + \frac{C_m}{C_L + C_m}\right)exp(-t/T_{M1}). \qquad (49)$$

$V_0(t)$ reaches V_{DD} at $t = t_1$, which is found from Eq. (49) as $t_1 = T_{M1}\log[1 + (C_m/(C_L + C_m))]$. After $t = t_1$ the circuit equation is

$$(C_L + C_m)\frac{dV_0(t)}{dt} - C_m\frac{dV_1(t)}{dt} = B[V_{DD} - 2V_1(t)]. \qquad (50)$$

By substituting $V_1(t)$ from Eq. (46) we get

$$\frac{dV_0(t)}{dt} = -\frac{1}{T_{M2}}[V_0(t) + V_X] \qquad (51)$$

$$\text{where } T_{M2} = \frac{C_LC_1 + C_1C_m + C_mC_L}{2BC_m} \qquad V_X = \frac{C_LC_1 + C_1C_m - C_mC_L - 3C_m^{\,2}}{2C_m(C_L + C_m)}V_{DD}.$$

and the solution satisfying the initial condition $V_0(t_1) = V_{DD}$ is

$$V_0(t) = (V_{DD} + V_X)exp[(t_1 - t)/T_{M2}] - V_X. \qquad (52)$$

$V_0(t)$ reaches zero when $t = t_2$. Time t_2 is found as $t_2 = t_1 + T_{M2}\log[1 + (V_{DD}/V_X)]$. The $V_1(t)$ waveform is constructed from the $V_0(t)$ waveform using Eq. (46): We see that the waveform for $V_1(t)$ is drawn by compressing $V_0(t)$ by a factor $C_m/(C_1 + C_m)$ and shifting by

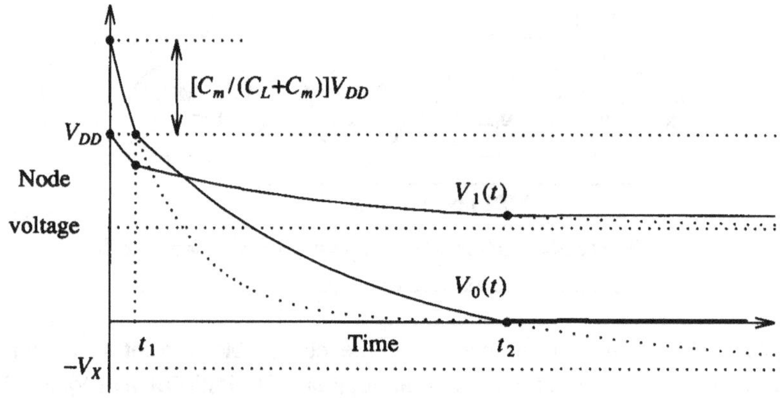

Figure 29 Node waveform of Miller aftereffect

$[C_1(C_L+C_m) - C_m^2]/[(C_1 + C_m)(C_L + C_m)]V_{DD}$. After $t=t_2$, $V_0(t)=0$, and $V_1(t)=V_1(t_2)$ (independent of time). Figure 29 shows the $V_1(t)$ and $V_0(t)$ waveforms.

3.7 Positive Miller Effect

To apply positive feedback through the Miller effect, we require an even number of cascaded inverters between the input and output. Except for the simplest case of *zero* inverters (which is a source follower), the analysis becomes quite complex. We study a few typical cases beginning with the Miller effect of a source follower circuit.

3.7.1 Miller Effect of Source Follower Circuit

We analyze the source follower circuit of Fig. 30. Capacitance C_1 and C_2 are the gate-to-source and gate-to-drain capacitances of the NFET, respectively. C_1 and C_2 depend on gate-to-source/drain voltages, but we neglect this dependence and use the average capacitances (Section 1.7). C_0 is the load capacitance, and $V_{THN}=0$. The circuit equations are

$$-C_1\frac{dV_0(t)}{dt}+(C_1+C_2)\frac{dV_1(t)}{dt}=\frac{V_{DD}-V_1(t)}{R} \tag{53a}$$

$$(C_0+C_1)\frac{dV_0(t)}{dt}-C_1\frac{dV_1(t)}{dt}=B_N[V_1(t)-V_0(t)]. \tag{53b}$$

Using transformation

$$\phi_0(t)=V_0(t)-V_{DD} \quad \phi_1(t)=V_1(t)-V_{DD}$$

we have

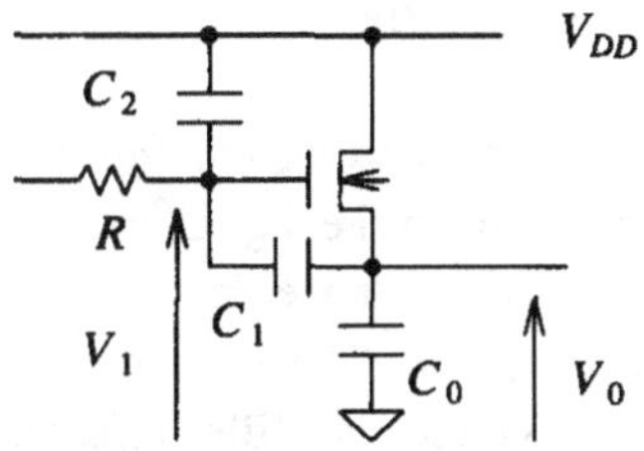

Figure 30 Miller effect of source follower circuit

$$-C_1\frac{d\phi_0}{dt}+(C_1+C_2)\frac{d\phi_1}{dt}=-\frac{\phi_1}{R}\quad (C_0+C_1)\frac{d\phi_0}{dt}-C_1\frac{d\phi_1}{dt}=B_N(\phi_1-\phi_0). \tag{54}$$

Assuming time dependence of the form ϕ_0, $\phi_1 \approx exp(St)$, parameter S is given by

$$S=\alpha\ (+\ \text{value}),\quad \beta\ (-\ \text{value})$$

where

$$\alpha,\beta=\frac{-[[(C_0+C_1)/R]+B_NC_2]\pm\sqrt{[(C_0+C_1)/R]+B_NC_2]^2-(4B_N/R)(C_0C_1+C_1C_2+C_2C_0)}}{2(C_0C_1+C_1C_2+C_2C_0)}$$

In the limit of small (B_NR) (fast input switching, or insignificant Miller effect)

$$\alpha=-\frac{B_N}{C_0+C_1}\quad \beta=-\frac{1}{C_TR}\quad \text{where}\quad C_T=C_2+\frac{C_0C_1}{C_0+C_1}.$$

In the limit of large (B_NR) (slow input switching, or significant Miller effect)

$$\alpha=-\frac{1}{C_2R}\quad \beta=-\frac{B_NC_2}{C_T(C_0+C_1)}.$$

We have

$$\phi_0(t)=Aexp(\alpha t)+Bexp(\beta t) \tag{55a}$$

$$\phi_1(t)=\frac{\alpha C_1A}{(C_1+C_2)\alpha+(1/R)}exp(\alpha t)+\frac{\beta C_1B}{(C_1+C_2)\beta+(1/R)}exp(\beta t). \tag{55b}$$

Coefficients A and B are determined from the initial conditions

$$V_0(0)=V_1(0)=0\quad \text{or}\quad \phi_0(0)=\phi_1(0)=-V_{DD}$$

and we obtain the following solution

$$V_1(t)=\phi_1(t)+V_{DD}=V_{DD}-\frac{V_{DD}}{\alpha-\beta}[\alpha(1+\beta C_2R)exp(\alpha t)-\beta(1+\alpha C_2R)exp(\beta t)].$$

In the limit of large (B_NR), $1+\alpha C_2R\approx 0$ and consequently $V_1(t)\approx V_{DD}[1-exp[-t/(C_2R)]]$. In this case there is complete cancellation of the gate-source capacitance C_1. Capacitance C_2 shows up as input capacitance. In the limit of small (B_NR), $V_1(t)\approx V_{DD}[1-exp[-t/(C_TR)]]$ and both input capacitances C_1 and C_2 load the input.

3.7.2 Multistage Positive Miller Effect

We study multi-stage positive Miller effect using the equivalent circuit of Fig. 31. The two-stage cascaded inverter input is driven from the signal source, V_I, which makes a low-to-high transition at time $t=0$ and has internal resistance R. The first and the second inverters have load capacitances C_2 and C_0, respectively. The output node of the second inverter is connected to the input node of the first inverter via Miller effect capacitance C_m. Capacitance C_1 represents the input capacitance of the first-stage inverter. We use the collapsible current generator FET model and assume that the FET threshold voltages are zero. Furthermore we assume that all the inverters of Fig. 31 are pull-up/pull-down symmetrical (NFET and PFET have same beta).

The approximate waveform for the positive Miller effect can be determined by assuming a simplified model of the two-stage cascaded inverter circuit. $V_1(t)$ increases with time following a time-constant process if the output node of the second stage inverter is held at ground potential by MN2. MN2 is assumed to be a large NFET. We simplify the problem by assuming that the output node of the second-stage inverter makes an instantaneous low-to-high transition at time T_D after V_1 reaches V_{GSW} at time t_1. T_D is the cumulative delay of the two cascaded inverters. We find t_1 as $t_1=(\log 2)R(C_1+C_m)$. We then have

$$V_1(t)=V_{DD}[1-exp[-t/R(C_1+C_m)]] \quad 0<t<t_2 \quad t_2=t_1+T_D\,. \tag{56a}$$

At $t=t_2$ there is a step increase in node voltage V_0 from 0 to V_{DD} due to the switching of the second-stage inverter. Then node voltage V_1 increases by $\Delta V_1=[C_m/(C_1+C_m)]V_{DD}$, and therefore

$$V_1(t)=V_1(t_2)+\Delta V_1+[V_{DD}-V_1(t_2)-\Delta V_1][1-exp[-(t-t_2)/R(C_1+C_m)]] \quad t>t_2\,. \tag{56b}$$

Before the step change in V_0 at t_2, MN2 loses the capability to absorb current from node 1 through capacitance C_m, and V_0 begins to increase. We neglected this effect. The V_1 waveform is shown in Fig. 32, where we assumed that $V_1(t_2)+\Delta V_1<V_{DD}$. Figure 32 shows that

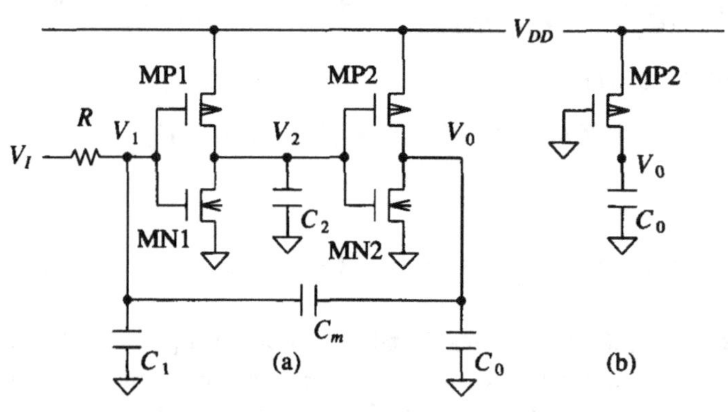

Figure 31 Analysis of positive Miller effect

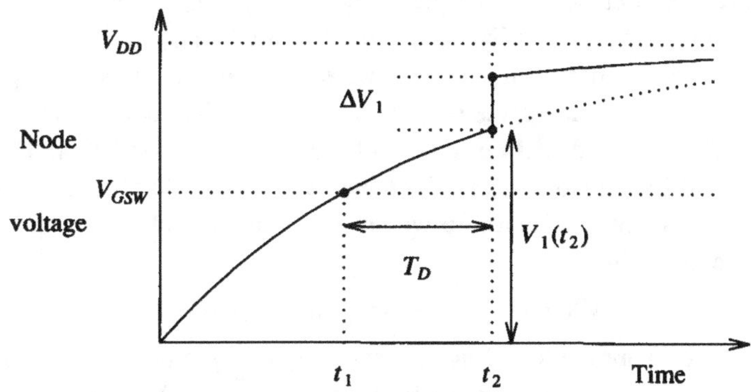

Figure 32 Positive Miller effect voltage waveform

instead of a plateau as in the conventional Miller effect, a step increase in V_1 occurs. The process of positive Miller effect is represented by the microstate sequence of Fig. 33. The microstate sequence was constructed as follows. The events that are caused by the input pull-up transition are (MN1: N → S) and (MP1: T → S). After that, node 2 pull-down begins. Then MP1 goes into the nonconducting region. When the event MP1: S → N takes place depends on the current from the step-function input signal source, and on the current through capacitance C_m. Therefore event (MP1: S → N) can be anywhere in the later part of the microstate sequence. The first effect of node 2 pull-down is (MP2: N → S), which is followed by (MN2: T → S). After the second event, node 0 pull-up begins. Events (MP2: S → T) and (MN2: S → N) must be after event (MN2: T → S), and event (MN1: S → T) must be after event (MN2: S → N) if the FETs are enhancement mode. The order of microstate events is determined if the FET threshold voltages are positive, even if they are infinitesimally small. When we assume $V_{THN} = V_{THP} = 0$, for problem simplification, we determine the order of microstate events in the following way. Since the second-stage inverter may switch either statically or dynamically, event MP2: S → T can be earlier or later than event MN2: S → N. The positive Miller effect starts at event (MN2: T → S) and ends at (MP2: S → T).

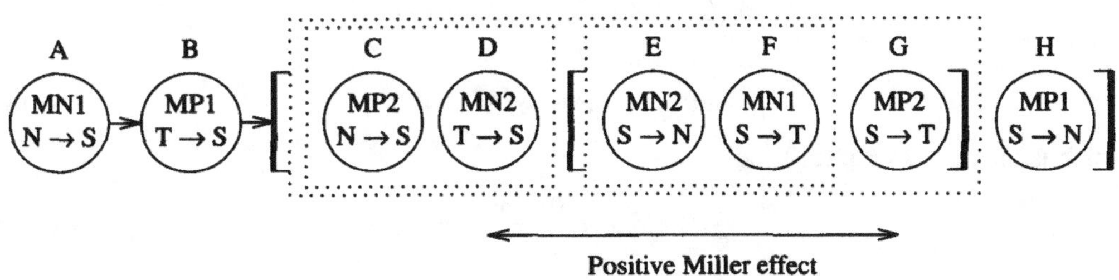

Figure 33 Microstate sequence of positive Miller effect

Now we give more detailed analysis of the positive Miller effect. After signal source V_I makes the low-to-high transition, MN1 conducts: the FET changes from the nonconducting region to the saturation region. Since capacitances C_m and C_0 make a capacitive voltage divider that responds without delay, the output node voltage V_0 may be driven up. Since MN2 is in the triode region, the induced current can be drained to ground if MN2 is large, but if MN2 is small, V_0 is driven up to a positive voltage. We assume that MN2 is large enough that V_0 is held at ground potential. Therefore we use the following initial condition immediately after turn-on of the input signal.

$$V_1(+0)=0 \quad V_2(+0)=V_{DD} \quad V_0(+0)=0 \tag{57}$$

where $V_i(+0)$ $(i=1,2,3)$ are the initial node voltages *after* V_I is turned on. Until V_{GSW} of the first inverter is reached, only V_1 changes with time. V_2 stays at V_{DD} and V_0 stays at ground. Node 1 voltage $V_1(t)$ is given by

$$(C_1+C_m)\frac{dV_1(t)}{dt}=\frac{V_{DD}-V_1(t)}{R} \tag{58}$$

and this is valid if MN2 stays in the triode region which requires

$$\frac{V_{DD}}{R}\cdot\frac{C_m}{C_1+C_m}<B_2V_{DD} \quad \text{or} \quad B_2R\frac{C_1+C_m}{C_m}>1\,.$$

We assume that this condition is maintained. This is the condition required to maintain output node 0 as a *virtual* ground at the beginning of node 1 pull-up. During this initial phase Eq. (58) can be solved to obtain

$$V_1(t)=V_{DD}[1-exp[-t/R(C_1+C_m)]] \qquad V_2(t)=V_{DD} \qquad V_0(t)=0 \tag{59}$$

and this solution is valid until $t=t_1$, which is reached when $V_1(t_1)=V_{DD}/2$. Time t_1 is given by

$$t_1=R(C_1+C_m)\cdot\log 2\,.$$

At time t_1, MP1 changes from the triode region to the saturation region, and V_2 begins to fall. Until $V_2(t)$ reaches the switching threshold voltage of the second inverter, only V_1 and V_2 change with time. The circuit equations during that time are

$$(C_1+C_m)\frac{dV_1(t)}{dt}=\frac{V_{DD}-V_1(t)}{R} \quad \text{and} \quad C_2\frac{dV_2(t)}{dt}=-2B_1[V_1(t)-V\frac{DD}{2}] \tag{60}$$

and they are solved subject to the initial condition

$$V_1(t_1)=V_{DD}/2 \qquad V_2(t_1)=V_{DD}$$

as

$$V_1(t)=V_{DD}[1-exp[-t/R(C_1+C_m)]] \tag{61a}$$

which is the same as Eq. (59) and

$$V_2(t)=V_{DD}-\frac{B_1V_{DD}}{C_2}(t-t_1) \tag{61b}$$

$$-\frac{2B_1V_{DD}R(C_1+C_m)}{C_2}[exp[-t/R(C_1+C_m)]-exp[-t_1/R(C_1+C_m)]]$$

$$V_0(t)=0. \tag{61c}$$

This set of solutions is valid when $t_1<t<t_2$, where t_2 is the time at which MN2 leaves the triode region, including the current from node 1 through capacitance C_m. We require

$$B_2V_2(t)\geq B_2[V_{DD}-V_2(t)]+C_m\frac{dV_1(t)}{dt}.$$

If the term $C_mV_1'(t)$ is small, t_2 is given by

$$t_2=R(C_1+C_m)\theta(1/2,\chi)\text{ where }\chi=\frac{C_2}{4B_1R(C_1+C_m)}+\frac{1+\log 2}{2}.$$

Until $t=t_2$ is reached, node 0 remained virtual ground. At time $t=t_2$ NFET MN2 moves from the triode region to the saturation region and V_0 begins to increase. Then capacitance C_m couples the increasing voltage of node 0 to node 1, thereby accelerating the input node transition. When all the three nodes become active, the circuit equations are

$$C_1\frac{dV_1(t)}{dt}=-C_m\frac{d}{dt}[V_1(t)-V_0(t)]+\frac{V_{DD}-V_1(t)}{R} \tag{62a}$$

$$C_2\frac{dV_2(t)}{dt}=-2B_1[V_1(t)-\frac{V_{DD}}{2}] \tag{62b}$$

$$C_0\frac{dV_0(t)}{dt}=-C_m\frac{d}{dt}[V_0(t)-V_1(t)]-2B_2[V_2(t)-\frac{V_{DD}}{2}]. \tag{62c}$$

Since Eqs. (62a)–(62c) are complex, the rest of the analysis is carried out using a computer program. Guidelines for writing programs of microstate sequence analysis were given in

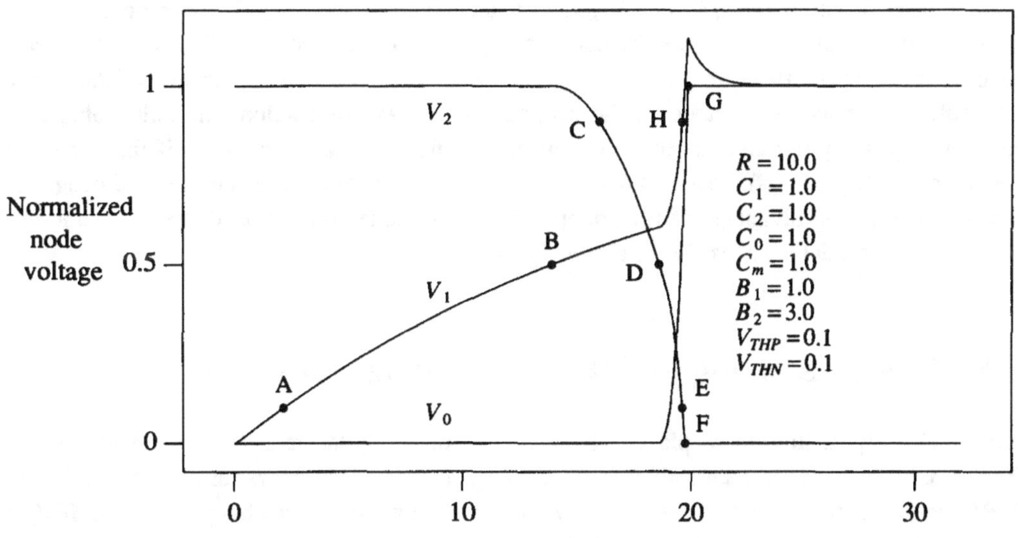

Figure 34 Positive Miller effect node waveforms

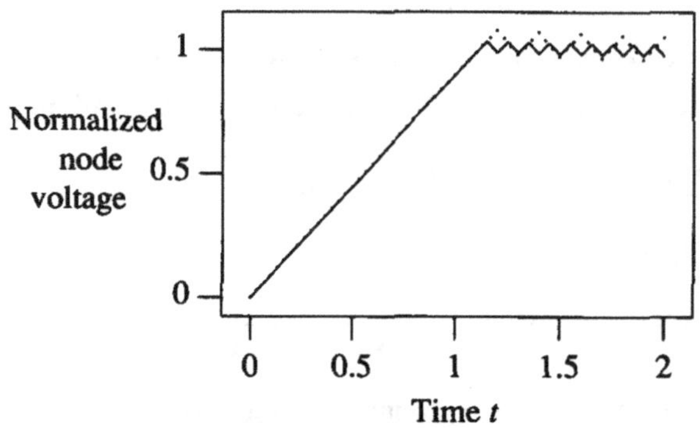

Figure 35 Instability in numerical analysis using the collapsible current generator model

Sections 1.17 and 2.13. A typical numerical result is shown in Fig. 34. In this Figure, microstate events of Fig. 33 are referred to using the same alphabetic symbols A–H. The V_1 waveform is much the same as that derived using the simple theory and shown in Fig. 32.

In numerical analysis of microstate sequences, the following problem must be handled properly. With reference to the analysis of the simplified circuit of Fig. 31(b), a straightforward computing algorithm might be

$$V_0 \to V_0 + (I_P/C_0)\Delta t$$

where the PFET current, I_P, is modeled as

$$I_P = B_P(V_{DD} - V_{THP}) \ (V_0 < V_{DD}) \quad I_P = -B_P(V_0 - V_{THP}) \ (V_0 > V_{DD})$$

since the source/drain identification of MP2 are different for $V_0 > V_{DD}$ and $V_0 < V_{DD}$. A discrete change of current, $2B_P(V_{DD} - V_{THP})$ appears if V_0 crosses V_{DD}. If the program that is written using the algorithm is run, the node voltage oscillates as shown in Fig. 35. The period of oscillation equals twice the integration time step. This is obviously an unrealistic result. The following may be included in the program to prevent oscillation. If node voltage V_0 reaches V_{DD}, the program computes the current that flows out from node 0. If the current is less than $B_P(V_{DD} - V_{THP})$, node 0 is *fused* to V_{DD}. This extra provision is often confusing, and it is obviously a disadvantage of the collapsible current generator model if used in numerical analysis. The model is primarily for closed-form analysis.

3.8 Charge Sharing—Transverse Signal Flow

Undesirable signal flow takes place even within a single digital gate [06]. An example is a NORA CMOS dynamic gate shown in Fig. 36(a). Before logic operation, the clock is low, MP1 conducts, input signals A and B are low, and output node 3 is precharged to V_{DD}. If V_2 is equal to or higher than $V_{DD} - V_{THN}$, NFET MN3 is nonconducting even if signal B makes low-to-high transition: output node 3 is isolated from the pull-down chain of the NFETs, and the

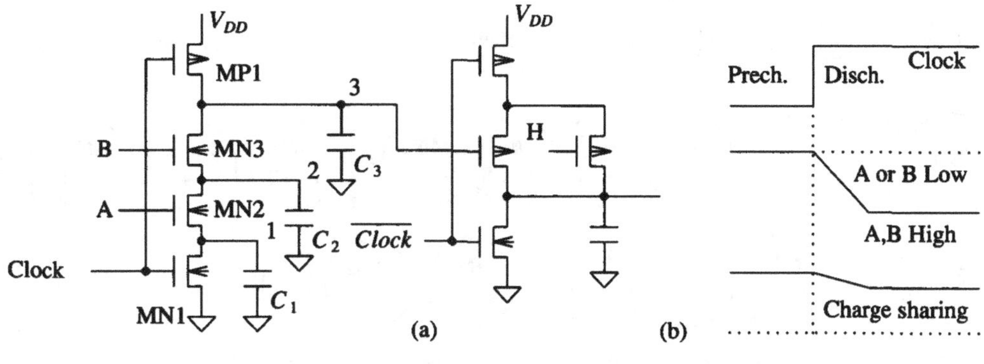

Figure 36 Dynamic NAND2 gate charge sharing

default high-logic level of node 3 remains intact. If V_2 is lower than that, however, node 3 is connected to node 2: the charge of node 3 is transferred to node 2. Voltages V_1 and V_2 depend on details of the precharge process. In which order and how fast the input signals A and B had stabilized at the low-logic level affects the precharge of internal nodes. If signal B stays high during the early precharge phase, V_2 is precharged close to $V_{DD} - V_{THN}$. If signal B is low at the beginning of precharge, V_2 can be close to ground. Intermediate cases are possible. V_2 is uncertain, and it can be significantly lower than the maximum, $V_{DD} - V_{THN}$.

At time $t=0$, clock makes a low-to-high transition, and the dynamic gate goes into the discharge (evaluate) phase. MP1 turns off, and MN1 (the ground switch) turns on. Signals A and B may make low-to-high transitions after that. Microstate sequence I of Fig. 37 shows the transition from precharge to discharge state, if both signals A and B remain at a low level. If internal nodes 1 and 2 are fully charged to $V_{DD} - V_{THN}$ and if both input signals A and B make a low-to-high transition simultaneously with the clock transition, MN2 and MN3 conduct and node 3 is pulled down. The pull-down transition can be represented by microstate sequence II of Fig. 37. This is a *skeleton* sequence (Section 1.16), which does not include possible *null* sequences that complicates the evolution of the circuit. The order in which the three NFETs MN1, MN2, and MN3 go into the triode region in this skeleton sequence depends on FET scaling and capacitances. If at least one of the signals A or B remains low, however, output node 3 should remain high, but this is not always the case. The microstate sequence of this case has three variations. If both input signals A and B remain low, microstate sequence I of Fig. 37 applies. If either B or A is low, microstate sequences shown in III and IV of Fig. 37 applies, respectively. If input signal B remains low, output node 3 is not affected (sequence III). As signal A makes a low-to-high transition, node 2 is pulled to ground by MN1 and MN2, and that is all. If input signal A remains low, and signal B makes a low-to-high transition, and if *internal* node 2 voltage at the end of the precharge phase is less than $V_{DD} - V_{THN}$, the node 3 charge is shared with node 2 as soon as signal B makes its transition, and the potential of node 3 decreases (sequence IV). We noted that voltage equalization does not take place if V_2 is equal or higher than $V_{DD} - V_{THN}$. Otherwise V_3 decreases. If V_2 is lower than

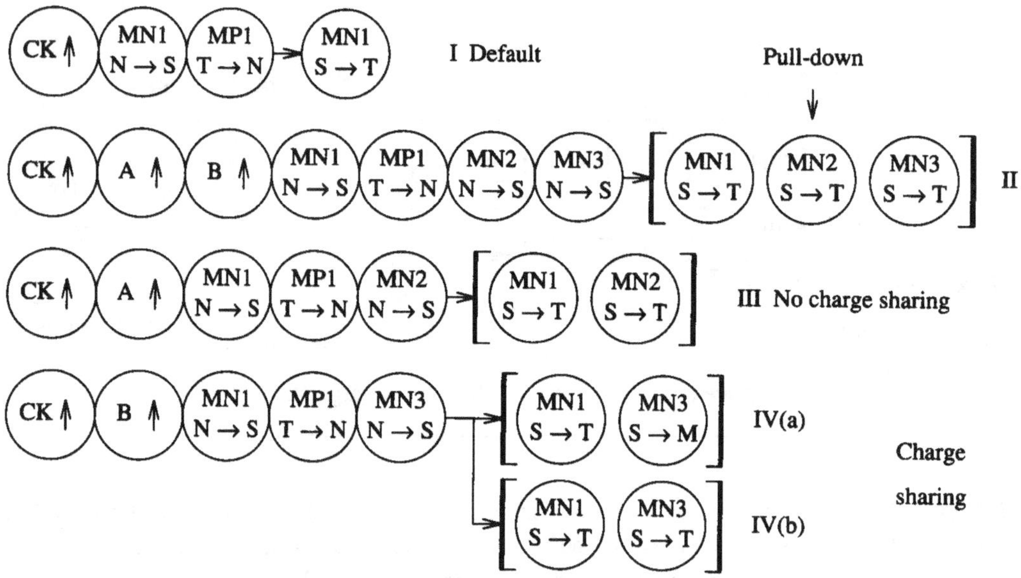

Figure 37 Dynamic NAND2 gate microstate sequence

$V_{DD} - V_{THN}$, but not significantly lower, and if the C_2 is small, the microstate sequence IV(a) of Fig. 37 shows that NFET MN3 ends up in the marginally conducting state, M, when equilibrium is reached. If capacitance C_2 is large, and the initial voltage is low, MN3 may go into the triode region, thereby equalizing V_2 and V_3, as shown by microstate sequence IV(b).

This microstate sequence shows clearly that the huge number of possible microstate sequences is the fundamental cause of the difficulty in understanding circuit failures. Circuit failure does not occur for *most* microstate sequences like I, II, and III of Fig. 37, but there are a few *unfortunate* ones like IV(a) and (b) that can lead to circuit failures. It has been difficult to consider all possibilities of circuit operation in the process of new circuit invention. To discover the existence of the rare *unfortunate* cases that may result in failures requires systematic study of all modes of circuit operation, and microstate sequences are useful for such analysis. The study of microstate sequence is supplemented by quantitative analysis and simulation of the circuit. The analysis helps to reduce the number of possible microstate sequences, and it gives a clear idea about how serious is the problem.

We begin with static analysis of charge-sharing. If the charge originally stored in capacitance C_3 of Fig. 36 is shared with capacitance C_2, there are two possible final microstates. In the first microstate, NFET MN3 is in the triode region with V_3 and V_2 equal. In the second microstate, MN3 is in the marginally conducting region, and node 3 voltage is higher than that of node 2. If V_2 was zero and V_3 was V_{DD} before charge-sharing, V_2 and V_3 reach equilibrium at voltage V_C given by $V_C = [C_3/(C_2 + C_3)]V_{DD}$. MN3 is in the triode region if

$$V_{DD} - V_C > V_{THN} \quad \text{or} \quad \frac{C_2}{C_2 + C_3} V_{DD} > V_{THN} \tag{63}$$

is satisfied. If Equation (63) is satisfied, the microstate sequence IV(b) of Fig. 37 is followed, if not, microstate sequence IV(a) is followed. The microstate sequence may terminate in several different states, depending on the design parameters and initial conditions of the circuit.

3.9 Charge Sharing

Although the mechanism of charge-sharing is intuitively obvious, accurate quantitative analysis of charge-sharing is not simple, because the many secondary effects that were neglected in the simple model of the last section must be included. In this section, the charge-sharing problem is solved in closed form by including the three most significant secondary effects. They are:

1) The effects of nonlinear (voltage-dependent) node capacitance
2) The effects of FET threshold voltage, and
3) The effects of capacitive induction from FET gates.

1) Nonlinear capacitance effects

The capacitance of a FET diffused source or drain island or a PN junction diode depends on the voltage applied to the junction. The small-signal capacitance of an abrupt PN junction is given by

$$C(V) = C_0 \left(\frac{V_{dif}}{V + V_{dif}} \right)^{1/2} \tag{64}$$

where C_0 is the zero-bias small-signal capacitance, V_{dif} is the diffusion potential, and V is the applied voltage. Charge stored in the PN junction capacitance $Q(V)$ as a function of the terminal voltage V is given by

$$Q(V) = \int_0^V C(V)dV = 2C_0\sqrt{V_{dif}}(\sqrt{V + V_{dif}} - \sqrt{V_{dif}}) \tag{65}$$

With reference to the equivalent circuit model of Fig. 38, a *voltage-independent* capacitance C_1 is charged to voltage V_0. The switch SW that models a FET is closed at time $t = 0$. After some time, potential equilibrium between the two capacitors is reached at the final voltage V_f. The FET (that was modeled by switch) is in the triode region when equilibrium is reached. This final voltage, V_f, is determined from the law of conservation of charge as

$$Q(V_f) + C_1 V_f = C_1 V_0 \quad \text{or} \quad 2C_0\sqrt{V_{dif}}(\sqrt{V_f + V_{dif}} - \sqrt{V_{dif}}) + C_1 V_f = C_1 V_0$$

which is solved as ($V_f < V_0$)

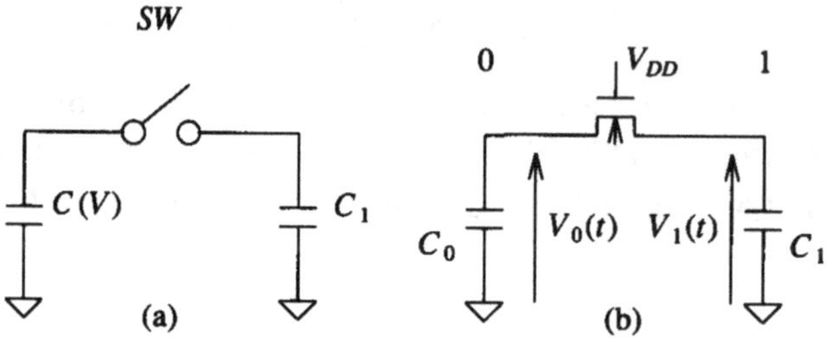

Figure 38 Charge-sharing, guide for analysis

$$V_f = V_0 - 2\left(\frac{C_0}{C_1}\right)\left[[(1+\frac{C_0}{C_1})^2 V_{dif}^{\;2} + V_0 V_{dif}]^{1/2} - (1+\frac{C_0}{C_1})V_{dif}\right]. \quad (66a)$$

In the limit of $C_0 \to \infty$

$$V_f \to \frac{C_1}{C_0 + C_1} V_0 \to 0 \quad (66b)$$

This is the case where the voltage change of the PN junction capacitor is so small that the effects of the voltage dependence is insignificant. In the limit as $V_0 \to \infty$

$$V_f \to V_0[1 - 2(C_0/C_1)(V_{dif}/V_0)^{1/2}] \quad (66c)$$

and the effects of charge-sharing decrease, although very slowly (note $1/\sqrt{V_0}$ dependence of the capacitance). This is because the PN junction capacitance decreases with increasing PN junction voltage. This analysis has the following practical significance. With reference to Fig. 36, node capacitance C_2 consists primarily of source/drain diffused island capacitance of MN2 and MN3. If C_2 was originally discharged, it takes a significant amount of charge to bring V_2 from 0 to the maximum voltage $V_{DD} - V_{THN}$. If charge-sharing is to be suppressed by increasing C_3, how much capacitance should be added must be determined including the voltage dependence of capacitance C_2.

From Eq. (66a) it is convenient to define parameter σ by

$$\sigma = \left(\frac{C_0 + C_1}{C_1}\right)\left(\frac{V_{dif}}{V_0}\right).$$

Parameter σ is a measure of the degree of charge sharing. Equation (66a) is rewritten as

$$\frac{V_f}{V_0} = f(\gamma, \sigma) \quad \text{where} \quad f(\gamma,\sigma) = 1 - 2\gamma[(\sigma^2 + \frac{1}{1+\gamma}\sigma)^{1/2} - \sigma] \quad (67)$$

and where $\gamma = C_0/C_1$. We find

$$f(\gamma,\sigma) \to \frac{1}{1+\gamma} \quad (\sigma \to \infty) \qquad f(\gamma,\sigma) \to 1 - 2\frac{\gamma\sqrt{\sigma}}{\sqrt{\gamma+1}} \quad (\sigma \to 0).$$

Figure 39 shows $(V_f/V_0) = f(\gamma, \sigma)$ for typical values of γ and σ.

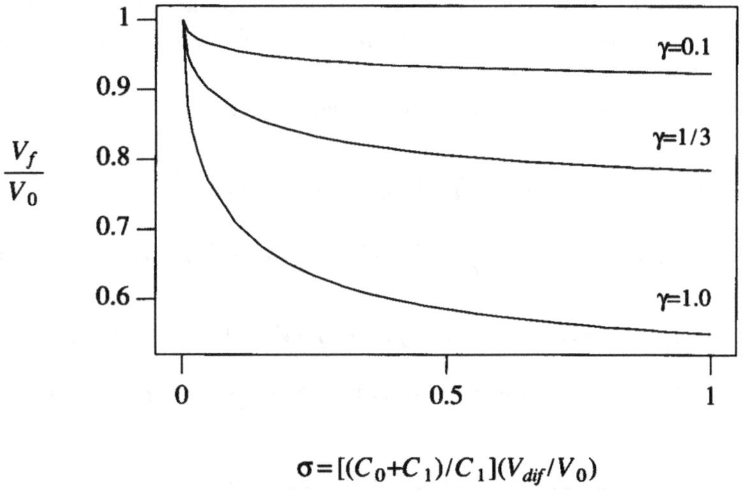

Figure 39 Effects of charge-sharing

---

2) Effects of FET threshold voltage

The switch shown in Fig. 38(a) is an oversimplified FET equivalent circuit. A FET is different from an ideal switch. In Fig. 38(b), capacitance C_1 is originally charged to $V_1(0)=V_{DD}$, and capacitance C_0 is completely discharged [or $V_0(0)$ is zero] and the NFET was turned off. At time $t=0$ the NFET gate is driven to V_{DD}, thereby turning the NFET on. Current flows through the NFET, and capacitance C_0 is charged by current from capacitance C_1.

The circuit equations are

$$C_1 \frac{dV_1(t)}{dt} = -I_N[V_0(t), V_1(t)] \quad C_0 \frac{dV_0(t)}{dt} = I_N[V_0(t), V_1(t)] \tag{68}$$

where node voltages $V_0(t)$ and $V_1(t)$ satisfy,

$$C_1V_1(t) + C_0V_0(t) = C_1V_{DD}$$

or the sum of the charges of the two capacitors is independent of time. We assume that $V_1(0)=V_{DD}$ and solve the problem subject to this initial condition. Both capacitances are assumed to be voltage-independent. I_N is the NFET current given as a function of voltages $V_0(t)$ and $V_1(t)$. If the collapsible current generator model is used,

$$I_N = 0 \text{ if } V_0 > V_{DD} - V_{THN}(V_0) \quad = B_N[V_{DD} - V_0 - V_{THN}(V_0)] \text{ otherwise}$$

where V_{THN} is a function of source voltage V_0 (back-bias effect). The back-bias effect makes the problem complex. We neglect this effect for simplicity and assume that

$$V_{THN}(V_0) = const\ (= V_{THN}).$$

Then we obtain the solution as

$$V_0(t) = (V_{DD} - V_{THN})[1 - exp[-(B_N/C_0)t]] \tag{69}$$

Node 0 voltage exponentially approaches $V_{DD} - V_{THN}$. Note that this solution assumes that charge-sharing is not extensive, and the two nodes never fuse. Condition $C_0 V_{DD} / (C_0 + C_1) < V_{TH}$ is required. The case where the condition is not satisfied and the nodes fuse is left for the reader.

Let us solve the same problem using the gradual channel FET model as well. I_N is given by

$$I_N = 0 \text{ if } V_{DD} - V_0 < V_{THN}(V_0) \quad (\textit{Nonconducting})$$

$$I_N = B_{GN}[V_{DD} - V_0 - V_{THN}(V_0) - \frac{1}{2}(V_1 - V_0)](V_1 - V_0) \text{ if } V_{DD} - V_{THN}(V_0) > V_1 \quad (\textit{Triode})$$

$$I_N = \frac{B_{GN}}{2}[V_{DD} - V_0 - V_{THN}(V_0)]^2 \text{ if } V_{DD} - V_{THN}(V_0) < V_1 \quad (\textit{Saturation}).$$

We assume $V_{THN}(V_0) = const = V_{THN}$. Immediately after time $t=0$, the NFET is in the saturation region, and therefore,

$$C_0 \frac{dV_0(t)}{dt} = \frac{B_{GN}}{2}[V_{DD} - V_0(t) - V_{THN}]^2$$

This equation is solved by separating the variables as

$$V_0(t) = (V_{DD} - V_{THN})\frac{t}{t + t_0} \quad \text{where} \quad t_0 = \frac{2C_0}{B_{GN}(V_{DD} - V_{THN})}. \tag{70}$$

Since $V_1(t) = V_{DD} - (C_0/C_1)V_0(t)$, the NFET is in the saturation region until $(C_0/C_1)V_0(t) > V_{THN}$. By solving this inequality for $t = t_S$ we obtain

$$t_S = \frac{C_1 V_{THN}}{C_0(V_{DD} - V_{THN}) - C_1 V_{THN}} t_0.$$

If $t > t_S$ the NFET is in the triode region. The circuit equation becomes

$$C_0 \frac{dV_0}{dt} = B_{GN}[V_{DD} - V_0 - V_{THN} - \frac{1}{2}(V_1 - V_0)](V_1 - V_0). \tag{71}$$

The initial condition is $V_0(t_S) = (C_1/C_0)V_{THN}$ and from Eq. (66) $V_1(t) - V_0(t) = V_{DD} - [1 + (C_0/C_1)]V_0(t)$. We then have

$$\frac{dV_0}{(V_A - V_0)(V_B - V_0)} = \frac{B_{GN}}{2C_0}\left(1 - \frac{C_0^2}{C_1^2}\right)dt \tag{72}$$

where

$$V_A = \frac{C_1}{C_0 + C_1}V_{DD} \quad \text{and} \quad V_B = \frac{C_1}{C_1 - C_0}(V_{DD} - 2V_{THN})$$

and the solution satisfying the initial condition is

$$V_0(t) = \frac{V_B - V_A f \exp[(t - t_S)/t_e]}{1 - f \exp[(t - t_S)/t_e]} \tag{73}$$

where

$$f = \frac{V_B - (C_1/C_0)V_{THN}}{V_A - (C_1/C_0)V_{THN}} \quad \text{and} \quad \frac{1}{t_e} = \frac{B_{GN}(V_B - V_A)}{2C_0}\left(1 - \frac{C_0^2}{C_1^2}\right).$$

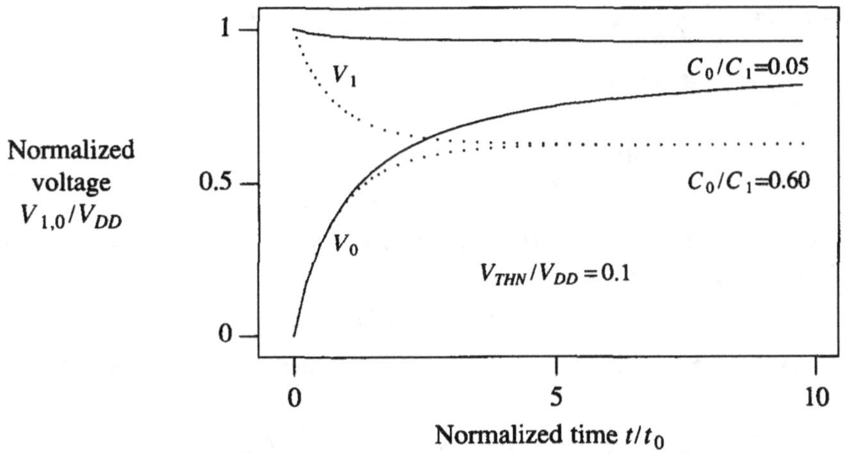

Figure 40 Charge-sharing

Figure 40 shows node voltage versus time. If $C_0(V_{DD}-V_{TH})<C_1V_{TH}$, the NFET stays in saturation. Then the capacitor voltages never equalize: the NFET turns off before the source voltage and the drain voltage equalize (solid curves, $C_0/C_1=0.05$). Otherwise the capacitor voltages equalize: first the NFET moves into the triode region, and it becomes equivalent to a linear resistor (dotted curves). In Fig. 40, $V_{THN}/V_{DD}=0.1$.

3) Capacitive coupling from FET gate

Since the FET gate is capacitively coupled to the source and to the drain, a gate voltage change induces a voltage change in the source and to the drain nodes. The capacitive coupling mechanism must be included in the quantitative analysis of charge sharing. Figure 41 shows the equivalent circuit of charge-sharing, including the gate capacitance coupling. Capacitance C_0 is completely discharged, and capacitance C_1 is charged to V_{DD} before the charge is shared with C_0. The gate voltage makes a step increase from 0 to V_{DD} at time $t=0$. Immediately after the transition, node voltages are

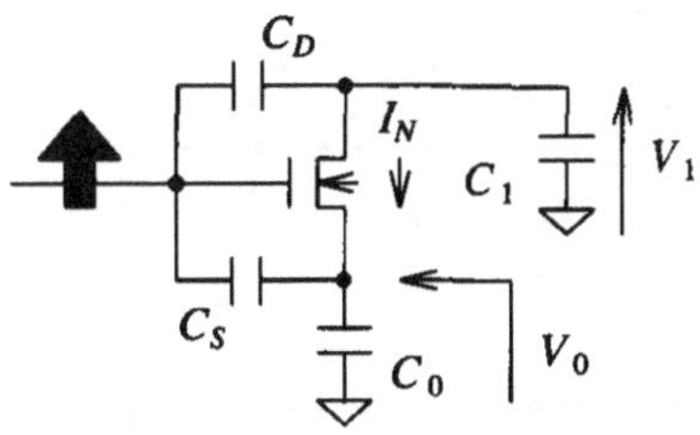

Figure 41 Charge-sharing analysis

$$V_0(+0) = \Delta V_0 = \frac{C_S}{C_0 + C_S} V_{DD} \quad V_1(+0) = V_{DD} + \Delta V_1 = V_{DD} + \frac{C_D}{C_1 + C_D} V_{DD} \tag{74}$$

where $V_i(+0)$ $(i = 0, 1)$ are the initial node voltages *after* the voltage step. The circuit equations are

$$(C_0 + C_S)\frac{dV_0(t)}{dt} = I_N \quad (C_1 + C_D)\frac{dV_1(t)}{dt} = -I_N\,. \tag{75}$$

We use the collapsible current generator model and we have $I_N = B_N[V_{DD} - V_{THN} - V_0(t)]$. An integral of Eq. (75) satisfying the initial condition is

$$(C_0 + C_S)V_0(t) + (C_1 + C_D)V_1(t) = (C_1 + 2C_D + C_S)V_{DD}$$

Using this special solution, the solution of Eq. (75) that satisfies the initial condition is found as

$$V_0(t) = (V_{DD} - V_{THN}) - (V_{DD} - V_{THN} - \Delta V_0)exp(-t/t_0) \tag{76a}$$

$$V_1(t) = V_{DD} - \Delta V_A + (\Delta V_A + \Delta V_1)exp(-t/t_0) \tag{76b}$$

where

$$\Delta V_A = -\frac{(C_S + C_D)V_{DD} - (C_0 + C_S)V_{THN}}{C_1 + C_D} \quad \text{and} \quad t_0 = \frac{C_0 + C_S}{B_N}\,.$$

Node waveforms $V_0(t)$ and $V_1(t)$ for the case where the two nodes fuse are plotted in Fig. 42. $V_0(t)$ increases and $V_1(t)$ decreases with time, and $V_0(t)$ equals $V_1(t)$ at $t = t_1$. Time t_1 is found by solving

$$V_0(t_1) = V_1(t_1)\,. \tag{77}$$

At time $t = t_1$ NFET goes into the triode region and $V_0(t)$ and $V_1(t)$ remain unchanged after that. We have

$$V_0(t) = V_1(t) = V_0(t_1) \quad t > t_1\,.$$

If Eq. (77) has no solution, the FET ends up in the marginally conducting state. Node voltage $V_0(t)$ tends to $V_{DD} - V_{THN}$ and $V_1(t)$ somewhat higher than that, as $t \to \infty$.

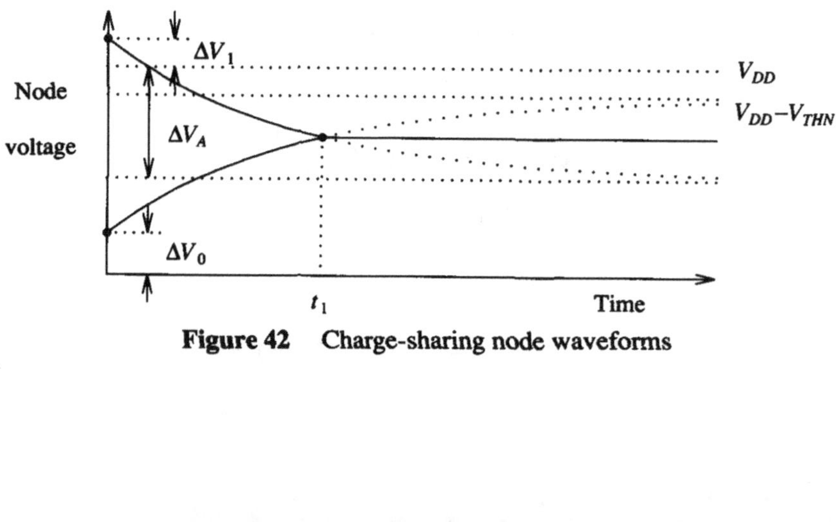

Figure 42 Charge-sharing node waveforms

3.10 Prevention of Charge-sharing

We summarize techniques used to prevent charge-sharing. Since charge-sharing takes place predominantly in CMOS dynamic logic circuits, use of CMOS static logic circuit resolves many problems. If CMOS dynamic circuits must be used the following techniques are available.

1) Quick fix of charge-sharing problem

If charge-sharing failures occur, it is desirable to suppress the problem temporarily using the minimum photomask change (the metal level mask only). If this is possible, chip diagnosis proceeds to the next step. The simplest cure is to add extra capacitance to the gate's output node. If the output node capacitance holds enough charge, effects of the charge-sharing can be reduced to the point that the circuit works normally. Capacitance can be made between the metal level and silicon substrate. Unfortunately capacitance between metal and silicon substrate is not high because of thick composite oxide. Capacitance is typically 0.3 pF for a 100 microns square patch (about the size of a bonding pad). Therefore significant area is required to fabricate the required capacitor. Whether or not this is feasible depends on the structure of the layout: metal-level wiring must be modified to make space for the capacitor in the vicinity of the gate output node, or a connecting wire must be routed to an open area where the capacitor is placed. According to the author's own experience, layout changes such as this are usually feasible with determined effort, and the temporary fix creates a working chip. How much capacitance should be added depends on the loading of the node. If the charge-shared dynamic gate is buffered by a static inverter, doubling the load capacitance reduces the effect of charge-sharing quite significantly. A temporary fix by adding capacitance must be verified by using circuit simulation including voltage dependence of the capacitance.

2) Internal node precharge

The most reliable method of preventing charge-sharing is to precharge the internal nodes of the dynamic gates, as shown in Fig. 43(a) [11]. A small PFET, MP1, is used to precharge parasitic capacitance C_X to V_{DD}. The clock of MP1 is the same as the clock used to control the dynamic gate. This technique is reliable, since internal node X is precharged to V_{DD}, which is higher than the minimum voltage, $V_{DD} - V_{TH}$, necessary to prevent charge-sharing. The internal node can be precharged using a NFET source-follower circuit. This scheme has the advantage of not overcharging the internal node, but the scheme requires a complementary clock, and the downgoing clockedge terminating the internal node precharge removes some charge from the node. The problem of these two methods is that the internal nodes get additional capacitive loading from the pull-up FET.

A gate's internal node precharge may be carried out by turning on the NFETs of Domino CMOS dynamic gates while gate precharge is in progress. The circuit shown in Fig. 43(b) works as follows. During precharge (CK is low), the internal node precharge signal, INP, is pulled down for a short time. NFET MNX conducts, and node X capacitance, C_X, is

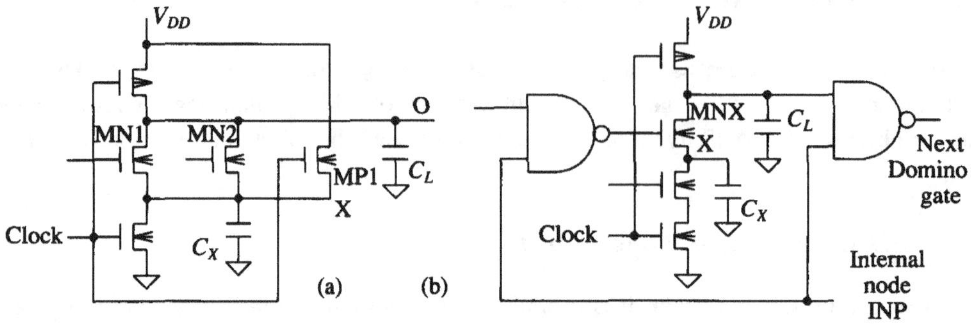

Figure 43 Internal node precharge

precharged to $(V_{DD} - V_{THN})$. Pull-down delay of the gate does not suffer from internal node precharge, because precharge is just enough to prevent charge-sharing. The pull-up transition of the NAND gate is fast, and therefore increase in delay time is small. It is unnecessary to turn on all the NFETs: in a multi-input NOR gate, for example, only one NFET need be turned on. Proper choice of the NFET prevents extra delay in the critical path. The design problem of this circuit is that effects of capacitive coupling from the NFET [MNX in Fig. 43(b)] gate that is turned on and off must be taken into consideration. Because of this capacitive coupling, V_X is reduced below $V_{DD} - V_{THN}$ after the INP pulse goes high. The effect is significant if C_X is small. There is another version of Domino CMOS that is resistant to charge-sharing. In this scheme, the ground switch is turned off completely (precharge phase) and turned on completely (discharge phase). The precharge PFET, however, is turned on completely for precharge, but turned off only to a low-conducting state during the discharge phase, and the PFET supplies a small current to the output node. This circuit is called *Zipper CMOS*.

3) *Weak* pull-up of the output node

Yet another version of Domino CMOS is as follows. In the clocked dynamic inverter circuit shown in Fig. 44(a), a small PFET, MP1, that is always conducting holds the output node at V_{DD} in the default precharged state, and it prevents charge-sharing or noise induction. Suppose that input signal A is low, and the inverter is clocked into the discharge state. An upgoing noise spike V_N arrives at terminal A, MNA conducts, and load capacitor C_L loses charge. This lost charge is compensated by current from MP1. A *passive* load like MP1 does not allow pull-down of the output node all the way to ground. Therefore sensitivity to ground-bus noise is compromised. There is a complementary relationship between the tolerance limit of high-logic-level noise and the tolerance limit of low-logic level noise. Suppose that input A is at a low-logic level and a small noise voltage V_N arrives. We assume that ground switch MN0 is so large that it presents negligible series resistance. If V_N is small

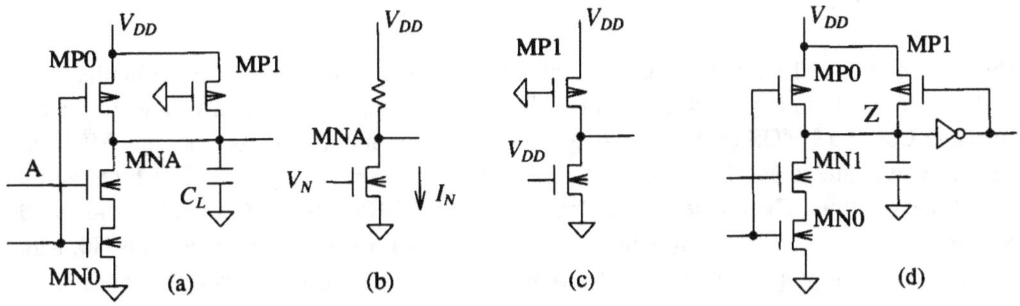

Figure 44 Preventing charge-sharing

[equivalent circuit of Fig. 44(b)], MNA is in saturation, and MP1 is in the triode region. We use the long-channel, low-field FET model (since the collapsible current generator model does not give required accuracy for this problem). MP1 can be approximated by a linear resistor $[B_{GP1}(V_{DD}-V_{THP})]^{-1}$. Then the maximum voltage drop of the output node due to noise V_N, ΔV_H, is given by

$$\Delta V_H = \frac{1}{B_{GP1}(V_{DD}-V_{THP})}\frac{B_{GNA}}{2}(V_N - V_{THN})^2 \approx \frac{1}{2}\left(\frac{B_{GNA}}{B_{GP1}}\right)\frac{(V_N - V_{THN})^2}{V_{DD}} \tag{78}$$

where B_{GP1} is beta of MP1 and B_{GNA} is that of MNA.

Let the voltage of the output node in the low state be V_L [equivalent circuit of Fig. 44(c)]. MP1 is in saturation, and MNA and MN0 are both in the triode region. V_L is determined from the equation

$$\frac{B_{GP1}}{2}(V_{DD}-V_{THP})^2 = B_{GNA}(V_{DD}-V_{THN}-\frac{1}{2}V_L)V_L \tag{79}$$

where the ground switch MN0 is so large that its series resistance is negligible. Solving Eq. (79) for V_L assuming that (B_{GP1}/B_{GNA}) is small

$$V_L = \frac{1}{2}\left(\frac{B_{GP1}}{B_{GNA}}\right)\frac{(V_{DD}-V_{THP})^2}{V_{DD}-V_{THN}} \approx \frac{1}{2}\left(\frac{B_{GP1}}{B_{GNA}}\right)V_{DD}$$

and therefore from Eq. (78) and (79)

$$V_L \Delta V_H = \frac{1}{4}(V_N - V_{THN})^2 . \tag{80}$$

Equation (80) shows that there is a complementary relationship between the reductions of margins of logic 0 and logic 1 voltages that is independent of the FET sizes. The circuit of Fig. 44(a) does not improve the gate's noise immunity, but the margins for the two logic levels are balanced and optimized.

4) Active load control

The weak output pull-up that was discussed in the last paragraph is extended in circuits shown in Fig. 44(d) [11]. In Fig. 44(d) a small PFET, MP1, is gated by the output voltage of the non-inverting Domino CMOS gate. If the output from the Domino CMOS gate is high, MP1 turns off. If input signal to MN1 is low, the NFET chain is nonconducting, node Z is high, and the output node is low. Then node Z is pulled up by PFET MP1. Node Z is held at V_{DD} by the conducting PFET, MP1. The problem of this circuit is that the size ratio of FETs MN0, MN1, and MP1 must be properly chosen: otherwise the circuit fails to pull down. Furthermore the positive feedback applied through the inverter and MP1 resists the pull-down of node Z during the early phase of the transient, and that increases the delay.

3.11 Response of Node to Injected Current

In this section we consider the responses of digital circuit nodes to small-signal injected currents. We consider this problem using both the collapsible current generator and the linear resistor FET models. There are three problems. (1) The node into which current is injected may change configuration due to the injected current; (2) the node accumulates the effects of the injected currents, both of which are considered using the collapsible current generator model; and (3) the node does not change configuration by current injection, and the node's impedance level is low. This last case is considered using the linear resistor model, since the collapsible current generator model is not adequate.

3.11.1 Node Configurational Change

Figure 45(a) shows the pull-down NFET chain of a NAND2 gate, consisting of NFETs MNA and MNB. Gates of MNA and MNB are biased to voltages V_{GA} and V_{GB}, respectively. We determine how nodes A and B respond to injected current I_0. This analysis is relevant to digital nodes that are biased close to the critical point of changing configuration. If $I_0 = 0$, nodes A and B are at ground potential, and this is the initial condition at time $t = 0$. For time $t > 0$, current I_0 is turned on and is independent of time. We define

$$I_{Amax} = B_A(V_{GA} - V_{THN}) \quad I_{Bmax} = B_B(V_{GB} - V_{THN}) \tag{81}$$

where I_{Amax} and I_{Bmax} are the maximum currents of MNA and MNB, respectively, if their sources are grounded. Using the parameter values I_{Amax} and I_{Bmax}, the problem can be classified into the following three cases, with reference to Fig. 46.

A) $I_{Amax} > I_0$ and $I_{Bmax} > I_0$. Both nodes A and B remain at ground potential. MNA and MNB are conductive enough that current I_0 is drained to ground without developing voltage [region 3 and 4 of Fig. 46].

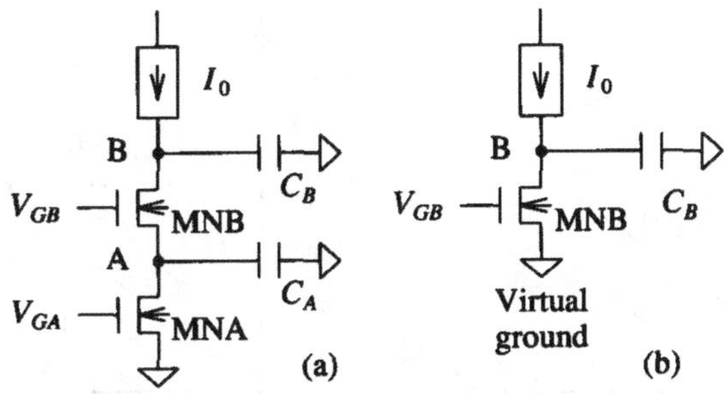

Figure 45 Analysis of response of digital node to injected current

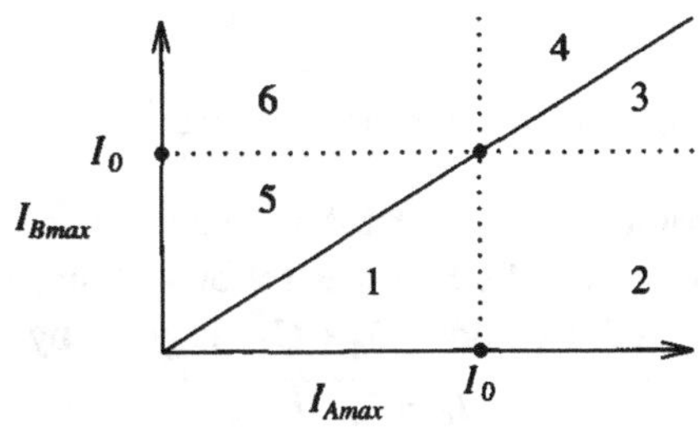

Figure 46 Classifying the node response problem

B) $I_{Amax} > I_{Bmax}$. We may assume that $I_{Bmax} < I_0$ since otherwise case (A) applies [region 1 and 2 of Fig. 46]. MNB carries the maximum current I_{Bmax}, while node A is held at ground potential. Since NFET MNA is able to drain current I_{Bmax} to ground without developing drain-source voltage, node A is a *virtual* ground. We have

$$V_A(t)=0\,. \tag{82}$$

The equivalent circuit of Fig. 45(a) changes to that of Fig. 45(b). Node B voltage increases with time as

$$V_B(t)=\frac{1}{C_B}(I_0-I_{Bmax})t\,. \tag{83}$$

Equation (83) holds until current generator I_0 collapses at some voltage (since we consider that I_0 is generated by a FET or similar signal source). The node waveforms are shown in Fig. 47(a).

C) $I_{Amax} < I_{Bmax}$. We may assume that $I_{Amax} < I_0$ [case 5 and 6 of Fig. 46]. Immediately after $t=0$ MNB carries more current than MNA. Therefore MNA goes into the saturation region immediately, but MNB remains in the triode region. Node A and B are fused, and their voltages change by

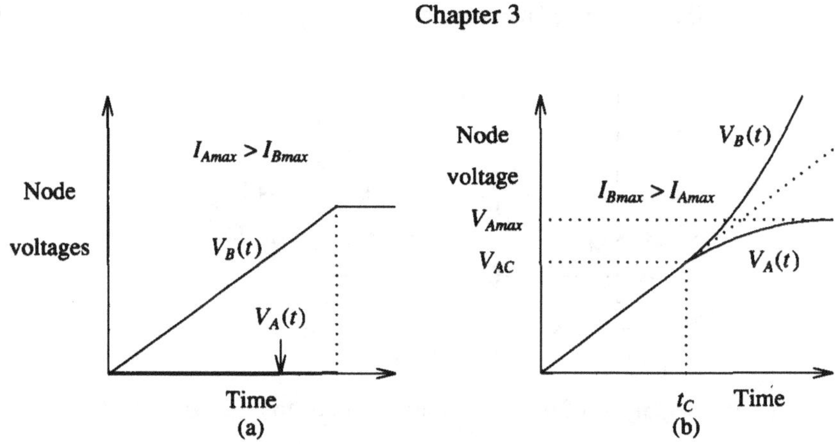

Figure 47 Response of digital nodes

$$V_A(t) = V_B(t) = \frac{I_0 - I_{Amax}}{C_A + C_B}\, t\,. \tag{84}$$

Since V_A increases, the maximum current of MNB decreases and at some time t_C, MNB goes into saturation as well. Time t_C can be found as follows. Since nodes A and B are fused, I_C charging the capacitance of the fused node $(C_A + C_B)$ is given by

$$I_C = I_0 - I_{Amax}$$

and capacitors C_A and C_B split the current I_C by the ratio of the capacitance values. The I_{CA} is given by

$$I_{CA} = \frac{C_A}{C_A + C_B}\, I_C\,.$$

Therefore I_B is $I_{CA} + I_{Amax}$. This current must be less than the maximum current allowed for MNB. Therefore the condition for nodes A and B to remain fused is

$$I_{CA} + I_{Amax} = \frac{C_A I_0 + C_B I_{Amax}}{C_A + C_B} \le I_{Bmax} - B_B V_A(t_C)$$

(we note that this expression is the same as the general equation derived in Section 2.14). We have

$$t_C = \frac{(C_A + C_B) I_{Bmax} - C_B I_{Amax} - C_A I_0}{B_B (I_0 - I_{Amax})}\,. \tag{85}$$

For $t > t_C$ the nodes A and B are separate nodes. Then the circuit equations change to

$$C_A \frac{dV_A(t)}{dt} + I_{Amax} = I_{Bmax} - B_B V_A(t) \quad C_B \frac{dV_B(t)}{dt} = I_0 - I_{Bmax} + B_B V_A(t)\,. \tag{86}$$

Equation (86) is written in the form

$$\frac{dV_A(t)}{dt} = \frac{B_B}{C_A}\,[V_{Amax} - V_A(t)]$$

where $V_{Amax}=(I_{Bmax}-I_{Amax})/B_B$. This equation is solved subject to the initial condition at $t=t_C$ given by

$$V_A(t_C)=V_B(t_C)=\frac{(C_A+C_B)I_{Bmax}-C_B I_{Amax}-C_A I_0}{B_B(C_A+C_B)}$$

as

$$V_A(t)=V_{Amax}-\frac{C_A}{C_A+C_B}\frac{\Delta I}{B_B}exp[-(B_B/C_A)(t-t_C)] \tag{87a}$$

$$V_B(t)=\frac{\Delta I}{C_B}(t-t_C) \tag{87b}$$

$$-\frac{C_A}{C_B}\frac{C_A}{C_A+C_B}\frac{\Delta I}{B_B}[1-exp[-(B_B/C_A)(t-t_C)]]+V_{Amax}-\frac{C_A}{C_A+C_B}\frac{\Delta I}{B_B}$$

where $\Delta I=I_0-I_{Amax}$. Equation (87a) and (87b) are plotted schematically in Fig. 47(b).

3.11.2 Linear Node Response

Next we analyze the problem of small-signal node response using the linear equivalent resistor FET model. This analysis is relevant to small current injection to digital nodes that have settled to the steady state. Figure 48(a) shows the equivalent circuit of a typical CMOS static node in the steady state. $V(t)$ is perturbed by an external noise voltage, $V_N(t)$. The node is connected either to V_{DD} by a PFET or to V_{SS} by an NFET, and the node is originally in steady state. In this equivalent circuit, C_m is the coupling capacitance, C_L is the load capacitance of the node excluding C_m, and R_D is the equivalent resistance of the FET [$R_D=1/B_G(V_{DD}-V_{THN})$ in the gradual channel, low-field FET model]. This equivalent circuit can also be used for a dynamic CMOS gate in the pull-down state. If the dynamic gate is in the default pull-up state, however, the output node is floating and this equivalent circuit does not apply. Then resistance R_D is missing (or very high).

The circuit equation satisfied by node voltage $V(t)$ is given by

$$(C_L+C_m)\frac{dV(t)}{dt}+\frac{V(t)}{R_D}=C_m\frac{dV_N(t)}{dt}. \tag{88}$$

This equation can be solved by the standard method to obtain

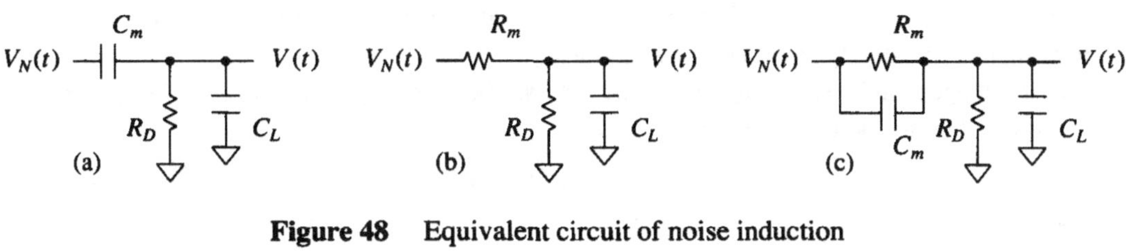

Figure 48 Equivalent circuit of noise induction

$$V(t) = exp[-t/R_D(C_L + C_m)] \tag{89}$$
$$\times [V(0) + \frac{C_m}{C_L + C_m} \int_0^t exp[\theta/R_D(C_L + C_m)] \frac{dV_N(\theta)}{d\theta} d\theta]$$

This equation can be partially integrated as

$$V(t) = V_A(t) + V_B(t) + V_C(t) \tag{90}$$

where

$$V_A(t) = [V(0) - \frac{C_m}{C_L + C_m} V_N(0)]exp[-t/R_D(C_L + C_m)] \qquad V_B(t) = \frac{C_m}{C_L + C_m} V_N(t)$$

$$V_C(t) = -\frac{C_m}{C_L + C_m} \frac{1}{R_D(C_L + C_m)} \int_0^t exp[-\xi/R_D(C_L + C_m)] V_N(t - \xi) d\xi .$$

This equation can be interpreted as follows: the first term represents discharge of the node with time constant $R_D(C_L + C_m)$ to the *initial* node voltage. The second term is the *instantaneous* capacitive coupling effect from the external node, with $C_m/(C_L + C_m)$ as the capacitive voltage division ratio. The integrand of the third term is the effect of capacitive coupling at time ξ before, which has been attenuated with time constant $R_D(C_L + C_m)$: the integral is the sum over all past contributions.

The response of a CMOS dynamic gate node that is high impedance, the floating node condition, is obtained by taking the limit $R_D \to \infty$. This gives

$$V(t) \to V(0) + \left(\frac{C_m}{C_L + C_m}\right)[V_N(t) - V_N(0)] . \tag{91}$$

A comparison of Eq. (90) and Eq. (91) shows that the effects of small-amplitude capacitive coupling is a *transient* for static nodes, but is a direct effect in floating nodes. Therefore floating nodes are more noise-sensitive. The difference in response between static nodes and floating nodes reflects the organization of this book. Floating nodes are considered in the group of abnormal signal flow, which is primarily discussed in this chapter. Problems of static nodes are considered as noise and are discussed in chapter 4. This classification is based on the viewpoint that the effects on floating nodes are permanent, and they are therefore a *digital signal* arrived at the node that has permanently changed the state. Another difference between abnormal signal flow and noise is that in the former the source of signal and its destination are within the same circuit (same chain of logic gates), and their time evolution can be determined from a single set of circuit equations, even though the equations may be complex. The analysis that assumes small signal allowed separate consideration of V_N and V. In general they are determined consistently from a set of equations.

If voltage is induced through resistance R_m as shown by the circuit of Fig. 48(b), the voltage induced on the node is given by

$$V(t) = V_A(t) + V_B(t)$$

where

$$V_A(t)=V(0)exp[-\frac{R_m+R_D}{C_LR_mR_D}t] \quad V_B(t)=\frac{1}{C_LR_m}\int_0^t \exp[-\frac{R_m+R_D}{C_LR_mR_D}\xi]V_N(t-\xi)d\xi$$

and if the voltage is induced through a parallel combination of resistance R_m and capacitance C_m as shown by the circuit of Fig. 48(c)

$$V(t)=V_A(t)+V_B(t)+V_C(t)$$

where

$$V_A(t)=[V(0)-\frac{C_m}{C_L+C_m}V_N(0)]exp[-\frac{(R_m+R_D)}{(C_L+C_m)R_mR_D}t] \quad V_B(t)=\frac{C_m}{C_L+C_m}V_N(t)$$

and

$$V_C(t)=-\frac{1}{(C_L+C_m)R_m}[\frac{C_m(R_m+R_D)}{(C_L+C_m)R_D}-1]\int_0^t exp[-\frac{R_m+R_D}{(C_L+C_m)R_mR_D}\xi]V_N(t-\xi)d\xi .$$

Figure 49 shows plots of the individual waveforms for the three cases, where the solid curves show a *scaled* source voltage waveform, $V_N(t)/10$. In Fig. 49(a), the capacitively induced voltage is plotted by open circles, and resistance induced voltage is plotted by closed circles. Resistance induced noise voltage is essentially an integration of the source waveform over time, since realistically $R_m \gg R_D$ can be assumed; R_m and C_L work as an integrator. Capacitively induced noise voltage is somewhere between the noise waveform itself and its time derivative: If R_D is large, C_m and C_L work as a capacitive voltage divider and if R_D is small, C_m and R_D work as a differentiator. Figure 49(b) shows several cases of operation of resistive-capacitive coupling circuit. If $R_mC_m = R_DC_L$, an attenuated, but exact replica of the source waveform is generated at the node. If R_mC_m is smaller than R_DC_L, the node waveform is close to time integration of the source waveform, since R_m is still much larger than R_D, and C_m becomes negligible; R_m and C_L work as integrator. If R_mC_m is larger than R_DC_L, the node waveform is more like a derivative of the source waveform, since C_m and the small R_D work as differentiator.

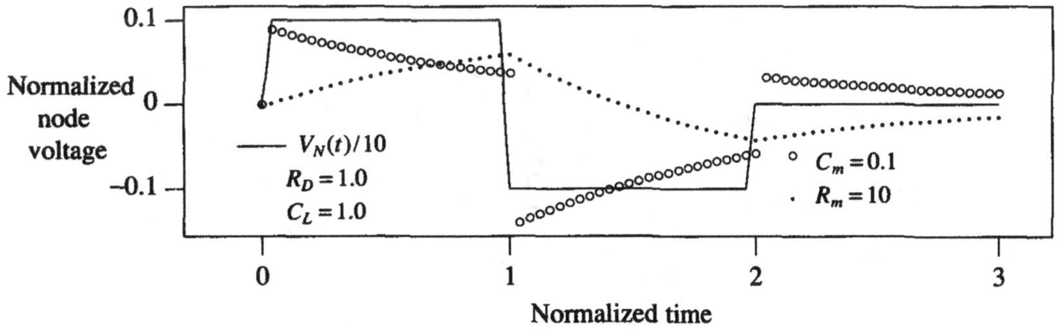

Figure 49(a) Node waveforms of induced noise

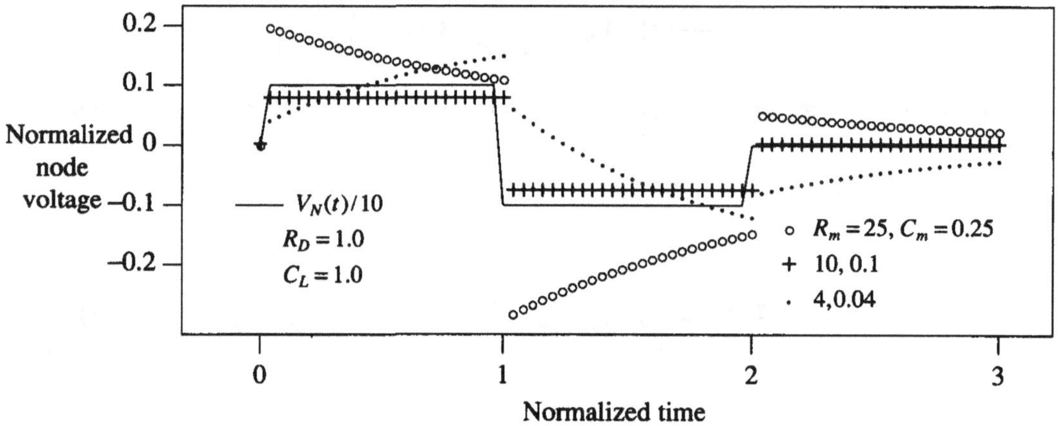

Figure 49(b) Node waveforms of induced noise

3.11.3 Accumulation of the Effects

Anomalous currents flow into a digital circuit node, and the currents generate spurious node voltages. The spurious voltages control the currents flowing into the second node that has the capability of accumulating small currents for a long time. The structure of the second node can be static as shown in Fig. 50(a), dynamic as shown in Fig. 50(b), or a hybrid of the two as shown in Fig. 50(c). Since circuit failure takes place at the second node, where the small effects accumulate, we study the voltage waveforms of the second node for the given first node waveform, $V(t)$. The response of the static node to time-dependent input voltage is determined by generalizing the theory of Section 2.8: The integral of the circuit equation is

$$V_0(t) = V_{DD} - \frac{B_N + B_P}{C_L} \int_{t_0}^{t} [V(\theta) - V_{GSW}]d\theta \quad (t \geq t_0)$$

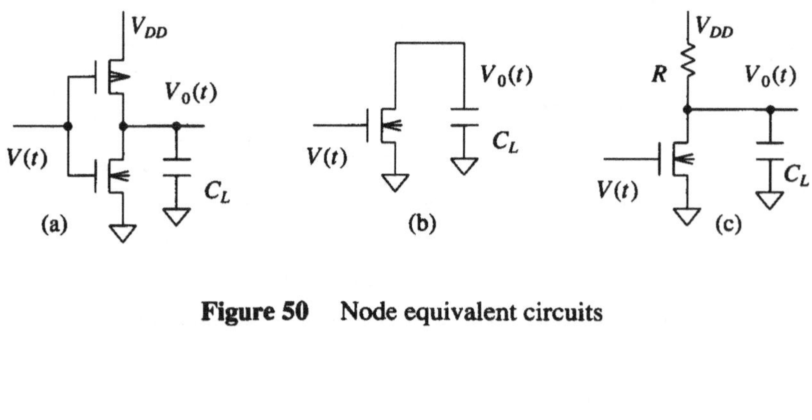

Figure 50 Node equivalent circuits

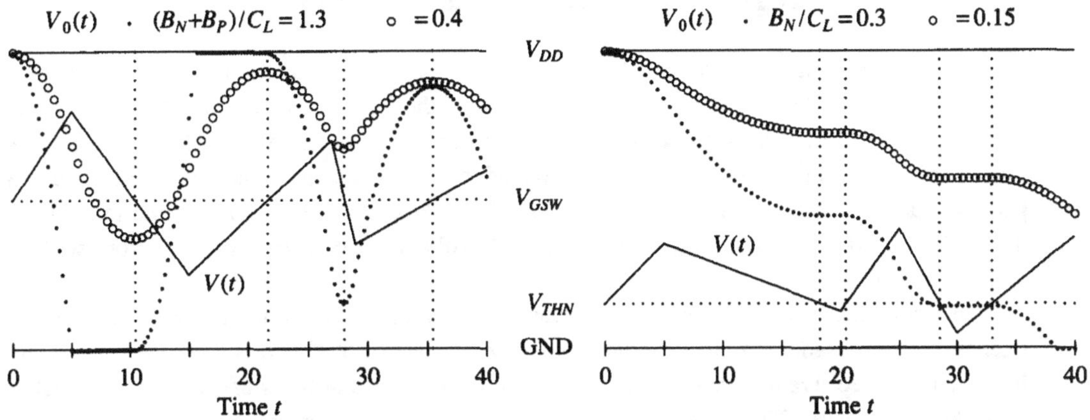

(a) Static inverter (b) Dynamic inverter

Figure 51 Static and dynamic node responses

where $V_{GSW}=(B_PV_{DD}-B_PV_{THP}+B_NV_{THN})/(B_N+B_P)$ is the switching threshold voltage of the inverter of Fig. 50(a). The output node of the inverter was originally at V_{DD}, and the integral is taken from time t_0. At time $t=t_0$, the integrand $V(t)-V_{GSW}$ becomes positive for the first time. If $V_0(t)$ becomes zero or V_{DD} later, $V_0(t)$ is clamped at that steady-state voltage. If $V_0(t)$ is clamped at zero $V_0(t)$ continues to develop later as follows: after the next moment t_1 when $V(t)-V_{GSW}$ becomes negative, $V_0(t)$ starts increasing from zero by a new integral form

$$V_0(t)=-\frac{B_N+B_P}{C_L}\int_{t_1}^{t}[V(\theta)-V_{GSW}]d\theta \quad (t\geq t_1).$$

A similar relation holds if $V_0(t)$ is clamped at V_{DD}. Figure 51(a) shows the response of a static node to $V(t)$ shown by the solid curve [closed circles for large, and open circles for small $(B_N+B_P)/C_L$].

The response of the dynamic node shown in Fig. 50(b) is given by a similar formula

$$V_0(t)=V_{DD}-\frac{B_N}{C_L}\int_{t_0}^{t}[V(\theta)-V_{THN}]d\theta .$$

The dynamic node was originally charged to V_{DD}. The integral is computed as follows: If the integrand $V(\theta)-V_{THN}$ is negative, the integrand is set to zero. Figure 51 shows the response of the dynamic node to the input node voltage waveform shown by the solid curve [closed circles for large, and open circles for small (B_N/C_L)].

If a hybrid node has a pull-up resistor as shown in Fig. 50(c), the integral of the circuit equation is computed in the same way.

3.12 Signal Stagnant at Gate-Metastable State

An interesting circuit failure that is due to anomaly in signal flow is the metastable state of a latch. The metastable state takes place if the effect of the input signal to the latch is too close to the decision threshold. Because of insufficient information retained in the latch, the latch is unable to decide which way to settle for an extended period of time. The metastable state is a peculiar state of a data storage circuit from which it takes an unusually long time to settle to the final state. If the latch circuit is complex, node voltages may even oscillate before settling at the final state.

In a completely synchronous logic circuit this problem occurs only at the maximum frequency limit. By reducing clock frequency, the *setup* time of the latch is increased. Then the latch acquires decisive information before the latch's input circuit turns off, and the metastable state does not occur. The metastable state matters essentially in *asynchronous* logic circuits. In asynchronous logic circuits, the output driver of the subsystem that sends data out and the input latch that receives the data have different clocks. Two different clocks have in general different frequencies. Even if *averaged* frequencies of the two are the same, variations of the frequencies about the same average can be different. Then there is no relationship between the clockedges of the two different clocks. With reference to Fig. 52(a) and (b), the clock edge, CKR [curve 1 of Fig. 52(b)], of chip B may have any timing relationship with respect to the clock edge, CKD of chip A that sends the signal out (curves 2–4). CKD can be earlier than CKR, at almost the same time as CKR, or later than CKR, as shown by curves 2, 3, and 4 of Fig. 52(b), respectively. If CKD is earlier than CKR as shown by curve 2, the latch captures stable (low logic level) data. If CKD is later than CKR, the latch captures stable (high logic level) data. If the two clocks are almost synchronous as shown by curve 3, the input data waveform 3′ provides confusing information to the latch of chip B. This is the case of metastability [12].

In recent years, the clock frequency of state-of-the-art digital data transmission systems ranges from 100 to 500 MHz. If signals are transported from one circuit pack to the other by

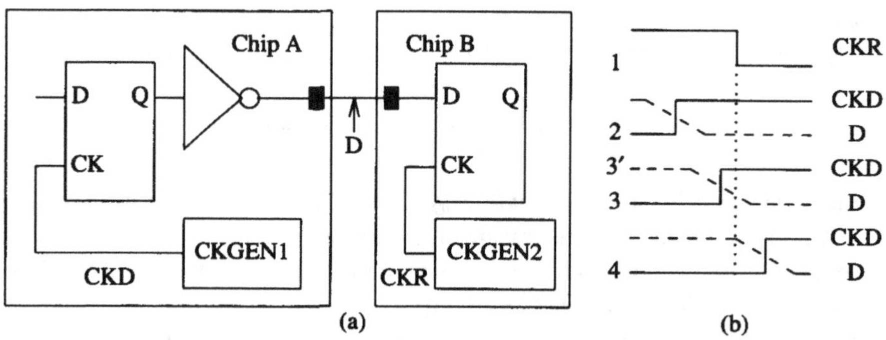

Figure 52 Metastable state

transmission media like a twisted pair of wires or coaxial cables, and if the size of the system is larger than about 1 meter, more than one bit of data is contained within the length of the transmission media (if propagation velocity is 1/2 ft/nsec, 1 bit at 600Mb/s signal occupies 3/4 ft). Even if all the active devices (ICs) of the system are clocked by perfectly synchronized clocks, the timing relation between the clock and the input data at the end of the transmission line is uncertain. Furthermore the timing relation may depend on environmental conditions like temperature. Metastable state may occur in such systems. The metastable state makes assembly of large, high-speed systems quite difficult. There are several solutions to this fundamental problem, but unfortunately all of them are costly. They are:

1) assemble system compactly and trim transmission delays accurately using some kind of advanced system assembly technology (Section 7.7);
2) send clock and data together, using matched transmission lines;
3) clock is recovered from signal.

A latch's response to clock and data is characterized by setup times. Figure 53(a) shows a realistic input circuit, and Fig. 53(b) shows clock and data waveforms. Suppose that data D makes a step function transition [solid curves 3 and 4 of Fig. 53(b)]. With reference to *internal* clock CKR that controlls the D-latch [curve 1 of Fig. 53(b)] and data D (curves 3 and 4), the input latch works as follows: a logic-high level is latched if logic-high data settles at least time T_S earlier than the clockedge (curve 3), but logic-low level is latched if data settles after T'_S (curve 4). The critical time T_S is called the setup time. Between T_S and T'_S, the latch's final state is uncertain. The state of the latch in this condition is metastable state. If input data makes a high-to-low transition, the setup times will be generally different. This definition of setup time is simple because the definition involves the CMOS latch circuit alone. Setup time of an integrated circuit is often defined with reference to *input* clock CKI of Fig. 53(a), rather than internal clock, CKR. In this case the setup time [T''_S in Fig. 53(b)] could be even negative, as shown by curve 2 of Fig. 53(b). If data D makes multiple transitions as shown by the

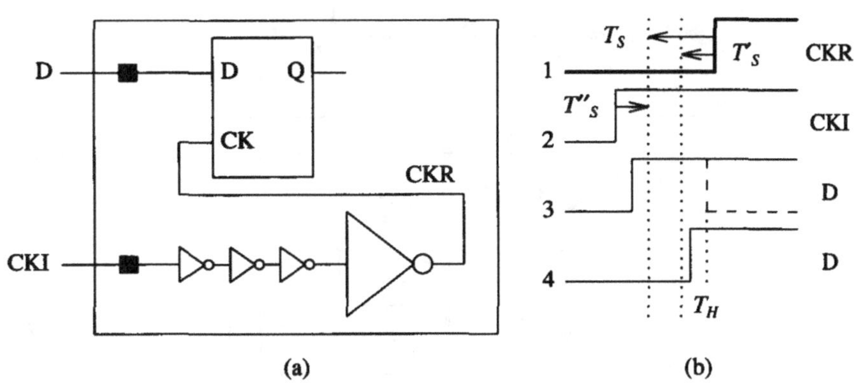

Figure 53 Details of setup time definition

dashed curve 3 of Fig. 53(b), the data D must be held until the *hold* time T_H. The latch captures the correct data only if the setup and hold time requirements are met. As will be discussed in Section 5.2, computation of these parameters is quite complex.

The requirement that the input data must have been stabilized at least by the setup time before the clock edge is equivalent to saying that the latch circuit must have acquired sufficient *internal* information to determine the future state when the connection of the data input path is cut off. How much information is required can be analyzed as follows. Metastability of a latch can be modeled using the equivalent circuit shown in Fig. 54. The two-stage cascaded inverters are originally biased close to the switching threshold. Since circuit behavior close to this bias point is of interest, metastability can be studied using a small-signal equivalent circuit. The circuit equations are

$$C\frac{dV_1(t)}{dt}+g_mV_0(t)=\frac{V_{DD}-V_1(t)}{R} \qquad C\frac{dV_0(t)}{dt}+g_mV_1(t)=\frac{V_{DD}-V_0(t)}{R} \tag{92}$$

where we used the collapsible current generator model of FET, with zero threshold voltage. The logic levels are defined as $V_H=V_{DD}$ and $V_L=0$ if $g_mR>1$ (Section 2.2). We assume the gain g_mR is larger than 1. Following the theory of Section 2.2, the unstable bias point that exists between V_L and V_H, V_{GSW} is determined by solving the equation

$$-\frac{1}{R}(V_{DD}-V_{GSW})+g_mV_{GSW}=0 \quad \text{as} \quad V_{GSW}=\frac{V_{DD}}{1+g_mR}.$$

Let $V_1(t)$ and $V_0(t)$ be written as

$$v_1(t)=V_1(t)-V_{GSW} \qquad v_0(t)=V_0(t)-V_{GSW}$$

and the circuit equations for $v_1(t)$ and $v_0(t)$ are

$$C\frac{dv_1}{dt}+\frac{1}{R}v_1+g_mv_0=0 \qquad C\frac{dv_0}{dt}+\frac{1}{R}v_0+g_mv_1=0. \tag{93}$$

At time $t=0$, v_0 and v_1 take the initial values $v_0(0)$ and $v_1(0)$, respectively. The solutions of Eq. (93) are sums of two terms of the form $exp(\alpha t)$, where two α's are given by

$$\alpha_0=-\frac{1}{CR}+\frac{g_m}{C} \qquad \alpha_1=-\frac{1}{CR}-\frac{g_m}{C}$$

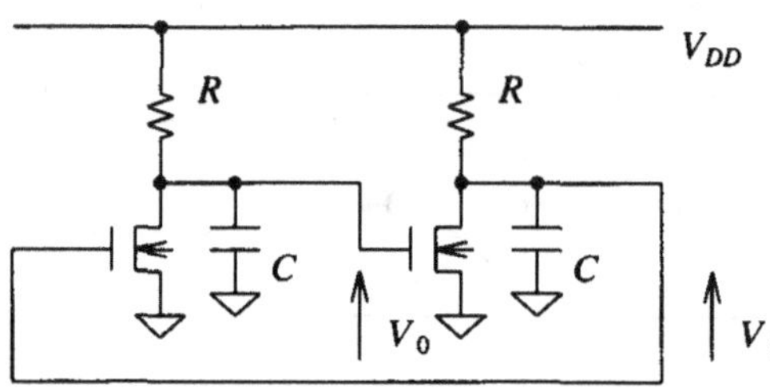

Figure 54 Equivalent circuit of latch in metastable state

where $\alpha_0 > 0$ and $\alpha_1 < 0$ if we assume gain $g_m R > 1$ (that is required for stable gate operation, Section 2.2). Solutions of Eq. (93) are

$$v_0 = A_{00} exp(\alpha_0 t) + A_{01} exp(\alpha_1 t) \quad v_1 = A_{10} exp(\alpha_0 t) + A_{11} exp(\alpha_1 t) . \tag{94}$$

Equation (93) requires that

$$A_{10} = -\frac{g_m}{C\alpha_0 + (1/R)} A_{00} = -A_{00} \quad A_{11} = -\frac{g_m}{C\alpha_1 + (1/R)} A_{01} = A_{01} . \tag{95}$$

A_{00} and A_{01} are the only two independent constants of integration and are determined from the initial condition as

$$A_{00} = \frac{1}{2} [v_0(0) - v_1(0)] \quad A_{01} = \frac{1}{2} [v_0(0) + v_1(0)] .$$

We note that $A_{01} = A_{11}$. Equations (94) and (95) show that the *mode* of node voltage change that simultaneously increases or decreases v_0 and v_1 decays exponentially with time like $exp(\alpha_1 t)$. The mode of changes in which node 1 voltage increases while node 0 voltage decreases (or vice versa) evolves with time like $exp(\alpha_0 t)$. α_0 can be positive or negative, depending on the value of $g_m R$. If gain $g_m R < 1$ the second mode decays also, but if $g_m R > 1$ as we assumed before, $\alpha_0 > 0$ and the second mode grows. This is the case where the latch is able to store digital data (Section 2.2). The latch settles if the second mode is excited. Suppose that the initial node voltage of the latch $V_0(0)$ and $V_1(0)$ are

$$V_0(0) = V_{GSW} + \Delta V_0 \quad V_1(0) = V_{GSW} + \Delta V_1$$

Then amplitude of the first and the second modes are given by

$$A_{00} = \frac{1}{2} (\Delta V_0 - \Delta V_1) \quad A_{01} = \frac{1}{2} (\Delta V_0 + \Delta V_1) .$$

Therefore if $\Delta V_0 = \Delta V_1$, the amplitude of the increasing mode is zero, and the latch goes into a metastable state. In this case, $v_0(t)$ and $v_1(t)$ decrease with time maintaining $v_0(t) = v_1(t)$, and the state of the latch approaches the metastable bias point. Unless some other external influence like noise perturbs the latch circuit, the circuit stays in the metastable state. As far as the present linearized theory is concerned, the metastability requirement $\Delta V_0 = \Delta V_1$ does not imply that ΔV_0 and ΔV_1 should both be small deviations from voltage V_{GSW}. In reality if ΔV_0 and ΔV_1 are not small, the nonlinear effects that were neglected in the basic circuit equation Eq. (93) emerge, and the neglected effects become crucial in determining the nature of critical phenomena like metastability.

There is an interesting viewpoint about the initial condition to the metastable state. Suppose that the latch is in the transmitting state, and the input data changes. Then assuming realistic signal rise/fall times of the signal nodes $\overline{Q}$ and Q of the latch have one inverter delay and therefore the two node voltages are always different, except when they cross-over. Then the initial condition of the metastable state is rather well known at the crossover point, and this is practically the only case when metastability takes place. This interesting viewpoint was raised by Miss France Herringer during her stay in AT&T's Bell Labs. in Fall, 1989.

The number of inversions within the closed loop of inverters that applies positive feedback need not be 2, which is the minimum. Any even number of inversions apply positive

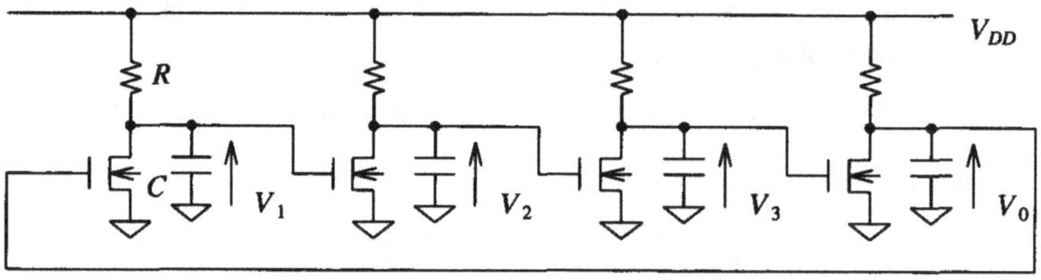

Figure 55 Four-stage cascaded inverters in a closed loop of latch

---

feedback. How a metastable state occurs in such *multistage* latches is interesting. Figure 55 shows a chain of four-stage cascaded identical inverters. The *unstable* equilibrium point V_{GSW} is given by $V_{GSW} = V_{DD}/(1 + g_m R)$ as before. If small voltage deviations, $v_i = V_i(t) - V_{GSW}$, are used to write the circuit equations we have,

$$C\frac{dv_1}{dt} + \frac{v_1}{R} + g_m v_0 = 0 \quad C\frac{dv_2}{dt} + \frac{v_2}{R} + g_m v_1 = 0 \tag{96a}$$

$$C\frac{dv_3}{dt} + \frac{v_3}{R} + g_m v_2 = 0 \quad C\frac{dv_0}{dt} + \frac{v_0}{R} + g_m v_3 = 0. \tag{96b}$$

The equations are solved subject to the initial conditions $v_i = v_i(0)$ at $t = 0$. The solutions are the sum of the terms of the form $exp(\alpha_i t)$ ($i = 0, 1, 2,$ and 3) where α_i's are found as

$$\alpha_i = \frac{1}{C}(-\frac{1}{R} + \chi_i) \quad i = 0, 1, 2, 3 \tag{97}$$

where $\chi_0 = g_m$, $\chi_1 = -g_m$, $\chi_2 = g_m i$ and $\chi_3 = -g_m i$. If the solutions are written as

$$v_i = \sum_{j=0}^{3} A_{ij} exp(\alpha_j t). \tag{98}$$

We have four independent constants of integration, A_{00}, A_{01}, A_{02}, and A_{03} and the other coefficients A_{ki} are given by

$$A_{ki} = [-\frac{g_m}{C\alpha_i + (1/R)}]^k A_{0i} \quad (0 \le i, k \le 3).$$

By substituting α_i's into A_{ki}'s we may rewrite Eq. (98) as

$$\begin{bmatrix} v_0 \\ v_1 \\ v_2 \\ v_3 \end{bmatrix} = \begin{bmatrix} A_{00} & A_{01} & A_{02} & A_{03} \\ -A_{00} & A_{01} & iA_{02} & -iA_{03} \\ A_{00} & A_{01} & -A_{02} & -A_{03} \\ -A_{00} & A_{01} & -iA_{02} & iA_{03} \end{bmatrix} \begin{bmatrix} exp(\alpha_0 t) \\ exp(\alpha_1 t) \\ exp(\alpha_2 t) \\ exp(\alpha_3 t) \end{bmatrix}. \tag{99}$$

Among the four α's, α_0 is the only one that may have positive real part. The solution for $v_0(t)$ is written as

$$v_0(t) = \mathrm{Re}\,[A_{00} \exp[(g_m R - 1)t/CR] + A_{01} \exp[-(g_m R + 1)t/CR] \tag{100}$$

$$+A_{02} \exp(-t/CR)(cos\frac{g_m}{C}t + i\sin\frac{g_m}{C}t) + A_{03} \exp(-t/CR)(cos\frac{g_m}{C}t - i\sin\frac{g_m}{C}t)].$$

Solutions for $v_1(t)$, ..., $v_3(t)$ may be written down similarly. We used complex coefficients A_{02} and A_{03}. If the initial condition provides $A_{00}=0$, a metastable state occurs. Since A_{02} and A_{03} may be chosen as complex conjugate

$$A_{02}+A_{03}=2\,\mathrm{Re}(A_{02}) \quad A_{02}-A_{03}=2i\,\mathrm{Im}(A_{02})$$

We have

$$v_0(0)=A_{01}+2\,\mathrm{Re}(A_{02}) \quad v_1(0)=A_{01}-2\,\mathrm{Im}(A_{02})$$

$$v_2(0)=A_{01}-2\,\mathrm{Re}(A_{02}) \quad v_3(0)=A_{01}+2\,\mathrm{Im}(A_{02})$$

The initial condition that results in the metastable state of the four-stage cascaded closed loop circuit is given by

$$v_0(0)=v_1(0)=v_2(0)=v_3(0)=const1 \;\; (=A_{01})$$

and $A_{02}=A_{03}=0$. Then the node voltages decay with time constant $(-\alpha_1)^{-1}=CR/(1+g_mR)$. This condition can be relaxed, however, as follows. If $v_0(0)=const1+const2$ and $v_2(0)=const1-const2$ and the others are $const1$, the metastable state takes place. The effects of the initial node 0 voltage cancel the effects of the initial node 2 voltage. Similarly if $v_1(0)=$ $const1$ + $const3$ and $v_3(0)$ = $const1$ − $const3$ and the others are $const1$, metastable state takes place. In this case the effects from node 1 and node 3 cancel. The most general condition is coexistence of all the three conditions, as shown schematically in Fig. 56. If coefficients $Re(A_{02})$ and $\mathrm{Im}(A_{02})$ are not zero, the node response is decaying oscillation [13].

It is useful to estimate how often the metastable state is expected to take place in an asynchronous system. This development follows the work by Veendrik [14]. We consider the simple latch circuit shown in Fig. 54. From Eq. (94) differential mode node voltage $v_D(t)=v_0(t)-v_1(t)$ grows exponentially with time, as

$$v_D(t)\approx const\times\exp(\alpha_0 t) \quad \text{where} \quad \alpha_0=(g_mR-1)/CR. \tag{101}$$

If $v_D(0)=\pm V_N$, the latch takes approximately time t_M to settle (assuming the exponential growth all the way), where t_M is given by

$$t_M=\frac{1}{\alpha_0}\log\frac{V_{DD}}{V_N}.$$

If the differential node voltage, $v_D(0)$, is in the range $V_N>v_D(0)>-V_N$, the minimum duration of the metastable state is given by t_M. Growth of the differential voltage with time is shown schematically in Fig. 57(a). At the time of latch closure, $v_D(0)$ may take any value in

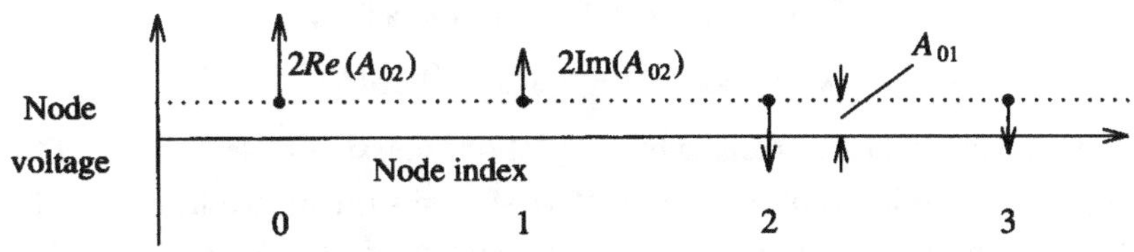

Figure 56 Initial node voltage for metastability of chain

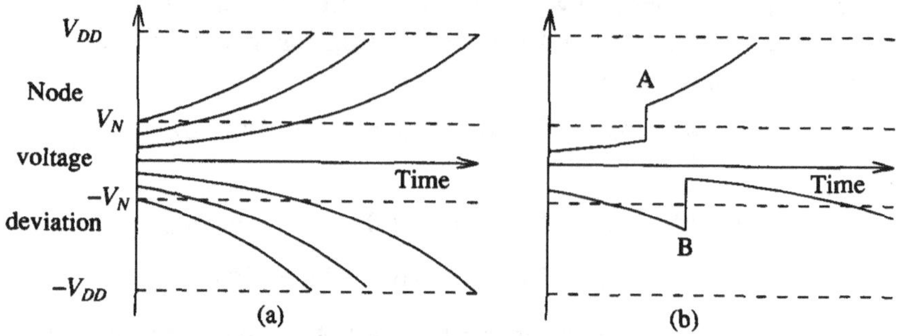

Figure 57 Metastable state node waveforms

the range $-V_{DD} < v_D(0) < V_{DD}$. We do not know, however, the probability weight. At high latch clock frequencies where the latch node voltages change quite often, we may assume equal probability for any values in the range as a crude approximation. Then the probability P that $|v_D(0)|$ is less than a given value V_N is

$$P = V_N/V_{DD}.$$

This is the probability that metastable state lasts longer than $t_M = (1/\alpha_0)\log(V_{DD}/V_N)$. We have a direct relationship between t_M and P as

$$P = \exp(-\alpha_0 t_M). \tag{102}$$

We need to investigate the assumptions of this theory. There is a range of differential node voltages within which the latch circuit is susceptible to the essential noise (we consider noise due to the fundamental physical mechanisms, like thermal noise). The range is shown by the pair of dashed lines at $\pm V_N$ in Fig. 57(b). Estimate of V_N due to thermal noise (Section 4.2) is $V_N = 1\ mV$ at room temperature. Veendrik proposed that the metastable state is independent of noise because of the following mechanism. As shown in Fig. 57(b), noise may terminate a metastable state (curve A) and noise may create a metastable state (curve B). Since the initial value of v_D is distributed uniformly over the range an equal number of terminations (curve A) and of creations (curve B) occur. As a consequence, probability is not affected by noise. This argument appears convincing. Indeed experimental evidences supporting the idea exists. Since there is no other convincing argument, a probability estimate of metastability based on this model is the best available at present. I wish to raise a few questions to this view in the following, however. We solve the probability formula for V_N as

$$V_N = (V_{DD}/2)\exp(-\alpha_0 t_M) = (V_{DD}/2)P$$

We consider a probability that is realistic in a digital transmission system that is of the order of 10^{-12} (1 metastable state in about 5 hours in a 50 Mb/sec system), we have $V_N = 2 \times 10^{-12}\ V$. Is this a rational magnitude of the initial differential voltage in ordinary CMOS D-latch? V_N is many orders of magnitude smaller than the noise voltages developed to the nodes by fundamental thermodynamic mechanisms.

I think the following two points must be clarified.

1) The number of metastable states that terminate by noise, and the number of metastable states that are created by noise must be well-balanced, so that the effects of input signal that is left in the latch circuit at the time of latch closure determines the latch's behavior on a statistical basis. Thermal noise at the node will be of the order of millivolts. If V_N is 9 orders or more of magnitude less than that of the intrinsic thermal noise, it is not immediately clear whether or not the assumption is well satisfied.

2) As we will see in chapter 4, noise induced on digital nodes from nearby switching nodes is several orders of magnitude higher than the fundamental noise such as thermal noise. The *induced* noise is not an entirely random phenomenon. Then the probability estimate should include the capacitive induction effect and the mode of operation of the entire chip, and a simple expression of probability cannot be a good approximation. The two points raised above indicate that the role played by noise must be considered in more detail. This problem will be discussed in a future paper.

Metastability is a problem to circuit designers. The issue must be resolved before full capability of submicron CMOS is unleashed, but all the schemes proposed so far seem to have some fundamental problems. The general scenario is that elimination of one metastable state creates the other. Multiple latching of input data has the effect equivalent to increasing the time assigned to a single latch, and therefore, no improvement. At present, the best circuit design strategy would be *not* to solve, but to *get around* metastability problems.

3.13 Process Variations

Wafer processing creates two types of interconnection failures: wires are short-circuited or open-circuited. A CMOS gate fails in various ways by these two simple mechanisms. This is the subject of circuit *testability*, and we do not go into the details. Process parameter variations, or inaccuracy in estimating the device parameters, result in similar failures. In the following sections we consider signal flow interruption by this mechanism. The common, and often not obvious, mechanism of failures is the wrong setting of the circuit operating point, which is caused by device parameter variations. Process variation of MOS device parameters is notoriously wide. Parameter variation affects circuits through two different mechanisms. The first is the effects on DC characteristics of gates, and the second is the effects on delay characteristics. We consider the DC problems here.

If a CMOS gate is in the steady-state, either the NFET or PFET is completely turned off. Because of this, a CMOS gate is *ratioless*, or CMOS gates work independent of the FET sizes. The output node voltage when the gate reaches steady state is either 0, or V_{DD}. This is an outstanding advantage of CMOS over any other technology. CMOS gates tolerate wide process variations. Even if the switching threshold voltage V_{GSW} varies, CMOS gates generate 0 or V_{DD} if the input voltage is V_{DD} or 0, respectively. Therefore the problems of process variation are encountered only if the range of logic voltage swing is restricted, as in the case of TTL to

CMOS or ECL to CMOS input buffers, and the process variation brings V_{GSW} outside the range of the input voltage swing. We consider failure problems of these circuits in this section.

The threshold voltage of a CMOS inverter depends on the *ratio* of drive capabilities of the pull-up and the pull-down FETs. If we assume the collapsible current generator FET model, the input (V_1)-output (V_0) characteristic of a CMOS inverter is given by,

$$V_D = V_{DD} \ (V_1 < V_{GSW}) \quad V_D = 0 \ (V_1 > V_{GSW})$$

where

$$V_{GSW} = \frac{B_P(V_{DD} - V_{THP}) + B_N V_{THN}}{B_N + B_P}. \tag{103}$$

Where the two regimes cross over, the V_0 versus V_1 curve is vertical. V_{GSW} is the switching threshold voltage of the CMOS inverter. The inverter has high voltage gain at $V_1 = V_{GSW}$. Therefore the circuit is used to interface small-amplitude input signals from the outside to large amplitude CMOS logic circuits on the chip. Stability of the switching threshold voltage is the major concern. V_{GSW} depends on the ratio of the transconductance parameters B_P and B_N as shown in Eq. (103), and the ratio is determined by process conditions. Equation (103) shows that $V_{GSW} \to V_{DD} - V_{THP}$ if $B_P \to \infty$, and $V_{GSW} \to V_{THN}$ if $B_N \to \infty$. In the vicinity of the respective infinities of B_P and B_N, V_{GSW} depends weakly on B_P and B_N, respectively. In this condition the size of the *smaller* FET determines the switching threshold voltage. Therefore we consider both ratios $\beta = B_N/B_P$ and its inverse, $\bar{\beta} = B_P/B_N$ such that the ratio that is less than 1 is meaningful. Voltages are normalized to V_{DD} as $v_{GSW} = V_{GSW}/V_{DD}$, $v_{THP} = V_{THP}/V_{DD}$ and $v_{THN} = V_{THN}/V_{DD}$. We have

$$v_{GSW} = \frac{1 - v_{THP} + \beta v_{THN}}{1 + \beta} \quad \text{or} \quad = \frac{(1 - v_{THP})\bar{\beta} + v_{THN}}{1 + \bar{\beta}}. \tag{104}$$

We need to know the sensitivity of v_{GSW} to variations of β and $\bar{\beta}$. We have

$$\frac{\partial v_{GSW}}{\partial \beta} = -\frac{1 - v_{THP} - v_{THN}}{(1+\beta)^2} \qquad \frac{\partial v_{GSW}}{\partial \bar{\beta}} = \frac{1 - v_{THP} - v_{THN}}{(1+\bar{\beta})^2} \tag{105}$$

by substituting β and $\bar{\beta}$ from Eq. (104) we get

$$\frac{\partial v_{GSW}}{\partial \beta} = -\frac{(v_{GSW} - v_{THN})^2}{1 - v_{THP} - v_{THN}} \qquad \frac{\partial v_{GSW}}{\partial \bar{\beta}} = \frac{(1 - v_{THP} - v_{GSW})^2}{1 - v_{THP} - v_{THN}}. \tag{106}$$

If we set

$$|\partial v_{GSW}/\partial \beta| = |\partial v_{GSW}/\partial \bar{\beta}|$$

and solve for v_{GSW}, we obtain $v_{GSW0} = (1 - v_{THP} + v_{THN})/2$, which means $\beta = 1$. The absolute values $|\partial v_{GSW}/\partial \beta|$ and $|\partial v_{GSW}/\partial \bar{\beta}|$ are plotted versus v_{GSW} in Fig. 58. In this figure the irrelevant part of the curves are indicated by dotted curves. If either FET is much larger than the other, the switching threshold voltage is effectively set close to the threshold voltage of the larger FET, and its variation is determined by the smaller FET. In this case, sensitivity of the switching threshold voltage to linear beta ratio variation gives a misleading impression. What

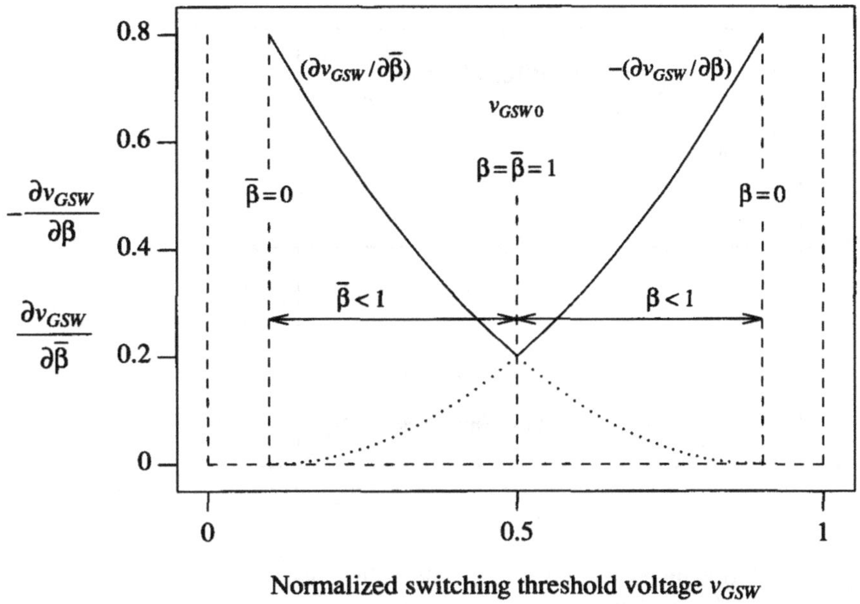

Figure 58 Sensitivity of switching threshold voltage on beta ratio

can be controlled by the wafer fabrication process is the conductivity ratio of PFET and NFET, which is of the order of unity. Then we must refer to the smaller one of $\beta = B_N/B_P$ or $\bar{\beta} = B_P/B_N$. Sensitivity of the switching threshold voltage of a CMOS inverter V_{GSW} against β or $\bar{\beta}$ defined in this way has a minimum at $v_{GSW0} = (1 - v_{THP} + v_{THN})/2$.

Sensitivity of v_{GSW} against variation of FET threshold voltages, v_{THN} and v_{THP}, are derived as

$$\frac{\partial v_{GSW}}{\partial v_{THN}} = \frac{\beta}{1+\beta} = \frac{1 - v_{THP} - v_{GSW}}{1 - v_{THP} - v_{THN}} \tag{107a}$$

$$\frac{\partial v_{GSW}}{\partial v_{THP}} = \frac{-1}{1+\beta} = - \frac{v_{GSW} - v_{THN}}{1 - v_{THP} - v_{THN}}. \tag{107b}$$

The absolute values $|\partial v_{GSW}/\partial v_{THN}|$ and $|\partial v_{GSW}/\partial v_{THP}|$ are plotted versus v_{GSW} in Fig. 59. Sensitivity to v_{THN} is unity if $v_{GSW} = v_{THN}$, and it is zero if $v_{GSW} = 1 - v_{THP}$. If the switching threshold voltage is close to the threshold voltage of the NFET, the sensitivity is high. At $v_{GSW0} = (1 - v_{THP} - v_{THN})/2$ the sensitivities against v_{THP} and v_{THN} variations are balanced, and this is the most stable condition of DC operation of the inverter.

If the switching threshold voltage goes outside the range of the input voltage swing, the inverter fails to switch, and the signal flow is interrupted. The simple analysis of this section applies only to inverters, since in multi-input gates inclusion of the back-bias effect (Section 1.3) is required. Multi-input gates are, however, seldom used as input buffers.

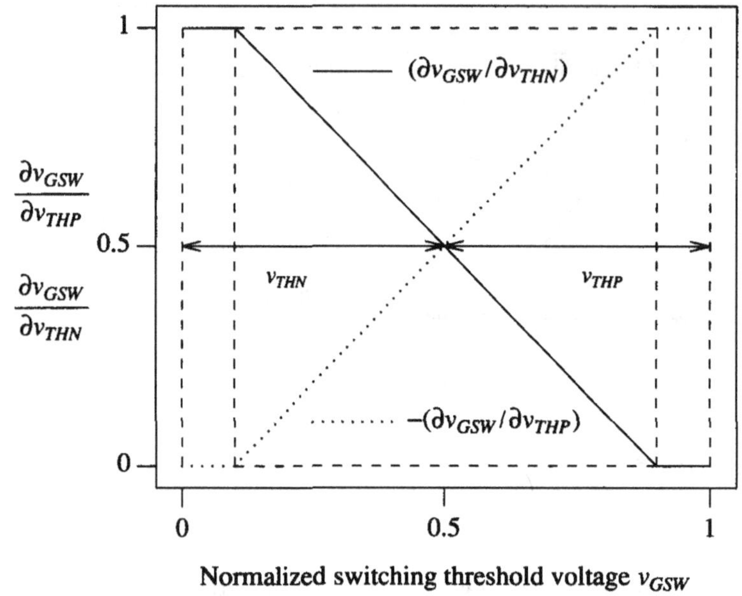

Figure 59 Sensitivity of switching threshold voltage on FET threshold voltage

3.14 Circuits Resistant to Process Variation

Process variations affect the DC operating point of circuits, thereby creating catastrophic signal interruption. The problem must be controlled by using circuit design techniques. In this section we summarize techniques of desensitizing circuits to process variation. Within a single chip, characteristics of two NFETs (or two PFETs) are well matched. Channel currents of two identically laid out NFETs within a 1cm square chip are usually matched within 2–5%. The same is true for two PFETs, but currents of PFETs and NFETs never match closely. Therefore circuits that use NFETs (or PFETs) only are significantly less sensitive to process variation. The circuit shown in Fig. 60(a) has switching threshold voltage that is only

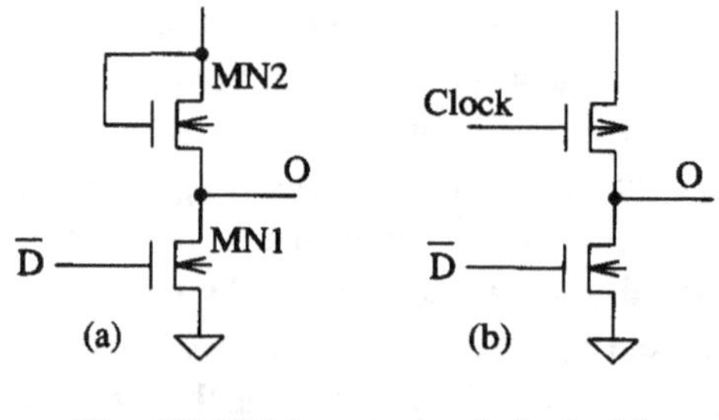

Figure 60 NMOS inverter and clocked inverter

weakly dependent on variations of NFET process parameters: If NFET MN1 has higher current, NFET MN2 has higher current as well, thereby maintaining the ratio of current that determines the switching threshold. One way of eliminating the effects of PFETs from circuits is to turn the PFETs off at the time of operation. A dynamic inverter shown in Fig. 60(b) has the PFET turned off after precharge. The inverter switches if input signal level ($\bar{D}$) is higher than the NFET threshold voltage. Process sensitivity is limited to variation in V_{THN} as far as DC steady state characteristics are concerned. Use of dynamic circuits are effective in reducing process sensitivity of certain circuits.

The differential amplifier circuit shown in Fig. 61(a) is practically immune to process variation. The amplifier switches at V_{REF}, which can be set practically independent of process parameters (either by dividing power supply voltage in a fixed ratio, or by using a reference voltage source). FETs MN1 and MN2, and MP1 and MP2 are closely matched (laid out as symmetrically as possible). The voltage swing at node O is not from 0 to V_{DD}. If node O is buffered by a CMOS inverter, CMOS logic levels are generated. Differential amplifier circuits can be used within positive feedback loops, so that small amplitude input signals are first captured and then amplified to full CMOS voltage levels and latched.

The circuit shown in Fig. 61(b) switches at threshold voltage V_{REF}, which is derived using divider resistors R_0 and R_1 (the resistor *ratio* can be made independent of process variations). MN0 and MN1, and MP0 and MP1 are matched. The equal switching threshold voltage of the two inverters is controlled by adjusting the local power supply voltage of the inverters (node V'_{DD}) by using a large inverter made from MPA and MNA. The circuit applies negative feedback to the second inverter so that the switching threshold voltage of the matched input inverter is kept at V_{REF}.

The amplifier shown in Fig. 61(b) is a single-ended amplifier. A differential amplifier can be designed using the same feedback circuit. In the circuit shown in Fig. 62, two input voltages V_A and V_B are averaged by a resistive averaging network consisting of resistors R_A and R_B, and the averaged voltage sets the switching threshold voltage of the amplifier. Let variations of voltages of nodes A, B, and C be ΔV_A, ΔV_B, and ΔV_C. They are related by

---

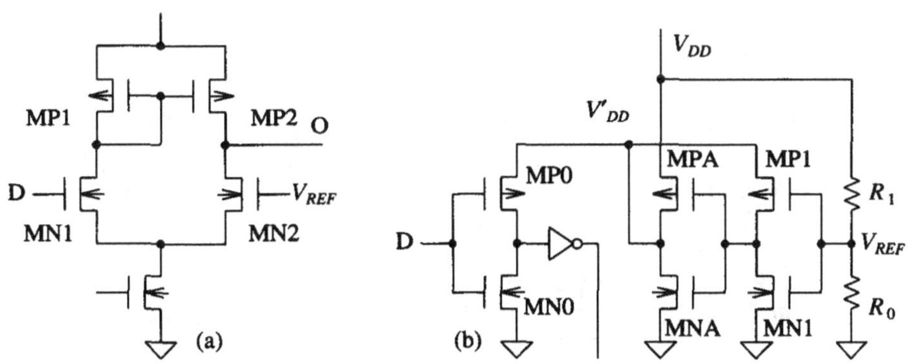

Figure 61 Differential amplifier and feedback technique

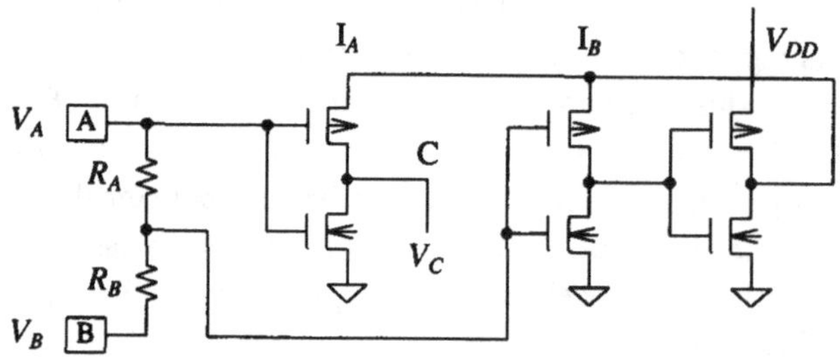

Figure 62 An alternative differential amplifier circuit

$$\Delta V_C = -G \cdot [\Delta V_A - \frac{1}{2}(\Delta V_A + \Delta V_B)] = -\frac{G}{2} \cdot (\Delta V_A - \Delta V_B)$$

and therefore the amplifier works effectively as a differential amplifier, although the circuit has a single-ended structure.

FET threshold voltage is dependent on the integrated dose of ionizing radiation, or aging of the chip. If threshold voltage must be maintained within a narrow range during the lifetime of the system, the best way is to apply negative feedback control using the back-bias effect. This amounts to controlling the substrate voltage to adjust the FET threshold voltage by the back-bias effect. Figure 63 shows NFETs (MNA-MNN) within the same P-tub that have matched characteristics, even under radiation dose. Sources of the NFETs are connected to ground, which is DC-wise isolated from the P-tub. A matched dummy NFET, MN1, is driven by small current generator I_0 to measure its threshold voltage. The measured threshold voltage is compared with the desired threshold voltage by an operational amplifier, and negative feedback is applied to the substrate voltage. All the matched NFETs within the same substrate have the desired threshold voltage [15].

As is shown by the few examples shown here, analog feedback control of digital circuit is powerful in reducing process sensitivity of digital circuits. Analog circuit techniques will be used increasingly in future, high-performance CMOS VLSI chips.

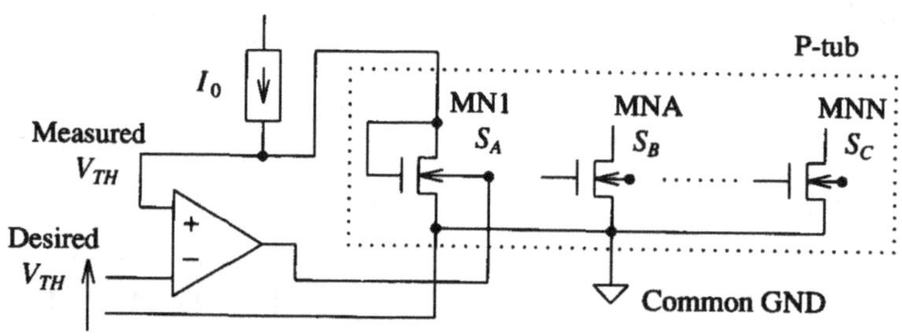

Figure 63 Feedback stabilization of FET threshold voltage

References

[01] Hakim, S. S. *Feedback circuit analysis*. New York: John Wiley & Sons, 1966.

[02] Shoji, M. *CMOS digital circuit technology*. Englewood Cliffs, NJ: Prentice-Hall Inc., 1987.

[03] Shoji, M. "Theory of transverse extension of Gunn domains." *J. Appl. Phys*. 41 (February, 1970): 774–78.

[04] Tomizawa, K., Kawashima, M., and Kataoka, S. "New logic functional device using transverse spreading of high-field domain in N-type GaAs." *Electronics Letters* 7 (May, 1971): 239–40.

[05] Ketchum, D. J., and Alvarez, E. C. *Pulse and switching circuits*. New York: McGraw Hill, 1965.

[06] Glasser, L. A., and Dobberpuhl, D. W. *The design and analysis of VLSI circuits*. Reading, MA: Addison-Wesley Publishing Co., 1985.

[07] Lamdan, T., and Aharon, M. "A circuit for high-speed carry propagation in MSI-FET technology." *The radio and Electronic Engineer* 46 (July, 1976): 337–41.

[08] Hwang, I. S., and Fisher, A. L. "Ultrafast compact 32-bit CMOS adders in multiple-output Domino logic." *IEEE J. of Solid-state circuits* 24 (April, 1989): 358–69.

[09] Cutler, P. *Semiconductor circuit analysis*. New York: McGraw-Hill, 1964.

[10] Shoji, M. "Negative capacitance bus terminator for improving the switching speed of a microcomputer databus." *IEEE J. Solid-State Circuits* vol. SC-20 (August, 1985): 828–32.

[11] Pretorius, J. A., Shubat, A. S., and Salama, C. A. T. "Charge redistribution and noise margins in Domino CMOS logic." *IEEE Trans. on Circuits and Systems* vol. CAS-33 (August, 1986): 786–93.

[12] Mead, C. A., and Conway, L. A. *Introduction to VLSI systems*. Reading, MA: Addison-Wesley Publishing Co., 1980.

[13] Chaney, T. J., and Molnar, C. E. "Anomalous behavior of synchronizer and arbiter circuits." *IEEE Trans. on Computers* vol. C-22 (April, 1973): 421–22.

[14] Veendrick, H. J. M. "The behavior of flip-flops used as synchronizers and prediction of their failure rate." *IEEE J.Solid-State Circuits* vol. SC-15 (April, 1980) 169–76.

[15] Kubo, M., Hori, R., Minato, O., and Sato, K. "A threshold voltage controlling circuit for short channel MOS integrated circuits." *ISSCC70 digest* (February, 1976): 54–55.

4 Noise Phenomena in Digital Circuits

4.1 Introduction

In the last chapter we discussed mechanisms of circuit failure that are caused by anomalous signal flow within a single digital circuit: such failures occur even if the circuit is isolated from any other circuit on the same chip. There are other failures that originate from interaction of the circuit with undesirable environment. Three types of electrical interactions are possible: the first is *induced* noise effects among the wiring in the upper structural layers on silicon; the second is the influence exercised from the conducting silicon substrate below the *active* silicon layer, and the third is coupling among circuits through global connections, especially through V_{SS} and V_{DD} *power* buses. In this chapter these three failure mechanisms, which are essentially digital circuit *noise*, are studied.

Noise in a digital CMOS VLSI chip is predominantly due to coupling from other digital nodes, and not caused by *intrinsic* noise generated by FETs or by any other active devices. In spite of that, digital noise is essentially an unpredictable, random phenomenon. The randomness originates from the complexity of the circuits on a chip. The problems of anomalous signal flow that were discussed in the last chapter did not include randomness or unpredictability. In this chapter we establish that the effects of intrinsic FET noise are negligible, and we then investigate *induced* noise. We then study the mechanisms of power-bus noise. The last part of the chapter deals with the noise originating in the substrate.

4.2 Noise of CMOS ICs

The noise in an *ideal* FET that is free from contact defects, trapping centers in the space-charge layer, etc., is the thermal noise of the channel resistance. Noise voltage is distributed along the length of the channel. Suppose that the FET gate is held at a constant DC potential.

Then the noise voltage generated within the channel effectively modulates the *constant* gate voltage. The FET becomes equivalent to a noise-free FET, whose gate is driven by the *effective* noise voltage source. Therefore, the FET generates noise voltage by *amplifying* its own internal thermal noise. MOSFET noise generated by this mechanism was analyzed by van der Ziel, Jordan and Shoji [01] [02] [03]. We review this theory briefly. Figure 1 shows the cross-section of a MOSFET. Coordinate x is set up such that $x=0$ at the source's edge and $x=L$ at the drain's edge of the channel. The gradual channel, low-field model is used in the following analysis. At point P, a distance x away from the source, the DC potential (*not* including noise) is $V_0(x)$. $V_0(x)$ satisfies equation

$$I = B_G L[V_G - V_{TH} - V_0(x)] \frac{dV_0(x)}{dx} \quad \text{where } B_G = \frac{\varepsilon_{OX}\mu W}{T_{OX} L} \tag{1}$$

where I is the DC channel current, V_G is the gate voltage relative to the grounded source, and the other parameters have their conventional meanings defined in chapter 1. In the following analysis we assume that the FET is driven by a current source I connected to the drain, and we determine the noise voltage developed at the drain terminal. Equation (1) is integrated as

$$Ix + const. = \frac{B_G L}{2}[(V_G - V_{TH})^2 - [V_G - V_{TH} - V_0(x)]^2] \tag{2}$$

and since $V_0(0)=0$ (condition at the source end of channel), $const.=0$. Therefore

$$V_G - V_{TH} - V_0(x) = [(V_G - V_{TH})^2 - \frac{2I}{B_G L}x]^{1/2}$$

$$= \sqrt{\frac{2I}{B_G L}}(x_{\max} - x)^{1/2} \quad \text{where } x_{\max} = \frac{B_G L}{2I}(V_G - V_{TH})^2 .$$

This relationship is plotted in Fig. 2 (curve A). Small segment of MOSFET channel centered at point P is shown in the inset of Fig. 1. This small segment of channel having length Δx has resistance ΔR which is given by

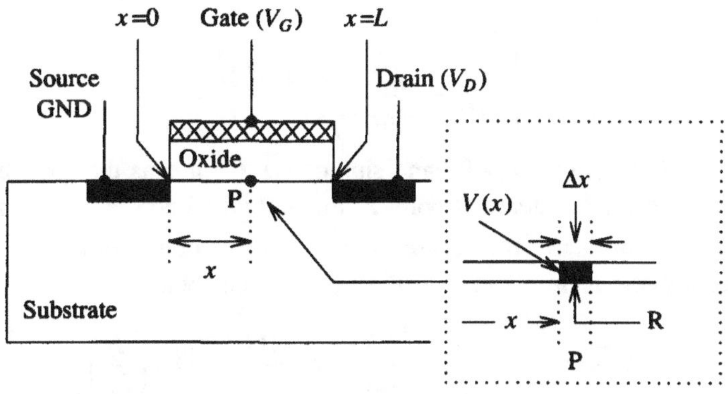

Figure 1 Analysis of MOSFET thermal noise

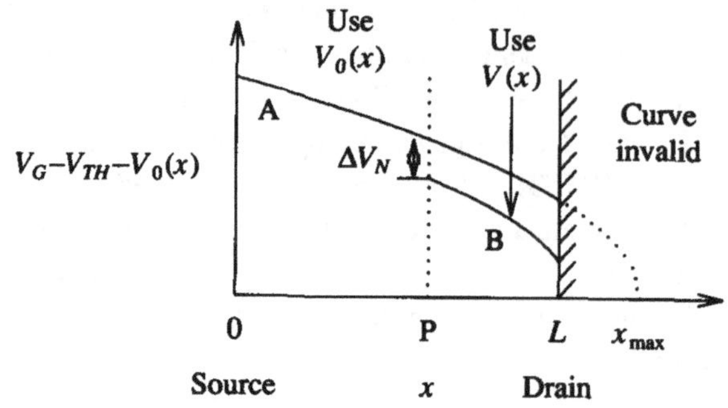

Figure 2 Guide to thermal noise analysis

$$\frac{\Delta R}{\Delta x} = \frac{T_{OX}}{\varepsilon_{OX}\mu \cdot [V_G - V_{TH} - V_0(x)] \cdot W} = \frac{1}{B_G L} \frac{1}{\sqrt{(V_G - V_{TH})^2 - (2I/B_G L)x}} \,. \tag{3}$$

Resistance ΔR generates conventional *thermal* noise voltage ΔV_N given by the formula

$$\overline{\Delta V_N{}^2} = 4kT(\Delta R)\,\Delta f \tag{4}$$

where Δf is the bandwidth of noise measurement, and T is the absolute temperature.

Because of noise voltage, ΔV_N, a voltage step of magnitude ΔV_N is created at point P. To the right-hand side of point P, curve A of Fig. 2 is discontinuously pushed down to curve B. To the right-hand side of P we need to use $V(x)$ instead $V_0(x)$. $V(x)$ is determined from Eq. (2) by re-selecting the integration constant to satisfy $V_0(x) + \Delta V_N = V(x)$ at point P. If ΔV_N is small, we find

$$const. = B_G L\sqrt{(V_G - V_{TH})^2 - (2I/B_G L)x}\,\Delta V_N \,.$$

The drain voltage change due to the noise is given by

$$\delta(\Delta V_D) = V(L) - V_0(L) = \chi \Delta V_N \tag{5}$$

where

$$\chi = \left[\frac{(V_G - V_{TH})^2 - (2I/B_G L)x}{(V_G - V_{TH})^2 - (2I/B_G)}\right]^{1/2}$$

and where notation $\delta(\Delta V_D)$ on the left-hand side of Eq. (5) means the contribution from the small section of the channel centered at point P. Since thermal noises from different locations of the channel are not correlated, the square average of drain noise voltage, $\overline{\Delta V_D{}^2}$, is given by integrating the contributions from all the sections of the channel as

$$\overline{\Delta V_D{}^2} = 4kT\Delta f \int_0^L \chi^2\left(\frac{\Delta R}{\Delta x}\right)dx = \frac{4kT\Delta f G_0}{G_D{}^2} f\left(\frac{I}{I_S}\right) \tag{6}$$

where

$$f(\theta) = \frac{2}{3}\frac{1}{\theta}[1 - (1-\theta)^{3/2}]$$

and where $G_0 = B_G(V_G - V_{TH})$ is the channel conductance at zero drain voltage, $G_D = (\partial I_D/\partial V_D) = B_G(V_G - V_{TH} - V_D)$ is the drain conductance, and $I_S = (B_G/2)(V_G - V_{TH})^2$ is the saturation current of the FET.

Equation (6) gives open-circuit drain noise voltage (the FET has a current generator as the load). The square average of short-circuit noise current, $\overline{\Delta i_D{}^2}$, is given by multiplying Eq. (6) by $G_D{}^2$ as

$$\overline{\Delta i_D{}^2} = 4kTG_0 f\left(\frac{I}{I_S}\right)\Delta f\,. \tag{7}$$

Figure 3 shows normalized square averaged noise current, $\overline{\Delta i_D{}^2}/\overline{\Delta i_D{}^2(0)}$, versus normalized drain current, I/I_S. At zero drain voltage, the channel resistance generates conventional thermal noise of the channel conductance (note that $G_D = G_0$ if $I = 0$). As drain current increases, square averaged noise current decreases gradually, and finally reaches 2/3 of the zero drain current value at saturation.

We use Eq. (7) to estimate the intrinsic FET noise voltage developed at a *digital* circuit node. We consider a symmetrical inverter as shown in Fig. 4(a), where the input node voltage is $V_{DD}/2$. Since the inverter is symmetrical, the output node voltage is also $V_{DD}/2$. Both the PFET and NFET are in saturation. We estimate the noise voltage at the output node of the inverter using the equivalent circuit of the output node shown in Fig. 4(b). This is the small-signal equivalent circuit of the inverter whose load impedance consists of the drain resistance of the FETs and the parasitic node capacitance. Resistance R_D is the drain resistance of the FET in the saturation region, defined by

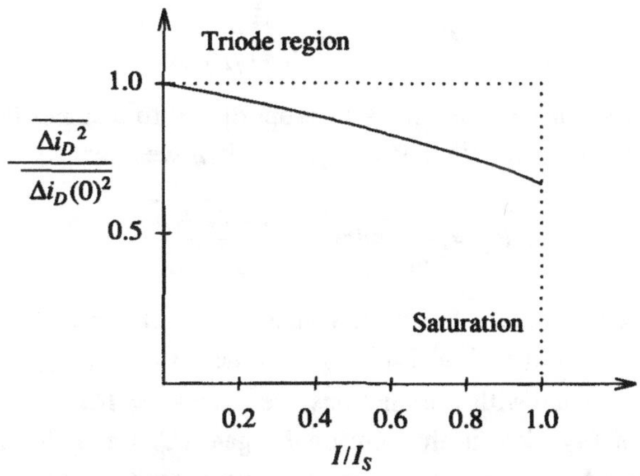

Figure 3 Thermal noise of MOSFET

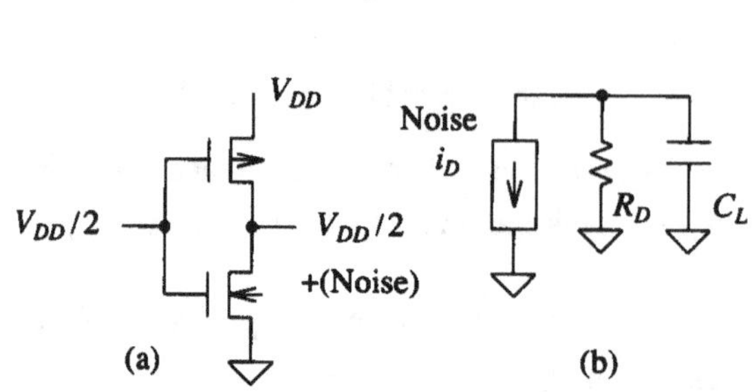

Figure 4 Thermal noise voltage analysis

$$R_D = \left(\frac{\partial I_D}{\partial V_D}\right)^{-1} [V_G = V_{DD}/2, V_D (= V_{DD}/2) > V_{DMAX} = V_G (= V_{DD}/2) - V_{TH}].$$

R_D is estimated based on the physical mechanism that the effective channel length L of FET decreases with increasing drain voltage V_D (Early effect) [04]. If V_D is higher than the saturation voltage V_{DMAX} $(= V_G - V_{TH})$, the *extra* drain voltage $|V_D - V_{DMAX}|$ is sustained by the space-charge layer that is created at the drain's edge of the channel. If $V_D < V_{DMAX}$ the space-charge layer does not exist. If $V_D > V_{DMAX}$, the length of the space-charge layer ΔL is approximately

$$\Delta L = \left[\frac{2\varepsilon_S(V_D - V_{DMAX})}{qN_A}\right]^{1/2}$$

where N_A is the acceptor concentration in the substrate, and the field at the drain's edge of the channel is assumed to be parallel to the silicon surface. At the drain's edge of the channel, length ΔL of the semiconductor is completely depleted of positive holes and the space-charge layer sustains voltage $V_D - V_{DMAX}$. The effective FET channel length decreases from L to $L - \Delta L$. Using this simple model, the R_D is calculated as

$$R_D = \frac{\Delta V_D}{\Delta I} = \frac{\Delta V_D}{I(\Delta L/L)}$$

where we note that drain current I is inversely proportional to channel length in the gradual-channel, low-field model of FET. When $V_D - V_{DMAX} \approx V_{TH}$ we have

$$R_D \approx \frac{\phi}{G_0} \quad \text{with} \quad \phi = \frac{2\sqrt{V_P V_{TH}}}{V_G - V_{TH}} \tag{8}$$

where $V_P = qN_A L^2/2\varepsilon_S$ is the drain-to-source voltage required to *punch-through* the FET, and $G_0 = B_G(V_G - V_{TH})$. Remember that Eq. (8) assumes $V_D - V_{DMAX} \approx V_{TH}$. If $N_A = 2\times 10^{16} cm^{-3}$, $\varepsilon_S = 10^{-12} F/cm$ and with L in microns, we have $V_P = 16L^2(V)$.

Capacitance C_L of Fig. 4(b) is the sum of the gate (C_G) and the drain diffused island (C_D) capacitances, given by $C_L = C_G + C_D$, and the two capacitances are approximately equal in CMOS circuits. This approximate relationship originates from the structural consistency of the integrated circuit structure, as discussed in Section 1.9. Gate capacitance C_G is given by $C_G = \varepsilon_{OX} LW/T_{OX}$, and therefore the total node capacitance C_L is estimated to be $C_L \approx 2C_G = 2\varepsilon_{OX} LW/T_{OX}$. Noise current generator i_D generates a white (frequency independent) noise current. The noise voltage developed at the node, however, depends on the node's frequency-dependent impedance. Impedance of the parallel combination of R_D and C_L decreases at high frequency. If we consider the -3db frequency of the R_D-C_L circuit of Fig. 4(a) as the *bandwidth* of noise (Δf), we have

$$\Delta f = \frac{\mu_N (V_G - V_{TH})}{2L^2} \frac{1}{\phi}$$

and therefore

$$\overline{\Delta V_N{}^2} = 4kTR_D{}^2 G_0 \cdot \frac{2}{3} \Delta f \quad \text{or} \quad \overline{\Delta V_N} = \left(\frac{4\phi}{3}\right)^{1/2} \left(\left(\frac{kT}{q}\right) \cdot \left(\frac{qT_{OX}}{\varepsilon_{OX} WL}\right)\right)^{1/2}. \tag{9}$$

Equation (9) shows that a crude estimate of thermal noise voltage amplitude of a digital node is the geometrical average of the thermal voltage $(kT/q) = 25mV$ (at room temperature) and the voltage developed at the gate capacitance when a single electronic charge is placed on the gate capacitor $[q/(\varepsilon_{OX} WL/T_{OX})]$, multiplied by a nondimensional factor of the order of unity, $\sqrt{4\phi/3}$ [05]. Since $V_P \approx 10\,V$, $V_{TH} \approx 0.5\,V$ and $V_G \approx 3\,V$, $\phi \approx 1.5$. Capacitance of a 1 μm by 3 μm gate whose gate oxide thickness is 150 angstroms is 0.007 pF. If a single electron is placed on the capacitor, the voltage developed across the terminals is 1.6×10^{-19} $(coulomb)/7 \times 10^{-15}\,(farad) = 2.3 \times 10^{-5}\,V$. Therefore

$$\overline{\Delta V_N} = 1.5 \times \sqrt{2.5 \times 10^{-2} \times 2.3 \times 10^{-5}} \approx 1\ millivolt\,.$$

This is the estimate of thermal noise at CMOS digital circuit nodes.

We attempt to compare the thermal noise with other noise, especially with noise induced by capacitive coupling [06]. The noise mechanism due to mutual capacitance coupling between nodes is schematically shown in Fig. 5(a). Two logic nodes A and B interact through coupling capacitance C_{AB} and the equivalent circuit is shown in Fig. 5(b). Nodes A and B have area S, and the insulating oxide thickness is T_{FOX}. We have

$$C_A = C_B = \frac{\varepsilon_{OX} S}{T_{FOX}} \quad \text{and} \quad C_{AB} \approx \frac{\varepsilon_{OX} S}{a}$$

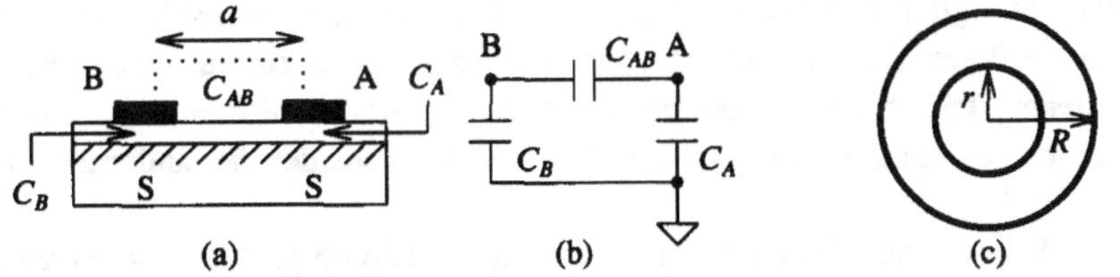

Figure 5 Induced noise voltage analysis

where a is the distance between nodes A and B. Capacitance estimates may have uncertainty factors of the order of 1, but we tolerate this to obtain a crude estimate of coupling (Section 4.4). When node A switches from 0 to V_{DD}, the voltage of *floating* node B increases due to capacitive coupling by

$$\Delta V_B = \frac{C_{AB}}{C_B + C_{AB}} V_{DD} \approx \frac{C_{AB}}{C_B} V_{DD} = \frac{T_{FOX}}{a} V_{DD} . \tag{10}$$

If $V_{DD} = 5\,V$, $T_{FOX} = 1\mu$m, and $a = 30\mu$m, (typical distance between two gates), induced noise voltage is 150mV. This estimate shows that induced noise voltage is from 2 to 3 orders of magnitude larger than the FET thermal noise voltage.

Integrated circuit nodes interact through electromagnetic induction as well. A simple estimate of the coupling strength is as follows. Suppose that there are two concentric loops of conductor on an integrated circuit chip, having radius r and R, where $r \ll R$. If current I flows through the outer loop, the magnetic field created at the center of the loop is given by $I/2R$. The voltage $|V|$ developed on the inner loop is then given by

$$|V| = \pi r^2 \frac{\mu_0}{2R} \left| \frac{dI}{dt} \right| . \tag{11}$$

We use $\mu_0 = 4\pi/10^7$ *(henry/meter)*, $r = 100\,\mu$m, $R = 1000\,\mu$m, and $|dI/dt| = 100\,(mA)/nsec$ giving $|V| = 2\,mV$. Since the parameter values are selected to represent the maximum realistic inductive coupling, this effect is approximately equal to thermal noise. This estimate is not for the case where the chip is in a strong external high frequency field.

Unless the capacitively induced noise is *shielded* completely, thermal noise and inductive noise need not be included in our primary noise evaluation. Special cases, where thermal noise must be considered, include linear small-signal amplifiers and the analysis of metastable states of bi-stable circuits.

4.3 General Theory of Induced Noise

An order-of-magnitude estimate that was given in the last section showed that digital circuit noise is primarily caused by capacitive coupling (electrostatic induction) from other circuit nodes in the vicinity. *Induced* noise in CMOS circuits depends on how the digital circuit nodes are excited. Therefore, induced noise is not strictly a random phenomenon. Practically, however, induced noise may be considered as a random phenomenon because of the number and complexity of digital circuits on a chip. Since digital noise is caused by switching of digital nodes, noise voltage is not as small as random noise in linear circuits caused by thermal or shot noise effects. The pseudorandom nature of digital circuit noise and its quite significant amplitude are the two features of digital circuit noise. We consider the randomness of induced noise first.

Figure 6(a) shows the wiring channel of a CMOS VLSI chip. Metal wires are fabricated on insulating oxide layer. Let us consider noise induced on the shaded wire from the nearby wires 1, 2, 3, ..., . We assume that the shaded wire is floating. Dotted lines show schemati-

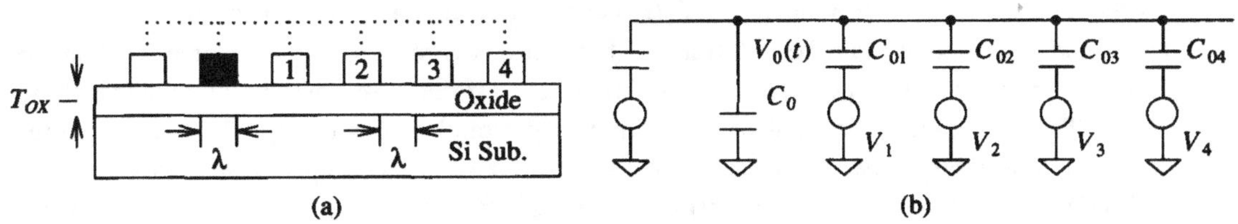

Figure 6 Randomness of induced noise phenomenon

cally the electric field lines that couple the conductors. In scaled-down CMOS, the width and gap between the conductor wires equals the feature size of the CMOS technology, λ. The capacitance of the shaded conductor to the grounded substrate, C_0, is given by

$$C_0 = f_0 \frac{\varepsilon_{OX}\lambda}{T_{OX}} \quad \text{(per unit length)} \tag{12a}$$

where f_0 is a nondimensional factor of the order of unity (typically 2–3, accounting for fringing effects). Coupling capacitance between the shaded wire and the i-th wire, C_{0i}, is approximated crudely by

$$C_{0i} = f \frac{\varepsilon_{OX}\lambda}{2i \cdot \lambda} = \frac{f\varepsilon_{OX}}{2i} \tag{12b}$$

where $(2i \cdot \lambda)$ is the approximate distance between the two conductors, and where f is a numerical factor of the order of unity (Section 4.4). $V_i(t)$ represents the node voltage of the i-th conductor switching between ground and V_{DD}. The induced noise $V_0(t)$ is then given by

$$V_0(t) = \frac{1}{C_0 + \sum_{i=1}^{N} C_{0i}} \sum_{i=1}^{N} C_{0i} \sum_{time} \Delta V_i$$

where ΔV_i is the change of $V_i(t)$. $\Delta V_i = +V_{DD}$ if a low-to-high transition takes place, and $\Delta V_i = -V_{DD}$ if a high-to-low transition takes place. The sum over time means the sum of ΔV_i over all past transitions. Before the first transition, we assume $V_0(t) = 0$. Since ΔV_i alternates $+V_{DD}$ and $-V_{DD}$, the sum over time of ΔV_i is $+V_{DD}$ and 0 or $-V_{DD}$ and 0, depending on whether the first transition is low to high or high to low, respectively. If the first transition is low to high, we set $sgn(i) = +1$ and if the first transition is high to low, $sgn(i) = -1$. Then we have

$$V_0(t) = \frac{1}{C_T} \sum_{i=1}^{N} C_{0i} V_{DD}\, sgn(i) F_i(t) = \frac{V_{DD}}{\sum_{i=1}^{N} (1/2i) + (f_0/f)(\lambda/T_{OX})} \sum_{i=1}^{N} \frac{sgn(i) F_i(t)}{2i} \tag{13}$$

where

$$C_T = C_0 + \sum_{i=1}^{N} C_{0i}$$

is the total capacitance of node 0 and $F_i(t)$ is a function that alternates between +1 and 0. In principle this function can be determined if the chip's design and details of excitation are known. In practice this is completely impractical, even if the most powerful simulator is available. There is another mechanism that increases the randomness of digital noise. Equation (13) contains a very slowly converging sum over index i. Induced noise from a large number of nodes add and subtract, and the residue becomes almost certainly random (noise voltage is not positive definite; more discussion will be given in Section 6.16). $V_0(t)$ is very much like random noise. There is yet another mechanism that increases the randomness of induced noise. Charges induced on circuit nodes dissipate through conducting paths consisting of FET channels to either V_{DD} or ground. The time required for dissipation, or the time constant of the conductor system's *relaxation* process, is different from node to node, and that depends on the details of circuit excitation also. Since every node is different in every possible way, the noise gets additional randomness. If there are a small number of predominant noise sources (or synchronization of many noise sources), however, $V_0(t)$ is not random. Analysis of this case is simpler than the random noise, since only several predominant noise sources are included. Induced noise, whether random or regular, can be analyzed using the following general theory.

Next we develop a general theory of induced noise, based on electrostatics using the following general model. Figure 7 shows the schematic of a CMOS static logic circuit that consists of capacitively coupled conductors. Each circle represents a circuit node connected by a CMOS FET, represented by a resistor, either to ground or to V_{DD}. Some nodes may be floating ($R_i = \infty$). There are $N+1$ nodes 0, 1, 2, ..., N with node 0 at ground potential. At time $t = -0$ (just before $t = 0$), node voltages are known as,

$$V_0(-0)\,(=0)\ \text{, and}\ V_1(-0), \ldots, V_N(-0) \quad (\text{either } 0 \text{ or } V_{DD})\,.$$

$V_1(-0) \cdots V_N(-0)$ are the steady-state node voltages. Charges stored at each node, $Q_i(-0)$ ($i = 0, 1, \ldots, N$) are given by [07]

$$Q_i(-0) = C_{i0} \cdot 0 + \sum_{j=1}^{N} C_{ij} V_j(-0)$$

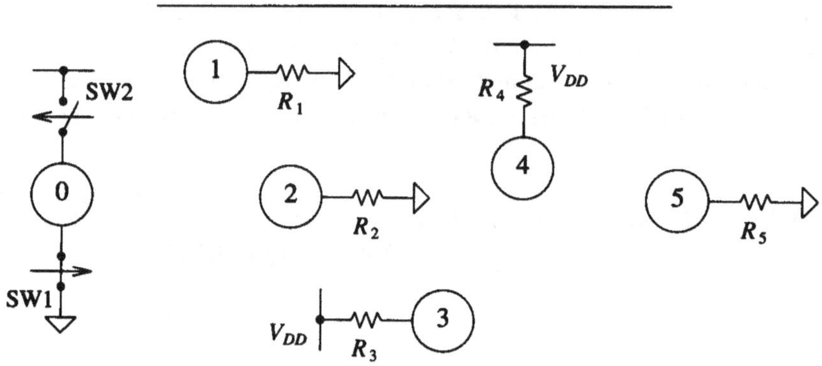

Figure 7 General theory of induced noise

where the term related to node 0 is separated out for convenience. C_{ij} is a set of capacitance coefficients of the conductor system. The reciprocity theorem of electrostatics requires that

$$C_{ij} = C_{ji} \quad (0 \le i, j \le N).$$

At $t=0$, V_0 changes abruptly from 0 to V_{DD}. Immediately after $t=0\,(t=+0)$, the voltage of nodes 1, 2, ..., N, $V_i(+0)$ is determined by solving a set of simultaneous equations

$$Q_i(-0) = C_{i0}V_{DD} + \sum_{j=1}^{N} C_{ij}V_j(+0) \quad (i=1,2,\ldots,N)$$

where $V_j(+0)\,(j=1,2,\ldots,N)$ are the conductor voltages immediately after the node 0 transition. These equations provide the initial conditions to the circuit equation of the multi-conductor system.

After the stepwise change of node voltages, they take some time to return to their respective steady state voltages. The charge on conductor i at time t, $Q_i(t)$, and its voltage, $V_i(t)$, are still related by

$$Q_i(t) = C_{i0}V_{DD} + \sum_{j=1}^{N} C_{ij}V_j(t) \tag{14a}$$

and the rate of change of $Q_i(t)$ is the current through resistance R_i to the voltage source (0 or V_{DD})

$$\frac{dQ_i(t)}{dt} = -\frac{V_i(t) - V_i(-0)}{R_i} \tag{14b}$$

where resistance R_i models FET channel. Therefore we have

$$\sum_{j=1}^{N} C_{ij}\frac{dV_j(t)}{dt} = -\frac{V_i(t) - V_i(-0)}{R_i}. \tag{15}$$

These simultaneous equations can be solved by the standard method [08]. The solution is a sum of terms of the form, $exp(St)$ as

$$V_i(t) = V_i(-0) + \sum_{j=1}^{N} A_{ij}exp(S_jt) \quad (i=1,2,\ldots.N). \tag{16}$$

The term $V_i(-0)$ is in this formula because voltages return to their value before the conductor 0 transition. The set of coefficients A_{ij} can be uniquely determined from the initial conditions at $t=+0$. Examples of noise analysis in this chapter are special cases of this generalized theory.

4.4 Magnitude of Induced Noise

High component density is achieved by packing FETs and interconnects side by side in a VLSI chip. There is capacitive coupling between devices and wires that are placed closely together, through which a voltage change of one circuit node affects the voltages of other nodes. Induced noise depends critically on layout. The worst-case structure in which two levels of interconnect overlap completely as shown in Figs. 8(a) and (b) is susceptible to induced noise.

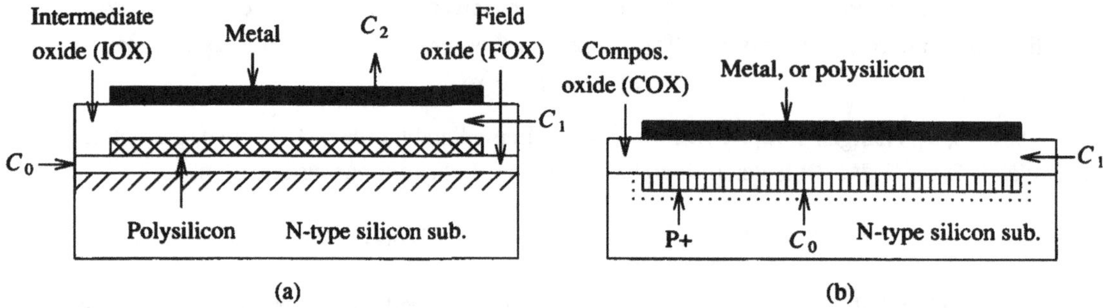

Figure 8 Structures susceptible to induced noise

Figure 8(a) shows metal-to-polysilicon capacitive coupling. We estimate magnitude of induced noise using the worst case. Polysilicon-to-substrate capacitance, C_0, and metal-to-polysilicon coupling capacitance, C_1, are identified in the figure. Metal also has capacitance to the substrate or to ground, C_2. This capacitance is small, since few electric field lines emerge from the top surface of the metal layer and reach the substrate. Switching of a metal wire from ground to V_{DD} induces voltage V_P on the *floating* polysilicon wire

$$V_P = V_{DD} \frac{C_1}{C_0 + C_1} \approx V_{DD} \frac{T_{FOX}}{T_{IOX} + T_{FOX}} \tag{17}$$

where T_{FOX} and T_{IOX} are thickness of field and intermediate oxides, respectively, and where the floating polysilicon node had ground potential before switching. Assuming $T_{IOX} \approx 2T_{FOX}$, $V_P \approx V_{DD}/3$.

Capacitively induced voltage from a polysilicon wire to a floating metal wire, V_M, is estimated using $V_M = V_{DD}[C_1/(C_1 + C_2)]$. Since C_2 is quite small, $V_M \approx V_{DD}$ and the induced noise voltage is quite significant. Coupling from a first-level metal wire to a second-level metal wire is similarly significant because a second-level metal wire has small capacitance to silicon substrate. Figure 8(b) shows capacitive coupling to P+ diffused wire from metal wire above. Equation (17) gives the noise estimate using capacitance C_0 and C_1 defined in Fig. 8(b). An advantage of diffused wires is that if the top surface is *shielded* by providing a grounded metal feature above, capacitive coupling to the diffused wires can be reduced quite significantly: grounded (or held at V_{DD}) silicon substrate or tubs on the three sides of the diffused wire reduces electrical interaction between the wires significantly. Furthermore, the space-charge layer separating diffused wire from the substrate has high capacitance.

As can be observed from these examples, the magnitude of induced noise depends on the ratio of the coupling capacitance to the node's total capacitance. The coupling capacitance is determined by the mechanical structure of the node, and the structure of other nodes. The coupling capacitance is determined by layout. The node's total capacitance depends strongly, however, on the processing technology. Bulk CMOS ICs are built on a conducting silicon substrate. Active components are fabricated on the surface of the conducting substrate, and components are isolated from the substrate by reverse-biased PN junction or oxide insulation. This structure strongly resembles traditional electronic circuits built on a PC board or on metal chassis, where the PC board or chassis works as a built-in ground reference plane. This historically proven and technically sound structure will prevail in the foreseeable future. As technology has progressed, however, several alternative structures have emerged that may be used for special purposes. SOI (silicon on insulator) technology is an example. Another recent trend is to provide more than one level of highly conductive metal interconnects. Up to three or four levels of metal interconnects are possible. Two adjacent metal layers have wires that are, most of the time, laid out perpendicular to each other. The topmost metal layer is often used to route power and ground supplies. This structure provides the second conducting substrate above the active devices.

SOS (SOI) CMOS devices are fabricated on an insulating substrate, or on silicon dioxide grown above the silicon. This structure has the advantage of reduced parasitic capacitance for FET drain-source islands and for interconnects. Smaller capacitance allows higher switching speed: SOS CMOS circuits are typically two times faster than comparable bulk CMOS circuits. GaAs MESFET devices have a similar structure, except that the insulator is GaAs itself. Figures 9(a) and (b) show interconnects of bulk and SOS CMOS ICs, respectively. Wiring capacitance of the central wire, C, is the sum of three components, C_C, C_{CA}, and C_{CB}, where C_{CA} and C_{CB} are capacitances of wire C to the two adjacent wires. Capacitances to wires further away may have a small effect, but they are neglected here. C_C is the capacitance of wire C to the substrate in bulk CMOS. In SOS CMOS, the physical meaning of capacitance C_C is not immediately clear. C_C is the sum of capacitances of wire C to all the conductors that exist *nearby*, except to wires A and B. All these conductors are, however, farther away from C than wires A and B, and therefore C_C is smaller than C_{CA} and C_{CB}. It is often negligible. In bulk CMOS ICs, C_C is larger than C_{CA} and C_{CB} by a significant factor. In SOS ICs C_C is often negligible compared with C_{CA} and C_{CB}.

The effectiveness of capacitances C_{CA} and C_{CB} as load capacitance to the driver of conductor C depends on the activity of signals on wires A and B. If signals A and B do not switch (wires A and B stay at constant potentials), C_{CA} and C_{CB} simply add to C_C. The total capacitance, C_T, is then given by

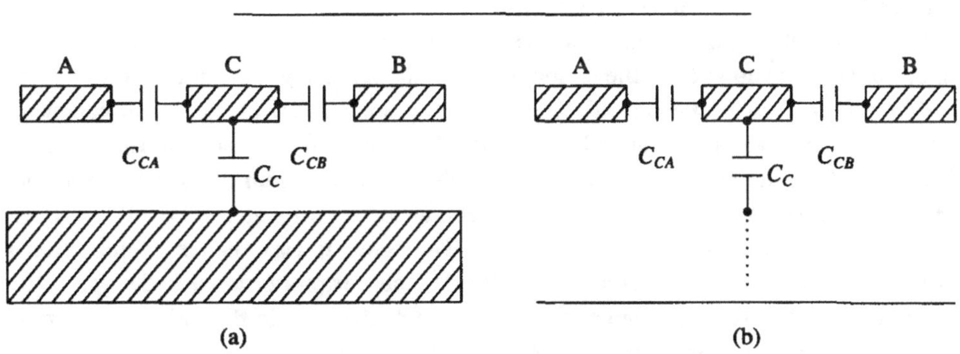

Figure 9 Schematic of interconnect

$$C_T = C_C + (C_{CA} + C_{CB}) . \tag{18}$$

If node C pulls up, and simultaneously the drivers of nodes A and B pull down, nodes A and B pull node C down through C_{CA} and C_{CB}. Since the node C driver pulls node A and B up simultaneously, the effects of capacitances C_{CA} and C_{CB} double. The total capacitance of node C_T is then given by

$$C_T = C_C + 2(C_{CA} + C_{CB}) . \tag{19}$$

If node C pulls up, and simultaneously nodes A and B pull up, the three nodes A, B, and C do not interact. Then the total capacitance of node C, C_T, is

$$C_T = C_C . \tag{20}$$

Equation (18) is typical, Eq. (19) is pessimistic, and Eq. (20) is an optimistic estimate.

C_T of a bulk CMOS circuit node under typical conditions may be considered as a constant capacitance, since $C_C \gg C_{CA}, C_{CB}$. In the case of SOS CMOS, $C_C \ll C_{CA} + C_{CB}$, and therefore the node capacitance depends essentially on the state changes of the nearby nodes. This effect has an important practical consequence. Timing simulation for design verification of gate array or standard cell devices are conventionally carried out using a gate-level timing simulator. A conventional gate-level timing simulator determines signal delays using capacitance values extracted from the layout. Capacitance contributed from interconnects is the most important part in wiring-intensive devices like standard cell devices. The capacitance is connected between the node and ground (or V_{DD}). Inclusion of the effects of the capacitances that couple different logic nodes is complicated (increase in the data size, the computation time and difficulties in the coupling capacitance determination). Since capacitive coupling between nodes is dominant in SOS devices, the structure of a gate-level timing simulator must be fundamentally altered to include the effects of the capacitive coupling. Existing gate-level timing simulators are unable to determine accurately the gate delay of SOS devices. A similar problem exists in GaAs MESFET ICs.

Yet another problem of SOS/SOI devices is the following. CMOS dynamic logic circuits use charge stored in the gate's output node capacitance. Since SOS devices do not have well-defined capacitance assigned to nodes, SOS/SOI CMOS dynamic logic circuits cannot be designed unless stable capacitance is added. Since this is cumbersome, SOS CMOS devices are usually designed using static logic circuits.

Dependence of coupling capacitance between two parallel wires on the distance between them, c, can be understood from the simple physical model of Fig. 10. In Fig. 10(a) the electric field lines are shown schematically. Conductor A is positively charged, and conductor B and the silicon substrate are grounded. We study the electric flux emanating from conductor A. If the two wires are close together or $c \to a$, where a is the width of the wire, most of the electric flux originating from the top surface of conductor A terminates at conductor B. Conductors A and B form a parallel plate capacitor. The coupling capacitance is proportional to a/c, where a is the area of conductor per unit length, and c is proportional to the gap between the electrodes. If the distance is large, however, most of the electric flux originating from the top surface of conductor A terminates on the grounded substrate. The electric flux in region 2 of Fig. 10(b) adds to conductor A-substrate capacitance. Only the small portion of the flux

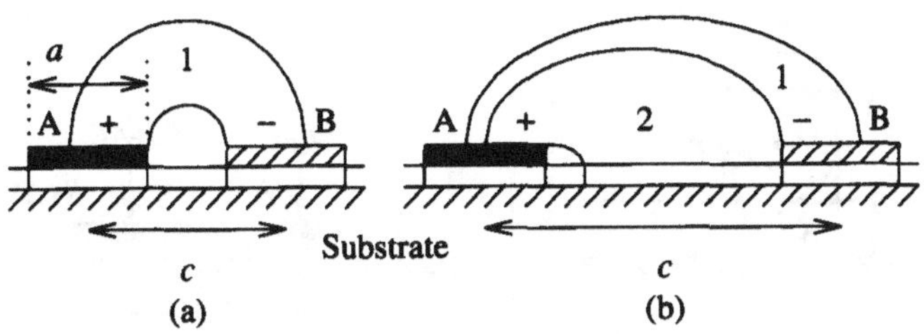

Figure 10 Coupling capacitance versus distance

originating from the top surface and proportional to a/c contributes to mutual coupling. Then we have

$$C_{AB} \approx a/c \quad (c \to 0) \quad C_{AB} \approx (a/c)^2 \quad (c \to \infty).$$

The coupling capacitance between two conductor wires whose sizes are comparable to their separation must be determined numerically. Figure 11 shows the coupling capacitance between two rectangular conductors versus distance. The dimensions are given in the inset. The capacitance per unit length (perpendicular to the page) is independent of the scale. Within the cross section only relative dimensions matter. Relative dielectric constants of oxide and nitride are 4 and 7, respectively. Within the range of the D shown, dependence of coupling capacitance on D changes from proportional to the inverse of the distance to the square of the inverse of the distance, as was predicted.

CMOS circuits like bistable latches and differential amplifiers are sensitive only to the *difference* in charges induced to the two *complementary* nodes. We give a simple theory to estimate the differential induced charge. In Fig. 12 a conductor having a square cross section of $W \times W$ is centered at point O. The conductor switches between the two CMOS logic levels.

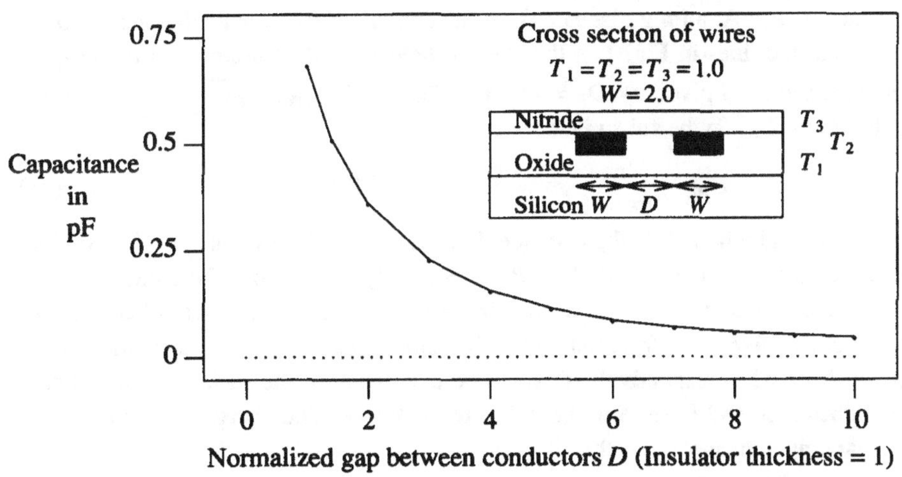

Figure 11 Line-to-line coupling capacitance

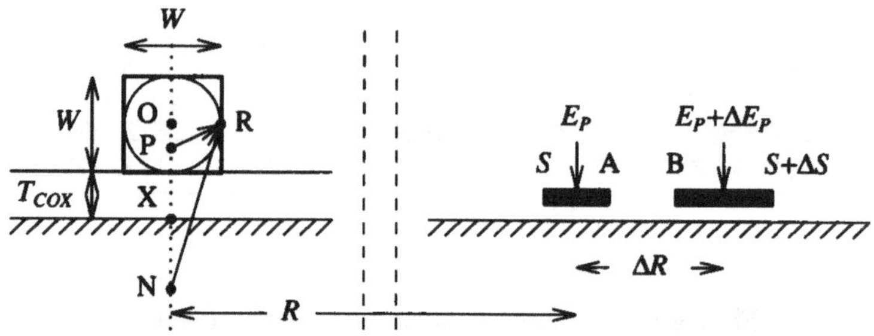

Figure 12 Differential noise induction analysis

The conductor is distance T_{COX} (thickness of the underlying oxide) above the grounded silicon substrate. The square conductor is approximated by a cylinder shown in Fig. 12. The field of the cylindrical conductor and the grounded silicon substrate can be equivalently described by two *image* line charges $\pm q$ per unit length, placed at P and N. The equipotential surfaces of the pair of line charges are *confocal* cylinders. The potential at arbitrary point Q (not shown in Fig. 12) is given by

$$\phi(Q) = -[q/(2\pi\varepsilon)]\log(r_P/r_N) \tag{24}$$

where r_P and r_N are the distances to point Q from line charges P and N, respectively. For the cylinder to best approximate the square conductor, we find that the location of line charge or distance $\overline{XP}$ ($=\overline{XN}$) should be $\overline{XP} = \sqrt{T_{COX}(T_{COX}+W)}$. The capacitance per unit length of the conductor is given, approximately, by

$$C = \frac{2\pi\varepsilon}{\log\dfrac{\sqrt{T_{COX}+W}+\sqrt{T_{COX}}}{\sqrt{T_{COX}+W}-\sqrt{T_{COX}}}}.$$

When this conductor is held at V_{DD}, charge on the conductor q is given by $q = CV_{DD}$ (per unit length). At a long distance from the conductor (schematically shown to the left of the pair of dashed lines in Fig. 12), the electric field is well approximated by a dipole field with dipole moment D_P given by $D_P = q \cdot 2\overline{XP} = 2CV_{DD}\sqrt{T_{COX}(T_{COX}+W)}$. The dipole creates an electric field, E_P, on the substrate surface

$$E_P = \frac{D_P}{2\pi\varepsilon R^2} \quad \text{where} \quad D_P = q \cdot 2\overline{XP} \tag{25}$$

is the diople moment, and where R is the distance defined in Fig. 12. Let us consider conductors A and B, which are distance R and $R+\Delta R$, respectively, away from the switching conductor. Conductors A and B model a pair of the circuit nodes, such as cross-coupled nodes of a flip-flop, that are sensitive to differential capacitive induction. The fields at conductors A and B are E_P and $E_P+\Delta E_P$, respectively, directed perpendicular to the surface of the silicon. The areas of conductors A and B are S and $S+\Delta S$, respectively. The charge induced on conductors A and B are given, respectively, by

$$Q_A = \varepsilon E_P S \quad Q_B = \varepsilon(E_P+\Delta E_P)(S+\Delta S) = Q_A + \varepsilon(\Delta E_P S + E_P \Delta S).$$

The difference in charge is $\Delta Q = Q_B - Q_A = \varepsilon(\Delta E_P \cdot S + E_P \cdot \Delta S)$. Assuming the dipolar field of the conductor line

$$\Delta Q = \frac{D_P S}{\pi R^2}\left(-\frac{\Delta R}{R} + \frac{\Delta S}{2S}\right). \tag{26}$$

If conductors A and B are identical, $\Delta S = 0$ and therefore $\Delta Q = -(D_P S/\pi R^3)\Delta R$. The differential charge induced on a pair of *symmetrical* nodes is inversely proportional to the third power of the distance.

We give an example of exact field analysis that is relevant to capacitive coupling of the topmost conductor, like second-level metal. This analysis clarifies some ambiguity in the definition of capacitance on chip also. Figure 13 shows the cross section of a long, rectangular capacitor fabricated on a silicon substrate. The long, rectangular shape is realistic in modeling interconnects. The structure and approximate shapes of the electric field lines are shown in Figs. 13(a) and (b), respectively. A first-order approximation for capacitance C (to the substrate) is given by

$$C = [\varepsilon_0(2aL)]/T_{FOX} \tag{27}$$

where T_{FOX} is the field oxide thickness and L is the conductor length. In order to improve the capacitance approximation, capacitance contributed from the fringing field at the edge of the capacitor, C_F (due to field lines like α) and capacitance contributed from the upper surface C_U (field lines like β, γ and δ) must be added to the parallel plate capacitance of Eq. (27). C_F is an integral part of the capacitor. C_F is called a *fringing* capacitance. The field lines associated with this capacitance exist in the vicinity of the capacitor edge. The qualitative difference between $C + C_F$ and C_U is as follows: If the upper surface is filled with a different dielectric, $C + C_F$ does not change, but C_U does. We are now interested in capacitance C_U, the upper surface capacitance.

We solve the electrostatic field problem of the conductors shown in Fig. 13(b), where a long, rectangular metal strip has width $2a$ and potential V_0. The region outside the strip is grounded. We calculate the charge, Q, on the upper surface using

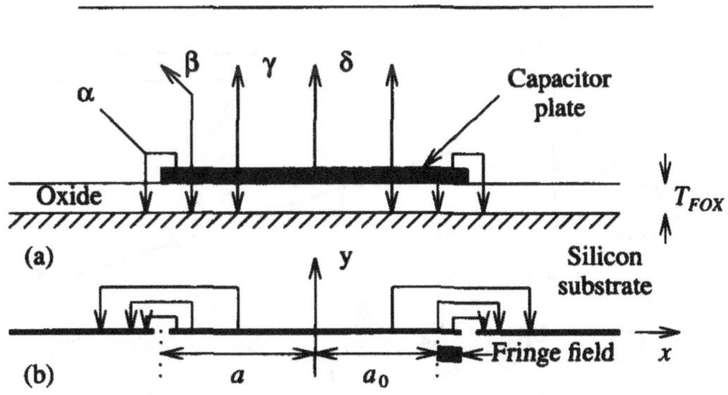

Figure 13 Capacitive coupling through top surface

$$Q = \int_{-a_0}^{a_0} \sigma(x)dx \tag{28}$$

where $\sigma(x)$ is the surface charge density (that is symmetrical) and a_0 ($< a$) is the effective width of the top surface of the capacitor. It is rational to assume, from geometrical consideration, $a - a_0 \approx T_{FOX}$.

We use the x-y coordinates shown in Fig. 13(b), and the potential $\phi(x, y)$. We have

$$\sigma(x) = \varepsilon_0 \left| \frac{\partial \phi(x, y)}{\partial y} \right|_{y=0} \tag{29}$$

where ε_0 is the dielectric constant of the material above (we assume vacuum), and where $\phi(x, y)$ satisfies the Laplace equation

$$\frac{\partial^2 \phi}{\partial x^2} + \frac{\partial^2 \phi}{\partial y^2} = 0 \tag{30}$$

This equation can be solved by substitution of $\phi(x, y) = X(x)Y(y)$ and separating variables. We find $\phi(x, y) = exp(-\mu y)\cos(\mu x)$ and where μ is a separation constant. The general solution that satisfies the boundary condition is

$$\phi(x, y) = \int_0^\infty f(\mu) exp(-\mu z)\cos(\mu x)d\mu$$

where arbitrary function $f(\mu)$ is determined from the boundary condition

$$\phi(x, 0) = V_0 \ (-a < x < a) \ \ \phi(x, 0) = 0 \ (|x| > a)$$

as

$$\phi(x, y) = \frac{2V_0}{\pi} \int_0^\infty \frac{\sin(a\mu)}{\mu} exp(-\mu y)\cos(\mu x)d\mu. \tag{31}$$

The charge density on the surface of the strip, $\sigma(x)$, is given by $\sigma(x) = \varepsilon_0 |\partial\phi(x)/\partial y|_{y=0}$. Charge within a strip of width a_0 and the capacitance C_U contributed from that part is given, respectively, by

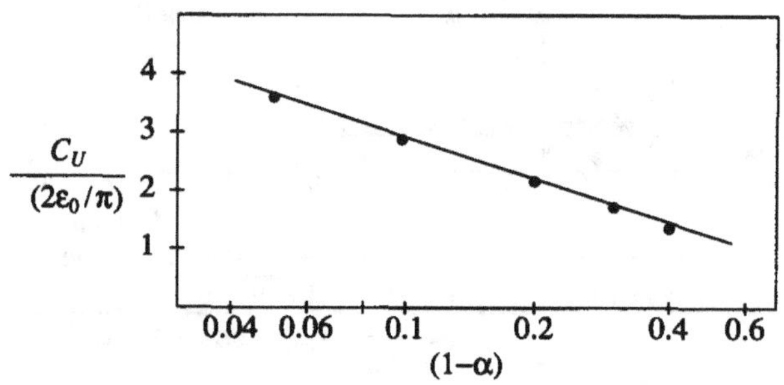

Figure 14 Top-surface coupling of capacitor

$$Q = \int_{-a_0}^{a_0} \sigma(x)dx \quad \text{and} \quad C_U = \frac{Q}{V}_0 = \frac{2\varepsilon_0}{\pi} \int_{-\alpha}^{\alpha} d\xi \int_0^{\infty} \sin\theta \cos(\xi\theta)d\theta \tag{32}$$

where $\alpha = a_0 / a$ and $\xi = x / a$. The *improper* integral can be evaluated in closed form as

$$C_U = [(2\varepsilon_0)/\pi] \log[(1+\alpha)/(1-\alpha)]. \tag{33}$$

Figure 14 shows $C_U / (2\varepsilon_0 / \pi)$ versus $1 - \alpha$, where $\alpha = a_0 / a$.

4.5 Metal-to-Metal Induced Noise

We saw that the induced noise pulse height is determined by the ratio of coupling capacitance to node capacitance [Eq. (17)]. We will show that the width of the induced noise pulse is determined by how fast the node is able to charge or discharge. The node returns to the original, steady-state voltage after some time. In a tristated node that is unable to charge or discharge, the effect of induced noise is permanent.

The most common inter-node capacitive coupling in a CMOS VLSI chip is between two metal wires. If two wires are side by side, significant coupling capacitance exists between the two wires. In addition to the coupling capacitance, each wire has its own capacitance to ground, or to V_{DD}. The mechanism of metal-wire-to-metal-wire induced noise is essentially capacitive voltage division followed by resistance-capacitance *relaxation* process, represented by the equivalent circuit shown in Fig. 15 where node A is driven by a low internal resistance, time-dependent voltage source, $V_1(t)$ [09]. Nodes A and B are coupled by capacitance C_1. Node B is initially at one of the steady-state CMOS logic levels (V_{SS} or V_{DD}) through a driver, whose internal resistance is approximated by a linear resistance, R_0. If the node B driver is in the steady state, and if the amplitude of the induced noise pulse is small, a linear resistor is a good approximation. C_0 is the capacitance of node B to ground and to all the other capacitively coupled nodes. Given $V_1(t)$, we determine $V_0(t)$. Suppose that $V_1(t)$ is zero until time $t=0$ and $V_0(t)$ is zero also. At $t=0$, $V_1(t)$ begins to increase. $V_0(t)$ begins to increase immediately, because capacitive voltage division has no delay. $V_1(t)$ increases with time, approaching

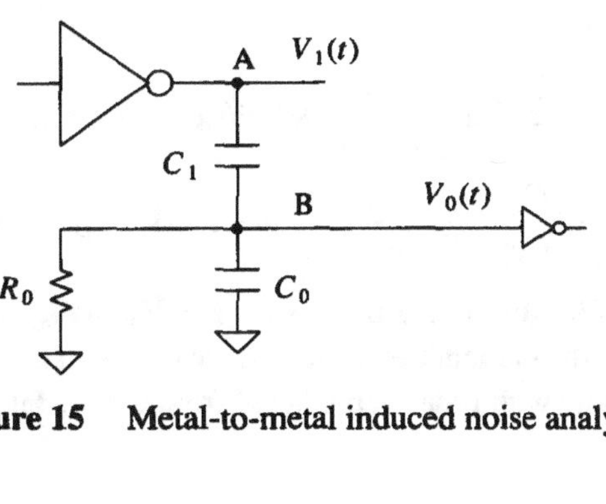

Figure 15 Metal-to-metal induced noise analysis

V_{DD} as $t \to \infty$. As $V_0(t)$ increases, current through resistance R_0 drains the induced charge to ground. $V_0(t)$ reaches a maximum and then decreases. For large t $V_0(t)$ approaches zero. If t is small, $V_0(t)$ is determined by capacitive division of $V_1(t)$, and if t is large, by the discharge process of the total capacitance $(C_0 + C_1)$ through resistance R_0.

The circuit equation satisfied by $V_0(t)$ is

$$C_0 \frac{dV_0(t)}{dt} + \frac{V_0(t)}{R_0} = C_1 \frac{d}{dt}[V_1(t) - V_0(t)]$$

which is solved as

$$V_0(t) = \frac{C_1}{C_0 + C_1} exp(-t/T_D) \int_0^t exp(\theta/T_D) V_1'(\theta) d\theta \tag{34}$$

$$\text{where} \quad T_D = (C_0 + C_1) R_0$$

We assume that $V_1(t) = 0$ if $t < 0$, and $V_1(t) > 0$ if $t > 0$. Then we have,

$$V_0(t) = \frac{C_1}{C_T}[V_1(t) - \frac{exp(-t/T_D)}{T_D} \int_0^t exp(\theta/T_D) V_1(\theta) d\theta] \tag{35}$$

where $C_T = C_0 + C_1$ is the total capacitance of node B. Equation (35) shows, in general, that

$$V_0(t) \to \frac{C_1}{C_T} V_1(t) \quad (t \to 0). \tag{36}$$

This is because the integral in Eq. (35) is proportional to t^2 in the limit as $t \to 0$. Equation (36) gives the induced noise voltage immediately after the noise source pulse is turned on. In the special case, $V_1(t) = 0$ $(t \le 0)$ and $V_1(t) = V_{DD}$ $(t > 0)$ Eq. (35) can be integrated to obtain

$$V_0(t) = (C_1/C_T) V_{DD} exp(-t/T_D) \quad \text{(waveform1)}. \tag{37}$$

In another special case,

$$V_1(t) = 0 \ (t < 0) \text{ and } V_1(t) = V_{DD}[1 - exp(-t/t_A)] \ (t > 0).$$

Then

$$V_0(t) = \frac{C_1}{C_T} V_{DD} \frac{T_D}{t_A - T_D}[exp(-t/t_A) - exp(-t/T_D)] \quad \text{(Waveform 2)}. \tag{38}$$

In the limit as $t \to \infty$

$$\to \frac{C_1}{C_T} V_{DD} \frac{T_D}{t_A - t_D} exp(-t/t_A) \quad (t_A > T_D)$$

$$\to \frac{C_1}{C_T} V_{DD} \frac{T_D}{T_D - t_A} exp(-t/T_D) \quad (T_D > t_A)$$

Equations (37) and (38) are normalized using $v_0 = V_0(t)/V_{DD}$ and $v_1 = V_1(t)/V_{DD}$ and are plotted in Fig. 16. Generally, the induced voltage increases with time, attains a maximum, and then decreases exponentially with time, as predicted from the model.

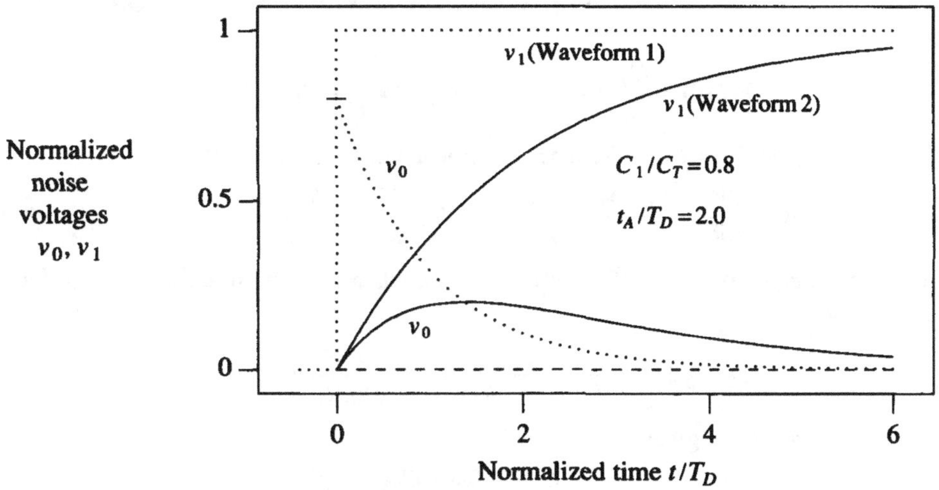

Figure 16 Metal-to-metal induced noise waveform

4.6 Metal-to-Polysilicon Induced Noise

Induced noise from a metal wire to a resistive polysilicon wire can be analyzed using the equivalent circuit shown in Fig. 17. Since the potential of a resistive polysilicon wire depends on location, a single variable description of the polysilicon wire voltage (as used in the previous section) is not adequate. Inclusion of the entire potential *profile* of the polysilicon wire, however, makes the problem complex. In the equivalent circuit of Fig. 17, a compromise was taken: the potential profile on the polysilicon wire is represented by two voltage variables, $V_0(t)$ and $V_1(t)$, at the two ends of the wire.

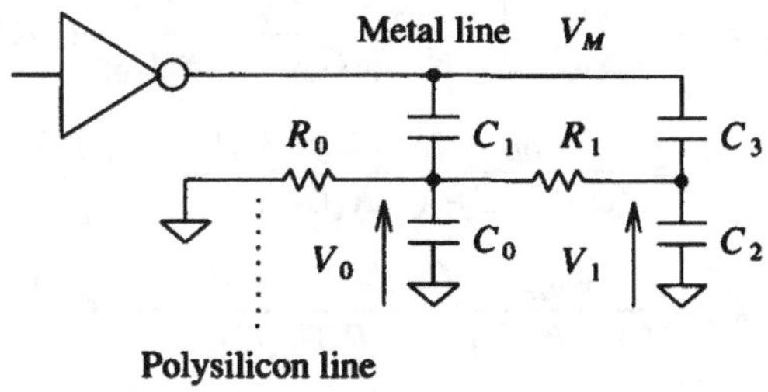

Figure 17 Metal-to-resistive-polysilicon induced noise analysis

At time $t=0$, the metal wire is pulled up from ground to V_{DD} within a negligibly short time. After pull-up

$$V_0(+0)=\frac{C_1}{C_{01}}V_{DD} \quad \text{and} \quad V_1(+0)=\frac{C_3}{C_{23}}V_{DD}$$

where $C_{01}=C_1+C_0$ and $C_{23}=C_2+C_3$. The circuit equations for time $t>0$ are

$$C_{01}\frac{dV_0}{dt}=\frac{V_1-V_0}{R_1}-\frac{V_0}{R_0} \qquad C_{23}\frac{dV_1}{dt}=-\frac{V_1-V_0}{R_1} \tag{39}$$

The solutions of the equations are of the form $exp(-St)$, where S is the solution of the following *secular* equation

$$C_{01}C_{23}S^2-\left(\frac{C_{23}}{R_0}+\frac{C_{01}+C_{23}}{R_1}\right)S+\frac{1}{R_0R_1}=0 \tag{40}$$

whose solutions, α and β, are given by,

$$\alpha\,(+),\,\beta\,(-)=\frac{\dfrac{C_{23}}{R_0}+\dfrac{C_{01}+C_{23}}{R_1}\pm\sqrt{\Delta}}{2C_{01}C_{23}}$$

$$\text{where}\quad \Delta=\left(\frac{C_{23}}{R_0}-\frac{C_{01}+C_{23}}{R_1}\right)^2+\frac{4C_{23}{}^2}{R_0R_1}.$$

The node voltages are given by,

$$V_0(t)=A_0exp(-\alpha t)+B_0exp(-\beta t) \qquad V_1(t)=A_1exp(-\alpha t)+B_1exp(-\beta t) \tag{41}$$

where

$$\frac{A_1+B_1-(A_0+B_0)}{R_1C_{23}}=\alpha A_1+\beta B_1 \qquad -\frac{A_1+B_1}{R_1C_{01}}+\frac{(R_0+R_1)(A_0+B_0)}{R_0R_1C_{01}}=\alpha A_0+\beta B_0.$$

The solution must satisfy the initial conditions

$$V_0(0)=A_0+B_0=(C_1/C_{01})V_{DD} \qquad V_1(0)=A_1+B_1=(C_3/C_{23})V_{DD}.$$

From the requirements, the coefficients A_0, B_0, A_1, and B_1 are determined as follows:

$$A_0=\frac{V_{DD}}{(\alpha-\beta)C_{01}}\left(-\frac{C_0C_3-C_1C_2}{R_1C_{01}C_{23}}+\frac{C_1}{R_0C_{01}}-\beta C_1\right)$$

$$B_0=\frac{V_{DD}}{(\alpha-\beta)C_{01}}\left(\frac{C_0C_3-C_1C_2}{R_1C_{01}C_{23}}-\frac{C_1}{R_0C_{01}}+\alpha C_1\right)$$

$$A_1=\frac{V_{DD}}{(\alpha-\beta)C_{23}}\left(\frac{C_0C_3-C_1C_2}{R_1C_{01}C_{23}}-\beta C_3\right)$$

$$B_1=\frac{V_{DD}}{(\alpha-\beta)C_{23}}\left(-\frac{C_0C_3-C_1C_2}{R_1C_{01}C_{23}}+\alpha C_3\right).$$

Equation (41) is plotted in Fig. 18. We give to the normalized variables $v_0(t)=V_0(t)/V_{DD}$, $v_1(t)=V_1(t)/V_{DD}$ the following parameter values: $R_0=R_1=1.0$, $C_0=C_2=1.0$ and

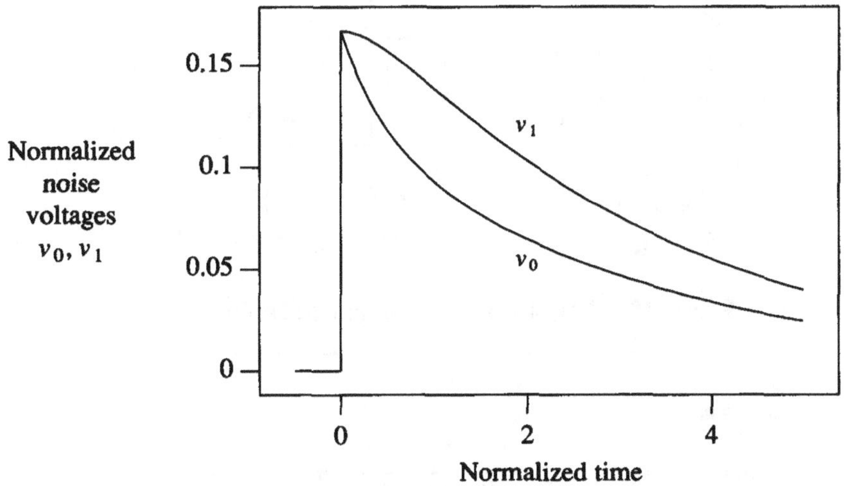

Figure 18 Metal-to-polysilicon induced noise

$C_1 = C_3 = 0.2$. The step increase in noise voltage at $t=0$ is because the V_M makes a step increase at that moment. If $V_M(t)$ increases continuously, $V_1(t)$ and $V_0(t)$ immediately after $t=0$ are approximated by

$$V_0(t) = (C_1/C_{01})V_M(t) \quad V_1(t) = (C_3/C_{23})V_M(t).$$

In Fig. 18, the waveforms are similar to those of metal-to-metal induced noise, but the decay time constant of the waveform depends on the location on the polysilicon wire. $V_0(t)$ (near the grounded end) decays faster than $V_1(t)$ (far from the grounded end). The initial rate of decrease of $V_1(t)$ is zero. This effect significantly stretches the noise pulse at the end of the polysilicon line.

4.7 Polysilicon-to-Metal Induced Noise

The polysilicon wire is described by two voltages, V_0 and V_1, as before. In order to solve this problem in closed form, we use the following approximation. We first assume that the metal wire is grounded as shown in Fig. 19(a), and we find the current I_N from the polysilicon wire to ground. We then compute V_N across the parallel connection of the driver's internal resistance R_D and capacitance C_L that exist between the metal wire and ground. This parallel combination is assumed to have a small impedance to the current injected from the polysilicon wire, and therefore is assumed to have little effect on the current. The equivalent circuit is shown in Fig. 19(b).

After node I of Fig. 19(a) is pulled up from ground to V_{DD} within a negligibly short time, node voltages V_0 and V_1 satisfy circuit equations

$$(C_0 + C_1)\frac{dV_0}{dt} = \frac{V_{DD} - V_0}{R_0} - \frac{V_0 - V_1}{R_1} \quad (C_0 + C_1)\frac{dV_1}{dt} = \frac{V_0 - V_1}{R_1} \tag{42}$$

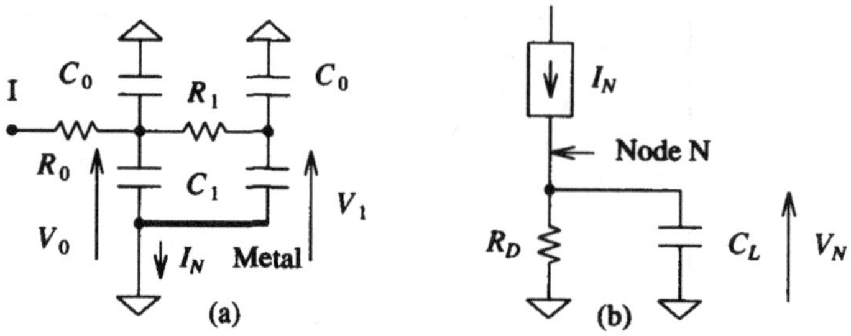

Figure 19 Polysilicon-to-metal induced noise

and the initial conditions are

$$V_0(0)=V_1(0)=0 \quad V_0'(0)=\frac{V_{DD}}{(C_0+C_1)R_0} \quad \text{and} \quad V_1'(0)=0.$$

Assuming $V_0(t)$ and $V_1(t)$ are the sums of the terms of the form $exp(-St)$, parameter S satisfies

$$S^2-\frac{1}{(C_0+C_1)}(\frac{1}{R_0}+\frac{2}{R_1})S+\frac{1}{R_0R_1(C_0+C_1)^2}=0. \tag{43}$$

Two roots of this equation, α and β, satisfy

$$\alpha+\beta=\frac{1}{C_0+C_1}(\frac{1}{R_0}+\frac{2}{R_1}) \quad \alpha\beta=\frac{1}{R_0R_1(C_0+C_1)^2}.$$

$V_0(t)$ and $V_1(t)$ satisfying the initial conditions are

$$V_0(t)=V_{DD}+Aexp(-\alpha t)+Bexp(-\beta t) \tag{44}$$

$$V_1(t)=V_{DD}+Cexp(-\alpha t)+Dexp(-\beta t)$$

where coefficients A, B, C and D are given by

$$A=\frac{1-\beta(C_0+C_1)R_0}{(C_0+C_1)R_0(\beta-\alpha)}V_{DD} \quad B=\frac{-1+\alpha(C_0+C_1)R_0}{(C_0+C_1)R_0(\beta-\alpha)}V_{DD}$$

$$C=-\frac{\beta}{\beta-\alpha}V_{DD} \quad D=\frac{\alpha}{\beta-\alpha}V_{DD}.$$

By using this solution, the total current to ground is found as

$$I_{total}=(C_0+C_1)\frac{d}{dt}[V_0(t)+V_1(t)]$$

$$=-\frac{\alpha V_{DD}}{\beta-\alpha}[\frac{1}{R_0}-2\beta(C_0+C_1)]exp(-\alpha t)+\frac{\beta V_{DD}}{\beta-\alpha}[\frac{1}{R_0}-2\alpha(C_0+C_1)]exp(-\beta t)$$

and the current injected to the metal wire, $I_N(t)$ is given by

$$I_N(t)=\left(\frac{C_1}{C_0+C_1}\right)I_{total}=X_\alpha exp(-\alpha t)+X_\beta exp(-\beta t) \tag{45}$$

$$X_\alpha = -\left(\frac{C_1}{C_0+C_1}\right)\left(\frac{\alpha V_{DD}}{\beta-\alpha}\right)\left(\frac{1}{R_0} - 2\beta(C_0+C_1)\right)$$

$$X_\beta = \left(\frac{C_1}{C_0+C_1}\right)\left(\frac{\beta V_{DD}}{\beta-\alpha}\right)\left(\frac{1}{R_0} - 2\alpha(C_0+C_1)\right).$$

Using these results, $V_N(t)$ that is developed across the parallel combination of capacitance C_L and resistance R_D is found using the equivalent circuit of Fig. 19(b). The circuit equation

$$C_L \frac{dV_N(t)}{dt} + \frac{V_N(t)}{R_D} = I_N(t) \tag{46}$$

is solved subject to the initial condition $V_N(0)=0$ to obtain

$$V_N(t) = \frac{1}{C_L}\left[\frac{X_\alpha[exp(-\alpha t) - exp(-t/C_L R_D)]}{(1/C_L R_D) - \alpha} + \frac{X_\beta[exp(-\beta t) - exp(-t/C_L R_D)]}{(1/C_L R_D) - \beta}\right] \tag{47}$$

Figure 20 shows V_N. The node voltages are normalized using $v_0 = V_0(t)/V_{DD}$ and $v_1 = V_1(t)/V_{DD}$. The parameter values are: $C_0 = 1.0$, $C_1 = 0.5$, $R_0 = 1.0$, $R_1 = 1.0$, $C_L = 5.0$ and $R_D = 0.2$. The noise voltage waveform is similar to that of the metal-metal or the metal-polysilicon induced noise. Although the driver's end of the polysilicon line was pulled up instantly, $V_N(t)$ increases continuously with time at $t=0$. This is because the stepwise increase of the voltage at the driver's end of the polysilicon line is attenuated by the resistance and capacitance of the polysilicon line.

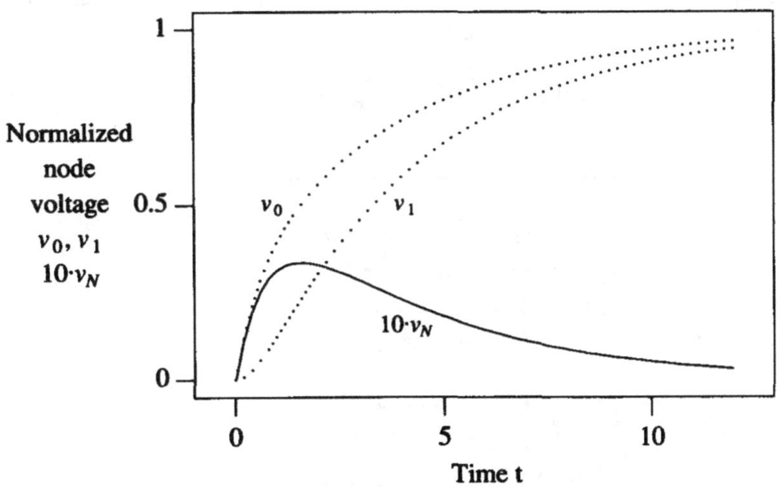

Figure 20 Polysilicon-to-metal induced noise

4.8 Polysilicon-to-Polysilicon Induced Noise

Polysilicon-wire-to-polysilicon-wire induced noise can be analyzed using the simplified equivalent circuit shown in Fig. 21. The four node voltages V_{A0}, V_{A1}, V_{B0}, and V_{B1} satisfy the following circuit equations.

$$C_0 \frac{dV_{A0}}{dt} + C_1 \frac{d}{dt}(V_{A0} - V_{B0}) + \frac{V_{A0} - V_{A1}}{R_1} = \frac{V_{DD} - V_{A0}}{R_0} \tag{48}$$

$$C_0 \frac{dV_{A1}}{dt} + C_1 \frac{d}{dt}(V_{A1} - V_{B1}) = \frac{V_{A0} - V_{A1}}{R_1}$$

$$C_0 \frac{dV_{B0}}{dt} + C_1 \frac{d}{dt}(V_{B0} - V_{A0}) + \frac{V_{B0}}{R_0} + \frac{V_{B0} - V_{B1}}{R_1} = 0$$

$$C_0 \frac{dV_{B1}}{dt} + C_1 \frac{d}{dt}(V_{B1} - V_{A1}) = \frac{V_{B0} - V_{B1}}{R_1}$$

A symmetrical pair of polysilicon lines was assumed. The four equations are transformed using

$$V_{A0} + V_{B0} = \psi_0 \quad V_{A1} + V_{B1} = \psi_1 \quad V_{A0} - V_{B0} = \psi_2 \quad V_{A1} - V_{B1} = \psi_3$$

to

$$C_0 \frac{d\psi_0}{dt} = \frac{V_{DD}}{R_0} - (\frac{1}{R_0} + \frac{1}{R_1})\psi_0 + \frac{\psi_1}{R_1} \qquad C_0 \frac{d\psi_1}{dt} = \frac{\psi_0 - \psi_1}{R_1}$$

$$(C_0 + 2C_1)\frac{d\psi_2}{dt} = \frac{V_{DD}}{R_0} - (\frac{1}{R_0} + \frac{1}{R_1})\psi_2 + \frac{\psi_3}{R_1} \qquad (C_0 + 2C_1)\frac{d\psi_3}{dt} = \frac{\psi_2 - \psi_3}{R_1}.$$

Assuming that ψ_0 to ψ_1 are sums of terms of the form $exp(-St)$, the first two equations containing ψ_0 and ψ_1 give the following secular equation

$$C_0^2 S^2 - C_0(\frac{1}{R_0} + \frac{2}{R_1})S + \frac{1}{R_0 R_1} = 0. \tag{49}$$

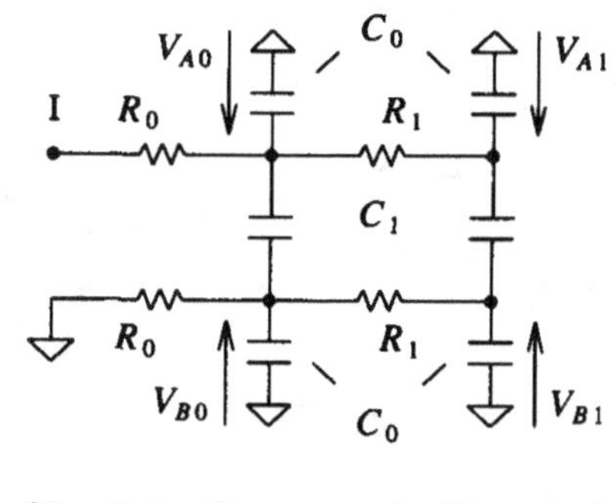

Figure 21 Polysilicon-to-polysilicon induced noise

If the two solutions are α and β, they satisfy

$$\alpha+\beta=\frac{1}{C_0}\left(\frac{1}{R_0}+\frac{2}{R_1}\right)\quad \alpha\beta=\frac{1}{R_0R_1C_0^2}.$$

For the second two circuit equations

$$(C_0+2C_1)^2S^2-(C_0+2C_1)\left(\frac{1}{R_0}+\frac{2}{R_1}\right)S+\frac{1}{R_0R_1}=0. \tag{50}$$

Here C_0 of the first secular equation is replaced by C_0+2C_1. If the solution of this equation is γ and δ,

$$\frac{\alpha}{\gamma}=\eta\quad \frac{\beta}{\delta}=\eta\quad \text{where}\quad \eta=1+2\frac{C_1}{C_0}\ (>1).$$

The solution that satisfies the initial conditions

$$V_{A0}(0)=V_{B0}(0)=V_{A1}(0)=V_{B1}(0)=0$$

$$V_{A0}(\infty)=V_{A1}(\infty)=V_{DD}\quad V_{B0}(\infty)=V_{B1}(\infty)=0\ \text{ or}$$

$$\psi_0(0)=\psi_1(0)=\psi_2(0)=\psi_3(0)=0\quad \psi_0(\infty)=\psi_1(\infty)=\psi_2(\infty)=\psi_3(\infty)=V_{DD}$$

is given by

$$\psi_0=V_{DD}+A_0exp(-\alpha t)-(V_{DD}+A_0)exp(-\beta t) \tag{51}$$

$$\psi_1=V_{DD}+A_1exp(-\alpha t)-(V_{DD}+A_1)exp(-\beta t)$$

$$\psi_2=V_{DD}+A_2exp[-(\alpha/\eta)t]-(V_{DD}+A_2)exp[-(\beta/\eta)t]$$

$$\psi_3=V_{DD}+A_3exp[-(\alpha/\eta)t]-(V_{DD}+A_3)exp[-(\beta/\eta)t].$$

The solution must satisfy

$$\left[\frac{dV_{A0}}{dt}\right]_{t=0}=\frac{V_{DD}}{R_0}\frac{1}{C_T}\quad \left[\frac{dV_{B0}}{dt}\right]_{t=0}=\frac{C_1}{C_0+C_1}\left[\frac{dV_{A0}}{dt}\right]_{t=0}$$

where $C_T=C_0+[C_0C_1/(C_0+C_1)]$ is the total capacitance of node N_{A0}. We have

$$A_0=\frac{V_{DD}}{\beta-\alpha}\frac{1-\beta R_0C_0}{R_0C_0}\quad V_{DD}+A_0=\frac{V_{DD}}{\beta-\alpha}\frac{1-\alpha R_0C_0}{R_0C_0}.$$

Using parameters A and B defined by

$$A=\frac{V_{DD}}{\beta-\alpha}\left(\frac{1}{R_0C_0}-\beta\right)\quad B=-\frac{\beta V_{DD}}{\beta-\alpha}$$

We obtain

$$V_{A0}(t)=V_{DD}+\frac{A}{2}[exp(-\alpha t)+exp[-(\alpha/\eta)t]]-\frac{V_{DD}+A}{2}[exp(-\beta t)+exp[-(\beta/\eta)t]] \tag{52}$$

$$V_{B0}(t)=\frac{A}{2}[exp(-\alpha t)-exp[-(\alpha/\eta)t]]-\frac{V_{DD}+A}{2}[exp(-\beta t)-exp[-(\beta/\eta)t]]$$

$$V_{A1}(t)=V_{DD}+\frac{B}{2}[exp(-\alpha t)+exp[-(\alpha/\eta)t]]-\frac{V_{DD}+B}{2}[exp(-\beta t)+exp[-(\beta/\eta)t]]$$

$$V_{B1}(t) = \frac{B}{2}[exp(-\alpha t) - exp[-(\alpha/\eta)t]] - \frac{V_{DD}+B}{2}[exp(-\beta t) - exp[-(\beta/\eta)t]]$$

$V_{A0}(t)$ through $V_{B1}(t)$ are normalized using $v_{XX}(t) = V_{XX}(t)/V_{DD}$ and are plotted in Fig. 22. Parameters are: $R_0 = 1.0$, $R_1 = 1.0$, $C_0 = 1.0$ and $C_1 = 0.3$. Noise pulse height and width depend on complex interaction of the capacitive voltage division and resistance-capacitance relaxation process.

4.9 Power-bus Noise

When a CMOS gate switches, a current spike is generated in the power-bus. Current spikes from many gates are summed to generate the *average* DC power supply current of the chip. The power supply current of CMOS VLSI chip contains, however, current spikes that cannot be averaged out: large spikes originate from switching of exceptionally large CMOS drivers like output drivers, and from many small gates that charge or discharge their load capacitances simultaneously. The spikes are too narrow and too large to be averaged out. Such current spikes generate power-bus noise that disturbs normal operation of other circuits on the chip [09].

The mechanism of power-bus noise generation and its effects can be understood from Fig. 23(a). The circuits on the chip are enclosed in the dotted box. If the output buffer discharges external load capacitance C, such as is done by PC board wires, a discharge current spike shown by arrow S is generated. This current is through the on-chip power-bus, the chip's bonding wire, and the package traces, all of which have series inductance and resistance. The current spike, S, generates a noise voltage, V_N, between the chip's ground-bus and

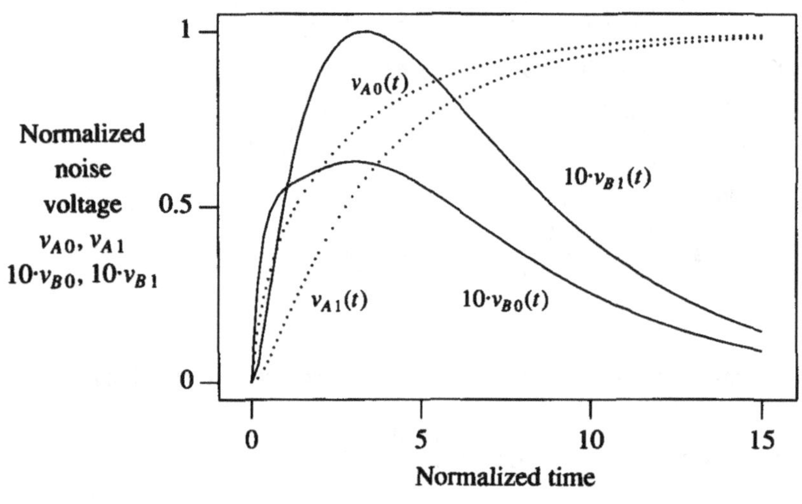

Figure 22 Polysilicon-to-polysilicon induced noise

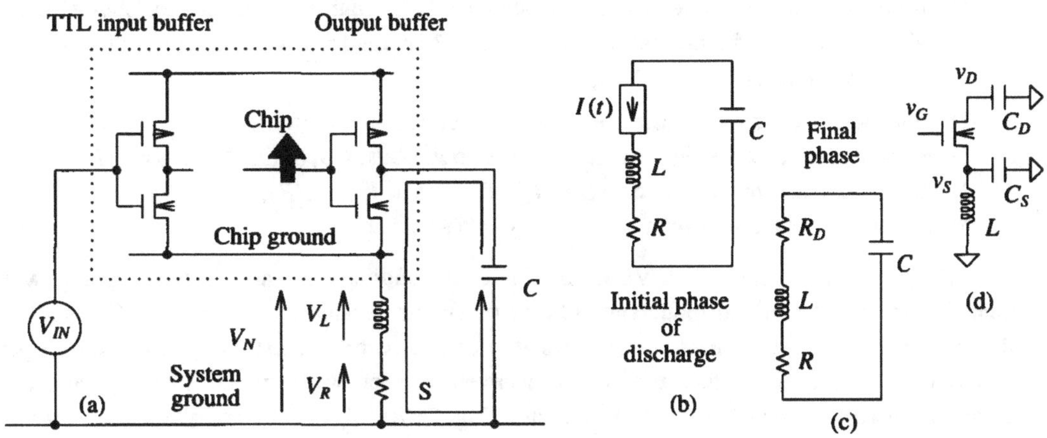

Figure 23 Effects of digital noise

the system's ground, which consists of resistance voltage drop, V_R, and inductance voltage drop, V_L. Suppose that the input buffer of the same chip is driven by external signal source V_{IN}. Since the chip's ground-bus voltage level is higher than that of the system ground by V_N, the voltage that drives the input buffer is $V_{IN} - V_N$, rather than V_{IN}. If the switching threshold voltage of the input buffer, V_{GSW}, is

$$V_{IN} > V_{GSW} \quad \text{and} \quad V_{IN} - V_N < V_{GSW}$$

the input buffer *considers* a high-input logic level as if it were a low-input logic level [we note that the input buffer has no *intelligence* to recognize the guardbanded logic voltage levels (Section 2.2)]. If the chip strobes in the wrong data value, circuit failure takes place.

With reference to Fig. 23 the mechanism of power-bus noise is analyzed as follows. At the beginning of the output driver pull-down transition, the loop that includes the package inductance is equivalent to the circuit shown in Fig. 23(b). The FET impedance is higher than the impedance of L and R. The FET current discharges capacitance C. Noise generated on the power-bus is determined by the NFET current working effectively as a time-dependent current generator. This process establishes current in the loop. Capacitor discharge continues, and the drain-source voltage of the NFET decreases. As the NFET moves out of the saturation region, the NFET internal resistance decreases, and the equivalent circuit of the loop changes to the one shown in Fig. 23(c). The NFET in the triode region is equivalent to resistance R_D. By this time a significant current is established in the loop. The later phase of power-bus noise is that the energy stored as current in the inductance is dissipated by the resistance, and the power-bus voltage returns to zero, either monotonically or oscillating.

The power-bus noise waveform depends on time constants of the circuit [10]. The five time constants of special interest are:

1) discharge time constant of load capacitance C through the channel resistance of the NFET, t_D (this is the product of the *equivalent* channel resistance of the FET, R_{EQ}, and load capacitance C, $t_D = CR_{EQ}$);
2) inductance-capacitance time constant, t_{LC}, defined as $t_{LC} = \sqrt{LC}$;
3) resistance-inductance time constant of the power-bus, t_{LR}, defined as $t_{LR} = L/R$;
4) resistance-inductance time constant, t'_{LR}, defined as $t'_{LR} = L/R_D$;
5) resistance-capacitance time constant, $t_{RC} = R_D C$.

The power interconnect of a VLSI chip includes bonding wires and on-chip metal power-bus. Let the resistance and inductance of any power interconnect be R and L. Current that flows through the interconnect is $I(t)$, and its time derivative is $dI(t)/dt$. Noise developed across the interconnect is the sum of noise developed across the inductance, V_L, and that developed across the resistance, V_R, as shown in Fig. 23(a). V_R and V_L are given, respectively, by

$$V_R(t) = RI(t) \qquad V_L(t) = L\frac{dI(t)}{dt}$$

and their ratio $V_L(t)/V_R(t)$ is

$$\frac{V_L(t)}{V_R(t)} = \frac{L}{R}\frac{1}{I(t)}\frac{dI(t)}{dt}.$$

Since the current spike width of the capacitance discharge through the FET is t_D, $|dI(t)/dt| \approx I(t)/t_D$. Then

$$\frac{V_L(t)}{V_R(t)} \approx \frac{t_{LR}}{t_D}. \tag{53}$$

Equation (53) shows that ratio of the inductive to the resistive noise of the power interconnect is essentially the ratio of t_{LR} to t_D. $|V_L/V_R| > 1$ for the power-bus noise of the I/O frame, and $|V_L/V_R| < 1$ for the noise within the chip.

L, R_D, R, and C of Fig. 23(a) form a series-resonant circuit. The series-resonant circuit is critically damped if

$$R + R_D = 2\sqrt{L/C} \quad \text{or} \quad t_{RC} \approx 2t_{LC} \quad \text{if} \quad R \ll R_D \quad (\text{or } t_{LR} \gg t'_{LR}). \tag{54}$$

If $t_{RC} < 2t_{LC}$, the LC tuned circuit goes into damped oscillation every time the output driver switches.

The power-bus of a CMOS VLSI chip consists of on-chip metal power-bus and bonding wires that connect the on-chip power-bus to package power terminals. Bonding wire is 1–2 mil diameter aluminum wire, whose ends are bonded to the chip's bonding pad and to the package terminal. A typical bonding wire in a conventional VLSI package is 1–1.5mm long. Resistance R and inductance L of a bonding wire having length $l(cm)$ is approximately,

$$R\,(ohms) = \frac{l}{\pi\sigma a^2} = 0.58l\,(cm) \qquad L\,(nH) = \frac{\mu_0 l}{2\pi}[\log(2b/a) + 0.25] \tag{55}$$

where $\mu_0 = 4\pi/10^9$ (*henry/cm*) is the magnetic permeability, a is the radius of the bonding wire, b is the distance to the ground conductor that serves as the current return path, and

$\sigma = 0.352 \times 10^6$ *mho/cm* is the conductivity of aluminum. In the resistance estimate, 1 mil diameter wire was assumed. The inductance-resistance time constant of the wire is L/R. If $a = 12.5$ *microns* (1 mil wire) and $b = 3000$ *microns*, $t_{LR} = 22$ *nanosec*. For electrical transients that complete within a time shorter than the time constant, a bonding wire works effectively as inductance. In typical, present high-performance CMOS ICs, the time required for an output driver to complete switching under the worst case (fastest) condition is 1 nsec or less. Then the bonding wire works practically as a pure inductance.

An on-chip power-bus is aluminum microstrip line deposited on composite oxide formed on the silicon surface. The silicon surface is doped to about $2 \times 10^{16}\ cm^{-3}$, and its conductivity $\sigma_{P,N}$ is in the range $1 - 10$ *mho/cm*. The dielectric relaxation time of silicon is given by ε_S/σ_P and ε_S/σ_N, respectively, for P-type and N-type substrate, where ε_S is the dielectric constant of silicon ($\varepsilon_S \approx 1.04 \times 10^{-12}\ F/cm$). The time constants are in the range of $0.1 - 1$ picoseconds. Since the switching time of scaled-down CMOS circuits is of the order of 1 nanosec, which is 1000 times longer than the dielectric relaxation time, the silicon bulk remains electrically neutral.

Penetration of the magnetic field into bulk silicon can be estimated from the skin depth. A sinusoidal electromagnetic wave having angular frequency $\omega = 2\pi f$, where f is the frequency in Hertz, penetrates to depth λ given by [07]

$$\lambda_{P,N}(\omega) = \sqrt{\frac{2}{\omega \mu_0 \sigma_{P,N}}} \quad \text{where} \quad \mu_0 = 1.25 \times 10^{-8} (henry/cm) \qquad (56)$$

A switching transient lasting 1 nanosecond contains a significant Fourier component at about angular frequency $\omega = 1/\tau = 10^9 radian/sec$. Then $\lambda_{P,N}(10^9) = 0.4/\sqrt{\sigma_{P,N}}$ where $\sigma_{P,N}$ in mho/cm. If doping is $2 \times 10^{16}/cm^3$, $\sigma_P = 1.2$mho/cm, $\sigma_N = 3.0$mho/cm, $\lambda_P(10^9) = 0.37$cm and $\lambda_N(10^9) = 0.23$cm. In twin-tub CMOS technology, a heavily doped substrate exists underneath relatively lightly doped P- and N-tubs. The substrate conductivity is higher than the conductivity assumed above by 2 orders of magnitude. Then λ_P and λ_N are one order of magnitude smaller: we expect the range of $\lambda_{P,N}$ from 200 to 4000μm. The skin depth is still equal to or larger than typical silicon chip thickness, which is about 500 μm. The back side of conventional silicon chip is contacted by metal, or directly attached to the metal pedestal of the package. This metal stops penetration of the magnetic field. As we saw, a high frequency electric field cannot exist in silicon bulk, but a high frequency magnetic field can. Therefore, as far as inductive properties of on-chip interconnects are concerned, the silicon substrate can be ignored but the metal backing can not. On-chip power-bus inductance is estimated from the formula [11]

$$L = 2.0 \times \log(8\frac{b}{W} + 4\frac{W}{b}) \ (nH/cm)$$

where W is the width of the power-bus, and b is the thickness of the silicon substrate. The back side of the silicon substrate has a metal layer, which serves as the current return path. The term, $8b/W$, can be explained by assuming that the magnetic field penetrates the entire thickness of the substrate. This formula assumes that there is *no* current return path other than the metal layer on the backside of silicon. If V_{SS} and V_{DD} buses are placed side by side and carry the supply and the return currents in opposite directions, the inductance is less. Effective inductance depends on current paths on the chip, which makes inductance estimates quite difficult.

Power-bus resistance is found by counting the number of squares as

$$R = \rho/W \quad (ohms/cm) \tag{57}$$

where ρ is the sheet resistance of metal. If we use $W = 100\ \mu m$, $b = 500\ \mu m$, $T_{OX} = 1\ \mu m$ (single-level metal) and $\rho = 0.04\ ohms/square$, $L = 7.2\ nH/cm$ and $R = 4.0\ ohms/cm$, and $t_{LR} = 1.8$ nanoseconds. This is only an approximation, since the effects of the eddy current loss (Section 1.9) are not included.

The power-bus capacitance to the substrate is given by

$$C = \frac{\varepsilon_{OX} W}{T_{OX}} = 0.345 \frac{W}{T_{OX}} \quad (pF/cm) \tag{58}$$

We have $C = 34.5\ pF/cm$, $t_{LC} = 0.50\ nsec/cm$ and $t_{RC} = 0.14\ nsec/cm^2$ assuming the same power-bus dimensions.

To understand power-bus noise waveforms in real FET circuits, we analyze the equivalent circuit shown in Fig. 23(d). The NFET is modeled using the simplified high-field model of Section 1.7 that gives drain current, I_D, as a function of the three terminal voltages V_D (drain), V_G (gate), and V_S (source) as

$$I_D = I_{Dmax} \frac{(V_{DD} + V_A)(V_G - V_S)(V_D - V_S)}{V_{DD}^2 (V_D - V_S + V_A)} \quad \text{where} \quad I_{Dmax} = B_{HN} \frac{V_{DD}^2}{V_{DD} + V_A}$$

Time and voltage are normalized to $T_D = C_D V_{DD}/I_{Dmax}$ and V_{DD}, respectively. Numerical results are shown in Fig. 24. The discharge process proceeds as follows. As the NFET turns on, load capacitance C_D discharges through the inductance at the NFET source. The NFET source is pulled up and the source and drain voltages eventually equalize. This is the point of changeover from the equivalent circuit of Fig. 23(b) to that of Fig. 23(c). After this, the circuit goes into damped oscillation, and the rate of decay of the oscillation depends on the FET channel resistance. Figures 24(a)–(d) use input waveform v_G, shown by the dashed curve, $C_S = C_D$, $V_A = 0.2V_{DD}$ and several values of $\Lambda = 1/L$. If the inductance increases (smaller Λ), the power-bus goes into significant damped oscillation. The circuit oscillates even if the inductance is reduced, since capacitance $C_S \neq 0$.

Power-bus noise reflects certain structural details of the circuit. Figure 25(a) shows current when the CMOS inverter switches. The CMOS inverter capacitive load can be split into two parts. Capacitance C_P is connected between the output node of the inverter and V_{DD}, and capacitance C_N between the output node and V_{SS}. If wiring from the output node of the inverter is routed on a grounded P-tub, the wiring capacitance contributes to C_N, and if it is on an N-tub, to C_P. If the inverter input pulls up rapidly, current flows through the FETs and the capacitive loads as shown by the broken line segments of Fig. 25(a). Since MN1 conducts, capacitance C_P charges up, and the charge-up current flows from V_{DD} to V_{SS}. This current comes from, and returns to the power supply. Capacitance C_N discharges through the closed

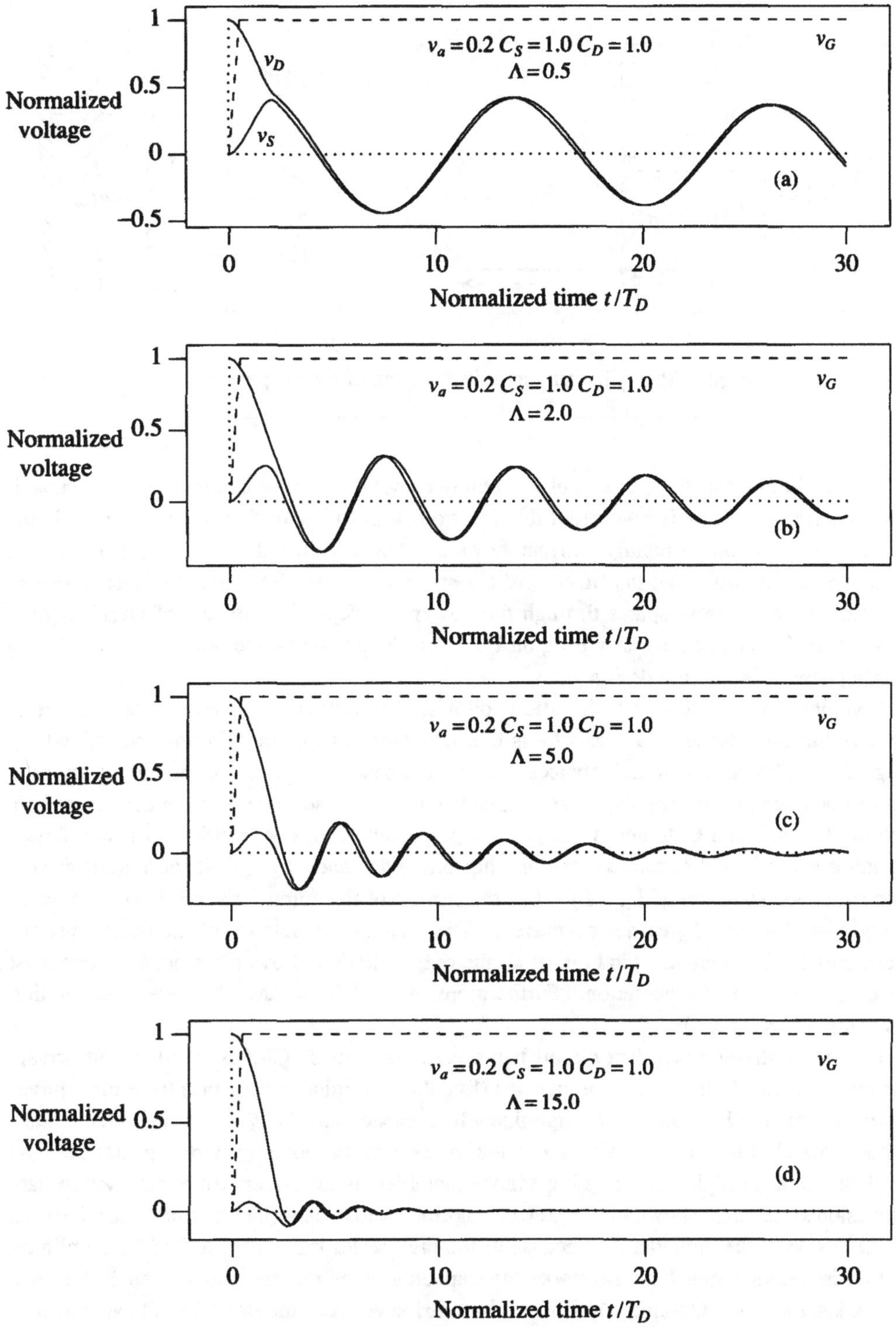

Figure 24 Power-bus noise waveforms

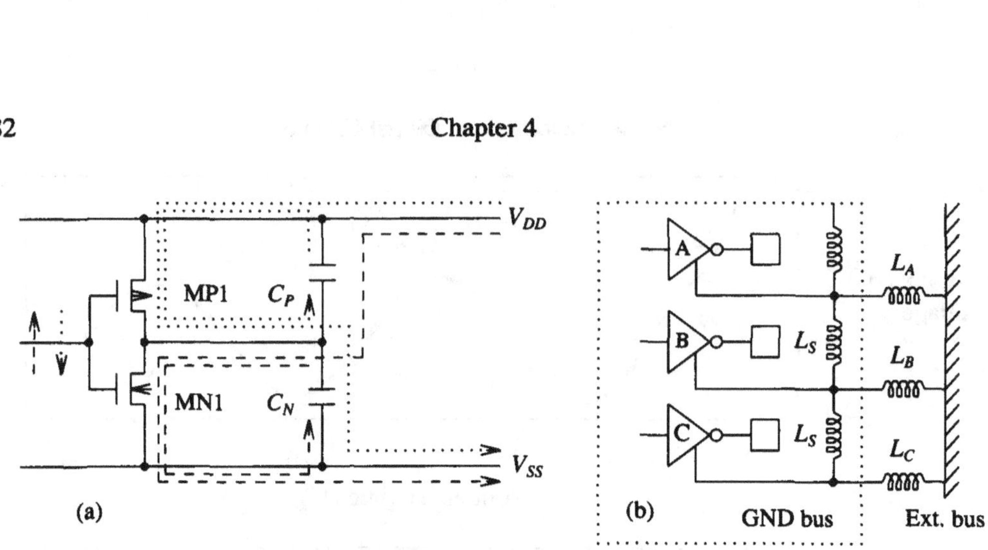

Figure 25 Charging and discharging of load capacitors

local loop including MN1 and C_N. This current does not involve the power supply. If capacitance C_P were zero, there is no current flowing from V_{DD} to V_{SS} during this transition. If the inverter input pulls down rapidly, current flows as shown by the dotted line segments: the roles played by the MP1 and the MN1, and C_P and C_N reverse. This detail of circuit operation indicates how current spikes through the power-bus depend on details of circuit layout. The details include percentage of wiring on P-tub and N-tub, which are often unknown during the early phase of VLSI chip design.

A similar issue of current distribution follows. The output-driver ground-bus circuit of Fig. 25(b) has inductance parasitics. L_S is the inductance of the on-chip power-bus, and L_A through L_C are bonding wire inductances. How the discharge currents of the output driver B are distributed among the bonding wires depends on how the set of the output drivers switch. If output drivers A and C do not switch, discharge current from output driver B is distributed to bonding wires A and C through the on-chip bus inductance L_S. If all the output drivers switch simultaneously, and if $L_A = L_B = L_C$, the current of the output driver B is conducted by bonding wire B alone. L_S carries no current. The effective inductance of the power-bus circuit depends on the *mode* of switching of all the output drivers. Power-bus noise theory must include the mode of I/O operation. Furthermore, mutual inductance between the bonding wires complicates the analysis.

The output-driver equivalent circuit has several variations. CMOS VLSI output drivers designed for relatively low clock frequencies (like those in microprocessors) have the equivalent circuit shown in Fig. 26(a). The significant load capacitance is *off-chip* (the dotted rectangle shows the chip boundary): it exists at the *far* end of the bonding wire inductance. The equivalent circuit of high-frequency impedance-matched output drivers (like that used in data transmission systems) is shown in Fig. 26(b). Significant load capacitance is *on-chip*: it exists at the *near* side of the inductance. Because of the high series resistance, R_S, of the amplitude reduction-impedance matching network, the capacitance at the far end of the inductance becomes insignificant. On-chip capacitance is contributed from the drain island capacitances of the output driver FET and from the bonding pad.

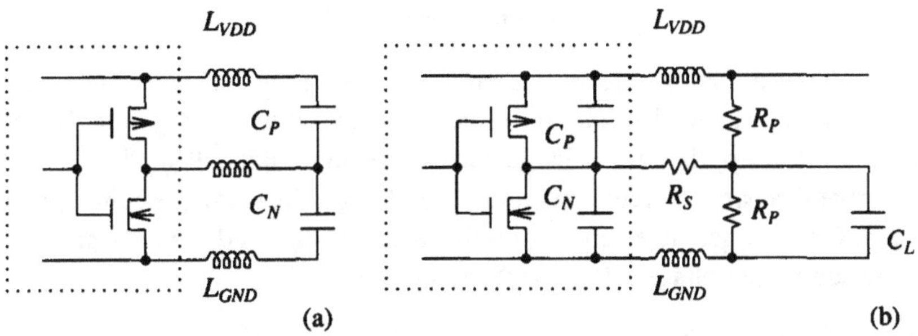

Figure 26 Equivalent circuit of output driver for different loads

4.10 Discharge Current Waveform of CMOS Gates

We consider the source of power-bus noise, which is capacitance discharge current through the FET. The discharge current waveform is schematically shown by the solid curve of Fig. 27(a). It begins to increase at time 0, reaches the maximum, I_{max}, at $t = t_{max}$, and then decreases. The current becomes 0 at $t = t_S$. The rise or fall time, t_S, is longer than the switching delay time t_D. They are approximately proportional. Therefore t_S is practically determined by the chip design requirements, like bus drive delay time. The area underneath the curve is the charge stored in the external capacitance, C_L, if overlap current is neglected. We have

$$Q = C_L V_{DD} = \int_0^{t_S} I(t)dt \tag{59}$$

and, therefore, the area is primarily determined by the design requirements also. Noise developed across the resistance, R, of the power-bus is proportional to $I(t)$, and noise developed across the inductance, L, is proportional to $dI(t)/dt$. Noise voltage, $V_N(t)$, is given by the sum of the two terms as

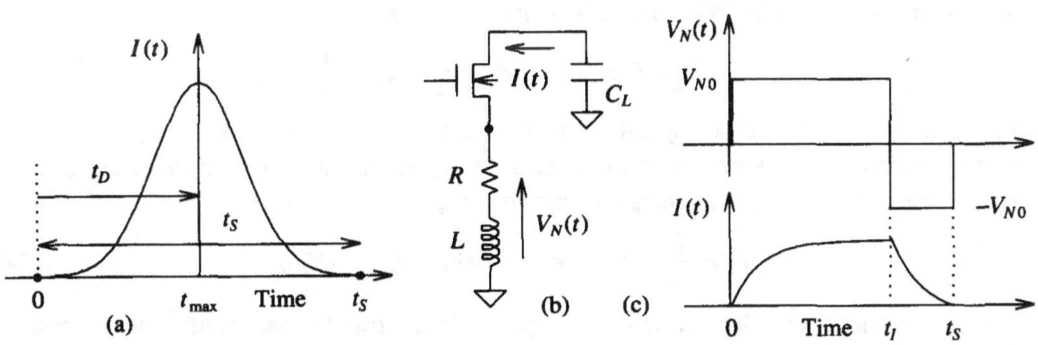

Figure 27 Optimum discharge current waveforms of output driver

$$V_N(t) = L\frac{dI(t)}{dt} + RI(t).$$

The optimum output driver waveform has the given width, encloses the given area, and has minimum $|V_N(t)|$. The equivalent circuit of the pull-down driver and ground-bus is shown in Fig. 27(b). We seek the optimum discharge current $I(t)$ minimizing $|V_N(t)|$.

Consider the discharge current waveform $I(t)$ of Fig. 27(c). Suppose that the initial phase of transition at $0 < t < t_I$ generates constant noise voltage, V_{N0}, and the later phase at $t_I < t < t_S$ generates constant noise voltage, $-V_{N0}$. $I(t)$ should satisfy

$$L\frac{dI_1(t)}{dt} + RI_1(t) = V_{N0} \quad (0 < t < t_I) \qquad L\frac{dI_2(t)}{dt} + RI_2(t) = -V_{N0} \quad (t_I < t < t_S). \tag{60}$$

These equations are solved subject to the *initial* condition, $I_1(0) = 0$, and the *final* condition, $I_2(t_S) = 0$ to obtain

$$I_1(t) = \frac{V_{N0}}{R}[1 - exp[-(R/L)t]] \qquad I_2(t) = -\frac{V_{N0}}{R}[1 - exp[(R/L)(t_S - t)]] \tag{61}$$

Time t_I when the current starts decreasing is given by solving the equation

$$I_1(t_I) = I_2(t_I) \quad \text{as} \quad t_I = \frac{L}{R}\log\frac{exp[(R/L)t_S] + 1}{2}.$$

The total charge $Q = C_L V_{DD}$ that must be drained to ground, is given by

$$Q = \int_0^{t_I} I_1(t)dt + \int_{t_I}^{t_S} I_2(t)dt.$$

After some algebra we have

$$V_{N0} = \frac{(R^2/L)Q}{2\log\frac{exp[(R/L)t_S] + 1}{2} - \frac{R}{L}t_S} \tag{62}$$

where $Q = C_L V_{DD}$. In normalized form this becomes

$$\frac{V_{N0}}{V_{DD}} = \frac{C_L R}{t_S} f(x) \quad \text{where } f(x) = \frac{x}{2\log\frac{exp(x) + 1}{2} - x} \qquad x = \frac{t_S}{t_{LR}} \tag{63}$$

and where $t_{LR} = L/R$ is the inductance-resistance time constant.

$$\lim_{x\to\infty} f(x) = 1 \quad \text{and} \quad \lim_{x\to 0} f(x) = \frac{4}{x}$$

Function $f(x)$ is plotted in Fig. 28. Let us consider two special cases. Suppose that the power-bus impedance is pure resistance. Then the optimum output driver conducts constant current during the period of discharge, as shown in Fig. 29(c). We have

$$I(t) = \frac{Q}{t_S} \qquad V_N(t) = \frac{RQ}{t_S} \quad \text{where } Q = C_L V_{DD}. \tag{64}$$

$V_N(t)$ is shown in Fig. 29(d). This limit is applicable if a resistive semiconductor substrate is used as part of the ground connection. The optimum discharge waveform when the power-bus connection is purely inductive is the equilateral triangle shown in Fig. 29(a). We have

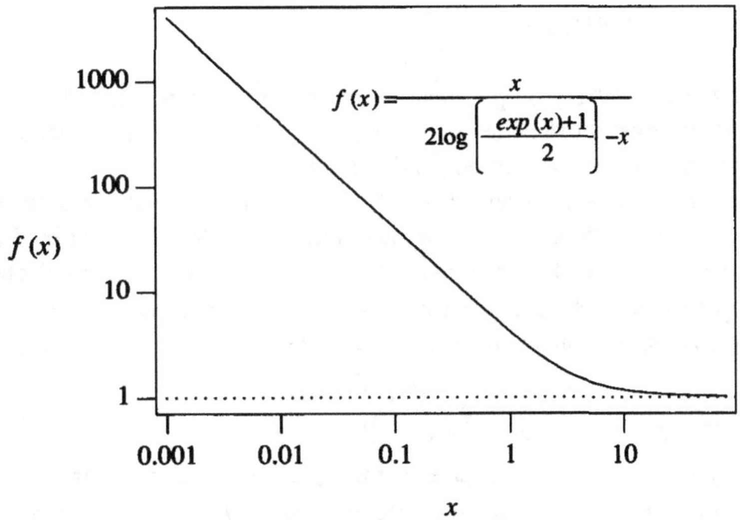

Figure 28 Function $f(x)$

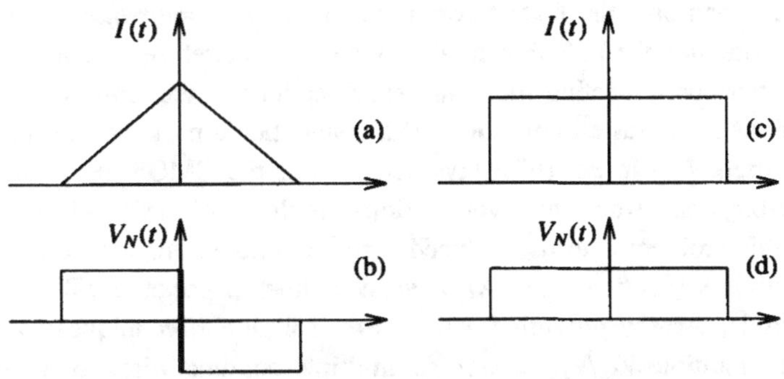

Figure 29 Minimum noise V_N and discharge current $I(t)$. (a), (b) for $R = 0$ and (c), (d) for $L = 0$

$$t_{max} = \frac{t_S}{2} \qquad \left|\frac{dI(t)}{dt}\right|_{max} = \frac{2I_{max}}{t_S} \qquad Q = CV_{DD} = \frac{1}{2} t_S I_{max} .$$

By eliminating I_{max} from the second and the third equations [10],

$$\left|\frac{dI(t)}{dt}\right|_{max} t_S^2 \geq 4Q = 4CV_{DD} . \tag{65}$$

This is a complementary relationship that is useful for output driver design. From this equation,

$$V_N(t)_{\max} = L\left|\frac{dI(t)}{dt}\right|_{\max} \geq \frac{4CLV_{DD}}{t_S^2} \tag{66}$$

and the power-bus noise is inversely proportional to the square of the rise or fall time, t_S. A properly designed output driver satisfies the limit of Eq. (65) within 20–30%. Equation (66) is a good estimate of power-bus noise, and of output driver design.

Suppose that the FET internal resistance is much higher than the impedance of the power-bus. Let the number of simultaneously switching output drivers be N_0, the number of ground connections be N_g, the inductance of each bonding wire be L_b, and the effective inductance of the chip power-bus to each driver be L_i. Then V_N, the noise voltage developed by simultaneous pull-down of N_0 drivers, is estimated by using dI/dt of Eq. (65) as

$$\frac{V_N}{V_{DD}} \geq 4\left(\frac{T_{LC}}{t_S}\right)^2 \tag{67}$$

where $T_{LC}{}^2 = [(N_0/N_g)L_b + L_i]C_L$ is the square of the *effective* LC time constant. The normalized power-bus noise is proportional to the square of the ratio (T_{LC}/t_S). Time constant T_{LC} is determined primarily by packaging. Using a pin grid array package on a PC board, T_{LC} is estimated as $T_{LC} = [6 \times 2(\textit{nanohenry}) \times 100(\textit{picofarad})]^{\frac{1}{2}} = 1\ \textit{nanosec}$. Then $V_N = 240$ *millivolt* if $t_S = 10$ *nanosec*, which is less than the FET threshold voltage. If this is true for the slowest process conditions, the fastest condition might be $t_S = 4\ \textit{nanosec}$. Then $V_N = 1.51$ *volts*, which is more than the FET threshold voltage. It is therefore difficult to design the output driver simultaneously meeting the delay specifications at the slowest conditions and the noise specifications at the fastest condition. If a conventional package like a pin grid array or a chipcarrier is used, T_{LC} in Eq. (67) stays constant, as the CMOS technology scales down. Then the power-bus noise would be a very serious problem in VHSIC VLSI circuits. A rational solution to this problem is to use reduced logic amplitude, impedance-matched, constant-current output drivers (Section 7.7). An advanced flipchip package will reduce T_{LC} by an order of magnitude, thereby providing a long-term solution. Techniques to ease the output driver noise, like multiple V_{SS}/V_{DD} bondings, multiple bonding wires to single bonding pad, and separating the output driver power-bus from the internal power-bus are already used. The I/O protocol can also provide an effective and immediate solution: when a chip is outputting data, no input of the chip should be strobing data in.

If the noise voltage is so large that the drain/source PN junctions conduct, the noise voltage is clamped. However, the resultant injection of minority carriers is undesirable because it increases latchup susceptibility. Minority carriers may spread out over a wide area, and they may discharge precharged circuit nodes.

4.11 Power-bus Response

In Section 4.9 we analyzed the ground-bus noise of an isolated output driver. If more than one output driver is integrated into an output driver *frame*, the analysis becomes complex, because any NFET of the output driver can be in any of three different states. They are: (1) the output driver drives the off-chip load and dumps current to the ground-bus; (2) NFET of the output driver is conducting, but it is in the quiescent state, and the channel resistance of the NFET dampens oscillation excited on the ground-bus; and (3) the NFET is turned off and the driver has no effect on the bus. (1), (2), and (3) correspond to S, T, and N regions of the microstate of the driver NFET, respectively.

We consider a high-speed output driver frame having one ground-bus bonding wire for each output driver. Such output driver frames are used in ultra high-frequency CMOS chips, but the analysis that assumes this simple structure has general validity if a single equivalent output driver is used to approximate several output drivers of a low-speed output buffer frame. Since the I/O frame of a VLSI chip includes many output drivers, and each driver may be in any of the three states, the number of possibilities is huge. The problem has no typical or simple case. Perhaps the *cleanest* case is that the ground-bus is infinitely long and the output drivers at all the locations connect load capacitance to the bus in the *quiescent* state, except for the one driver that discharges the load capacitance. The FET is assumed to have high internal resistance, and the ground-bus noise can be computed using the optimum discharge current waveform, which was derived in the last section, and conventional linear circuit theory. The analysis requires the power-bus impedance.

The impedance of an infinitely long power-bus can be determined as follows. With reference to Fig. 30(b), the impedance looking into the infinitely long transmission line of the ground-bus (arrow sign) Z satisfies the following recursive relationship for Z

$$Z = Z_S + \frac{Z_P Z}{Z_P + Z} \tag{68}$$

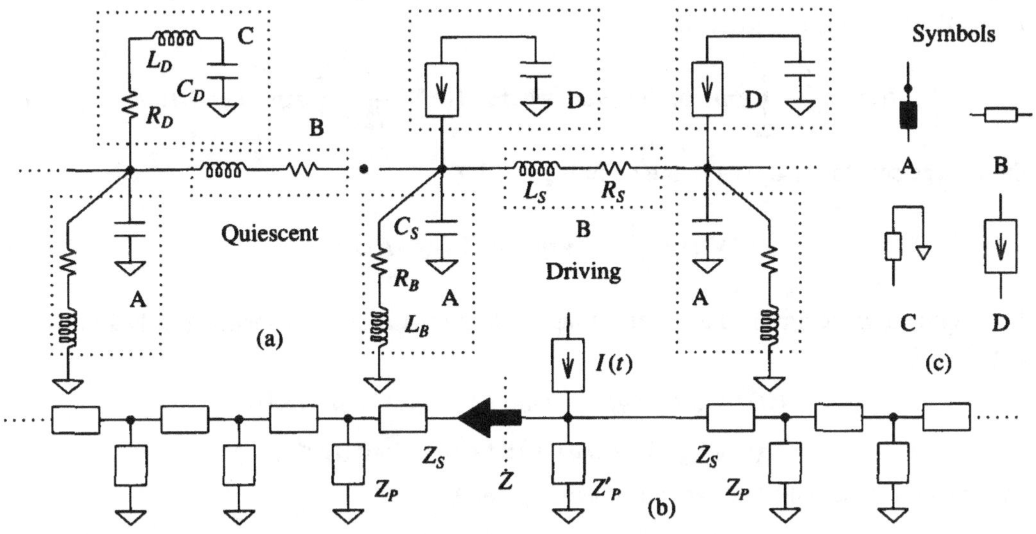

Figure 30 Model of power-bus section

where Z_S is shown by box B of Fig. 30(a), and Z_P by the parallel combination of boxes A and C. This equation is solved for impedance Z as

$$Z = \frac{1}{2}(Z_S + \sqrt{Z_S^2 + 4Z_S Z_P}) \tag{69}$$

where

$$Z_S = R_S + i\omega L_S \quad Z_P = \frac{1}{i\omega C_S + \dfrac{1}{R_B + i\omega L_B} + \dfrac{1}{R_D + i\omega L_D + \dfrac{1}{i\omega C_D}}}$$

and where the square root of Eq. (69) is a complex argument multivalued square root function, since Z_S and Z_P are complex numbers. A *branch* of the multivalued function is chosen such that $Z \to Z_S$ in the limit as $\omega \to \infty$, and $Z \to R_0$ (> 0) in the limit as $\omega \to 0$. The function must vary continuously in the range $0 \le \omega < \infty$. The resistance R_0 is given by

$$R_0 = \frac{1}{2}(R_S + \sqrt{R_S^2 + 4R_S R_B})$$

Figure 31 shows impedance Z versus angular frequency ω for typical parameter values given in Fig. 32. Marker dots on the curve start at $\omega = 1$ (unit is 10^9 radian/sec) at the lower left side, with increment $\Delta\omega = 1$ to $\omega = 28$. The units of ω are chosen to conform to convenient choice of units for inductance, capacitance, resistance, and time in nanohenrys, nanofarads, ohms, and nanoseconds, respectively.

The driving point impedance to the current generator $I(t)$ of Fig. 30(b), Z_{DP}, is given by

$$Z_{DP} = \frac{ZZ'_P}{Z + 2Z'_P}$$

where Z'_P is the impedance presented by dotted box A of Fig. 30(a). If the current waveform $I(t)$ is Fourier-transformed as [12]

$$I(t) = \frac{1}{\sqrt{2\pi}} \int_{-\infty}^{+\infty} exp(i\omega t) I(\omega) d\omega \quad \text{where} \quad I(\omega) = \frac{1}{\sqrt{2\pi}} \int_{-\infty}^{+\infty} exp(-i\omega t) I(t) dt \tag{70}$$

the voltage developed at the driving point is given by

$$V(t) = \frac{1}{\sqrt{2\pi}} \int_{-\infty}^{+\infty} exp(i\omega t) Z_{DP}(\omega) I(\omega) d\omega \tag{71}$$

If the current waveform is the isoceles triangle discussed in the last section (dashed curve of Fig. 32),

$$I(t) = 0 \; (t < -t_H) \quad = I_{\max}[1 + (t/t_H)] \; (-t_H \le t < 0)$$

$$I(t) = I_{\max}[1 - (t/t_H)] \; (0 \le t < t_H) \quad = 0 \; (t_H \le t),$$

where $t_H = t_S/2$, and the Fourier component is given by

$$I(\omega) = \sqrt{2/\pi}\, t_H I_{\max} \frac{1 - \cos(\omega t_H)}{(\omega t_H)^2}. \tag{72}$$

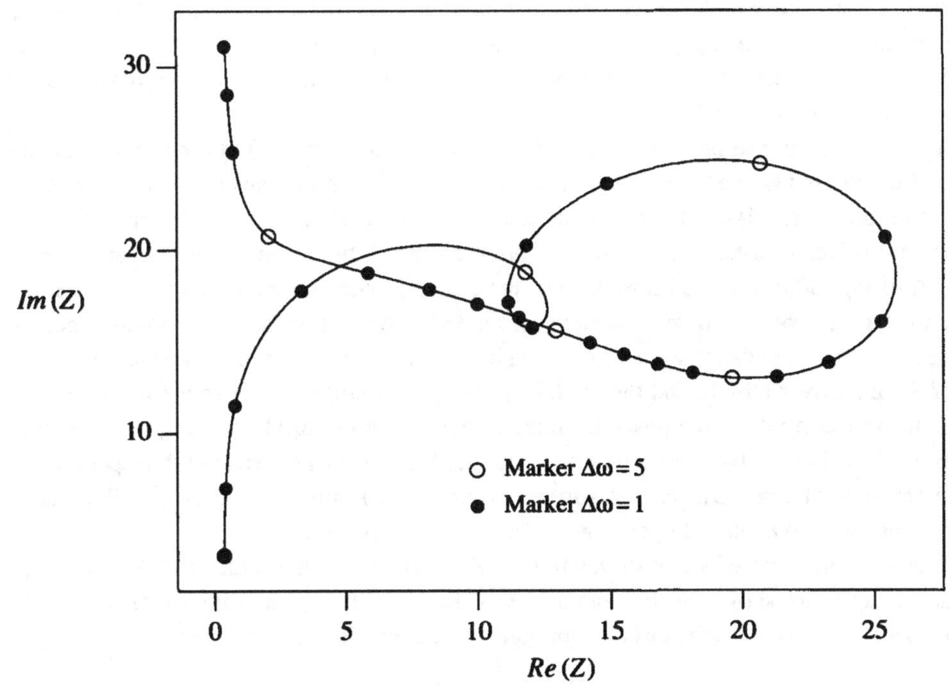

Figure 31 Impedance of power-bus

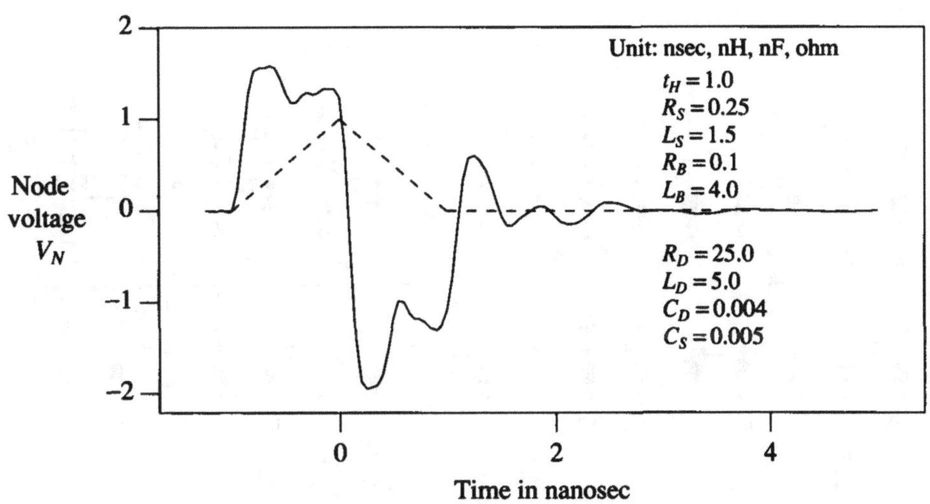

Figure 32 Noise voltage waveform

If the Fourier transform Eq. (71) is carried out numerically, we obtain the results shown in Fig. 32. We assumed $t_H = 1$ and $I_{max} = 1$. The impedance of the power-bus is primarily inductance, and therefore the isoceles triangle current waveform is differentiated once to approximate the node voltage waveform crudely.

The ground-bus noise problem in an infinitely long and *uniform* I/O frame was relatively simple. If the drivers take arbitrary state, however, the problem becomes quite complex, even in a short ground-bus. Figure 33 shows possible states of single-driver, two-driver, and three-driver output driver frames. The three symbols used in the figure are defined in Fig. 30(c). Symbol A of Fig. 30(c) models impedance of the bonding wire and the package power terminal. The P-type silicon substrate is securely grounded. The end of the power interconnect is connected to a low-impedance V_{SS} source. Symbols C and D represent the passive and active states of the the driver NFETs and the load, respectively. A single-driver frame has three different states reflecting the three possible states of the output driver (1–3 of Fig. 33). A two-driver frame has six possible states, assuming that the frame is symmetrical with respect to the two drivers (4–9 of Fig. 33). A three-driver symmetrical frame has eleven possible states (10–20). They are the *modes* of operation of the output driver frame.

We analyze the case of a two-driver frame. A schematic of the frame and definitions of the node voltages are shown in the dotted box of Fig. 33. If the simplified high-field FET model of Section 1.7 is used, the circuit equations for the two-driver frame are

$$C_S \frac{dV_A}{dt} = I_A - \frac{1}{L_B}\int_0^t V_A dt - \frac{1}{L_S}\int_0^t (V_A - V_B) dt \qquad (73)$$

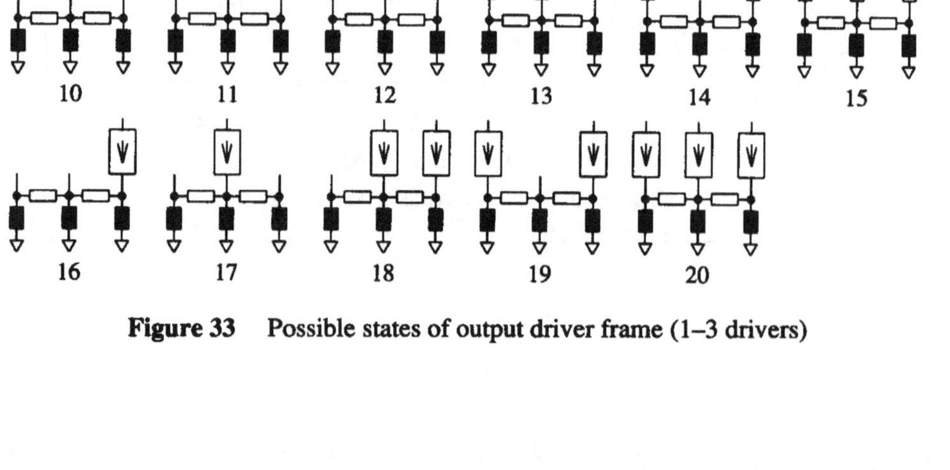

Figure 33 Possible states of output driver frame (1–3 drivers)

$$C_S \frac{dV_B}{dt} = I_B - \frac{1}{L_B}\int_0^t V_B dt + \frac{1}{L_S}\int_0^t (V_A - V_B)dt$$

$$C_D \frac{dU_A}{dt} = -I_A \quad C_D \frac{dU_B}{dt} = -I_B$$

$$I_A = I(U_A, V_{GA}, V_A) \quad I_B = I(U_B, V_{GB}, V_B)$$

$$I(U, V_G, V) = I_{Dmax} \frac{(V_{DD} + V_A)(V_G - V)(U - V)}{V_{DD}^2(U - V + V_A)}$$

where V_A and V_B are the ground-bus voltages at the locations of driver A and B, respectively, and where $I_{Dmax} = B_H V_{DD}^2/(V_{DD} + V_A)$ (Section 1.7). U_A and U_B are the output node voltages of driver A and B, respectively, and I_A and I_B are the currents of driver A and B, respectively. The set of circuit equations was solved numerically and the results are shown in Figs. 34(a)–(c). Parameter values were: $C_S = C_D = 4\ pF$, $I_{Dmax} = 0.04$, $V_A = 1.0$, $V_{DD} = 5.0$, $L_B = 2.5\ nH$ and L_S varies from $L_B/2$ (strong driver-driver coupling) to $2L_B$ (weak coupling). Figure 34(a) shows node A (solid curve) and node B (dots) waveforms when both drivers discharge the load capacitance C_D synchronously. Since the electrical transient is symmetrical with respect to nodes A and B, the two curves are identical. The damped oscillation period is estimated by considering the limit as $L_S \to \infty$ and assuming that the NFET has small internal impedance. Then $T_P = 2\pi\sqrt{L_B(C_S + C_D)} \approx 0.9$ nanosec. Since two output drivers discharge the load capacitance synchronously, the power-bus noise is high ($V_{Amax} \approx 0.36\ volts$).

In Fig. 34(b) only one driver (driver A) discharges. Since the two bonding wires share the current, the maximum noise is less ($V_{Amax} \approx 0.2\ volts$). Driver B has a high input and the NFET of driver B is turned on all the time. The three plots of Fig. 34(b) show the node voltages for successively increasing on-chip ground-bus inductance L_S, starting from $L_S = L_B/2 = 1.25\ nH$ to $L_S = 2L_B = 5\ nH$. As the inductance of the on-chip power-bus increases, the discharge current is carried more and more by the bonding wire at the driver's location, and therefore V_{Amax} increases and V_{Bmax} decreases. The period of damped oscillation is close to that estimated for Fig. 34(a). The two node voltages approach the same waveform after a few

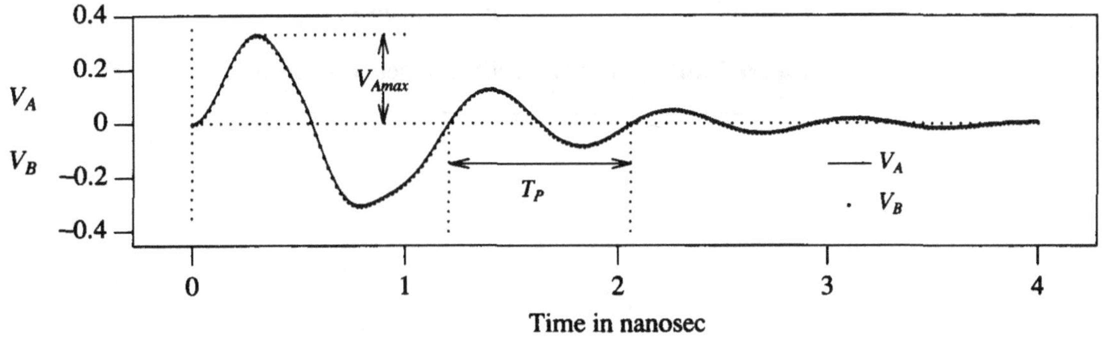

Figure 34(a) Drivers A and B driving

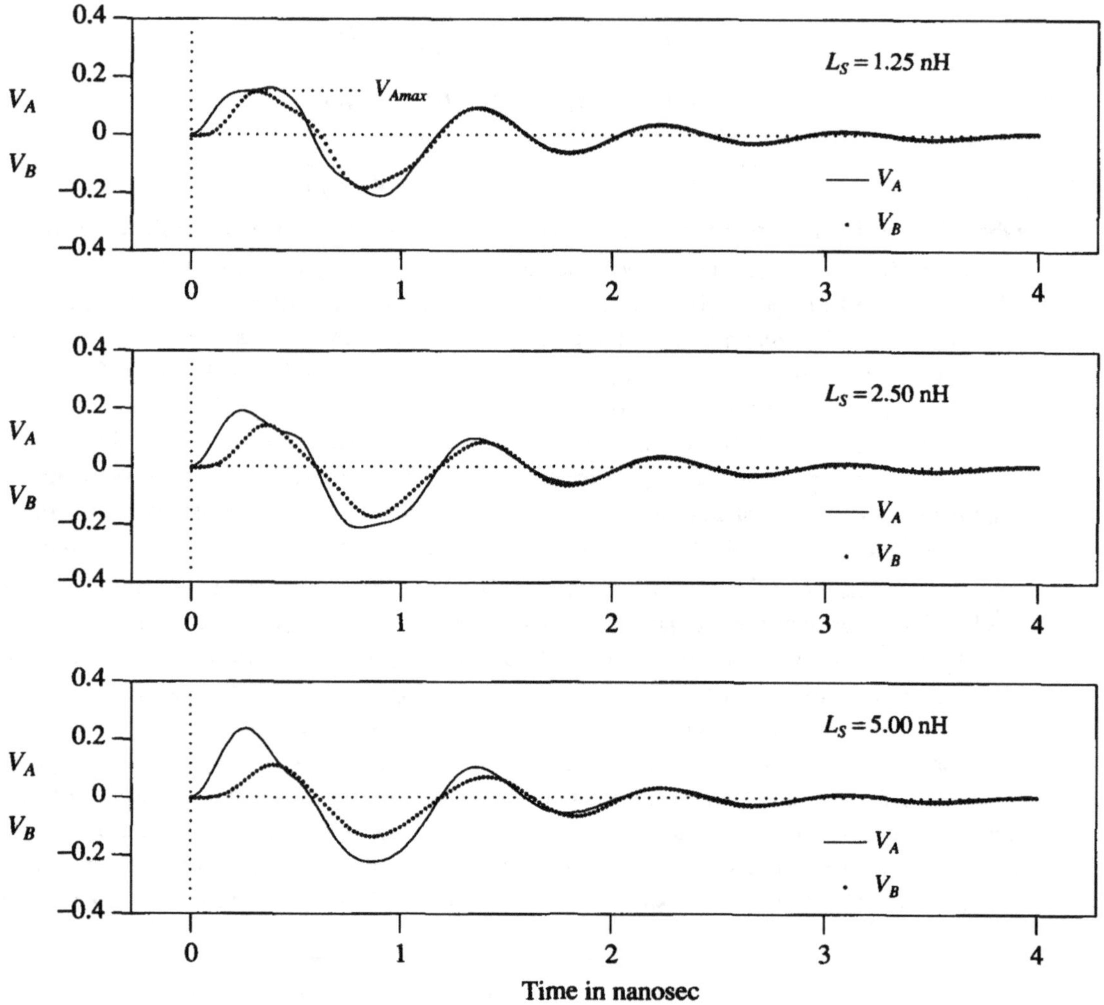

Figure 34(b) Driver A driving, driver B conducting

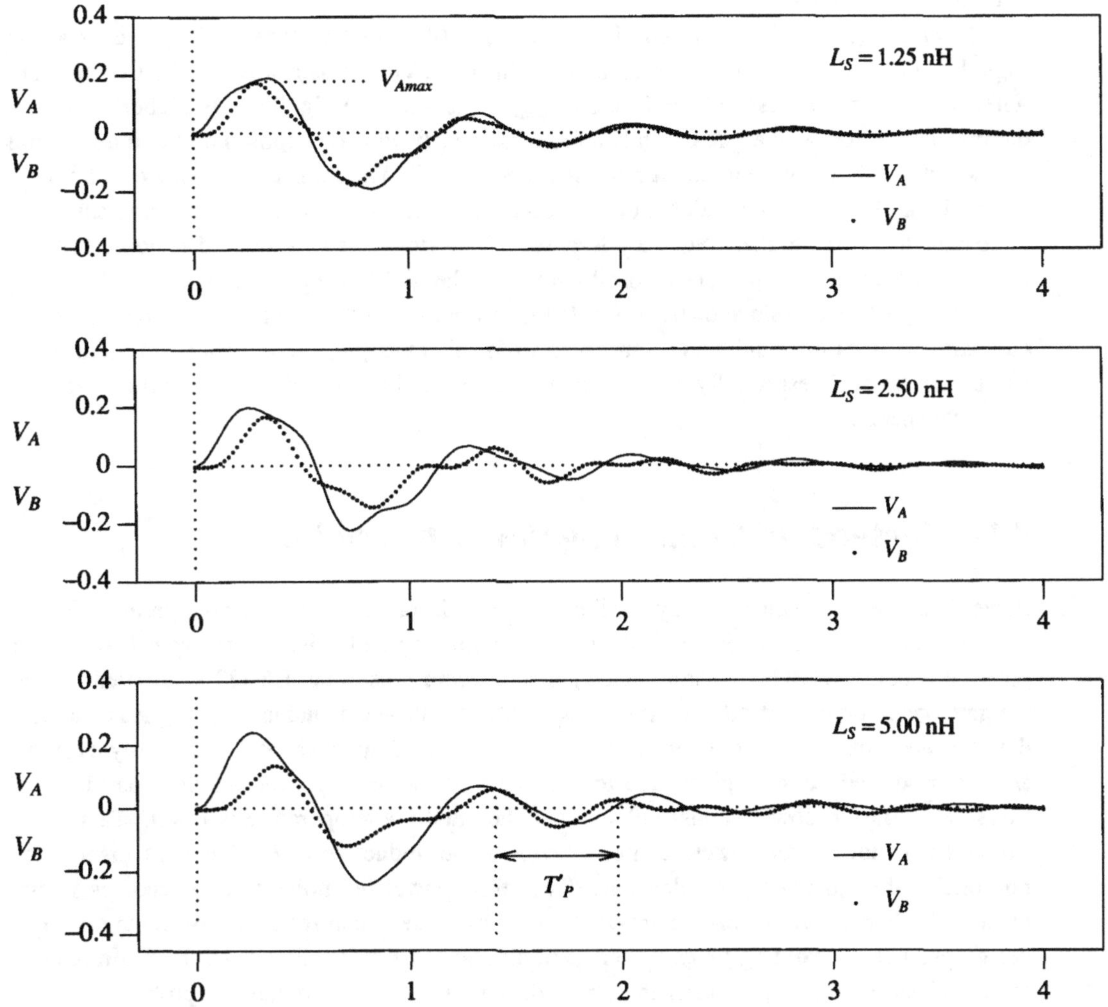

Figure 34(c) Driver A driving, driver B turned off

cycles of damped oscillation. This is because of the universal equipartition law of energy among loosely coupled oscillators: since the resonant circuits at node A and at node B become identical at the final phase of the discharge, electrical transient energy is shared equally by the two resonant circuits [13].

In Fig. 34(c) the driver at node B is turned off (the input voltage of the driver is a low logic level and the NFET is not conducting). Since the current is shared by the two bonding wires, the maximum noise voltage is less ($V_{Amax} \approx 0.2\ volts$). A significantly higher frequency damped oscillation with approximately a 0.6 nanosec period is superposed on the fundamental frequency oscillation having about a 0.9 nanosec period. The higher frequency oscillation is due to the $L_B - C_S$ tuned circuit at node B. Since this circuit does not have effective damping, the oscillation at the higher frequency lasts for a long time. After about 1.5 nanosec the predominant component of power-bus oscillation is at the higher frequency. The two node voltages are asynchronous since the two *oscillators* at nodes A and B are different, and the law of equipartition does not hold. As is observed from this example, the electrical transient in an I/O buffer frame is essentially a problem of normal modes, and of the equipartition law of electrical energy.

4.12 Methods of Controlling Power-bus Noise

Power-bus noise is generated by well-synchronized, fast-switching large circuit. Output drivers, on-chip memories, PLAs, and CMOS dynamic logic circuits are the typical sources of power-bus noise. CMOS dynamic gates generate more noise than CMOS static gates, since the gate operation is controlled by the clock. CMOS static-gate chains are designed to use all the allocated timing budget. Current spikes are distributed approximately uniformly over the entire period, and current spikes sum to a smooth DC current. Power-bus noise is higher if FETs switch faster, power supply voltage is higher, and the temperature is lower. This is the *worst-case* condition for power-bus noise. Noise in an inductive power-bus is inversely proportional to the square of gate delay time. Therefore, power-bus noise rapidly becomes significant as discharge current spikes narrow. Power-bus noise is controlled by the common-sense design practice of avoiding *precise* overlap of discharge current spikes [14]. By shifting non-critical clock edges, and by inserting small delays, many power-bus noise problems can be eliminated.

In this section we discuss various methods of controlling power-bus noise. First we consider a single output driver. We saw that the optimum discharge current waveform of a CMOS output driver and inductive power-bus is an isoceles triangle. $|dI/dt|$ at the leading edge and at the trailing edge must be equalized. Figure 35(a) shows the circuit diagram of a CMOS output driver. We consider that input signal D makes an instantaneous high-to-low transition. How fast discharge current, I_D, of the output stage increases depends on how fast node X is pulled up. The slope of the leading edge of the discharge current waveform shown in Fig. 35(b) is determined from the size of pull-down NFET MN1, and from the scaleup ratio of the first and the second inverters (size ratio of MP0, MN0 to MP1, MN1). The discharge current increases

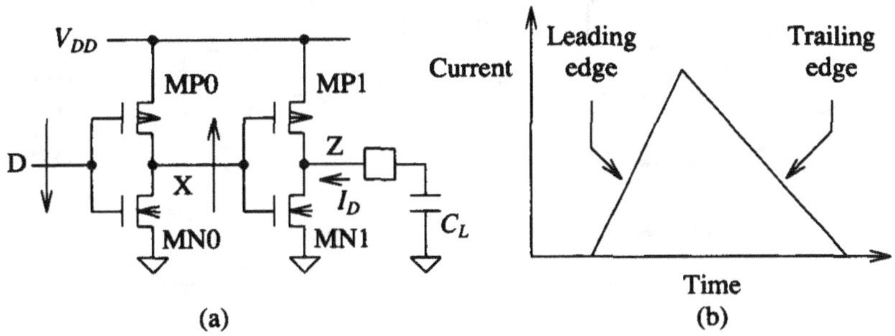

Figure 35 CMOS output driver and discharge waveform

with time, reaches a maximum, and then decreases. The rate of decrease of the discharge current is determined by how fast capacitance C_L discharges through the channel resistance of MN1. The slope of the trailing edge is determined by the conductivity of MN1 (in case of pull-up, MP1). The problem of output driver optimization consists in adjusting the sizes and scaleup ratio of the two stages to attain an isoceles triangle discharge waveform.

Equation (66) shows that $|dI/dt|$ is inversely proportional to the square of the rise/fall time t_S, which is proportional to the driver delay time. The driver delay time varies over a wide range, as process conditions, temperature, and power supply voltage vary. Cumulative variation of some CMOS technologies can be as much as a factor of 3. Then power-bus noise varies, $3^2=9$ times. For the slowest condition, driver delay must be less than a certain maximum, and for the fastest condition, power-bus noise must be less than a maximum. It is difficult to bring these two parameters simultaneously under control, if such a wide range of variation exists. Power-bus noise would be brought under control if a chip's parameters have a smaller range. Several circuit techniques have been proposed to suppress the wide range.

The fundamental solution is the application of analog feedback control techniques. Although there can be many different implementations, the basic concept is demonstrated

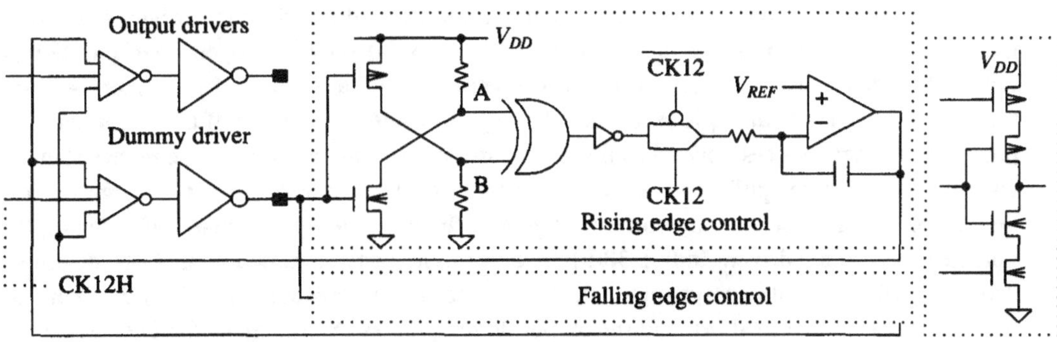

Figure 36 Control of output buffer switching time

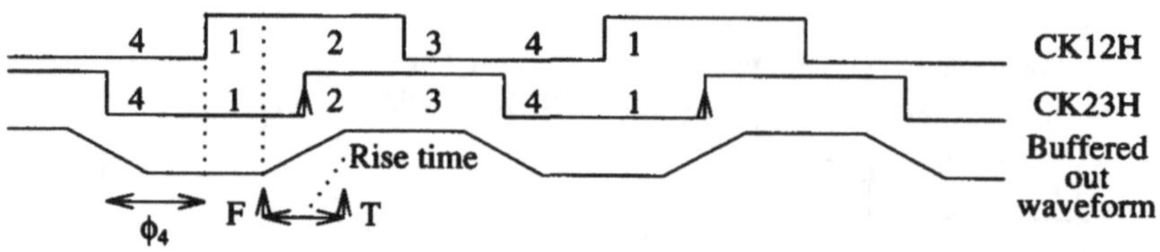

Figure 37 Waveform of risetime control

using the circuit shown in Fig. 36, and the waveforms shown in Fig. 37. Devices within a single CMOS VLSI chip track. In the circuit of Fig. 36, all the output drivers and one dummy driver have identical layout, load, and internal electrical control. If the rise or fall times of the dummy driver are controlled to be equal to the maximum specification by negative feedback control, all drivers have rise and fall times equal to the maximum specification: In Fig. 36, the circuit is clocked by a pair of two-phase primary clock pulses, CK12H and CK23H (Section 5.4). The chip goes through four phases 1, 2, 3, and 4 within one machine cycle. CK12H is sent out from the chip by the dummy output driver, which has capacitive load equal to other drivers. CK12H, CK23H, and the waveform of the buffered output of CK12H (delayed) are shown in Fig. 37. The rise time of the buffered output CK12H (interval between the two vertical arrows) is measured using the circuit shown in Fig. 36. Before the up-going transition of the buffered CK12H, the output of the exclusive OR gate has a logic-low level, since both inputs A and B are high. When the buffered up-going clockedge of CK12H arrives, node A pulls down. The output of the exclusive OR changes to a high level. When the tail of the up-going edge arrives, one pulse rise time later, the PFET turns off, and node B pulls down by the resistor. Both inputs A and B are at a low level. The exclusive OR output returns to a low level. Therefore, the pulse at the output of the exclusive OR gate has width equal to the pulse rise time. The output pulse is averaged, and an operational amplifier applies negative feedback, based on the average voltage, to maintain constant rise time. The three-input control gate circuit of Fig. 36 is shown in the dotted box to the right, and it will be discussed in Section 5.4. Rise or fall time is now independent of variations in the chip's processing, temperature and environmental variations.

Since $|dI/dt|$ is proportional to the charge that must be drained by the driver, power-bus noise can be reduced by reducing the charge. Figure 38(a) shows an open-drain pull-down-only output driver that has external pull-up resistor to a voltage source V_{BB}, which is lower than V_{DD}. Since $|dI(t)/dt|_{\max}$ is proportional to V_{DD} [Eq. (66)], reduction of the maximum output driver voltage reduces power-bus noise proportionally. This bus drive scheme has standby current. The external pull-up resistor can be replaced by a clocked bus precharge device, thereby eliminating the standby current. Figure 38(b) shows another scheme of reducing the voltage of the output driver. In Fig. 38(b) the transmission gate is turned on first. If the output driver is pull-up/pull-down symmetrical, the external bus is precharged to $V_{DD}/2$. Then the transmission gate is turned off, and the output driver charges or discharges the bus. Since the

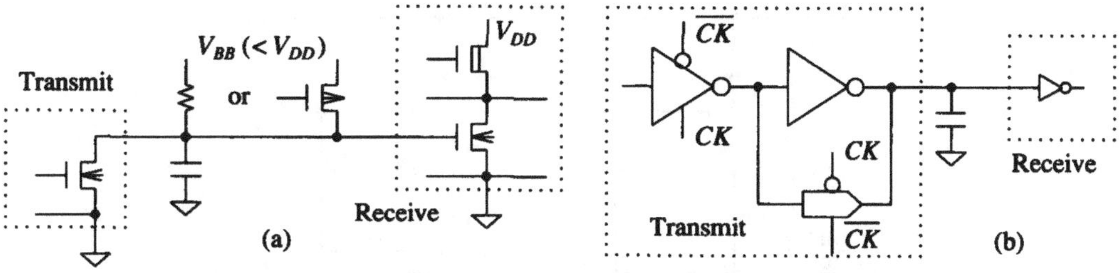

Figure 38 Techniques of reducing power-bus noise generation

output driver begins to drive from $V_{DD}/2$ to either V_{DD} or V_{SS}, the circuit works as if the power supply voltage were halved [15].

A widely practiced method of power-bus noise control is to separate the power-bus of the output drivers from the power-bus that feeds the internal logic circuits. *Noisy* power-buses are allowed to bounce as output drivers send data out. The *quiet* power-bus is used strictly to feed the internal logic circuits. In bidirectional I/O buffers, the output drivers and the input buffers must be supplied from different power-buses. Noisy and quiet power-buses are connected separately to the system's power supply terminals available in the package or PC board, as shown by the equivalent circuit of Fig. 39. In Fig. 39 inductance symbols represent bonding wires. This scheme works if power-bus noise is less than the PN junction turn-on voltage. Then noisy power-buses can be completely isolated from the substrate. If the noisy bus voltages bounce within the limit that PN junction is not forward-biased, no noise voltage is induced on the quiet bus. If voltage swing of the noisy power-bus is higher than that, the scheme has the risk of minority carrier injection into the substrate from accidentally forward-biased PN junction of source/drain islands.

CMOS PLAs, static and dynamic memories, and ROMs (read-only memories) include large, well-synchronized dynamic circuits. Figure 40(a) shows the equivalent circuit of a con-

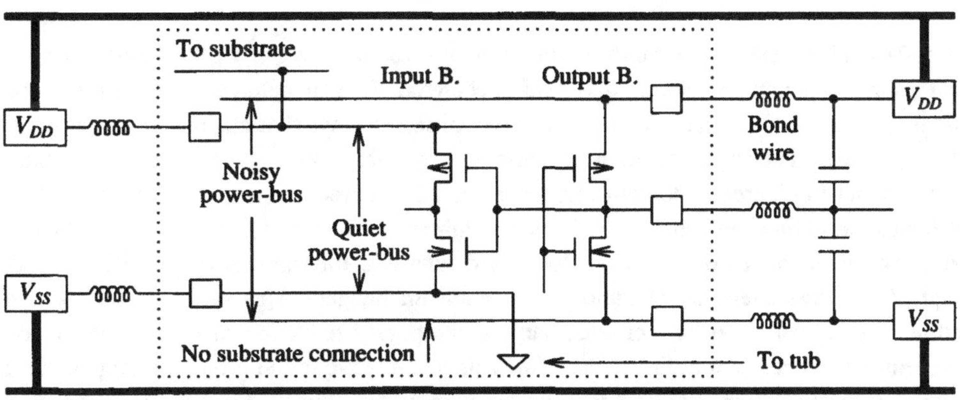

Figure 39 Connection of output driver

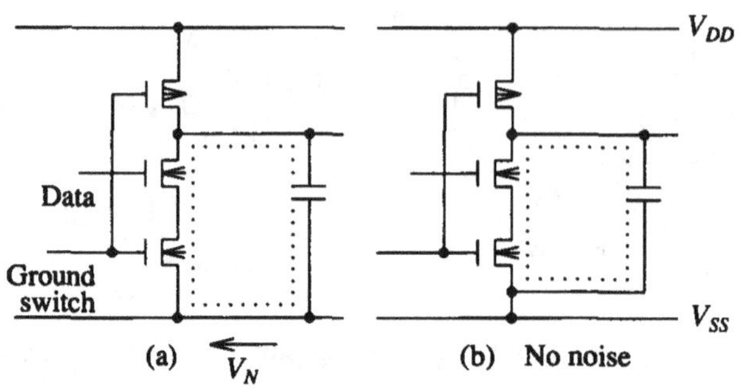

Figure 40 One-point grounding

ventional CMOS dynamic PLA. Load capacitance between the output node and V_{SS} (P-tub) is discharged through the FETs and the ground switch. Since the discharge current path includes a section of power-bus, noise that is generated on the power-bus section (V_N) affects the rest of the circuit [Fig. 40(a)]. If the load capacitance is first connected to the source of the ground switch, and then connected to the V_{SS} bus, the discharge current loops within the PLA, and the noise spike affects no external circuit [Fig. 40(b)]. To implement this in a PLA layout, the PLA ground-bus must be separated, and the P-tub area where the code FETs are placed must be securely grounded to the *local* PLA ground-bus. This layout technique helps reduce power-bus noise during PLA discharge. Power-bus noise during PLA precharge cannot be avoided. To reduce precharge noise, it is possible to precharge the entire PLA circuit at a controlled rate. It is also possible to precharge the PLA decoder and PLA ROM at different clock edges, or shifted clock edges. Then the precharge current spikes of the decoder and the ROM never overlap. It is possible to reduce precharge noise still further by splitting the decoder into several sections and precharging them sequentially.

4.13 Effects of Noise

Power-bus noise creates uncertainty in the *reference* voltage level of CMOS logic. Even if the input voltage to a gate or CMOS IC is held unchanged, the gate behaves as if the input voltage changed, if V_{SS} or V_{DD} changes. There are two different ways that changes take place. If on-chip capacitance is charged through the on-chip power-bus, V_{DD}, the bus voltage of the chip decreases and V_{SS} increases temporarily. The net effect is reduction of the power supply voltage for the duration of the chargeup. This is a *differential* mode of power supply voltage variation. When an off-chip capacitive load is discharged through on-chip NFET pull-down driver, V_{SS} increases temporarily, and V_{DD} remains unchanged. This variation may be considered as a superposition of two effects, namely, temporary reduction of power supply voltage and temporary increase of both V_{SS} and V_{DD} voltages relative to the ground voltage to the chip (voltage of the PC board ground-bus at the location of the chip). This latter is *common* mode of power supply voltage variation. It is therefore convenient to consider that power-bus noise consists of the two components-common mode V_{COM} and differential mode V_{DIF}. They are defined by

$$V_{COM} = \frac{V_{SS} + V_{DD}}{2} - \frac{V_{DD}(0)}{2} \qquad V_{DIF} = V_{DD} - V_{SS} - V_{DD}(0) \tag{74}$$

The differential mode power-bus noise modulates the gate's response time. If a current surge takes place, the power supply voltage decreases as a whole and the CMOS circuit fails, because gate delay time increases with decreasing power supply voltage. This problem will be discussed in Section 5.17. Effects of common mode power-bus noise are quite different. If the input voltage to a gate is fixed, and if the gate is subjected to the chip's power-bus noise whose common mode component V_{COM} is not zero, the gate responds as if there were an induced noise at the input node, having amplitude, $-V_{COM}$. In the following analysis we consider common mode power-bus noise only, and we do not worry about the origin or the source of noise: power-bus and induced noise mechanisms are not distinguished.

We consider two methods of evaluating the effects of noise on digital circuits. The first is to use static gate characteristics, and the second is to use dynamic gate response [16]. The first method is convenient for wide noise spikes, and the second method must be used for narrow noise spikes. The spike width that separates the two regimes is about the gate delay time. Since power-bus noise is quite narrow and the width of induced noise spikes is of the order of the gate delay time, the second method is more realistic than the first.

A) Static noise criteria of static CMOS gate (noise margin) [17]

Figure 41(b) shows the static characteristic of the CMOS inverter shown in Fig. 41(a). If V_1 changes from 0 (V_{SS}) to V_{DD}, the output voltage V_0 changes from V_{DD} to 0. There are two points, P and Q, on the static characteristic curve where the slope of the tangent to the curve is unity. At input voltage V_{1L}, output voltage is V_{OH}, and at input voltage V_{1H}, output voltage is V_{OL}. Within the input voltage range $V_{1L} < V_1 < V_{1H}$, the slope is greater than unity. When the inverter is biased at operating point P, the inverter has unity small-signal gain. If input voltage V_1 consists of a DC component V_{1L} and small variation ΔV_1

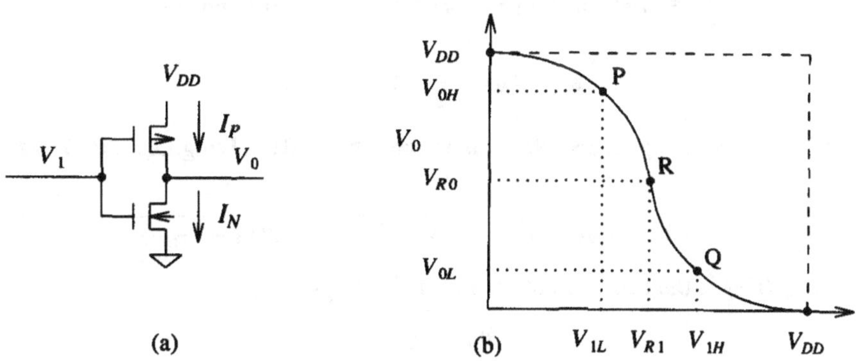

Figure 41 Definition and analysis of noise margin

$$V_1 = V_{1L} + \Delta V_1 \tag{75a}$$

V_0 is

$$V_0 = V_{0H} - \Delta V_1 . \tag{75b}$$

The inverter works as a small-signal unity gain inverter. When the inverter is biased at operating point R, between point P and Q, Eqs. (75a) and (75b) change to

$$V_1 = V_{R1} + \Delta V_1 \text{ and } V_0 = V_{R0} - A\Delta V_1$$

where $A > 1$. The inverter works as voltage amplifier. Operating points P and Q are the two ends where the inverter changes from small-signal active amplifier to passive attenuator. From this observation, we may conclude that an up-going or down-going noise spike whose height is less than V_{1L} or $V_{DD} - V_{1H}$, respectively, does not disturb logic operation of the gate that is settled at the respective steady states. This conclusion is independent of the width of the noise spike, as long as the noise spike is wide. This *conservative* upper limit of noise is called the *noise margin* [18].

Voltages V_{1L} and V_{1H} can be found using the simple FET characteristic as follows. PFET current I_P and NFET current I_N of a CMOS static inverter shown in Fig. 41(a) must be equal on the static characteristic as

$$I_P = I_N . \tag{76a}$$

If the collapsible current generator model is used, $I_P = B_P(V_{DD} - V_1 - V_{THP})$ and $I_N = B_N(V_1 - V_{THN})$. We have

$$V_{1L} = \frac{B_P(V_{DD} - V_{THP}) + B_N V_{THN}}{B_N + B_P} \text{ and } V_{DD} - V_{1H} = \frac{B_N(V_{DD} - V_{THN}) + B_P V_{THP}}{B_N + B_P} .$$

This is the switching threshold voltage and its complement, since the static characteristic is discontinuous. In a symmetrical inverter, $B_N = B_P = B$ and $V_{THN} = V_{THP} = V_{TH}$. Then $V_{1L} = V_{1H} = V_{DD}/2$. This is the maximum noise margin attainable. The model based on the collapsible current generator is too simplified to yield useful information.

If the low-field, gradual-channel FET model is used, we have

$$I_P = B_{GP}[V_{DD} - V_1 - V_{THP} - \frac{1}{2}(V_{DD} - V_0)](V_{DD} - V_0) \tag{76b}$$

$$\text{and } I_N = \frac{B_{GN}}{2}(V_1 - V_{THN})^2$$

where PFET is in triode region and NFET is in saturation. By solving Eqs. (76a) and (76b) for V_0,

$$V_0 = V_1 + V_{THP} + \sqrt{(V_{DD} - V_1 - V_{THP})^2 - \beta(V_1 - V_{THN})^2}$$

where $\beta = B_{GN}/B_{GP}$ is the beta ratio of the FETs. By requiring

$$(\partial V_0/\partial V_1) = -1$$

and solving for V_1

$$V_{1L} = (V_{DD} - V_{THP} - V_{THN}) f(\beta) + V_{THN} \quad \text{where} \quad f(\beta) = \frac{3}{2\sqrt{\beta(\beta+3)} + \beta + 3} \tag{77}$$

and

$$V_{1L} = \frac{3}{8} V_{DD} - \frac{3}{8} V_{THP} + \frac{5}{8} V_{THN}$$

if $\beta = 1$. If $V_{THP} = V_{THN} = V_{TH}$ is also assumed

$$V_{1L} = \frac{3}{8} V_{DD} + \frac{1}{4} V_{TH} .$$

We note that since $V_{TH} < V_{DD}/2$, $V_{1L} < V_{DD}/2$. V_{1H} can be found from Eq. (77) by substitution of

$$(V_{1L} - V_{THN}) \rightarrow (V_{DD} - V_{1H} - V_{THP}) \quad \text{and} \quad \beta \rightarrow (1/\beta)$$

as

$$V_{DD} - V_{1H} = (V_{DD} - V_{THP} - V_{THN}) f(1/\beta) + V_{THP} .$$

The V_{GSW} is given as (Section 2.4)

$$V_{GSW} = \frac{V_{DD} - V_{THP} + \sqrt{\beta} V_{THN}}{1 + \sqrt{\beta}} .$$

Then

$$V_{GSW} - V_{1L} = g(\beta)(V_{DD} - V_{THP} - V_{THN}) \tag{78}$$

$$\text{and} \quad V_{1H} - V_{GSW} = g(1/\beta)(V_{DD} - V_{THP} - V_{THN})$$

where

$$g(\beta) = \frac{2\sqrt{\beta(\beta+3)} + \beta - 3\sqrt{\beta}}{[1 + \sqrt{\beta}][2\sqrt{\beta(\beta+3)} + \beta + 3]} .$$

If $\beta = 1$, $g(\beta) = 1/8$. Therefore even using the gradual channel, low-field model, the CMOS static gate has a quite wide noise margin, close to the theoretical limit of $V_{DD}/2$. Noise margin is the most conservative criteria of a gate's noise sensitivity. This criteria is, however, excessively conservative for narrow noise pulses. Since the width of digital noise pulse is usually comparable to gate delay time, analysis of real noise sensitivity should include effects of noise pulse width, as well as noise pulse height.

Noise margin characterizes noise sensitivity of a single stage. Effects of noise to a cascaded inverter chain is quite complex. A special-case analysis is given in Section 4.14. Noise exists practically everywhere on a VLSI chip in operation, but the circuits pick up noise at only certain sensitive locations. Noise makes entry where impedance level is matched, and where there is gain. To the entry point of noise, the single-stage analysis is realistic.

B) Dynamic noise criteria of CMOS static gates

The second and more realistic method of characterizing the effects of digital noise on circuit operation is to study gate's response to noise pulse shown in Fig. 42(b), which has width T_N and height V_{1max}. We determine the parameters that characterize the response to the noise pulse, like curves A and B of Fig. 42(b). Especially we need to know width T_R and height V_R of the noise response. We consider the noise effects on a single gate. If many gates are cascaded, the noise effects are complex, as shown by the simplest example in Section 4.14. The following analysis holds approximately if the input noise pulse is substituted by the sum of the noise directly induced to the node and the noises induced to the other nodes and amplified before reaching the node.

This analysis is carried out using the collapsible current generator FET model. Channel current, I_N, is given by

$$I_D = I_N = B(V_1 - V_{TH}) \quad V_O > 0 \quad V_O = 0 \text{ if } I_D < I_N$$

and PFET has the same characteristic. We assume that the inverter shown in Fig. 42(a) is symmetrical.

We assume that the input noise voltage is a square pulse defined by [Fig. 42(b)],

$$V_1(t) = 0 \ (t < 0) \quad V_1(t) = V_{1max} \ (0 < t < T_N) \quad V_1(t) = 0 \ (T_N < t)$$

The microstate sequence of an inverter subjected to a noise pulse is shown in Fig. 43. In this sequence, circles connected by thick bars are equivalent to contacting circles. If V_{1max} is less than V_{THN}, no response is generated (sequence END). If V_{1max} is more than V_{THN} but is less than V_{GSW}, MN1 turns on, and then turns off (Sequence Z). Again, no transient is generated in the circuit. If $V_{1max} > V_{GSW}$ ($= V_{DD}/2$) but $V_{1max} < (V_{DD}/2) + (C_L V_{DD}/2BT_N)$, the switching transient is shown by sequence A of Fig. 43. MP1 goes into saturation region, but NFET MN1 does not have time to go into the triode region. Responding to the noise pulse, a down-going triangular pulse shown by dotted waveform A of Fig. 42(b) is generated at the output. If $V_{1max} > (V_{DD}/2) + (C_L V_{DD}/2BT_N)$, NFET MN1 goes into the triode region before the noise pulse turns off: the pulse has a flat bottom [dashed curve B of Fig. 42(b)]. If V_{1max} is close to $V_{GSW} = V_{DD}/2$, PFET MP1 never turns off: microstate sequence B represents this case. If V_{1max} is close to V_{DD}, MP1 turns off, and sequence C represents the switching.

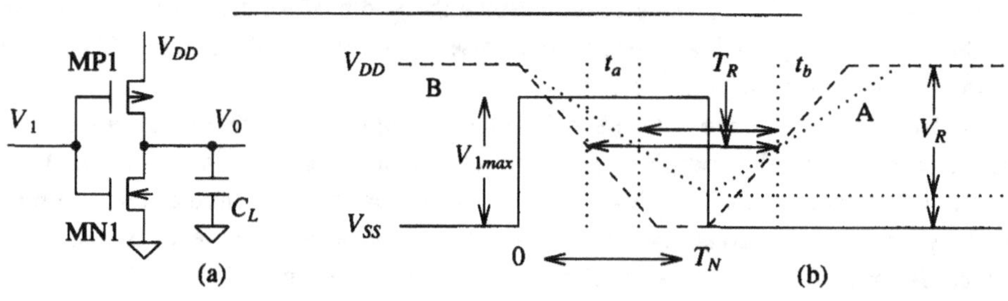

Figure 42 Effects of digital noise

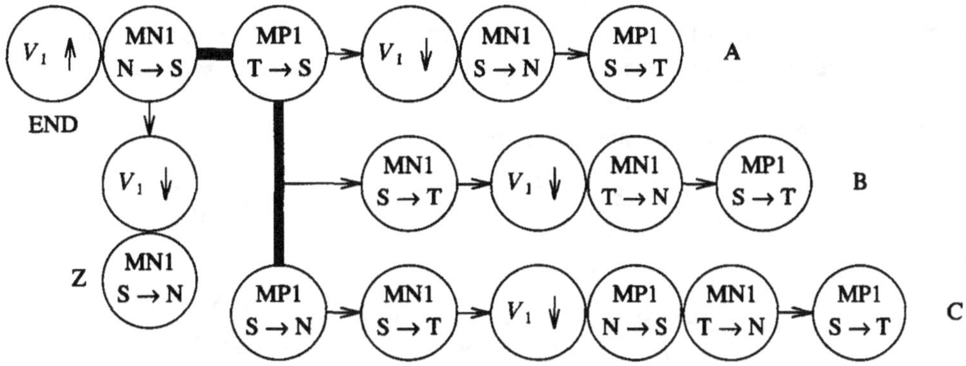

Figure 43 Microstate sequence of noise response

We consider cases of microstate sequences A and B (noise pulse is not excessively high). After the noise pulse starts, the output node voltage decreases with time as

$$V_0(t) = V_{DD} - \frac{2B[V_{1max} - (1/2)V_{DD}]}{C_L} t \qquad V_{1max} > \frac{V_{DD}}{2}. \tag{79}$$

$V_0 = (1/2)V_{DD}$ at $t = t_a$, where t_a is given by

$$t_a = \frac{C_L V_{DD}}{4B[V_{1max} - (1/2)V_{DD}]}.$$

At $t = T_N$, $V_0(t)$ reaches the minimum, $V_{\min}$, given by

$$V_{\min} = V_{DD} - \frac{2B[V_{1max} - (1/2)V_{DD}]}{C_L} T_N$$

if $V_{1max} < (1/2)V_{DD} + [(C_L V_{DD})/(2BT_N)]$, and $V_{\min} = 0$ otherwise. In the first case, the waveform does not have a flat bottom, and for $t > T_N$,

$$V_0(t) = V_{\min} + \frac{B(V_{DD} - V_{TH})}{C_L}(t - T_N). \tag{80}$$

Time t_b, when $V_0 = V_{DD}/2$ is reached again, is given by

$$t_b = T_N + \frac{C_L[(V_{DD}/2) - V_{\min}]}{B(V_{DD} - V_{TH})}.$$

By substituting $V_{\min}$ we obtain the down-going pulse width at $V_{DD}/2$ as

$$T_R = t_b - t_a = \frac{(2V_{1max} - V_{TH})[[V_{1max} - (1/2)V_{DD}]T_N - (C_L V_{DD}/4B)]}{(V_{DD} - V_{TH})[V_{1max} - (1/2)V_{DD}]} \tag{81}$$

if $(1/2)V_{DD} + (C_L V_{DD}/2BT_N) > V_{1max} > (1/2)V_{DD} + (C_L V_{DD}/4BT_N)$ [the second half of inequality means that $V_{\min} < (1/2)V_{DD}$] and

$$T_R = T_N + \frac{C_L V_{DD}}{2B}\left(\frac{1}{V_{DD} - V_{TH}} - \frac{1}{2V_{1max} - V_{DD}}\right) \tag{82}$$

if $V_{1max} > (1/2)V_{DD} + (C_L V_{DD}/2BT_N)$. In this case the response pulse has a flat bottom. Output pulse width at $V_{GSW} = V_{DD}/2$ is a good measure of how much the effects of noise is transmitted to the later stage (Section 2.2). From these results, the noise sensitivity of a CMOS static inverter can be determined as follows.

Equations (81) and (82) are normalized using

$$\tau_N = \frac{T_N}{(C_L/B)} \quad \tau_R = \frac{T_R}{(C_L/B)} \quad v_{1max} = \frac{V_{1max}}{V_{DD}} \quad v_{TH} = \frac{V_{TH}}{V_{DD}}$$

to obtain

$$\tau_N = \tau_R - 0.5\left[\frac{1}{1 - v_{TH}} - \frac{1}{2v_{1max} - 1}\right] \text{ if } \tau_N > \frac{0.5}{v_{1max} - 0.5}$$

$$\tau_N = \frac{\tau_R(1 - v_{TH})(v_{1max} - 0.5) + 0.5(v_{1max} - 0.5v_{TH})}{2(v_{1max} - 0.5)(v_{1max} - 0.5v_{TH})} \text{ if } \tau_N > \frac{0.25}{v_{1max} - 0.5}$$

Figure 44 shows a two-dimensional, equal-height, contour map of the normalized width of the inverter output pulse as a function of the normalized input noise pulse height $v_{1max} = V_{1max}/V_{DD}$, and normalized input noise pulse width, $\tau_{1max} = T_N/(C_L/B)$. If $V_{1max} < V_{DD}/2$ or, in general, if input noise is less than V_{GSW}, the inverter attenuates noise pulse. If the noise pulse height is higher than $V_{DD}/2$, or, in general, if input noise pulse is higher than V_{GSW}, the inverter amplifies and generates approximately full-height noise pulse. If v_{1max} and τ_N are under the dotted curve A, the noise pulse at the output does not reach the critical height $V_{DD}/2$. If v_{1max} and τ_N are above the dotted curve B, the pulse at the inverter's output has a flat bottom (or, the output node voltage drops to 0). Contours for the parameter values that give output pulse width 0.5, 1.5 and 4.5 $\times(C_L/B)$ are shown by the three solid curves. If input noise pulse height is close to $V_{GSW} = V_{DD}/2$, the noise pulse width must be

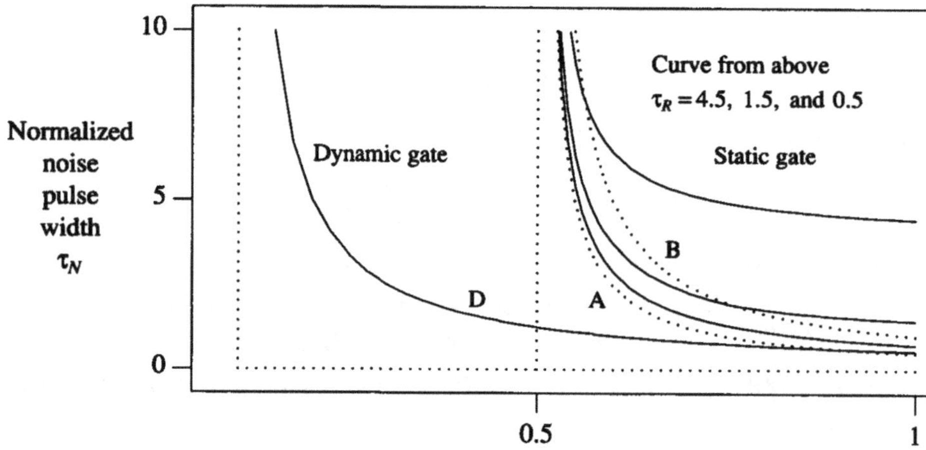

Figure 44 Response of inverter to noise pulse

quite wide to influence the circuit. In the wide pulse limit, the noise margin discussed before provides an alternative characterization of noise sensitivity, but in real operation of the circuit, noise sensitivity should be characterized by dynamic criteria, as discussed here.

C) Effects of noise in dynamic circuits

Effects of noise on dynamic circuits are estimated in the following closed-form analysis. The dynamic gate equivalent circuit is shown in Fig. 45(a). Before the noise pulse arrives, MN1 is turned off, the output node of the dynamic gate is precharged to V_{DD}, and the node holds total charge $C_L V_{DD}$. The charge is lost through MN1 if a noise pulse, $V_1(t)$, arrives at the gate of MN1. The output node voltage time dependence, $V_0(t)$, is determined from the circuit equation,

$$C_L \frac{dV_0(t)}{dt} = -B[V_1(t) - V_{THN}] \tag{83}$$

which is solved as,

$$V_0(t) = V_{DD} - \frac{B}{C_L} \int_0^t [V_1(t) - V_{THN}]dt \tag{84}$$

$$= V_{DD} - \frac{B}{C_L}(V_{1max} - V_{THN})T_N \quad \text{if} \quad (t > T_N)$$

V_{1max} and T_N are the height and the width of the noise pulse, respectively. Figure 45(b) shows the output node waveforms. If the decrease of the output node voltage is more than the PFET threshold voltage, the dynamic circuit that is connected to the gate output is affected. In this case we have,

$$(v_{1max} - v_{THN})\tau_N = v_{THP} \tag{85}$$

where $v_{1max} = V_{1max}/V_{DD}$, $v_{THN} = V_{THN}/V_{DD}$, $v_{THP} = V_{THP}/V_{DD}$ and $\tau_N = T_N/(C_L/BV_{DD})$. There is a *complementary* relationship between noise pulse height and width. The relationship is shown by curve D in Fig. 44. If we ask whether or not the response will reach height $V_{DD}/2$ to

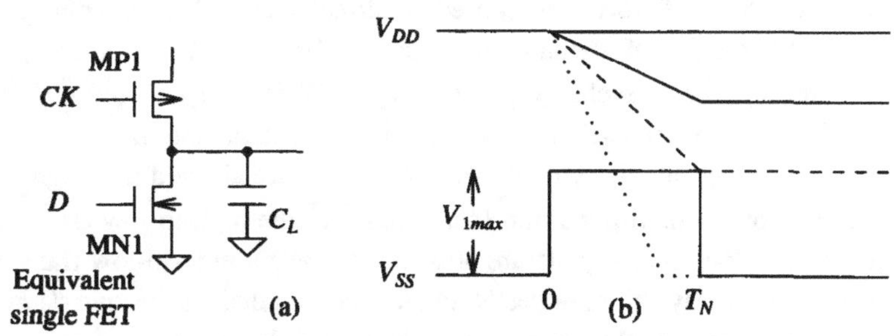

Figure 45 Noise effects of CMOS dynamic gate

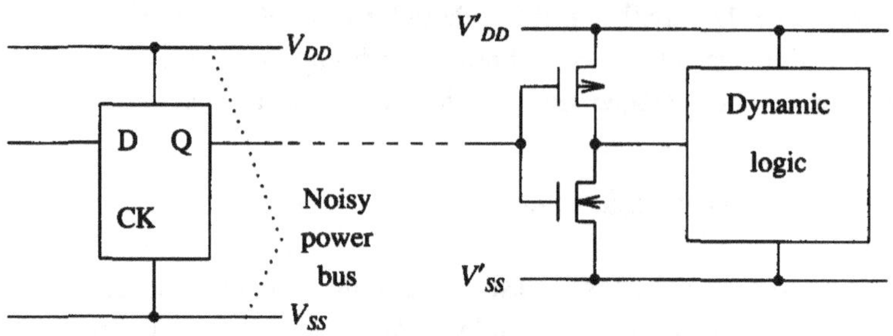

Figure 46 Minimizing the effects of power-bus noise

affect a static gate, v_{THP} on the right-hand side of Eq. (85) should be replaced by 1/2. In dynamic gates an *infinitely* wide output pulse is generated.

These results were derived using the collapsible current generator FET model. The conclusions are translated to the case of more realistic FET model as follows. Analysis in this section showed that a symmetrical CMOS static inverter is immune to noise whose amplitude is less than a maximum, which is close to $V_{DD}/2$. CMOS dynamic gates are many times more sensitive to noise than static gates. CMOS static inverters can be used together with a CMOS dynamic gate to increase noise immunity as the following example shows. Since a PLA generates significant noise, microprocessor control signals generated by a PLA should not be introduced directly to sensitive datapath destinations that have many dynamic gates. One way to eliminate the noise effect is to buffer the control signal from a noisy PLA source by an inverter that shares power and ground-bus with the dynamic gates at the destination, as shown in Fig. 46. Unless the PLA power-bus noise is higher than $V_{DD}/2$, the static buffer prevents PLA power-bus noise from reaching the dynamic logic gate at the destination.

4.14 Cascaded Tristated Gate Chain

NORA (No-Race) CMOS dynamic logic circuits [19] have many failure problems in a VLSI circuit environment. Circuit failures are caused by directly cascaded dynamic gates. Figure 47 shows a simple NORA CMOS dynamic logic circuit that ORs all the input signals A_1, B_1, A_2, ..., . At the time of gate's precharge-predischarge, CK is held low and $\overline{CK}$ is held high. Nodes N_1, N_3, N_5, ... are predischarged to ground potential, and nodes N_2, N_4, ... are precharged to V_{DD}. During that time, inputs A_1, B_1, A_2, ... are all held at a high level. When clock CK makes the low-to-high transition (and clock $\overline{CK}$ the high-to-low transition simultaneously) some or all of the input signals are allowed to make a high-to-low transition. If any of the inputs A_i $(i = 1-6)$ stays high, nodes N_i to N_6 maintain their respective *default* states set during precharge-predischarge: all these nodes are high impedance. In the case when A_1 or B_1 is high, all the nodes $N_1 - N_6$ are tristated, alternatively, to low and high voltage levels (N_1,

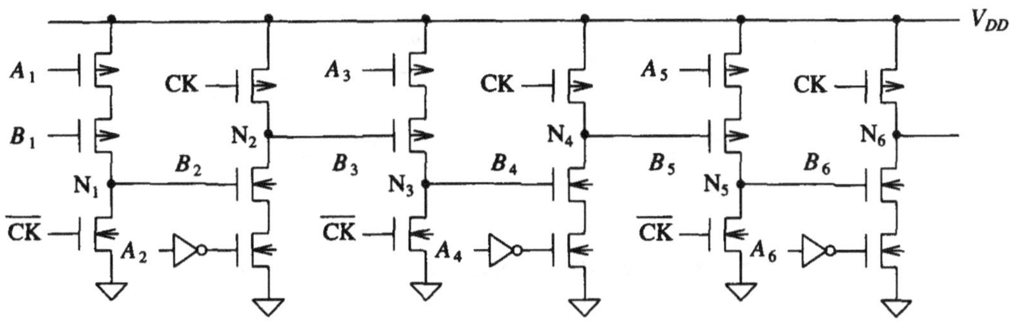

Figure 47 NORA dynamic logic circuit

N_3, ... are at ground potential, and N_2, N_4, ... are at V_{DD}). A circuit having many tristated nodes that may interact is sensitive to external influence like noise. This problem is a generalization of the problem of the last section.

A cascaded chain of tristated gates has unidirectional voltage gain, by which a signal induced on a node is amplified before reaching the next node. There are two cases in which the effects from many stages accumulate. The first is a long chain of cascaded and tristated gates through which a signal travels unidirectionally, and the second is a loop of tristated gates in which a signal circulates many times. The second case will be discussed in Section 6.5. We study a long chain of tristated gates in this section. Figure 48 shows a simplified equivalent circuit of a long chain of tristated gates. Series-connected conducting FETs are replaced by a single equivalent FET. Nodes N_1, N_3, ... are originally at ground potential, and nodes N_2, N_4, ... are at V_{DD} (immediately after precharge/predischarge). We analyze the case where the input node to the chain is affected by noise. The input node is pulled down to $V_{DD} - V_I$ at time $t = 0$. The microstate sequence of this process is shown in Fig. 49. The gate chain begins discharging from the left end. Turn-on transitions (N $\rightarrow$ S) of FETs take place in sequence. It is not obvious, however, which FET arrives at the triode region first. The bracket [...] of the microstate sequence indicates that the order in which the final triode regions are reached cannot be determined from the circuit structure alone. We assume the FET collapsible current generator

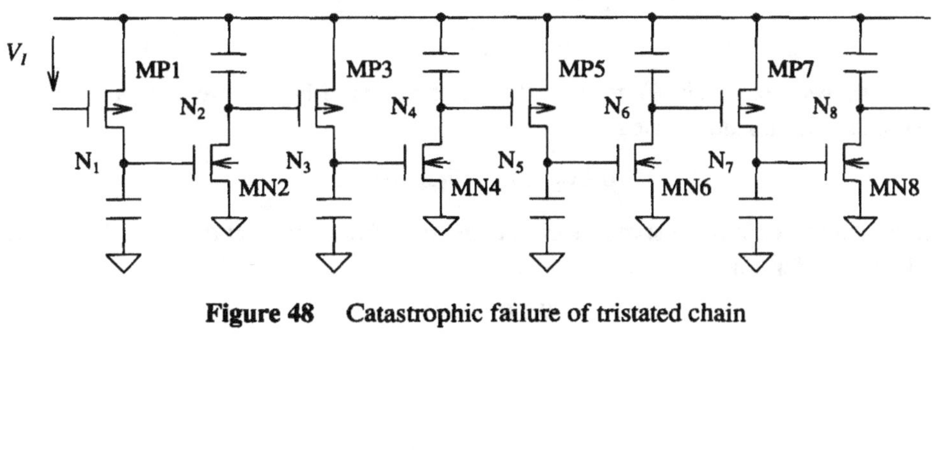

Figure 48 Catastrophic failure of tristated chain

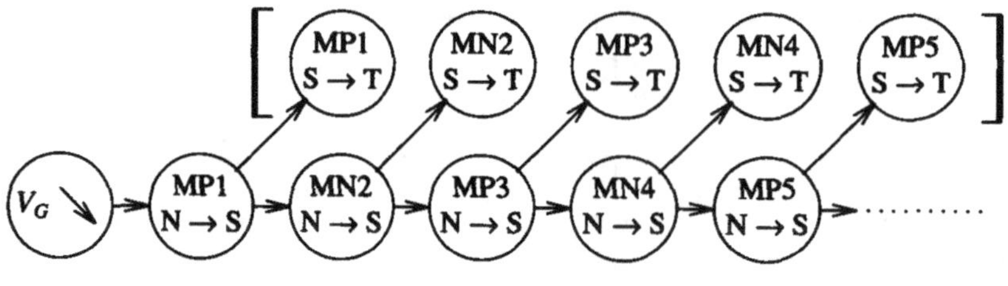

Figure 49 Microstate sequence of NORA gate chain

model with zero threshold voltage, and that PFETs and NFETs have same transconductance. This special case can be solved in closed form. With reference to Fig. 48, voltages of even index number nodes (N_{2n}), $V_{2n}(t)$ and of odd index number node (N_{2n+1}), $V_{2n+1}(t)$ satisfy circuit equations

$$C\frac{d}{dt}[V_{DD} - V_{2n}(t)] = BV_{2n-1}(t) \qquad C\frac{dV_{2n+1}(t)}{dt} = B[V_{DD} - V_{2n}(t)] \tag{86}$$

which are integrated subject to the following initial conditions:

$$V_{2n}(0) = V_{DD} \qquad V_{2n+1}(0) = 0$$

to obtain

$$V_{2n}(t) = V_{DD} - \frac{B}{C}\int_0^t V_{2n-1}(t)dt \qquad V_{2n+1}(t) = \frac{B}{C}\int_0^t [V_{DD} - V_{2n}(t)]dt \tag{87}$$

Using $V_0(t) = V_{DD} - V_I$, this recurrence equation can be solved in a closed form. Using the normalization

$$\frac{B}{C}t = \theta \quad \frac{V_{2n}(t)}{V_{DD}} = v_{2n}(t) \quad \frac{V_{2n+1}(t)}{V_{DD}} = v_{2n+1}(t) \quad \frac{V_I}{V_{DD}} = v_I \tag{88}$$

we obtain

$$v_0(\theta) = 1 - v_I \quad v_1(\theta) = v_I\theta \quad v_2(\theta) = 1 - \frac{1}{2}v_I\theta^2 \tag{89}$$

$$v_3(\theta) = \frac{1}{2}\frac{1}{3}v_I\theta^3 \quad v_4(\theta) = 1 - \frac{1}{2}\frac{1}{3}\frac{1}{4}v_I\theta^4$$

$$\cdots\cdots\cdots\cdots\cdots$$

Figure 50 shows $v_1(\theta)\ldots, v_6(\theta)$, for $v_I = 0.1$. The results of Fig. 50 are valid until any of the following inequalities are violated:

$$0 \leq v_{2n}(\theta), \qquad v_{2n+1}(\theta) \leq 1 \tag{90}$$

Figure 50 shows that $v_3(\theta)$ first moves out of the range from 0 to 1, at $\theta = 3.92$. Therefore the analysis is valid for time

$$0 \leq \theta \leq 3.915\,.$$

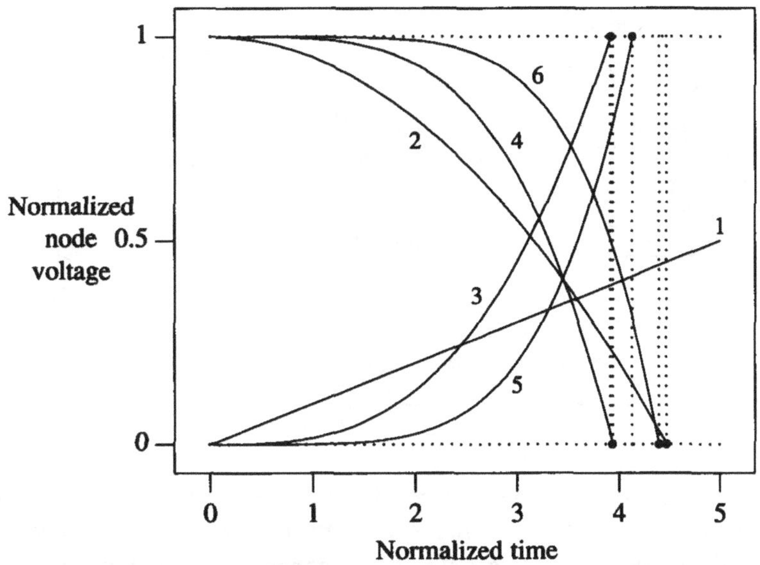

Figure 50 Node voltage versus time

That the maximum range of voltages specified by Eq. (90), is exceeded by $v_3(\theta)$ first [and not by $v_1(\theta)$ or $v_2(\theta)$] means that the third stage reaches the wrong logic level first, and therefore a circuit failure occurred there first. The first stage to fail depends on two factors. Since V_I is amplified as the signal propagates through the cascaded stages, later stages (large n) are likely to fail first. Since it takes time for the signal to propagate from one stage to the next, earlier stages (small n) are also likely to fail first. By a compromise of the two conflicting mechanisms, the cascaded circuit fails at an intermediate location first.

4.15 Input Noise

The gate delay time of scaled-down CMOS technology is short, typically 1 nanosecond or less. Therefore only a few nanosecond wide noise pulse generates a propagating signal in logic gate chains. This high sensitivity has the following consequences. Application of CMOS is not limited to very high clock frequency systems. Because of its low power advantage, CMOS chips are used in systems whose clock frequency is low (often less than 100kHz). Such slow CMOS devices can be fabricated using the old CMOS technology, whose feature size is 5–10 microns and whose typical gate delay times are 10 nanoseconds or more. Large feature size CMOS technologies were widely used until the mid-1970s. Because of the rapid progress in CMOS technology, such obsolete technologies were scrapped by the mid-1980s. A designer has no other choice than to use the current, scaled-down CMOS technology of

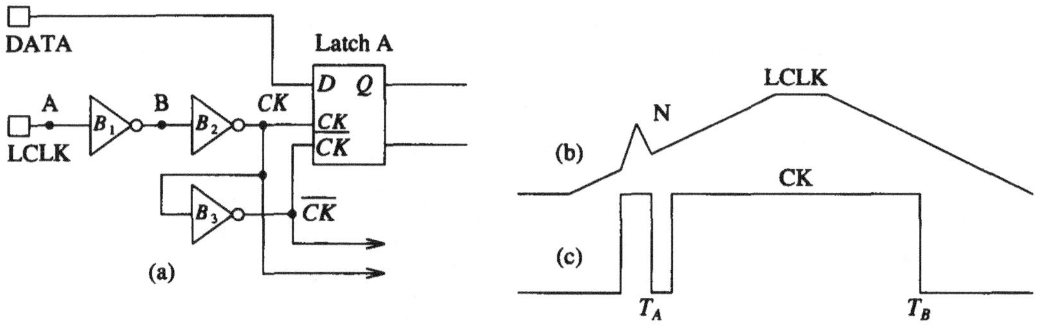

Figure 51 Noise of input circuits

about 1 micron feature size to design even slow CMOS devices. Furthermore, even slow chips can be LSI or VLSI, and the smaller feature size produces much lower power dissipation.

If a slow CMOS chip is designed using fast CMOS technology, a new digital noise problem emerges. The problem can be understood from the circuit shown in Fig. 51. Latch A captures signal DATA when LCLK makes a high-to-low transition. Suppose that the chip was designed using fast, scaled-down CMOS technology, and the chip was driven by a slow LCLK pulse as shown in Fig. 51(b). The LCLK pulse is generated outside the chip, where noise induction is inevitable. When noise N is induced on the LCLK waveform while it is making a slow transition, the fast input buffer B_1 is able to respond to the noise spike: noise N is amplified to generate a glitch at node B, and after further buffering by inverters B_2 and B_3, clock pulse CK shown in Fig. 51(c) is generated. Latch A strobes DATA in at time T_A instead of at the correct time T_B, and the chip goes one clock period ahead, with wrong input data. This results in a malfunction of the circuit [20].

If the amplitude of noise N is small, this problem can be analyzed using the equivalent circuit model shown in Fig. 52(a). Since the input voltage changes very slowly, the inverter is DC-biased at the highest gain operating point when the input noise pulse arrived. We consider

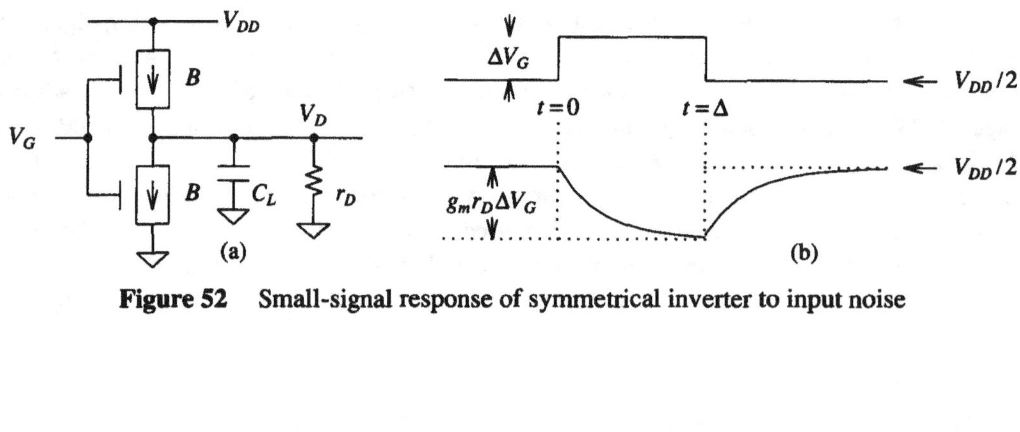

Figure 52 Small-signal response of symmetrical inverter to input noise

a symmetrical inverter where the PFET and NFET are modeled by using a single current generator whose transconductance g_m is given by

$$g_m = \left(\frac{\partial I_D}{\partial V_G}\right)_{NFET} (V_G = V_D = V_{DD}/2) - \left(-\frac{\partial I_D}{\partial V_G}\right)_{PFET} (V_G = V_D = V_{DD}/2) = 2B_G$$

In Fig. 52(a) we used load resistance, r_D, that represents the drain resistance of the FETs and C_L, that is the capacitive load. The circuit includes r_D since the signal is slow and therefore the low-frequency gain of the inverter matters. The circuit equation is written using small variations of node voltages v_G and v_D defined by $v_G = V_G - (V_{DD}/2)$ and $v_D = V_D - (V_{DD}/2)$ as

$$C_L \frac{dv_D}{dt} + \frac{v_D}{r_D} + g_m v_G = 0. \tag{91}$$

This equation is solved to obtain

$$v_D(t) = -\frac{g_m}{C_L}\int_0^t v_G(\theta)exp[-(t-\theta)/C_L r_D]d\theta + v_D(0)exp(-t/C_L r_D). \tag{92}$$

Suppose that input noise waveform $v_G(t)$ is given by

$$v_G(t) = 0 \quad t < 0 \quad = \Delta V_G \quad 0 < t < \Delta \quad = 0 \quad \Delta < t.$$

Then Eq. (92) becomes

$$v_D(t) = 0 \quad t < 0 \tag{93}$$

$$v_D(t) = -\frac{g_m \Delta V_G}{C_L}\int_0^t exp[-(t-\theta)/C_L r_D]d\theta$$

$$= -(g_m r_D)\Delta V_G[1 - exp(-t/C_L r_D)] \quad 0 < t < \Delta$$

$$v_D(t) = v_D(\Delta)exp(-t/C_L r_D) \quad \Delta < t.$$

Equation (93) is plotted in Fig. 52(b). This result is interpreted as follows. In the steady state, the inverter output node voltage change is $-G\Delta V_G$ where $G = g_m r_D$ is the gain of the inverting amplifier. The output voltage, v_D, approaches $-G\Delta V_G$, and then returns to $V_{DD}/2$ after the pulse is turned off, with the same time constant $C_L r_D$.

Whether or not the input waveform of Fig. 52(b) produces a *full glitch* pulse is a large-signal circuit problem that can be resolved using the same assumptions, and by the following graphical method. We consider two cascaded *symmetrical* inverters as shown in Fig. 53. Input voltage V_A, shown in Fig. 54(a), is a superposition of a monotonically and slowly increasing waveform and noise that arrives at about the time when the input voltage reaches the switching threshold voltage, $V_{DD}/2$. The inverter's output node waveform, V_B of Fig. 54(a), is determined by solving the equation

$$C_B \frac{dV_B(t)}{dt} = -2B[V_A(t) - \frac{1}{2}V_{DD}] \tag{94}$$

$$\text{or} \quad V_B(t) = V_{DD} - \frac{2B}{C_B}\int_{t_1}^t [V_A(\theta) - \frac{1}{2}V_{DD}]d\theta$$

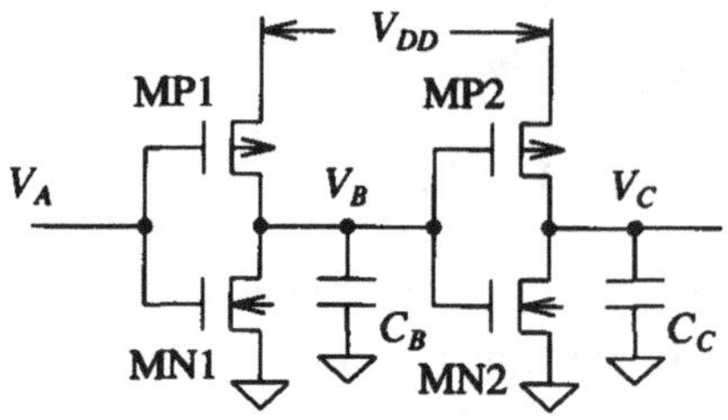

Figure 53 Large-signal analysis of input noise

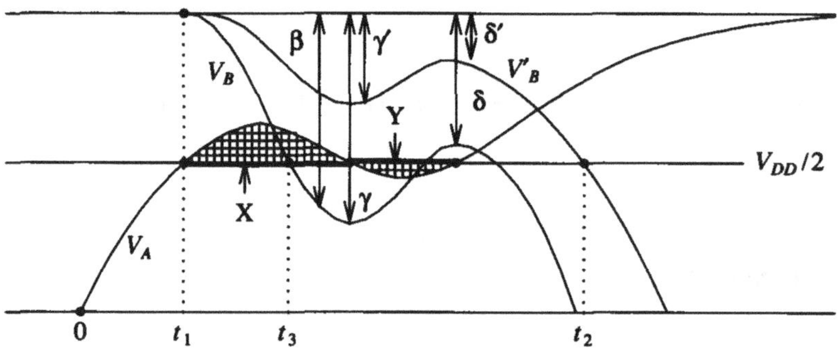

Figure 54(a) Noise waveform and glitch

where B is the transconductance of the FETs and where t_1 is the time when $V_A(t)$ reaches the switching threshold voltage $V_{DD}/2$ for the first time, as shown in Fig. 54(a). Using Eq. (94) curve $V_B(t)$ of Fig. 54(a) is drawn as follows: the length of arrow β (after t_1) equals the shadowed area surrounded by arrow β, curve $V_A(t)$, and straight line $V = V_{DD}/2$, multiplied by scale factor $(2B/C_B)$. Figure 54(a) shows the case when area X (above $V_{DD}/2$) is significantly larger than area Y (below $V_{DD}/2$). Then $V_B(t)$ looks like curves V_B or V'_B of Fig. 54(a).

$$\text{Length of } \gamma, (\text{or } \gamma') = (2B/C_B)(\text{area } X)$$

$$\text{Length of } \delta, (\text{or } \delta') = (2B/C_B)(\text{area } X - \text{area } Y)$$

We note that the length of arrow δ (at the voltage maximum) is not zero because area X is larger than area Y. The magnitude of parameter $2B/C_B$ determines whether the node waveform is like V_B (large $2B/C_B$, or fast inverter) or V'_B (small $2B/C_B$, or slow inverter). The node B waveform drives the second-stage inverter, which has the same switching threshold voltage. If $2B/C_B$ is small (V'_B), the second-stage inverter switches at $t = t_2$, and in this case the effect of noise is only to increase delay of the cascaded inverter. No glitch is generated. If $2B/C_B$ is large, the minimum of V_B (arrow γ) is larger than $V_{DD}/2$. The second-stage inverter begins to switch at the front edge of noise pulse, t_3. If the length of arrow δ is longer than $V_{DD}/2$, the second-stage inverter pulls up continuously: the delay time of cascaded inverters decreases due to the noise, and no glitch is generated. Distortion of the switching waveform

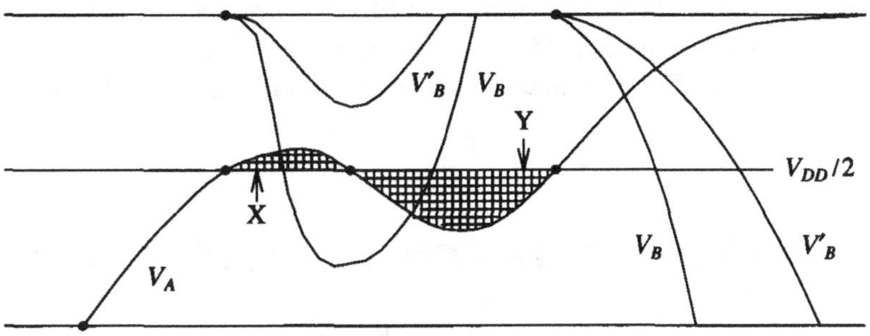

Figure 54(b) Noise waveform and glitch

(plateau of voltage waveform) is significant if $2B/C_B$ (or delay of inverter) is in the intermediate range. In this case a glitch is not generated except for rather unusual circumstances.

The noise waveform, V_A, shown in Fig. 54(b) has area X smaller than area Y. In this case, a glitch is always generated if $2B/C_B$ is increased (or the inverter delay is reduced). Because area Y is larger than area X in Fig. 54(b), *separation* of the front end from the *main* transition (that follows later) is assured. Therefore, if capacitive loading is small or the transconductance is high, the spurious front-end transition is separated and a glitch is generated. In the case of Fig. 54(a) the front end of the noise pulse is still an integral part of the main transition, and therefore a separate pulse is not likely to be generated. The cases of Fig. 54(a) and 54(b) show essentially two different regimes.

Microstate sequence A of Fig. 55 is *normal* static switching of an inverter. Sequence B has an extra component, the *null* sequence in the microstate sequence (shown by the double arrow). The null sequence indicates not only non-monotonic change of node voltages but also existence of a glitch pulse that may cause failure.

To prevent input noise effects, at least one of the following measures must be taken.

1) The customer is requested to provide clean, noise-free signals to the chip.

2) A noise filter (low-pass filter) is provided at the input of the chip. It is possible to design the noise filter to serve simultaneously as an ESD (electrostatic discharge) protection circuit. Then a single circuit serves two useful purposes.

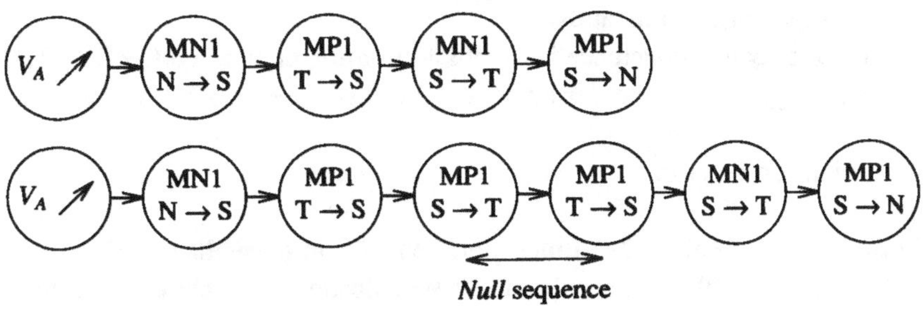

Figure 55 Microstate sequence of input noise

3) Response time of the input buffer is deliberately increased so that the input buffer is unable to respond to a short noise pulse. Channel length of the first-stage FETs is increased to slow down the response. This technique simultaneously minimizes the effects of process variation.

4.16 How To Reduce Noise Effects of Digital Circuits

From the discussions of the previous sections, we learned that there are several generic design techniques for reducing noise effects in digital circuits. They are summarized in this section.

I) Minimization of noise

Amplitude of induced noise depends on the ratio of coupling capacitance to node capacitance. The ratio must be kept small. Insulating substrate CMOS technology is undesirable for this reason. In multilevel metal technology, noise induction from the first-level metal to the second-level metal is significant, because second-level metal has a small capacitance of its own. Sensitive nodes in the second-level metal should not be parallel and overlapping wide first level metal wires. Sensitive nodes should be confined within narrow area to reduce capacitive coupling from other nodes.

The width of induced noise spikes is proportional to the product of the total node capacitance and the resistance of the node to any low-impedance DC power supply level (V_{SS} or V_{DD}). Since capacitance, being an *essential* parasitic, cannot be reduced below a certain limit, the resistance of the node must be reduced. Long interconnect wires, especially of second-level metal, should not be tristated. Such logic signals should be buffered at intermediate points by static CMOS buffers and then the buffered signals are retransmitted through the rest of the interconnect.

Noise sources and sensitive nodes should be electrostatically isolated. Circuit nodes that are an especially strong noise generator, or especially sensitive to noise, can be shielded electrostatically. Power- and ground-buses that generate large power-bus noise must be separated from the rest of the power-bus system.

Synchronous precharge-discharge of large dynamic gates must be carefully controlled, so that not all currents overlap. By adding small delays between precharge/discharge processes, noise can be reduced quite significantly.

Reduction of package inductance is essential in future, high-speed CMOS VLSI systems. Packaging technology based on solder bump and area pads (Section 7.7) is the most desirable.

II) Desensitization of circuits

CMOS dynamic gates are about 3–5 times more sensitive to noise than CMOS static gates. If CMOS static gates are able to meet the delay specifications, the chip should be assembled using CMOS static gates alone.

CMOS dynamic gate circuits must be designed carefully. Practically, the output node of any dynamic gate should be buffered by a static inverter. This leads to the domino CMOS circuit configuration. NORA dynamic gates are vulnerable to noise effects.

Noise that is induced from outside the chip must be filtered before being introduced to the chip. The elimination of input noise is essential in slow CMOS circuits fabricated using fast CMOS technology.

There is an essential limit in circuit desensitization: the sum of noise margins to logic-low and logic-high levels is less than V_{DD}. In order to increase the limit still further, hysteresis must be introduced in the switching characteristic. This solution is discussed in the next section.

4.17 Switching of Schmitt Trigger Circuits

In the last section we saw that a CMOS static gate is sensitive to up-going noise spikes higher than V_{GSW}, and to down-going noise spike larger than $V_{DD} - V_{GSW}$. The sum of the two noise pulse heights is fixed at V_{DD}. If V_{GSW} is set high, sensitivity to an up-going noise spikes decreases, but sensitivity to down-going noise spike increases. Conventional design policy is to set V_{GSW} at about $V_{DD}/2$ so that the two sensitivities are balanced.

There is a way to reduce noise sensitivity further. We use a gate that switches at different threshold voltages for up-going and down-going transitions. If the switching threshold for an up-going transition [$V_{GSW}(U)$] is higher than the switching threshold for a down-going transition [$V_{GSW}(D)$], the gate has high noise immunity. This kind of gate is called a Schmitt trigger.

There are several versions of Schmitt trigger circuits. The circuit shown in Fig. 56(a) is perhaps the simplest circuit and is used frequently [21]. This circuit consists of a CMOS inverter shown in Fig. 56(b) plus an NFET and a PFET. If the extra FETs, MN3 and MP3, are much smaller than MN1, MN2, MP1, and MP2, the circuit of Fig. 56(a) switches at the same

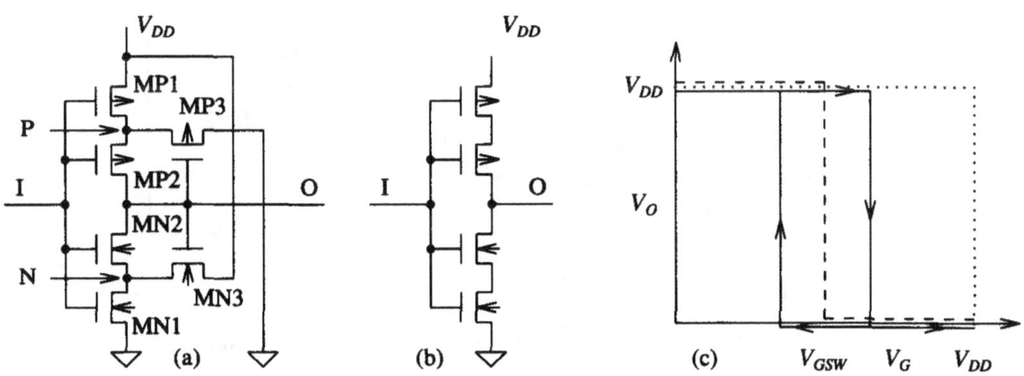

Figure 56 Schmitt trigger circuit and characteristics

V_{GSW} as the inverter of Fig. 56(b). If FETs MN3 and MP3 have comparable conductivities as the other FETs, V_{GSW} for up-going transitions becomes higher than the switching threshold voltage of the inverter of Fig. 56(b), and the switching threshold voltage for down-going transition becomes lower than that of the inverter. If the FET collapsible current generator model is used, static switching characteristics of the two circuits are as shown in Fig. 56(c). The dashed curve shows the static characteristic of the inverter of Fig. 56(b). The solid curves show static characteristics of the Schmitt trigger circuit of Fig. 56(a), and the arrows show directions of voltage changes. The static characteristic has *hysteresis*.

We study the process of Schmitt trigger circuit switching. Suppose that at time $t=0$, input voltage V_G is 0, and the voltage increases gradually. We consider the static switching process. For simplicity, we assume MN1 and MN2 have the same size, as do MP1 and MP2, and we neglect the back-bias effect. The initial condition of the Schmitt trigger circuit is given by

$$V_G=0 \quad V_O=V_{DD} \quad V_P=V_{DD} \text{ and } V_N=V_{DD}-V_{THN}. \tag{95}$$

Until $V_G \geq V_{THN}$ the circuit remains in the initial state (output high). When V_G exceeds V_{THN}, MN1 begins to draw current but MN2 is not conducting yet. The currents of MN1, I(MN1), and of MN3, I(MN3), must be equal. Using node voltage V_N of Fig. 57(a) we write

$$I(MN1)=I(MN3) \text{ where } I(MN1)=B_N(V_G-V_{THN}) \tag{96}$$

$$I(MN3)=b_N(V_{DD}-V_N-V_{THN}).$$

We note that V_O is still V_{DD}. We solve for V_N from Eq. (96) as

$$V_N=(V_{DD}-V_{THN})-\frac{B_N}{b_N}(V_G-V_{THN})$$

Since V_N decreases with increasing V_G, MN2 turns on when

$$V_G>V_N+V_{THN} \text{ or } V_G>\frac{V_{DD}+(B_N/b_N)V_{THN}}{1+(B_N/b_N)}=V_{GC1}$$

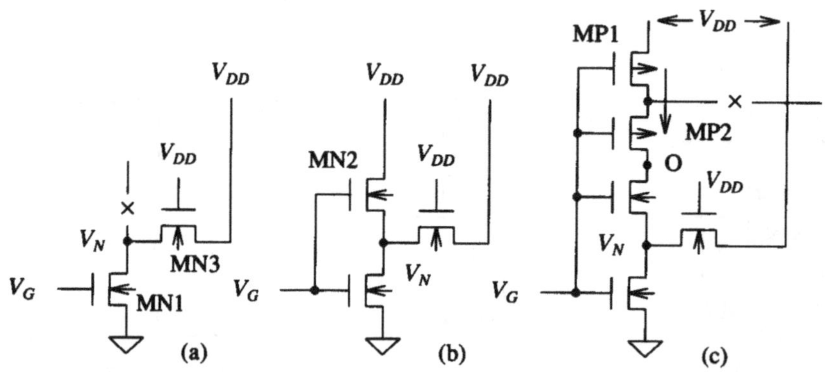

Figure 57 Initial phase of pull-down transition of Schmitt trigger

is reached. At this input voltage, node N voltage V_{NC1} is

$$V_{NC1} = \frac{b_N}{B_N + b_N}(V_{DD} - V_{THN}) \ (>0)$$

and therefore MN1 has not yet reached the triode region. After MN2 turns on, V_N is determined from

$$I(MN2) + I(MN3) = I(MN1) \tag{97}$$

where $I(MN2) = B_N(V_G - V_N - V_{THN})$. We have

$$V_N = \frac{b_N}{B_N + b_N}(V_{DD} - V_{THN}) = V_{NC1}\,.$$

V_N is now independent of V_G. The current of MN2 is given by

$$I(MN2) = B_N[V_G - V_{THN} - \frac{b_N}{B_N + b_N}(V_{DD} - V_{THN})]\,.$$

The chain of the two PFETs MP1 and MP2 is able to carry the maximum current, $I(MP^*)$, given by $I(MP^*) = B_P(V_{DD} - V_G - V_{THP})$. If the current limit is not exceeded, the PFET chain develops no voltage (triode region). If that limit is exceeded, the PFET chain develops voltage that initiates switching. This limit is given by equating $I(MN2)$ to $I(MP^*)$ as

$$V_G > V_{GC2} \tag{98}$$

where

$$V_{GC2} = \frac{1}{B_N + B_P}\left[B_P(V_{DD} - V_{THP}) + B_N[V_{THN} + \frac{b_N}{B_N + b_N}(V_{DD} - V_{THN})]\right].$$

If this limit is exceeded, either MP1 or MP2 moves into saturation, thereby developing voltage across its source and drain. Which FET develops voltage first depends on details like FET size and the back-bias effect. If MP1 and MP2 have the same size, as we assumed, MP2 goes into saturation first. Then the gate voltage of PFET MP3 decreases, and MP3 turns on. Then MP1 goes into the saturation region because it must supply current to both MP2 and MP3. Then MP2 turns off rapidly. The effect is regenerative: there is no more stable state beyond the second critical point. V_{GC2} is the switching threshold voltage of the Schmitt trigger for an up-going transition. The microstate sequence of the Schmitt trigger is shown in Fig. 58. When we repeat the same analysis for a down-going transition, we find the critical voltage

$$V_{GC2} = \frac{1}{B_N + B_P}\left[B_N V_{THN} - B_P V_{THP} + \frac{B_P(B_P V_{DD} + b_P V_{THP})}{B_P + b_P}\right]. \tag{99}$$

We define the switching threshold voltage of the inverter shown in Fig. 57(c):

$$V_{GSW} = \frac{B_P V_{DD} - B_P V_{THP} + B_N V_{THN}}{B_N + B_P}\,.$$

We then have the following results:

$$V_{GC2}(Up) - V_{GSW} = \frac{B_N b_N}{(B_N + B_P)(B_N + b_N)}(V_{DD} - V_{THN}) \ (>0) \tag{100a}$$

and

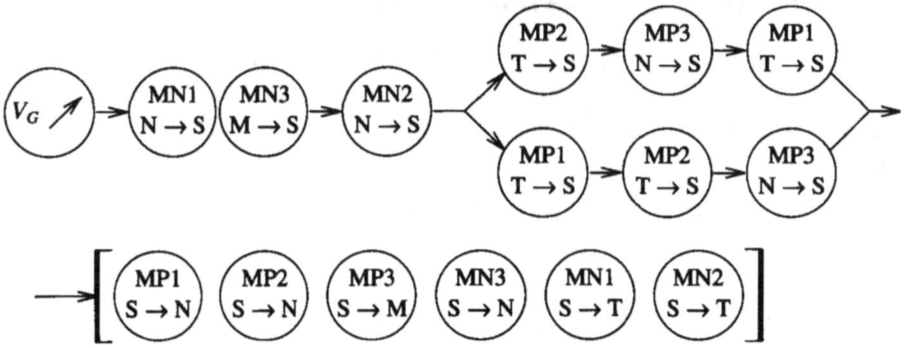

Figure 58 Schmitt trigger pull-down transition microstate sequence

$$V_{GSW} - V_{GC2}(Down) = \frac{B_P b_P}{(B_N + B_P)(B_P + b_P)}(V_{DD} - V_{THP}) \ (>0) . \tag{100b}$$

Therefore, the switching threshold for an up-going transition is higher than that for a down-going transition. The difference between the switching thresholds and the switching threshold voltage of the reference inverter is proportional to the transconductance of MN3 and MP3 if they are both small, as

$$V_{GC2}(Up) - V_{GSW} \approx o(b_N) \quad V_{GSW} - V_{GC2}(Down) \approx o(b_P) . \tag{100c}$$

4.18 Voltage Swing of the Substrate

In this section we consider the effects of a conducting silicon substrate. We consider injection of *majority* carriers only. Minority carrier injection changes circuit connectivity drastically. Therefore, minority carrier issues will be discussed in chapter 6. Not all on-chip power- or ground-buses have the same potential all of the time, and the potential difference forces current flow into the substrate. Capacitive coupling to the substrate injects current also. The first problem is how to determine the substrate potential distribution. We assume DC and a simple uniform substrate structure first.

Figures 59(a) and (b) show current I injected into the substrate at point A. The substrate has thickness l, uniform conductivity σ, and the bottom surface is either glued to a nonconducting floor of the package cavity [Fig. 59(a)] or bonded to a metal layer inside the package that is connected to the appropriate power supply voltage level [Fig. 59(b)]. This analysis is valid for bulk CMOS chips. We assume that the chip size (horizontal dimension) is much larger than thickness l. This problem of field analysis can be solved using the image method [22]. We use coordinate system (x,y) defined in Fig. 59(a). If the bottom of the substrate is glued to a nonconducting floor of the package, image current sources ($2I$) are placed at points

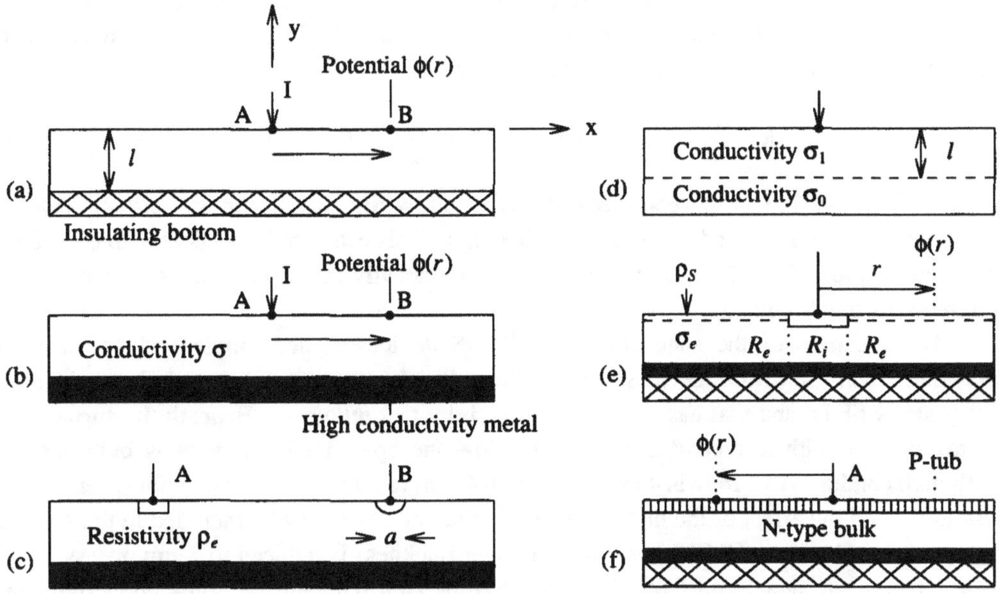

Figure 59 Current injection into substrate

$(x=0,\ y=2ln)$ where $n = ..., -2, -1, 0, 1, 2, ...$ to satisfy the boundary conditions. As contributions from all the image current sources are summed, we obtain potential $\phi(r)$ as,

$$\phi(r)=\frac{I}{4\pi\sigma l}\,\chi_{A,B}(r/2l)+const \text{ where } \chi_A(\rho)=\sum_{n=-\infty}^{\infty}\left(\frac{1}{\sqrt{\rho^2+n^2}}-\frac{1}{|n|}\right). \tag{101}$$

If the chip is bonded to the package cavity metal floor [Fig. 59(b)], image current sources $(-1)^n \cdot 2I$ are placed at the same locations, and we obtain

$$\chi_B(\rho)=\sum_{n=-\infty}^{\infty}(-1)^n\left(\frac{1}{\sqrt{\rho^2+n^2}}-\frac{1}{|n|}\right) \tag{102}$$

In Eqs. (101) and (102), the constant of ρ, $1/|n|$, is subtracted from each term of the infinite series to guarantee convergence. If $1/|n|$ is subtracted, the terms in the sum tends to zero like $1/n^2$ in the limit as $n\to\pm\infty$, and the convergence of the sum is guaranteed. The term with $n=0$ does not include $1/|n|$. Function $\chi_A(\rho)$ decreases with increasing ρ. If $\rho\to\infty$ the current density within the substrate becomes uniform, and the problem is solved in closed form as $\chi_A(\rho)\to -2\log\rho$, which requires properly choosing the additional constant of Eq. (101) to match to the reference potential level. We note that $\chi_B(\rho)$ has the limit:

$$\lim_{\rho\to\infty}\chi_B(\rho)=2\sum_{n=1}^{\infty}\frac{(-1)^{n-1}}{n}=2\log 2=1.38\,. \tag{103}$$

Figure 60 shows functions $\chi_A(\rho)$ and $\chi_B(\rho)$ versus ρ. Function $\chi_B(\rho)$ has an application in estimating resistance of a grounded contact applied to the substrate. We approximate a small square contact A of Fig. 59(c) by a semi-spherical contact B that has diameter a. Contact resistance, R, equals the voltage developed on the semi-spherical surface referenced to the grounded high-conductivity metal if unit current is forced into the contact. We have

$$R=\frac{1}{a\sigma}\,\chi(a/4l)\quad\text{where}\quad \chi(\rho)=\frac{\rho}{\pi}\,[\chi_B(\rho)-2\log 2]\,. \tag{104}$$

Figure 60 shows $10\cdot\chi(\rho)$ versus ρ ($10\times$ is to allow use of a common vertical scale). If $\sigma=1\ mho/cm$, $a=1\ \mu m$ and $l=10\ \mu m$, $R=3\ Kohm$. This estimate is applicable to contacts on moderately doped P- or N-tubs grown directly on a heavily doped substrate that serves as the reference (ground) voltage.

The substrate of the state-of-the-art CMOS devices is not simple and uniform bulk material. Twin-tub CMOS devices have a diffused surface layer (tub) that works as the doped substrate of FETs, and that has medium conductivity ($\approx 1\ mho/cm$). Beneath the surface is an epitaxial layer with low conductivity, and below the epitaxial layer there is bulk substrate with high conductivity. In twin-tub CMOS of 1–2 μm feature size, the epitaxial layer is 10–20 μm thick, and the effect of the high resistivity epitaxial layer must be included in the substrate analysis. In submicron CMOS, the epitaxial layer thickness is reduced to 5 μm or less, to prevent latchup. In submicron CMOS, P- and N-tubs almost touch the high-conductivity bulk substrate. The CMOS substrate structure is significantly different for different CMOS technologies.

We study the effects of highly conducting bulk substrate under the pure epitaxial layer shown in Fig. 59(d). Let the conductivity of the pure epitaxial layer be σ_1, that of the highly doped substrate be σ_0, and thickness of the epitaxial layer be l. The doped substrate is assumed to be very thick, and the bottom of the substrate is grounded (zero potential). The potential field within the epitaxial layer is determined by placing image current sources and

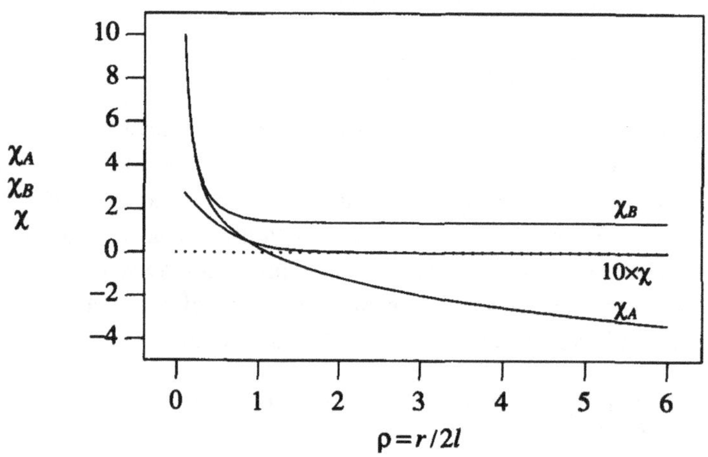

Figure 60 Substrate response function

assuming that the entire space has conductivity σ_1. We use the same coordinate system as defined in Fig. 59(a) and the image current sources as follows:

At $x=0$ and $y=0$, a current source $2I$ (current I is injected to the lower half-space).

At $x=0$ and $y=\pm 2li$ (where i is a positive integer), a current source $\sigma_f{}^i \cdot 2I$ with parameter σ_f given by

$$\sigma_f = (\sigma_1 - \sigma_0)/(\sigma_1 + \sigma_0) .$$

The epitaxial layer surface potential at distance r from the point of current injection $\phi(r)$ is given by

$$\frac{\phi(r)}{(I/4\pi\sigma_1 l)} = \frac{1}{\rho} + 2 \sum_{n=1}^{\infty} \frac{\sigma_f{}^n}{\sqrt{n^2+\rho^2}} \quad \text{where } \rho = \frac{r}{2l} . \tag{105}$$

Potential $\phi(r)$ for: Case C, highly conducting substrate, Case U, uniform substrate (substrate and epitaxial layer are the same material), and Case I, insulating substrate are shown in Fig. 61. In a highly conducting substrate, the potential drops to zero quite rapidly.

Since the bulk substrate in epitaxial CMOS devices has very low resistivity (of the order of 0.01 ohm cm), the substrate may be considered as a constant voltage reference for practical purposes. Assuming that the bulk substrate is held at 0 potential, we study the effects of a medium conductivity surface layer (tub), at the top of the pure epitaxial layer. The following analysis refers to Fig. 59(e). The medium conductivity surface layer above the low conductivity epitaxial layer is characterized by sheet resistance ρ_S (*ohms/square*), and the low conductivity epitaxial layer is characterized by bulk conductivity σ_e (*mho/cm*^{2}). Conductivity σ_e is defined as the current that flows across the epitaxial layer having unit area, if the potential difference between the medium conductivity surface and the high conductivity substrate is main-

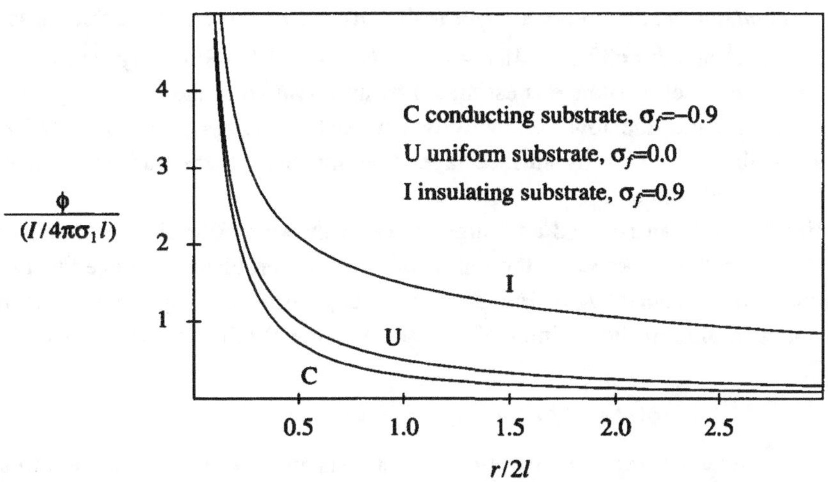

Figure 61 Response function of substrate

tained at 1 volt. The field has axial symmetry around the point of the current injection. The substrate surface potential $\phi(r)$ and current within the medium conductivity surface layer $I(r)$ are determined from the current conservation law

$$-2\pi r\left(\frac{1}{\rho_S}\right)\frac{d\phi(r)}{dr}=I(r) \quad \text{and} \quad \frac{dI(r)}{dr}=-2\pi r\sigma_e\phi(r). \tag{106}$$

By eliminating $I(r)$, the potential profile $\phi(r)$ is found from the equation

$$\frac{d^2\phi(r)}{dr^2}+\frac{1}{r}\frac{d\phi(r)}{dr}-k^2\phi(r)=0 \ \text{ where } k^2=\rho_S\sigma_e. \tag{107}$$

In the limit as $r \to 0$, $\phi(r) \approx -(\rho_S I/2\pi)\log r$, and in the limit as $r \to \infty$, $\phi(r) \to 0$. The solution of Eq. (107) satisfying the boundary conditions is given by using the modified Bessel function K_0 as [23]

$$\phi(r)=(\rho_S I/2\pi)K_0(kr) \quad K_0(\zeta) \to \sqrt{\pi/2\zeta}\,exp(-\zeta)\ (\zeta \to \infty)$$

$$K_0(\zeta) \to -\log(\zeta/2)\ (\zeta \to 0).$$

Let us apply this result to compute contact resistance. If a circular contact has diameter a, its contact resistance, R, is given by the following formula

$$R=\frac{R_i R_e}{R_i+R_e} \quad \text{where} \quad R_e=\frac{\phi(a/2)}{I(a/2)}=-\frac{\rho_S}{2\pi}\frac{2}{ka}\frac{K_0(ka/2)}{K_0'(ka/2)} \quad \text{and} \quad R_i=\frac{4}{\pi a^2\sigma_e} \tag{108}$$

or

$$R=\left[\frac{\rho_S}{\sigma_e}\right]^{1/2}\frac{1}{\pi a}G(ka/2) \quad \text{where} \quad G(x)=\frac{1}{(x/2)+[K_1(x)/K_0(x)]} \tag{109}$$

where we used the relationship $K_0'(x)=-K_1(x)$. With reference to Fig. 59(e), R_i is the resistance of the epitaxial layer directly underneath the contact, and R_e is the resistance of the surface layer and the epitaxial layer outside. The two component resistances are effectively in parallel. If $\rho_S = 2\ Kohm/square$, the epitaxial layer resistivity is $5\ ohm\ cm$ and the thickness is $10\ \mu m$, $\sigma_e = 200\ mho/cm^2$ and $k = 632/cm$. If $a = 1\ \mu m$, $ka/2 = 0.032$. Since $\sqrt{\rho_S/\sigma_e} = 3.16$, $R = 1.16\ Kohms$. If the contact resistance is estimated by approximating the contact by semi-spherical contact buried inside the low conductivity epitaxial layer, $R \approx \rho_e/\pi a \approx 16\ Kohms$. The effects of the medium conductivity surface layer that spreads out the current from the small contact area is quite significant.

If current is injected into an isolated and large P-tub as shown in Fig. 59(f), the current spreads out quickly within the thickness of the thin P-tub, and the problem is reduced to two-dimensional current flow. If contact A of Fig. 59(f) is a long distance away from the P-tub boundary, the potential profile in the vicinity of contact A using P-tub sheet resistance ρ is found as

$$\phi(r)=-(\rho I/2\pi)\log r + const. \tag{110}$$

Since the potential is slowly varying at $r \to \infty$, the contact resistance cannot be assigned to an isolated contact. We need to consider the entire P-tub, into which currents are injected and

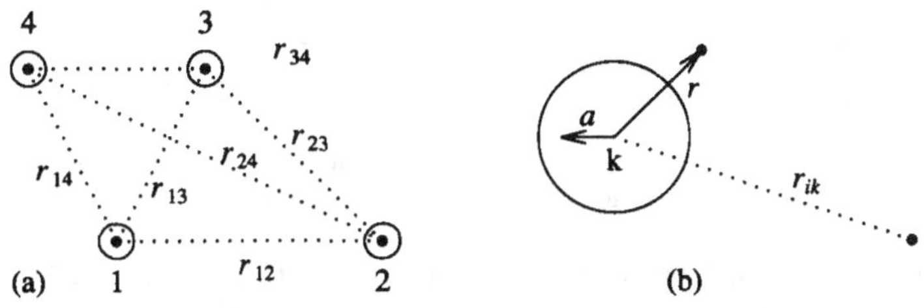

Figure 62 Analysis of DC current distribution in isolated tub

extracted at many locations. The sum of the currents is zero. Current distribution within an isolated tub can be determined as follows. With reference to Fig. 62(a), contacts 1, 2, 3, ... have potentials V_1, V_2, V_3, ... and carry currents I_1, I_2, I_3, ..., respectively. Current is positive if it flows into the tub. Currents must satisfy the conservation law

$$\sum_{i=1}^{N} I_i = 0 \tag{111}$$

where N is the number of contacts. We assume that contacts can be approximated by a circular shape, which has radius a_i, and a_i is much smaller than the distance between the contacts, r_{ik}, as shown in Fig. 62(a). The potential within the tub is the sum of the potentials contributed by the currents from each of the N contacts. Each contact contributes potential

$$\text{Contribution of contact } i = -\frac{\rho I_i}{2\pi} \log r_i \tag{112}$$

where r_i is the distance between the particular location where potential is measured and contact i. We assume the tub boundaries are far away from the contacts. We have

$$V(\vec{r}) = -\frac{\rho}{2\pi} \sum_{i=1}^{N} I_i \log r_i + V_\infty \tag{113}$$

where vector $\vec{r}$ specifies the location within the tub where $V(\vec{r})$ is measured (r_i is the distance from contact i to the specified location $\vec{r}$). V_∞ is the tub potentuial far away from the contacts [note Eq. (111)]. This equation can be approximated as follows. Since the distance between the contacts is much larger than the size of the contacts,

$$V(\vec{r}) = -\frac{\rho}{2\pi} (I_k \log r + \sum_{i \neq k} I_i \log r_{ik}) + V_\infty \tag{114}$$

if location $\vec{r}$ is in the vicinity of contact k (where r is the distance to location $\vec{r}$ from the center of contact k). Since the potential on a circle centered at the k-th contact with radius a_k is V_k, we have a set of simultaneous equations

$$V_k = -\frac{\rho}{2\pi} (I_k \log a_k + \sum_{i \neq k} I_i \log r_{ik}) + V_\infty$$

together with Eq. (111) the set of equations are written in matrix form as

$$\begin{bmatrix} 0 & 1 & 1 & 1 & 1 \\ -2\pi/\rho & \log a_1 & \log r_{12} & \cdots & \log r_{1N} \\ -2\pi/\rho & \log r_{21} & \log a_2 & \cdots & \log r_{2N} \\ \cdots & \cdots & \cdots & \cdots & \cdots \\ -2\pi/\rho & \log r_{N1} & \log r_{N2} & \cdots & \log a_N \end{bmatrix} \begin{bmatrix} V_\infty \\ I_1 \\ I_2 \\ \cdots \\ I_N \end{bmatrix} = -\frac{2\pi}{\rho} \begin{bmatrix} 0 \\ V_1 \\ V_2 \\ \cdots \\ V_N \end{bmatrix}. \tag{115}$$

By solving this set of simultaneous equations, the current carried by each contact is found from the set of given contact voltages. If current I_i is known, the potential distribution within a tub can be determined from Eq. (113). If the tub area is small and the effects of the tub boundary cannot be neglected, or if there is area that is held at a fixed potential (like a large *area* contact), standard techniques of images from electrostatics can be applied.

An example is given in Fig. 63. Figure 63(a) shows a horizontal V_{SS} bus having three P-tub contacts A, B, and C, and a vertical V_{SS} bus having a continuous *area* contact. The two buses are not directly connected by high conductivity metal. We determine the current distribution within the P-tub when the two V_{SS} buses have different potentials. This problem has application to the analysis of an on-chip power-bus system interconnected via resistive crossunders. Figure 63(b) shows equipotential contours within the P-tub if the horizontal V_{SS} bus is at unit potential above the grounded vertical V_{SS} bus. In Fig. 63(b) $\overline{\text{OA}}=1$, $\overline{\text{OB}}=2$ and $\overline{\text{OC}}=3$ and the radius of each contact is 0.1. Potential was determined by placing three image current sources to the left side of the vertical axis: the set of six real and image current sources maintain the vertical axis at zero potential. Two dotted curves α and β show flow lines that are the boundaries of currents originating from contacts A, B and C. The current from contact A flows below flowline α. Contact C is the farthest away from the vertical V_{SS} bus, but cur-

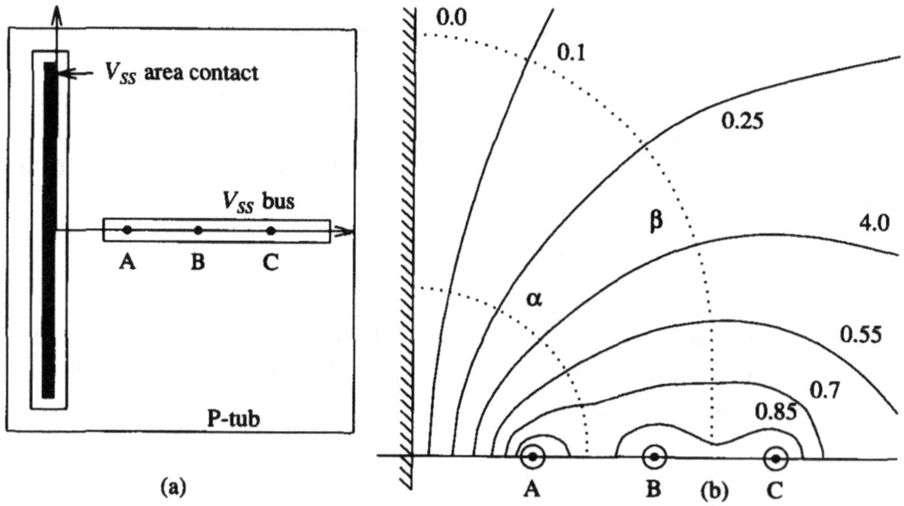

Figure 63 Current distribution within *isolated* P-tub

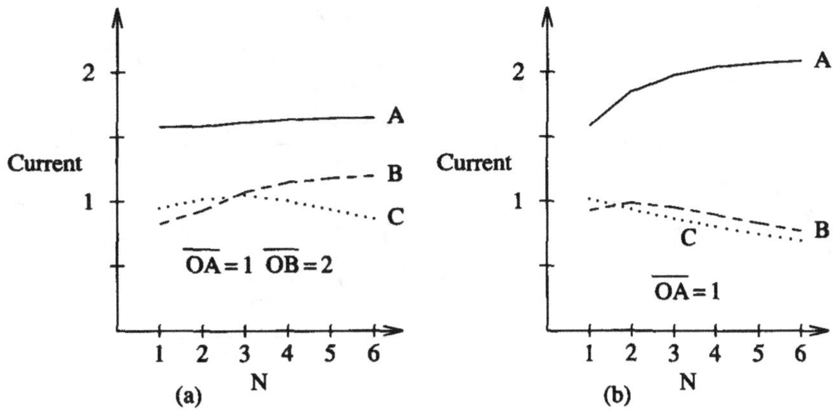

Figure 64 Contact current versus contact distance

rent originating from contact C is able to spread out over the entire area above flowline β. Therefore current from contact C decreases very slowly if contact C moves away from the vertical V_{SS} bus. Figure 64 shows current carried by contacts A, B, and C versus distance of the contacts from the vertical V_{SS} bus where N is a parameter specifying the distances. In Fig. 64(a) we have $\overline{OA}=1$ and $\overline{OB}=2$, and $\overline{OC}=3\ (N=1), =4\ (N=2), =8\ (N=3), =16\ (N=4), = 32\ (N=5), =64\ (N=6)$. In Fig. 65(b) we have $\overline{OA}=1$, $\overline{OC}=2\overline{OB}$, and $\overline{OB}=2^N$. We see that decrease of current by moving contacts away from the vertical V_{SS} bus is quite slow. This means that wherever a contact is, it carries approximately the same DC current.

4.19 Dynamic Electrical Phenomena in P-tub

In N-substrate CMOS technology, P-tubs are isolated by reverse-biased PN junctions from the global N-type substrate. In modern twin-tub CMOS technology, N-substrate consists of N+ bulk substrate and lightly doped N-type epitaxial layer, into which P-tubs and N-tubs are implanted and diffused. The vertical structure of an isolated P-tub is shown in Fig. 65(b). We study electrical transients in the P-tub. The P-tub can be represented by the equivalent circuit shown in Fig. 65(a). We assume that the P-tub is a rectangular strip, and we consider electrical phenomena that are one-dimensional: the potential depends on time t and coordinate x, but not on y. In the equivalent circuit of Fig. 65(a), R_0 is the P-tub sheet resistance (ρ_P *ohms/square*) divided by width W, R_1 is the resistivity of the epitaxial layer (ρ_E *ohms* $\cdot$ *cm*) times thickness of the undepleted portion of the epitaxial layer (l_E *cm*) divided by W, and $C(F/cm)$ is the capacitance per unit area of the reverse-biased PN junction C_{PN} times W, as

$$R_0 = \frac{\rho_P}{W} \qquad R_1 = \frac{\rho_E l_E}{W} \qquad C = C_{PN} W .$$

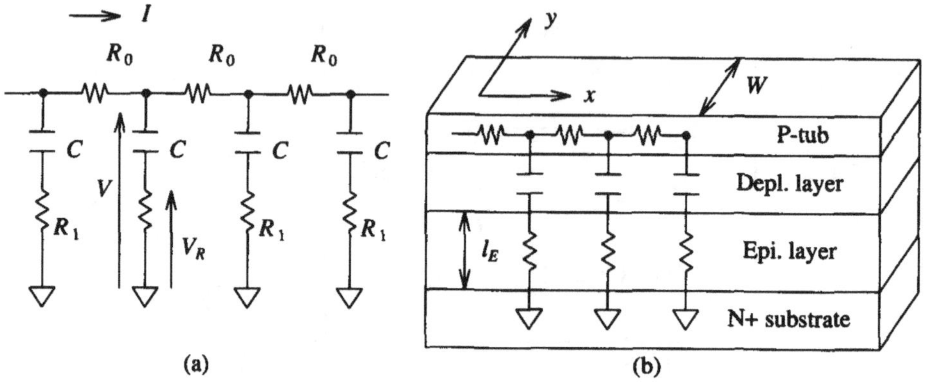

Figure 65 P-tub equivalent circuit

Voltage dependence of the PN junction capacitance is neglected. The resistance of the epitaxial layer that is in series with the capacitance is significant in 1–2 μm CMOS, but it is less significant in submicron CMOS: In submicron CMOS the epitaxial layer is quite thin.

Electrical transients of the equivalent RC chain circuit of Fig. 65(a) can be determined from the following set of circuit equations:

$$V_R = -R_1 \frac{\partial I}{\partial x} \qquad \frac{\partial I}{\partial x} = -C \frac{\partial}{\partial t}(V - V_R) \qquad \frac{\partial V}{\partial x} = -R_0 I$$

with I, V, and V_R defined in Fig. 65(a). If I and V_R are eliminated from the equations, we obtain

$$D \frac{\partial^2 V}{\partial x^2} + \lambda \frac{\partial^3 V}{\partial^2 x \partial t} = \frac{\partial V}{\partial t} \tag{116}$$

where $D = 1/R_0 C$ and $\lambda = R_1/R_0$. This equation reduces to an ordinary diffusion equation if $\lambda = 0\,(R_1 = 0)$. In this limit, the P-tub works effectively as an RC low-pass filter circuit. This case has been analyzed in detail in chapter 1.

In the other limit, R_1 and C are large. This is for a rapid transient in the P-tub. The first term of Eq. (116) can be neglected. The equation can be integrated once by time t as

$$\lambda \frac{\partial^2 V}{\partial x^2} - V = V_0 \tag{117}$$

where V_0 is the constant of integration by time t. One solution of this equation is

$$V = -V_0 + A exp(x/\sqrt{\lambda}) + B exp(-x/\sqrt{\lambda}) \tag{118}$$

where A and B are constants of integration by x. Since the potential must be bounded, a solution that is valid in the limit as $x \rightarrow \infty$ must have $A = 0$. This solution means that the P-tub voltage converges to $-V_0$ exponentially, with characteristic length $\sqrt{\lambda} = \sqrt{R_1/R_0}$. For a rapid transient, the P-tub works effectively as a resistive potential divider. This potential profile is established immediately after pulse current is injected into the P-tub. This exponential potential profile is established within a very short time, and the profile becomes the *initial condition* to the following evolution governed by the equation that includes all terms of Eq. (116).

The other special case of Eq. (116) is that $\partial V/\partial t$ is small compared with the second term on the left-hand side. Then,

$$\frac{\partial^2}{\partial x^2}(DV+\lambda\frac{\partial V}{\partial t})=0 \quad \text{or} \quad \frac{\partial V}{\partial t}=-\frac{D}{\lambda}(V-\frac{Ax+B}{D}) \tag{119}$$

where A and B are constants of integration by x. This equation is solved to obtain

$$V=Kexp[-(D/\lambda)t]+\frac{Ax+B}{D} \tag{120}$$

where K is constant of integration with respect to t and where $D/\lambda=1/R_1C$. This is the case where a DC potential profile is established within the P-tub, and as far as the time dependent change is concerned, everywhere is the same.

The general solution of the partial differential equation, Eq. (116), is found as follows. We consider the response of the substrate to a current pulse injected at location $x=\xi$ at time $t=\tau$. The substrate response to arbitrary excitation can be found by using the superposition principle. We first solve

$$\frac{\partial V_0}{\partial t}-D\frac{\partial^2 V_0}{\partial x^2}-\lambda\frac{\partial^3 V_0}{\partial x^2\partial t}=N\delta(x-\xi)\delta(t-\tau) \tag{121}$$

where N is a normalization factor. We use the Fourier transform as follows [24][25].

$$V_0(x,\xi,t,\tau)=\frac{1}{4\pi^2}\int_{-\infty}^{\infty}\int_{-\infty}^{\infty}V_0(k,\omega)exp[i(k(x-\xi)-\omega(t-\tau))]dk\,d\omega$$

$$\delta(x-\xi)\delta(t-\tau)=\frac{1}{4\pi^2}\int_{-\infty}^{\infty}\int_{-\infty}^{\infty}exp[i(k(x-\xi)-\omega(t-\tau))]dk\,d\omega\,.$$

By substituting this integral into the partial differential equation we get

$$V_0(k,\omega)=\frac{-1}{i\omega-Dk^2+i\lambda\omega k^2}$$

and therefore

$$V_0(x,\xi,t,\tau)=-\frac{N}{(2\pi)^2}\int_{-\infty}^{\infty}exp[ik(x-\xi)]dk\int_{-\infty}^{\infty}\frac{exp[-i\omega(t-\tau)]}{i\omega-Dk^2+i\lambda\omega k^2}d\omega \tag{122}$$

where the second integral is rewritten as a complex integral of variable ω as

$$\frac{1}{1+\lambda k^2}\int_{-\infty}^{\infty}\frac{exp[-i\omega(t-\tau)]}{i\omega-p(k)}d\omega$$

where $p(k)=Dk^2/(1+\lambda k^2)$. The complex integrand has a pole at $\omega=-ip(k)$ as shown in Fig. 66, and the residue at that pole is $-i\cdot\exp[-Dk^2(t-\tau)/(1+\lambda k^2)]$. With reference to Fig. 66, the integral along the upper arc tends to zero at large radius if $t<\tau$. Since there is no singularity inside, the integral is zero. When $t>\tau$ the integral along the lower arc tends to zero, and since there is a pole inside, the integral is $-2\pi\exp[-Dk^2(t-\tau)/(1+\lambda k^2)]$. Therefore

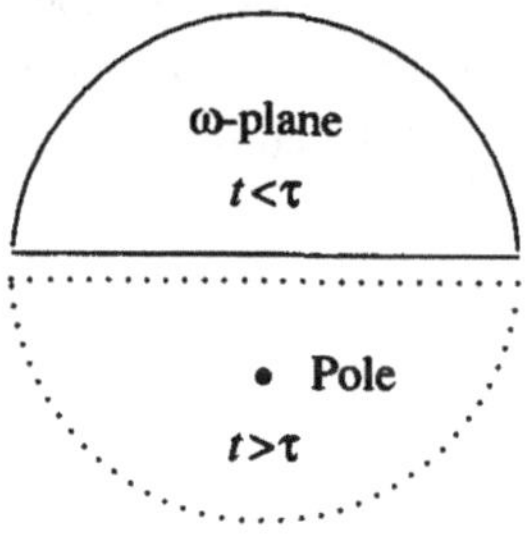

Figure 66 Path of integration

$$V_0(x,\xi,t,\tau)=\frac{N}{2\pi}\int_{-\infty}^{\infty}\frac{\cos k(x-\xi)}{1+\lambda k^2}\,exp[-\frac{Dk^2}{1+\lambda k^2}(t-\tau)]dk \tag{123}$$

$$=\frac{N(x-\xi)}{\pi\lambda}\int_0^{\infty}\frac{\cos\theta}{\theta^2+a^2}\,exp[-\frac{D(t-\tau)}{\lambda}\frac{\theta^2}{\theta^2+a^2}]d\theta$$

where $a^2=(x-\xi)^2/\lambda$.

Although the derivation of Eq. (123) is not elementary, it is easy to show that V_0 of Eq. (123) satisfies Eq. (121), by direct substitution. V_0 of Eq. (123) has the following limits. In the limit as $\lambda\to\infty$, the exponential factor within the integral tends to unity, and we have

$$V_0(x,\xi,t,\tau)\to\frac{N(x-\xi)}{\pi\lambda}\int_0^{\infty}\frac{\cos\theta}{\theta^2+a^2}d\theta=\frac{N}{2\sqrt{\lambda}}exp[-(x-\xi)/\sqrt{\lambda}]\,.$$

This is a special case given previously by Eq. (118). In the limit as $\lambda\to 0$

$$V_0(x,\xi,t,\tau)\to\frac{N}{\pi(x-\xi)}\int_0^{\infty}\cos\theta\,exp[-\frac{D(t-\tau)\theta^2}{(x-\xi)^2}]d\theta=\frac{N}{2\sqrt{\pi D(t-\tau)}}exp[-\frac{(x-\xi)^2}{4D(t-\tau)}]$$

and this is the solution of the diffusion equation to which Eq. (116) reduces in the limit as $\lambda\to 0$. If λ has intermediate values, V_0 of Eq. (123) can be computed numerically. Figure 67 shows $(\pi/N)\times V_0(x,\xi,t,\tau)$ versus normalized distance $(x-\xi)/\sqrt{\lambda}$, for several values of normalized time $t/(\lambda/D)$. Immediately after pulse current injection, the potential profile within the P-tub is established by resistive voltage division. The profile has exponential dependence on x as given by Eq. (118) (curves A and B). The profile becomes the initial condition for the subsequent development by the *diffusion* effect, which produces potential profiles like curves C and D. The general solution of Eq. (116) is given by

$$V(x,t)=\int d\tau\int d\xi V_0(x,\xi,t,\tau)U(\xi,\tau) \tag{124}$$

where $U(\xi,\tau)$ is a function that specifies injected current that is distributed both in time and in location. $V_0(x,\xi,t,\tau)$ is the Green's function of Eq. (116).

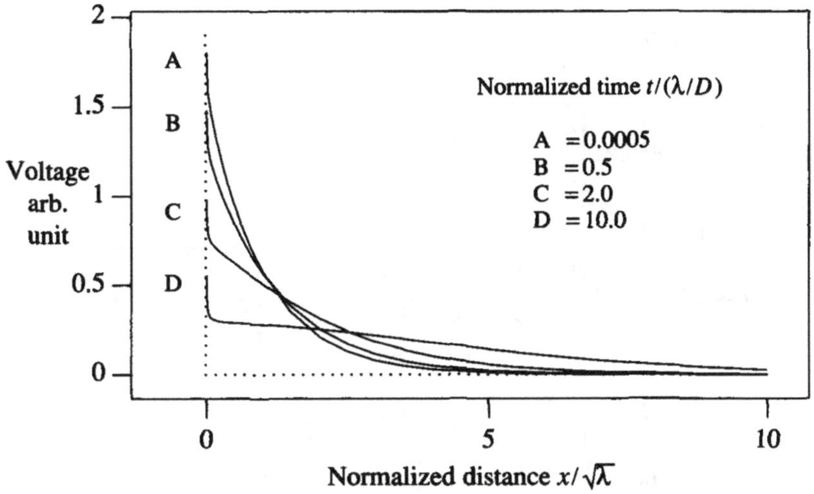

Figure 67 Green function of P-tub substrate response

4.20 Substrate and Tub Potential Fluctuation

The substrate (or tub) of bulk CMOS is connected to V_{SS} or V_{DD} power-bus, but substrate contacts are often sparse. The area of substrate or tub far away from the contacts may have significant potential fluctuation. Since FETs are imbedded in the substrate or in the tub, and the sources of the FETs are connected directly to the power- or ground-bus, deviation of substrate or tub potential affects the devices. Let us consider an NFET fabricated on a P-tub. The NFET source is connected to the ground-bus. If the substrate potential is below ground potential, back-bias is applied to the NFET, and the FET threshold voltage increases. NFET current decreases and pull-down delay of the FET increases. This results in reduction in the maximum operating frequency of the circuit. If the substrate potential is above ground, more drastic circuit failure may take place in CMOS dynamic logic circuits like PLAs. Since the NFET threshold voltage decreases, NFETs weakly turn on, even at zero gate voltage. In CMOS dynamic gates with many pull-down NFETs, the sum of the leakage currents can be high enough to destroy a logic-high level held by charge stored in the output node capacitance.

This problem takes place in CMOS dynamic PLAs, shown schematically in Fig. 68(a). Figure 68(a) shows the decoder section of a PLA, fabricated on a single large P-tub. In the figure, the P-tub is connected to the ground-bus along the right edge. Before a logic operation, the clock pulls down, and precharge PFETs pull the word lines up. In some PLA designs, the drain nodes of the ground switches are also precharged to prevent charge sharing and to speed

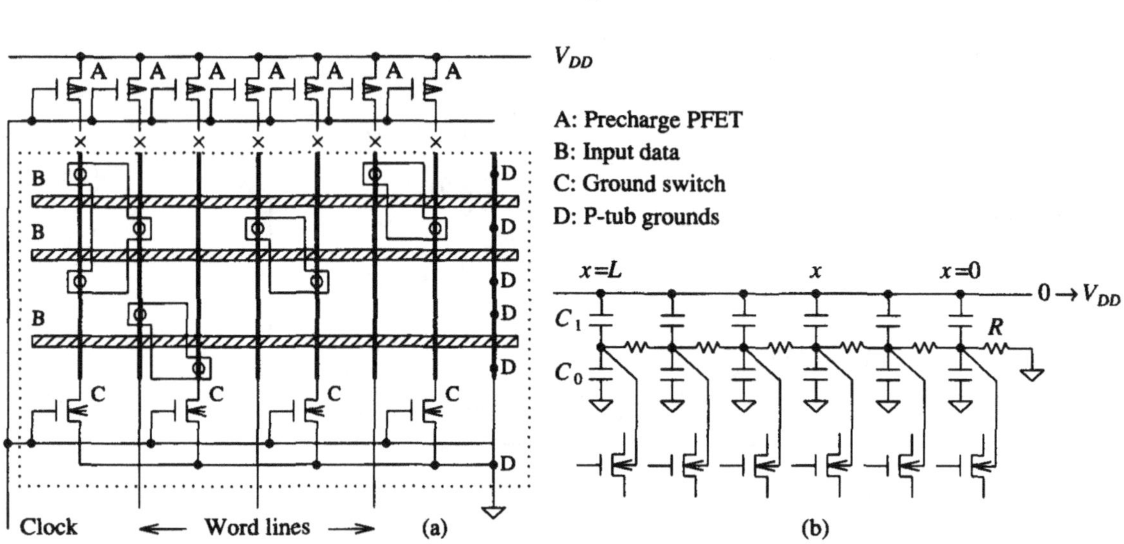

Figure 68 CMOS dynamic PLA schematic

up precharge. As word lines are pulled up, source/drain islands of the code NFETs are also charged up. The word line to P-tub capacitance C_1 consists of the wordline wiring capacitance and the source/drain island capacitances of the code NFETs. The P-tub has capacitance to the constant potential N-type global substrate. The equivalent circuit of the word line, the P-tub and the global N-substrate, is shown in Fig. 68(b). C_1 is the total capacitance between P-tub and a word line, and C_0 is the P-tub to substrate capacitance. We neglect the voltage dependence of capacitance C_0. R is the series resistance of the P-tub. C_0, C_1, and R are the average values per unit length perpendicular to the word line of the PLA. Our problem is to analyze transients of the equivalent RC chain circuit of the P-tub, in response to a stepwise voltage increase of the word lines.

If we take coordinate x in the direction perpendicular to the word lines as shown in Fig. 68(b), the equation satisfied by the time-dependent potential profile at location x at time t, $V(x,t)$ is

$$\frac{\partial V(x,t)}{\partial t} = \frac{1}{R(C_0+C_1)} \frac{\partial^2 V(x,t)}{\partial x^2} . \tag{125a}$$

Equation (125a) is solved subject to the initial condition

$$V(x,0) = V_N = \frac{C_1 V_{DD}}{C_0+C_1} \tag{125b}$$

and the boundary condition

$$\frac{\partial}{\partial x} V(x,t)|_{x=L} = 0 \quad \text{and} \quad V(0,t) = 0 . \tag{125c}$$

The solution satisfying the boundary conditions is the sum of the terms of the form

$const \cdot exp(-\alpha t)\sin(\beta x)$. The parameters are determined from the initial and the boundary conditions as

$$V(x,t) = \frac{4V_N}{\pi} \sum_{n=1}^{\infty} \frac{exp[-[1/R(C_0 + C_1)](\pi/2L)^2(2n-1)^2 t]}{2n-1} \sin[(\pi/2L)(2n-1)x] \quad (126)$$

Figure 69 shows $V(x,t)/V_N$ versus x/L for several typical values of normalized time $\tau = t/[L^2 R(C_0 + C_1)]$.

We attempt a rough numerical estimate of the substrate noise effects. Since the N-type substrate is 1–2 orders of magnitude more highly doped than P-tub surface ($2 \times 10^{16}/cm^3$), C_0 is estimated using the conventional PN junction capacitance formula as $C_0 = 6 \times 10^{-9}\ F/cm^2$. This can be a slight overestimate, but we do not attempt to elaborate. NFETs in CMOS dynamic PLA have approximately size $2\lambda \times 4\lambda$, where λ is the feature size of the CMOS technology. Each FET has capacitance to substrate estimated as

$$C(FET - SUB) = C_A \cdot 8\lambda^2 + C_P \cdot 12\lambda$$

$$\text{where } C_A = 3 \times 10^{-4}\ pF/\mu m^2 \quad C_P = 3 \times 10^{-4}\ pF/\mu m$$

If $\lambda = 1\ \mu m$ $C(FET - SUB) = 6 \times 10^{-3}\ pF$/FET. Since the PLA pitch is about $4\lambda \times 4\lambda$, there can be $10^8/16\lambda^2$ NFETs. If 10% of the possible sites are occupied by FETs, there are 6×10^5 FETs per 1 centimeter square. Then $C_1 = 6 \times 10^5 \times 6 \times 10^{-3} = 3.6 \times 10^{-9}\ F/cm^2$. We have

$$V_N = 1.8\ V \quad \text{and} \quad \text{minimum time constant} = R(C_0 + C_1)(2L/\pi)^2 = 1.8\ nsec$$

assuming $R = 2\ Kohm/square$ and $L = 150\ \mu m$. The numerical estimate shows that violent substrate noise can exist within CMOS dynamic PLAs. High-speed PLA design must include

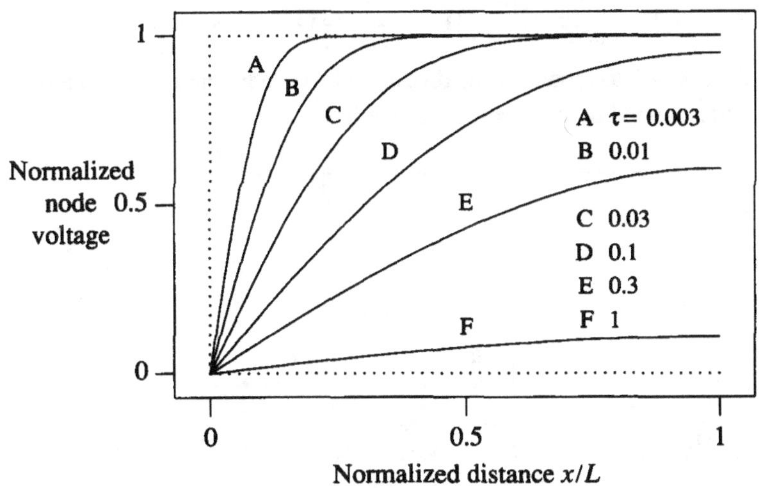

Figure 69 PLA substrate noise

evaluation of the substrate noise, and of P-tub grounding methods. The example shows that the distance between the P-tub contact must be less than about 100 μm.

Figure 70(a) shows a cross section of metal or polysilicon wire fabricated on oxide that insulates the wire from the P-tub. Capacitance between the wire and the underlying P-tub is in series with the P-tub to N-substrate junction capacitance. Since the P-tub has high sheet resistance, not all junction capacitance is effectively in series with the oxide capacitance C_{OX}. We determine the effective series capacitance and the substrate potential variation. Impedance of the metal wire above the oxide to the substrate is given by a series combination of the oxide capacitance C_{OX} and the impedance of the P-tub. The P-tub impedance is given by this parallel combination of impedance $Z_0/2$ (distributed RC chain circuit on both sides of the wire) and the impedance of capacitance C_{PS}, which is the capaictance of the portion of the P-tub directly under the metal wire. For narrow wires, the series capacitance is much larger than the capacitance between the wire and the silicon substrate.

We consider a metal wire of long rectangular shape. The P-tub below the wire can be modeled using the RC chain circuit of Fig. 70(b). The problem is one-dimensional. Series resistance R of the P-tub, and capacitance C of the P-tub to the N-substrate are both per unit length in the direction perpendicular to the wire. We need to determine impedance Z_0 of the infinite length uniform RC chain circuit (that extends on the side of metal wire). Since the impedance looking into the RC chain circuit at the left end, and that looking into the RC chain at the first node to the right are the same, we have

$$Z_0 = \frac{1}{i\omega C\lambda + \dfrac{1}{R\lambda + Z_0}} \tag{127a}$$

where λ is the length of the single RC link. This equation can be solved for Z_0, and the physically meaningful solution is identified as

$$Z_0 = \frac{R\lambda}{2}\left[\left(1 + \frac{4}{i\omega CR\lambda^2}\right)^{1/2} - 1\right] \to \frac{R}{\sqrt{2\omega CR}}(1-i) \quad (\lambda \to 0). \tag{127b}$$

Here we took the limit as $\lambda \to 0$ to represent the distributed RC chain circuit accurately.

If we use admittance defined by $Y(\omega) = 1/Z_0(\omega)$ we have

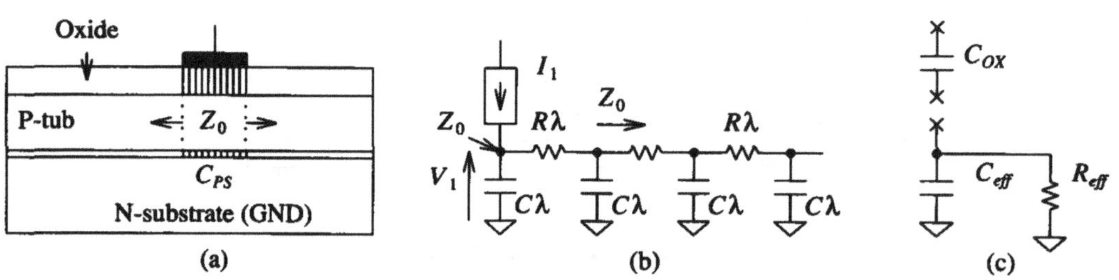

Figure 70 Effectiveness of capacitance on P-tub parasitic

$$Y(\omega) = Y_R(\omega) + iY_I(\omega) \quad \text{and} \quad Y_I(\omega) = i\omega C_{eff}(\omega) \tag{128}$$

where $C_{eff}(\omega)$ is the effective capacitance of the RC chain circuit. Using Eq. (127b) we have

$$Y_R(\omega), Y_I(\omega) \rightarrow \left(\frac{\omega C}{2R}\right)^{1/2} \quad \text{and} \quad C_{eff}(\omega) \rightarrow \left(\frac{C}{2\omega R}\right)^{1/2}$$

The effective capacitance increases with decreasing angular frequency ω. The RC chain circuit is represented by a parallel combination of effective resistance $R_{eff} = R\sqrt{2/\omega CR}$ and effective capacitance $C_{eff} = C\sqrt{1/2\omega CR}$. This means that all the resistance and capacitance within distance about $\sqrt{1/\omega CR}$ from the driving point is lumped at the driving point. The point on the RC chain about distance $\sqrt{1/\omega CR}$ away from the driving point is effectively grounded. If $C = 1 \times 10^{-8}\ F/cm^2$, $R = 2\ kohms/square$ and $\omega = 10^9\ radians/sec$, $\sqrt{1/\omega CR} = 0.7 \times 10^{-2} =$ 70 μm. Since this is much larger than typical metal wire width, the P-tub to substrate capacitance in series with the metal wire to P-tub capacitance is ineffective. If the metal wire is several hundred microns wide, the effects of the series capacitance become discernible.

We study the voltage response $V_1(t)$ of the RC chain impedance, Z_0, to a square current pulse, $I_1(t)$, having height I_{max} and duration $2t_H$. The current pulse is a superposition of the Fourier components as

$$I_1(t) = \frac{2I_{max}}{\pi} \int_0^{\infty} \frac{\sin(\omega t_H)\cos(\omega t)}{\omega}\, d\omega\,.$$

Each Fourier component $\cos(\omega t)$ generates response

$$R_{eff}\cos(\omega t) + \frac{1}{\omega C_{eff}}\sin(\omega t) \quad \text{where} \quad R_{eff} = \frac{R}{\sqrt{2\omega CR}} \quad C_{eff} = \left[\frac{2C}{\omega R}\right]^{1/2}$$

from Eq. (127b). By adding the component responses we obtain

$$V_1(t) = V_P \int_0^{\infty} \frac{\sin\theta[\sin(\tau\theta) + \cos(\tau\theta)]}{\theta^{3/2}}\, d\theta$$

$$\text{where} \quad V_P = \frac{2}{\pi}\left[\frac{R}{2C}\right]^{1/2} I_{max}\sqrt{t_H} \qquad \theta = \omega t_H \qquad \tau = \frac{t}{t_H}\,.$$

Figure 71 shows the plot of normalized time $V_1(t)/V_P$ versus normalized time $\tau = t/t_S$. The solid curves show the best fit to the numerical results assuming the simple forms $V_1(t) \approx const\sqrt{\tau + 1}$ (curve A) and $V_1(t) \approx const/\sqrt{\tau - 1}$ (curve B). The square-root time dependence is a feature of the uniform long RC chain circuit. As for curve A, charge $Q = I_{max} \cdot \Delta t$ is injected into the chain during time interval Δt, and during that time capacitance within distance proportional to $\sqrt{\Delta t}$ is charged up. Therefore, node voltage increases proportional to $\sqrt{\Delta t}$. After the pulse is turned off, the injected charge is distributed approximately uniformly over the capacitances within distance proportional to $\sqrt{\Delta t}$, and therefore node voltage decreases as $const/\sqrt{\Delta t}$ in the limit as $t \rightarrow \infty$.

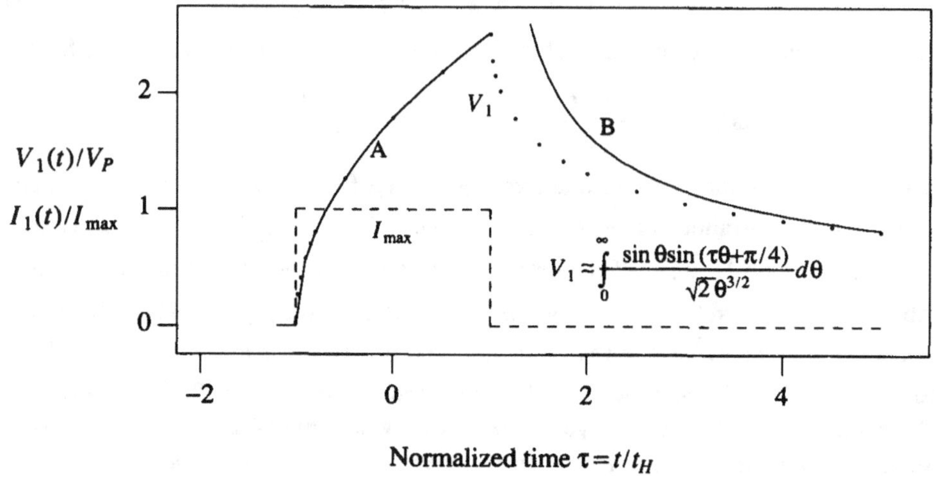

Figure 71 Response of P-tub to injected current pulse

These results are consistent with the discussion in the author's book *CMOS digital circuit technology*, chapter 3, where effective capacitance is set proportional to the square root of the switching delay time. The more detailed analysis of this section shows that the frequency-dependent capacitance is paralleled with resistance, which presents equal impedance.

The series resistance of on-chip parasitic capacitance (parasitic of parasitic!) may create quite significant uncertainty in estimating CMOS output driver delay time. The CMOS driver of an impedance-matched transmission line has the equivalent circuit shown in Fig. 72(a) (more discussion in Section 7.7). Because of the resistive voltage divider/impedance-matching network consisting of resistors R_S and R_L, capacitance to the right side of resistor R_S has practically no effect on delay. Delay of the driver is determined by the capacitances C_0 and C that are mostly on-chip (bonding pad and drain island capacitances of the driver FETs). Figure 72(b) shows a simplified equivalent circuit of the output driver that is modeled using the collapsible current generator. The delay time of the circuit is determined from the circuit equations

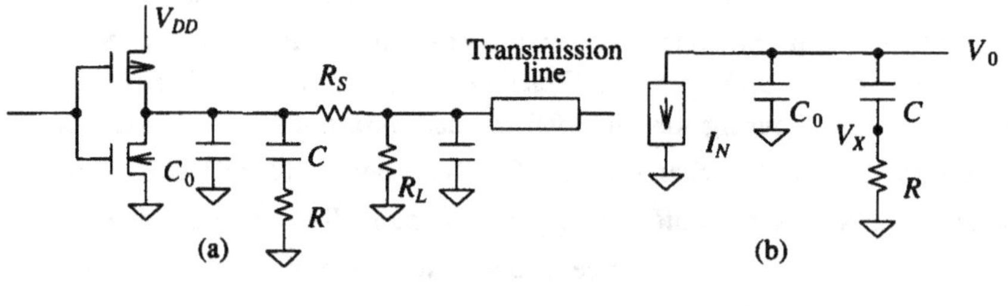

Figure 72 Analysis of resistance-loaded transmission line driver

$$C\frac{d}{dt}(V_0 - V_X) = \frac{V_X}{R} \quad \text{and} \quad \frac{V_X}{R} + C_0\frac{dV_0}{dt} + I_N = 0 \tag{129}$$

that are solved subject to the initial condition $V_0(0) = V_{DD}$ $V_X(0) = 0$ to obtain

$$V_0(t) = R\left(\frac{C}{C + C_0}\right)^2 I_N\left[exp(-\frac{C + C_0}{CC_0R}) - 1\right] + V_{DD} - \frac{I_N}{C + C_0}t. \tag{130}$$

Delay time, T_D, is determined by requiring that $V_0(T_D) = V_{DD}/2$, as

$$T_D = T_0 \cdot (\zeta\gamma) \cdot \theta(-\gamma, 1 - \frac{1}{2\zeta}) \text{ where } \gamma = C_0/C \quad \zeta = \left(\frac{C}{C + C_0}\right)^2 \frac{RI_N}{V_{DD}} \tag{131}$$

where we use θ-function of chapter 2 and where $T_0 = (C + C_0)V_{DD}/I_N$. Figure 73 shows T_D/T_0 versus normalized resistance ζ. If normalized resistance $RI_N/V_{DD} \approx 1$, the response time decreases quite significantly if C_0/C is small. $RI_N/V_{DD} \approx 1$ can be reached in realistic 1 μm CMOS circuits. This mode of operation indicates that *bulk* CMOS is operating like SOI CMOS. The substrate and tub are not conductive enough to provide solid ground voltage reference. In performance comparison between SOS CMOS and bulk CMOS, it is important to note that bulk CMOS devices occasionally work like SOS CMOS devices (Section 7.4).

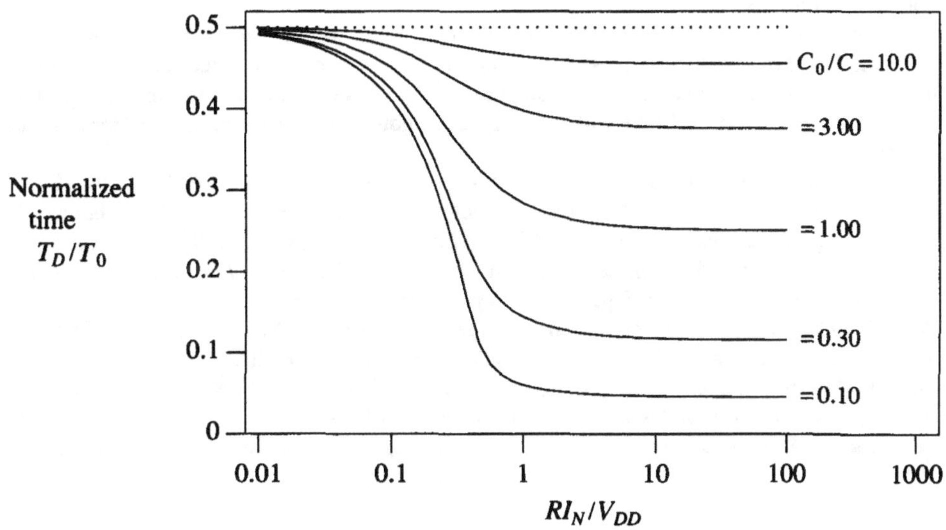

Figure 73 Delay time versus series resistance to the load resistor

References

[01] Ziel, A. van der. "Thermal noise in field-effect transistors." *Proc. IRE* 50 (August, 1962): 1808–12.

[02] Jordan, A. G., and Jordan, N. A. "Theory of noise in metal oxide semiconductor devices." *IEEE Trans. on Electron Devices* vol. ED-12 (March, 1965): 148–56.

[03] Shoji, M. "Analysis of high-frequency thermal noise of enhancement mode MOS fielf-effect transistors." *IEEE Trans. on Electron devices* vol. ED-13 (June, 1966): 520–24.

[04] Richman, P. *MOS field-effect transistors and integrated circuits.* New York: John Wiley and Sons, 1973.

[05] Ziel, A. van der. "Noise in VLSI" in *VLSI handbook.* N. G. Einspruch, ed. Orlando, FL: Academic Press, 1985. [Equation (9) and van der Ziel's result are essentially the same, although a few specific assumptions were made].

[06] Wallmark, J. T. "Noise spikes in digital VLSI circuits." *IEEE Trans. on Electron Devices* vol. ED-29 (March, 1982): 451–58.

[07] Stratton, J. A. *Electromagnetic theory.* New York: McGraw-Hill, Inc., 1941.

[08] Ross, S. L. *Introduction to ordinary differential equation.* New York: John Wiley & Sons, 1966.

[09] Shoji, M. *CMOS digital circuit technology.* Englewood Cliffs, NJ: Prentice Hall, 1988.

[10] Shoji, M. "Reliable chip design method in high performance CMOS VLSI." *ICCD86 digest* (October, 1986): 389–92.

[11] Hasegawa, H., Furukawa, M., and Yanai, H. "Properties of microstrip line on Si-SiO_2 system." *IEEE Trans. on Microwave Theory and Techniques* vol. MTT-19 (November, 1971): 869–81.

[12] Franklin, P. *Fourier methods.* New York: McGraw Hill, 1949.

[13] Mandel, F. *Statistical physics.* New York: John Wiley and Sons, 1988.

[14] Shoji, M. "Electrical design of BellMac-32A microprocessor." *ICCC82 digest* (September, 1982): 112–15.

[15] Wada, T., Shinohara, H., and Kayano, S. "A low noise data-out buffer in CMOS SRAM." *Japanese Telecommunication Society Meeting Digest* (1986): 2–229.

[16] Lohstroh, J. "Static and dynamic noise margins of logic circuits." *IEEE J. Solid-State Circuits* vol. SC-14 (June, 1979): 591–98.

[17] Professor F. Rosenberger gave me valuable comments on the issues of noise margin. In the original text, I tried to avoid using noise margin because it is related to the mechanism of information transfer between the gates in very complicated way. I consider that noise margin is still not included in the theoretical framework properly, but significant improvements were made by his comments.

[18] Carr, W. N., and Mize, J. P. *MOS/LSI design and application.* New York: McGraw Hill, 1972.

[19] Goncalves, N. P., and Man, H. J. de. "NORA: a racefree dynamic CMOS technique for pipelined logic structures." *IEEE J. Solid-State Circuits* vol. SC-18 (June, 1983): 261–68.

[20] Yamasaki, H., Adachi, S., Yamasaki, N., and Wakasa, Y. *Noise reduction techniques in electronic circuits and systems.* (in Japanese). Tokyo: Ohm Publishing Co., 1981.

[21] Wang, N. *Digital MOS integrated circuits.* Englewood Cliffs, NJ: Prentice-Hall Inc., 1989.

[22] Seely, S., and Poularikas, A. D. *Electromagnetics.* New York: Marcell Dekker Inc., 1979.

[23] Pipes, L. A. *Applied mathematics for engineers and physicists.* New York: McGraw-Hill, 1946.

[24] Ahlfors, L. V. *Complex analysis.* New York: McGraw Hill, 1966.

[25] Seeley, R. T. *An introduction to Fourier series and integrals.* New York: W. A. Benjamin Inc., 1966.

5 Circuit Failures Due to Timing Problems

5.1 Introduction

There is a group of circuit failure mechanisms that are caused by timing or clocking of circuits. An example of the failure mode is as follows. If the timing of a circuit is forced from an externally generated clock signal, the time allowed by the clock for the circuit may not be enough to carry out the required data processing. The chip fails at the maximum clock frequency. Any chip suffers from this generic failure mechanism. In properly designed circuits, the only way to remove this failure mechanism is to reduce the clock frequency. This failure mechanism can be generalized as follows. One signal (primary clock) is split into two, one of which generates a signal, and the other generates a (delayed) clock. Data released from a storage circuit controlled by the primary clock is the *signal*. The primary clock is delayed by the clock processing circuit to generate the second clock signal. The *signal* and the *clock* go through different chains of logic circuits, and they are later *combined* when the processed data are captured at the destination storage circuit that is controlled by the second clock. Relative delays of the two (or more, in general) signals determine whether or not the circuit fails. This failure is independent of the primary clock frequency. This failure mechanism is often called a "RACE" condition, because the *signal* and the *clock* compete, and which of them arrives at the destination first determines whether or not the circuit fails. In synchronous circuits this mechanism includes failure at the frequency limit, if the logic chain and the *external* clock generator are considered together.

These timing failures originate from a single fundamental process: splitting the clock signal into two, processing separately, and recombining the two signals later. The definition sounds unusually specific. Why is the definition of timing failures so specific? Let us use a single standard clock delay circuit and many different signal processing circuits. We determine whether or not the signal that is latched into the destination is correct. What we did here was to determine whether or not the time required for signal processing is more (fail) or less (pass) than the standard clock delay time. The process of splitting the clock and recombining it is the most fundamental way of measuring time. Since time is a *relative* variable that has no intrinsic zero reference, we need this complex procedure, and that is why the failure mechanism sounds unusually specific. We recall that the failure mechanisms due to anomalous signal flow and noise studied in chapters 3 and 4 were rather vaguely defined. Charge, which was the key parameter of the circuit failures in chapters 3 and 4 is an *absolute* parameter that has well-defined zero. Therefore, the failure depends on a simple criterion, namely, how much charge is accumulated (or lost). We begin this chapter with a review of the mechanisms of splitting and recombining a clock, or the mechanism of sequential data processing.

5.2 Basics of CMOS Logic Circuit Clocking

Theoretically any logic operation can be carried out by integrating enough gates into a single combinational logic circuit, and by driving the logic circuit inputs with data to be processed. This scheme is, however, impractical since a astronomical number of gates are required to execute even simple arithmetic operations. The scheme turns out to be quite uneconomical. A more practical method is to use a logic circuit block consisting of a group of gates again and again, by feeding the logic circuit with previously processed data and logic circuit *reconfiguration* control data, thereby processing the data sequentially.

This type of signal processing requires control of sequencing, usually performed with a *clock*. The clock determines a time interval within which one specified set of logic operations must be completed by the logic circuit. At the end of the period, the processed data are temporarily stored in a set of storage circuits (latches). The logic circuits are reconfigured to reflect the needs of the next step processing, and the previously processed data are returned to the inputs of the logic circuits. The entire circuit is synchronized by the clock. The general structure of the processing circuit is shown in Fig. 1(a).

Latches have two states. In one state, they accept data and pass it to their output, and in the other state, they retain existing data and ignore new data. The two states are controlled by the clock logic level. Figure 1(b) shows the most commonly used CMOS D-latch circuit. The

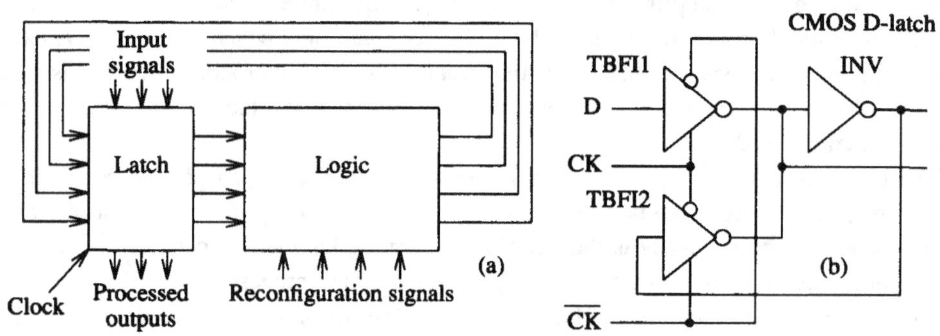

Figure 1 Sequential processing circuit and D-latch

circuit of Fig. 1(b) consists of two tristatable inverters and one plain inverter. When *CK* is high and $\overline{CK}$ is low, data *D* is inverted by tristatable inverter TBFI1 and $\bar{Q}$ is driven to the complement $\bar{D}$. Since tristatable inverter TBFI2 is disabled, TBFI1 is the only driver of node $\bar{Q}$. $\bar{Q}$ is inverted by INV, and output node *Q* is driven to the same logic level as *D*. For *D* to propagate through the latch, a time that equals the sum of the delay times of TBFI1 and INV is required.

When *CK* makes a high-to-low transition, and simultaneously $\overline{CK}$ makes a low-to-high transition, TBFI1 is disabled and TBFI2 is enabled. The latch *closes* the data input path and secures the existing data. TBFI2 must adjust voltages of the two internal nodes *Q* and $\bar{Q}$, and a steady state must be established. If node $\bar{Q}$ and *Q* have been in the steady state reflecting data $\bar{D}$ and *D*, respectively, before the clock transition, the time is zero. If nodes $\bar{Q}$ and *Q* are close to, but have not yet reached, the steady state reflecting the input data, the latch takes as long as the delay time of TBFI2 to reach the steady state. It was pointed out that the time required to store data in a latch by applying positive feedback through TBFI2 determines critically whether or not a circuit fails by RACE condition [01] (Section 5.7). An important point is that the data storage circuit requires time just to *store* data. In a high-speed sequential data processing system, the storage circuit delay time limits the ultimate performance of the circuits. The time required to store data is conventionally characterized by the setup time. To store data, the voltage representing the data must stabilize the setup time before the edge of the clock that closes the latch [conventionally *falling* edge (Section 3.12)]. The characterization is convenient for logic operation analyses. The problem of this method is that setup time depends on how the data and the clock voltages drive the latch. If the setup and hold times are used, they are *not* constants.

In order to reduce latch circuit delay, the CMOS dynamic latch circuit shown in Fig. 2, which has only one inversion per latch, is often used. A transmission gate *samples* the input voltage (signal D) when *CK* is high and $\overline{CK}$ is low. The sampled voltage level is stored in the capacitor at the output of the transmission gate. The capacitance need not be intentionally added: stray capacitance like wiring capacitance and the input capacitance of the CMOS buffer

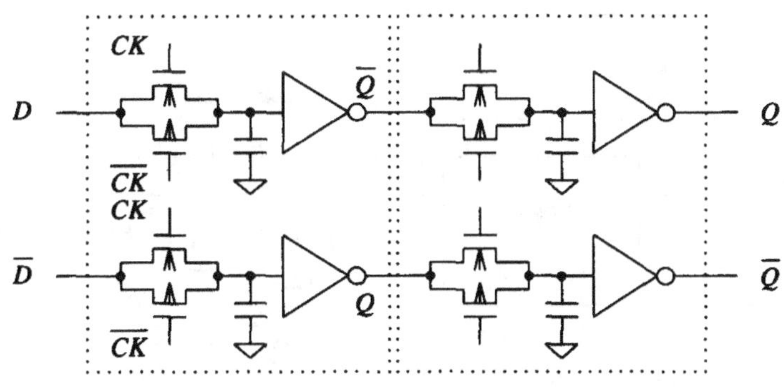

Figure 2 CMOS dynamic D-latch circuit

is often sufficient. The sampled signal is buffered by the inverter to restore logic swing. Since this latch includes only one inversion, only the $\bar{Q}$ is available. This restriction is often inconvenient. Therefore a *dual* rail scheme, which stores data D and its complement $\bar{D}$ in a pair of the dynamic latches, is used. Figure 2 shows the cascaded two stages of latches in the master-slave latch configuration.

In the dynamic latch of Fig. 2, the transmission gate and the storage capacitor have signal delay. We note that memory is created by signal delay. Then there is the question, What has delay of logic circuits to do in the sequential data processing? or can we use the delay for temporary data storage [02]? In order to complete one step of *sequential* processing, the latch data given to the logic circuit (source data) must remain unchanged until the processing is completed. At that time, the output data overwrite the latch data. This simple processing scheme has the following subtle details. The delay time of signals from the input to the output of the logic circuit when the circuit carries out the most complex (longest propagation delay) operation is called the critical path delay. The critical path delay determines the maximum operating frequency. Another, more subtle, point is that the delay time of some short logic paths of the circuit must be maintained longer than a certain lower limit required by the response time of latches, for all the possible conditions of process, temperature, and power supply voltage, if some special means, like the master-slave timing (Section 5.3) is not used. This is a design problem that is, in principle, satisfiable, but because of complexity in design it is difficult to achieve. Some logic signal paths within a complex logic subsystem can be very short, and the logic delay time can be shorter than the time required for the signal to propagate through the latch's internal circuit and to complete the change of state.

With reference to Fig. 3(a), the latch symbol represents a bank of *single* D-latches [such as that shown in Fig. 1(b)]. The arrows connecting the latch and the logic circuit block represent multiple connections. With reference to the waveforms shown in Fig. 3(b), let the delay of logic signals through the logic block (from node N_1 to node N_0) be T_{LO}, propagation time of signal through the latch in the transmitting state (from node N_0 to node N_1) be T_{LA}, the clock pulse width be T_C, and the clock period be T_P. We made a crucial assumption that the data and the clock waveforms are crisp, and therefore the delay times can be defined clearly. T_{LA} is the delay time of a signal through the inverter and tristatable inverter of the latch. T_{LA}

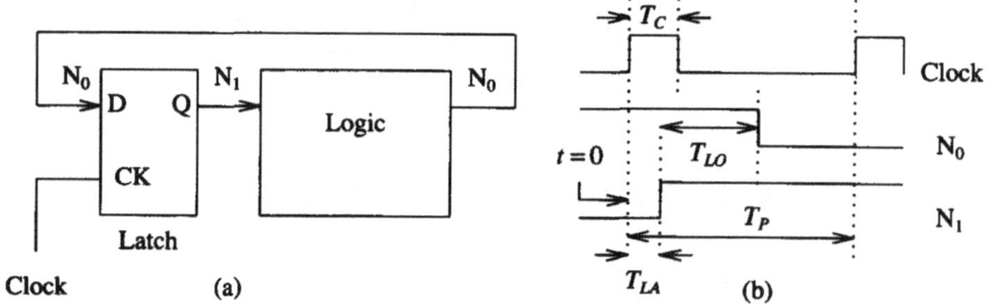

Figure 3 A model of sequential circuit

is longer than the setup time, but the relation of magnitude between the two cannot be determined simply (Sections 3.12 and 5.6). We keep the relation as an open issue, and we consider the cases for the minimum (in an ideally fast latch, 0) and the maximim, T_{LA}. Since there are many different signal paths, the delay time T_{LO} has a wide range. We seek the requirement for sequential processing to be carried out successfully. Suppose that a positive clock edge arrived at $t = 0$, and the latch transmits data. Suppose that the data stored in the latch and the data transmitted from the input are different. At time $t = T_{LA}$ the transmitted data arrives at the output terminal (N_1) of the latch, and the logic operation begins. The logic circuit sends back the processed data to the latch (node N_0) at time $t = T_{LA} + T_{LO}$. This time must be absolutely earlier than the trailing edge of the *next* clock pulse, which occurs at time $t = T_P + T_C$. If the setup time is the minimum, 0,

$$T_{LA} + T_{LO} < T_P + T_C \tag{1}$$

Here the right-hand side of the inequality is $T_P + T_C$ rather than T_P, because shifting the gates to balance delay time and optimize the performance is now a well-established design technique. If the setup time is the maximum, T_{LA},

$$T_{LA} + T_{LO} < T_P + T_C - T_{LA} \quad \text{or} \quad 2T_{LA} + T_{LO} < T_P + T_C \tag{2}$$

This is one necessary condition, namely, that signal processing must be completed within the available time. Since the first condition is the optimum case, the violation of it creates failure.

At time $t_1 = T_{LA} + T_{LO}$, the processed data arrives at the input of the latch (node N_0). Let us consider a short logic path. If the latch is still transmitting, the processed signal appears at the output of the latch (node N_1) at time $t_2 = 2 \cdot T_{LA} + T_{LO}$ and at the output of the logic circuit (node N_0) at $t_3 = 2 \cdot (T_{LA} + T_{LO})$. Because the data input to the logic circuit are generally different, this output data are different from the correctly processed data at the output of the logic circuit at time t_1 $(= T_{LA} + T_{LO})$. The short, closed logic path modifies the input data to the long logic paths that share the input data. How this modification happens will be discussed in Section 5.7. If the clock makes a transition after time t_2, it can be guaranteed that the latches have stored wrong source data for the *current* clock period (note that the maximum setup time was assumed). Therefore a *sufficient* condition for logic error to occur is

$$t_2 \, (= 2 \cdot T_{LA} + T_{LO}) < T_C \, . \tag{3}$$

If we require that a latch needs steady input data maintained at the latch input only at the time of closure (short setup time), we have

$$t_1 \, (= T_{LA} + T_{LO}) < T_C$$

as the sufficient condition of circuit failure. Although Eqs. (1)–(3) can be violated by reducing T_C (and thereby making the circuit operative), T_C cannot be arbitrarily short. Since we assumed that the waveforms are crisp, if

$$T_C > T_{LA} \tag{4}$$

or if input data stabilized at least T_{LA} before the negative clock edge, the data signal has propagated through the latch when the down-going clock edge turned the input off. The latch captures the correct data. T_{LA} is the upper limit of the latch setup time. In practice if the delays satisfy

$$T_{LA} + T_{LO} > T_C \quad T_C > T_{LA} \quad \text{and} \quad 2T_{LA} + T_{LO} < T_P + T_C \tag{5}$$

for any signal path in the logic circuit, a single D-latch and logic circuit shown in Fig. 3 are able to process signal sequentially. From Eq. (5) such design requires reducing T_C to the minimum, just enough for the latch to capture data. The operation of this circuit can be interpreted as follows: Since the minimum delay is required for any logic operation, logic circuit (*not* including the latch) effectively *stores* data temporarily, while executing the sequential logic operation. This mode of operation is feasible if every logic path through the logic block is designed to have at least the specified minimum delay conforming to Eq. (5). We reiterate that the most fundamental assumption of the argument is that the delay times must be defined without ambiguity, or the node waveform must be crisp: Node N_0 and N_1 waveforms must have rise (or fall) times that are much less than the delay times T_{LO} and T_{LA}. Delay times cannot be determined clearly from slowly rising/falling waveforms, since the delay time depends on the logic threshold voltage (Section 2.2). Gates of the logic chain are operating in dynamic switching mode. The dynamic switching mode has memory, and the memory effect is an integral part of this sequential logic operation. If such design is feasible, the logic circuit has high performance. It may be that such a design is required in future ultra-high performance CMOS ICs. High performance is achieved by minimizing delay for latch (storage of state between cycles).

We consider a possible application of the single-latch timing scheme of Fig. 3. A shift register is a circuit that shifts data by one location every clock cycle. A standard shift register circuit is a cascade of unit cells consisting of static D-latches, shown in Fig. 4(a). Two cascaded D-latches, 1 and 2, are clocked by CK and $\overline{CK}$, respectively. When CK is high, latch 1 transmits input data and latch 2 retains the previous data. When CK makes a high-to-low transition, latch 2 transmits, and latch 1 retains the data. Then latch 2 receives the data latch 1 acquired when CK was high. The data are shifted from latch 2 to the first latch of the next stage when CK makes another low-to-high transition.

If a conventional CMOS D-latch circuit shown in Fig. 1(b) is used in the shift register circuit, the shift register unit cell includes four gates in the signal path (four gate delays, or four *inversions*). One way to decrease the number of inversions for circuit speedup is to use $\overline{Q}$ output of the latch. $\overline{Q}$ output does not have strong capacitance drive capability. If CMOS

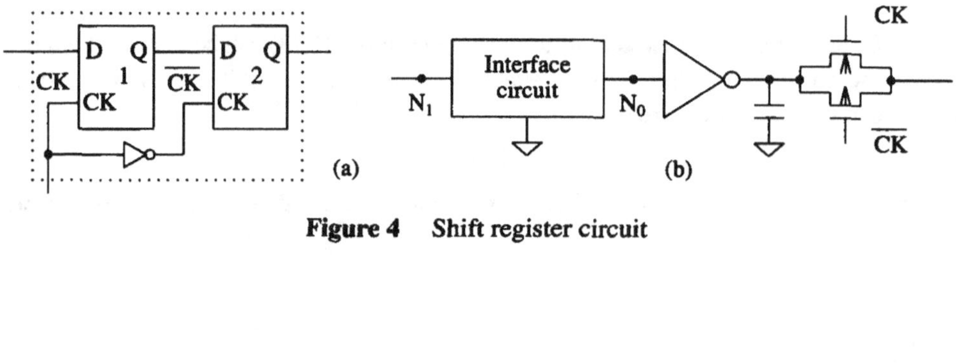

Figure 4 Shift register circuit

dynamic D-latch as shown in Fig. 2 is used, the cell includes a little more than two gate delays (note that the transmission gates add delay), but the output has strong drive capability. We ask, Is there any way to design a high-speed shift register cell that includes only one inversion? This question has fundamental importance in CMOS digital circuit technology. If fast shift register is available, fast serial to slow parallel conversion is possible, and the area of application of relatively slow CMOS expands greatly.

Figure 4(b) shows a candidate circuit that we will study. The first latch of Fig. 4(a) was replaced by the interface (delay) circuit. The crucial point is that the delay can be less than an inverter delay, and it is adjustable. As we saw before, node N_0 waveform must be delayed to node N_1 waveform, but certain waveform requirements must be met. The node waveforms are shown in Fig. 5(a). We note that delay time defined at the logic gate's threshold voltage is often misleading. The node N_0 waveform A0 makes the inverter and transmission gate operate gradually and simultaneously, such that the delay time definition becomes ambiguous. The circuits as a whole operate more or less simultaneously and the desired function of sequential processing is never carried out successfully. We need to have a crisp switching waveform as shown in waveform B0 and B1 of Fig. 5(a). This requirement means physically the following: In Section 2.8 we saw that an inverter (and generally any *enabled* CMOS gate) has two different modes of switching operation, static switching and dynamic switching. The gradual waveforms of node, curves A0 and A1 of Fig. 5(a), are the waveforms for static switching, and the crisp waveforms of curves B0 and B1 are the waveforms for dynamic switching. Dynamic switching involves essentially delay or *memory* effect in the gate circuit, which is required for the operation of the shift register. Static switching does not have the required memory effect. A circuit that has memory is required to substitute for the first D-latch of the shift register.

Since an inverter that is switching dynamically has memory effect, at least one stage of a shift register should be replaced by an inverter, as shown in Fig. 4(b). The stage provides power gain to regenerate signal. What sort of passive delay circuit generates a crisp waveform shown by node waveform B of Fig. 5(a)? An RC delay circuit shown in Fig. 5(b) is not adequate. A cascaded two-stage RC low-pass circuit also has difficulties of gradual voltage change at the output node. In order to suppress voltage change at the output node N_0 immediately following the input node (N_1) voltage change, inductance can be inserted in series, as shown in Fig. 5(c). The inductance supresses the initial change, but later when the signal begins to change toward the final, steady-state voltage level, the inductance forces current into the output node and the output node voltage increases rapidly. If this idea is extended to the

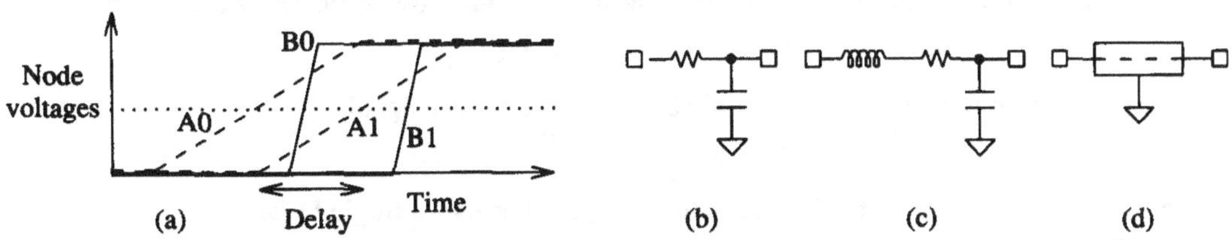

Figure 5 Node waveforms and interface circuit

limit, the ideal circuit is a segment of a transmission line shown in Fig. 5(d), that is either series or parallel terminated (Section 7.8). It is well known that a transmission line can be used as memory. Use of inductance or transmission line on chip will be acceptable in future, very high-speed ICs. Transmission line delay can be taylored accurately by adjusting the length. The process variation can be minimized, since the parameters are mostly process-independent. A shift register circuit shown in Fig. 4(b) that has some form of transmission line segment will be attractive for multiplexer applications for high-speed digital transmission systems.

In an ordinary CMOS circuit design, meeting the exhaustive minimum delay time requirements of logic gate chains under all realistic conditions is difficult. Therefore sequential circuits are almost always built using two latches, with one being called the master latch and the other the slave latch. The two latches never transmit data simultaneously. If one latch transmits data, the other blocks data. To achieve this mode of operation, the two latches are driven by clocks that are shifted from each other. There are several different schemes for clocking sequential circuits of this basic structure, and this subject will be discussed in the next section.

5.3 Master-Slave Data Transfer Timing

In order to carry out sequential data processing reliably, two latches are cascaded and clocked such that both latches are never simultaneously transparent [03]. Direct implementation of this scheme is schematically shown in Fig. 6(a). The latches are driven by clock pulses CKM (master latch clock) and CKS (slave latch clock) that do not overlap each other. The two-phase non-overlapping master-slave timing shown in Fig. 6(a) uses two independent clock pulses that open the respective latches when the clock pulses are at high logic level. There is nonzero interval between master latch close and slave latch open, and slave latch close and master latch open. The *dead zones* T_2 and T_4 in timing diagram of Fig. 6(a) guarantee stable operation of the sequential logic circuit. The circuit tolerates small clock skew that might exist within the chip. Logic circuits can be placed between slave latch and master latch, or between master latch and slave latch or both, depending on the processing needs. We define four time intervals T_1, T_2, T_3 and T_4 as *phases* 1, 2, 3, and 4, respectively, within a single clock cycle. Clock period T_P is given by

$$T_P = T_1 + T_2 + T_3 + T_4 .$$

The maximum time available for logic block M to carry out signal processing is given by

$$T_M = T_1 + T_2 + T_3 - 2 \times T_{LA} \tag{6}$$

and the maximum time available for logic block S to carry out signal processing is

$$T_S = T_3 + T_4 + T_1 - 2 \times T_{LA} \tag{7}$$

where T_{LA} is the time required for signal to propagate through a single D-latch.

Edge-triggered master-slave timing is a special case of nonoverlapping two-phase timing: one of the two dead zones in the nonoverlapping timing is collapsed. This circuit must be

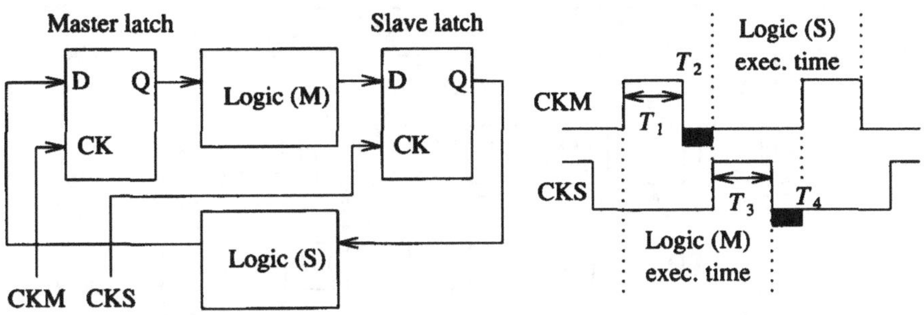

(a) Nonoverlapping 2-phase master-slave timing

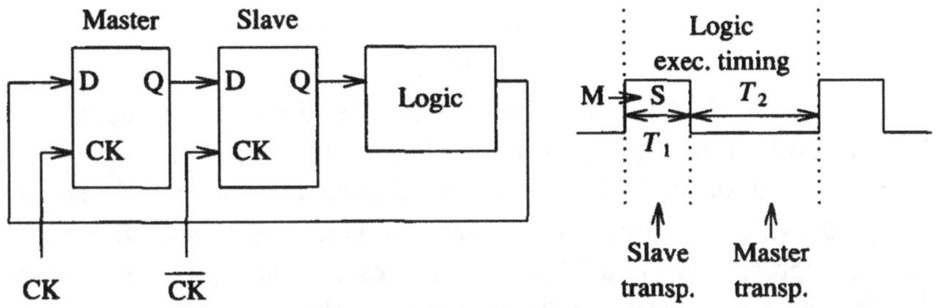

(b) Edge-triggered master-slave timing

Figure 6 CMOS VLSI timing scheme

designed carefully. CK and $\overline{CK}$ must be synchronized well, and the rise-fall times must be short. Short rise-fall times are required to eliminate ambiguity in assuring synchronization. The two CMOS D-latches are placed side by side and metal wires are used to cross-connect the CK and $\overline{CK}$ to the master and slave latches. Then the time for signal to propagate through the two latches can be made more than the small clock skew, and therefore the circuit operates stably. This scheme requires only one clock (two polarities), and therefore the scheme is convenient for high clock-frequency CMOS circuits. The time available for logic circuit operation is

$$T_L = T_1 + T_2 - 2 \times T_{LA} = T_P - 2 \times T_{LA}$$

where $T_1 + T_2 = T_P$ is the clock period.

The objective of both timing schemes of Fig. 6 is to keep the connectivity-wise closed circuit loop always open electrically. Electrically closed signal path has a property that may be compared to *will* of humans (Section 6.3). If independent *will* exists in a subordinate system, the system will not execute the expected function. The timing schemes are to suppress the independent will of the circuit and to bring the circuit entirely under the designer's control.

It is important to establish a guideline for selecting proper clock frequency. Figure 7(a) shows the path of signal propagation during one clock period of a sequential processing cir-

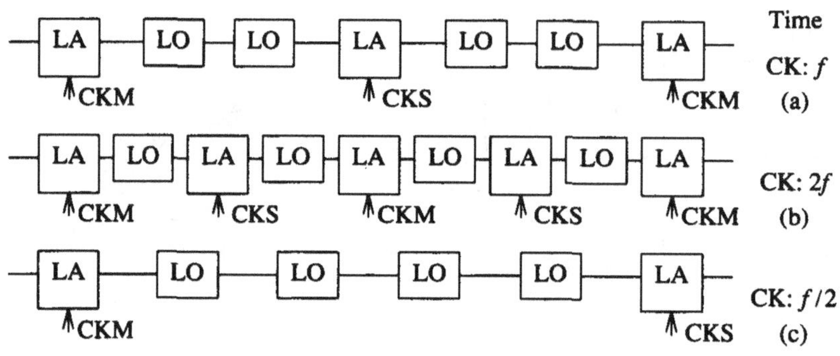

Figure 7 Selection of basic clock frequency

cuit. LA stands for latches, and LO for logic circuits. Let the clock frequency be f (Hz). This logic circuit could have been designed using clock at $2f$ (Hz) as shown in Fig. 7(b), or using clock at $f/2$ (Hz) as shown in Fig. 7(c). Both designs have advantages and disadvantages. The design shown in Fig. 7(c) has advantages in large-scale, *random* data processing subsystems like microprocessors. In a VLSI microprocessor, a huge number of signals must be combined within one basic cycle, and the resulting complex logic gates make long chains. Since a CMOS static D-latch is approximately equivalent to two cascaded gates, it is undesirable to insert unnecessary D-latches, thereby further increasing the length of the already long logic chain. The fundamental requirement of such systems is that data processing is carried out within the time allowed. For such applications, long and unclocked time intervals allow flexibility in timing optimization. This point is clear from the following example.

Let us consider the logic circuit shown in Fig. 8(b). Suppose that the time slot B, immediately *upstream* of a latch that is clocked by CK, is tight in timing budget, but the time slot A′ that is immediately *downstream* of the latch has timing margin. Then gates A and B are shifted to downstream of the latch, and extra latches are added as shown in Fig. 8(a) to capture the input data for the *shifted* gates. If all the logic gates belonged to a single, long timing slot

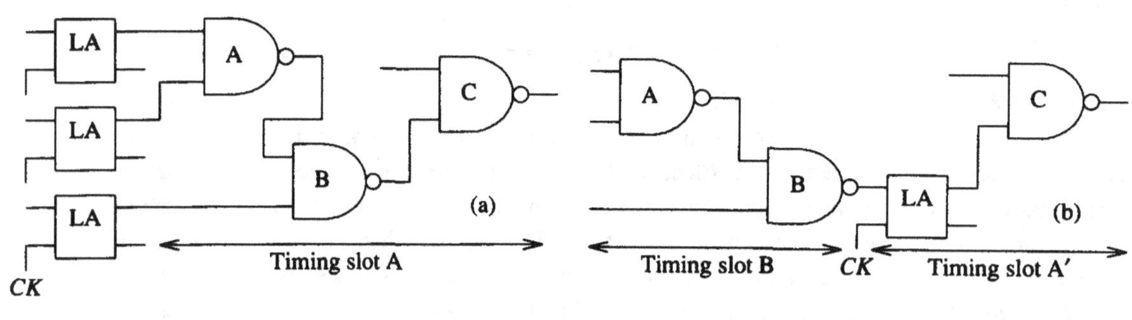

Figure 8 Timing optimized logic design

at a lower clock frequency, the design would have been easier. A single, long time slot allows more processing to be carried out than many short timing slots separated by clocked latches. We note that latches cannot be removed arbitrarily from the high-frequency circuit of Fig. 7(b) to create a timing slot long enough to carry out complex signal processing. Because gate delay time varies with temperature, power supply voltage, and processing conditions, it is impossible to set both the minimum and the maximum delay limits. For instance, to start processing at the first clock period and to capture the answer in the third clock period all the time is difficult. Unless the required processing is completed within the allocated maximum time of one clock cycle for the worst-case delay conditions, there is no way to synchronize the entire sequential processing. In a VLSI or ULSI circuit, it is difficult to distribute skew-free clocks, especially at high clock frequencies. Selection of the lowest clock frequency compatible with operating principles of the system makes logic and circuit optimization easy. In microprocessors, it is often necessary to have more clockedges than available from the basic clock. This requirement is not a justification to choose a higher frequency clock: rather, the extra clockedges should be generated using analog circuit design techniques, as we will see in the next section. Microprocessor clock frequency should be chosen as low as the data-bus and address-bus transfer requirements permit. To operate I/O circuits at a higher frequency clock than the internal logic is one way to optimize microprocessor performance.

Timing design using the highest clock frequency [shown in Fig. 7(b)] has advantages in high-frequency serial data processing circuits, such as used in digital data switching and transmission systems. The clock frequency of such systems is much higher than that of VLSI microprocessors (150MHz versus 20MHz). With such high clock frequency, maintenance of crisp signal waveforms that have precise duration is the most important design consideration. Every time data are captured by a latch, the waveform is restored. Figure 9 shows how a pulse waveform deteriorated by signal processing (the up-going pulses get narrow) is restored by clocking. In high-frequency serial data processing systems, the logic operations required between two clocking points are not extensive. Furthermore, the number of clocking points in the processing system can be adjusted. This flexibility originates from the fact that in a long-haul data transmission system, the delay of one extra clock period spent within the terminal equipment is practically negligible. As we saw, clock frequency selection depends on whether the circuit is used in a microprocessor system or in a data-transmission system. The difference between these logic systems is fundamental, and it will be further discussed in Section 7.6.

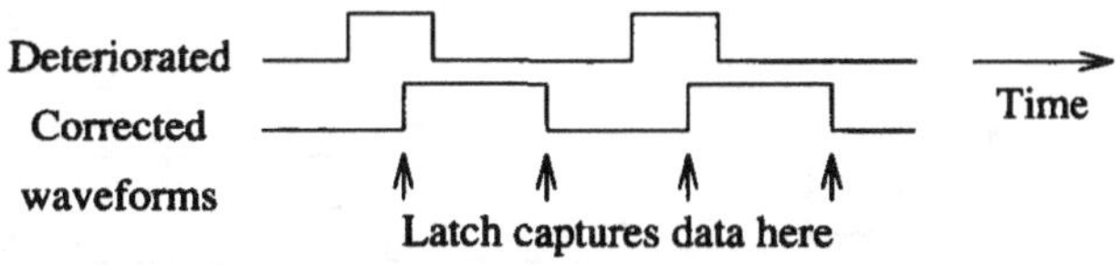

Figure 9 Correction of deteriorated waveform by reclocking

5.4 Clock Signal Generation

Master-slave timing described in the last section requires a precision clock signal. In this section methods of generating CMOS clocks are discussed. The clock signal source for present high-performance data-processing systems is almost always a crystal-controlled oscillator, which is used in order to maintain high frequency stability. A small-amplitude sinusoidal signal from a crystal oscillator is amplified to CMOS logic levels. Crystal clock generators of CMOS or TTL logic levels are commercially available as packaged devices. We consider the problems of *reshaping* the crystal clock signal having unspecified waveform to generate clock signals for specific CMOS circuit applications.

A clock provides basic timing information to the chip, but how much information a clock signal contains depends on the clock waveform *precision.* If a clock waveform is not specified at all, the clock signal contains only the basic period information (clock contained in the pseudo-random data signal of digital transmission system is an example). Since only period is known, the clock phase must be adjusted to gate the serial data securely. If either rising or falling edge is specified in time, clock signal contains information about *beginning of period,* as well. If both rising and falling edges are specified, one extra time point within one period of clock can be specified in addition. Information of period plus two time points are the maximum information a *binary* clock signal can contain.

It is possible to include more than the three basic parameters in a single clock waveform by using complex waveforms and by maintaining high waveform precision (for example, by using pseudo-ternary waveforms). This method is not practical for CMOS, because clock circuits must be designed to have the minimum delay and short rise-fall times. If complex waveforms are used, on-chip circuits that decode clock waveforms become complex and slow. Therefore more than one clock signal, or clock signals at higher harmonics frequencies, are used to provide more than two timing edges per cycle to the chip. An example is shown in Fig. 10. In Fig. 10 a pair of precision fundamental frequency clocks CK23 and CK34 provide a total of four clock edges within one cycle. The primary clock signals, CK23 and CK34, are *decoded* to four phase clocks CK1, CK2, CK3, and CK4. It is possible to use one fundamental frequency clock and one twice the fundamental frequency clock to create four clock edges. This scheme is less precise, however, than the scheme using the two fundamental frequency clocks. It is easier to attain high precision by using only one clock frequency. The clocking

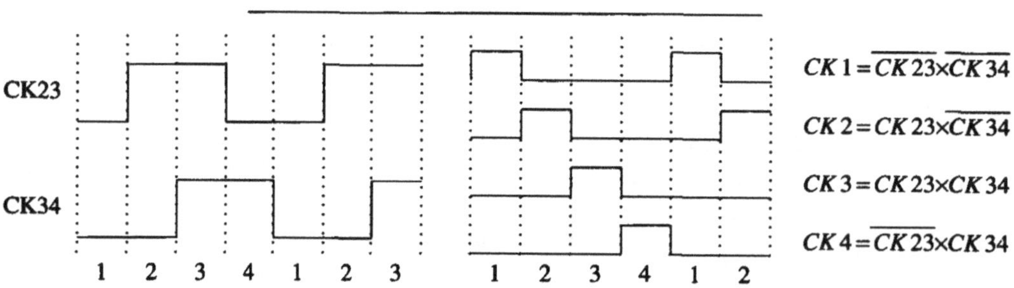

Figure 10 Clock signals in four-phase clocking system

scheme of Fig. 10 requires a symmetrical (50% duty cycle) clock pulse, and a second symmetrical clock pulse that is delayed from the first clock pulse by a specified fraction of the basic clock period. We will review techniques to generate such clock signals next.

It is possible to obtain square, 50% duty cycle clock pulses by counting down a higher frequency clock signal [04]. Figure 11 shows a simple countdown circuit called a Moebius counter. Suppose node N_1 is high and node N_0 is low. When up-going clock edge A arrives, the N_1 signal is transmitted to node N_0 through latch 0, and therefore node N_0 makes a low-to-high transition. Latch 1 retains the previous data. When down-going edge B arrives, the N_0 data are transmitted to node N_1 inverted (note $\bar{Q}$ output), and therefore node N_1 makes a high-to-low transition. Latch 0 retains the previous data. A pair of half-frequency clock pulses that are shifted by the width of the input clock pulse are generated. Node N_0 and N_1 waveforms both have 50% duty cycles. This pair of clocks can be used as the primary clocks of Fig. 10, thereby generating four clock edges within one fundamental period. If four *equally spaced* clock edges within a period are required, four times the fundamental frequency clock is generated first, and it is counted down twice using the cascaded countdown circuit shown in Fig. 11. The waveforms at the nodes $N_0 - N_3$ are shown in Fig. 11.

One problem of the circuit of Fig. 11 is that the states of the latch 0 through 3 immediately after turn-on of power supply voltage are unknown. If only one clock signal is used, there is no problem. If more than one waveform is used to maintain proper clockedge timing relationships, however, the latches must be set to the proper initial state before clocking starts.

If many countdown circuits are cascaded, delay between the primary clock edge and the clock edge of the counted-down clock pulse (T_D of Fig. 11) becomes significant. If both the high-frequency primary clock and the low-frequency (counted-down) clocks are used for timing, the relative delay cannot exceed a certain small limit. To resynchronize all the clocks, the low-frequency clock is *re-timed* by using the *sampling* circuit of Figs. 12(a) or (b).

In Fig. 12, N_A is the *delayed* low-frequency clock pulse. If the pulse is *sampled* by a transmission gate driven by the primary, high-frequency clock, the clockedges are re-aligned

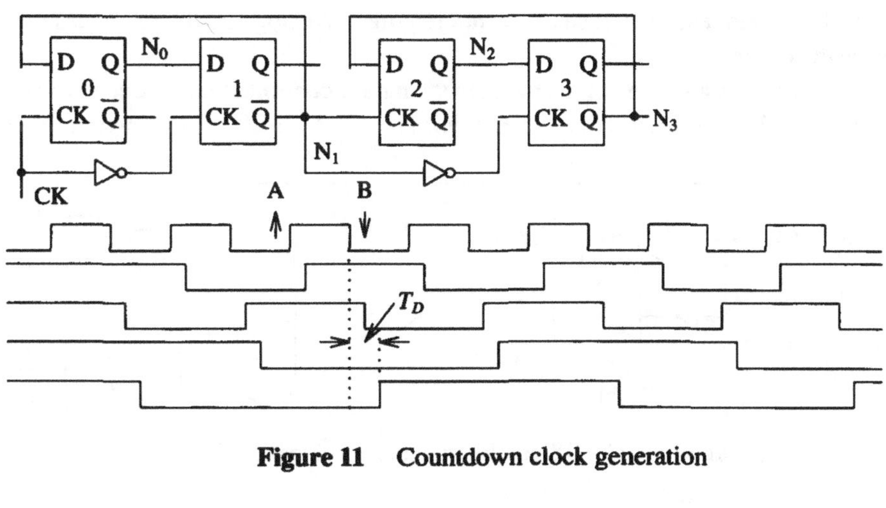

Figure 11 Countdown clock generation

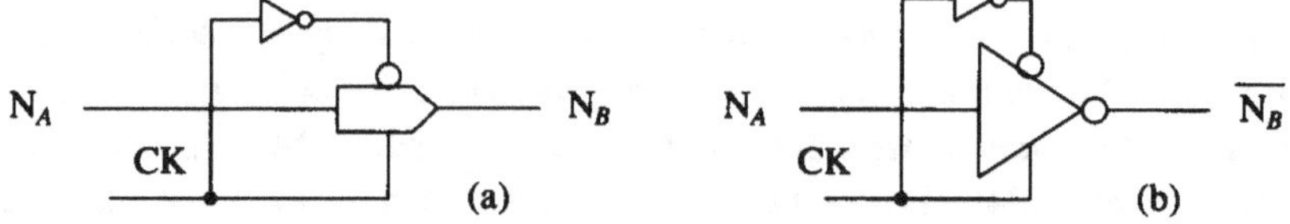

Figure 12 Retiming circuit

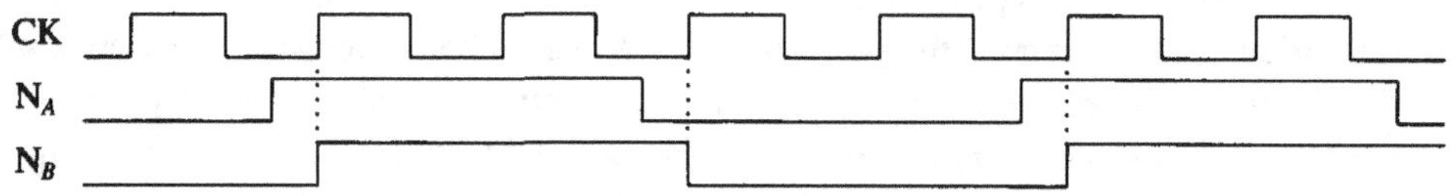

Figure 13 Retimed clock waveform

within the delay time of the transmission gate. The waveforms are schematically shown in Fig. 13.

The transmission gates and tristatable inverters of Fig. 12 require a pair of complementary clocks. CMOS latches require a pair of complementary clock signals also. Figure 14 shows various circuits that generate a pair of CMOS clocks, CK and $\overline{CK}$, from single clock source CK. The circuit of Fig. 14(a) is simple and fast, but $\overline{CK}$ is delayed from CK by one inverter delay time. The circuit of Fig. 14(b) allows the generation of a complementary, skew-free pair of clocks, but the delay is relatively long. The circuit of Fig. 14(c) uses positive feedback, but the NAND gate in the circuit delays clock.

Generation of a 50% duty cycle clock by countdown is widely used, and it is reliable. The problem of this scheme is availability and cost of high-frequency crystal-controlled clock sources. Crystal-controlled oscillators with frequencies higher than about 30MHz are hybrid bipolar integrated circuits that are expensive. It is desirable for cost reasons that two times or four times the basic frequency clock source is not required. Analog circuit techniques are useful to achieve the objective.

Analog circuits that are useful for clock reshaping are controlled delay circuits and duty cycle (or pulse width) control circuits. The circuit shown in Fig. 15(a) has digital input D1

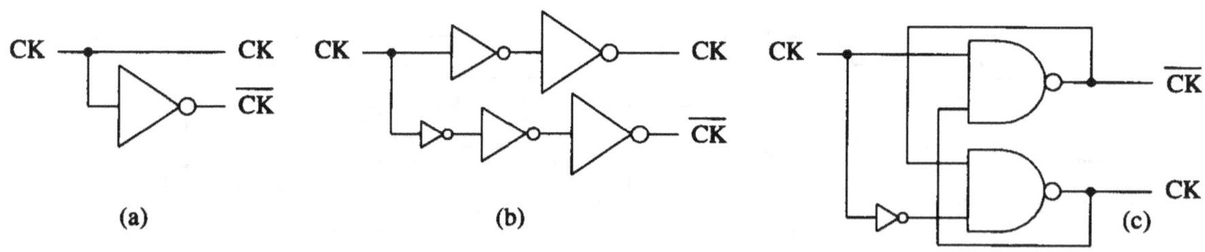

Figure 14 CMOS complementary clock generation

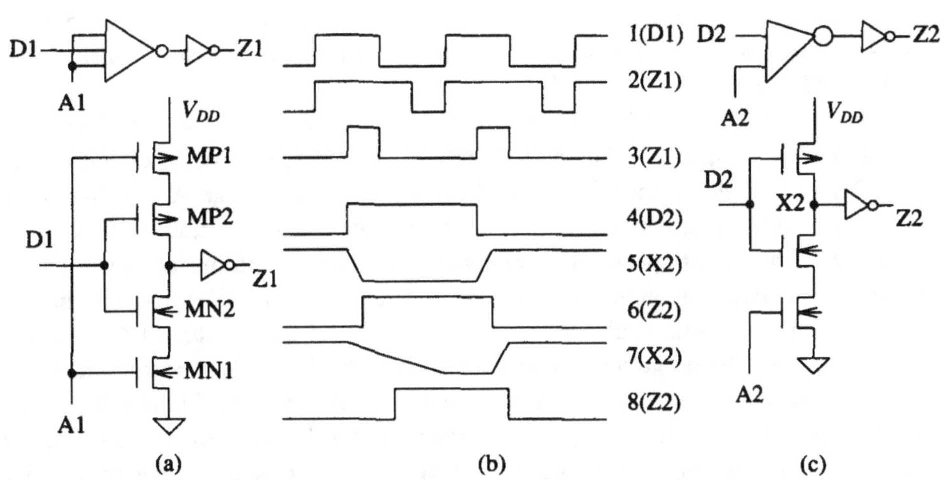

Figure 15 Clockpulse shaping circuit

and analog control input A1. If the analog control voltage is high, NFET MN1 has high conductance, and PFET MP1 has low conductance. Then the pull-down delay of the controlled inverter is short, but pull-up delay is long. If the analog control voltage is low, however, pull-down delay is long and pull-up delay is short. The inverter that buffers the output of the controlled inverter reshapes the signal to obtain a crisp clock signal. The input waveform D1, and output waveforms Z1 for high and low control voltages of the analog input A1, are shown in Fig. 15(b) waveforms 1, 2, and 3, respectively. This circuit can be used in a negative feedback loop as shown in Fig. 16(a). The circuit generates a 50% duty cycle (symmetrical) clock pulse from the source pulse, which may have any duty cycle (typically 25–75%). Figure 16(b) shows the internal node waveforms. The feedback control works as follows. If input clock CK is a train of narrow up-going pulses, the node N_0 waveform is narrow up-going pulses also. The N_0 waveform is averaged by resistance R, capacitance C, and the operational amplifier, and a high analog control voltage is generated (note that the negative input of the operational amplifier is driven by the averaged voltage). Then the controlled delay circuit pulls

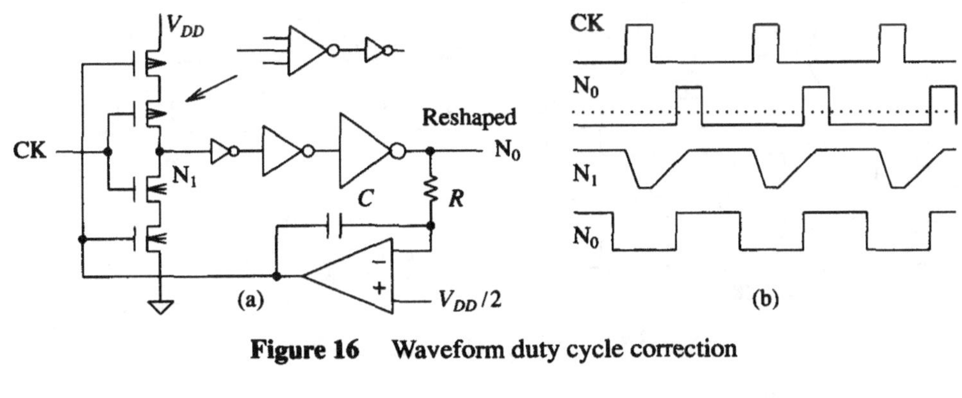

Figure 16 Waveform duty cycle correction

down rapidly and pulls up slowly. Narrow up-going pulse generates wider down-going pulse at node N_1. If the gain of the operational amplifier is high, the duty cycle of pulse at node N_0 is forced to 50% [05].

The circuit of Fig. 15(c) delays input digital signal D2, and the delay time is controlled by the analog voltage at input A2. The circuit consists of a controlled delay circuit and buffer. Pull-up delay time of this circuit is practically independent of voltage A2. Pull-down delay time depends on the analog control voltage: if the control voltage decreases, pull-down delay time increases. The circuit operation is understood from the waveforms of Fig. 15(b). If the control voltage is high, both pull-up and pull-down delays are short. The up-going input pulse at D2 (curve 4) and the output pulse at Z2 (curve 6) have the same width. If the control voltage is low, however, the up-going transition of input pulse D2 is delayed significantly more than the down-going transition, and a narrower pulse is generated at Z2 (curve 8). If an even number of the bare control circuit [Fig. 15(c) without buffer] is cascaded and controlled by the same analog voltage, the input up-going and the input down-going edges are equally delayed and pulse D2 is shifted in time, without significantly altering the pulse width. In order to maintain the pulse width, a fast noninverting buffer should be inserted between consecutive stages of the delay control gates to recover crisp switching waveform.

The circuit of Fig. 17(a) is an example of an analog circuit that is used to generate an extra clock edge, Y, between two clock edges X and Z. Clock edges X and Z are generated by *input* clock CK. CK is delayed by the controlled delay circuit shown within the dotted rectangle of Fig. 17(a). The original clock, CK, and the delayed clock, DCK, are compared and *DC* voltages proportional to time interval $\overline{YZ}$ (voltage A) and $\overline{XY}$ (voltage B) are generated. This is carried out by first generating a positive-going pulse having widths $\overline{YZ}$ and $\overline{XY}$ and then averaging the pulse voltages using circuits that have long RC time constants. The two *DC* voltages proportional to $\overline{YZ}$ and $\overline{XY}$ are compared using the differential amplifier, and negative

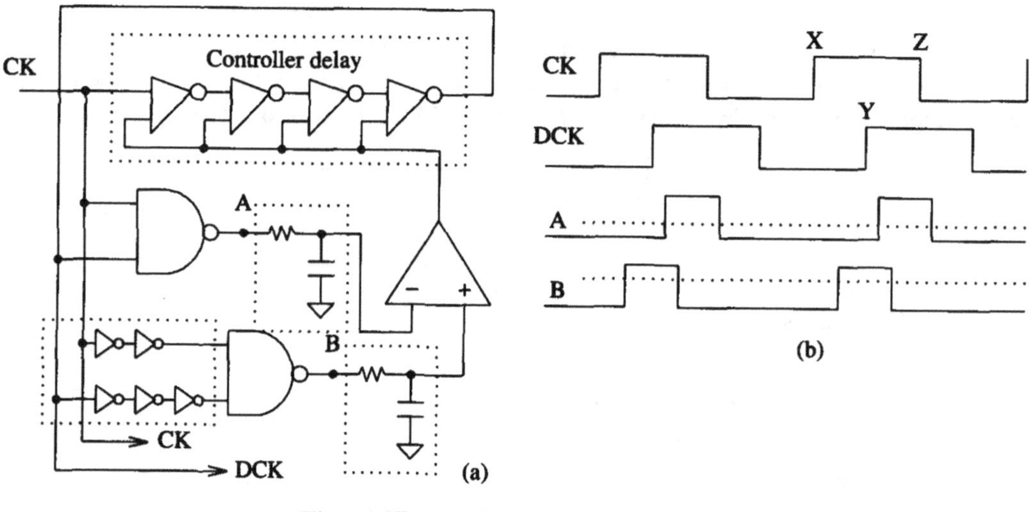

Figure 17 Analog clockedge insertion

feedback is applied to the controlled-delay circuit. Clock edge Y is set so that $\overline{XY} = \overline{YZ}$, if the gain of the differential amplifier is high. The two sets of cascaded inverters within the dotted box must have their delays matched under all the conditions. How to design such a matched delay circuit will be discussed later in Sections 5.14 and 5.15.

5.5 Clock Signal Distribution

Generation of skew-free clock signals is only a part of clock subsystem design. The clock signals must be delivered to the destinations. The objective of clock distribution circuit design is twofold: (1) to distribute a skew-free primary clock all over the VLSI chip, and (2) to design the clock post-processing circuits that interface the *primary* clocks to the final destination with minimum skew and delay.

Clock signals are distributed by most conductive metal wires. The signal delay through unloaded metal wire is the lower bound of the clock delay. The metal wire has capacitive (C), resistive (R), and inductive (L) parasitics, all of which contribute to clock delay. If a chip is a (cm) square, the clock-bus length is about $2a$ (cm). Realistic commercial VLSI chip size is about 1 (cm) square. The delay of 2 (cm) long, unloaded metal wire is shown in Fig. 18 [06]. The vertical axis is the delay in nanoseconds, and the horizontal axis is the width of the clock-bus in microns. From Fig. 18, a 20-micron-wide clock-bus (considered very wide) has unloaded delay of 0.5 nanosec. In a VHSIC circuit whose clock frequency is 50–100MHz the primary clock distribution delay must be included in chip design and in chip timing specifica-

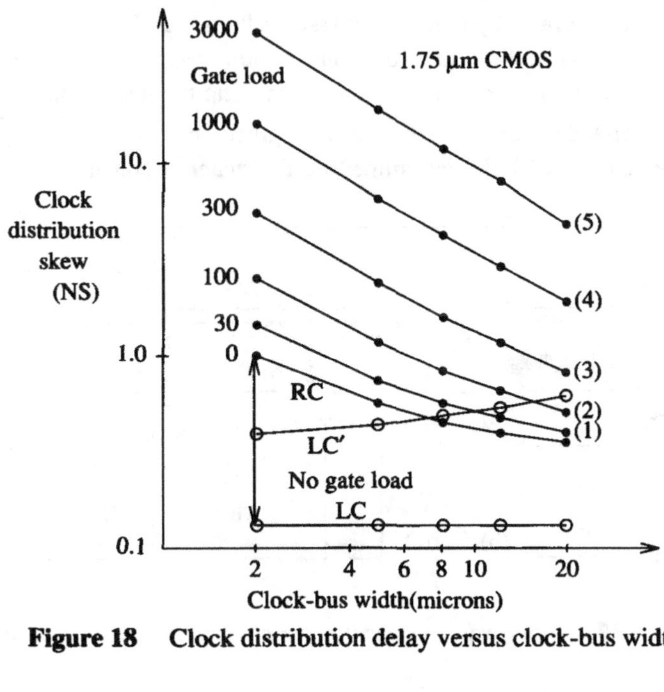

Figure 18 Clock distribution delay versus clock-bus width

tions. Real delay time is significantly more than this estimate, since the destination circuits load the clock-bus. Curves 1–5 of Fig. 18 show the delay when the clock-bus is loaded by CMOS gates. The additional capacitive load is from minimum size 1.75 micron CMOS inverters whose gate capacitance is 0.04 picofarad. If there are 10–30K gates in the random logic, 1–3K gates will receive clock signals directly. Then the delay will be as high as 10 nanosec. Curve LC′ shows the clock delay due to inductance-capacitance parasitics including silicon skin effect. Because of the skin effect of silicon, propagation velocity of signal along the metal wire can be as much as several times less than the velocity of light (curve LC).

An electrical signal propagates along a microstrip line that is fabricated on silicon and silicon dioxide dielectric significantly slower than the velocity of light within silicon dioxide. Figure 19(a) shows a cross section of a microstrip line fabricated on highly conductive metal. The wave propagates along the microstrip line in the direction indicated by the thick arrow. The horizontal arrows within the conductors indicate directions of local conduction currents, + and − signs show the charges accumulated by the currents, and the vertical arrows indicate displacement currents that are proportional to the rate of change of the vertical electric field. Currents create magnetic fields, whose directions are shown by the arrowhead and the arrowtail. The electric field in the space between the conductor is in the vertical direction, and the magnetic field is perpendicular to the page. The fields are perpendicular to each other, and they are both perpendicular to the direction of propagation. The mode of propagation is that of a transverse electromagnetic (TEM) wave. The propagation velocity of the wave c is given by $c = 1/\sqrt{\varepsilon\mu}$ where ε is the dielectric constant and μ is the magnetic permeability of the silicon dioxide insulator. This is the velocity of light within the dielectric material.

This propagation mode is modeled by the equivalent LC ladder circuit shown in Fig. 19(b). Capacitance C represents the portions of the microstrip line centered at locations A, B, C, ... (shown by the boxes drawn by dotted lines) and inductances L_1 and L_2 model the section of the microstrip line connecting the locations of the capacitances, AB1, AB2, BC1, BC2, ... Both inductance and capacitance are per unit length values. The wave propagation velocity along the equivalent LC ladder circuit is given by $c = 1/\sqrt{(L_1 + L_2)C}$. Since the velocity equals the velocity of light determined by the material constants ε and μ of silicon

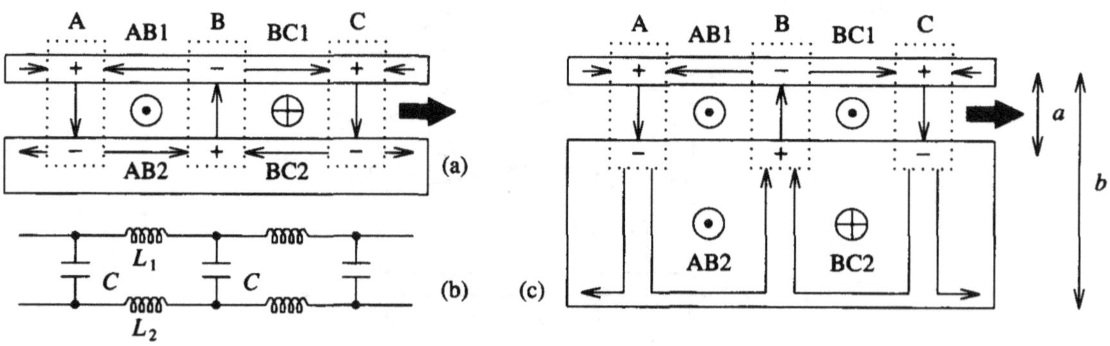

Figure 19 Slow-wave propagation in microstrip line on silicon

dioxide insulator, the capacitance and the inductance of the equivalent microstrip line satisfy the relationship $(L_1 + L_2)C = \varepsilon\mu$.

This conclusion is based on the assumption that the current and the charge are distributed within the thin surface layer of the conductors. As for the charge, this condition is well satisfied even in the silicon substrate. How fast the excess charge in a semiconductor decays depends on the dielectric relaxation time ε/σ, where σ and ε are conductivity and dielectric constant of silicon, respectively. If $\sigma = 1\ mho/cm$, $\varepsilon/\sigma = 1\ psec$. This time is much shorter than typical transients of CMOS circuits, which are 100 picosec or more at present. Charge can exist only on the semiconductor surface. As for current, the assumption of a thin surface current is satisfied in metal, which has high conductivity. In doped silicon substrate, however, the conductivity is in the range of 1–100 mho/cm, as compared to $10^6 mho/cm$ in metal. The *skin depth*, the measure of penetration of current into a conductor, is given by $\Lambda = \sqrt{2/\omega\mu\sigma}$ where ω is the angular frequency of the signal. Λ is in the range 4–0.4 mm if $\omega = 10^7 - 10^9\ rad/sec$. The skin depth is more than the silicon substrate thickness. Therefore the field and current associated with the wave is modified from that of Fig. 19(a) to that of Fig. 19(c). The magnetic field penetrates into the semiconductor substrate and current flows deep inside. Then inductance L_2 in the equivalent circuit increases over that of an ordinary transmission line approximately by a factor of $\log_e(b/a)$ (b/a is the ratio of the distance b to the *nearby* conductor that carries return current, to the thickness of the insulator a). Since capacitance of the microstrip line section remains the same, but inductance of the line increases, the propagation velocity decreases. If we use $b = 500\ \mu$m (thickness of silicon substrate) and $a = 1\ \mu$m (thickness of insulating oxide), the propagation velocity is reduced by a factor of $\sqrt{\log_e(b/a)} = 2.5$. This mode is called *slow wave* of signal propagation [07]. Whether or not slow wave propagation is actually excited on a clock-bus depends on the structure of the clock-bus. If a large and parallel metal structure like a power/ground-bus exists nearby, slow wave mode may not be excited.

Clock signals at different locations on the clock-bus have different delays, and that causes clock skew. Except at very high clock frequencies, the clock-bus delay is due to the wiring resistance. The primary clock-bus must be designed using continuous metal wire, and high-resistance crossovers should be avoided. In a case when a crossover is absolutely required, the number of squares of the crossover must be reduced so that the crossover resistance is comparable to or less than the primary clock metal bus resistance itself. A large number of contacts to the crossover level must be used, since contact resistance from metal to the less conductive crossover level is quite significant. The primary clock-bus wire width should be properly chosen. The clock-bus should not be allowed to couple with other circuits that switch simultaneously. If the primary clock-bus is adjacent to a signal wire that switches simultaneously, capacitive coupling creates uncertainty in clock-bus loading, thereby creating unpredictable clock skew. For this reason, the primary clock-bus is often routed between V_{DD} and V_{SS} power-buses. This structure provides current return path for faster signal propagation.

Clock distribution delay is optimized by using the distributed clock system shown in Fig. 20(a) [08]. In this distributed clock system, a buffered system clock signal is distributed over the entire chip through the primary clock-bus. Every time the clock-bus crosses over other cir-

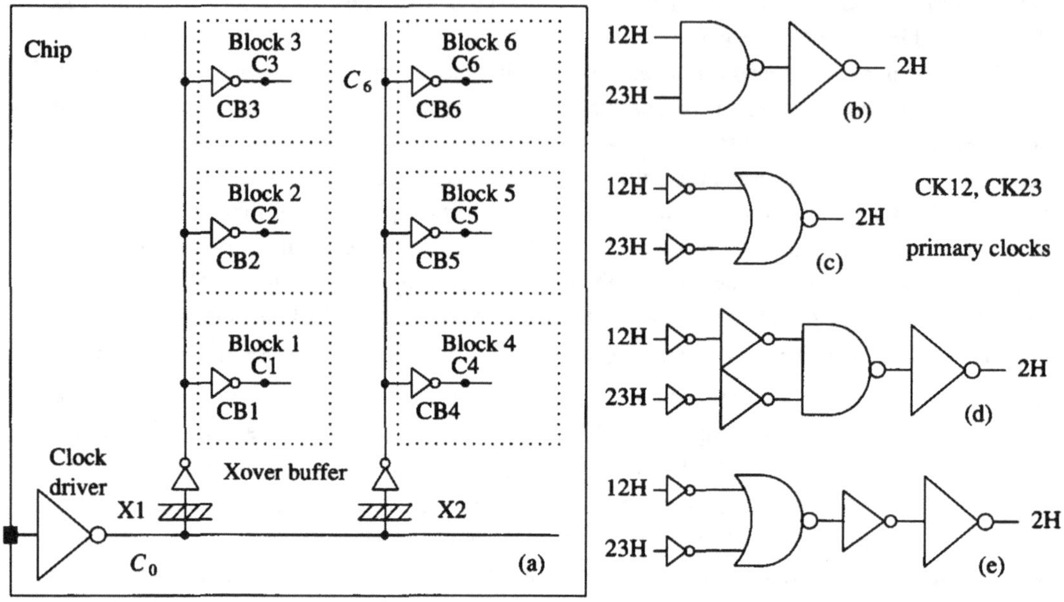

Figure 20 Distributed clock system

cuit features, a clock buffer is provided at the far end of the crossover. Clock edges at destination nodes C1–C6 are synchronized by adjusting the FET sizes of the crossunder buffers and of the clock signal generator circuit at the individual logic blocks, CB1, CB2, ..., CB6. Clock buffers CB1–6 of Fig. 20(a) can be clock decoders. The scheme therefore accommodates nonoverlapping two-phase clock systems (Fig. 10). From flexibility in delay compensation, the distributed clock system is the most promising choice in VHSIC.

Clock decoder and clock enable-disable circuits of the distributed clock system should be designed to equalize the delays of all clock circuits. Since these circuits are *local* on a VLSI chip, clock decoding and conditioning circuits do not have long subsequent distribution delay. Matching clock delays, however, is not a simple matter. To achieve a high degree of matching the *precision delay matching* technique that is discussed in Sections 5.14 and 5.15 is the most effective. Since there can be large variation in clock loads, the number of buffer stages and the types of gates must be optimized. The primary clock-bus must be driven strong enough to tolerate the clock load uncertainty at the early chip design phase. Figures 20(b)–(e) show various choices of local clock decoder logic. Primary clocks are CK12H and CH23H, and the logic circuits generate CK2H. If CK2H is heavily loaded, a simple circuit like Figs. 20(b) or (c), designed using large FETs, gives the shortest delay. If the load is light and if delay must be added for skew control, however, a circuit like Figs. 20(d) or (e), which has two stages of buffer, is preferred. If this circuit is used, gates should be designed using small FETs, so that clock-bus loading is minimized.

If the clock period is reduced to about 20 times the clock distribution delay, adjusting the primary clock delay to all the destinations becomes too complex. If the clock frequency of a CMOS VLSI chip is higher than about 100 MHz, the scheme called H-clock distribution method shown in Fig. 21 can be used. In this clock distribution scheme, clock signal paths from the clock source (location of clock buffer) to any destination A, B, C, ..., Z are *electrically* all the same. The structure is a *fractal* constructed from "H"-character [09]. Clock signals are available only from the leaves of the *tree*. It is illegal to tap the clock signal at an arbitrary location like X. The destinations must all have the same type of local clock buffer, so that the clock distribution tree is loaded equally at all destinations. If any destination is not used, a dummy buffer must be placed there.

We need to consider clocking problems of multi-chip subsystems. If many chips are clocked by the same system clock, and if the chips exchange signals that are timed by the clock, the chips' *internal* clock edges must be well synchronized. Since the chips are processed under different process conditions, the internal clock delays are not equal, even if the clock circuit designs are identical. If clock delay is T_D for the slowest chip, clock skew that is as much as $T_D/2$ must be tolerated when the fastest and the slowest chips communicate. If internal clock edges of chips are different due to skew, the processing time budget is reduced, and system performance deteriorates. The best strategy is to reduce the internal clock delay T_D to the minimum, and to introduce a selection procedure to match sets of chips.

The system clock is brought into the chip's clock buffer through an ESD (electrostatic discharge) protection circuit consisting of a diffused resistor and a pair of reverse-biased PN junctions. The circuit, having the structure of a low-pass RC filter, delays the clock signal. In order to provide sufficient protection for harsh environments, several nanoseconds of delay must be tolerated in the ESD protection circuit. Since the clock buffer must drive all the internal clock loads, the buffer is large, and large gate capacitance of its FETs adds to the capacitance of ESD protection circuit. The clock buffer must be large enough to produce rapid internal clock signal transitions. Realistically, the *effective* fanout ratio of the clock driver to clock load should be about 3, or less. Selection of the number of the clock buffer stages simultaneously minimizing delay and transition time is an important design consideration. The clock input stage should be designed using the largest possible FETs within the limit of tolerable

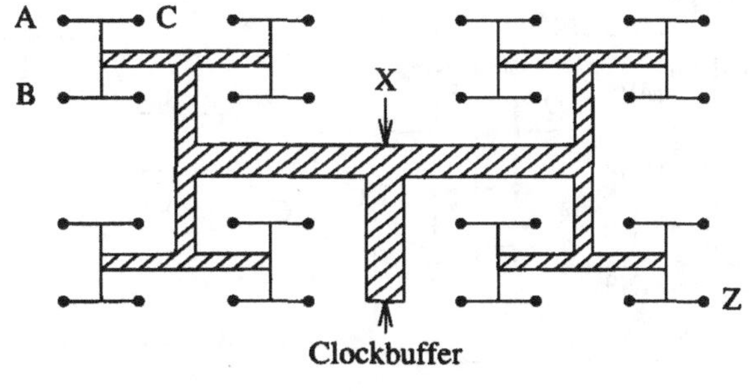

Figure 21 H-clock distribution scheme (clock-bus)

input-protection circuit delay. We regard clocks as AC *power supply* to the chip, rather than an ordinary input *signal*. Series resistance in the clock signal path has serious performance impacts.

A perfectly synchronized clock system does not always provide the best performance. It is impossible, however, to assign a balanced timing budget to the logic circuit by adjusting individual clock edges, since the number of clock loads in a VLSI circuit is huge, amounting to 10–15% of the number of random gates. The best strategy is to design a well-synchronized clock system, and then, for a small number of critical paths, the timing budget is balanced by shifting clock edges.

Before ending this section, I will give a new viewpoint on sequential logic circuits. In a sequential processing system, all the signals become available as the system clock ticks. The primary clock is split to many destinations, and they are combined later, as the logic answer is captured at the destinations. From this viewpoint, an entire sequential logic circuit operates by first splitting and then recombining the primary clocks. In this sense the clock is truly an AC power supply.

5.6 D-latch Operations

The CMOS D-latch that is shown in Fig. 22(a) consists of two tristatable inverters and one plain inverter. With reference to curve 2 of Fig. 22(b), if *CK* is low and $\overline{CK}$ is high, the first tristatable inverter tristates. The second tristatable inverter drives, and together with the inverter, it makes a positive feedback loop that is equivalent to two cascaded inverters in a closed loop. The circuit stays in one of the two possible states into which it was forced. Forcing the state of the latch is accomplished by the first tristatable inverter, when the latch is transmitting (or transparent). When the input data D changes, the interval between the time when the input data changes for the last time and the time when the clocks close the latch may

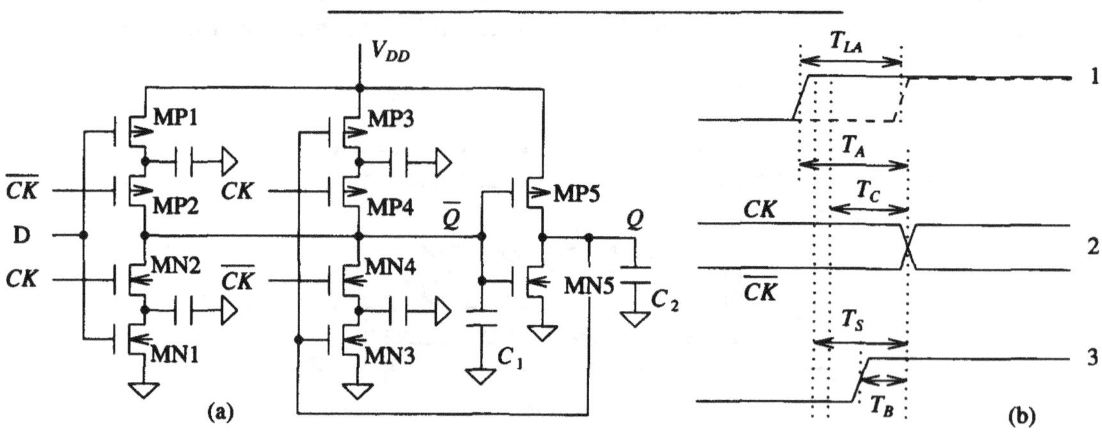

Figure 22 CMOS D-latch and definition of setup time

be too short to *capture* the input data at the clockedge. In Fig. 22(b), clock waveforms and two input waveforms are shown. The data signal D (curve 1) of Fig. 22(b) makes a low-to-high transition at time T_A earlier than the clock transition. Time T_A is long enough to introduce the high logic level into the latch circuit: after the clock transition, a logic-high level is stored in the latch. The second data waveform (curve 3) of Fig. 22(b) shows an input data transition that occurs at time T_B earlier than the clock transition. In this case the transition is too late to introduce a logic-high level into the latch circuit. Logic-*low* data are stored in the circuit when the circuit settles to the steady state. Obviously there is a boundary between the two regimes. This boundary is shown by time T_C. If input data change before that time, the latch captures the following data: If input data changed after that, the latch captures the earlier data.

Suppose that the latch is in a noise-free environment. If the transition of the input data happens exactly at the T_C before the clock transition, the latch is unable to settle: At the moment when the latch *closes*, the latch has not acquired enough *internal* information to determines the final state. This conclusion, however, is not realistic. Rather, it takes a long time for the latch to settle to the final state, and which state the latch will eventually settle to depends on many unpredictable real effects like noise. This phenomenon is called metastability (Section 3.12). The metastable state is the *grey* boundary between the two *definite* settling conditions of the latch. T_C and the other timing parameters are related as follows. Curve 1 of Fig. 22(b) includes node D (solid) and node Q (dashed) waveforms. If transparent, the signal takes time T_{LA} to propagate from D to Q. After T_{LA} the $\bar{Q}$ voltage is lower than V_{GSW} of the inverter, and the Q voltage is higher than V_{GSW}. If the clocks close the latch after T_{LA}, the latch captures the high level, even if small noise exists, because $\bar{Q}$ is one gate delay time earlier than Q, and in this state the latch has already acquired sufficient information to go to the direction. The setup time, T_S, that guarantees storage of the data in a realistic condition [10] is therefore shorter that T_{LA}. Then we have

$$T_C < T_S < T_{LA}.$$

The definition of T_S was given assuming certain clock and data signal waveforms (such parameters as the rise and the fall times of the signals and the clocks). If waveforms are specified differently, the T_S will be different. Definition of the setup time T_S requires yet another consideration on the effects of noise. It is difficult to make a clean statement of what the setup time T_S of latch is. Often I have a reservation to use setup time, because its definition is not simple and clear. Here we consider its lower bound, T_C, which is more clearly defined theoretically than T_S. In this section we study how T_C is determined using the simplest model.

During the transparent phase, the second tristatable inverter is tristated. The input circuit of the latch is shown in Fig. 23(a). In order to make the analysis simple, we assume that capacitances of the internal nodes X and Y are negligible. We use the FET collapsible current generator model with zero threshold voltage. Transconductance of both FETs of the first tristatable inverter are B_0, those of the second, B_1, and those of the inverter, B_2. We assume that input signal D makes an instantaneous low-to-high transition. Immediately after this transition, MN1 turns on and MP1 turns off, and since the internal node Y is assumed to have negligible capacitance, the series-connected PFETs have no effect on the operation. The microstate

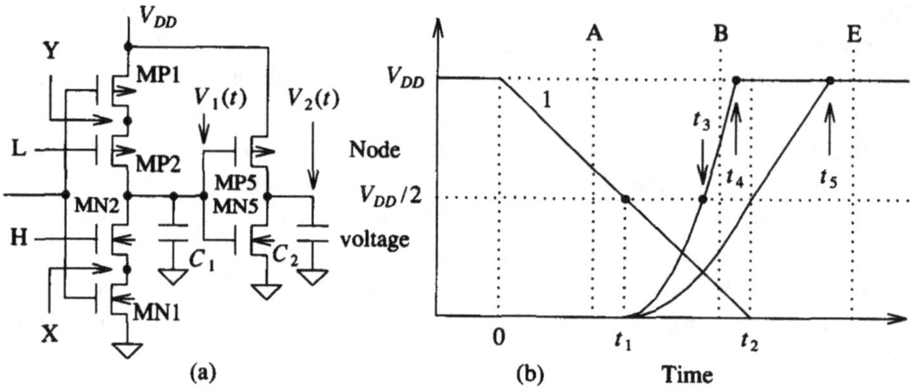

Figure 23 Analysis of latch setup time—I

---

sequences of the D-latch are shown in Fig. 24. The two events of sequence I involving PFET MP2 (MP2: T → S and MP2: S → M) do not influence any other events. The gates of MN1 and MN2 are both at V_{DD} after the signal D transition. There are two possible states of the NFET chain consisting of MN1 and MN2, but only one of MN1 or MN2 can be in the triode region. If MN2 is in the triode region, MN1 is not. Then V_X is positive, and therefore MN2 current is less than MN1 current. V_X decreases within a negligibly short time and settles at ground. MN1 goes into the triode region. Therefore, the only stable circuit configuration immediately after the up-going input signal D transition is with MN2 in the saturation region and MN1 in the triode region. Within a short time after the transition, series Z of microstate sequence II is completed. We assumed that MN1 and MN2 have the same size, and node X has negligibly small capacitance (how fast part Z of the microstate sequence is completed is proportional to the capacitance of node X). Microstate change MP5: N → S follows immediately because $V_{THP}=0$ was assumed. Since node X becomes a *virtual* ground, $V_1(t)$ is determined by current through MN2. This current discharges C_1.

$$C_1 \frac{dV_1(t)}{dt} = -B_0 V_{DD} \tag{8}$$

which is solved subject to the initial condition $V_1(0)=V_{DD}$ to obtain

$$V_1(t) = V_{DD} - \frac{B_0}{C_1} V_{DD} t .$$

This is shown by curve 1 of Fig. 23(b). It takes time, $t_1 = C_1/(2B_0)$, to reach the switching threshold voltage of the symmetrical inverter consisting of MP5 and MN5, $V_{DD}/2$. The inverter is driven by the linear ramp waveform. V_1 reaches zero at time $t_2=2t_1$. After $t>t_2$, $V_1(t)=0$. The pull-up transition of the inverter begins at $t=t_1$, and the waveform $V_2(t)$ at the output node of the inverter (node 2) is given by solving the circuit equation

$$C_2 \frac{dV_2(t)}{dt} = B_2[V_{DD} - V_1(t)] - B_2 V_1(t)$$

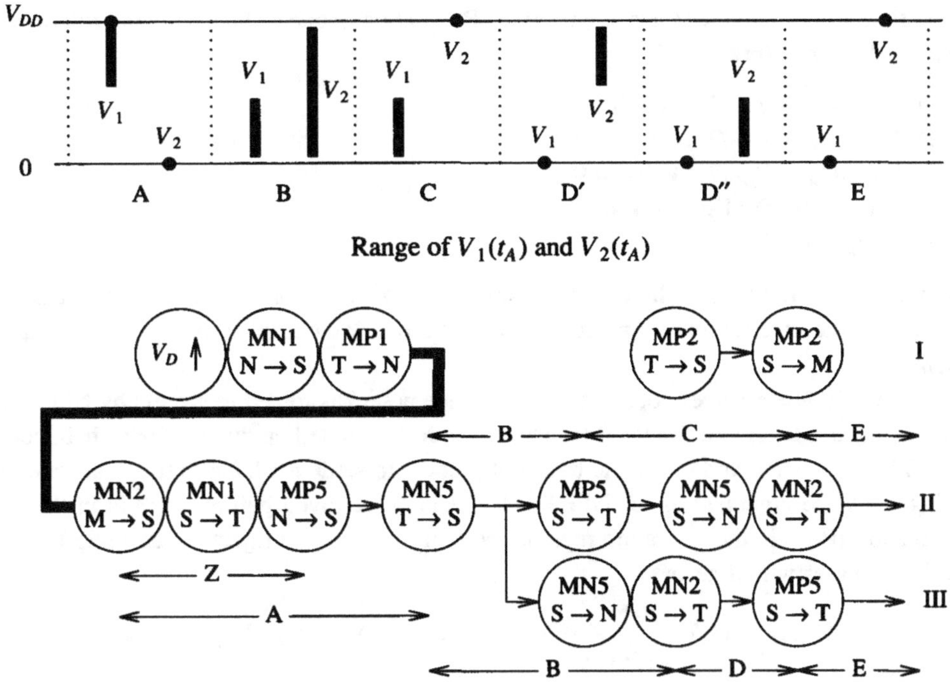

Figure 24 Microstate sequence of CMOS D-latch

subject to the initial condition $V_2(t_1)=0$ to obtain

$$V_2(t)=V_{DD}\frac{B_0B_2}{C_1C_2}(t-t_1)^2. \tag{9}$$

At time t_3, $V_2(t)$ reaches $V_{DD}/2$ and at time t_4, $V_2(t)$ reaches V_{DD}. Times t_3 and t_4 are given, respectively, by

$$t_3=t_1+\sqrt{(C_1C_2)/(2B_0B_2)}\quad t_4=t_1+\sqrt{(C_1C_2)/(B_0B_2)}.$$

Depending on which of t_2 or t_4 is larger, there are two cases. If $t_4<t_2$, Eq. (9) is valid for $t_1\le t\le t_4$. For $t>t_4$, $V_2(t)=V_{DD}$. If $t_4>t_2$, Eq. (9) is valid only for $t_1\le t\le t_2$ and after t_2 we have

$$V_2(t)=V_2(t_2)+\frac{B_2}{C_2}V_{DD}(t-t_2)$$

until $t=t_5$, where $t_5=(C_2/B_2)+t_2-(B_0/C_1)(t_2-t_1)^2$. At time $t=t_5$, node voltage $V_2(t)$ reaches V_{DD}, and for $t>t_5$, $V_2(t)=V_{DD}$. Although the consideration of individual cases is cumbersome, the procedure is clear. In any event, we know V_1 and V_2 at the moment when the pair of clock signals at $t=t_A$ turn the input tristatable inverter off and turn the second tristatable inverter on. Let V_1 and V_2 at that moment be $V_1(t_A)$ and $V_2(t_A)$, respectively. Depending on the time slot of the microstate sequence A, B, C, D or E, one of the following *initial* conditions is established (Figure 24):

A) $V_1(t_A)>V_{DD}/2 \quad V_2(t_A)=0$
B) $0<V_1(t_A)<V_{DD}/2 \quad 0<V_2(t_A)<V_{DD}$
C) $0<V_1(t_A)<V_{DD}/2 \quad V_2(t_A)=V_{DD}$
D) $V_1(t_A)=0 \quad 0<V_2(t_A)<V_{DD}$
E) $V_1(t_A)=0 \quad V_2(t_A)=V_{DD}$

The equivalent circuit after the clock transition is shown in Fig. 25. The second tristatable inverter and the inverter make a positive feedback loop. We consider each case of initial conditions.

Case (A): Node 1 does not affect node 2, since node 2 is forced to ground by NFET MN5 that is in the triode region. $V_2(t)=0$ for $t>t_A$. Node 1 is pulled up by the PFET chain consisting of MP3 and MP4, and this is the transient process to settle the latch to the final state. Following the same argument as for NFETs MN1 and MN2 given before, MP3 is in the triode region and MP4 is in the saturation region, and current B_1V_{DD} is injected into node 1. Therefore V_1 increases with time as

$$V_1(t)=V_1(t_A)+\frac{B_1V_{DD}}{C_1}(t-t_A)\ (t_A<t<t_{CA}) \quad V_1(t)=V_{DD}\ (t>t_{CA}) \tag{10}$$

where

$$t_{CA}=t_A+(C_1/B_1V_{DD})[V_{DD}-V_1(t_A)]$$

Waveforms are shown in Fig. 26 by symbols "A". The latch settles at state $Q=Low$ $[V_2(\infty)=0]$.

Case (C): This is the *mirror image* of case (A). Node 1 does not affect node 2, and $V_2(t)=V_{DD}$ $(t>t_A)$. Node 1 voltage decreases with time as

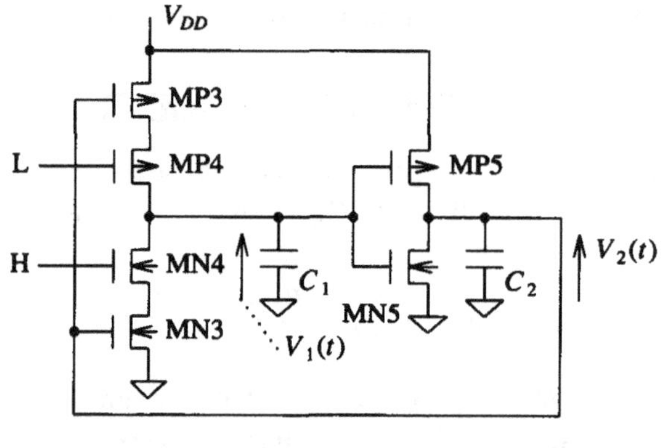

Figure 25 Analysis of setup time—II

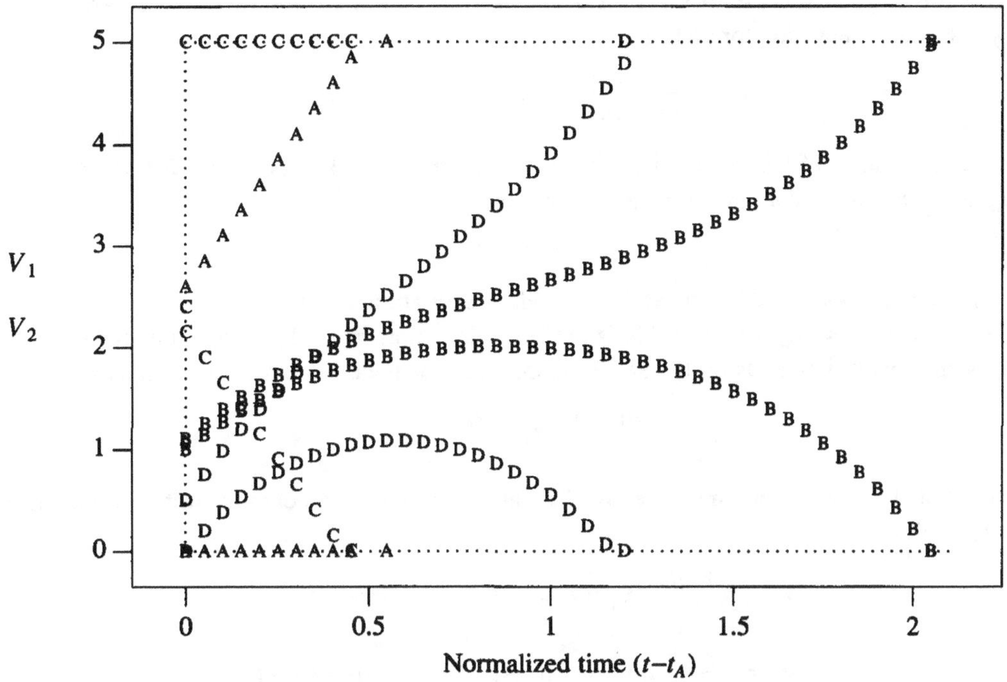

Figure 26 Node waveforms of latch

$$V_1(t)=V_1(t_A)-\frac{B_1V_{DD}}{C_1}(t-t_A)\,(t_A<t<t_{CC})\quad V_1(t)=0\,(t>t_{CC}) \tag{11}$$

where

$$t_{CC}=t_A+(C_1/B_1V_{DD})V_1(t_A).$$

$V_1(t)$ and $V_2(t)$ are shown in Fig. 26 by symbols "C". The latch settles at state $Q=\text{High}$ $[V_2(\infty)=V_{DD}]$.

Case (E): This case requires no further study. The latch has settled already at $Q=\text{High}$ $[V_2(\infty)=V_{DD}]$.

Case (D): This case consists of two subcases. In the first subcase (D′) $V_{DD}>V_2(t_A)>V_{DD}/2$ and V_2 increases with time as

$$V_2(t)=V_2(t_A)+\frac{B_2V_{DD}}{C_2}(t-t_A)\,(t_A<t<t_{CD})\quad V_2(t)=V_{DD}\,(t>t_{CD}) \tag{12}$$

where

$$t_{CD}=t_A+(C_2/B_2V_{DD})[V_{DD}-V_2(t_A)]$$

The latch settles at $Q=\text{High}$ $[V_2(\infty)=V_{DD}]$ at $t=t_{CD}$. The other subcase, (D″), $[0<V_2(t_A)<V_{DD}/2]$, and case (B) require special consideration.

Case (B) and subcase of (D″): If $V_1(t)$ and $V_2(t)$ are both in the range between 0 and V_{DD}, the circuit equation for node 2 is

$$C_2 \frac{dV_2(t)}{dt} = 2B_2[\frac{V_{DD}}{2} - V_1(t)]. \tag{13}$$

If the capacitance of the internal node is negligible, the current of NFETs MN3 and MN4 in Fig. 25, I, is independent of $V_1(t)$, and is given by

$$I = B_1 V_2(t)$$

If $V_1(t) > V_{DD} - V_2(t)$, MN3 and MN4 are both in saturation, and both carry the same current $B_1 V_2(t)$. If $V_1(t) < V_{DD} - V_2(t)$, MN4 is in the triode region, and MN3 carries current $B_1 V_2(t)$. The same conclusion holds for the PFET chain. Then the node 1 equation is written as

$$C_1 \frac{dV_1(t)}{dt} = 2B_1[\frac{V_{DD}}{2} - V_2(t)] \tag{14}$$

The set of circuit equations are solved by the standard method subject to the initial condition at $t = t_A$ to obtain

$$V_1(t) = \frac{V_{DD}}{2} + X_1 exp[\lambda(t - t_A)] + Y_1 exp[-\lambda(t - t_A)] \tag{15}$$

$$V_2(t) = \frac{V_{DD}}{2} - \mu X_1 exp[\lambda(t - t_A)] + \mu Y_1 exp[-\lambda(t - t_A)]$$

where

$$\lambda = 2\left(\frac{B_1 B_2}{C_1 C_2}\right)^{1/2} \quad \mu = \left(\frac{C_1}{C_2}\frac{B_2}{B_1}\right)^{1/2}. \tag{16}$$

and where

$$X_1 = \frac{1}{2}\left[V_1(t_A) - \frac{1}{\mu}V_2(t_A) - \frac{V_{DD}}{2}(1 - \frac{1}{\mu})\right]$$

$$Y_1 = \frac{1}{2}\left[V_1(t_A) + \frac{1}{\mu}V_2(t_A) - \frac{V_{DD}}{2}(1 + \frac{1}{\mu})\right]$$

Typical cases of (B) and (D) are shown by symbols "B" and "D" in Fig. 26, respectively.

The initial conditions $V_1(t_A)$ and $V_2(t_A)$ can be represented by a point on the V_1-V_2 coordinate plane within the region, $0 \le V_1, V_2 \le V_{DD}$. It is convenient to track the node voltage evolution on the V_1-V_2 *phase* plane, as we did in Section 2.13. Time t is eliminated from Eq. (15) to obtain

$$\frac{[V_1(t) - (V_{DD}/2)]^2}{4X_1Y_1} - \frac{[V_2(t) - (V_{DD}/2)]^2}{4X_1Y_1\mu^2} = 1$$

This is a set of hyperbolas centered at $V_1 = V_{DD}/2$, $V_2 = V_{DD}/2$ as shown in Fig. 27. With increasing time, the phase point (V_1, V_2) moves on the hyperbola in the direction indicated by the arrows. Starting from any initial condition, $[V_1(t_A), V_2(t_A)]$, the node voltage evolution

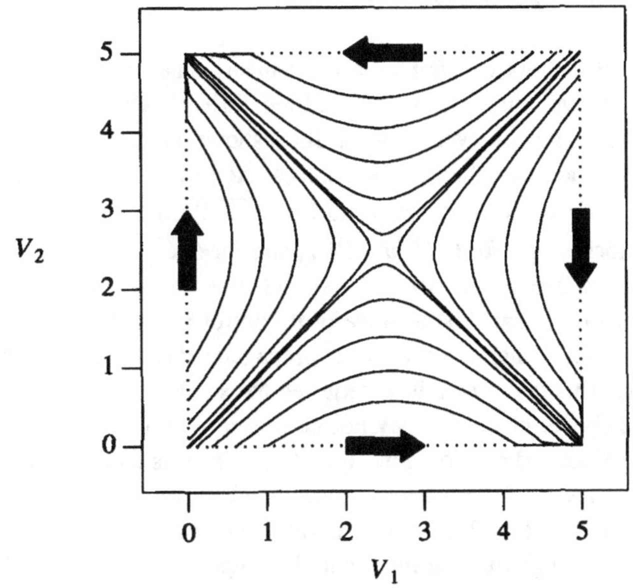

Figure 27 Phase trajectories of settling latch

can be easily understood. If a point on a hyperbola moves out of the region $0 < V_1, V_2 < V_{DD}$, the phase point moves along the edge of the square. In case (B) the phase point moves first inside the square and then along the edge, and in case (D″) only along the edge. The boundary point between the hyperbola and the edge corresponds to the microstate transition S → T of a FET in the latch circuit.

Equation (16) shows that the time constant for latch settling is related to the time constants of the component inverting gates by a simple relationship. The two inverters have time constants $t_1 = C_1/2B_1$ and $t_2 = C_2/2B_2$. They are the delay times of symmetrical inverter responding to a digital input signal making rapid transition. In a feedback circuit, the two time constants are combined to produce one *growing* and one *decaying* time constant. In Eq. (15) we note

$$t_{LATCH} = \frac{1}{2}\left(\frac{C_1 C_2}{B_1 B_2}\right)^{1/2} \approx \frac{1}{2}\frac{1}{2}\left(\frac{C_1}{B_1} + \frac{C_2}{B_2}\right) = \frac{t_1 + t_2}{2}$$

since geometrical and arithmetic averages of two numbers are close together if they have approximately the same magnitude. This means that t_{LATCH} is the sum of the delay times t_1 and t_2 multiplied by a factor of the order of unity. This demonstrates a fundamental character of a latch circuit: until a signal makes one round trip of the closed loop, no significant state change takes place.

5.7 Data Flow-through Problem

Sequential data-processing circuits can fail by a mechanism called data flow-through. This happens in a cascaded gate chain that makes a closed loop, through which a signal goes around more than once during a single clock cycle. Figure 28(a) shows a pair of CMOS D-latches: M (Master) and S (Slave) that are cascaded in the configuration of edge-triggered master-slave timing system. Two pairs of clocks $CK1$ ($\overline{CK1}$) and $CK2$ ($\overline{CK2}$) drive the pair. In ideal conditions, we use the same clock signal for $CK1$ and $\overline{CK2}$, and for $\overline{CK1}$ and $CK2$. A pair of complementary clocks are skew-free if the clock signal and its complement cross $V_{DD}/2$ at the same time, and the transition is fast. This waveform control guarantee that if one latch is transparent, the other is closed. Then a signal never propagates directly from node N_1 to node N_0 and back to node N_1. This is the condition required for stable sequential data processing. The conditions specified above, however, may not be met in real VLSI chips. Two pairs of clock waveforms could be skewed as shown in Fig. 28(b). There is an interval, T_0, from time t_M to t_S, when the two latches are simultaneously transparent.

The box labeled "Logic" in Fig. 28(a) shows multiple chains of combinational logic. If a logic chain is enabled (observing from the input and the output, the chain is *logically* equivalent to cascaded inverters), there is a closed loop that consists of the logic chain and the two D-latches. If T_0 is much longer than a typical gate delay, the signal is able to go around the loop more than once. Then the circuit may break into oscillation, as will be shown in Section 6.3 (odd number of inversions) or may settle at the state determined from initial conditions. In either case, the circuit fails to work correctly. This problem is called data flow-through across the master-slave latches. If a signal completes the loop once, it is often sufficient to create circuit failure.

Figures 29 and 30 show simple sequential logic circuits, and a schematic timing diagram of the circuits is shown in Fig. 31. In this timing diagram, painted bars show the periods of time when the respective latch is closed, and open sections show the periods when the latch is

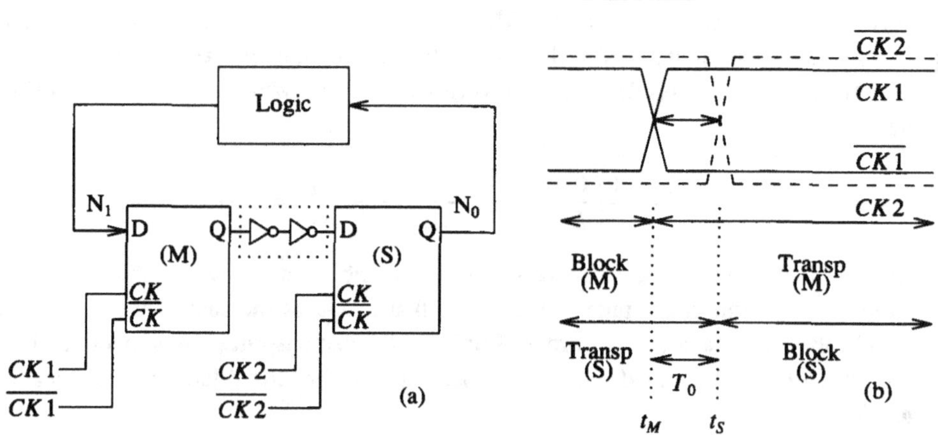

Figure 28 Data flow-through conditions

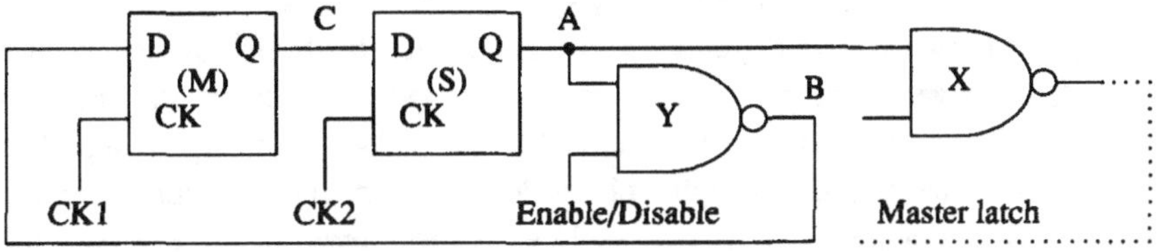

Figure 29 Data flow-through (failure)

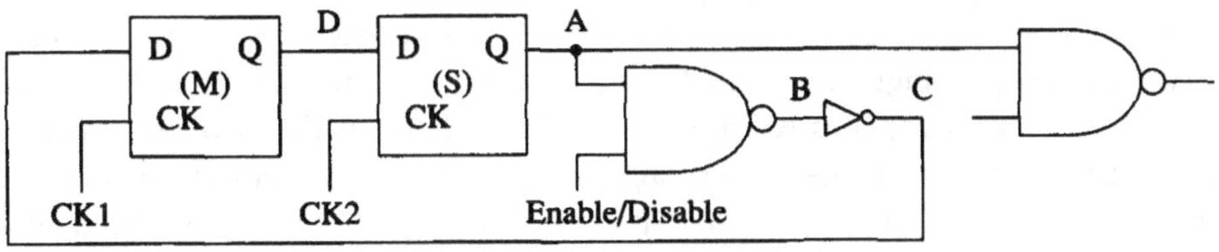

Figure 30 Data flow-through (no failure)

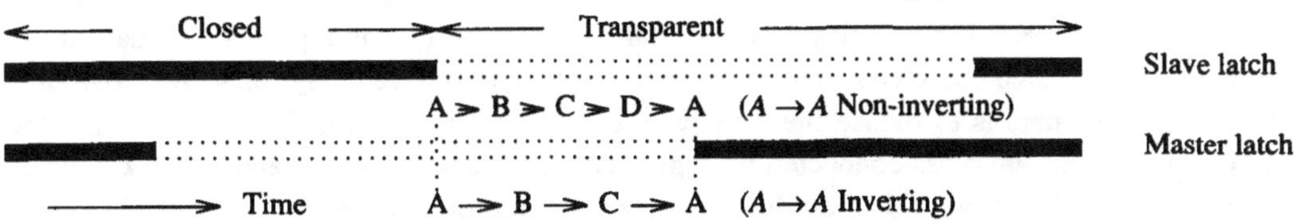

Figure 31 Open window in timing

transparent to the input signal. If the circuit is designed properly, the period of time when the master latch is transparent precedes the period of time when the slave latch is transparent, and there is no overlap of the two transparent periods. This requirement may not be satisfied in a VLSI chip. Signal processing of the subsystem shown in Fig. 29 is intended to be carried out during the period of time when the slave latch is closed and is retaining the previously captured data. The logic gate chain starting with gate X is long, and the signal processing takes most of the period of time when the slave latch is closed. The logic circuit includes a short chain of logic gates consisting of the single gate Y, and gate Y shares an input data with gate X. Suppose that node A of Fig. 29 became high at the time when the slave latch was clocked into the transparent state. This is the correct input data to the logic circuits. Gate Y is enabled, and Node B pulls down. At that time the master latch is still transparent and node C pulls down. The slave latch is still transparent, and therefore node A pulls down. Node A now has low logic level. At that time, the master latch closes, and the loop is broken. The slave latch, however, acquired wrong data: logic chain Y as well as logic chain X get the wrong input data. This sequence of events is shown schematically in Fig. 31 by sequence

A → B → C → A (inverting). If NAND2 gate A is disabled, the logic error does not take place. The circuit failure is data-dependent. Let us consider the logic circuit shown in Fig. 30. In this circuit, the signal that propagates along the short logic chain starting with gate A returns to node A while maintaining the same Boolean value. Even if the signal propagates through the master-slave latches through a closed loop, no logic error occurs. The sequence of events is shown by sequence A → B → C → D → A (non-inverting) in Fig. 31. In this case, the short loop including the two latches, NAND2 gate and inverter became a latch circuit that processed, captured, and stored correct data.

It is interesting to note that failure due to the data flow-through problem may be considered as a circuit configurational change (Section 2.3) induced by overlapping master and slave latch clocks. A circuit that includes an even number of inversions applies DC positive feedback in a closed-loop configuration, and the circuit becomes a latch. A circuit that includes an odd number of inversions applies AC positive feedback. In effect the loop circuit is converted into a ring oscillator, and as soon as the ring oscillator begins to oscillate, the digital circuit fails (this is of course an extreme case). Although this is a valid alternative classification of the failure, we consider this problem as a timing failure.

The data flow-through problem in an edge-triggered master-slave latch timing system can be alleviated by inserting delay between the master (M) and slave (S) latches, such as that shown in Fig. 28(a). The delay element is the two-stage cascaded inverters in the dotted box between the latches. If this delay is more than clock skew (T_0 of Fig. 28), the data flow-through is eliminated completely, since the latches themselves have delays as well. Obviously what was done here is to reduce the frequency of the ring oscillation. If the master and the slave latch clocks are cross-connected by highly conductive metal wires, and if clock rise/fall times are short, insertion of a small delay between the master and the slave latches eliminates the data flow-through problem practically completely. There is yet another technique, which will be discussed in the next section, to make this delay insertion more effective.

To maintain short rise-fall time clocks is crucial in preventing data flow-through. Whether or not the clock signals to the master and to the slave latches overlap can be determined only if the times of clock transitions are accurately determined. They are the times when the clock signal voltages cross V_{GSW}. Because V_{GSW} of multi-input gates like a tristatable inverter depends on both clock signal voltage and the input data voltage, V_{GSW} has *essential uncertainty* (Section 2.6). If rise/fall times of the clock signal are not fast, the clock edges for the master and for the slave latch clocks cannot be determined accurately. In such a *fuzzy* clock, whether or not an overlapping period of transparency exists cannot be determined accurately.

The best way to prevent data flow-through is to adopt the two-phase, nonoverlapping master-slave latch timing of Fig. 10. One fundamental cycle of the chip is divided into four phases, phase 1, phase 2, phase 3, and phase 4. The master latch is transparent only during phase 3 and the slave latch is transparent only during phase 1 (or vice versa). Existence of two wide *dead* phases, phase 2 and phase 4, guarantees that the two latches are never transparent simultaneously, even if significant clock skew exists. This version of two-phase, nonoverlapping clock system is the safest choice.

5.8 Data Flow-through and Logic Polarity

Data flow-through of CMOS circuits can be partially alleviated by selecting the proper logic polarity. The technique can be explained using the following examples. With reference to Fig. 32, the clock pull-down (downward arrow 1) precharges the output node of the dynamic NAND2 gate. We consider that the dynamic logic circuit and the tristatable inverter make a master-slave latch. This is the issue of data flow-through, since the dynamic logic circuit may be considered as the combination of logic gate and data storage circuit consisting of the output node capacitor. Pull-up of the output node by PFET, MP1, (upward arrow 2) destroys the *processed* data that has been retained by node 2. If the pull-up transition is transmitted to the tristatable inverter, output node 3 pulls down (arrow 3), and the data stored there is destroyed as well.

If the dynamic gate and the tristatable inverter share the same (down-going) clock as shown in Fig. 32 (note that there is no inverter between MP1's gate and MN1's gate), however, tristatable inverter pull-down can be prevented, since MN1 turns off as soon as MP1 turns on. Such *sharing* of clock signals often eliminates data flow-through [11].

A similar technique works for the cascaded CMOS static D-latches shown in Fig. 33. Latch A and latch B make an edge-triggered master-slave pair. Suppose at first that CKA and $\overline{CKB}$ are low and $\overline{CKA}$ and CKB are high, and therefore latch A is closed and latch B is transmitting. Suppose that input D is high, and latch A has $\bar{Q} = high$ and $Q = low$. When $\overline{CKA}$ makes a high-to-low transition, the state of latch A is unaffected. The state of the latch is changed only when CKA makes a low to high transition. When this transition takes place the input signal at D pulls node $\bar{Q}$ of latch A down (arrow 1), node Q of latch A up (arrow 2) and node D of latch B (arrow 3) down. Pull-up by MP2 is blocked, however, by the series-connected PFET, MP1, that receives an up-going clock signal $\overline{CKB}$ that is synchronized to CKA. Therefore, flow-through of data through the two cascaded latches is prevented, even if

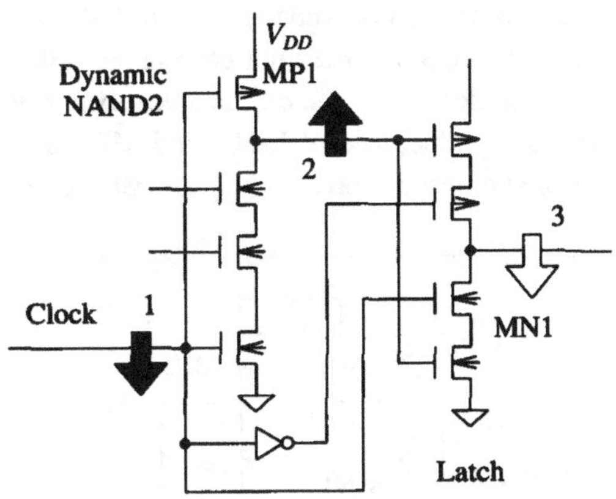

Figure 32 Conditions of data flow-through and signal transition polarity

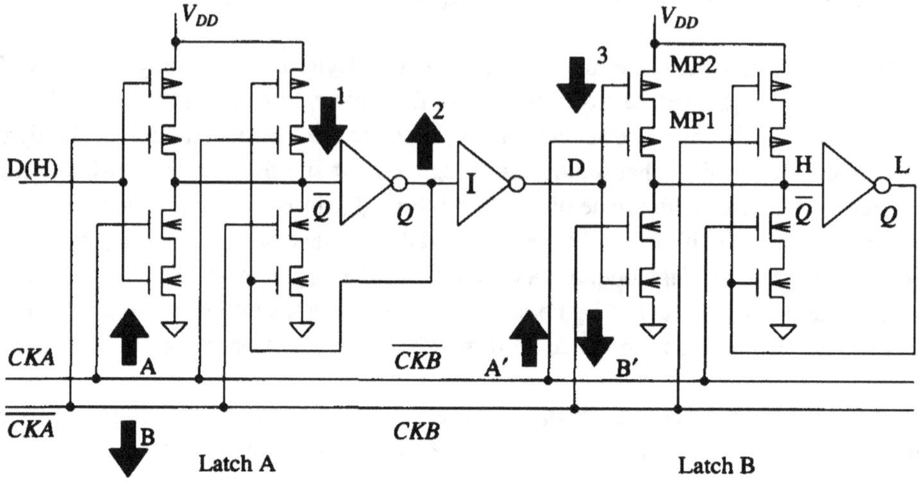

Figure 33 Transition polarity and data flow-through

CKA and $\overline{CKA}$ (or $\overline{CKB}$ and CKB) are not perfectly synchronized. For this preventive mechanism to work, inverter I is crucial. The inverter provides proper polarity at the input of the slave latch. If inverter I is missing, CKA and $\overline{CKA}$ must be well synchronized in addition to CKA and $\overline{CKB}$, and $\overline{CKA}$ and CKB. It is important to note that the technique works only if the rise/fall times of the clock signal are significantly shorter than typical *loaded* delay time of the inverter.

It may be surprising that the data flow-through problem depends on the seemingly unrelated number of inversions between the master and slave latches. This peculiar effect originates from a symmetry property of complementary circuits. We show a general and illustrative example. Figure 34 shows a two-stage directly coupled complementary amplifier. If load resistors R_N and R_P change arbitrarily, V_{OUT} changes even if V_{IN} is held constant. If R_N and R_P changes are correlated, however, V_{OUT} may not change, as follows. This is a large-signal DC analysis. The FETs are modeled using the collapsible current generator model with zero threshold voltage. Let the transconductance of MNA and MPA be B_N and B_P, respectively. We find that input voltage V_{IN} and output voltage V_{OUT} are related by

---

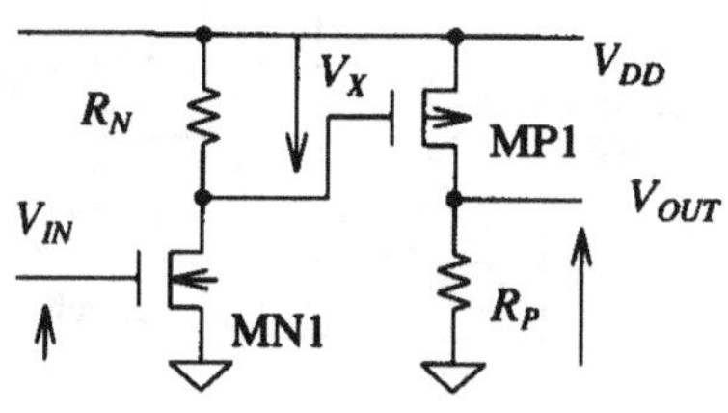

Figure 34 Symmetry of complementary circuit

$$V_{OUT} = B_P R_P V_X = (B_P B_N)(R_P R_N) V_{IN} . \quad (17)$$

This equation shows that the output voltage of the amplifier remains unchanged if $R_P R_N$ is constant: R_P and R_N may vary, but as long as their product remains unchanged, the output voltage remains unchanged. In the circuits of Fig. 32 the clock signal causes approximately this kind of variation of the effective resistance of the *load* devices MP1 (corresponding to R_N of Fig. 34) and MN1 (corresponding to R_P of Fig. 34). R_N decreases and R_P increases after the clock transition. Ideally node 3 voltage does not change by clock transition.

The examples of Figs. 32–34 suggest a useful concept of circuit design. The cause of circuit parameter variation need not be voltage variation. If we want a circuit that is tolerant of process parameter variation, we design a circuit that is deliberately susceptible to the process parameter variation at several key locations of the signal path, and we optimize the circuit such that the cumulative effect of the parameter variation vanishes. This design concept provides significant flexibility in complementary circuits like CMOS. An example of this design will be discussed in Sections 5.14 and 5.15.

5.10 Hazard

Circuit failures that are due to recombination of split signals may be classified into two groups. If the difference in the delay times of the two (or more) signals is significantly more than typical rise/fall time of the signal, an extra pulse could be generated in the logic circuit. This is known as a *hazard*. The other is that the difference in delays is comparable to the rise/fall times. In this case, signal waveforms deform (creating a plateau or a hump), and many circuit delay parameters become uncertain.

When two input signals are combined by a gate, the output waveform depends on how the transition edges of the two input signals are aligned. Alignment matters in the following way. Figure 35(a) shows a CMOS gate and Fig. 35(b) shows the waveforms for input signals A and B and output signal Z. Signal A was originally high [state I of Fig. 35(b)]. Signal A makes a high-to-low transition, and in final state II, is low. Signal B is low in the initial state I, makes a low-to-high transition, and is high in the final state. Transitions of the two signals

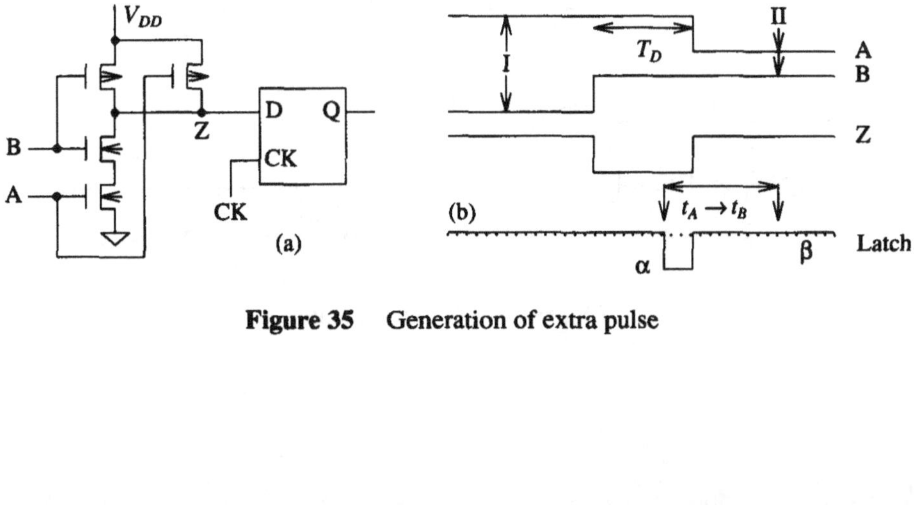

Figure 35 Generation of extra pulse

occur at time T_D apart. There is an interval of time when both input signals have a high logic level. If the gate has a short response time, the waveform of output node Z would be as shown by the third curve of Fig. 35(b). There is a down-going pulse having width T_D. What was described here is normal operation of NAND2 gate. Output of the logic gate is strobed into the destination latch at a specified time: Depending on when the answer is captured [in Fig. 35(b) two possibilities, a long slot from time t_A to t_B, and narrow slot at t_B are shown], the output waveform of the latch would be as shown by curves α (solid) or β (dotted), respectively. If the latch clock is narrow at t_B, the isolated pulse at node Z never goes through the latch. We recognize that a properly clocked latch prevents an isolated pulse of Fig. 35(b) propagating from the gate. The clocked latch has an important function of *cleaning-up* and reshaping signal waveforms. If the clocked latch is removed, the effect of the down-going pulse spreads out, and it may cause logic circuit failures.

The down-going pulse was generated because the NAND2 gate output was pulled down by the first input signal transition (B) and then pulled up by the second input signal transition (A). The cumulative effect of the two input signal changes was to maintain the output node at the same (high logic level), which is the correct answer. Whether or not the output node stays at the high logic level *all the time* depends on the details of the timing of the two input transitions. Since the pulse width equals the interval between the two transitions, the extra pulse is not generated if the transition edges of the input signals A and B are exactly coincident, or if the down-going edge of the signal A is earlier than the up-going edge of signal B. In a more complex gate than a NAND2, more than two signals may change within a single clock cycle, and the signals may change any time within the cycle. Then more than one *extra pulse* can be generated. If more than one signal changes state within a single clock period, generation of extra pulse is inevitable. This problem is caused because N-input logic gate carries out $2^N \to 2$ mapping: If more than one variable changes, the process of switching depends on the order the variable changes.

If several gates are assembled into a logic circuit, even the change of a single input signal may generate an extra pulse, or glitch. An example is the *glitch* generator circuit shown in Fig. 36. A circuit of this type generates complex output waveforms (for example, see Fig. 40), even if the circuit is quite simple. Then, because of the misleading advantage of generating complex waveforms by a simple circuit, the circuit is quite often *misused*.

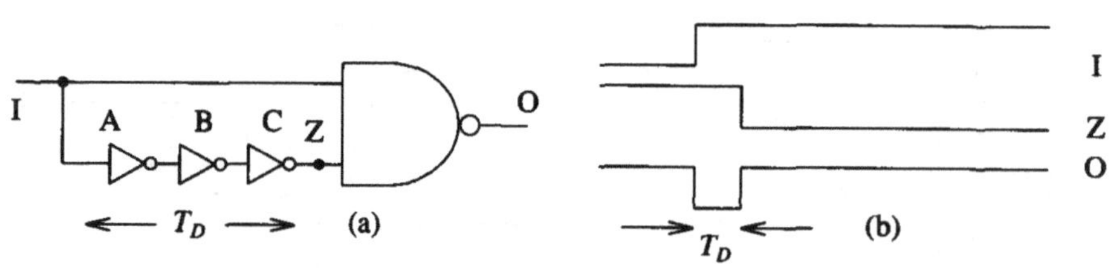

Figure 36 Glitch generator

In Fig. 36(a), delay of the NAND2 gate is assumed to be negligible. Cascaded inverters have delay and input signal I is delayed by T_D. When signal I and its delayed version Z are NANDed, a down-going glitch having width T_D is generated every time input signal I makes a low-to-high transition. Conventionally the problem of Fig. 36 is called *hazard* if the extra pulse generated by the circuit is not desirable. In this example, switching of a single input signal generates the glitch. This point is ambiguous, however, since signals A and B of Fig. 35(a) may be regarded as generated from a *single* clock signal that drives their respective source latches. Therefore we cannot take the criterion of *single* input change too literally.

From these examinations the following conclusion is drawn. The extra pulse or glitch generation appears strange because two levels of observations are applied to a logic circuit, one being more precise than the other. What is considered to be a *simultaneous* signal transitions to *coarse* observation (logic operation) is not simultaneous for *precise* observation (circuit level observation that is able to resolve time of the order of a gate delay), and what is considered to be a *single* transition of logic variables for coarse observation is really a sequence of voltage transients to a precise observation. A logic circuit consisting of several gates may be considered as a single equivalent logic gate for coarse observation, but it is a complex FET-level circuit for precise observation. As such, the phenomena of glitch generation is deeply rooted in the way the logic circuits are implemented.

Figure 37 shows the microstate sequence of the glitch generator shown in Fig. 36(a). As we saw in the example, hazards are a gate-level phenomenon, and therefore we need not describe FET-level details. The abbreviated symbols of the gate level for the microstate sequence have been introduced in Section 1.16. In Fig. 37 the direct response of the NAND gate to the up-going transition of input signal I is shown by branch S, and the delayed action through the chain of inverters by branch D. The relative timing between the events that belong to the different branches affects glitch generation.

We study several examples of glitch generation, or hazard, to obtain more familiarity with interesting effect. Figure 38 shows generalization of the circuit already shown in Fig. 36(a): the second input of the NAND2 gate may have an even or odd number of inverters. If the number of inversions is odd, a glitch is generated following the up-going edge of signal A, as

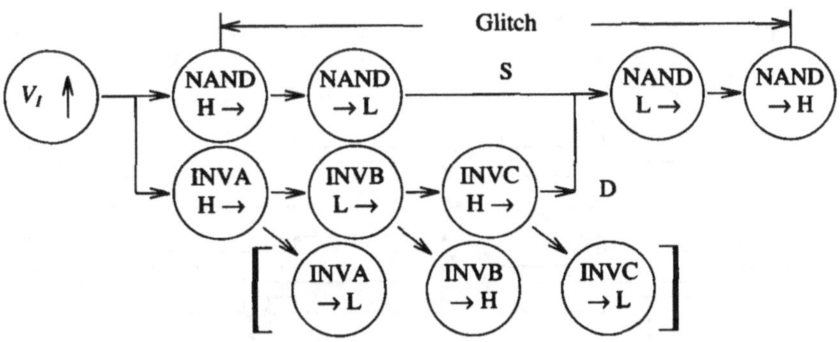

Figure 37 Microstate sequence of glitch generator

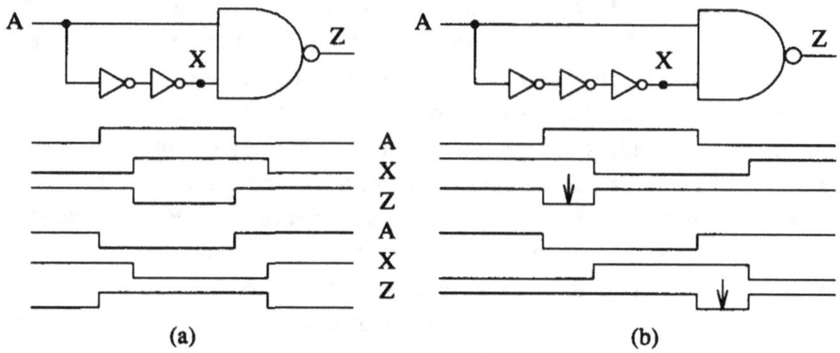

Figure 38 Extra pulse generation by NAND2 gate

shown by the arrows (this is the case of Fig. 36). If the number of the inversion is even, however, the width of the down-going output pulse at node Z is less than the width of the up-going input pulse A by the delay time of the cascaded inverters. If the input pulse is down-going, the width of the up-going output pulse increases. This mode of operation of the similar circuit as a pulse width modulator has not been utilized, but the circuit appears useful. There are many subsystems that require this function.

Figure 39 shows a similar circuit that has a NOR2 gate instead of a NAND2 gate. A glitch is generated following down-going edges of input signal A, in contrast to the up-going edge of the input signal.

The circuits of Fig. 40 are cascaded parity circuits that are driven by successively more delayed versions of a single input signal, A. If the parity circuit has N-inputs, N state transitions are generated. If N is even, the state of the output node Z before the input signal transition and the state a long time after the transition are the same. If N is odd, however, the initial and the final output states are different, and glitches are superposed to the step function of the output voltage. This type of glitch, shown in Fig. 40(a), is called a *dynamic* hazard (since the

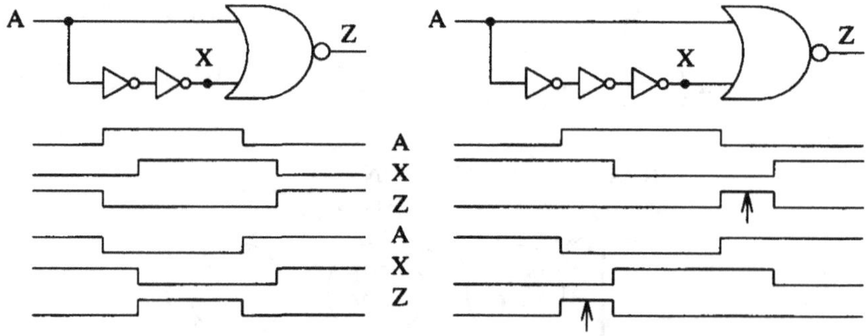

Figure 39 Extra pulse generation by NOR2 gate

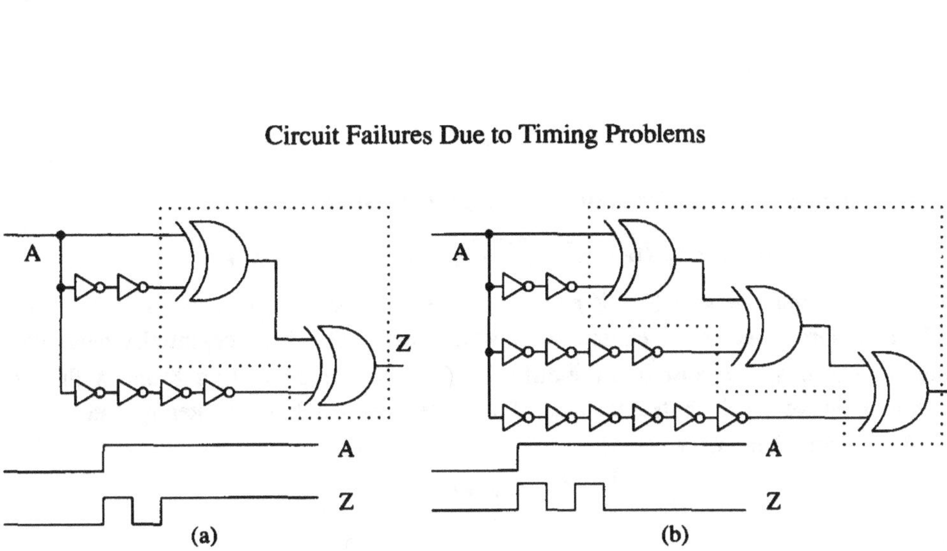

Figure 40 Hazard, second example

output makes transition), and those of Figs. 38, 39 and 40(b) are called *static* hazard (since the output stays unchanged).

5.11 Signal Recombination Creating Hazards

In high-performance CMOS systems, hazards are sometimes used to generate complex logic signals, especially clocks. An example is the circuit shown previously in Fig. 36: The circuit is convenient to generate isolated clock pulses. Whether or not a hazard is harmful depends on the designer's intention. If a hazard is not wanted, however, it is necessary to have a systematic method of detecting them. Whether or not a hazard exists depends on logic design, as well as circuit design. A classic example is shown in Fig. 41. The circuit works as follows: the NOR of signals W and X (signal A) and the NOR of signals $\overline{X}$ and Y (signal B) are generated first, and then the two signals A and B are NORed together. If signals W and Y are both low, the two NOR gates are enabled—they are equivalent to inverters—and the circuit is essentially equivalent to the single shot circuit of Fig. 36(a). The circuit has hazard. Our problem is how to detect the hazard *systematically*. The following analysis follows closely the book published by McCluskey [12].

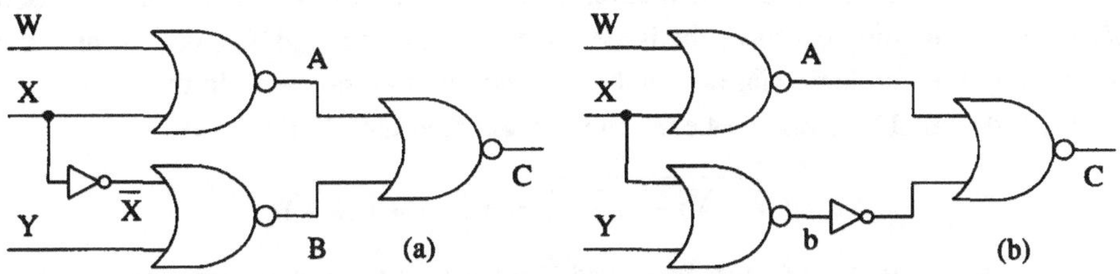

Figure 41 Hazard detection by logic design analysis

The output signal of the circuit of Fig. 41(a) is expressed as

$$C(\text{general}) = \overline{\overline{(W+X)} + \overline{(\bar{X}+Y)}} = W\bar{X} + WY + X\bar{X} + XY\,. \tag{18}$$

In this Boolean equation, variables X and $\bar{X}$ were regarded as *independent* variables. They do not take complementary values except in the steady state. This is an essentially new detail of McCluskey's analysis, because it is the difference in delays in generating $\bar{X}$ and X that create the glitch. First we consider steady states of the gate using Eq. (18). In steady state, variables X and $\bar{X}$ take complementary values.

$$X\bar{X} = 0$$

and therefore

$$C(\text{steady state}) = W\bar{X} + WY + XY\,. \tag{19}$$

We have

$$C(\text{general}) = C(\text{steady state}) + X\bar{X}\,. \tag{20}$$

We seek the condition when the *extra* term $X\bar{X}$ in Eq. (20) generates a high logic level temporarily during the signal transition. Since $X\bar{X}$ may become 1, which makes the difference, the steady state output of the circuit C (steady state) must remain at 0. We seek a pair of steady states of input variables, one of which is the initial and the other is the final state that satisfy the following criteria:

1) both states give 0 to output signal C after the steady states are reached;
2) the value of variable X contained in the first set of values of variables specifying the initial state is the complement of the value of X contained in the second set of values of variables specifying the final state;
3) all the other logic variables maintain the same values, before and after the transition.

We seek the initial and the final states. From Eq. (19) we find that $W=1$ is not acceptable, since we have to make the term $W\bar{X}$ vanish independent of $\bar{X}$. If $W=0$, then the term XY must be zero irrespective of X. Then we must have $Y=0$. Then the two steady states are

State 1: $W=0$, $X=1$, and $Y=0$
State 2: $W=0$, $X=0$, and $Y=0$

A glitch is generated if the transition between the two states 1 and 2 takes place and if the signal delay allows that both variables X and $\bar{X}$ become simultaneously 1 during the transient.

We point out that there is a certain ambiguity in identifying what is meant by variables X and $\bar{X}$, and this ambiguity confuses the issue. In the circuit of Fig. 41(a), both X and $\bar{X}$ can be clearly identified on the logic diagram, but this is not always the case. In the circuit shown in Fig. 41(b), node variable $\bar{X}$ does not exist in the logic diagram. In this case

$$C = \overline{\overline{(W+X)} + \overline{(\overline{\overline{X}}+Y)}} = (W\bar{Y})\bar{X} + \bar{Y}(X\bar{X})\,. \tag{21}$$

In order to have a hazard, $W\bar{Y}$ must be 0, and $\bar{Y}$ must be 1 in order to make term $X\bar{X}$ *visible*. Therefore we must have $W=0$ and $Y=0$. Although the output node of NOR2 gate B gives $\bar{X}$

under this restricted condition, in general the node represents logic variable $\overline{X+Y}$. Whether or not X and $\overline{X}$ exist explicitly on the logic diagram, or in which order one of them is generated from the other does not matter. Whichever the case, the existence of a term like $X\overline{X}$ in the Boolean expression (if other Boolean variables like W, Y, ..., are properly chosen) indicates that variable X was split, processed separately and then combined.

Figure 42(a) shows a circuit that has a *dynamic* hazard. In analyzing dynamic hazards, all the signals that are split from the single original signal [like signal X of Fig. 42(a)] must be labeled separately. The circuit generates signal C that has the following Boolean expression

$$C=\overline{X_1\overline{(X_2Y)}+\overline{\overline{X_3}Z}}=Z\overline{X_1}\,\overline{X_3}+(YZ)X_2\overline{X_3} \tag{22}$$

where X_1, X_2 and X_3 are the signals split from the original signal X. When steady state is reached, $X_2\overline{X_3}=0$, and therefore we have

$$C(\text{steady state})=Z\overline{X_1}\,\overline{X_3}\rightarrow Z\overline{X} \tag{23}$$

where all split signals tend to the same signal, or X_1, X_2, and $X_3 \rightarrow X$ as the steady state is reached. From Eq. (23), signal C makes a high-to-low transition as signal X makes a low-to-high transition. This transition should be *enabled.* Then we must have $Z=1$. In Eq. (22), the second term, $X_2\overline{X_3}$, is zero for both the initial and the final steady states that have different Boolean values, but it can be 1 during transition if $X_2=1$ and $X_3=0$ takes place. For this term to be *visible*, $YZ=1$ is required in Eq. (22). We then have

$$C=\overline{X_1}\,\overline{X_3}+X_2\overline{X_3}. \tag{24}$$

Let us consider timing of this dynamic hazard. Glitch generation occurs as follows. Suppose that X_1 makes a low-to-high transition at time $t=t_1$, X_2 at $t=t_2$, and X_3 at $t=t_3$ such that $t_1<t_2<t_3$. This means that signal path delays are longer in the order of path 1, 2, and 3. Then we have

$$C=1\ (t<t_1)\quad =0\ (t_1<t<t_2)\quad =1\ (t_2<t<t_3)\quad =0\ (t_3<t)$$

and a dynamic hazard is generated during time $t_2<t<t_3$. In order to have a dynamic hazard, the logic expression must have a term like $\overline{X_1}\,\overline{X_3}$ of Eq. (22) that switch state if the input signal changes state, and a second term of the form $X_2\overline{X_3}$ that creates a glitch during the switching process. In order for the temporary 1 of $X_2\overline{X_3}$ to be observable, $\overline{X_1}\,\overline{X_3}$ must have finished its

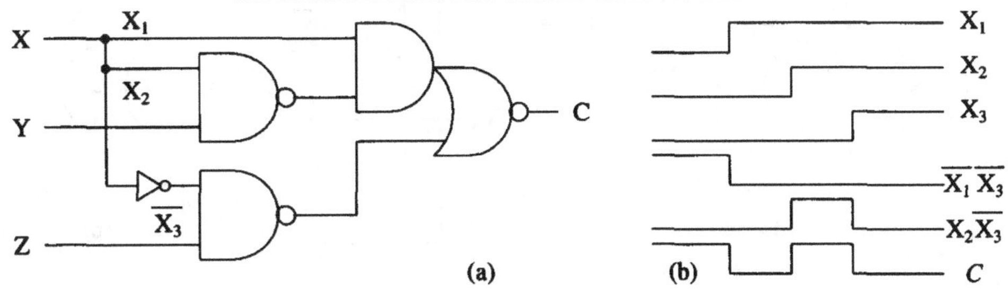

Figure 42 Dynamic hazard detection

transition before the glitch pulse is generated ($t_2 > t_1$). The node waveforms are shown in Fig. 42(b).

Figure 43 shows that the logic function of Fig. 41(a) can be implemented using a different configuration of gates. The circuit of Fig. 43(b) generates logic variable C given by

$$C = \overline{\overline{(W+X)} + \overline{(\overline{X}+Y)} + \overline{(Y+W)}} = (W+X)(\overline{X}+Y)(Y+W) \qquad (25)$$
$$= (Y+W)X\overline{X} + (W\overline{X} + WY + XY)$$

and the two logic circuits are identical in the steady states, but they are different during signal transitions. Because of the term $(Y+W)$ multiplied by $X\overline{X}$, $(Y+W)X\overline{X}$ remains at 0 if $Y+W=0$. Then there is no hazard. The circuit of Fig. 43 is called a *de-glitched* logic circuit. In logic gate circuits, certain difficulties that occur only in rare sets of input values can be resolved, case by case, by adding extra logic circuits, and this is an example. Several examples were discussed in Section 3.2 (diodes in logic gates) and in Section 3.3.

Signals X and $\overline{X}$ are regarded as independent signals in the hazard analysis, but how they vary with time and when they both become 1 depend on the details of the circuit. In Fig. 43(a) positive-going glitch pulse is generated if both signals A and B become low. If the delay of NOR2(A) is longer than that of the NOR2(B)-inverter combination, nodes A and B can be simultaneously low after the *down-going* transition of signal X. If the delay of NOR2(A) is shorter than that of the NOR2(B)-inverter combination, signals A and B become simultaneously low after the *up-going* edge of input signal X. As we observe from this example, the hazard can be controlled by adjusting gate delays. Figure 44 shows the timing diagram of the signals for the circuit shown in Fig. 43(a). Signal $\overline{X}$ is delayed with respect to signal X by delay time T_X (inverter delay time). Signal A is delayed with respect to signal X by T_A (NOR2 gate A delay time) and so forth. Since signal A and signal B can both be low, as shown by waveforms A and B of Fig. 44, an up-going glitch shown by waveform C is generated. If this circuit was redesigned so that the down-going edge of gate A is delayed as shown by broken curve α, the glitch would be eliminated.

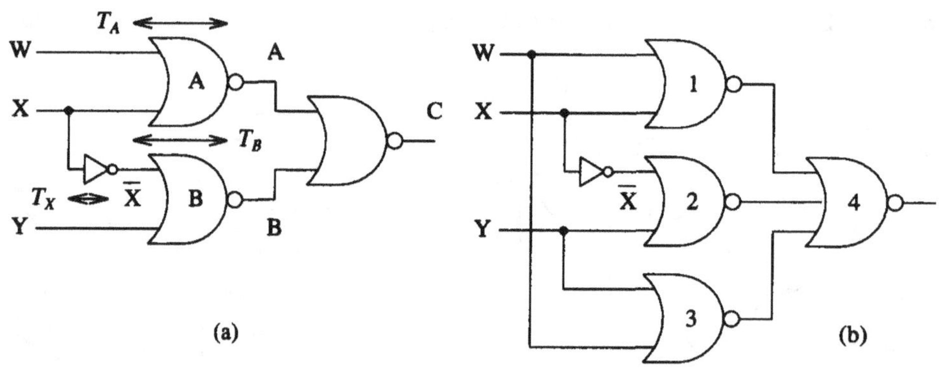

Figure 43 Hazard-free logic design

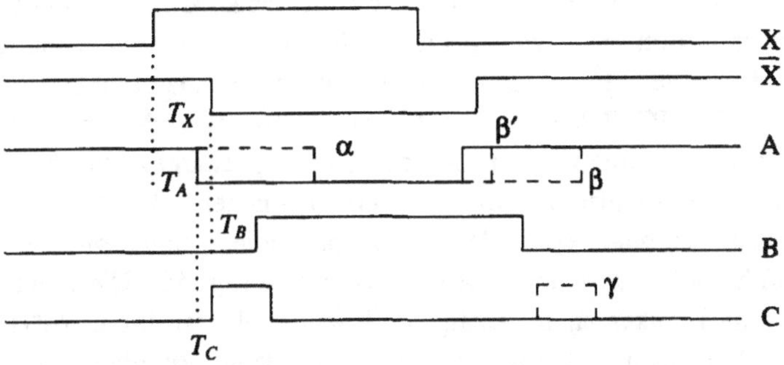

Figure 44 Elimination of hazard by timing design

A problem of this technique is that the up-going edge of gate A should not be delayed significantly. If the up-going edge is delayed as much (dashed curve β), the up-going glitch γ appears every time when signal X makes a high-to-low transition. Therefore gate A must be designed such that gate pulls down slowly (curve α) but pulls up rapidly (curve β'). Techniques of designing a gate having asymmetrical delays by adjusting FET sizes (large PFET and small NFET) are now well established. A simple design technique of eliminating glitches is to provide heavy capacitive load to gate C, so that gate C is unable to respond to narrow glitch pulse, but this is not a preferred method.

5.11.1 Internal Hazard

As we studied before, a hazard can be created if a logic variable and its time-shifted version are combined by a logic gate. If we generalize, the time-shifted version can be the *logic variable* stored in the internal node of the gate (Section 1.10). Then a hazard is observed even within a single inverting gate. Let us study an example. Figure 45(a) shows a static NAND2

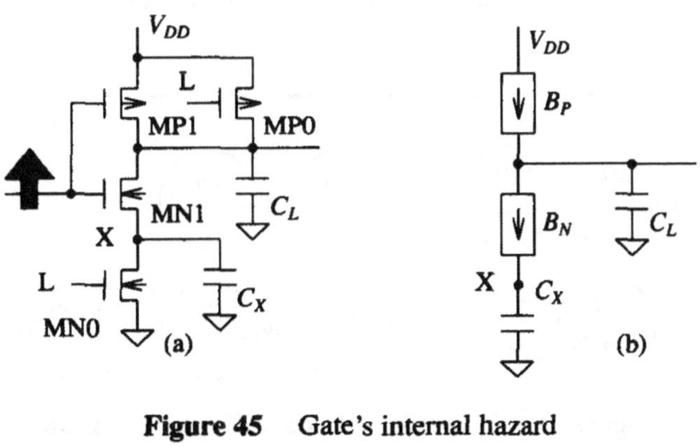

Figure 45 Gate's internal hazard

gate. Suppose that internal node X has large parasitic capacitance C_X, and C_X is completely discharged to ground. Both input voltages are logic low. If the input signal to MN1 makes a low-to-high transition, the output node is pulled down, as the charge of capacitor C_L is shared with capacitor C_X. The PFET MP0 may not be conductive enough to hold the output node at V_{DD} if MN1 conducts a large current. As capacitor C_X charges, this output node voltage recovers to V_{DD}, but before that, a down-going glitch is generated at the output node. This is quite similar to a conventional hazard. We call the phenomena an *internal* hazard.

The internal hazard microstate sequence is shown in Fig. 46. The input voltage to MN1 and MP1 [Fig. 45(a)] makes an instantaneous low-to-high transition. MN1 turns on, MP1 turns off, and MP0 moves from the triode region to the saturation region, in rapid succession. If MP0 remains in the triode region, a glitch is not generated: the no-glitch sequence of Fig. 46. The events that follow have two possibilities. If capacitance C_X is small, MP0 eventually returns to the triode region and MN1 approaches the marginally conducting region M (sequence A of Fig. 46). If MN1 is large and MP0 is small, however, MN1 may go into the triode region and the output node and internal node X fuse. As V_X increases, the fused node breaks apart and the FETs go into the same final state as sequence A (sequence B of Fig. 46). In this sequence we note that event MN1: T $\rightarrow$ S precedes event MP0: $S \rightarrow T$ because otherwise node X is charged up to V_{DD}. Event MN1: S $\rightarrow$ M is the last, since it takes an infinitely long time to reach the final, marginally conducting state.

The case of microstate sequence A can be analyzed as follows. The circuit equations are

$$C_L \frac{dV_0(t)}{dt} = B_P(V_{DD} - V_{THP}) - B_N[V_{DD} - V_X(t) - V_{THN}] \tag{26}$$

$$C_X \frac{dV_X(t)}{dt} = B_N[V_{DD} - V_X(t) - V_{THN}]$$

where we assume that FET threshold voltages are independent of the source voltages. The equations are solved subject to the initial condition

$$V_X(0)=0 \quad V_0(0)=V_{DD}$$

to obtain

$$V_X(t)=(V_{DD} - V_{THN})[1 - exp[-(B_N/C_X)t]]\,. \tag{27a}$$

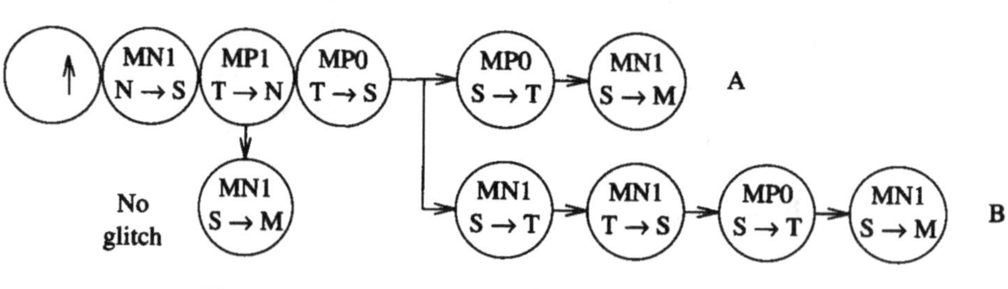

Figure 46 Microstate sequence of internal hazard

$V_X(t)$ increases monotonically with time, and approaches $V_{DD}-V_{THN}$. We obtain $V_0(t)$ satisfying the initial condition as

$$V_0(t)=V_{DD}+\frac{B_P}{C_L}(V_{DD}-V_{THP})t-\frac{C_X}{C_L}(V_{DD}-V_{THN})[1-exp[-(B_N/C_X)t]]\,. \qquad (27b)$$

If $t \to 0$

$$V_0(t)=V_{DD}-(1/C_L)(I_{Nmax}-I_{Pmax})t$$

For an internal hazard to take place, the coefficient of time, t, must be negative. We must have

$$I_{Nmax}>I_{Pmax} \text{ where } I_{Nmax}=B_N(V_{DD}-V_{THN}) \;\; I_{Pmax}=B_P(V_{DD}-V_{THP})$$

$V_0(t)$ decreases, attains a minimum, and then increases again. At time $t=t_1$, $V_0(t)=V_{DD}$ again. Time t_1 is determined as

$$t_1=(C_X/B_N)\theta(\kappa_1, 1)$$

where

$$\kappa_1=(I_{Pmax}/I_{Nmax})=B_P(V_{DD}-V_{THP})/B_N(V_{DD}-V_{THN})$$

and where $\theta(\kappa, \chi)$ is the θ-function defined in the Appendix of chapter 2. Waveforms $V_X(t)$ and $V_0(t)$ are schematically shown in Fig. 47(a).

In the case of sequence B of Fig. 46, the two curves $V_0(t)$ and $V_X(t)$ intersect at $t=t_2$, where t_2 is given by

$$t_2=(C_X/B_N)\theta(\kappa_2, \chi)$$

and where

$$\kappa_2=\frac{C_X}{C_X+C_L}\frac{I_{Pmax}}{I_{Nmax}} \qquad \chi=\frac{C_X(V_{DD}-V_{THN})-C_LV_{THN}}{(C_X+C_L)(V_{DD}-V_{THN})}\,.$$

At $t=t_2$ the two nodes fuse. The circuit equation changes to

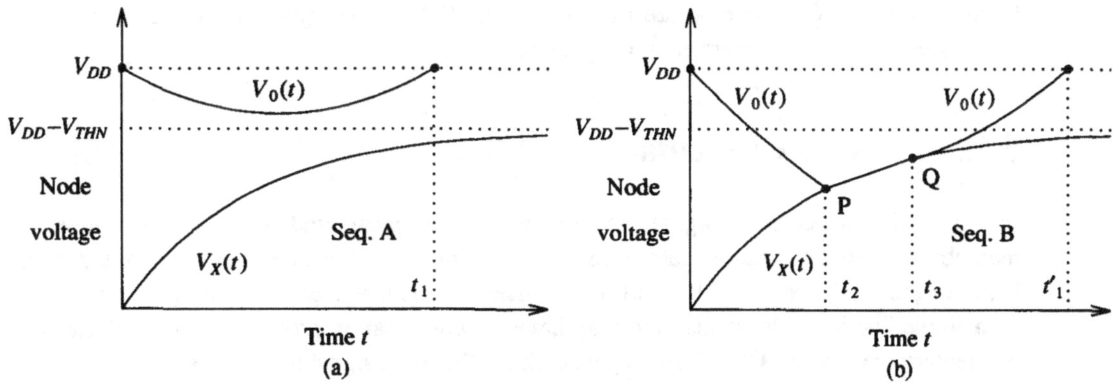

Figure 47 Internal hazard node waveforms

$$(C_X + C_L)\frac{dV_0(t)}{dt} = B_P(V_{DD} - V_{THP}) \quad V_0(t) = V_X(t)$$

subject to the initial condition at $t = t_2$. We have

$$V_0(t) = V_X(t) = V_0(t_2) + \frac{B_P}{C_X + C_L}(V_{DD} - V_{THP})(t - t_2)$$

With reference to Fig. 45(b), the current in MN1 equals $C_X dV_0(t)/dt$. If this current becomes higher than the maximum current for MN1 to stay in the triode region

$$\frac{C_X}{C_X + C_L} B_P(V_{DD} - V_{THP}) > B_N[V_{DD} - V_X(t) - V_{THN}]$$

the two nodes break apart. This takes place at time $t = t_3$ given by

$$t_3 = t_2 - \frac{C_X}{B_N} + \frac{C_X + C_L}{B_P}\,\frac{V_{DD} - V_{THN} - V_0(t_2)}{V_{DD} - V_{THP}}.$$

After $t > t_3$, the same solutions as in the interval $0 < t < t_2$ with different initial conditions apply. We have

$$V_0(t) = V_0(t_3) + \frac{B_P}{C_L}(V_{DD} - V_{THP})(t - t_3) \tag{28a}$$

$$-\frac{C_X}{C_L}(V_{DD} - V_{THN})[1 - exp[-(B_N/C_X)(t - t_3)]]$$

$$V_X(t) = V_0(t_3) + [V_{DD} - V_{THN} - V_0(t_3)][1 - exp[-(B_N/C_X)(t - t_3)]] \tag{28b}$$

At time $t = t_1'$, $V_0(t)$ reaches V_{DD}. Time t_1' is given by

$$t_1' - t_3 = \frac{C_X}{B_N}\theta(\kappa_1, \chi)$$

where

$$\kappa_1 = I_{Pmax}/I_{Nmax} \qquad \chi = \frac{C_L[V_{DD} - V_0(t_3)] + C_X(V_{DD} - V_{THN})}{C_X(V_{DD} - V_{THN})}.$$

Node waveforms for this case are plotted in Fig. 47(b). The $V_0(t)$ waveform shows that a down-going glitch was generated at the gate output.

5.12 Overlap Currents

In this section we consider signal *recombination* at the most fundamental level. The signal recombination that we studied before assumes two or more independent logic signal paths on a logic diagram. The problem that will be discussed here is that even a *single* signal represented by a single line in the logic diagram may have a recombination problem because of the complementary structure of CMOS static gates. Since the input signal to a CMOS static gate must control both the NFET and PFET, signal recombination takes place even within a single CMOS static gate. The simplest example is shown in Fig. 48. Note that load capacitance C_L

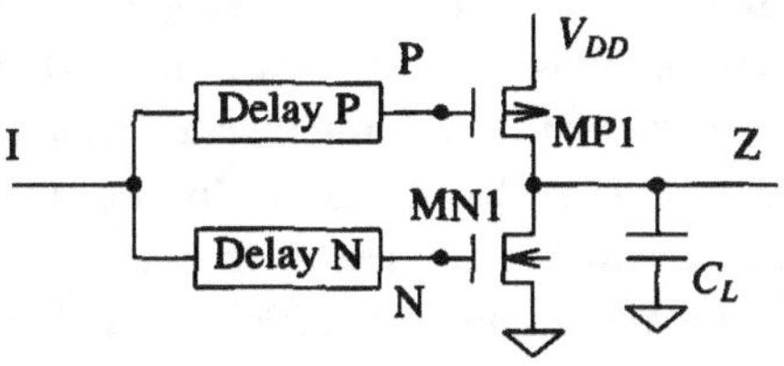

Figure 48 Signal arriving at the gates of NFET and PFET of single gate at different times

is assumed to exist between the output node and ground. Input signal I drives the gates, but the electrical signals at the FET gates can be different. Both FETs can be on or both can be off during switching. If both FETs are turned on simultaneously, V_{DD} and V_{SS} are directly connected via conducting FETs, and current flows through the circuit. This current is called *overlap current.*

Suppose that input signal I makes an instantaneous low-to-high transition, and the delay to the gate of MN1, Delay N, is shorter than delay to the gate of MP1, Delay P. As soon as the up-going signal arrives at node N, MN1 is driven into the saturation region, and we have current

$$I_N = B_N(V_{DD} - V_{THN})$$

through MN1. If this current is higher than the current of MP1 with its gate grounded, which is given by

$$I_P = B_P(V_{DD} - V_{THP}),$$

output node Z is pulled down. The microstate sequence in this case is shown by sequence A of Fig. 49. After MN1 pulls the output node to ground (MN1 goes into the triode region), current I_P ($I_P < I_N$ in the present case) flows from V_{DD} to V_{SS}, and this state continues until MP1 is turned off upon arrival of the up-going signal at node P. The interval when overlap current flows is indicated by the double arrow. Note that the node voltage transition takes place first. After output voltage transient is over, overlap current continues. If $I_N < I_P$, however, output

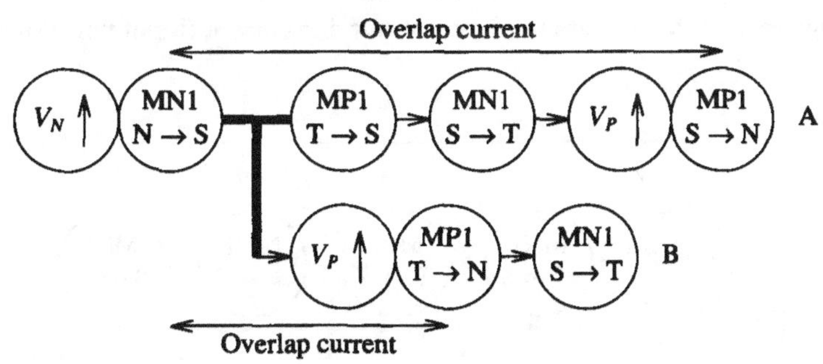

Figure 49 Microstate sequence of overlap current (input pull-up)

node Z remains at V_{DD} potential, since MP1 is able to supply enough current to hold the output node at V_{DD}. Overlap current remains until MP1 is turned off by the up-going input signal arriving at the PFET's gate, as shown by the microstate sequence B of Fig. 49. Pull-down switching of the output node begins at that moment. In this case the node voltage transition takes place after the overlap current is terminated. In either case, the overlap current is the smaller of I_P or I_N, the maximum currents of the PFET and NFET, respectively. It is convenient to define overlap current to carry the total *excess* charge over that stored in the load capacitance. Overlap current flows directly from V_{DD} to V_{SS} during the switching transient. In the pull-down transition of the Fig. 48 inverter, overlap current equals MN1 current minus the discharge current of capacitor C_L. In the other, equivalent interpretation overlap current equals the current through MP1 during the same transient. We assumed that load capacitance C_L of Fig. 48 is connected between the output node and ground. If C_L is split into two components, one of which is between the output node and ground, and the other to V_{DD}, the problem becomes more complex. We do not go into this in detail.

We consider that input signal I of Fig. 48 makes a high-to-low transition. The microstate sequence is shown in Fig. 50. MN1 turns off first since delay N is shorter than delay P. Until the down-going signal arrives at the gate of MP1, the output node is tristated (shown by double-ended arrow). With this polarity of input signal transition, there is no overlap current.

Even if NFET and PFET of an inverter receive the same signal with the same timing and waveform, significant overlap current exists if the input signal transition is slow. Figure 51 shows the microstate sequence of this case (microstate sequence for *static* switching). Suppose that input voltage $V_I(t)$ changes with time as

$$V_I(t)=0\ (t<0)\ \ =\alpha t\ (0<t<t_{IS}=V_{DD}/\alpha)\ \text{ and }\ =V_{DD}\ (t>t_{IS}).$$

During phase A of the microstate sequence of Fig. 51, MN1 is in the saturation region, carrying current $I_N=B_N[V_I(t)-V_{THN}]\ =\ B_N(\alpha t-V_{THN})\ \ (t_I<t<t_{IS})$, where $t_I=V_{THN}/\alpha$.

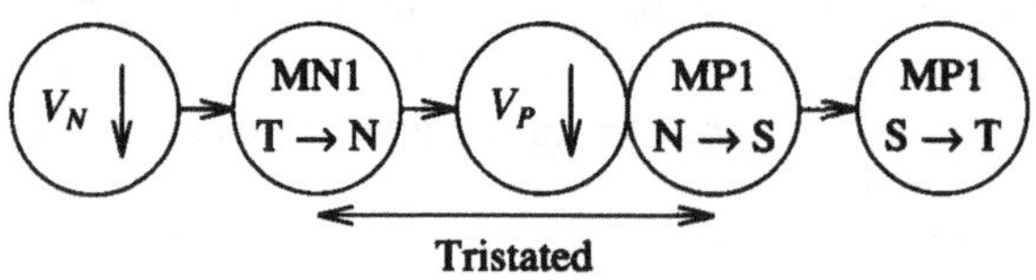

Figure 50 Microstate sequence of overlap current (input pull-down)

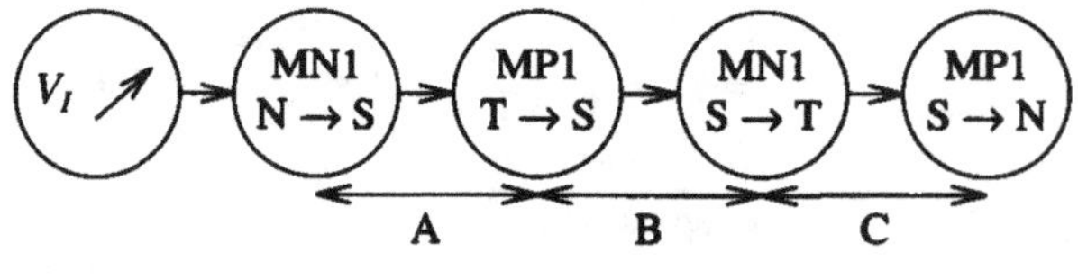

Figure 51 Microstate sequence of slow switching

The maximum current of MP1 during phase A is $I_P = B_P[V_{DD} - V_I(t) - V_{THP}] = B_P(V_{DD} - V_{THP} - \alpha t)$. During phase A, $I_P > I_N$. Then overlap current $I_O(t)$ is the wasted NFET current given by

$$I_O(t) = I_N(t) = B_N(\alpha t - V_{THN})$$

for $t_I < t < t_{SW}$, where t_{SW} is given by

$$t_{SW} = \frac{V_{GSW}}{\alpha} \quad \text{and where} \quad V_{GSW} = \frac{B_P(V_{DD} - V_{THP}) + B_N V_{THN}}{B_P + B_N}$$

is the switching threshold voltage of the inverter. During phase B, the inverter output voltage changes. The current discharging load capacitor C_L is given by

$$I_N(t) - I_P(t) = (B_N + B_P)\alpha(t - t_{SW})$$

and therefore the output voltage change is found by solving

$$C_L \frac{dV_O(t)}{dt} = -(B_N + B_P)\alpha(t - t_{SW})$$

to obtain

$$V_O(t) = V_{DD} - \frac{(B_N + B_P)\alpha}{2C_L}(t - t_{SW})^2 .$$

After switching begins, the current through MP1 is *wasted* as overlap current. Then

$$I_O(t) = I_P(t) = B_P(V_{DD} - V_{THP} - \alpha t) .$$

At time t_{END} the input voltage $V_I(t)$ reaches $V_{DD} - V_{THP}$, and the overlap current vanishes. This time t_{END} is given by

$$t_{END} = \frac{V_{DD} - V_{THP}}{\alpha}$$

The charge that is wasted as overlap current is given by

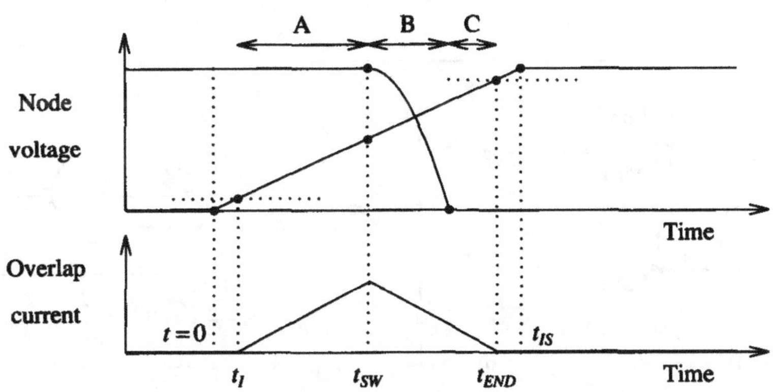

Figure 52 Node voltage and overlap current

$$Q_O = \frac{1}{2} I_P(t_{SW})(t_{END} - t_I) = \frac{1}{2\alpha} \frac{B_N B_P}{B_N + B_P} (V_{DD} - V_{THP} - V_{THN})^2 .$$

The ratio of stored charge to lost charge is

$$\frac{Q_O}{Q_C} = \frac{B_P}{B_P + B_N} \frac{t_{IS}}{2t_0} \left(1 - \frac{V_{THP} + V_{THN}}{V_{DD}}\right)^2 \qquad (29)$$

where $t_{IS} = V_{DD}/\alpha$ and $t_0 = C_L/B_N$. Wasted charge is proportional to the rise (or fall) time of the input signal, t_{IS}. If charge is wasted, DC power is wasted. To exploit the low-power advantage of CMOS fully, circuits that carry heavy overlap current must be avoided.

Circuits that suffer from overlap current are shown in Fig. 53. Figure 53(a) shows MP1 driven by a delayed signal through the resistive wiring. This happens if a large FET of an output driver is laid out using a meandering polysilicon gate. Figure 53(b) shows a Domino CMOS gate, whose precharge and discharge clocks are routed by different polysilicon wires through the tall structure of a microprocessor datapath. If CKD makes an up-going transition earlier than the up-going CKP, all the FETs are turned on, and overlap current exists. Figure 53(c) shows the function of a CMOS static NOR3 gate *synthesized* using a NAND3 gate and a single pull-up PFET (the gate is faster than a conventional NOR3 gate [13]). The gate has higher driving capability than a conventional NOR3 gate but suffers from overlap current.

Figure 54(a) shows a common data communication media (bus) that is driven by multiple tristatable drivers, only one of which is enabled at any time. The data on the bus goes to many destinations (shown by small inverters). If there is only one destination, the circuit is a multiplexer, and if there are multiple destinations, the circuit is a data-bus. Figure 54(b) shows a circuit diagram of two tristatable inverters driving a common data-bus. Control signals of the tristatable inverters C_1 ($\overline{C_1}$) and C_2 ($\overline{C_2}$) are generated so that no two tristatable inverters are enabled simultaneously. This design objective may not be achieved: If more than one of them are enabled, even for a short time, and if data D_1 and D_2 are not the same, current flows as shown in Fig. 54(b), and V_{DD} and V_{SS} buses are effectively short-circuited. This problem is called bus drive conflict. Since bus drivers usually have large FETs, their overlap currents are high. The high overlap currents increase power consumption and create large power-bus noise.

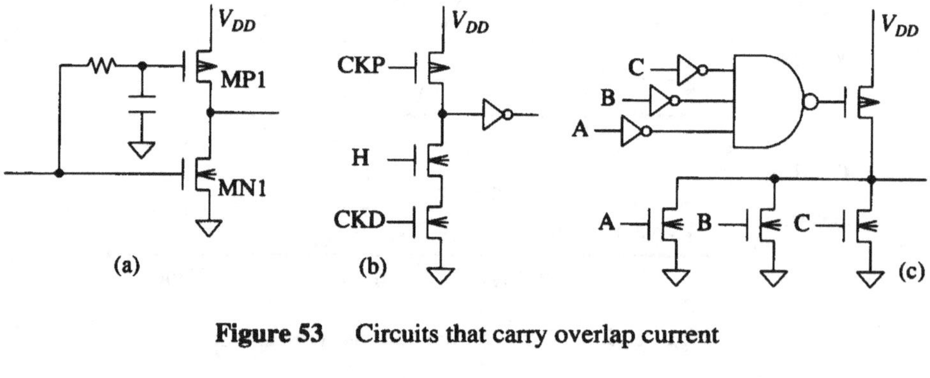

Figure 53 Circuits that carry overlap current

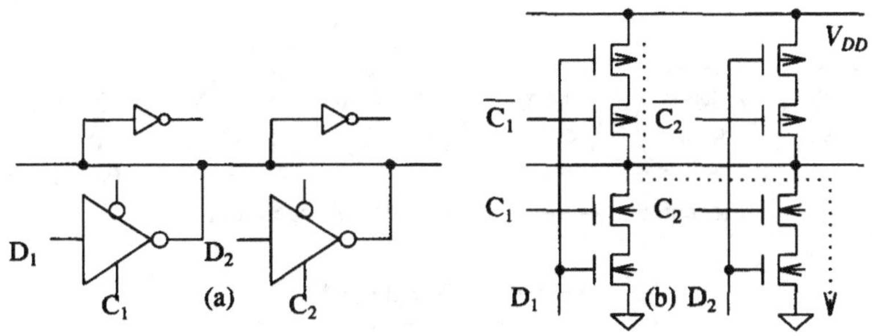

Figure 54 Multiplexer and data-bus

5.13 Failures of CMOS Dynamic Gates at the Maximum Frequency

CMOS dynamic gates have a peculiar failure mechanism at the maximum frequency of operation. This peculiar failure mode may be interpreted as the recombination of two signals that propagate along the *same* signal path at different times. The failure mode is as follows. A dynamic gate operates normally if both precharge and discharge processes work correctly. At the maximum operative frequency, the two processes interact. Figure 55(a) shows a CMOS dynamic inverter. With reference to Fig. 55(b), the precharge clock, CK, is low during period $-t_P<t<0$ and is high during period $0<t<t_N$. *Normal operation* of the dynamic inverter occurs at frequencies significantly lower than the maximum operating frequency. The node waveform of the normal operation is shown by the solid curve of Fig. 55(b), and its microstate sequence is shown in Fig. 56 (the first sequence). By the time precharge is terminated, the output node is at V_{DD} (MP0 reaches the triode region), and when discharge is completed the output node is at ground potential (MN0 reaches the triode region). Curve ABB′E in Fig. 55(b) shows this ideal operation of the dynamic inverter. The memory of previous transients is completely wiped out. We note that the pull-up and pull-down device going into the triode region

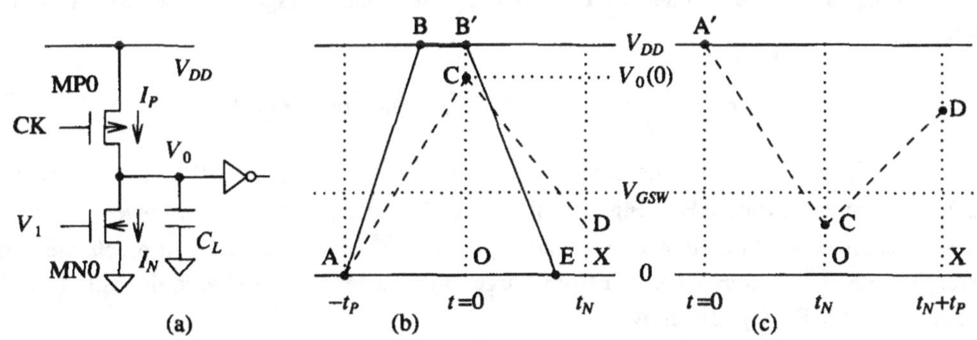

Figure 55 Dynamic gate failure at the frequency limit

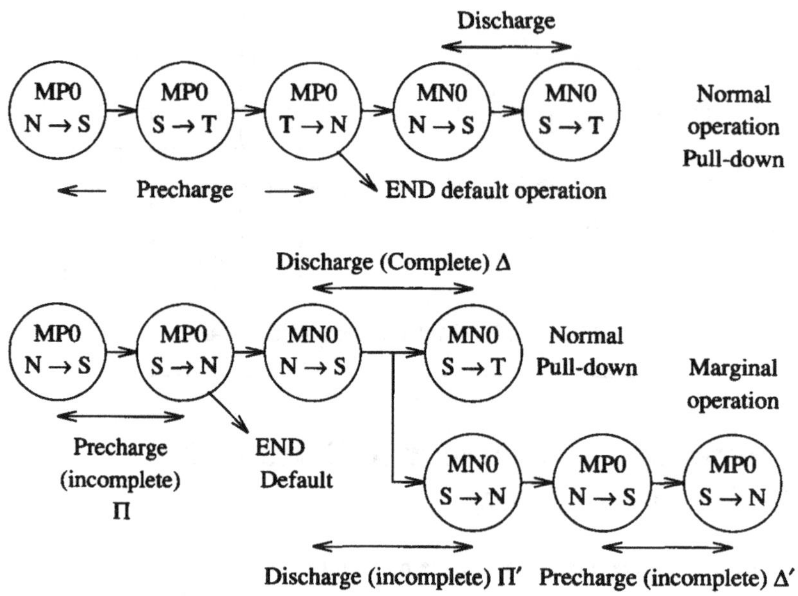

Figure 56 Microstate sequences of precharge-discharge interaction

indicates the *erasure* of the switching history. Although this is an advantage in modeling and not in real circuits, it is significant: We need not set an arbitrary criterion of pass-fail, at least, in the theory. A dynamic gate is used, however, almost always with a static gate that buffers its output. The Boolean value of the signal at the dynamic date's output node must be *interpreted*, as was discussed in Section 2.2. To do that we must include *marginal* operation of dynamic gate into consideration. The *marginal* operation sequence of Fig. 56 shows operation of CMOS dynamic gate subject to incomplete precharge and incomplete discharge. Since MP0 and MN0 never go into the triode region, the dynamic gate output node retains *memory* of past switching events, and that complicates the circuit operation. We study how incomplete precharge affects the dynamic gate discharge. Output node voltage, V_0, is zero at $t=-t_P$. We use the collapsible current generator FET model. The node voltage increases with time following

$$V_0(t)=\frac{I_P}{C_L}(t+t_P) \text{ where } I_P=B_P(V_{DD}-V_{THP}) \tag{30}$$

If the clock frequency is low, the output node is pulled up to V_{DD} by MP0, and at that time precharge completes [line $\overline{AB}$ in Fig. 55(b)]. If the clock frequency is high, however, the output node is pulled up to a voltage less than V_{DD} [line $\overline{AC}$ in Fig. 55(b)], and precharge stops there. The degree of *completeness* of precharge is characterized by charge deposited on the output node by MP0, Q_P, given by

$$Q_P=I_P t_P \quad (I_P t_P<C_L V_{DD}) \qquad Q_P=C_L V_{DD} \text{ (otherwise)}.$$

The final voltage reached during precharge must be higher than V_{GSW} of the destination. The destination is the inverter shown by the small logic symbol in Fig. 55(a). To *interpret* the dynamic gate operation, we assume that the destination gets the high level if $V_0 > V_{GSW}$ and the low level if $V_0 < V_{GSW}$ at the time when the operation result is required. Q_P must satisfy

$$Q_P > Q_C \quad Q_C = C_L V_{GSW} . \tag{31}$$

The pull-down NFET is turned on at time $t=0$. At time $t=t_N$ the logic value is required by the destination. In the process of discharge, MN0 drains charge given by

$$Q_N = I_N t_N \quad (I_N t_N < Q_P) \quad Q_N = Q_P \text{ otherwise}$$

where $Q_P = C_L V_0(0)$ [Fig. 55(b)], and where $I_N = B_N(V_{DD} - V_{THN})$. The charge MN0 drains, Q_N, is less than the charge deposited by MP0 before or $Q_P \geq Q_N$. The output node voltage must be less than V_{GSW}, or

$$Q_C > Q_P - Q_N \tag{32}$$

Since Q_P and Q_N are in the range between 0 and $Q_{max} = C_L V_{DD}$, a pass-fail condition of a dynamic gate can be represented within the square region shown in Fig. 57(a). The horizontal axis is Q_P and the vertical axis is Q_N. Since $Q_P \geq Q_N$, the region above the diagonal of the square region is irrelevant. A dynamic inverter normally operates within the rhombic region. To the left of the rhomboid, precharge is incomplete, and under the rhomboid, charge deposited by precharge is too much more than charge drained by discharge, and therefore the inverter has discharge failure. Figure 57(a) shows that incomplete precharge helps discharge (note that area closer to precharge fail requires less Q_N for normal operation). Similar analysis, using Fig. 55(c), can be repeated to study the effects of incomplete discharge to subsequent precharge. The result is summarized in Fig. 57(b). A dynamic inverter operates normally within the rhombic region. Below that region the inverter fails to discharge, and in the shaded region to the left, by incomplete precharge. Again, incomplete discharge helps precharge. In the region close to V_{GSW}, the effects of the destination gate delay cannot be neglected, and this simple analysis fails (pass-fail criteria is *interpretation*). The region is indicated by thick lines in Figs. 57(a) and (b).

A CMOS dynamic PLA consists of cascaded two-stage dynamic NOR gates as shown in Fig. 58(a). The first stage is called the *decoder* and the second stage the ROM. The clock

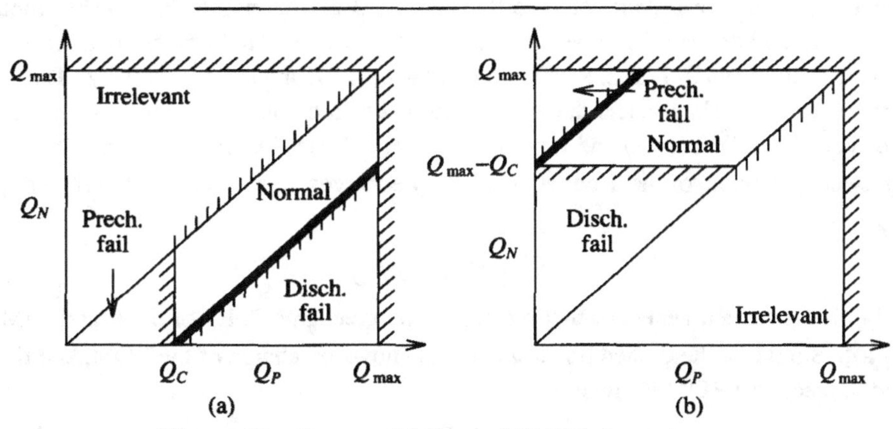

Figure 57 Range of failure of CMOS dynamic gates

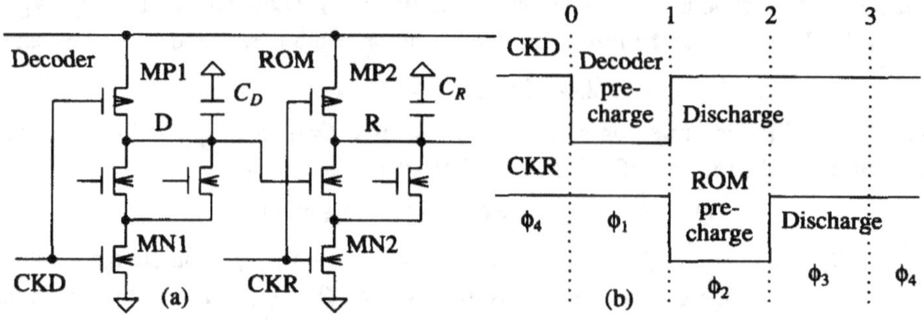

Figure 58 CMOS dynamic PLA

waveforms are shown in Fig. 58(b). If CKD is low, the decoder output D of Fig. 58(a) is precharged by MP1 to V_{DD}. When CKD makes a low-to-high transition, node D is discharged if any PLA data input is high. The dynamic gate executes the NOR function of the input signals. The second stage NOR gates work in the same way. The cascaded dynamic gates fail by either of the two distinct mechanisms at the frequency limit: incomplete precharge that fails at a high level, and incomplete discharge that fails at a low level. The two failure mechanisms of the two stages interact in a complicated way. It is not an easy matter to determine how to improve the frequency limit of a PLA design. A systematic method of studying PLA failure modes at the frequency limit is required.

We need a simple model of this complex problem: the circuit of Fig. 58(a) is modeled using the collapsible current generator model: CKD and CKR switch the FETs instantly from conducting to nonconducting states, and vice versa, and the width of the clock pulses is one phase long, in the four-phase master-slave timing. Therefore, MN1 and MN2 are ideal switches, with zero on-resistance, that turn on and off instantly when CKD and CKR make transitions.

C_D and C_R are the capacitive loads of nodes D and R, respectively. Clock waveforms CKD and CKR, and the instants of time 0, 1, 2, and 3 are defined in Fig. 58(b). Figure 59 shows simplified node waveforms versus time. The time axis is marked by four vertical lines 0, 1, 2, and 3. Intervals 0–1, 1–2, and 2–3 are phases ϕ_1, ϕ_2, and ϕ_3, respectively. The vertical axis represents node voltages, and the line segments drawn on the coordinate system are waveforms of node D (solid curves) and R (dashed curves) of Fig. 58. In Fig. 59(a), line segment $\overline{0Z}$ represents precharge of the decoder. Node D precharges up to V_{DD}, and the time required, $\overline{00'}$, is given by,

$$\overline{00'} = (C_D V_{DD})/I_{PD}$$

where I_{PD} is the current generator that represents the precharge PFET of the decoder [MP1 of Fig. 58(a)]. Similarly, the dashed line segment $\overline{1B}$ shows precharge of the ROM, and time $\overline{AB}$ required to precharge ROM is given by

$$\overline{AB} = (C_R V_{DD})/I_{PR}$$

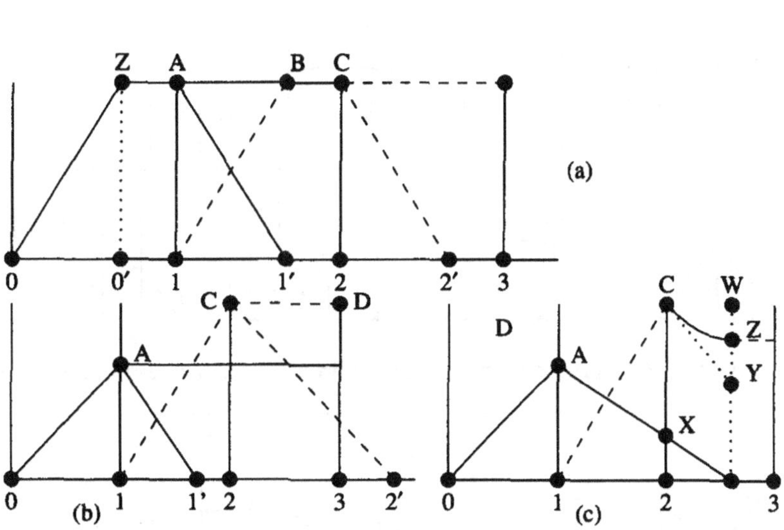

Figure 59 Schematic of discharge waveforms of PLA

where I_{PR} is the current from MP2 that precharges the ROM section of the PLA. Line segment $\overline{A1'}$ and $\overline{C2'}$ represent discharge of the decoder and the ROM, respectively. Discharge delay times $\overline{11'}$ (decoder) and 22′ (ROM) are given by using the decoder and the ROM discharge currents I_{DD} and I_{DR} as

$$\overline{11'} = \frac{C_D V_{DD}}{I_{DD}} \qquad \overline{22'} = \frac{C_R V_{DD}}{I_{DR}} .$$

By reducing the phase duration as shown in Fig. 59(b), the effects of increasing clock frequency can be studied. The slope of line segment $\overline{0A}$ of Fig. 59(b) remains the same as that of $\overline{0Z}$ of Fig. 59(a), and the slopes of line segment $\overline{A1'}$ of Fig. 59(a) and (b) remain the same. A similar relationship applies for ROM precharge. As for ROM discharge, the slope of segment $\overline{C2'}$ of Fig. 59(b) is less than that of segment $\overline{C2'}$ of Fig. 59(a) by voltage ratio $\overline{A1}/V_{DD}$. This is because the current generator representing the ROM NFET is proportional to the gate voltage (voltage of node D). We assumed, for simplicity, that the threshold voltages of NFETs are zero. In Fig. 59(c), discharge of the decoder is slow: discharge continues into interval $\overline{23}$ (Decoder clock CKD is kept turned on through phase 3). Then output node R loses charge by leakage current of incompletely turned off ROM NFET. In Fig. 59(c), the slope of segment $\overline{CY}$ is given by the slope of segment $\overline{C2'}$ of Fig. 59(a), multiplied by $\overline{2X}/V_{DD}$. Curve $\overline{CZ}$ is a parabola with symmetry axis $\overline{WZY}$ so that $\overline{WZ} = \overline{ZY}$. The graphical construction allows any PLA waveform to be drawn subject to the simplifying assumptions. This study is carried out by using a computer program that automatically draws the schematic node voltage waveforms shown in Figs. 60 and 61. We assume that if a node voltage is less than $V_{DD}/2$, the logic level is zero after buffering, and if a node voltage is more than $V_{DD}/2$, the logic level is one. Then a PLA may produce correct output at R, even if the internal node does not work properly. It is necessary to study a great variety of cases and get a general idea.

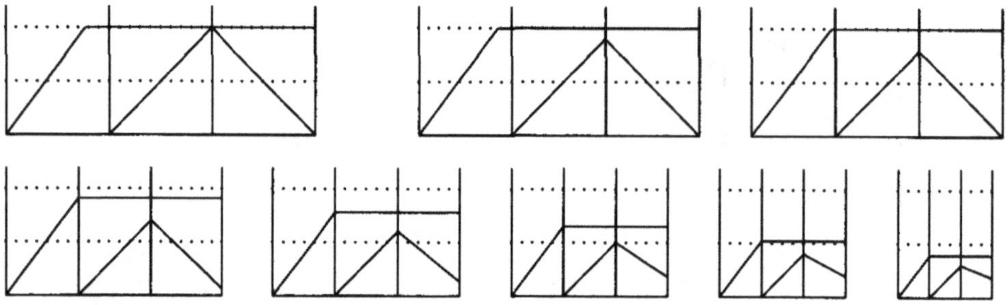

Figure 60 PLA failure at the frequency limit (decoder does not discharge)

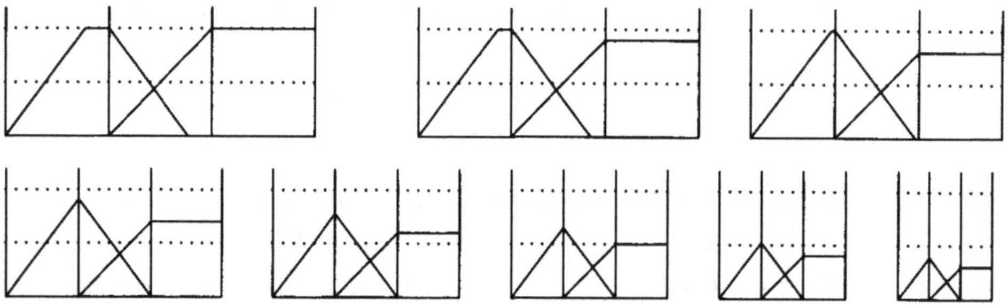

Figure 61 PLA failure at the frequency limit (decoder discharges)

Depending on the input data, the decoder may or may not discharge. In Fig. 60 the decoder does not discharge: the PLA does not fail even if the decoder output node fails. In Fig. 61 the decoder discharges: the PLA fails by insufficient precharge of the ROM. As we saw, failures are caused by improper precharge of ROM and improper discharge of decoder, but the failure mechanisms interact, and they are indeed quite complex. ROM precharge failure is aggravated if decoder discharge is slow: node R loses charge during the discharge phase of the ROM, since the decoder output node has not yet pulled down completely. This faulty discharge of the ROM is slowed down by insufficient precharge of the decoder. When ROM precharge is slow, ROM discharge is fast. The combination of these cases amounts to a complex failure mode at the PLA frequency limit. To understand this complex problem, the simplified model and the graphical method are quite useful.

5.14 Linear Amplifier and Inverter

Problems of CMOS circuit timing design are classified into two types. One type is resolved if signal delays of two logic paths can be matched robustly against process, temperature, and environmental variations. The other type requires the generation of absolute delay times, and these are more difficult. My experience shows that the second type occurs quite rarely. Therefore we consider the first type in this section. The technique we discuss in this section is a new FET scaling technique to match delays of two independent signal paths robustly, against process variations, etc. The design technique has many applications, including clock skew control, pulse width control, elimination of race conditions or hazards, and design optimization of tricky circuits like glitch generators. An interesting application of this technique to pseudo static gates will be discussed later.

In order to match the delays of two logic chains, the conditions of the two chains must be specified precisely. The logic chains must be *logically* equivalent to cascaded chains of inverters: For all the gates in the chains the voltage levels of the inputs other than the ones that switch must have been stabilized at their respective steady state logic levels that enable the gates, before the switching begins. The effects of past switching must have been completely wiped out: voltage levels of all the *internal* nodes of the gates must have been settled at their steady state voltages. We are able to consider the robust delay matching only subject to these restricted conditions.

Subject to the constraints, matching delays of two signal paths for a set of precisely specified process conditions, power supply voltage, and temperature is an easy design objective to achieve if an accurate circuit level simulator is available. To make the design robust against *variations* in process, etc., is the non-trivial part. In order to understand this problem, we need to consider a fundamental problem that is often overlooked in digital circuit theory: A CMOS inverter is different from a linear amplifier in an essential detail, and the difference is crucial in robust delay matching. We consider linear (small-signal) and digital (large-signal) operations of the same CMOS inverter. For small-signal linear operation, the DC operating point can be established using a large resistor R_B that is connected between input and output of the inverter, as shown in Fig. 62(a). If input node I is driven by a small-amplitude voltage source, gain G is given by

$$G = r_d(g_{mN} + g_{mP}) \tag{33}$$

where g_{mN} and g_{mP} are transconductances of NFET and PFET, respectively, at the DC bias point, and r_d is the load resistance (consisting primarily of the Early effect resistances of both

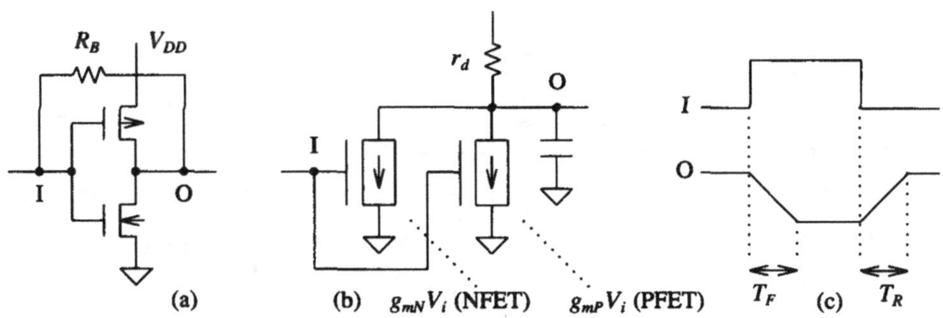

Figure 62 CMOS linear amplifier

FETs and R_B). The equivalent circuit of the small-signal amplifier is shown in Fig. 62(b). As the input voltage goes up, the NFET current increases and the PFET current decreases. Contributions from the PFET and NFET add, or the PFET and NFET work *jointly*. Therefore, the amplified waveform shown in Fig. 62(c) is symmetrical with respect to up-going and down-going transitions: rise time T_R and fall time T_F are the same. T_R and T_F remain the same as long as the signal amplitude is small, and the effective transconductance $g_{mN} + g_{mP}$ is independent of the signal amplitude. The ratio of transconductance of PFET and of NFET does not matter for *faithfully* scaled reproduction of the small-amplitude input waveform at the output node, because $\Delta I/\Delta V = constant$ for + or $- \Delta V$.

A CMOS digital inverter is shown in Fig. 63(a). If the input signal switches from 0 to V_{DD} in a negligibly short time, the NFET that was nonconducting turns on, and the PFET that was conducting turns off. The entire pull-down of the output node is carried out by the NFET alone. Similarly, if the input switches rapidly from V_{DD} to 0, the entire pull-up of the output node is carried out by the PFET. In dynamic operation of a digital inverter, pull-down is carried out by the NFET alone, and pull-up is carried out by the PFET alone: joint operation of the PFET and NFET, as in a linear amplifier, is nonexistent in the digital inverter. If the NFET has high drive capability but the PFET has low drive capability, a symmetrical input pulse is deformed as shown in Fig. 63(c) (note that $T_F \ll T_R$). A linear amplifier and a digital inverter reflect two extrema modes of operation of complementary circuits. Linear amplification is not only a small-signal operation but also a slow operation, to maintain waveform accuracy. Linear amplification does not use the entire switching capability of the FETs. Signals change gradually, and the NFET and PFET share the work of amplification. Digital operation is a fast operation that uses the device's switching capability to the limit to make a transition between the two separate voltage levels. Signals vary rapidly, and therefore PFET or NFET, but not both, execute pull-up or pull-down operations, respectively.

Since the PFET and NFET show up separately in digital operations, a special caution is required to desensitize the circuit against process-temperature-voltage variations. Process variation is characterized not by one parameter but by two parameters, one for NFETs and the other for PFETs. Such variation emphasizes the *difference* in the driving capabilities of PFETs and NFETs. The objective of this section is to study the mechanisms of delay matching of cas-

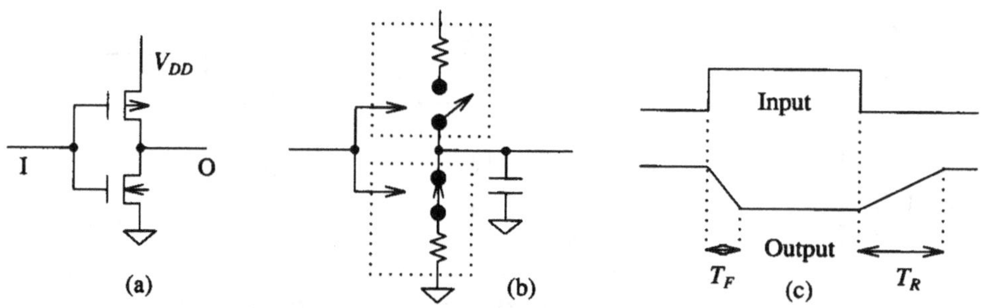

Figure 63 CMOS inverter

caded gate chains by going back to this fundamental viewpoint. We use gate delay models having various degree of sophistication, and we show the theoretical foundation of the robust delay-matched circuit design. The design guideline must be practical and include the following restriction. A designer does not have the freedom to specify all the device parameters: He is able to control only the *horizontal* dimensions of the device and interconnects or, alternatively, the device scaling and adjustment of interconnect.

Variation of process conditions is included in the analysis as follows. High-speed and low-speed process parameters are defined such that only 2% of the processed devices from the production facility have drive capabilities outside the higher and the lower specifications: 96% of the entire population should fall within the high and the low switching speed limits. A convenient parameter to define the FET drive capability is the time constant of a FET, T_D defined by [14],

$$T_D = (C_G + C_D)(V_{DD} / I_{sat}) \tag{34}$$

where C_G is the gate capacitance, C_D is the drain diffused island capacitance, and I_{sat} is the drain current that flows if both gate and drain voltages are set at the maximum of digital switching ($V_G = V_D = V_{DD}$). T_D is a parameter proportional to gate delay: typically five times the averaged T_D for the PFET and for the NFET is the delay of an inverter with fanout of three. Table 1 lists T_D, for three process conditions for twin-tub 1.75 and 0.9 micron CMOS processes. The time constants were determined using the ADVICE circuit simulator for minimum size FETs at 85 degrees centigrade.

Table 3: FET time constants (T_D)

Technology	1.75 μm process				0.9 μm process			
FET	NFET		PFET		NFET		PFET	
Unit	(psec)	(ratio)	(psec)	(ratio)	(psec)	(ratio)	(psec)	(ratio)
High current(H)	88.0	0.556	201.0	0.620	56.6	0.912	158.3	1.130
Typical(M)	158.0	1.000	324.0	1.000	62.0	1.000	139.5	1.000
Low current(L)	273.0	1.730	530.0	1.630	77.5	1.250	173.5	1.240

In Table 3, *ratio* means the ratio to typical processing. The drive capability is the inverse of the time constant. This result shows that the ratio of drive capability of NFET to PFET in 1.75 μm CMOS varies from 6.02(NH-PL) process to 0.73 for (NL-PH) process. In 0.9 μm CMOS, the variation is less, reflecting significantly improved process control. The variation is a worst-case limit, since parameters like NFET channel length and thin oxide thickness approximately track the respective PFET parameters. The realistic process parameter variation never occurs such that all the parameters specifying the drive capability of NFETs fluctuate in the direction to increase the drive capability, and simultaneously all the parameters of PFETs fluctuate in the direction to decrease the drive capability. There is positive correlations among the parameters. If the set of the fastest possible parameter values of one type FET were used together with the slowest possible parameter values of the other type FET, by neglecting the correlations, the set of parameter values would be unrealistically skewed. Still, the designer must be prepared that the device parameters could vary within remarkably wide range.

Figure 64 shows two chains of inverters whose delay is to be matched. Suppose that at time $t=0$ the signal at the input of the first inverter chain makes a low-to-high transition, and the signal at the input of the second chain makes a high-to-low transition. The signal transition polarities at all the nodes are shown by the arrows in Fig. 64. The delay of the first logic chain is the sum of the pull-down delay of NFET MN1 [$T_{NF}(1)$], delay of the resistive interconnect [$T_P(1)$], the pull-up delay of PFET MP2 [$T_{PF}(1)$], delay of the resistive interconnect [$T_P(2)$], ..., and so on. The delay of the first and the second chain, T_D and T_D' respectively, are given by [15],

$$T_D = T_{NF} + T_P + T_{PF} \tag{35}$$

where

$$T_{NF} = T_{NF}(1) + T_{NF}(2) + \ldots\ldots$$

$$T_P = T_P(1) + T_P(2) + \ldots\ldots \qquad T_{PF} = T_{PF}(1) + T_{PF}(2) + \ldots\ldots$$

and

$$T'_D = T'_{NF} + T'_P + T'_{PF} \tag{36}$$

where

$$T'_{NF} = T'_{NF}(1) + T'_{NF}(2) + \ldots\ldots$$

$$T'_P = T'_P(1) + T'_P(2) + \ldots\ldots \qquad T'_{PF} = T'_{PF}(1) + T'_{PF}(2) + \ldots\ldots$$

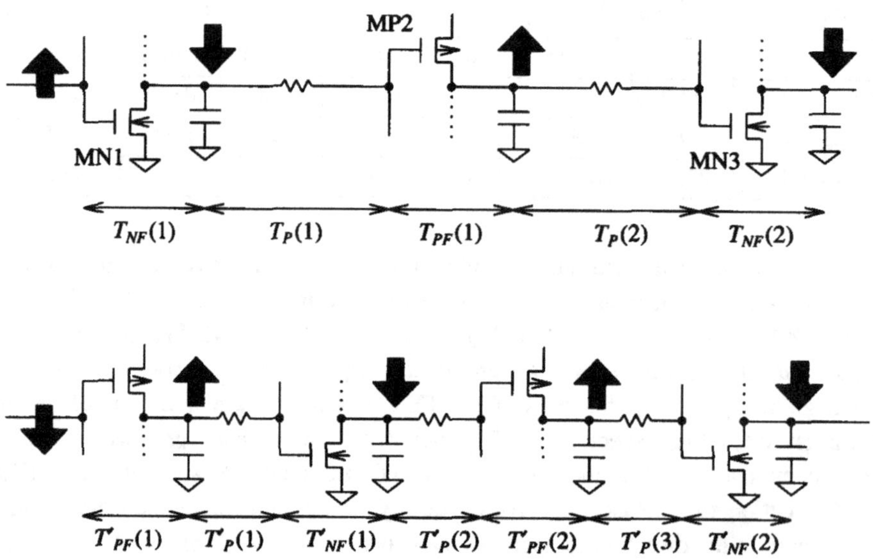

Figure 64 Detailed balance scheme of delay

Equations (35) and (36) are good approximations of the delay times of the chains if the signal waveform distortion caused by heavy loading and by Miller effect is small. Equations (35) and (36) lead to an important practical conclusion. When the process conditions vary, T_{NF} and T'_{NF} vary approximately by the same factor (λ_{NF}), and so do T_{PF} and T'_{PF} (by a factor λ_{PF}) and T_P and T'_P (by a factor λ_P). Then if Eqs. (35) and (36) are written for typical process conditions at typical temperature and power supply voltage, the delay times at an arbitrary set of process and environmental conditions are

$$T_D = \lambda_{NF} T_{NF} + \lambda_P T_P + \lambda_{PF} T_{PF} \tag{37}$$

and

$$T'_D = \lambda_{NF} T'_{NF} + \lambda_P T'_P + \lambda_{PF} T'_{PF} \,. \tag{38}$$

In order to design the two circuits to have equal delay for typical conditions, it is necessary to have

$$T_D = T'_D \tag{39}$$

for typical conditions. This design, however, does not remain skew-free if the process and environmental conditions vary. Instead, if we design a circuit such that

$$T_{NF} = T'_{NF} \quad T_P = T'_P \text{ and } T_{PF} = T'_{PF} \tag{40}$$

the delays are matched robustly against all the process-temperature-voltage variations. By Eqs. (39) and (40) the delays contributed by NFETs, PFETs and interconnects are *separately* matched. Even if NFET, PFET, and interconnect delays vary independently, T_D and T'_D remain matched. If component matching of Eq. (40) is exercised in design, the delays of the two chains are robustly matched, whatever the process variations may be. If only the total delays of the two chains T_D and T'_D are matched without paying attention to the component delays, the delay will not be matched if the conditions vary.

If the wiring delays are negligible, skew of the two signal paths is more sensitive to a *differential* variation of the drive capabilities of the NFET and the PFET (NFET drive capability is higher than that of the typical process, and PFET drive capability is *less* than that of the typical process, and vice versa). Any circuit generates relatively small skew for a uniform increase or decrease of drive capabilities of both PFET and NFET. Suppose wiring delays are neglected. Delays are determined by the size of all the FETs. If there are ten FETs in the circuit, there are ten adjustable FET sizes, and there are only two equations. There are many solutions to the problem. Among the many solutions, the best solution can be selected using the following criteria. These criteria are all to maintain good switching waveforms at all the nodes.

1) No single inverter should have significantly more delay than any other.
2) Inverters should be as symmetrical as possible to pull-up and pull-down.
3) The last stage of each chain should have about the same drive capability (for equal capacitance load).

The best way to design delay-matched chains is to determine approximate FET size by a simple algorithm and then determine precise FET size by iterating precise circuit simulation [16].

In this section, process variation was characterized by using one parameter for PFETs, and one for NFETs. This is still a simplification. Process variation specification may require more than one parameter for each type FET. This type of theory is developed in the next section.

5.14.1 Application of the Delay Matching Technique

The basic concept of robust delay matching has a number of applications, including the following example. Domino CMOS dynamic logic circuit is two to three times faster than the corresponding CMOS static logic circuit. Although Domino CMOS is a cascadable dynamic logic circuit, its application is limited to noninverting logic, unless *double rail* logic is used [14]. This restriction is often inconvenient. Directly cascaded CMOS dynamic logic circuits like CMOS PLAs are even faster than Domino CMOS logic circuits, and they do not have the *inversion* restriction. The speed advantage of directly cascaded CMOS dynamic logic circuits should be exploited to its full extent. An ideal solution is a mixture of directly cascaded dynamic logic and Domino CMOS as shown schematically in Fig. 65, where square boxes with bubbles represent CMOS dynamic logic gates, and boxes without bubbles represent non-inverting Domino CMOS gates. This circuit has not commonly been used because the circuit is sensitive to process, temperature, and voltage variations. This problem can be understood from the FET diagram shown in Fig. 66. Suppose that input signals A1 and A2 are both high, and input signal B2 is low. Nodes B1, C1, and D1 are precharged to V_{DD} before the logic operation is carried out. When CKA makes a low-to-high transition, NFETs MNA1, MNA2, and MNA3 all conduct, and node B1 (output node of gate A) is pulled down. Some time later MNB2 is turned off. CKA is delayed by two cascaded inverters to generate CKB. CKB makes a low-to-high transition, but the edge of this transition may be before, or may be after, the time MNB2 is turned off. If CKB makes its low-to-high transition after MNB2 is turned off, the cascaded gates work properly. If CKB makes its transition before MNB2 is turned off, however, node C1 (output node of gate B) is pulled down, and therefore gate B fails. From this observation, the delay of the cascaded inverters should be made long enough to guarantee that CKB is later than the MNB2 turnoff, and under the most extreme conditions of process, temperature, and voltage variation. This is the reason why the high performance advantage of CMOS dynamic logic cannot be fully exploited.

This difficulty can be removed by using the technique of precision delay matching. When clock CKA makes its low-to-high transition, the chain of three NFETs MNA1–MNA3

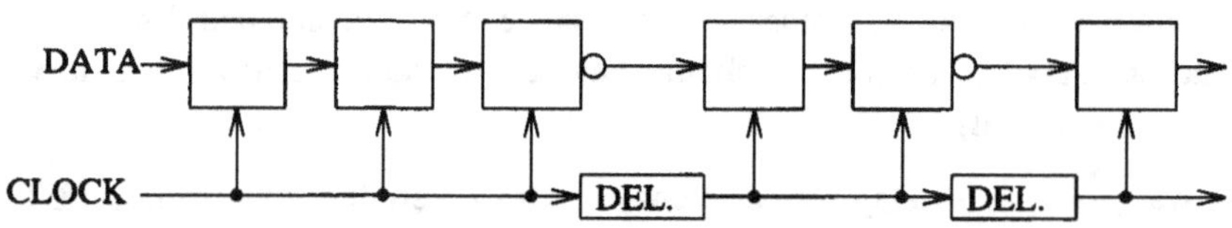

Figure 65 Mixed use of Domino CMOS and CMOS dynamic logic gates

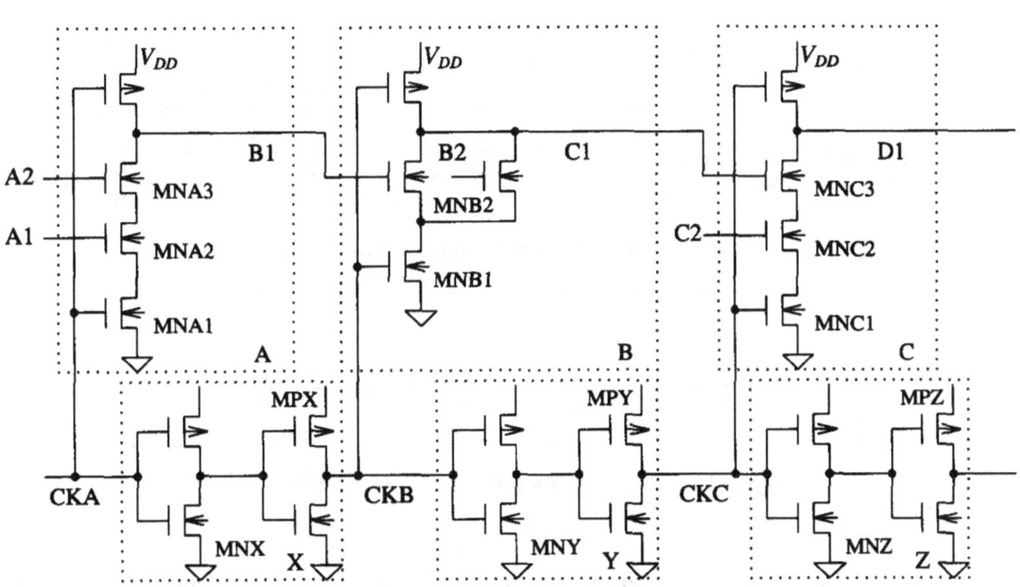

Figure 66 Matching delays of signal and clock

pulls node B1 down. How fast depends on the conductance and capacitance of the NFETs. Simultaneously with pull-down of node B1, MNX of the first stage buffer of the clock delay chain pulls the output node down. How fast the output node is pulled down depends on the NFET, in the same way as in the NFET chain. Then, if the delay of node B1 pull-down and of the MNX pull-down are ratioed properly, the circuit becomes robust against process variation. Delay of CKB relative to CKA has a component that is contributed from the pull-up delay of MPX as well. Therefore the basic idea of delay matching is to increase the component of clock delay contributed from MNX (or its equivalent, if the clock delay chain has more inverters). The pull-up delay of MPX is used as the safety margin. In this way, the clock delay necessary to guarantee safe operation of directly cascaded CMOS dynamic gates can be reduced to a tolerable minimum.

5.15 Mathematical Formulation of Precision Delay Matching

We study the robust delay matching method by using a more precise gate delay model than that in the last section. As an example, we consider the long inverter chain shown in Fig. 67, and we match the chain's delay time for an input *pull-up* transition and for an input *pull-down* transition. The concept of precision delay matching is applicable to pull-up and pull-down delays of a single gate chain. An important assumption for this problem is that the width of the input pulse is wide enough to reduce interaction between the two consecutive transitions. If the interaction is not negligible, we have a different problem, which will be discussed in

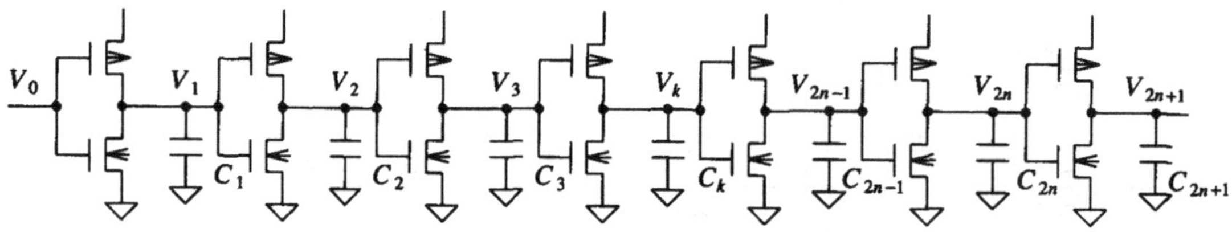

Figure 67 Pull-up/pull-down delay matching of a long inverter chain

Figure 68 Signal waveform in a long inverter chain

Section 6.15. Figure 68 shows node waveforms within the chain schematically. If the two delays are matched, the width of an isolated pulse that propagates along the chain is maintained against all the variations. The solution of this problem has a wide range of applications in the design of a digital transmission system. An intuitive solution of this problem is to cascade inverters that have a fixed PFET/NFET size ratio, and to scale the inverter chain up by a constant scaleup ratio, from the beginning to the end. The point of the following theory is to show that this is *not* the only solution: the existence of other solutions opens many new possibilities for robustly delay-matched design.

Node capacitance C_k of k-th node $(1 \le k \le 2n)$, where $2n$ is the number of cascaded inverters, is given by

$$C_k = (C_{GP}P_{k+1} + C_{GN}N_{k+1}) + (C_{DP}P_k + C_{DN}N_k) \tag{41}$$

where P_k and N_k $(1 \le k \le 2n+1)$ are sizes (widths) in μm of PFETs and NFETs, respectively, of the k-th inverter [the $(2n+1)$st stage is the final load of the inverter chain]. C_{GP} and C_{GN} are the gate capacitances of PFETs and NFETs, respectively, per unit width (pF/μm). C_{DP} and C_{DN} are the drain island capacitances of PFETs and NFETs, repectively, per unit width. The first parenthesis of Eq. (41) is the input capacitance of the $(k+1)$st inverter, and the second parenthesis is the output capacitance of the kth inverter. The Miller effect capacitances are not included.

Delays from the $(k-1)$st node to the kth node are given by

$$t_{kD} = f\,\frac{C_k V_{DD}}{I_N N_k} \quad \text{and} \quad t_{kU} = f\,\frac{C_k V_{DD}}{I_P P_k} \tag{42}$$

for node k pull-down and node k pull-up transitions, respectively. I_N and I_P are the NFET and PFET currents per unit device width, respectively, if the gate and drain are both biased to V_{DD}. Factor f is a nondimensional constant of the order of unity required to calibrate delays in real time units. Equation (42) assumes that the transition time of node voltage is short. Let us consider the cumulative delay from node 0 to node $2n$ of Fig. 67. Input pull-up delay (node 0 makes a low-to-high transition) $T_D(U)$ is given by

$$\frac{T_D(U)}{f\cdot V_{DD}}=\frac{C_1}{I_N N_1}+\frac{C_2}{I_P P_2}+\frac{C_3}{I_N N_3}+\cdots+\frac{C_{2n}}{I_P P_{2n}} \tag{43a}$$

and input pull-down delay (node 0 makes a high-to-low transition) is given by

$$\frac{T_D(D)}{f\cdot V_{DD}}=\frac{C_1}{I_P P_1}+\frac{C_2}{I_N N_2}+\frac{C_3}{I_P P_3}+\cdots+\frac{C_{2n}}{I_N N_{2n}}\,. \tag{43b}$$

After some algebraic rearrangement we have

$$\begin{aligned}\frac{T_D(U)}{f\cdot V_{DD}}&=\frac{C_{GP}}{I_N}\left(\frac{P_2}{N_1}+\frac{P_4}{N_3}+\ldots\right)+\frac{C_{GP}}{I_P}\left(\frac{P_3}{P_2}+\frac{P_5}{P_4}+\ldots\right)\\&+\frac{C_{GN}}{I_N}\left(\frac{N_2}{N_1}+\frac{N_4}{N_3}+\ldots\right)+\frac{C_{GN}}{I_P}\left(\frac{N_3}{P_2}+\frac{N_5}{P_4}+\ldots\right)\\&+\frac{C_{DP}}{I_N}\left(\frac{P_1}{N_1}+\frac{P_3}{N_3}+\ldots\right)+n\frac{C_{DP}}{I_P}+n\frac{C_{DN}}{I_N}+\frac{C_{DN}}{I_P}\left(\frac{N_2}{P_2}+\frac{N_4}{P_4}+\ldots\right)\end{aligned}$$

$$\begin{aligned}\frac{T_D(D)}{f\cdot V_{DD}}&=\frac{C_{GP}}{I_P}\left(\frac{P_2}{P_1}+\frac{P_4}{P_3}+\ldots\right)+\frac{C_{GP}}{I_N}\left(\frac{P_3}{N_2}+\frac{P_5}{N_4}+\ldots\right)\\&+\frac{C_{GN}}{I_P}\left(\frac{N_2}{P_1}+\frac{N_4}{P_3}+\ldots\right)+\frac{C_{GN}}{I_N}\left(\frac{N_3}{N_2}+\frac{N_5}{N_4}+\ldots\right)\\&+n\frac{C_{DP}}{I_P}+\frac{C_{DP}}{I_N}\left(\frac{P_2}{N_2}+\frac{P_4}{N_4}+\ldots\right)+\frac{C_{DN}}{I_P}\left(\frac{N_1}{P_1}+\frac{N_3}{P_3}+\ldots\right)+n\frac{C_{DN}}{I_N}\,.\end{aligned}$$

Since (C_{GP}/I_N), (C_{GN}/I_N), (C_{DP}/I_N), (C_{DN}/I_N), (C_{GP}/I_P), (C_{GN}/I_P), (C_{DP}/I_P) and (C_{DN}/I_P) are the process dependent parameters, we match the coefficients multiplied by the terms of $T_D(U)$ and $T_D(D)$. We have

$$(C_{GP}/I_N):\quad \frac{P_2}{N_1}+\frac{P_4}{N_3}+\cdots=\frac{P_3}{N_2}+\frac{P_5}{N_4}+\cdots \tag{44a}$$

$$(C_{GN}/I_N):\quad \frac{N_2}{N_1}+\frac{N_4}{N_3}+\cdots=\frac{N_3}{N_2}+\frac{N_5}{N_4}+\cdots \tag{44b}$$

$$(C_{DP}/I_N):\quad \frac{P_1}{N_1}+\frac{P_3}{N_3}+\cdots=\frac{P_2}{N_2}+\frac{P_4}{N_4}+\cdots \tag{44c}$$

The coefficients of (C_{DN}/I_N) are automatically matched. Furthermore

$$(C_{GP}/I_P):\quad \frac{P_3}{P_2}+\frac{P_5}{P_4}+\cdots=\frac{P_2}{P_1}+\frac{P_4}{P_3}+\cdots \tag{44d}$$

$$(C_{GN}/I_P):\quad \frac{N_3}{P_2}+\frac{N_5}{P_4}+\cdots=\frac{N_2}{P_1}+\frac{N_4}{P_3}+\cdots \tag{44e}$$

$$(C_{DN}/I_P):\quad \frac{N_2}{P_2}+\frac{N_4}{P_4}+\cdots=\frac{N_1}{P_1}+\frac{N_3}{P_3}+\cdots \tag{44f}$$

The coefficients of (C_{DP}/I_P) are automatically matched. Equations (44) are *indefinite* equations that have infinitely many solutions. If the number of stages of inverters, $2n$, is more than 4, there are enough degrees of freedom in selecting $2n$ unknowns $P_1, ..., N_{2n}$. Then for such a set of FET sizes the effects of process-temperature-voltage variations are canceled, and the inverter chain remains at a robustly equal delay for the two transition polarities. A computer program to search for the optimum FET sizes using this kind of algorithm has been reported.

Equations (44) are satisfied by an obvious solution to this problem, which was mentioned at the beginning of this section. If the PFET/NFET size ratio is η and the uniform scaleup ratio is κ, both sides of Eq. (44a) are equal at $\eta\kappa n$, and the equation is satisfied. All the other equations are satisfied as well. The point I wish to show by this development is that this is *not* the only solution: There are an infinite number of solutions that satisfy the requirement of robust delay matching. Robust delay matching can be obtained even if the total number of stages of the two signal paths are different, or even if the types of gates involved in the two chains are not the same.

Robust delay matching becomes more flexible if the number of process dependent parameters is reduced. If C_{GP} and C_{GN} are equal, Eqs. (44a) and (44b) are combined to obtain

$$\frac{P_2+N_2}{N_1}+\frac{P_4+N_4}{N_3}+\cdots=\frac{P_3+N_3}{N_2}+\frac{P_5+N_5}{N_4}+\cdots \tag{45}$$

Similarly Eqs. (44d) and (44e) are combined to obtain

$$\frac{P_3+N_3}{P_2}+\frac{P_5+N_5}{P_4}+\cdots=\frac{P_2+N_2}{P_1}+\frac{P_4+N_4}{P_3}+\cdots \tag{46}$$

and therefore the number of equations that must be satisfied by P_k and N_k is reduced to 4.

The theory of robust delay matching is simple and practical, but the effects of signal waveforms, especially the effects of rise/fall times on gate delay, are not included in this theory. Effects of slew rate of input signal on gate delay time is complex, because both the PFET and NFET contribute to pull-up or pull-down of the output node (Section 5.14). A satisfactory theory for this case has not yet been developed.

5.16 Dependence of Gate Delay on Input Signal Transition Time

The gate delay analysis of the last section assumed that the input signal rise/fall times are shorter than the time required for the output node to complete its transition. The process is the *dynamic* switching of the gate, which was discussed in Section 2.8. If the rise/fall time of the input signal increases, the switching process changes gradually from dynamic to static switching, and the change affects the delay time.

The delay time of a CMOS gate is the interval from the time when increasing (or decreasing) input voltage reaches $V_{DD}/2$ to the time when decreasing (or increasing) output voltage reaches $V_{DD}/2$. This definition has the advantage that cumulative delay of a chain of logic gates can be expressed as the sum of delays of the individual gates. This definition is convenient for circuit design, but a few curious problems emerge from this definition, because the *real* switching threshold voltage of gate is not always $V_{DD}/2$. The delay time of the gate may even become negative.

To understand this problem we need to analyze the inverter switching transient in closed form using a time-dependent input voltage. An analysis that uses a linear ramp input voltage was given in the author's book *CMOS digital circuit technology* [14]. The entire switching transient is split into several regions, and the solution of one region is connected to adjacent regions. Here we develop a simpler theory that gives the delay times with less effort. We consider the inverter shown in Fig. 69(a). MN1 and MP1 have transconductance B_N and B_P, respectively, and we assume that FETs have zero threshold voltage. The switching threshold voltage V_{GSW} of the inverter is given by

$$V_{GSW} = \frac{B_P V_{DD}}{B_N + B_P}. \tag{47}$$

We consider the input voltage waveform $V_1(t)$ shown in Fig. 69(b). Let the time t, when $V_1(t)$ equals the switching threshold voltage V_{GSW}, be chosen as 0. In Fig. 69(b) the case $V_{GSW} < V_{DD}/2$ is shown. Time $t = t_1$ when $V_1(t)$ equals $V_{DD}/2$ is then positive. If $V_{GSW} > V_{DD}/2$, t_1 is negative. We use a simple waveform for $V_1(t)$ so that t_1 can be obtained in closed form. One such choice is

$$V_1(t) = V_{GSW} + (V_{DD} - V_{GSW})\frac{t}{t + t_0}$$

where time t_0 is the measure of the input waveform rise time. We solve for t_1 as

$$t_1 = \left(1 - \frac{2V_{GSW}}{V_{DD}}\right)t_0 = \frac{1 - (B_P/B_N)}{1 + (B_P/B_N)}\,t_0. \tag{48}$$

At time $t = 0$, the inverter begins to switch. The switching process is described by the circuit equation

$$C_L \frac{dV_0(t)}{dt} = (B_N + B_P)[V_{GSW} - V_1(t)].$$

The circuit equation is solved subject to the initial condition $V_0(t) = V_{DD}$ as

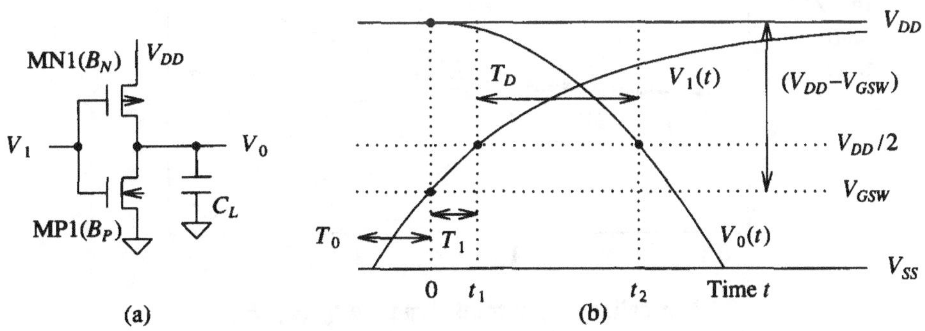

Figure 69 Analysis of dependence of delay on input slew rate

$$V_0(t) = V_{DD}[1 - (B_N/C_L)(t - t_0 \log \frac{t + t_0}{t_0})]. \tag{49}$$

The time when $V_0(t)$ equals $V_{DD}/2$, $t = t_2$ is given by

$$t_2 = t_0 \cdot \tau(\zeta) \quad \text{where} \quad \zeta = \tau(\zeta) - \log[1 + \tau(\zeta)]$$

is solved for $\tau(\zeta)$ to define the implicit function. The value of ζ that we use is given by $\zeta = (1/2t_0)(C_L/B_N)$. The implicit function $\tau(\zeta)$ can be approximated closely by

$$\tau(\zeta) = \sqrt{2\zeta} + 0.771\zeta$$

in the range $\zeta \leq 5$. The gate delay time T_D, which is defined in Fig. 69(b), is given by

$$T_D = t_2 - t_1 = (C_L/B_N)\left[\sqrt{\theta} + 0.386 - \frac{1 - (B_P/B_N)}{1 + (B_P/B_N)}\theta\right]. \tag{50}$$

Figure 70 shows plots of normalized delay time, $T_D/(C_L/B_N)$, versus normalized switching time, $\theta = t_0/(C_L/B_N)$, for several values of B_P/B_N.

If $V_{GSW} > (V_{DD}/2)$ (or $B_N < B_P$), $T_D \to +\infty$ as $\theta \to \infty$. If $V_{GSW} < (V_{DD}/2)$ (or $B_N > B_P$) $T_D \to -\infty$ as $\theta \to \infty$. The divergence originates from the last term in Eq. (50), or deviation of the switching threshold voltage of the inverter from the defined threshold voltage $V_{DD}/2$.

Equation (50) also shows divergence of a different kind. Weak divergence like $\theta^{1/2}$ at $(B_P/B_N) = 1$ is the consequence of the perfectly saturating collapsible current generator FET model. At the moment when the input voltage reaches $V_{GSW} = V_{DD}/2$, the output node is still at V_{DD}. The currents of the PFET and the NFET are equal, and therefore there is no excess current available to discharge the load capacitance. After that time, the difference in the currents increases linearly with time. The output node voltage V_0 decreases proportional to the square of the time. The time required to reduce the voltage by $V_{DD}/2$ is therefore proportional to the inverse of the square root of the input voltage slew rate.

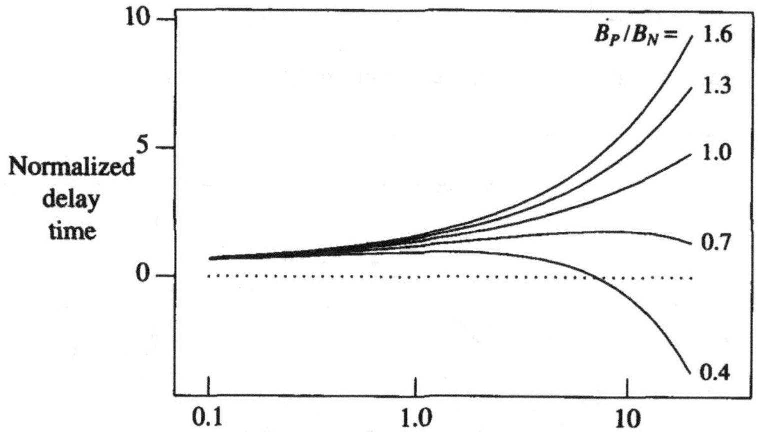

Normalized input transition time $t_0/(C_L/B_N)$

Figure 70 Delay versus input switching time

If current-voltage characteristics of the FETs are not discontinuous at zero drain voltage, the divergence proportional to $\theta^{1/2}$ does not exist. This is demonstrated as follows. We use the simplified FET current-voltage characteristic introduced in Section 1.7, and we consider a symmetrical inverter. The circuit equation is then

$$\frac{C_L}{B_H}\frac{dV_0}{dt}=f(V_{DD}-V_1,V_{DD}-V_0)-f(V_1,V_0) \text{ where } f(V_1,V_0)=\frac{V_1V_0}{V_A+V_0}.$$

We assume $V_1=\alpha t$ where α is the slew rate of the input voltage. Since we consider the limit of slow switching, solution of an implicit equation $V_0=\phi_0(V_1)$ obtained by setting $dV_0(t)/dt=0$

$$f\,[V_{DD}-V_1,V_{DD}-\phi_0(V_1)]=f[V_1,\phi_0(V_1)]$$

is the first approximation of the solution of the differential equation. Function $\phi_0(V_1)$ can be found in a closed-form as

$$\phi_0(V_1)=\frac{V_{DD}(V_{DD}-V_A-2V_1)+\sqrt{V_{DD}{}^2(V_{DD}-V_A-2V_1)^2+4V_{DD}V_A(V_{DD}-2V_1)(V_{DD}-V_1)}}{2(V_{DD}-2V_1)}$$

where $\phi_0(V_1)$ satisfies $\phi_0(0)=V_{DD}$, $\phi_0(V_{DD})=0$ and $\phi_0(V_{DD}/2)=V_{DD}/2$, since the inverter is pull-up/pull-down symmetrical. This formula gives the gate's static switching characteristics. The second approximation of the solution, $\phi_1(V_1)$, is found by approximating the time derivative using the first approximation as

$$f\,[V_{DD}-V_1,V_{DD}-\phi_1(V_1)]-f\,[V_1,\phi_1(V_1)]=(C_L/B_H)\phi_0'(V_1)$$

where the time derivative, $\phi_0'(V_1)$, is given by

$$\phi_0'(V_1)=\frac{d\phi_0(V_1)}{dt}=\frac{dV_1}{dt}\frac{d\phi_0(V_1)}{dV_1}=\alpha\frac{d\phi_0(V_1)}{dV_1}.$$

This equation can be solved in closed form: $\phi_1(V_1)$ becomes a solution of another quadratic equation

$$A_0\phi_1(V_1)^2+A_1\phi_1(V_1)+A_2=0 \tag{51}$$

$$\text{where } V_C=(C_L/B_H)\phi_0'(V_1) \quad A_0=V_{DD}-2V_1-V_C$$

$$A_1=-V_{DD}{}^2+2V_{DD}V_1+V_{DD}V_A+V_{DD}V_C$$

$$\text{and } A_2=(V_{DD}+V_A)V_AV_C-(V_{DD}-V_1)V_{DD}V_A$$

Figure 71 shows the static switching waveform $\phi_0(\alpha t)$ (closed circles) and typical *slow* dynamic switching waveform $\phi_1(\alpha t)$ (open circles). The dynamic waveform (approximation of the real waveform) is delayed with respect to the static waveform $V_0=\phi_0(V_1)$, and the difference is the finite delay time.

The inverter switching delay time can be approximated as follows. If we set $\phi_1(V_1)=V_{DD}/2$ and $V_1=(V_{DD}/2)+\Delta V_G$ in Eq. (51) we have

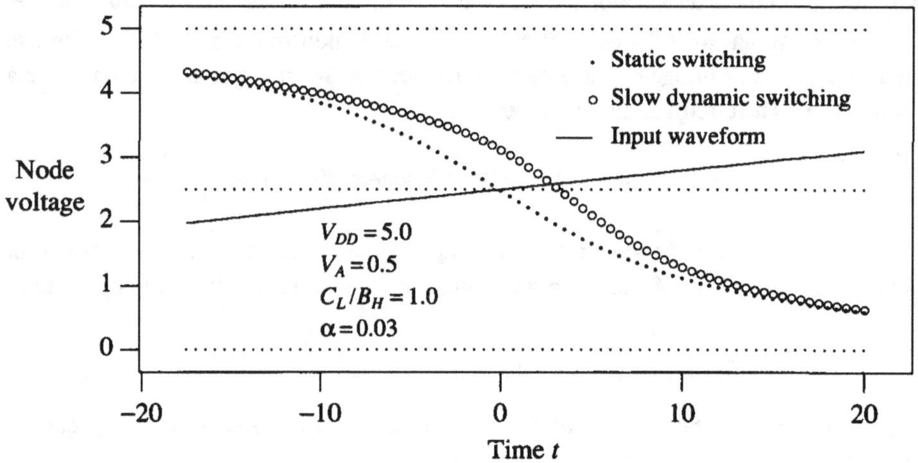

Figure 71 Static and almost static switching waveforms

$$\Delta V_G=-\frac{1}{2}\frac{C_L}{B_H}\alpha\left[\frac{d\phi_0(V_1)}{dV_1}\right]_{V_1=(1/2)V_{DD}}\left[\frac{\partial f}{\partial V_1}\right]^{-1}_{V_1=V_0=(1/2)V_{DD}}$$

and since $\Delta V_G=\alpha T_D$ where T_D is the delay time, the delay time in the limit as $\alpha \to 0$ can be derived. The delay time of the symmetrical inverter remains finite and is of the order of C_L/B_H.

5.17 Power Supply Voltage Variation

Differential mode power-bus noise, or power supply voltage variation, has the effect of modulating gate delays [17]. If the cumulative delay of a gate chain exceeds the timing budget due to this delay modulation, the circuit fails. This is the mechanism that was introduced in Section 4.12. The effects of power supply voltage variation on CMOS gate delay is understood from the following two extrema cases: the first is the effect of slow, global voltage variation, and the second is the effect of fast, local voltage variation. The two cases are schematically shown in Figs. 72(a) (global) and (b) (local), respectively.

5.17.1 Global Power Supply Voltage Variation

Let us consider the first case, slow global power supply voltage variation [14]. By this type of voltage variation, CMOS logic amplitude varies simultaneously with the power supply voltage, and this is the crucial point. Pull-down delay time of the inverter shown in Fig. 72(a) is determined using the simplified high-field FET model of Section 1.7. The collapsible current

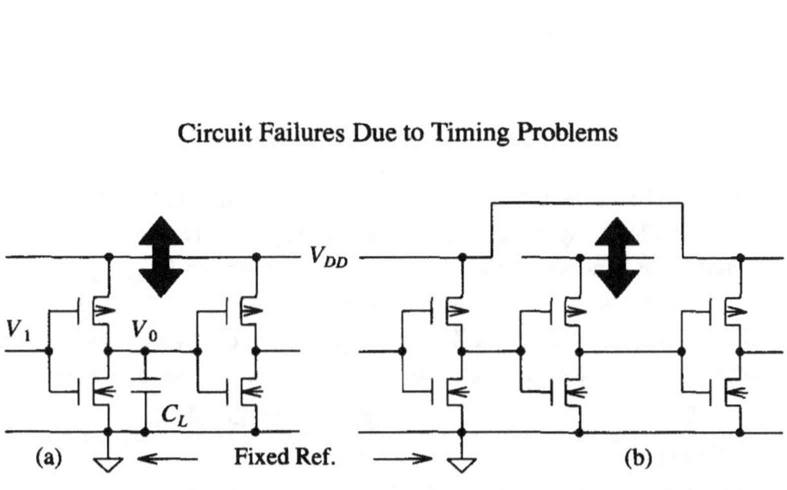

Figure 72 Effects of power supply fluctuation on gate delay

generator model is too simple to provide details. How FET drain current increases with increasing drain voltage is crucial.

In the circuit of Fig. 72(a), V_1 makes an instantaneous low-to-high transition at time $t=0$. Output node voltage $V_0(t)$ satisfies the circuit equation

$$C_L \frac{dV_0(t)}{dt} = -\frac{B_{HN}(V_{DD}-V_{THN})V_0(t)}{V_0(t)+V_A}.$$

The equation is solved subject to the initial condition $V_0(0)=V_{DD}$ to obtain

$$V_{DD}-V_0(t)+V_A\log[V_{DD}/V_0(t)]=(B_{HN}/C_L)(V_{DD}-V_{THN})t$$

Delay time T_D is found by setting $V_0(t)=V_{DD}/2$ to obtain

$$T_D=\frac{C_L}{B_{HN}}\frac{(V_{DD}/2)+V_A\log 2}{V_{DD}-V_{THN}} \tag{52}$$

$$\to\frac{1}{2}\frac{C_L}{B_{HN}}\frac{1}{1-(V_{THN}/V_{DD})}\ (V_A\to 0)\ \text{and}\ \to\frac{C_L}{B_{HN}}\frac{V_A\log 2}{V_{DD}[1-(V_{THN}/V_{DD})]}\ (V_A\to\infty).$$

In the limit as $V_A\to 0$ (the collapsible current generator model), T_D depends only on the ratio V_{THN}/V_{DD}. As long as (V_{THN}/V_{DD}) is small, the T_D is practically independent of the power supply voltage. In scaled-down 5-volt CMOS, inverter delay is indeed only weakly dependent on power supply voltage down to about 3 volts, where the effects of the term in the denominator of Eq. (52), including V_{THN}/V_{DD}, begin to show up, and the delay increases rapidly with further decrease of V_{DD}. In the limit as $V_A\to 0$, or in the collapsible current generator model, the current of a fully turned-on FET is proportional to the power supply voltage minus the FET threshold voltage. Since the charge that must be drained by the FET in the process of switching is proportional to the power supply voltage, the delay time, which is the ratio of the two, is independent of the power supply voltage. This conclusion is approximately valid in NAND/NOR gates as well. The rate of charge drained from the entire FET chain is controlled by the current of the single NFET that is closest to the grounded end of the NFET chain. The time to discharge the chain completely is the ratio of the stored charge to the current

$$[\sum_i C_i(V_{DD}-V_{THN})+C_LV_{DD}]/B_N(V_{DD}-V_{THN})=\text{function of }(V_{THN}/V_{DD})\text{ only}$$

where C_i and C_L are the capacitances of the i-th internal node of the NFET chain (from the grounded end) and the load capacitance of the chain, respectively (we neglect voltage dependence of the island capacitances). Delay depends only on the ratio V_{THN}/V_{DD}.

In the other limit, as $V_A \to \infty$ (the linear resistor model), T_D depends on the ratio V_{THN}/V_{DD} and V_{DD} itself. Different FET models give different power supply voltage dependence. For instance, the gradual-channel, low-field FET model gives delay time

$$T_D = \frac{C_L}{B_{HN} V_{DD}} \Delta(x) \quad \text{where} \quad \Delta(x) = \frac{2x}{(1-x)^2} + \frac{\log(3-4x)}{1-x} \quad \text{and } x = \frac{V_{THN}}{V_{DD}}$$

which contains both V_{DD} and V_{THN}/V_{DD}. In the gradual channel, low-field FET model, conductance is proportional to the power supply voltage. Since delay time is the ratio of the output node capacitance and the conductance, direct V_{DD} dependence remains. The difference in the voltage dependence in the two limits of the FET model is due to the difference in the degree of *activity* of the FET. The FET is active, or self-regulating, to the maximum extent if modeled using the collapsible current generator model. Only the gate voltage determines the current. The gradual-channel, low-field model FET is much less active, and both the gate and the drain voltages determine the current. The extra dependence of channel current on the drain voltage adds dependence of delay on V_{DD} itself. In both cases the ratio V_{THN}/V_{DD} is the second factor that determines the delay, and this is the essential factor in the power supply voltage dependence.

If we define nominal drain voltage, $V_{DD}(0)$, and normalize the parameters in Eq. (52) using $v_{DD} = V_{DD}/V_{DD}(0)$, $v_{THN} = V_{THN}/V_{DD}(0)$ and $v_A = V_A/V_{DD}(0)$

$$\frac{T_D}{T_D(0)} = \frac{1 - v_{THN}}{v_{DD} - v_{THN}} \frac{v_{DD} + 2\log 2 \cdot v_A}{1 + 2\log 2 \cdot v_A} \quad \text{where} \quad T_D(0) = \frac{C_L}{B_{HN}} \frac{[V_{DD}(0)/2] + \log 2 \cdot V_A}{V_{DD}(0) - V_{THN}}.$$

$T_D/T_D(0)$ versus v_{DD} is plotted in Fig. 73 for several values of v_A.

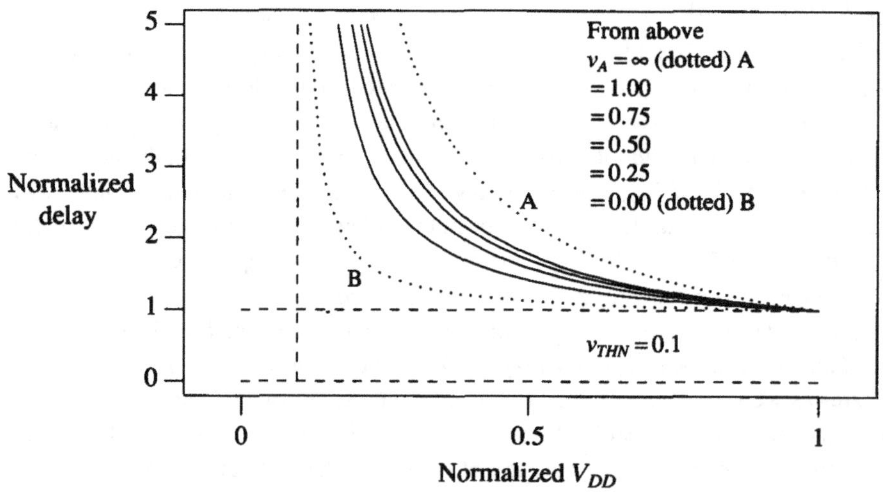

Figure 73 Inverter delay versus power supply voltage

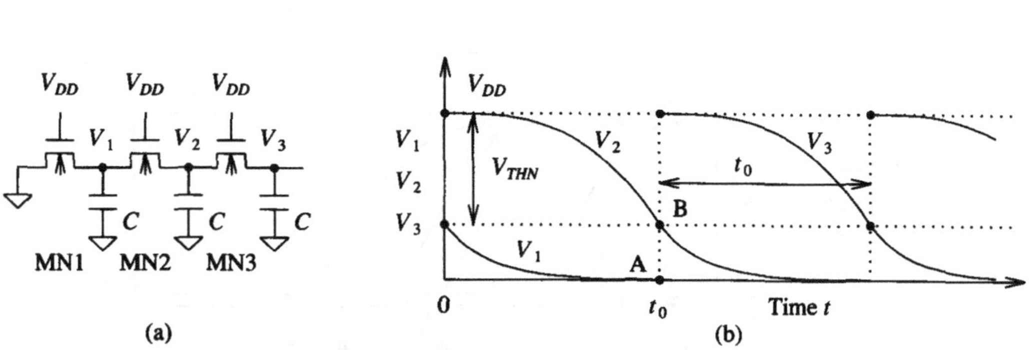

Figure 74 Sequential discharge of FET chain at low voltages

If a CMOS gate consists of series-connected NFETs, the power-supply voltage dependence becomes complex. We show that a FET chain discharges by two distinct modes at high ($V_{DD} \gg V_{THN}$) and at low ($V_{DD} \approx V_{THN}$) drain voltages. At high V_{DD} many or all the NFETs in the NFET chain of Fig. 74(a) conduct simultaneously. Since several consecutive sections of the NFET chain of Fig. 74(a) transfer charge to the left side simultaneously, the discharge process proceeds rapidly. Waveforms are shown in Fig. 75(a). Three to four NFETs are conducting simultaneously. At low V_{DD}, gate voltages of the NFETs are close to the threshold voltage V_{THN}. No two NFETs can be conducting simultaneously. The second NFET is unable to turn on until the first NFET pulls node 1 practically down to ground. Until node 2 is pulled down to ground, node 3 discharge does not begin. In this way the NFET chain discharges more sequentially than at higher voltages. Node voltage waveforms are shown in Fig. 75(b). Because of this effect, NAND gate pull-down delay time increases more rapidly than inverter delay time, with decreasing power supply voltage.

How this sequential FET chain discharge proceeds in the limit of low drain voltage is interesting. A similar problem of long NFET chain discharge was solved in Section 2.15 assuming that the threshold voltage V_{THN} is zero. Here we study the case $V_{DD} \approx V_{THN}$. In Fig. 74(a), nodes 1, 2, ... are all charged to V_{DD} before discharge begins. NFETs are modeled using the collapsible current generator model, and the threshold voltage of the NFET is independent of the source voltage. $\Delta V = V_{DD} - V_{THN}$ is small. Suppose that MN1 of Fig. 74(a) discharged node 1 to $V_1 = V_{DD} - V_{THN}$. MN2 conducts and V_1 is found from the circuit equation

$$C\frac{dV_1(t)}{dt} = -B_N(V_{DD} - V_{THN}) + B_N[V_{DD} - V_1(t) - V_{THN}] = -B_N V_1(t)$$

as

$$V_1(t) = (V_{DD} - V_{THN})exp[-(B_N/C)t]$$

$V_2(t)$ is determined from

$$C\frac{dV_2(t)}{dt} = -B_N[V_{DD} - V_1(t) - V_{THN}]$$

using the solution for $V_1(t)$, $V_2(t)$ is found as

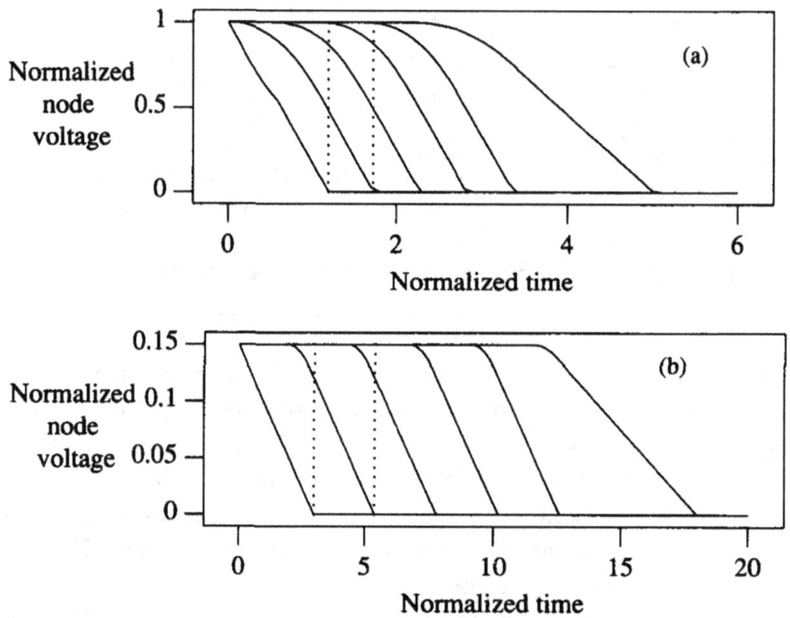

Figure 75 Sequential and nonsequential discharge process of NFET chain

$$V_2(t) = V_{DD} - \frac{B_N}{C}(V_{DD} - V_{THN})[t - \frac{C}{B_N}[1 - exp[-(B_N/C)t]]]\,.$$

In the limit as $t \to 0$, $V_2(t)$ may be approximated as follows:

$$V_2(t) \to V_{DD} - \frac{V_{DD} - V_{THN}}{2}\left(\frac{B_N}{C}\right)^2 t^2\,.$$

If $V_2(t_0) = V_{DD} - V_{THN}$, t_0 of Fig. 74(b) is determined as

$$t_0 \approx \frac{C}{B_N}\left(\frac{2V_{THN}}{V_{DD} - V_{THN}}\right)^{1/2} \gg \frac{C}{B_N}\,.$$

$V_1(t)$ decays with time constant C/B_N. Node 2 takes much longer to discharge from V_{DD} (original) to $V_{DD} - V_{THN}$ because of the factor $\sqrt{2V_{THN}/(V_{DD} - V_{THN})}$. By that time, V_1 has practically reached ground. Although theoretically V_1 decreases but never reaches ground, the *residual* voltage can be neglected. After $t > t_0$, V_2 is given by

$$V_2(t) = (V_{DD} - V_{THN})exp[-(B_N/C)(t - t_0)]$$

We have a *travelling* node voltage waveform, with per stage delay time t_0, as shown in Fig. 74(b). This analysis is only approximation, since FET model at small current is precise only if subthreshold current is included.

5.17.2 Local Power Supply Voltage Variation

We study the effects of fast and local power supply voltage variation shown in Fig. 72(b). If the ground voltage remains unchanged and V_{DD} changes, the gate's output node is not influenced if it is low. If it is high, however, the output node and the V_{DD} power-bus are connected by the PFET channel resistance, and V_{DD} voltage variation is coupled more or less directly to the gate output node.

The pull-up PFET is in the triode region. The equivalent circuit in this case is shown in Fig. 76(b). The linear resistor FET model is more realistic than the collapsible current generator model for this case. The channel resistance, R_D, is the small-signal resistance of the fully conducting channel [$R_D = 1/B_G(V_{DD} - V_{THP})$ in the gradual channel, low-field FET model]. The circuit equation is

$$(C_P + C_N)\frac{dV_0(t)}{dt} = C_P\frac{dV_{DD}(t)}{dt} + \frac{V_{DD}(t) - V_0(t)}{R_D}.$$

By using $V_0(t) = V_{DD}(0) + \Delta V_0(t)$ and $V_{DD}(t) = V_{DD}(0) + \Delta V_{DD}(t)$ the solution is given by

$$\Delta V_0(t) = \frac{C_P}{C_P + C_N}\Delta V_{DD}(t) + \frac{C_N}{R_D(C_P + C_N)^2}\int_0^t \Delta V_{DD}(\theta) exp[-(t-\theta)/R_D(C_P + C_N)]d\theta$$

This case was discussed before, in Section 3.12.

Suppose that the input voltage to the inverter makes an instantaneous high-to-low transition. Following the input node transition, the output node voltage increases with time. If V_{DD} bus noise occurs during the pull-up transition, the output node voltage changes accordingly. This effect modifies the gate delay time, thereby introducing circuit delay uncertainty.

The equivalent circuit for power-bus noise is shown in Fig. 76(a). If power-bus noise occurs during the pull-up transition, the circuit equation is written down from the equivalent circuit of Fig. 76(a) using the collapsible current generator model as

$$(C_P + C_N)\frac{dV_0(t)}{dt} = C_P\frac{dV_{DD}(t)}{dt} + B_P[V_{DD}(t) - V_{THP}]$$

We write $V_{DD}(t) = V_{DD}(0) + \Delta V_{DD}(t)$ [$\Delta V_{DD}(0) = 0$]. The circuit equation is solved subject to the initial condition $V_0(0) = 0$ as

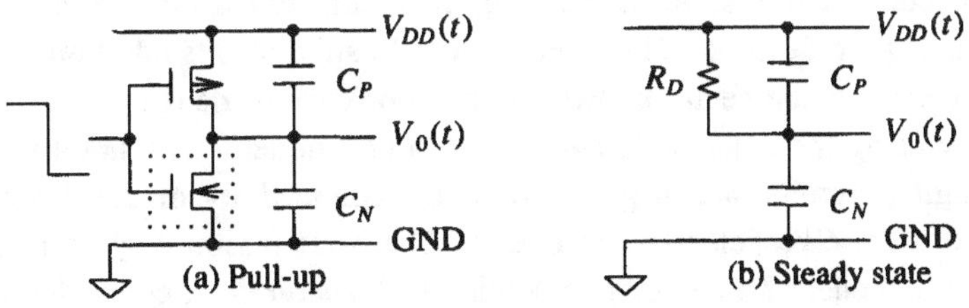

Figure 76 Effects of local power-bus noise

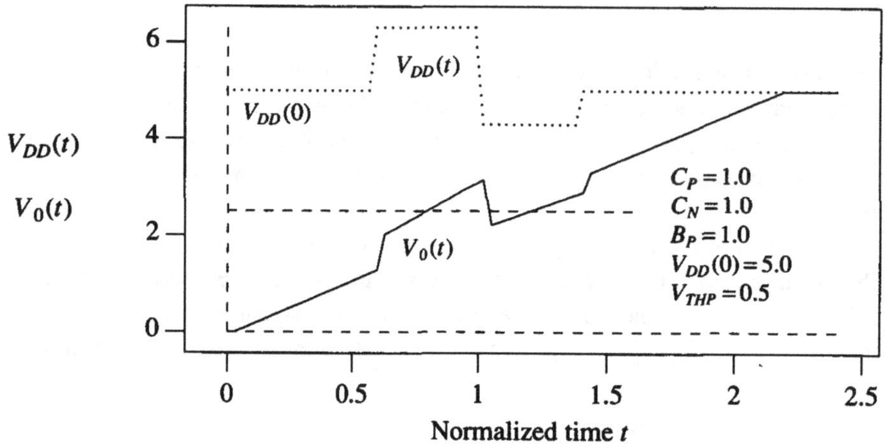

Figure 77 Power-bus noise and node waveform

$$V_0(t) = \frac{B_P}{C_P + C_N}[V_{DD}(0) - V_{THP}]t \tag{53}$$

$$+ \frac{C_P}{C_P + C_N}\Delta V_{DD}(t) + \frac{B_P}{C_P + C_N}\int_0^t \Delta V_{DD}(\theta)d\theta .$$

Figure 77 shows $V_{DD}(t)$ and $V_0(t)$ for typical parameter values.

5.18 General Theory of Digital Node Impedance

The basic requirement for understanding timing-related CMOS circuit failure mechanisms is that delay times and node waveforms of a CMOS circuit can be determined. They are determined from the capability of the node driving source and from the response characteristic of the circuit that *loads* the node. Driving sources in CMOS digital circuits are conducting FETs, whose characteristics can be derived from the basic semiconductor device physics. If more than one FET is connected in series or parallel, the characteristics of the *driving source* must be determined using the devices and circuit theory. In spite of this additional complexity, the properties of the driving source are known. In this book we made the ultimate driving source simplification by using the collapsible current generator model. This simplification made the properties of digital circuits that originate from the connectivity of the devices, rather than from the devices themselves, clearly observable. The basic reason why such simplification is possible is that the model of a device is essentially a *structureless*, controlled current or voltage source, whose strength is determined by the values of only a few terminal voltages and currents. This simplification is valid as long as the transit time of carriers within the device is significantly less than the switching delay time. The response characteristics of the circuits that load the node are, however, complicated. The impedance of the loading circuit depends on the values of many signals, both the present and past, because loading circuits are extensive, have internal structure, and can be an active circuit that includes feedback. In digital circuits, no special design techniques are used to make node impedance well behaved. The standard techniques used in analog circuits such as cathode follower and cascode are practically never used in digital circuits. Then digital circuit node impedance is far more complicated than analog circuit node impedance. Since there are significant differences in the physical nature between the driver and the load, it is useful to develop a new theory. In the author's opinion, switching transients of CMOS digital circuits has never been analyzed by separating the effects of the driving source and the characteristics of the load impedance. The objective of this section is to derive a model of node impedance that is compatible and that is convenient if used with the collapsible current generator model. Circuit loading a node is conventionally characterized as equivalent to *two-terminal* impedance. The impedance is complicated, however, because the impedance is not simply capacitive, resistive, inductive, or often not even *passive*. A quite general theory of two-terminal impedance is required. The theory presented here is intended as the first step toward this unexplored area.

We approach this problem as follows. If a CMOS digital circuit node is driven by a time-dependent current generator $I(t)$, voltage $V(t)$ is developed between the terminal and ground. Similarly if the same terminal is driven by a time-dependent voltage source $V(t)$, we have current $I(t)$. We do not specify *driving* waveforms $I(t)$ or $V(t)$ too closely: we assume $I(t)$ and $V(t)$ are *typical* charge-discharge current or node voltage, respectively. The relationship between $I(t)$ and $V(t)$, however, may have many varieties, but the relationships that practically exist in CMOS digital circuits are rather limited. By studying a few typical cases, we gain insight into the nature of the digital node impedance.

In Fig. 78(a) the relationship between $I(t)$ and $V(t)$ may be *approximated* by

$$I(t) \approx C\frac{dV(t)}{dt}. \qquad (54)$$

Then the circuit presents essentially a *capacitive* impedance to the driver. If $I(t)$ and $V(t)$ were related by

$$V(t) \approx RI(t) \qquad (55)$$

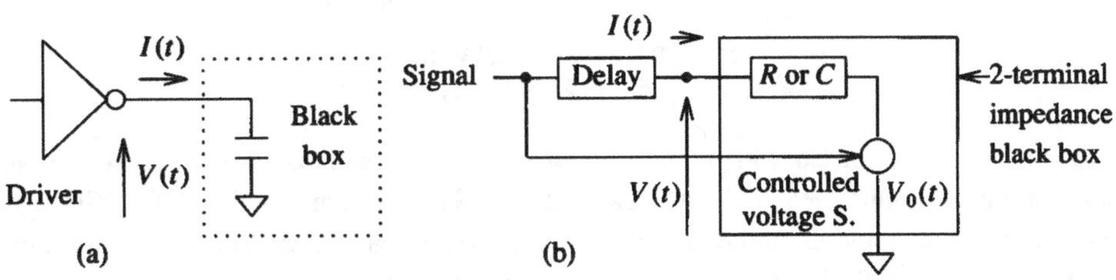

Figure 78 Generalized structure of two-terminal node impedance

the circuit within the box presents essentially a resistive impedance to the driver. These two examples are expected to be the most common in CMOS circuits because of the physical construction of the circuits.

Equations (54) and (55) can be generalized by including one more component that is connected to the other terminal of the resistor or the capacitor, and to ground. The component can be any kind of impedance or *controlled* voltage source, an example of which is shown in Fig. 78. If the series impedance is capacitance, the two-terminal characteristic of the black box is expressed by

$$I(t) \approx C \frac{d}{dt} [V(t) - V_0(t)] \tag{56}$$

and if the impedance is resistance,

$$I(t) \approx [V(t) - V_0(t)]/R \,. \tag{57}$$

This generalization reflects physical mechanisms of CMOS circuits well. Any circuit can be connected to the other end of the resistance or the capacitance. The capacitor or resistor receives excitation from voltage source $V_0(t)$ simultaneously, a short time before or after, or a long time before or after the voltage source $V(t)$. Suppose that the series impedance is resistance. The controlled voltage source $V_0(t)$ that follows $V(t)$ with zero delay has the effect of changing the resistance value. Response only a short time before or after can be represented by generalized capacitance, and response some significant time before or after can be represented by an *independent* voltage source. The simple combination of impedance and voltage source is able to represent a wide range of node impedances. In this sense, Fig. 77(b) shows a general equivalent circuit of a digital node.

To gain more insight, we assume a simple case

$$V_0(t) = G \cdot V(t - \tau)$$

or $V_0(t)$ is a constant *gain*, G, times the value of $V(t)$ a time τ earlier. Suppose that τ is small. We assume G and τ can be either positive or negative. We expand $V(t-\tau)$ into a power series

$$V(t-\tau) = V(t) - \frac{dV(t)}{dt}\tau + \frac{1}{2!}\frac{d^2V(t)}{dt^2}\tau^2 - \ldots..$$

with derivatives computed at time t. By substituting this into Eq. (57) we obtain

$$I(t) = \frac{1-G}{R}V(t) + \frac{G\tau}{R}\frac{dV(t)}{dt} - \frac{G\tau^2}{2R}\frac{d^2V(t)}{dt^2} + \frac{G\tau^3}{6R}\frac{d^3V(t)}{dt^3} - \frac{\tau^4}{24R}\frac{d^4V(t)}{dt^4} + o(\tau^5) \tag{58}$$

The physical meaning of Eq. (58) is that the two-terminal port appears as if it were a parallel combination of resistance $R/(1-G)$ and capacitance $G\tau/R$ for small values of τ. Since G and τ may take any value and sign, the effective resistance and capacitance can be positive or negative. If $G = 1$ and $\tau > 0$, the load appears as if it were a pure *capacitance*.

Similarly from Eq. (56)

$$I(t) \approx C\frac{d}{dt}[V(t) - G \cdot V(t-\tau)] = C(1-G)\frac{d}{dt}V(t) + C\tau\frac{d^2V(t)}{dt^2} - \frac{C\tau^2}{2}\frac{d^3V(t)}{dt^3} \qquad (59)$$

$$+\frac{C\tau^3}{6}\frac{d^4V(t)}{dt^4} - \frac{C\tau^4}{24}\frac{d^5V(t)}{dt^5} + o(\tau^5)$$

and the node impedance appears like capacitance having a different value. If $G=1$ the capacitance vanishes, and if $G>1$ the capacitance becomes negative.

We determine the condition required for the generalized two-terminal impedance to be similar to a conventional two-terminal impedance. We consider a typical digital node waveform $V(t)$ shown in Fig. 79. At time $t=t_{SW1}$, $V(t_{SW1})=V_{GSW}$, where V_{GSW} is the *typical* switching threshold voltage of a CMOS gate. Rise/fall times of the waveform are T_{RF}, which is the time required to make transition between the threshold voltages V_{THN} to $V_{DD}-V_{THP}$. Figure 79 shows two waveform possibilities of the internal voltage generator $V_0(t)$, which switches at delayed or advanced time $\pm t_{SW0}$. Obviously the simple impedance approximation works if

$$|t_{SW0} - t_{SW1}| \ll T_{RF}.$$

If the inequality is reversed

$$|t_{SW0} - t_{SW1}| \gg T_{RF}$$

then the internal voltage generator works effectively as an *independent* voltage source. It is interesting to study how the two distinct regimes change over.

We assume a simple Gaussian node waveform

$$V(t) = V_{DD} f_0[(t - t_{SW1})/t_R] \quad \text{where} \quad f_0(\theta) = \frac{1}{2} + \frac{1}{\sqrt{\pi}}\int_0^\theta exp(-u^2)du$$

and where t_R is the measure of the rise time of signal. Setting $t_{SW1}=0$ we have

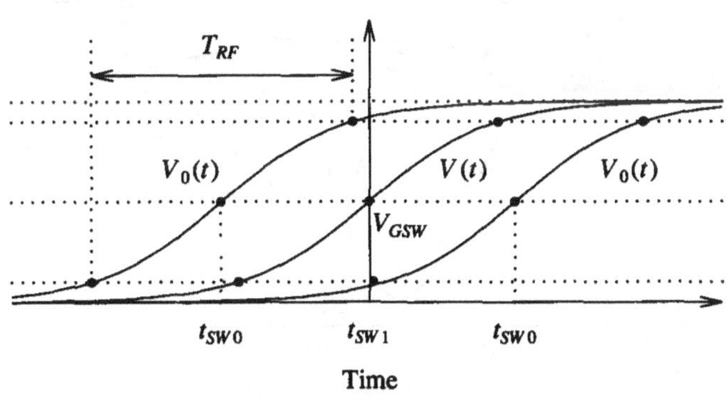

Figure 79 Digital node waveforms, rise/fall times

$$\frac{d^{(k)}V(t)}{dt^{(k)}} = \frac{V_{DD}}{t_R{}^k} f_k(t/t_R).$$

where

$$f_k(\theta) = \frac{1}{\sqrt{\pi}} P_k(\theta) exp(-\theta^2) \qquad P_{k+1}(\theta) = -2\theta P_k(\theta) + P'_k(\theta).$$

The first several terms are as follows.

$$P_1(\theta) = 1 \quad P_2(\theta) = -2\theta \quad P_3(\theta) = 4\theta^2 - 2 \quad P_4(\theta) = -8\theta^3 + 12\theta$$

$$P_5(\theta) = 16\theta^4 - 48\theta^2 + 12 \quad P_6(\theta) = -32\theta^5 + 160\theta^3 - 120\theta.$$

Figure 80 shows plots of $f_k(\theta)$ versus θ. In Eq. (58), $I(t)$ is the superposition of terms including $V(t)$ and its derivatives. We assume $G = 1$. Since the first term of Eq. (58) then vanishes, the dominant term is proportional to the first derivative, $dV(t)/dt$, which represents an isolated pulse shown by solid curve 1 of Fig. 81. If the next higher derivative $d^2V(t)/dt^2$ is added, the isolated pulse shifts on the time axis, as shown by dotted curve 2 of Fig. 81. If the third derivative is added, the width of the pulse increases or decreases as shown by solid curve 3. If an isolated pulse is characterized by its position in time and its duration, the two modes of variation of pulse are the essential ones.

Using Eq. (58), the circuit response is expanded in a power series of τ. If τ is small, the power series converges; by taking a few terms, a good approximation of the waveform can be obtained. Figure 81 shows the response for $\tau/t_R = 1.2$. Dots in the figure show the current computed exactly, and curves 1, 2, 3, 4, 5, and 6 are approximations by taking sum of first 1, 2, 3, 4, 5, and 6 terms, respectively, of Eq. (58). The first term (curve 1) represents essentially capacitive response. Obviously the successive approximation converges to the exact current.

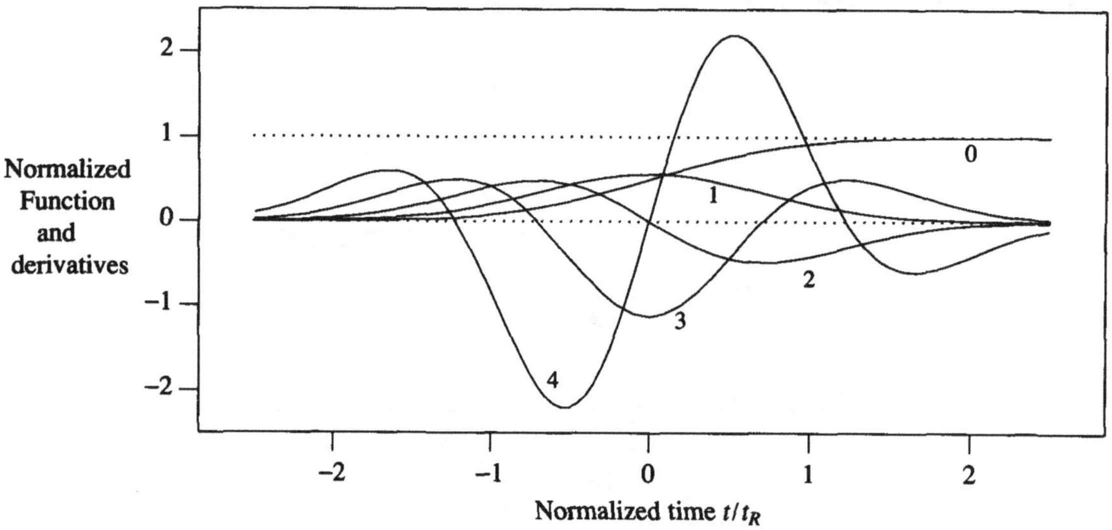

Figure 80 Node waveform and its higher derivatives

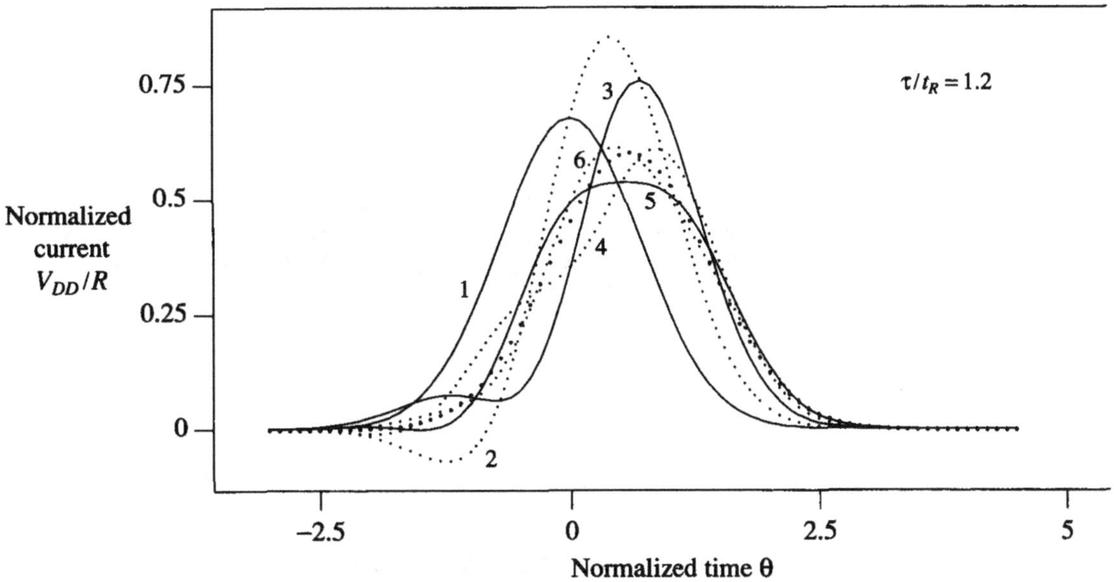

Figure 81 Approximation of node current

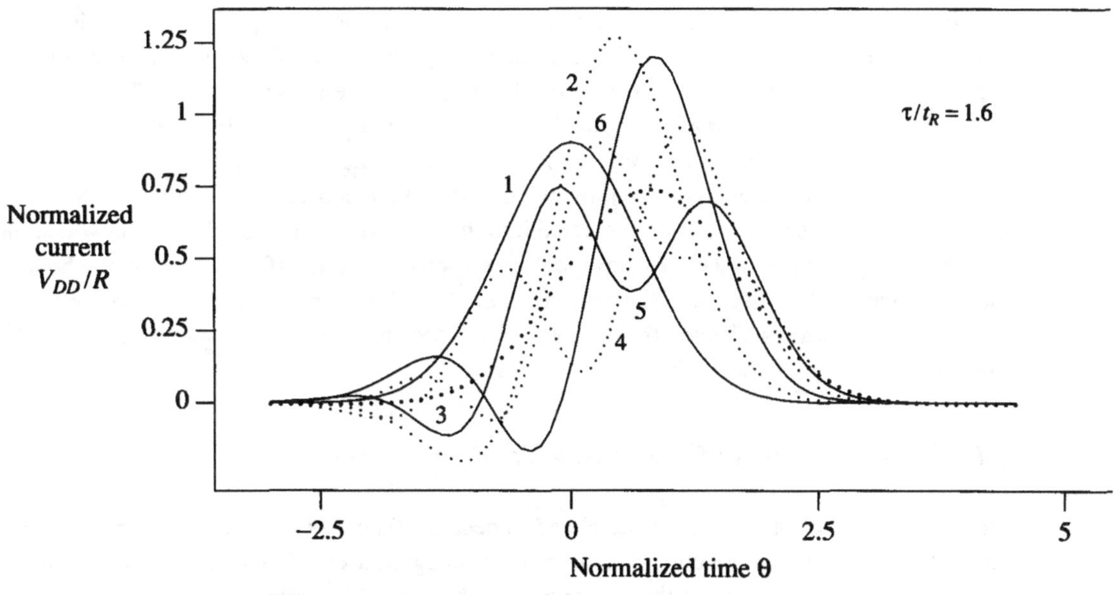

Figure 82 Approximation of node current

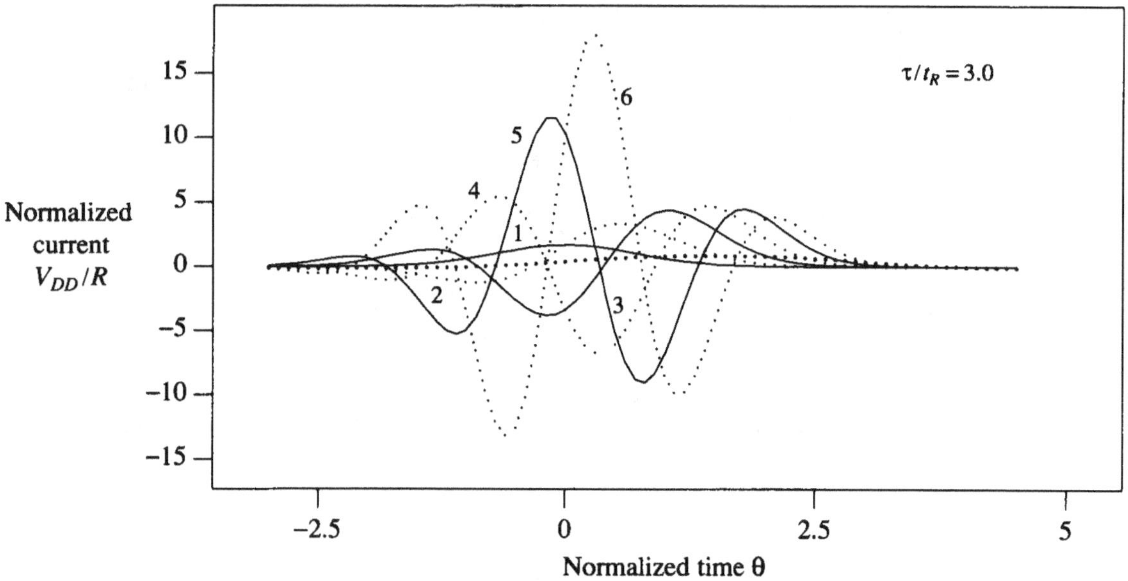

Figure 83 Approximation of node current

In Fig. 82 $\tau/t_R = 1.6$. Convergence gets worse with increasing delay time. Figure 83 shows response when $\tau/t_R = 3.0$. The successive approximation diverges, and therefore the impedance approximation fails. Delay time is too long to define impedance. The circuit is equivalent to a series combination of a resistance and an independent voltage source. The two regimes are identified clearly by convergence or by divergence of the infinite series. Convergence of the infinite series expansion assures that the two electrical phenomena are *coherent*. The generalized impedance can be separated into the two extrema cases: the case in which the internal voltage source delay (or advance) is so small that interference between the two coherent signals takes place, and the case in which the delay is too long to maintain the coherence. The latter case is simple: the problem is reduced to the backward signal flow from the independent voltage source to the driver.

5.18.1 Application of Generalized Node Impedance

An interesting application of the generalized impedance theory is as follows. We noted that a capacitor built on the conducting substrate of an integrated circuit has undesirable parasitic capacitance. The lower face of the capacitor plate has parasitic capacitance to the silicon substrate. The bulk CMOS substrate surface is held at either V_{SS} or V_{DD}. Figure 84(a) shows the capacitor cross-section in an MOS IC. Electrodes P_1 and P_2 are metal and polysilicide, respectively, and P_3 is the substrate (such as P+ on P-tub). We intend to use the capacitance between electrodes P_1 and P_2 [capacitance C_1 in the equivalent circuit of Fig. 84(b)], but

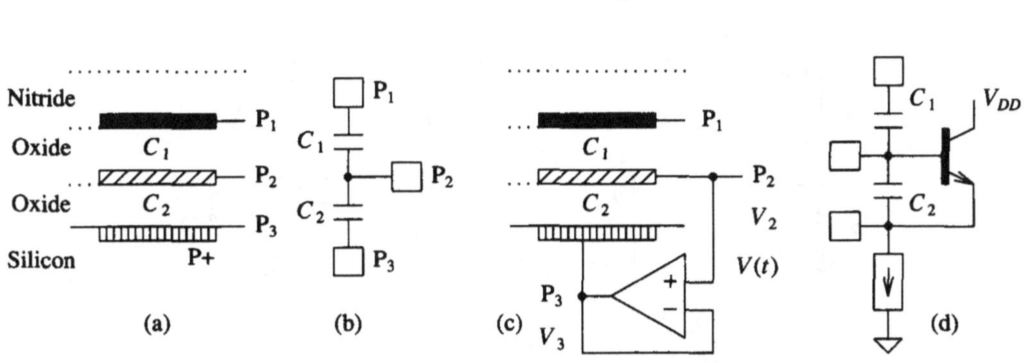

Figure 84 Substrate capacitance compensation

inevitably capacitance between electrodes P_2 and $P_3[C_2$ in Fig. 84(b)] is associated with C_1. If node P_2 is either grounded or driven from a signal source having low internal impedance, the parasitic capacitor C_2 causes no harm, but such use of capacitor C_1 is restrictive.

In Fig. 84(c), the potential of electrode P_3 is set by an active voltage follower. Since the voltage of electrode P_3 follows P_2, capacitance C_2 does not affect node P_2. A simple and fast voltage follower can be built using the emitter follower of a fast BJT available in BiMOS technology. The circuit is shown in Fig. 84(d). Suppose that V_2 of Fig. 84(c) is $V(t)$, and V_3 is $(1-\alpha)V(t-\tau)$, where α is error from unity gain and τ is a small delay of the voltage follower. Then the current I through capacitance C_2 is given by

$$I = C_2 \frac{d}{dt}[V(t)-(1-\alpha)V(t-\tau)] = C_2 \frac{d}{dt}[\alpha V(t)+(1-\alpha)\tau \frac{dV(t)}{dt}] \tag{60}$$

Equation (60) can be rewritten as

$$I = \alpha C_2 \frac{d}{dt}[V(t)+R\alpha C_2 \frac{dV(t)}{dt}] \tag{61}$$

and this equation means that the impedance due to C_2 on node P_2 is the series combination of *reduced* capacitance αC_2 in series with negative resistance $-R$ where

$$R = \frac{(1-\alpha)\tau}{\alpha^2 C_2}. \tag{62}$$

Let the current through capacitance C_2 when the voltage follower is turned off be I_0. Then

$$\frac{I}{I_0} = \alpha + i\omega\tau \tag{63}$$

and therefore $I \to 0$ if $\alpha \to 0$ and $\tau \to 0$. We need a fast and high transconductance device for the circuit. If $\alpha = 0.98$, $\tau = 50psec$, and $\omega = 45(MHz)\times 2\pi = 2.8\times 10^8/sec$, $|I/I_0| \approx 0.03$, or about 97% of the parasitic capacitance can be compensated. There are many practical difficulties in using this scheme, but such a new technique creates a future for circuit engineering. As an example of the difficulties, the circuit using an emitter follower requires careful considera-

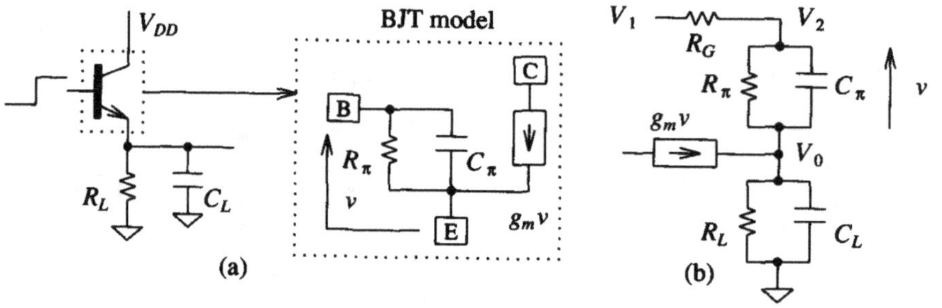

Figure 85 Emitter follower ringing

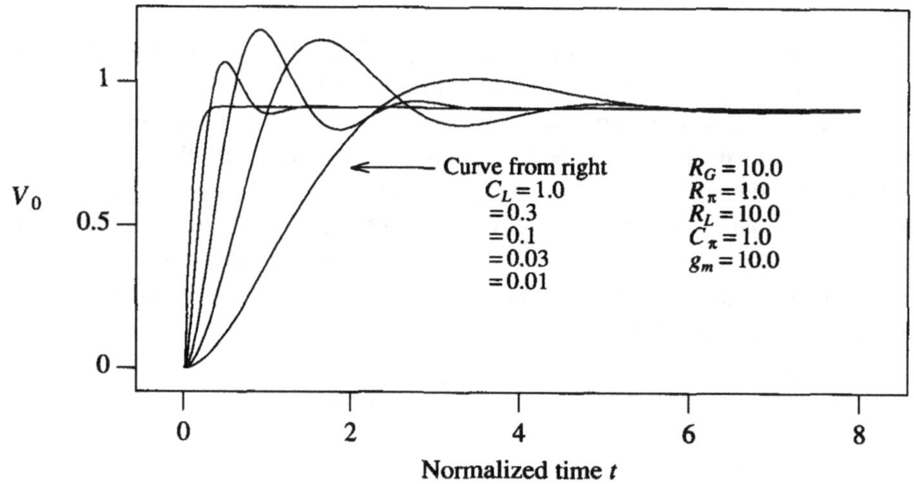

Figure 86 Response of emitter follower with capacitive load

tion of stability, since a capacitively loaded emitter-follower circuit has a singular ringing problem [18].

Figure 85(a) shows the emitter-follower circuit loaded with capacitor C_L, and Fig. 86 shows the switching transient waveforms of the output voltage. We note that ringing occurs only if the base-emitter time constant of BJT $R_\pi C_\pi$ is about the same order of magnitude as the load time constant $R_L C_L$. The BJT of Fig. 85(a) is modeled using the simple equivalent circuit shown in the dotted box of Fig. 85(a). The entire emitter-follower circuit becomes equivalent to the voltage divider circuit shown in Fig. 85(b). If V_1 (base voltage) makes a low-to-high transition, the wide-band voltage divider, consisting of R_π, C_π, R_L and C_L and satisfying $R_\pi C_\pi \approx R_L C_L$, divides the input pulse voltage, and it rapidly establishes voltage v, which is the emitter-base voltage of the BJT. The BJT, or its equivalent current generator $g_m v$, drives the emitter output node voltage up, and node voltage V_2 goes up above the pulse voltage V_1. A voltage overshoot is generated during this phase. Since V_2 becomes higher than pulse voltage V_1, R_G discharges the base node (capacitance C_π), and the emitter base voltage, v, of the BJT drops below the BJT conduction threshold voltage. Then current generator $g_m v$ turns off, the output node voltage falls, but the emitter-base voltage, v, is unable to recover until about time $R_G C_\pi$ or $R_\pi C_\pi$ elapses. Now the output node voltage V_0 undershoots. The output node voltage goes through a damped oscillation, as shown in Fig. 86. For this oscillation mechanism to work, the time constants $R_\pi C_\pi$ and $R_L C_L$ must be about the same. If either time constant is significantly larger than the other, the circuit does not go into damped oscillation. This is easily understood, since the circuit can be described by a single, first-order circuit equation, that has no oscillating solution.

Capacitance compensation may occur unintentionally in CMOS digital circuits that have positive feedback. An example is the capacitively coupled latch discussed in Section 6.5. Active capacitance compensation reduces node capacitance to a low level, and as a result the node may become noise sensitive.

References

[01] Fishburn, J. private communication

[02] The following idea is a significant deviation from the orthodoxy of the present digital circuit timing techniques. This discussion was inevitable, because I stand on the viewpoint that static and dynamic switching modes are fundamentally different, and the memory effect of dynamic switching mode can be used for circuit timing. Professor Rosenberger gave a detailed critical review on this contention, and the original text was modified quite significantly. I am grateful for his comments.

[03] Weste, N., and Eshraghian, K. *Principles of CMOS VLSI design.* Reading, MA: Addison-Wesley Publishing Co., 1985.

[04] Oberman, R. M. M. *Electronic counters.* Barnes and Noble Books, 1973.

[05] Krambeck, R. H., and Shoji, M. "Skew-free clock circuit for integrated circuit chip." U. S. patent No. 4,479,216.

[06] Shoji, M. "Reliable chip design method in high performance CMOS VLSI." *ICCD86 digest* (October, 1986): 389–92.

[07] Hasegawa, H., Furukawa, M., and Yanai, H. "Properties of microstrip line on Si-SiO_2 system." *IEEE Trans. on Microwave Theory and Techniques* vol. MTT-19 (November, 1971): 869–81.

[08] Shoji, M. "Electrical design of BellMac-32A microprocessor." *ICCC 82 digest* (September, 1982): 112–15.

[09] Tewksbury, S. K. *Wafer level integrated systems: implementation issues.* Boston, MA: Kluver Academic Publishers, 1988.

[10] Peatman, J. B. *Digital hardware design.* New York: McGraw Hill, 1980.

[11] Goncalves, N. P., and Man, H. J. de. "NORA: a racefree dynamic CMOS technique for pipelined logic structures." *IEEE J. Solid-State Circuits* vol. SC-18 (June, 1983): 261–68.

[12] McCluskey, E. J. *Logic design principles.* Englewood Cliffs, NJ: Prentice-Hall Inc., 1986.

[13] Shoji, M. "Synthetic CMOS static logic gates." U.S. patent No. 4,649,296.

[14] Shoji, M. *CMOS digital circuit technology.* Englewood Cliffs, NJ: Prentice-Hall Inc., 1987.

[15] Shoji, M. "Elimination of process-dependent clock skew in CMOS VLSI." *IEEE J. Solid-State Circuits* vol. SC-21 (October, 1986): 875–80.

[16] Argade, P. V. "Sizing an inverter with a precise delay: Generation of complementary signal with minimal skew and pulse width distortion in CMOS." *IEEE Trans. on Computer-aided Design* vol. 8 (January, 1989): 33–40.

[17] Wagner, K. D., and McCluskey, E. J. "Effect of supply voltage on circuit propagation delay and test applications." *ICCAD85 digest* (1985): 42–44.

[18] Hakim, S. S. *Feedback circuit analysis.* New York: John Wiley & Sons, 1966.

6 Essential Uncertainty in CMOS Circuits

6.1 Introduction

It is taken for granted that every detail of CMOS circuits on a silicon chip can be known, and every parameter of the circuit can be determined to the desired precision from properly arranged computer simulation, or experiments. This is not a *realistic* expectation. In this chapter we study circuit failures that originate from *essential unavailability* of information in CMOS circuits. It is natural to conclude that a circuit that is designed using wrong or uncertain assumptions could fail. Close examination of circuit operation reveals that ICs are really designed on uncertain design information. Recognition of this problem sets a limit to blind trust placed on computer simulation results. Problems that originate from circuit or circuit parameter uncertainties can be classified into three major groups.

1) Circuit connectivity is uncertain. A CMOS circuit on a silicon chip has inevitable active or passive parasitic components. It is conventional to neglect unimportant parasitics, but the criteria for neglecting parasitics are not entirely certain. If real parasitic components are included, circuit connectivity may change drastically. We may consider that one CMOS circuit coexists with the other, unwanted circuit, and if the second circuit is, by its own right, a *reliable* circuit, the second circuit takes over the *really desired* circuit under certain operating conditions. A drastic circuit failure occurs from this mechanism.

2) Circuit parameters and initial conditions are uncertain. Since CMOS gates are multi-input amplifiers, the circuit operation depends on the initial states and the details of the changes of all the input signals. A gate's transient characteristics depend on the entire past history of switching of the gate. Gates must be characterized in their working environment within a CMOS chip, and not as independent, isolated objects. Then it is impossible to carry out complete characterization, except simulating the entire chip by circuit level simulator, for all possible cases of circuit excitation. This is beyond the capability of any simulation. We must assume that *simulation is always incomplete*.

3) It is impossible to determine certain circuit parameters. Some circuit parameters (for example, relative delay of two gate chains) cannot be determined by simulation or numerical analysis, even in principle, under certain conditions. Even if simulation can be carried out, the accuracy of the result cannot be guaranteed. Furthermore certain device, interconnect, and circuit parameter values cannot be determined better than a certain accuracy limit, since a circuit built on a silicon chip is so complex.

We begin with the problem of circuit connectivity, or a circuit's configurational uncertainty that occurs when active and passive parasitics are included. One frequent scenario of this failure mechanism is shown schematically in Fig. 1. The circuit within box A has *directed* energy gain for the signal that flows from input to output (arrow). This gain is used to carry out desired signal processing. Circuit P is composed of unwanted parasitics that divert a fraction of the output signal back to the input. Because of the energy gain along the path of signal flow, even minute input-output coupling can create serious consequences: As the signal circulates around the loop, the amplitude of the signal increases with time, finally attaining the logic voltage swing at the output node. Since circuit P provides only minute coupling between the input and the output nodes of the gain stage, the structure of the integrated circuit that provides the small unwanted coupling misses the designer's attention easily.

This failure mechanism might be considered the same as the anomalous flow of signals or induced noise phenomena discussed in chapters 3 and 4, respectively. There is, however, a significant difference from the general small-signal flow mechanisms. In the mechanisms of anomalous signal flow or induced noise, the signal amplitude that flows through the parasitic path was always small. The node that receives the signal has a mechanism to accumulate the small effect, and the state of the circuit is finally upset. There is no signal flow from the node that receives the signal to the node that sends the signal. In our present mechanism the amplitude of signal in the parasitic closed path increases with time because the signal circulates, and the signal may reach a significant amplitude when the *steady state* is reached. In this sense an amplifier circuit is converted into an oscillator circuit. This *classical* case has a number of ramifications. Therefore we consider a new group of failures, called *circuit configurational uncertainty*. The circuit P of Fig. 1 need not be an electrical circuit; for instance, the circuit may include thermal or magnetic interactions. Such failures include parasitic signal paths that may be quite unobvious to the designer.

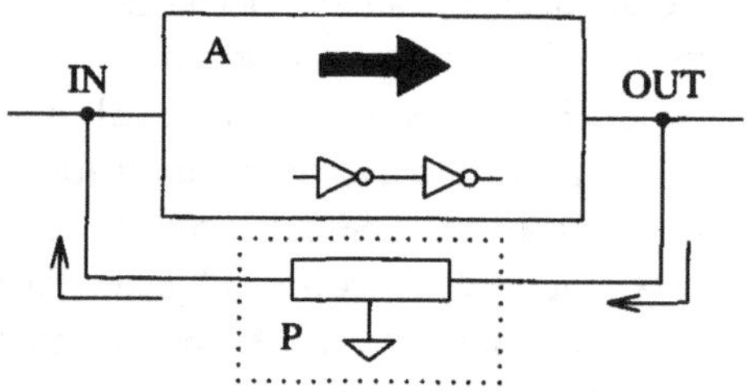

Figure 1 Mechanism of circuit configurational change

6.2 Current Distribution among Paralleled Devices

We begin with the simplest example of circuit configurational uncertainty. It will be remembered that the charge-sharing that was discussed in Section 3.9 occurs if many FETs are connected in series. Charge-sharing tends to make the potential profile of a series-connected FET chain more *uniform*. If identical active devices are connected in parallel, often the parallel combination behaves in a strange way. A rather remarkable phenomenon that creates *nonuniform* current distribution among paralleled devices occurs, and under certain conditions this nonuniform current distribution may even result in self-excited oscillation. The strange phenomena originate from the often overlooked device characteristics and parasitics. This is because a parallel combination may create a positive feedback mechanism that enhances spontaneously generated current unbalance between the paralleled devices. We saw in Section 2.14 that current distribution among paralleled devices is quite involved. Paralleling FETs or any other devices is a common practice if a single FET is unable to handle the required power, or unable to provide low impedance. High power is required, for example, in a switching power supply. We consider DC current distribution among paralleled devices.

Figure 2(a) shows parallel-connected, identical NFETs. At normal MOSFET operating temperature (0–125 degrees centigrade), the channel resistance of an NFET increases with increasing temperature. Suppose that MNA carries more current than MNB. MNA consumes more power, and the channel temperature becomes higher than that of MNB. Then the current of MNA decreases. This *negative* feedback effect tends to equalize current distribution among the two NFETs, so that the current is equally divided by the two NFETs. Figure 2(b) shows parallel-connected identical BJTs. Suppose that BJTA carries more current than BJTB. Power consumed by BJTA becomes higher, and therefore its emitter-base junction temperature becomes higher than that of BJTB. For the same base-emitter voltage of BJTA and BJTB, the current of BJTA becomes higher than the current of BJTB. Then more current flows through BJTA, and the effect is regenerative. Large DC current unbalance is established by this mechanism. This phenomenon is called thermal runaway. Both BJT and FET currents increase with increasing input voltage, but the temperature coefficients of the currents have different signs. If BJTs are paralleled, additional components like emitter series resistance are required

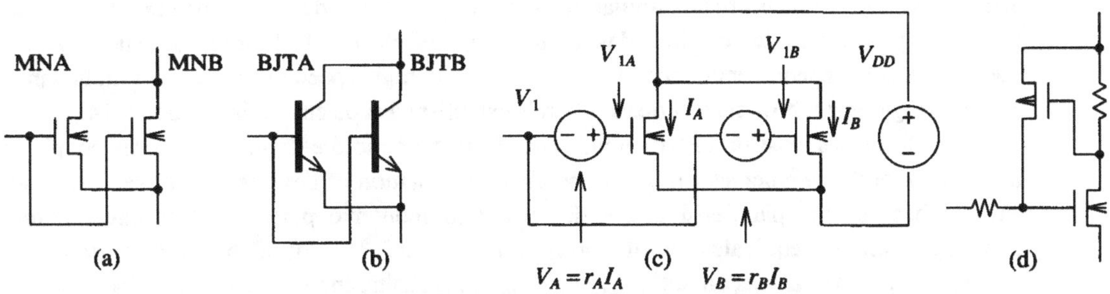

Figure 2 Paralleled devices and DC equivalent circuits

to control thermal runaway. Similar runaway phenomena may happen in JFETs at low temperatures, where carrier freezeout causes reversal in the temperature coefficient.

Thermal runaway can be understood qualitatively from the equivalent circuit model shown in Fig. 2(c). Dependent voltage source $V_A = r_A I_A$ is in series with gate input voltage V_1. The voltage source represents the effects of device heating by the channel current I_A. If I_A increases with increasing temperature, r_A is positive. Parameter r_A is a *transresistance* that characterizes the strength of the thermal coupling. Temperature rise is proportional to dissipated power. We assumed that the devices are driven by constant voltage source V_{DD}, and therefore power dissipation and temperature rise is proportional to the device current. We have

$$I_A = B_A(V_{1A} - V_{TH}) \quad I_B = B_B(V_{1B} - V_{TH}) \tag{1a}$$

where B_A and B_B are the two FET's transconductances, and

$$V_{1A} = V_1 + r_A I_A \quad V_{1B} = V_1 + r_B I_B . \tag{1b}$$

In Eqs. (1a) and (1b) we used the collapsible current generator FET model. I_A and I_B are found as

$$I_A = \frac{B_A}{1 - B_A r_A}(V_1 - V_{TH}) \quad I_B = \frac{B_B}{1 - B_B r_B}(V_1 - V_{TH})$$

and the ratio of the currents is given by

$$\frac{I_A}{I_B} = \frac{B_A(1 - B_B r_B)}{B_B(1 - B_A r_A)} . \tag{2}$$

In ordinary MOSFETs at normal operating temperatures, r_A and r_B are negative, since the negative temperature dependence of mobility determines FET current. Suppose that they somehow became positive. Then if $B_A r_A \approx 1$ or $B_B r_B \approx 1$, the ratio of the currents of the two FETs become sensitive to a small difference in the device parameters. The current of one device becomes so large that the device draws practically all the currents supplied from the source. Suppose that the two paralleled FETs are identical. Before power is turned on, temperature of one FET is higher than that of the other. Then that FET draws all the current after power is turned on. Thermal feedback converted the parallel combination effectively into a binary memory circuit. Although ordinary silicon MOSFETs do not have positive transresistance, it is possible to create *effective* positive transresistance using the circuit shown in Fig. 2(d). Paralleling two such circuits creates similar problems to paralleling devices. The current distribution among paralleled devices includes a wide range of electrical phenomena. Thermal runaway is a slow phenomenon. At the other limit of high speed where circuit inductance becomes significant, how current distribution is established is discussed in Section 2.14.

If inductances exist in a FET circuit, and if the gate and the drain inductances couple, a new circuit problem emerges [01]. The parallel combination of devices could even oscillate. This is because the paralleled active devices with inductive parasitics that have mutual coupling could be equivalent to the most *reliable* push-pull oscillator circuit, which is shown in Fig. 3(a). In this circuit, L_{DA} and L_{DB} are parasitic inductances of the drain wiring, and L_{GA} and L_{GB} are those of the gate wiring. The two groups of inductances couple mutually,

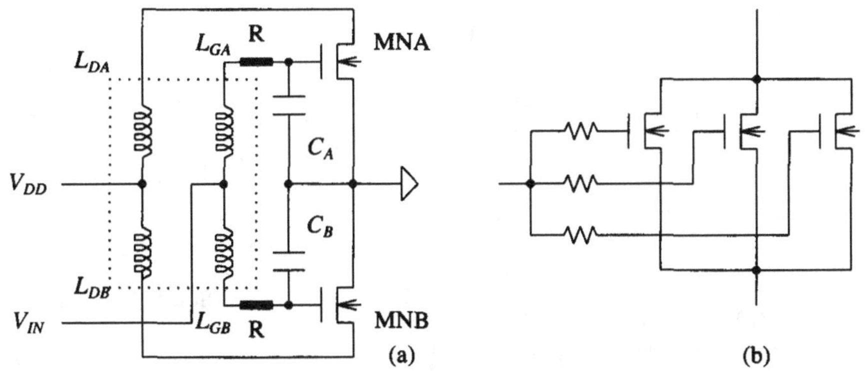

Figure 3 Parasitic oscillation of paralleled MOSFETs

thereby forming a transformer. The secondary wiring of the transformer is shunted by FET gate capacitances C_A and C_B, and therefore the gate circuit is tuned to the resonant frequency. This circuit is a FET version of the *tuned-grid, push-pull* vacuum-tube RF oscillator very popular until the 1950s [02][03]. Because transconductances of the two active devices add, oscillation starts very reliably. Coexistence of such a *reliable* parasitic circuit creates real problems. To eliminate oscillation, series resistances are inserted into the gate circuit, as shown in Fig. 3(b). The series resistances inserted at locations R in Fig. 3(a) reduce Q of the tuned secondary circuit of the transformer. Then voltage gain of the push-pull amplifier at the resonant frequency is reduced to the point that oscillation is unable to start.

6.3 Effects of Negative and Positive Feedback

Circuit oscillation takes place because a circuit that is intended as an amplifier changes configuration to an oscillator circuit. CMOS logic circuits are usually free from self-excited oscillation. The mechanism of oscillation must be understood, however, to avoid risky designs. In order for a self-excited oscillation to start, the circuit must have *small-signal* gain. A CMOS gate has high small-signal voltage gain only within narrow input voltage range, owing to the complementary-symmetry structure of enhancement mode FETs. If the operating point is set outside this narrow range, the circuit has no small-signal gain, and it is therefore stable.

Since a CMOS circuit has high DC voltage gain, a special technique is required to bias CMOS inverters within the narrow range having high gain. Some form of DC negative feedback maintains this operating point. The circuit shown in Fig. 4(a) keeps input node I and output node O of a CMOS inverter at the same DC potential. This DC *negative feedback* keeps the inverter stably at the high small-signal gain bias point. This is understood from the static characteristic. The bias point is determined from the intersection of the static characteristic of the inverter and the straight line with unit slope (input voltage = output voltage). Even if the device parameters vary, the bias point is maintained within the range having high small-signal

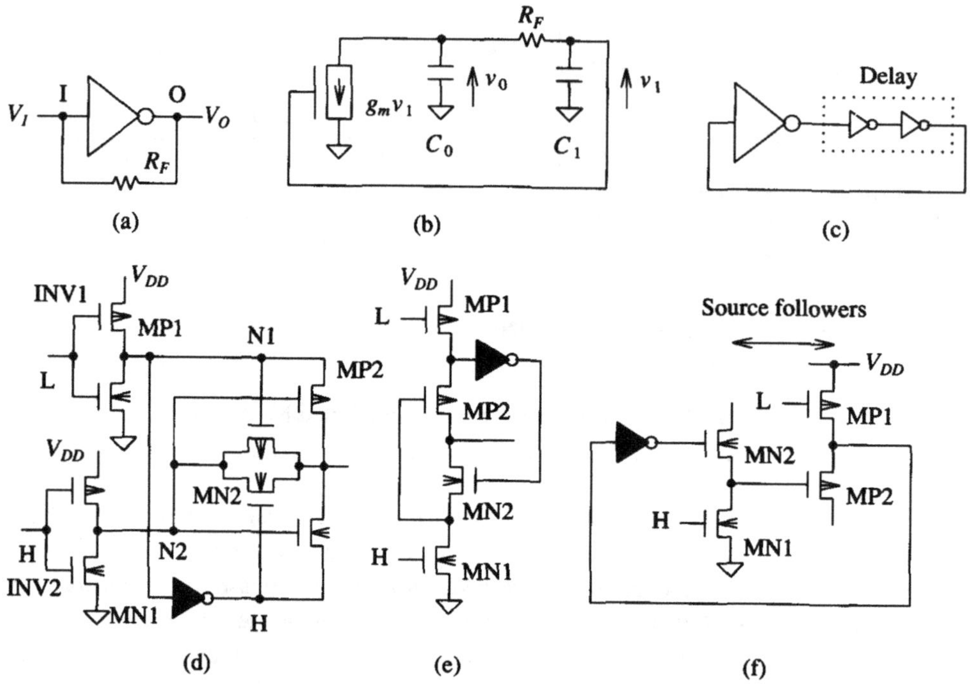

Figure 4 Negative feedback around one-stage inverting amplifier

gain. The circuit of Fig. 4(a) is often used to amplify the small-amplitude signal from a crystal controlled oscillator, or from a capacitively coupled signal source. For this DC biasing scheme to work, resistance R_F should be large, so that the stage that drives the input of the small-signal amplifier is not loaded by the feedback resistor. To feed signal, capacitive input signal coupling is used.

The circuit of Fig. 4(a) is stable if nothing except resistance R_F is connected between the input and the output of the inverter. The small-signal equivalent circuit of the inverter shown in Fig. 4(b) is described by the circuit equation

$$C_0 \frac{dv_0}{dt} = -g_m v_1 - \frac{v_0 - v_1}{R_F} \qquad C_1 \frac{dv_1}{dt} = \frac{v_0 - v_1}{R_F} \tag{3}$$

where g_m is the sum of the transconductances of the PFET and the NFET at the DC bias point. If the small-signal voltages are expressed as the sum of terms of the form $exp(-St)$, parameter S satisfies the following equation:

$$C_0 C_1 S^2 - \frac{C_0 + C_1}{R_F} S + \frac{g_m}{R_F} = 0$$

which is solved to obtain,

$$S = S_0 \pm S_1 \tag{4}$$

where

$$S_0 = \frac{C_0 + C_1}{2C_0 C_1 R_F} \qquad S_1 = \frac{1}{2C_0 C_1 R_F} \sqrt{(C_0 + C_1)^2 - 4C_0 C_1 g_m R_F}$$

and therefore the real part of S is always positive. Any small perturbation to the node voltage decays with time, and the stability of the DC bias point is guaranteed.

If a complex circuit includes an inverter whose input and output are connected by a noninverting coupling circuit, the inverter may set an undesirable stable operating point. An example is the six-FET exclusive OR gate circuit shown in Fig. 4(d) [04]. The circuit is simplified, as shown in Fig. 4(e), by eliminating a few FETs for convenience of explanation. In this simplified circuit, the input and output of the inverter (shown by the shadowed logic symbol) are connected obviously via two stages of the source followers. The inverter with the negative feedback may have an undesirable stable operating point. If the circuit settles at the operating point, it fails. If nodes N1 and N2 are strongly held by INV1 and INV2, respectively, the negative feedback path is cut off, and the circuit operates normally [04]. Negative feedback creates elementary *will* of the circuit, and if the *will* conflicts with the external drive, the circuit fails.

A stable bias point, however, is subject to the qualification that the inverter does not have any other delay. In Eq. (3) the term $g_m v_1$ is meant to have argument t, the time *at present*. If the device has internal delay τ, t is replaced by $t - \tau$ and we have

$$g_m v_1(t - \tau) = g_m v_1 - g_m \tau \frac{dv_1}{dt}. \tag{5}$$

The secular equation is rewritten as,

$$C_0 C_1 S^2 - \frac{1}{R_F}(C_0 + C_1 - g_m \tau)S + \frac{g_m}{R_F} = 0$$

and solutions S_0 and S_1 are rewritten as

$$S_0 = \frac{C_0 + C_1 - g_m \tau}{2C_0 C_1 R_F} \quad \text{and} \quad S_1 = \frac{1}{2C_0 C_1 R_F} \sqrt{(C_0 + C_1 - g_m \tau)^2 - 4C_0 C_1 g_m R_F}.$$

Then $S_0 < 0$ if $C_0 + C_1 < g_m \tau$. If simultaneously the term inside the square root becomes negative, a self-excited oscillation starts. If $C_0 + C_1 < g_m \tau$ and R_F is small, both S become real and negative. This case may be regarded as a relaxation oscillation. We do not go into detail here. The oscillation, called phase-shift oscillation, has been used conventionally to generate clean sine-wave oscillation [05]. In a conventional phase-shift oscillator circuit, the additional delay τ is created by using an RC phase-shift network that is connected between the output and the input of the amplifier. This arrangement is suitable for a sinusoidal oscillator, since RC delay circuits have high attenuation, which limits hard saturation of the amplifier. If the oscillation waveform does not matter, the delay circuit may be replaced by cascaded inverters as shown in Fig. 4(c). In order to obtain a *noninverted* delayed signal, an even number of cascaded inverters are required. From Fig. 4(c) we see that a cascaded *odd* number of inverters oscillate. This circuit is known as a *ring* oscillator. We note here that the ring oscillator belongs to the phase-shift oscillator class.

The problem of circuit oscillation may be reinterpreted as follows. A self-excited oscillation starts if the input and output of an *inverting* amplifier are connected by a passive noninverting feedback network and the passive network is DC-coupled. If the amplifier has small gain, the amplifier settles at a stable DC bias point, where the input and the output DC voltages of the amplifier become consistent. The bias point is, however, not held strong enough to assure stability under all circumstances. If voltage gain and delay of the inverting amplifier are larger than a certain limit, the bias consistency may not be maintained once a small deviation from the operating point occurs. Then the circuit becomes inconsistent. Since the input and the output voltages are never consistent at any other bias point of the amplifier, the amplifier keeps correcting the inconsistency. Self-excited oscillation is the evidence of this continuous correction. How frequent the correction takes place, or what is the oscillating frequency, depends on the delay of the signal.

We consider a simplified circuit model of cascaded inverting amplifier stages that consist of ideal amplifier and passive interstage components. We consider the stability of the circuit, if the chain is closed in a loop, by connecting the input and the output. For convenience we assume that zero DC input voltage to the inverting amplifier corresponds to zero DC output voltage, and positive input voltage corresponds to negative output voltage. For a feedback circuit to make the operation of the inverting amplifier chain inconsistent, the output voltage must be positive when input voltage is positive, and vice versa. This biasing inconsistency makes the output voltage go to negative polarity, and at that time the input voltage becomes negative also. This inconsistency of bias point generates the original inconsistency of positive-input, positive-output bias, and thereby it retains oscillation permanently. To create the permanent inconsistency using an inverting amplifier, it is necessary to shift the phase of the output voltage by 180 degrees before driving the input. This 180 degrees is in addition to inversion, which is another 180 degrees, but which is frequency independent. The extra phase shift depends on the frequency. Phase shift may be carried out in an interstage circuit, in interconnect, or by the internal circuit of the amplifier itself. In operational amplifiers, most of the phase shift takes place within the amplifier itself. If a signal that loops from input through the amplifier and through the interstage circuit back to the input again has cumulative frequency-dependent phase shift of $|\phi|$ degrees, and if $|\phi|$ is significantly less than 180 degrees, there is no possibility for the feedback amplifier to oscillate. A parameter called phase margin

$$\Delta = (180 - |\phi|)(\text{degrees}) \tag{6}$$

is positive. The circuit of a one-stage inverting amplifier has a phase margin of 90 degrees. The circuit is unconditionally stable.

A recent trend is to include analog control circuits in CMOS digital VLSI chips to improve performance or to tighten specifications. Spontaneous oscillation is always a designer's concern in an analog feedback control system. Figure 5(a) shows the schematic of an analog feedback control system that is used, for instance, in the clock duty cycle correction circuit that was discussed in Section 5.4. Figure 5(b) shows the electrical equivalent circuit of

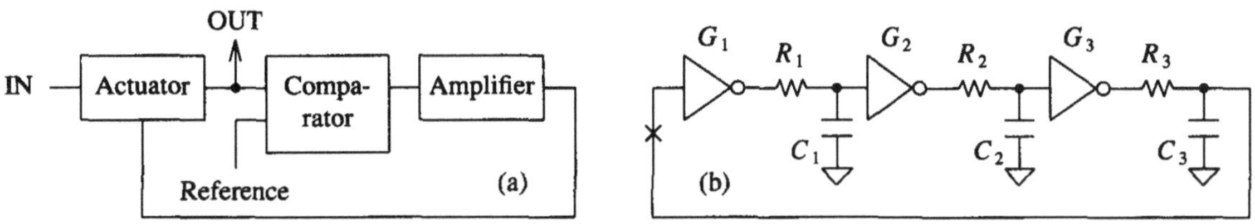

Figure 5 Analog feedback control system and its circuit model

the control loop. The inverting amplifiers are ideal (no delay). If this circuit is broken open at the X sign, the input-output transfer function $G(\omega)$ is given by

$$G(\omega)=\prod_{i=1}^{m}\frac{G_i}{1+i\omega R_iC_i}$$

where m is the number of cascaded inverter stages, and the phase angle (excluding inversion) in the limit as $\omega \to \infty$ is

$$|\phi| \to \frac{\pi}{2}\cdot m$$

Therefore *uncompensated* cascaded amplifiers of more than one stage lose phase margin at high frequencies.

Amplifier gain enters into this problem as follows. Even if the phase margin is zero or negative at some frequency, the feedback amplifier is unable to oscillate if the amplifier has no cumulative gain (amplifier and the feedback circuit combined, called *open loop* gain) at that frequency. Therefore, the stability condition can be rephrased as follows: If the phase margin is positive at the frequency where the open-loop gain is unity, the circuit is stable. If we say *phase margin* without qualification, we mean phase margin at the unity gain frequency. Practically, the minimum phase margin that includes some safety margin is about 45 degrees [06]. Figure 6 shows an example of the response of a clock duty cycle control circuit. The circuit has five cascaded phase-shift stages (two of them have gain), and therefore at high frequencies $|\phi| \to 5\pi/2$ (> 180 degrees). If the gain of the loop is higher than unity at the frequency where the phase angle is 180 degrees, the circuit breaks into oscillation. This control circuit was designed such that the time constant of one of the low-pass filter circuits of Fig. 6(b) is significantly greater than that of the others. Then the entire circuit is approximated by a one-stage RC low-pass filter circuit at low frequencies, whose maximum phase shift is 90 degrees. If the low-pass filter circuit is able to reduce gain of the loop below unity before the rest of the low-pass filter stages begin to contribute phase shift, there should be approximately 90 degrees of phase margin. The phase margin of the control circuit of Fig. 6 is about 59 degrees, as indicated by the arrow.

Analog control of digital circuits is a solution to many design problems of digital circuits. Figure 7 shows a schematic of an analog controlled digital circuit. The analog *regulator* is a precision circuit that has high small-signal gain. If the analog regulator and the digital switching circuit are fabricated on the same chip, the delay time of the regulator T_A is much longer

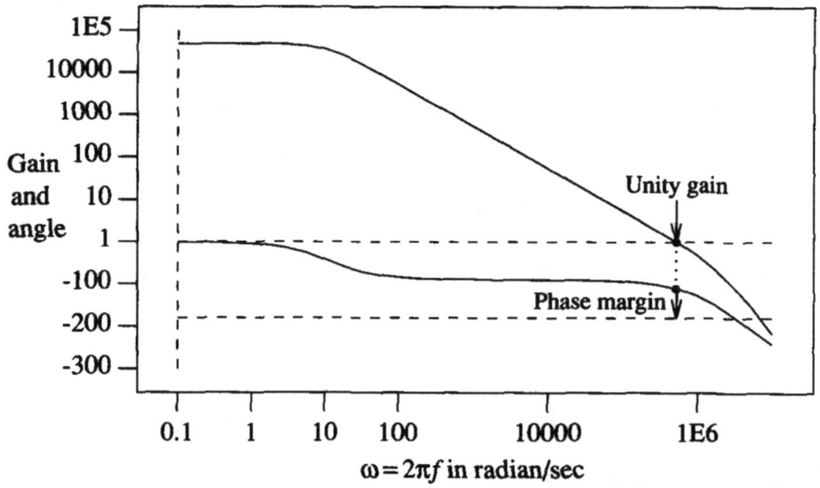

Figure 6 Gain and phase angle versus angular frequency

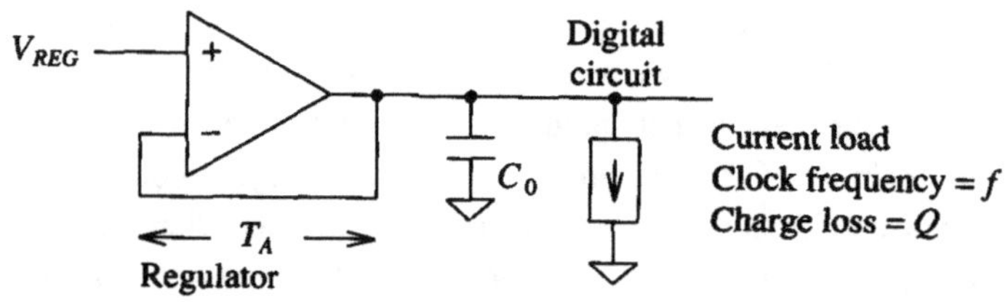

Figure 7 Analog circuit control of digital circuit

than that of the digital circuit, since the digital circuit exploits the device's switching capability to the limit, but the analog circuit requires relaxed specifications on the switching speed to obtain high gain and precision (Section 5.14). Suppose that the digital circuit of Fig. 7 is modeled using a pulse current generator that draws charge Q per cycle. Then the analog regulator is unable to respond until charge $Q \cdot fT_A$ is lost from the output node. The charge must be made available from the *load* capacitance of the regulator, C_0. Then

$$C_0 V_{REG} \gg QfT_A \quad \text{or} \quad C_0 \gg QfT_A / V_{REG}$$

is required for good voltage regulation. The important point is that the capacitance C_0 is an integral part of the control circuit and cannot be reduced below a certain minimum.

6.4 Ring Oscillator

A ring oscillator is an odd number of inverters connected in a ring as shown in Fig. 8. If the delay of each stage is T_D, a signal that starts at the input of the first inverter returns there, polarity inverted, after time T given by

$$T = (2N + 1)T_D$$

where $(2N + 1)$ is the number of stages. One period of oscillation is two times the round trip delay and therefore

$$\text{Frequency of oscillation} = 1/2T. \tag{7}$$

We study the mechanism of ring oscillation using the inverter microstate sequence. The pull-down transition microstate sequence of an inverter consists of four state changes. These are NFET (N → S), PFET (T → S), PFET (S → N) and NFET (S → T). The first two microstate changes come in that order, but the last two may have either order. The ring oscillator microstate sequence should be constructed by shifting and superposing the sequences of the successive cascaded stages. In doing so there are the following restrictions. The output node voltage of the first-stage inverter of Fig. 8 begins to change only after PFET MP0 changes from the triode to the saturation region. Therefore MP1 (N → S) follows MP0 (T → S).

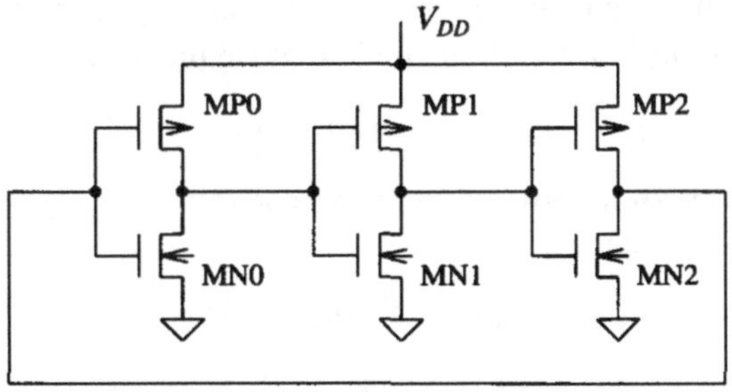

Figure 8 Ring oscillator

Furthermore, MN1 turns off before MN0 goes into the triode region, or MN1 (S $\rightarrow$ N) precedes MN0 (S $\rightarrow$ T). As for any other state changes, there is no reason to place one state change before or after the others. The microstate sequence in which MP0 (S $\rightarrow$ N) precedes MN0 (S $\rightarrow$ T) (and similar order maintained in all the other cascaded stages) is shown in Fig. 9. This is the case of *dynamic* switching of the inverters.

If the microstate change occurs in the reverse order for every inverter stage (static switching), however, the regular ring oscillator microstate sequence cannot be drawn, as shown in Fig. 10. In a ring oscillator, the FET that works as a load (MP0, for instance) must turn off before the *active* FET (MN0, for instance) completes the transition. This is reasonable: If the load FET is on after the active FET completed the node transition, delay is determined by the *input* signal, not by the delay time of the inverter. Without a clearly defined delay time assigned to each stage of the inverter, a ring oscillator is unable to find its oscillation frequency. This example shows that in certain circuits, microstate sequence is forced to follow a certain, structurally determined order. Regularity of microstate sequence (Fig. 9) does not immediately mean perfectly periodic oscillation. The microstate sequence represents quasi-periodic oscillations as well.

How easily a ring oscillation starts can be analyzed as follows. Figure 11 shows the small-signal equivalent circuit of a ring oscillator consisting of $(2N+1)$ cascaded CMOS inverter stages. The NFET symbol represents the *cumulative* transconductances of both the inverter NFET and PFET, and resistor R represents the resistive load due to the Early effect of the FETs. The circuit includes load capacitance, C_L, and Miller capacitance, C_m. Inclusion of the Miller capacitance is required to describe realistically small amplitude oscillations. Node voltages v_0, ..., v_{2N} ($v_{2N+1} = v_0$) represent small deviations from the equilibrium node voltage (V_S), or $V_i = V_S + v_i$ ($0 \le i \le 2N$). Node voltages v_{k-1}, v_k and v_{k+1} ($0 \le k \le 2N$) satisfy the following small-signal circuit equation

$$C_L \frac{dv_k}{dt} + C_m \frac{d}{dt}(v_k - v_{k-1}) + C_m \frac{d}{dt}(v_k - v_{k+1}) + g_m v_{k-1} = -\frac{v_k}{R}. \tag{8}$$

If time dependence of the form $exp(St)$ is assumed

$$(g_m - C_m S)v_{k-1} + [(C_L + 2C_m)S + \frac{1}{R}]v_k - C_m S v_{k+1} = 0$$

where v_k's are *complex* amplitude of the node voltages. We assume a travelling wave solution of this equation. The following mathematics is a close analogy to the analysis of lattice waves or phonons in solid-state physics, especially optical mode phonons [07]:

$$v_{k+1} = -\exp(i\theta)v_k \tag{9}$$

where $exp(i\theta)$ is the phase shift between two consecutive nodes. By using Eq. (9) iteratively, we find

$$v_k = (-1)^k exp(ik\theta)v_0$$

and since $v_{2N+1} = v_0$

$$[exp(i\theta)]^{2N+1} = -1 \text{ or } \theta = \pm \frac{\pi}{2N+1} l$$

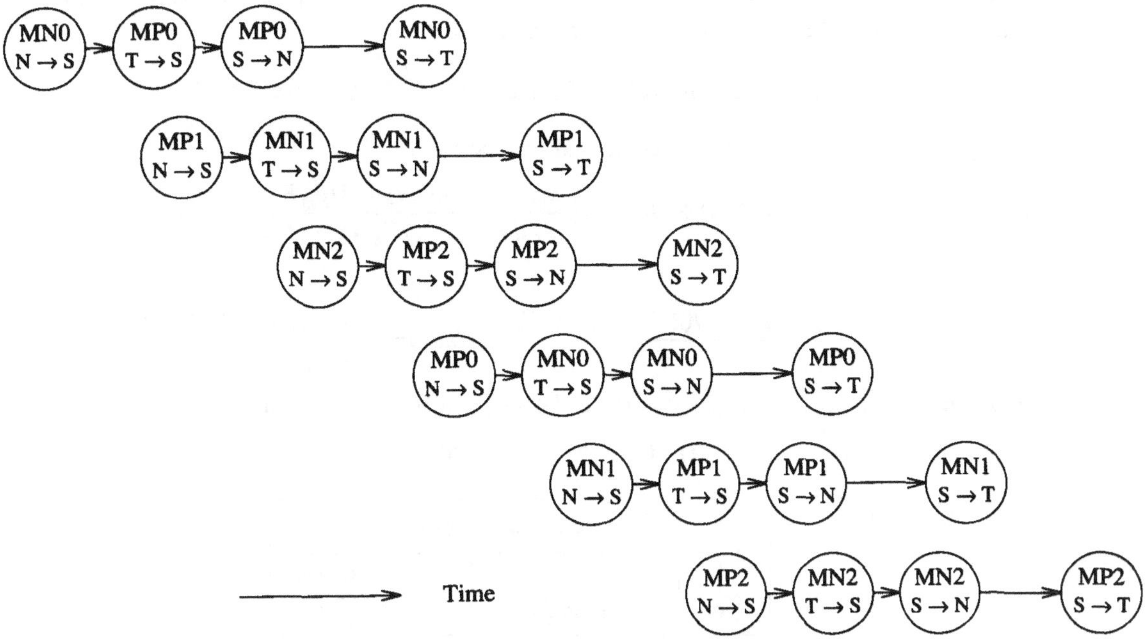

Figure 9 Microstate sequence of ring oscillator

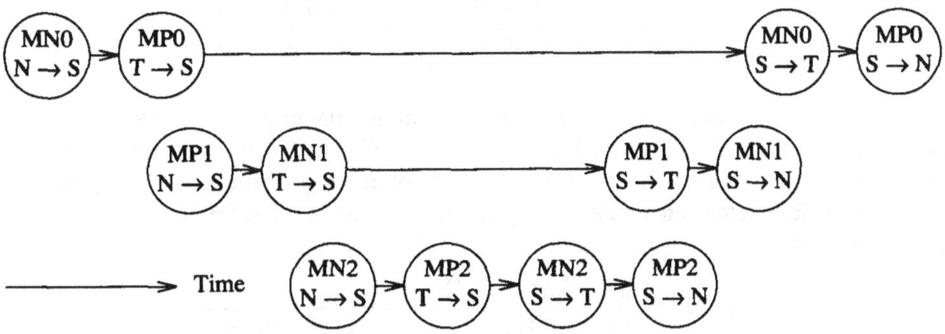

Figure 10 Microstate sequence of ring oscillator

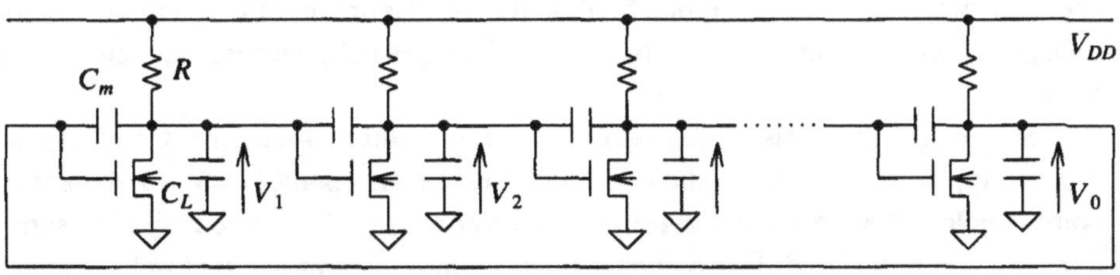

Figure 11 Analysis of ring oscillator including the Miller effect

where l is an integer that takes values $l = 1, 3, 5, \ldots, (2N+1)$. There are $(2N+1)$ *normal modes* $[l = (2N+1)$ counts only once]. We then have

$$(g_m - C_m S)exp(-i\theta) - [(C_L + 2C_m)S + \frac{1}{R}] - C_m S exp(i\theta) = 0$$

Using $\theta = \pm \pi l/(2N+1)$ in this equation and solving for S

$$S = Re(l) + i\, Im(l) = \left(\frac{g_m}{C_L}\right) \frac{exp(-i\theta) - (1/g_m R)}{1 + 2(C_m/C_L)(1 + \cos\theta)} \tag{10}$$

where

$$Re(l) = \left(\frac{g_m}{C_L}\right) \frac{cos[(\pi l)/(2N+1)] - (1/G)}{\Delta} \qquad Im(l) = -\left(\frac{g_m}{C_L}\right) \frac{\sin[(\pi l)/(2N+1)]}{\Delta}$$

and where $\Delta = 1 + 2(C_m/C_L)[1 + \cos(\pi l/(2N+1))]$ and $G = g_m R$. We have

$$v_k(l) = (-1)^k exp[ik[\pi l/(2N+1)]]v_0(l).$$

Table 1: Real/imaginary part of S		
l	$Re(l) \cdot (C_L/g_m)$	$Im(l) \cdot (C_L/g_m)$
1.00	0.4722	–0.1205
3.00	0.3816	–0.3538
5.00	0.1916	–0.5574
7.00	–0.1068	–0.6895
9.00	–0.4945	–0.6768
11.00	–0.8690	–0.4396

Table 1 shows the real and imaginary part of the complex parameter S versus the mode index l, assuming $G = 30$, $2N + 1 = 13$ and $C_m/C_L = 0.25$. The real part is positive for $l = 1$ and decreases with increasing l. The normal mode that corresponds to the largest positive real part of parameter S dominates, since this mode grows the fastest (That is for $l = 1$). Then the imaginary part is set at

$$|Im(1)| \approx \left(\frac{g_m}{C_L}\right) \frac{[\pi/(2N+1)]}{1 + 4(C_m/C_L)} = \frac{g_m}{C_L + 4C_m} \frac{\pi}{2N+1}$$

and this determines the angular frequency of the ring oscillation. We note that Miller capacitance, C_m, is twice as effective as simple load capacitance. Simple load capacitance is connected between the node and ground, while the other terminal of C_m is driven by the antiphase voltage. Since there are two C_m's for each node, $C_L + 4C_m$ becomes the effective load capacitance.

Small-signal analysis of ring oscillator is important to assure that the oscillation starts. If it does not, the node voltages should be close to the bias point $V_i = V_S$, and somehow the bias point is held. The most likely cause is low inverter gain, but noise may play some role. It is interesting to note that difficulty in starting oscillation is a problem closely related to metastability of D-latch (Section 3.12).

6.5 Capacitively Coupled Latches

A chain of tristated gates may have internal nodes that are coupled through parasitic capacitance. The coupling capacitance converts the cascaded gates into a capacitively coupled *latch*. The simplest circuit structure is shown in Fig. 12(a). If capacitance C_m does not exist, Fig. 12(a) is equivalent to a NORA dynamic logic circuit in the default state (Section 4.14). Adding C_m, the circuit is converted to a capacitively coupled latch. This circuit is analyzed using the FET collapsible current generator model, with zero FET threshold voltage. The circuit includes noise current generator, I_N. Before the noise current generator turns on, V_1, V_2, and V_0 are at their precharged-predischarged states,

$$V_1(-0)=0 \quad V_2(-0)=V_{DD} \quad \text{and} \quad V_0(-0)=0$$

where -0 in the argument means the state of the circuit at time $t<0$. At time $t=0$, I_N turns on momentarily, thereby depositing positive charge Q_T on C_1. Charges are defined in Fig. 12(b). From the law of charge conservation we have

$$Q_m+Q_0=0 \quad -Q_m+Q_1=Q_T \quad \text{and} \quad \frac{Q_1}{C_1}+\frac{Q_m}{C_m}=\frac{Q_0}{C_0}$$

By solving the simultaneous equations we determine

$$Q_0=\frac{Q_T}{C_1\Delta} \quad Q_m=-Q_0 \quad \text{and} \quad Q_1=\left(\frac{1}{C_0}+\frac{1}{C_m}\right)\frac{Q_T}{\Delta}$$

where $\Delta=(1/C_0)+(1/C_1)+(1/C_m)$. We have the initial voltages of the circuit at time $t=+0$ (immediately after I_N is turned off) given by

$$V_1(+0)=\left(\frac{1}{C_0}+\frac{1}{C_m}\right)\frac{Q_T}{C_1\Delta} \quad \text{and} \quad V_0(+0)=\frac{C_m}{C_0+C_m}V_1(+0) \tag{11}$$

From this initial state the microstate sequence evolves as shown in Fig. 13. Since I_N turns off before $t=+0$ we derive the relationship between $V_0(t)$ and $V_1(t)$ using $V_0(t) = V_1(t) + (Q_m/C_m)$ and $Q_1-Q_m=Q_T$ (Q_T is independent of time)

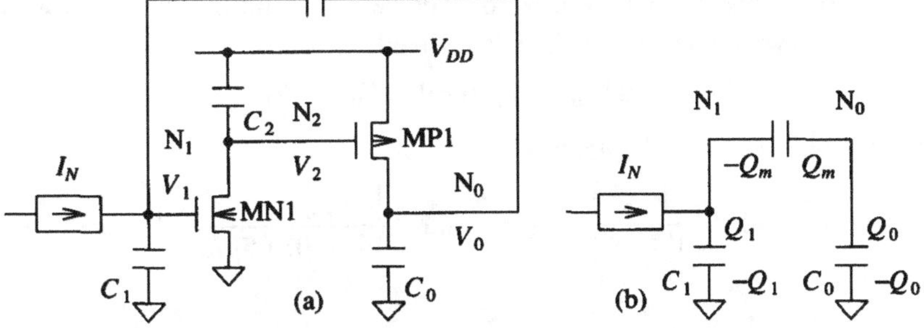

Figure 12 Capacitively coupled latch

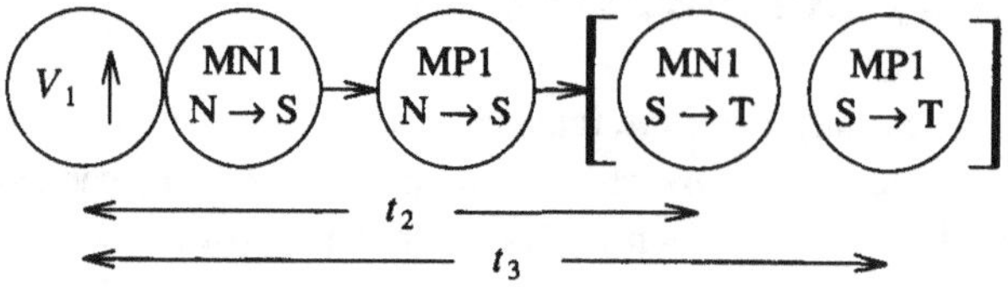

Figure 13 Microstate sequence of flipping capacitively coupled latch

$$V_0(t)=\frac{C_1+C_m}{C_m}[V_1(t)-\frac{C_0C_1+C_0C_m+C_1C_m}{(C_0+C_m)(C_1+C_m)}V_1(+0)]\,.$$

The circuit equations for $V_0(t)$ and $V_1(t)$ are

$$\frac{dV_0(t)}{dt}=\frac{C_1+C_m}{C_m}\frac{dV_1(t)}{dt}\quad\text{and}\quad [C_0+\frac{C_mC_1}{C_m+C_1}]\frac{dV_0(t)}{dt}=B_P[V_{DD}-V_2(t)] \tag{12a}$$

Equations for $V_1(t)$ and $V_2(t)$ are found as

$$C_X\frac{dV_1(t)}{dt}=B_P[V_{DD}-V_2(t)]\qquad C_2\frac{dV_2(t)}{dt}=-B_NV_1(t) \tag{12b}$$

where $C_X = (C_0C_1 + C_0C_m + C_1C_m)/C_m$. By assuming time variations of the form V_1, $V_2 \approx exp(St)$,

$$S_+=\alpha\quad S_-=-\alpha\quad\text{where}\quad \alpha=\left(\frac{B_NB_P}{C_XC_2}\right)^{1/2}.$$

We note that the time constant, $1/\alpha$, is the geometrical average of the time constants of the two cascaded stages. This is a characteristic feature of a feedback circuit, as we discussed in Sections 2.9 and 5.6. The solutions satisfying the initial conditions are

$$V_1(t)=V_1(+0)\cosh(\alpha t)\qquad V_2(t)=V_{DD}-\left(\frac{B_N}{C_2}\right)\frac{1}{\alpha}V_1(+0)\sinh(\alpha t) \tag{13a,b}$$

$$V_0(t)=\frac{C_1+C_m}{C_m}V_1(t)-\frac{C_X}{C_0+C_m}V_1(+0) \tag{13c}$$

We note that $V_0'(0)=V_1'(0)=0$. V_2 is V_{DD} at $t=+0$, and MP1 carries no current. Therefore time derivatives of V_1 and V_0 start from zero. The set of circuit equations cease to be valid when any of the following three inequalities is violated.

$$V_1(t)\le V_{DD}\qquad V_2(t)\ge 0\qquad V_0(t)\le V_{DD}$$

for $t\le t_1$, $t\le t_2$, and $t\le t_0$, respectively, where

$$t_1=\frac{1}{\alpha}\cosh^{-1}\left\{\frac{V_{DD}}{V_1(+0)}\right\}\qquad t_2=\frac{1}{\alpha}\sinh^{-1}\left\{\frac{V_{DD}}{V_1(+0)}\frac{\alpha}{(B_N/C_2)}\right\}\qquad\text{and}$$

$$t_0=\frac{1}{\alpha}\cosh^{-1}\left\{\frac{C_m}{C_1+C_m}\frac{V_{DD}}{V_1(+0)}+\frac{C_XC_m}{(C_0+C_m)(C_1+C_m)}\right\}$$

where $\cosh^{-1}(x)=\log(x+\sqrt{x^2-1})$ and $\sinh^{-1}(x)=\log(x+\sqrt{x^2+1})$. The time is the smallest of t_1, t_2 and t_0. If t_1 is the smallest, V_1 reaches the steady state first (high). If t_2 is the smallest, V_2 reaches the steady state (low) first, and if t_0 is the smallest, V_0 reaches V_{DD} first. In the circuit of Fig. 12, node N_1 is not connected to any source of charge, but in reality the node should have a PFET driver whose drain island PN junction conducts if $V_1 > V_{DD}+V_{BE}$, where V_{BE} is the turn-on voltage of PN junction diode. In the present analysis we set $V_{BE}=0$, and the voltage is clamped at V_{DD}. Figure 14 shows plots of node waveforms for the case when t_2 is the smallest. In this case, Eqs. (13a)–(13c) cease to be valid at $t=t_2$. At that time MN1 goes into the triode region and $V_2(t)=0$ after that time. Therefore

$$V_0(t)=V_0(t_2)+\frac{C_1+C_m}{C_m}\frac{B_P}{C_X}V_{DD}(t-t_2)$$

and

$$V_1(t)=\frac{C_m}{C_1+C_m}V_0(t)+\frac{C_XC_m}{(C_0+C_m)(C_1+C_m)}V_1(+0).$$

At time $t=t_0$, $V_0(t)=V_{DD}$, and after that the circuit reaches the steady state. After $t=t_2$, V_1 is determined by the capacitive voltage division ratio of C_m and C_1. In a capacitively coupled latch, some nodes may not reach the steady-state logic voltage level, 0 or V_{DD}, at the end of the transient. If one of such nodes drives a static gate, the gate carries current: significant leakage current may be created by this mechanism (Section 3.6).

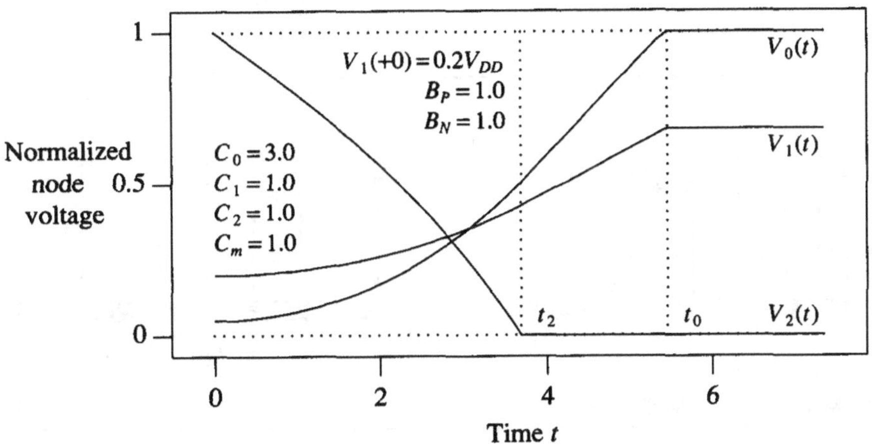

Figure 14 Node waveforms of capacitively coupled latch

6.6 Substrate DC Coupling

We consider the problems due to the silicon substrate. Ideally the substrate should provide perfectly insulating mechanical support and good heat-sinking to the integrated circuit above, and it should provide good electrical isolation between the components, but these are incompatible requirements. For the latter requirement to be satisfied, the circuit requires a solid ground reference, and the substrate should serve as the ground conductor. For that purpose the substrate is not always sufficiently conductive, as we saw in chapter 4. As for insulation, a reverse-biased PN junction is never a perfectly insulating barrier. The insulation breaks down dramatically if the substrate contains minority carriers.

FETs, resistors, and capacitors on a bulk CMOS chip are mutually isolated at each element's structural boundary by reverse-biased PN junctions when the chip operates normally. There are mechanisms that destroy the insulation of the PN junction. They are (1) injection of minority carriers to either side of the PN junction; (2) change of the type of semiconductor by external field, thereby removing the insulating PN junction itself; and (3) leakage current of poorly fabricated PN junctions. Mechanisms (1) and (2) are understood from the structure of the CMOS IC shown in Fig. 15. The figure consists of four sections, I, II, III, and IV, that show an NFET, PFET, N+ and P-tub resistor, and P+ resistor, respectively, that are isolated by reverse-biased PN junctions. They are the fundamental components used to build CMOS ICs, and they are directly imbedded in the conductive substrate. Since such direct contact does not exist in polysilicon resistors, they are not included in the figure. Polysilicon resistors are practically the only CMOS IC resistor that can be made free from *conductance* parasitics.

Section I of Fig. 15 shows NFET MN consisting of source island β, drain island δ, polysilicon gate electrode γ, channel under polysilicon gate γ, and P-tub contact α. In this figure we assume CMOS technology using an N-type substrate (N+ substrate and N-type epitaxial layer grown on top of the substrate body). P-tubs are made by diffusion into the N-type epitaxial layer. N+ source/drain diffused islands are biased positively relative to the grounded P-tub if the chip operates normally. Reverse-biased PN junctions under the diffused

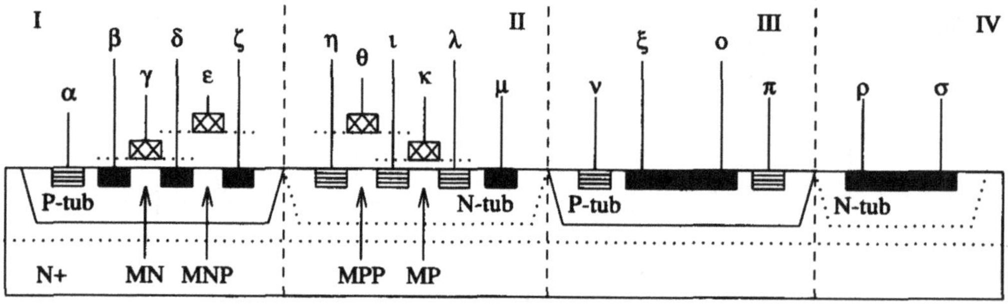

Figure 15 Structure of bulk CMOS IC

islands carry negligibly small current. If any of these N+ diffused islands are biased negatively with respect to the P-tub (or the P-tub is locally biased positively), the PN junction between the N+ diffused island and the P-tub conducts, and significant minority electrons are injected into the P-tub (and some minority holes are injected into the N+ island). Beneath the P-tub, there exists N-type substrate that is positively biased relative to the grounded P-tub. This parasitic component is an NPN transistor that consists of the three layers: the source N+ island as emitter (β), P-tub as the base (α), and N-substrate as collector (η). The P-tub thickness used in the state-of-art CMOS is around 5–10 microns. The NPN parasitic bipolar transistor has a reasonably thin base, high current gain, and the characteristic is good. All bipolar transistors of this type share collectors, which are the positively biased N-type substrate. Therefore the BJTs are called *substrate* transistors, and they have been frequently used in emitter follower circuits in linear bipolar technology. The condition in a CMOS circuit that turns any substrate NPN BJT on is a serious threat to normal operation of a bulk CMOS chip, since V_{DD} and ground are short-circuited through the device.

There is another NPN BJT structure, which consists of two independent N+ diffused islands as emitter (β) and collector (δ), and the P-tub as base (α). The current gain of this BJT is, however, poor because the thin and *unbounded* P-tub is unable to retain electrons injected from the emitter: they are absorbed mostly by the N-type substrate. This type transistor is practically unusable as a circuit component.

The P-tub surface is doped only to 3×10^{16} *acceptors/cm*^{3}. The doping level is set by the requirement of fabricating enhancement mode NFETs having the desired threshold voltage. This doping density is not high enough to prevent influence on the semiconductor surface from the electric field applied by the upper structural components of the chip like wiring. In Fig. 15, if the voltage applied to a polysilicon (or metal) wire, ε, that covers two nearby N+ diffused area δ and ζ inverts the surface of P-tub to N-type, the two diffused islands, δ, and ζ, are electrically connected; the CMOS structure may have a parasitic *thick oxide* FET associated with it. This effect, called *channel*, is a major problem of component insulation. To prevent channel, CMOS technologies use *channel stop* diffusion to cover the area that is not used for NFET channel. Channel stop diffusion is shallow P-type diffusion on P-tub (or shallow N+ diffusion on N-tub) whose surface concentration is high enough to prevent surface inversion under any electric field generated by wires above at *ordinary* operating voltages.

Section II of Fig. 15 shows PFET MP fabricated on an N-tub. An obvious BJT structure does not exist, but many effects similar to that discussed for NFETs exist for PFETs also. One significant difference is that all the N-tubs are connected together by the global N+ substrate. Then parasitic PNP devices have all their bases connected together by the N-substrate. There are basically three different types of parasitic PNP BJTs. In the first type, two independent P+ diffused islands work as emitter and collector. In the second type, two P-tubs work in a similar manner. In the third type, a P-tub and a P+ diffusion island work as collector and emitter, respectively. The last type parasitic PNP BJT (N substrate base) and the NPN substrate BJT (P-tub as base) have built-in connections to form a bipolar latch circuit, and a disastrous effect called latchup occurs if this circuit turns on. To control latchup, the bipolar latch circuit in the shadow must be degraded by degrading the parasitic BJTs and by adjusting the latch circuit parameters. Electrical phenomena originating from parasitic BJTs are very complex, since all P+ and N+ diffused islands and P-tubs effectively form weird *integrated* multi-emitter-multi-collector bipolar transistors. Section III of Fig. 15 shows an isolated P-tub and N+ diffused area fabricated on the P-tub as resistors. Section IV of Fig. 15 shows a P+ diffused area on an N-tub as a resistor. These components may initiate latchup, if any of them is biased in the forward direction.

6.7 Minority Carriers in CMOS Substrate

We need to review a few basic concepts of the physics of semiconductors [08]. Semiconductor substrate material used in CMOS ICs has resistivity less than about 10 ohm cm, or conductivity, σ, more than 0.1 mho/cm. The dielectric constant of silicon, ε_S, is approximately 1 pF/cm. If positive or negative mobile charge is created within the silicon bulk, the charge decays within several times the *dielectric relaxation time* T_D given by

$$T_D = \frac{\varepsilon_S}{\sigma} \leq \frac{1\,(pF/cm)}{0.1\,(mho/cm)} = 10\,psec\,. \tag{14}$$

Since the typical switching delay time of a 1 μm feature size CMOS gate is 1 nsec, silicon bulk remains practically in the charge-neutral state all the time. If an electric field exists within bulk substrate, current flows within the substrate, following Ohm's law. The current creates charge at PN junction or at the surface of the semiconductor, thereby reducing the field within the substrate.

Suppose that the substrate silicon is N-type. The mobile electron charge density $Q\,(coulomb/cm^3)$ is given in terms of ionized *donor* density, $N_D\,(1/cm^3)$, as

$$Q = qN_D$$

where q is the charge of an electron ($1.6\times10^{-19}\,coulomb$). Ionized donors and free electrons coexist, such that charge-neutrality is maintained everywhere within the semiconductor in steady state. Significant deviation from charge neutrality occurs only within small spatial dimensions of the order of the Debye length (Section 1.8)

$$l = \left(\frac{\varepsilon_S kT}{q^2 N_D}\right)^{1/2} = 2\times10^{-6}\,cm \ \text{ if } \ N_D = 2\times10^{16}\,(cm^{-3})\,. \tag{15}$$

The Debye length is derived by equating the electron's thermal energy with the electrostatic energy produced by violation of charge neutrality. Charge neutrality is violated in the channel of FETs and in the space-charge layer of PN junctions but in no other places in the CMOS substrate. If no electric field exists, majority electrons have Brownian motion within the boundary of the semiconductor crystal lattice, but in macroscopic scale they remain in the *same* place. They are so distributed that the charge is neutral everywhere if spatial dimensions larger than the Debye length are used to measure the field and the charge density.

If positive holes exist within the N-type bulk semiconductor, charge neutrality holds by including the positive hole charge. Positive holes and ionized donors behave similarly in many ways. We may consider this as if the donor density is increased locally by the hole density. The difference is that holes are able to move, and their density changes with time by generation or recombination processes. Then electron charge (that neutralizes hole charge) varies with time as well. If excess holes are injected into an N-type substrate, they annihilate sooner or later by the recombination process, but the process does not reduce the hole density to absolute zero: there is a certain lower limit, P_{min}, given by

$$P_{min} = \frac{n_i^2}{N_D} \tag{16}$$

where n_i is the *intrinsic* carrier density of the semiconductor ($n_i = 10^{10}\ cm^{-3}$ in silicon at room temperature). We consider that holes are injected (or generated) if their density is higher than P_{min}.

Minority holes more than the equilibrium density, P_{min}, may be injected or created locally within the N-type bulk. As soon as the excess holes are created, majority electrons surround the excess holes, and charge neutrality is reestablished within approximately a few times the dielectric relaxation time. If holes are created by ionizing radiation, electrons excited from the valence band to the conduction band exist in the immediate vicinity of holes that are created simultaneously, and the electrons provide neutralizing negative charge. If holes are *injected* from outside of the N-type bulk (for instance, from a positively biased and highly doped P+ contact formed on the N-type bulk region), the forward-biased PN junction is unable to supply electrons. Neutralizing electrons are supplied instead from the N+ contacts (Ohmic contacts) attached to the N-type bulk somewhere else. Without such contacts, it is impossible to bias the PN junction. This mechanism has the following details: Electrons within the N-type bulk make *local* motion like dominos falling into the region occupied by the holes, and within a few times the dielectric relaxation time electrical neutrality is reestablished by the electrons that are *sequentially* squeezed into the area. No electron physically moves across the entire semiconductor bulk. The balance of electrons needed are injected from the N-type (Ohmic) contact to the bulk.

Excess minority holes have a nonuniform distribution within the neutral N-type bulk. Suppose that there is no electric field. Two effects will follow. First, the excess holes recombine with electrons, and their density decreases. This process occurs while maintaining charge neutrality. The second effect is that excess holes spread out over a wide area, thereby approaching an essentially uniform excess hole density profile. This *diffusion* effect is caused by thermal motion of the holes. Since thermal motion is perfectly random, spreading of the holes occurs symmetrically: the center of mass of the holes remains unchanged with time, and the area occupied by the holes increases with time. Since holes spread out by diffusion, the density of the majority electrons must change also. The excess electron density that neutralizes holes spreads out. During the process of diffusion of the excess positive holes, charge neutrality within the N-type bulk is maintained. Since there is no net charge, Gauss' law of electrostatics dictates that there is no electric field. Therefore no external force works on the holes, except the random force of thermal fluctuation that shows up as the diffusion effect. Holes cannot move by any mechanism other than diffusion.

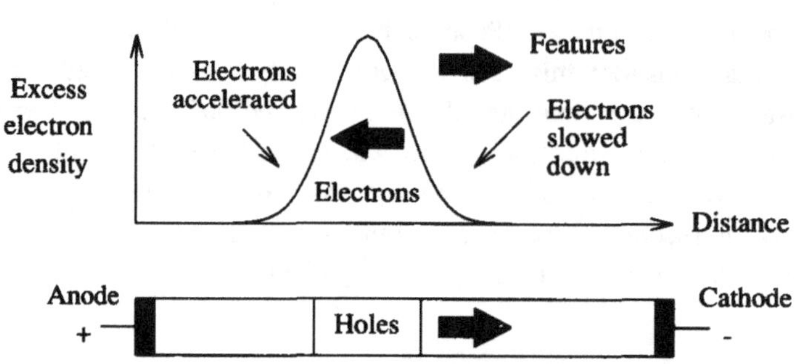

Figure 16 Motion of elecvtron density profile in long sample

When an electric field is applied, holes move in the direction of the field, and electrons move in the direction opposite to the field. Since *individual* holes move, the density profile (consisting of the excess holes and appearing as a spike in Fig. 16) moves in the direction of the field. This *drift* motion of holes takes place while maintaining charge neutrality. This means that the peak of electron density profile (consisting of the excess electrons that neutralize the excess holes) moves in the direction of the field also. The motion of the feature of electron density profile is in the *opposite* direction to the motion of the electrons themselves.

Figure 16 explains the mechanism of motion of the feature of the majority electron density profile (*spike*). The figure shows a long sample that contains excess holes that are located approximately in the middle. An electric field drives the spike of the excess holes to the right. The plot shows excess electron density versus distance along the sample. Electrons flow from the cathode to the anode. On the cathode side of the *spike* of electron density profile, electrons are slowed, and they accumulate there. On the anode side of the spike electrons are accelerated, and they deplete the region. Therefore the spike moves effectively to the right (to the cathode), that is, *opposite* to the direction of real electron flow. The electrons *stream* through the feature of the density profile.

If the electric field is absent, holes move only by process of diffusion. If a field exists, holes gain drift velocity. It is interesting to compare how effective are these two mechanisms of transport. In Fig. 17 we assume a one-dimensional hole distribution. The hole density is P_0 at $x=0$ and 0 at $x=x_0$. How to enforce this boundary condition will be discussed shortly.

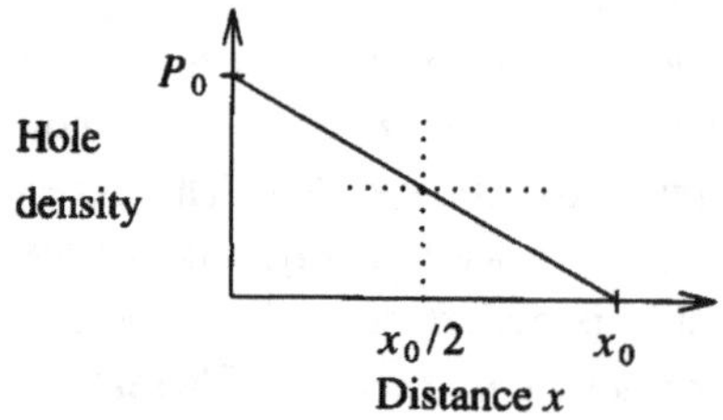

Figure 17 Hole distribution and diffusion current

Since the flux of hole flow is $-D_P dP(x)/dx$ where $P(x)$ is hole density profile, the hole's diffusion current is given by

$$J_{DP} = -qD_P \frac{dP(x)}{dx} = qD_P \frac{P_0}{x_0}.$$

If a field E (volts/cm) exists, conduction current at the halfway point $x_0/2$ of the profile is given by

$$J_{CP} = q\mu_P \frac{P_0}{2} E.$$

The ratio of the two currents is given by

$$\left| \frac{J_{DP}}{J_{CP}} \right| = 2\left(\frac{D_P}{\mu_P} \right) \frac{1}{x_0 E}. \tag{17}$$

If Einstein's relation, $D_P = (kT/q)\mu_P$, is used we have

$$\left| \frac{J_{DP}}{J_{CP}} \right| = 2 \frac{(kT/q)}{x_0 E} = 2 \frac{25\ mV\ at\ 300 deg.K}{potential\ drop} \tag{18}$$

where *potential drop* is the voltage developed by the field across the section of semiconductor having length x_0. This estimate shows that if a potential difference greater than twice the thermal voltage (25 mV) is developed within the semiconductor, conduction current predominates over diffusion current. This point is important in understanding certain aspects of latchup.

The effects of nonuniform donor density within N-type bulk on the motion and distribution of minority holes are important in understanding latchup. Figure 18(a) shows a piece of N-type semiconductor, the left half of which is doped to concentration N (*donors/cm*^{3}) and the right half to N_+ (*donors/cm*^{3}), respectively, where $N \ll N_+$. Since the right side is populated by many more mobile electrons than the left side, some electrons diffuse to the left side. Then the right side is positively charged, and the left side is negatively charged. The potential difference, $\Delta\phi$, equals the difference in the Fermi potential given by

$$\Delta\phi = \frac{kT}{q} \log \frac{N_+}{n_i} - \frac{kT}{q} \log \frac{N}{n_i} = \frac{kT}{q} \log \frac{N_+}{N} \tag{19}$$

where n_i is the intrinsic carrier density of silicon. If N_+/N is 10, 100 and 1000, $\Delta\phi$ is 58, 115 and 173 mV, respectively, at room temperature. This is a potential difference large enough to

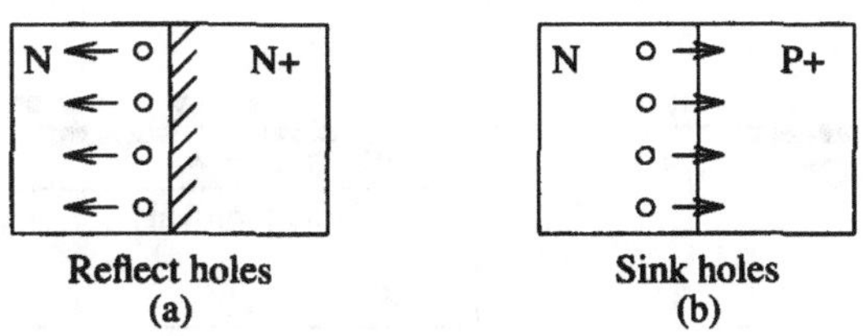

Figure 18 Nonuniformities within n-type bulk

influence hole flow. If the N+ region is doped much higher than the N region, holes are unable to penetrate into the N+ region. Holes are *reflected* at the high-low boundary, to the lower doping region. This effect is used in bipolar ICs. The *buried* layer underneath a BJT's isolation island prevents penetration of practically any holes that might be injected into the island. If a CMOS IC is fabricated using N on N+ substrate, minority holes are practically confined within N-type epitaxial layer, and they are reflected by the highly doped N+ substrate.

If the N+ region is replaced by a P-region, a PN junction is formed within the semiconductor bulk as shown in Fig. 18(b). Then a different effect takes place. Since the N-region is void of holes, and the P-region is void of electrons, both carriers diffuse into the other region. As a result, the N-type region is charged positively and the P-type region is charged negatively. The potential established across the PN junction boundary is significantly higher than that of N-N+ junction. The potential difference is close to the width of the forbidden band gap of silicon, about 1 volt, rather than several hundred millivolts. The potential difference is called the *diffusion potential* of a PN junction. This potential difference pulls minority holes at the edge of the N-type region into P-type region, and minority electrons of the P-type region into the N-type region. At the edge of a reverse-biased PN junction, the density of minority carriers is practically zero. This condition holds as long as the P-region of the PN junction is not biased positively, so that minority carrier injection does not take place. A zero-biased or reverse-biased PN junction works as an effective *sink* for minority holes in the N-type region, and for minority electrons in the P-type region.

In Fig. 19(a) a P-type contact, P_1, is forward-biased relative to the N-type epitaxial layer, and holes are injected. If the substrate were uniform N-type bulk, holes spread out to all directions from the point of injection by diffusion. When the substrate consists of epitaxial layer deposited on an N+ substrate, the injected holes are confined within the N-type epitaxial layer. One way of capturing the injected holes in the N-type epitaxial layer is to provide a PN junction contact, P_2, in the vicinity of contact P_1 that is reverse-biased and that sinks the injected

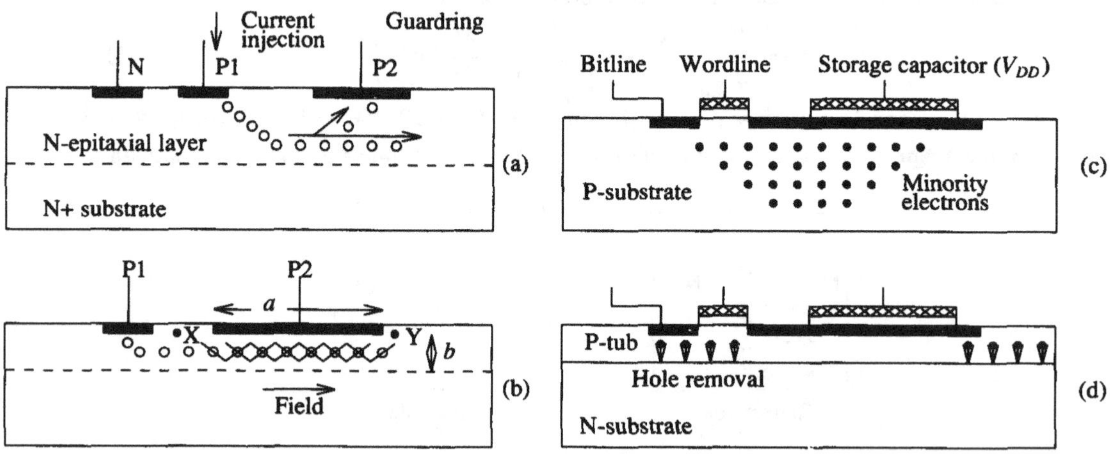

Figure 19 Transport of minority carriers by field and diffusion

holes. Within the narrow space between the P+ contact and the reflecting N+ substrate, injected holes move in a random walk, and once they come under the influence of the space-charge layer of the PN junction, they are pulled in. This random walk mechanism reduces the hole concentration approximately by the exponential of the aspect ratio a/b, $exp(a/b)$, as shown in Fig. 19(b). If the ratio is made large, the structure is effective in collecting minority holes. The mechanism works, however, only if there is a negligibly small transverse electric field in the channel. If there is a potential of the order of the thermal voltage, $kT/q = 25mV$ at room temperature, between point X and Y of Fig. 19(b), the holes are accelerated in the direction of diffusion, and the random walk mechanism does not remove holes effectively. Since it is practically impossible to control the multi-dimensional electric field within a CMOS substrate, a hole sink using reverse-biased PN junctions is not a practical method of minority carrier control. The standard method of controlling minority carriers in CMOS substrates is to connect the substrate to a highly conducting metal power-bus at many locations so that no transverse potential difference is established within the substrate: the better way of controlling minority carriers is *not* to inject minority carriers, rather than to collect already injected minority carriers.

Leakage currents of PN junctions do not significantly affect digital CMOS static logic circuits. CMOS dynamic logic circuits are, however, sensitive to leakage currents under certain circumstances. Figure 19(c) shows the cross section of a dynamic random-access memory cell (DRAM). The upper plate of the storage capacitor is a layer of polysilicon, and the lower plate is a diffused island fabricated on silicon. An NFET connects or disconnects the storage capacitor to the bit line, which controls cell access by the read/write circuit. The memory cell stores digital value 0 if the lower plate is at ground potential. Since the lower plate is a diffused area that is surrounded by grounded P-substrate, the *discharged* state is stable, and the state is held. When the cell stores digital value 1, however, the lower plate is charged to V_{DD} volts relative to the grounded P-substrate. Since the positively charged and floating N+ island is surrounded by the grounded P-substrate, the N+ island loses positive charge gradually by leakage current to the P-substrate. Digital value 1 is not held in the memory cell forever. A DRAM cell must be periodically *refreshed* (writing the same data as was read) to restore a well-defined logic voltage level. Discharge of the positively charged N+ island occurs because minority electrons in the P-substrate are collected by the diffused island, thereby neutralizing the positive charge there. This current is small, since the wide energy gap of silicon keeps minority electron density in the P-substrate low. When temperature is high, or when minority electrons are created in the P-substrate, however, leakage current increases, and discharge is accelerated.

When minority carriers are generated within the uniform bulk silicon of Fig. 19(c), they diffuse over distance λ_e given by

$$\lambda_e = \sqrt{D_e \tau_e} \tag{20}$$

where τ_e is minority electron lifetime, the average time before an excess minority carrier recombines. The electron life time is of the order of 1–10 microseconds and $D_e = 10(cm^2/sec)$. Then the range of diffusing electrons is about 100 microns, as estimated from Eq. (20). This estimate shows that once minority electrons are created within the P-type bulk, they diffuse out over a wide range, and they stay for a long time as compared with modern CMOS circuit clock periods, which range from 10 to 100 nanoseconds. In N-substrate CMOS technology, the P-tub vertical dimension is only 5 to 10 microns. This distance is an order of magnitude smaller than the diffusion length in bulk silicon. Then minority electrons are absorbed by the positively biased N substrate, and N-substrate CMOS technology is, therefore, effective in controlling minority electrons in the P-tub. The structure of the memory utilizing this effect is schematically shown in Fig. 19(d).

If a reverse-biased PN junction is used to isolate circuit components, and if excess minority carriers exist in the substrate, the isolated circuit nodes get leakage current. If a floating circuit node consists of both P+ and N+ diffused islands, the potential of the node after a long time is determined by which of N+ and P+ islands is leakier than the other. Leakage current is relevant to the design of source-follower circuits for voltage reduction, shown in Fig. 20. In this circuit, MN1 charges a large capacitor, C_1, to potential V_S (relative to grounded P-tub) determined by the implicit equation

$$V_S = V_{DD} - V_{THN}(V_S)$$

where $V_{THN}(V_S)$ is the functional relationship between the NFET threshold voltage and the source voltage (the back-bias effect). If capacitor C_1 is not charged originally, the node potential increases up to V_S, and as soon as this potential is reached, MN1 turns off (the marginally conducting state). This state, however, is unstable if leakage current exists. If extra current is injected into the output node, the potential increases above V_S and stays there. In order to hold the potential of the node close to V_S, a small leakage compensation FET, MP1, that is conducting all the time is added as shown in Fig. 20. MP1 should drain the leakage current I_l from MN1 and should hold the voltage close to, but slightly below V_S.

A CMOS dynamic NOR gate in a PLA suffers NFET leakage currents. Figure 21(a) shows a multi-input CMOS dynamic NOR gate used in a PLA. N NFETs are connected parallel to the output node of the dynamic gate. N can be as high as 100 in PLAs used for a microprocessor. Let the leakage current of a single NFET be I_L, and the channel current of a single fully conducting NFET be I_C. For integrity of dynamic gate operation, the state of the gate in which only one NFET is turned on and the state of gate in which all the NFETs are turned off must be clearly distinguishable. Then

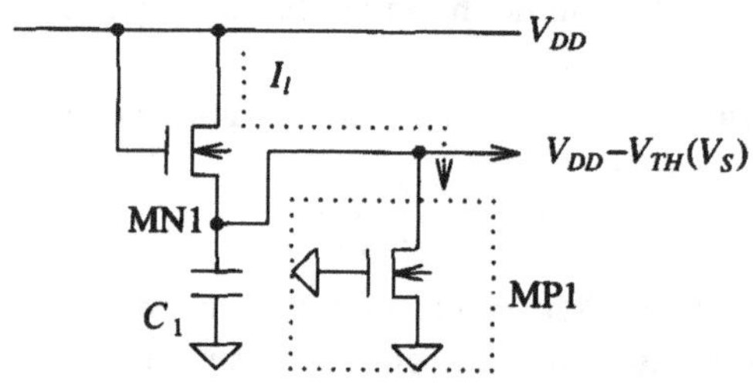

Figure 20 Leakage current effects in source-follower nodes

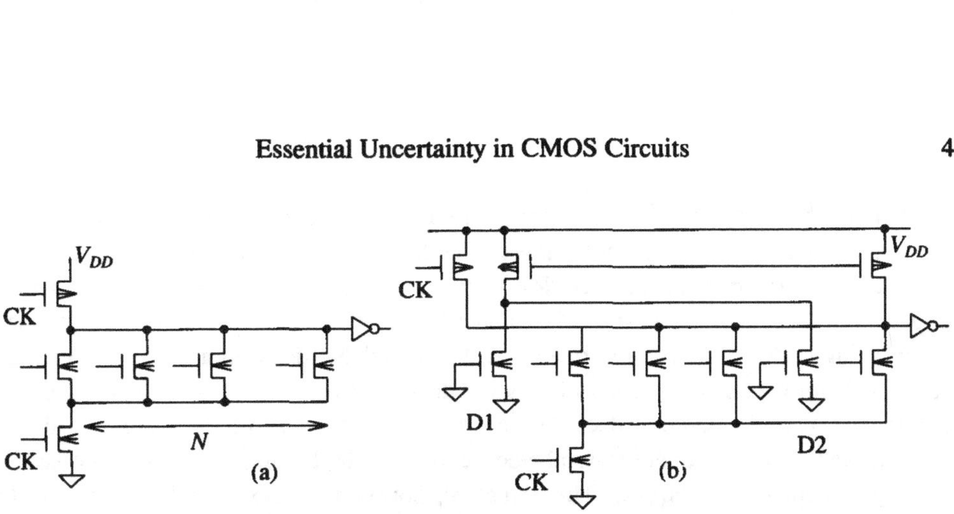

Figure 21 PLA leakage current compensation

$$I_C \gg N \cdot I_L \tag{21}$$

is required. This becomes a difficult requirement to satisfy, because N is quite large. The only obvious remedy is to reduce temperature, but that is not universally acceptable. An idea for intermediate solution to this problem could be compensating exactly the charge lost by leakage current. Figure 21(b) shows a design for a leakage-current-compensated CMOS dynamic NOR gate. A PLA multi-input dynamic NOR gate has several dummy NFETs like D_1 and D_2. The dummy NFETs have leakage current that reflects *local* conditions of the large, distributed PLA NOR gate. This is necessary, because a dynamic NOR PLA gate is often 1–2 mm long, and many environmental parameters like temperature and noise may not be uniform over the distance. The current is reflected back by a scaled PFET current mirror and returned to the output node of the dynamic gate. A similar idea of compensating noise reflecting local variation is used in folded bit-lines of memory.

Before concluding this section, I want to consider mechanisms of minority carrier injection into substrates and tubs. The most drastic cause is wrong chip design. Especially, it is not allowed to make a PN junction to the substrate or to the tub that can be forward-biased. Figure 22(a) shows an N+ to P-tub PN junction intended for use as a PN junction diode. This device does not work. Since the P-tub is on the N+ substrate, the P-tub-substrate (reverse-biased) PN junction works as the collector of an NPN transistor, and the structure becomes equivalent to

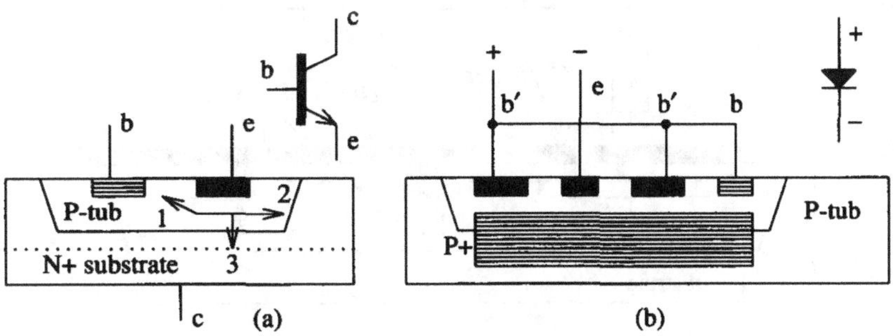

Figure 22 PN junction diode on CMOS IC

an NPN transistor that goes into the saturation region (high conductance, short-circuit state) as soon as the N+ to P-tub PN junction injects electrons. To make the structure of Fig. 22(a) work, it is necessary to retain electron flow 1 (from N+ diffused area e to P-tub contact b), but to suppress electron flow 2 (to P-tub edge) and 3 (to substrate). The established technique is to provide a ring of N+ diffused area b′ around the central N+ diffused area e as shown in Fig. 22(b). The ring captures electrons diffusing sidewise effectively. Furthermore, a deep diffused area (P+) is provided to reflect electrons diffusing directly to the substrate. This *buried* layer is also effective in reducing lateral electric fields under the ring b′, thereby increasing the efficiency of capturing electrons. This structure, however, cannot be fabricated in ordinary CMOS, unless extra process steps are added to produce the buried layer under P-tub.

Minority carriers in the substrate could be generated by thermal excitation, photoexcitation, and avalanche breakdown of the PN junction, in addition to the most obvious forward injection of the PN junction. There are many *circuit* operations in which minority carriers are pumped into the substrate. In Fig. 23, a large capacitor on an isolated P-tub is charged. This is an equivalent circuit of a large-scale circuit on a CMOS VLSI chip, such as CMOS dynamic PLAs and ROMs, that require precharge. Current flows as shown by the arrows, and the P-tub underneath the N+ contact is biased positively with respect to the N+ diffused island. Since the P-tub has high sheet resistance (1–2 kohms/square), a large transverse potential is established. Then electrons are injected into the P-tub from the N+ contact.

In Fig. 24, input node I is at ground and output node O at V_{DD}. At time $t=0$, V_1 makes a low-to-high transition in a short time. There is inevitable coupling capacitance between input I and output O by Miller capacitance, C_m. When the input makes its low-to-high transition, charge is injected into the output node from the input node, and the output node is momentarily driven above V_{DD}. Since the input transition occurs within a short time, V_0 is driven up to

$$V_{FF} = V_{DD} + \Delta V_{FF} \quad \text{where} \quad \Delta V_{FF} = \frac{C_m}{C_L + C_m} V_{DD} \,. \tag{22}$$

Since C_m/C_L can be as high as 1/3, ΔV_{FF} is $V_{DD}/4$. If $V_{DD} = 5\,V$, $\Delta V_{FF} = 1.25\,V$. Since the excess voltage is more than the threshold voltage of conduction of silicon PN junctions, the diffused drain island of the PFET is forward-biased, and minority holes are injected into the N-type substrate. The total charge injected into the substrate is estimated as

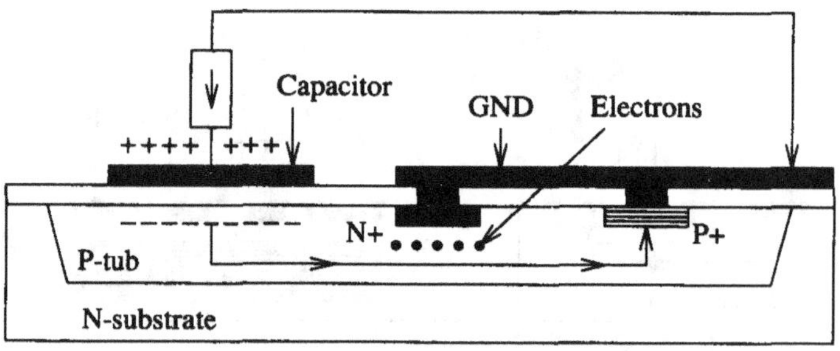

Figure 23 Substrate current and potential buildup

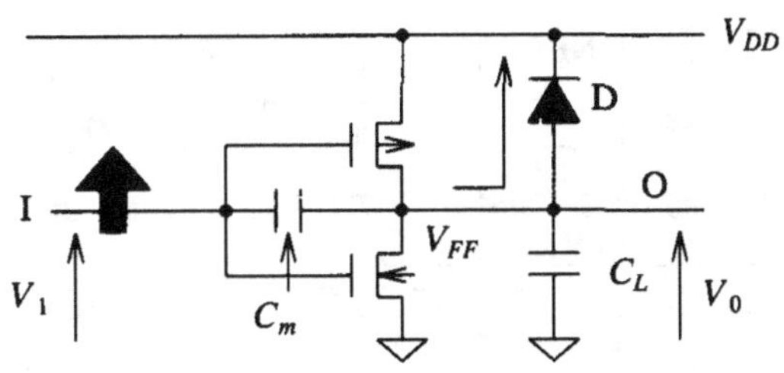

Figure 24 Minority carrier injection by feedforward

$$Q_I = (C_L + C_m)(\Delta V_{FF} - V_{BE})$$

where V_{BE} is the threshold voltage of conduction of PN junction of silicon. The charge stored in the output node is,

$$Q_T = (C_L + C_m)V_{DD}$$

and therefore

$$\frac{Q_I}{Q_T} = \frac{C_m}{C_L + C_m} - \frac{V_{BE}}{V_{DD}}. \tag{23}$$

If $C_L = 3C_m$ and $V_{BE} = 0.6\,V$, $Q_I/Q_T = 0.13$. This is the maximum minority carrier charge that is injected into the substrate and is not a small fraction. Capacitive coupling to floating circuit nodes is common in CMOS ICs, and the effect can be quite significant. Substrate or tubs of CMOS ICs are not a very *clean* environment to place devices and components. One effective way of cleaning up the substrate is to bias the P-tubs below 0, and the N-tubs above V_{DD} potential, using auxiliary power supplies.

6.8 Catastrophic Effects of Minority Carrier Injection into the Substrate

If a large minority carrier current is injected into the substrate, catastrophic failure of CMOS circuits can occur due to latchup [09]. CMOS devices may be permanently damaged by latchup. Latchup was once considered the fatal weakness of CMOS technology. Fortunately, latchup has been controlled effectively using epitaxial substrates and by many other processing and layout innovations. Using epitaxial substrates, CMOS technology established itself as the choice for VLSI technology. Latchup problems, however, may reappear in submicron CMOS. In submicron CMOS, latchup occurs in many ways. Latchup was once considered a problem of input/output (I/O) circuits only. In scaled-down high-speed CMOS, internal circuits may latchup as well. The following analysis assumes a CMOS output driver circuit as an example.

Figure 25 shows a CMOS output driver cross-section. Suppose that minority hole current is generated within the N-substrate. Hole current can be initiated, for instance, by forcing the

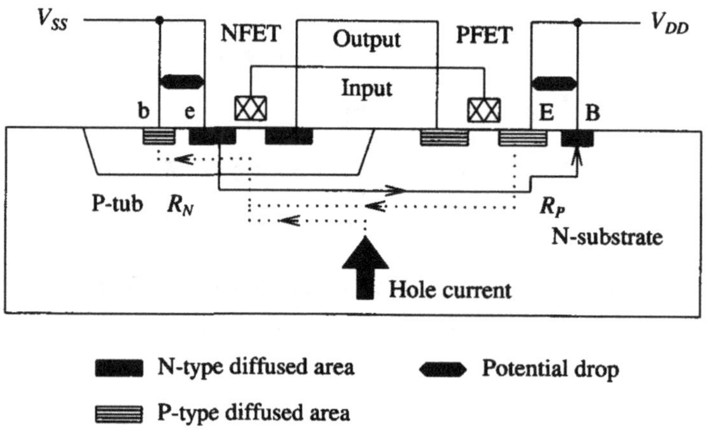

Figure 25 Mechanism of latchup

output node of the driver momentarily above V_{DD}. Since the drain island of a PFET is P-type and is highly doped, the current is predominantly carried by holes that are injected into the N-type substrate. The hole current (dotted line segments) flows to the P-tub, and then flows (within the P-tub) to the P-tub contact b. The hole current creates a potential between the P-tub contact b, and the NFET source contact, e. Then the N+ diffused island e on the P-tub is forward-biased. Since the N+ diffused island has high donor concentration, minority electrons are injected into the P-tub. The minority electrons diffuse and they are captured by the N-substrate. The electrons flow through the N-substrate (solid line segments) to the N-tub contact B. This electron current generates a potential between the N-tub contact B and the source contact E of the PFET. The forward-biased P-type diffused island E injects hole current into the N-substrate that takes over the role of the hole current originally generated in the N-substrate. The process is regenerative: even if the original hole current source is turned off, the process continues, and soon a large current exists between V_{DD} and V_{SS}. The mechanism of Fig. 25 can be represented by the equivalent circuit of Fig. 26. In Fig. 26(a), nodes of the equivalent circuit e, b, E, and B are identified by the same symbols as in Fig. 25. Resistance R_1 models the P-tub sheet resistance between contact e and the P-tub contact b. Resistance R_2 models the N-type bulk silicon resistance between contacts B and E. The circuit of Fig. 26(a) does not initiate latchup if the transistors have no initial current. Current that initiates latchup, such as the hole current of Fig. 25, is required. The current source is shown by I_{PP} or I_{NN} in Fig. 26(a) (only one is needed to start latchup).

DC analysis of the equivalent circuit of Fig. 26 is useful in understanding latchup. If R_1 and R_2 are both very large, the PNP1 collector current, I_P, becomes directly the NPN1 base current I_{BN}. The base current generates a collector current of NPN1 β_N times larger than I_{BN}.

$$I_N = \beta_N I_{BN} = \beta_N I_P \tag{24}$$

The collector current I_N becomes the base current of PNP1, I_{BP}. The current generated at the collector of PNP1 I'_P is given by

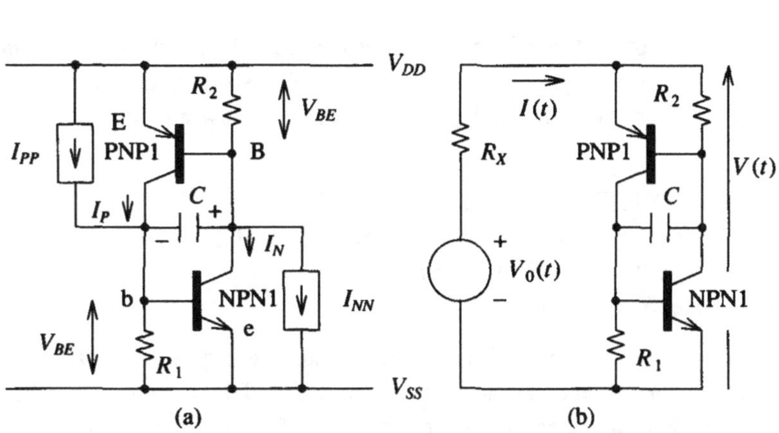

Figure 26 Dynamic response of CMOS latchup

$$I'_P = \beta_P I_{BP} = \beta_P \beta_N I_P$$

If $I'_P > I_P$ the positive feedback mechanism creates a catastrophic current increase. This happens if the product of the common-emitter current gain of the two bipolar transistors is more than unity, or

$$\beta_P \beta_N > 1\,. \tag{25}$$

If β_P and β_N are small, latchup does not occur. By doping the substrate with gold, or by reducing the operation temperature to 77 degrees Kelvin, latchup can be eliminated.

The equivalent circuit of CMOS latchup shown in Fig. 26 includes a P-tub to substrate parasitic capacitance C. C must be included to study dynamic effects of CMOS latchup. If a CMOS IC operates normally, capacitor C is charged to V_{DD}. The following dynamic characteristics of latchup can be derived from this equivalent circuit. Suppose that current pulse I_{PP} having width T_W excites the circuit of Fig. 26(a). The charge supplied to capacitor C is $I_{PP}T_W$. If NPN1 is turned on by the pulse, the charge must reduce the capacitor's terminal voltage by the base-emitter voltage of NPN1, V_{BE}. Then

$$CV_{BE} < I_{PP}T_W \tag{26}$$

or at least $T_W > CV_{BE}/I_{PP}$ is required. There is a lower limit to the current pulse width (or total charge) to initiate latchup.

Latchup analysis in the following will be carried out using the equivalent circuit of Fig. 26(b). The analysis was originally published by R. Troutman and H. Zappe [10]. Their theory is convenient because the analysis technique is, in many respects, similar to that of CMOS circuits discussed in this book. They studied how latchup is induced when power supply voltage $V_0(t)$ is increased. This model does not require an external current generator to initiate latchup, and therefore the analysis becomes especially simple. They assumed the simplest Ebers-Moll BJT model [11] having current-independent alpha and crisp turn-on characteristic of the base-emitter diode at the threshold voltage V_{BE}.

The mathematical procedure of the latchup problem can be understood by using the *modified* microstate sequence of Fig. 27. The concept of microstate sequence can be extended to complementary bipolar circuits: In the bipolar circuits the following correspondence to CMOS circuits exists:

NPN BJT ←→ NFET and PNP BJT ←→ PFET
Nonconducting state (N) ←→ Nonconducting region (N)
Conducting state (C) ←→ Saturation region (S)
Saturation state (S) ←→ Triode region (T)

where *saturation* refers to the high impedance conducting region of FETs, and to the low impedance pseudo *short-circuit* state in BJTs. A *saturated* BJT is more like a FET in the triode region than to a FET in the saturation region. Use of the term *saturation* in bipolar and in FET circuits is now conventional, however, and we must live with this confusing use. The identifiers indicating the states of a BJT, N, C, and S are defined using these terms.

The collector current versus collector voltage characteristic of a BJT has a crisp corner at the point where the collector current saturates (the point where the bipolar transistor moves out of *saturation* by increasing collector voltage). Therefore BJTs are better described by the collapsible current generator model than are MOSFETs. Dependence of collector current on base voltage is, however, exponential, and is highly nonlinear. We saw that scaled-down CMOS FETs have a linear gate-voltage drain-current relationship. The difference is minor, and the concept of microstate sequence can be extended to BJT circuits. In the equivalent circuit of Fig. 26(b) and in the microstate sequence of Fig. 27, we assume that resistor R_1 is larger than R_2. Then NPN1 turns on before PNP1. We consider this case only. The other case can be studied with minor modifications. The power supply voltage begins to increase from zero at time $t = 0$, and saturates at V_{DD} at time $t = t_P$, as shown in Fig. 28. Progress toward latchup occurs during the period $0 < t < t_P$. After t_P, current that flows through capacitor C decreases and no more progress toward latchup is made, unless the circuit had passed a critical point.

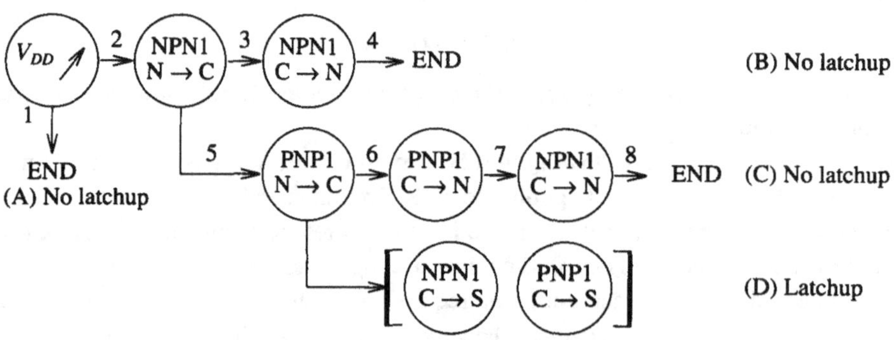

Figure 27 Microstate sequence of latchup

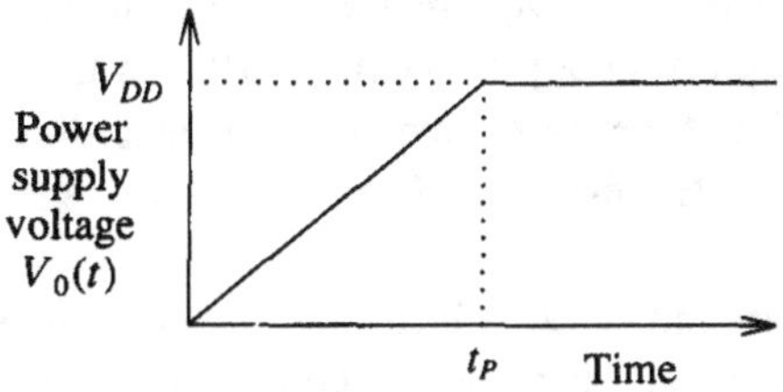

Figure 28 Time dependence of power supply voltage

Subject to these assumptions, the latchup process can be described as follows. With reference to the microstate sequences of Fig. 27 and the $V_0(t)$ waveform shown in Fig. 28, we consider first the case where capacitance C and the rate of increase of the power supply voltage

$$\frac{dV_0(t)}{dt} = \frac{V_{DD}}{t_P} = K$$

are both small. Capacitor C blocks DC current. The current through R_1 and R_2 is the displacement current of the capacitor C, which is small if K is small. Then the voltage developed across resistor R_1 is not enough to turn NPN1 on. Since NPN1 is off, even at $t = t_P$, the transient dies soon after $t = t_P$, keeping both BJTs off all the time, as shown by microstate sequence (A) of Fig. 27. In this case even the first event toward latchup, NPN1: N $\rightarrow$ C did not take place. We may classify the progress toward latchup by noting how many *events* toward latchup occur. Turn-on of NPN1 is the first event. The second event is turn-on of PNP1, and the third event is onset of saturation of either BJT. Latchup process can be understood by the *levels* reached, as shown in Fig. 29 (this classification was used by Troutman and Zappe: Fig. 29 is designed to show their basic ideas schematically). In this Figure the horizontal axis is time, and definitions of times t_1–t_4 and t_a–t_c are also given there. Triangular curve A of Fig. 29 shows that power supply turn-on is so slow, and capacitance C is so small that not even NPN1 turned on.

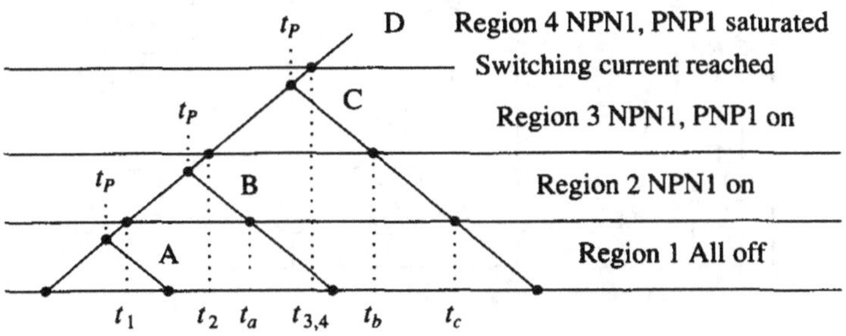

Figure 29 Definition of time parameters

If the rate of increase of power supply voltage K and capacitance C are increased, NPN1 turns on at time $t=t_1$. Latchup does not proceed any further if time $t=t_P$ is reached soon after that. Then NPN1 turns off at time $t=t_a$ and the transient dies soon after time t_a. This sequence of events is shown by the first line of microstate sequence (B) in Fig. 27 and by the triangular curve B of Fig. 29.

If parameters K and C are increased still further, PNP1 turns on at time $t=t_2$. This state is close to entering the latchup state, but there is still a possibility that time $t=t_P$ is reached before latchup. The sequence of events is shown by the triangular curve C of Fig. 29, and by the third microstate sequence (C) of Fig. 27. If this is the case, PNP1 turns off at $t=t_b$, NPN1 turns off at $t=t_c$, and the transient dies soon after that. If the latchup process proceeds further, however, the latchup state is established at time $t=t_3$ or t_4 of Fig. 29 (we use t_3 if the latchup state is established after $t=t_P$, and we use t_4 if the state is established before $t=t_P$). The third microstate sequence (D) of Fig. 27 shows that the latchup state was established.

Troutman and Zappe analyzed each case shown in the microstate sequence of Fig. 27 using the equivalent circuits shown in Fig. 30. Figure 30(a) shows the equivalent circuit that is valid before NPN1 turns on (region 1 of Fig. 29), Fig. 30(b) shows the equivalent circuit after NPN1 turned on but before PNP1 turns on (region 2 of Fig. 29), and Fig. 30(c) shows the equivalent circuit after both BJTs turned on. Figure 30(c) is the equivalent circuit that includes BJTs in saturation (strongly conducting state). In Fig. 30 $V_0(t)$ is the power supply voltage, V_{BE} is the base-emitter voltage necessary to turn BJTs on (we assume the same voltage for NPN1 and PNP1), and α_1 and α_2 are the grounded-base current gains of NPN1 and PNP1, respectively. In the equivalent circuit of Fig. 30(c), the $V(t)$ is higher than $2 \cdot V_{BE}$ immediately after turn-on of the BJTs (they are not in the saturation region yet). $V(t)$ decreases as the BJT current increases, and finally reaches $1 \cdot V_{BE}$ when the BJTs go into saturation. This voltage is less than the sum of the two voltage generators $2 \cdot V_{BE}$. This is because the current generators of Fig. 30(c) drive node N to a higher voltage than node P, thereby charging capacitor C in the *reverse* direction. Node P is close to ground and node N is close to V_{DD}. Current gains α_1 and α_2 are assumed independent of current. This is only an approximation (especially in BJT saturation region), but it makes closed-form analysis easy.

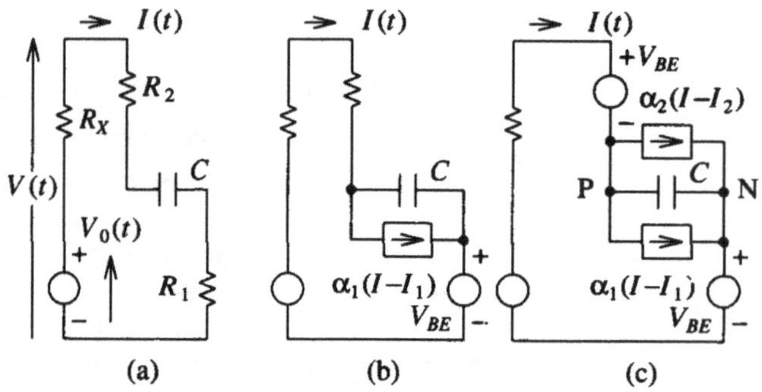

Figure 30 Analysis of latchup process

The analysis of the circuit is tedious but straightforward. Current parameters I_1 and I_2 are defined as follows:

$$I_1 = \frac{V_{BE}}{R_1} \qquad I_2 = \frac{V_{BE}}{R_2}$$

R_X is the series resistance that models the internal impedance of the power supply. The circuit equations satisfied by current $I(t)$, and the solution in each of the regions 1–3 of Fig. 29 are given as follows. From $I(t)$ all other voltages and currents can be determined.

Region 1.

$$C\frac{d}{dt}[V_0(t) - (R_X + R_2 + R_1)I(t)] = I(t) \tag{27a}$$

which is solved to obtain

$$I(t) = \frac{1}{R_X + R_2 + R_1}\int_0^t V_0'(\theta)exp[-(t-\theta)/\tau_1]d\theta \tag{27b}$$

where time constant $\tau_1 = C(R_X + R_2 + R_1)$.

Region 2.

$$C\frac{d}{dt}[V_0(t) - (R_X + R_2)I(t)] = I(t) - \alpha_1[I(t) - I_1] \tag{28a}$$

where I_1 is the current required to maintain the base-emitter voltage of NPN1. Equation (28a) is solved to obtain

$$I(t) = -\frac{I_1}{1-\alpha_1}[\alpha_1 - exp[-(t-t_1)/\tau_2]] + \frac{1}{R_X + R_2}\int_{t_1}^t V_0'(\theta)exp[-(t-\theta)/\tau_2]d\theta \tag{28b}$$

where time constant $\tau_2 = C(R_X + R_2)/(1-\alpha_1)$.

Region 3.

$$C\frac{d}{dt}[V_0(t) - R_X I(t)] = I(t) - \alpha_1[I(t) - I_1] - \alpha_2[I(t) - I_2] \tag{29a}$$

which can be solved to obtain

$$I(t) = -\frac{1}{1-(\alpha_1+\alpha_2)}[I_k - I_3 exp[-(t-t_2)/\tau_3]] \tag{29b}$$

$$+\frac{1}{R_X}\int_{t_2}^t V_0'(\theta)exp[-(t-\theta)/\tau_3]d\theta$$

where time constant $\tau_3 = R_X C/[1-(\alpha_1+\alpha_2)]$, and

$$I_3 = \alpha_1 I_1 + (1-\alpha_1)I_2 \quad \text{and} \quad I_k = \alpha_1 I_1 + \alpha_2 I_2$$

Algebraic operations to derive the closed-form results for given $V_0(t)$ are straightforward but are quite tedious. The results are shown with the identification number in the microstate sequence of Fig. 27 [(1-UP) means branch 1 of the microstate sequence, with $I(t)$ increasing].

[A] Small KC values

$$I(t)=KC[1-exp(-t/\tau_1)] \quad (0\le t\le t_P) \quad \tau_1=C(R_1+R_2+R_X) \tag{1-UP}$$

$$I(t)=KC[1-exp(-t_P/\tau_1)]exp[-(t-t_P)/\tau_1] \quad (t_P\le t). \tag{1-DN}$$

As $t\to\infty$, $I(t)\to KC$ in Eq. (1-UP). Then if $KC<I_1$ $(=V_{BE}/R_1)$, the transient never enters region 2, or NPN1 never turns on. The circuit is latchup free. If $KC>I_1$ NPN1 turns on at time $t=t_1$, which is given by

$$t_1=\tau_1 \log[KC/(KC-I_1)]$$

[B] Medium KC values

If $KC>I_1$, and if the transient moves into region 2, $I(t)$ is given as follows.

$$I(t)=KC[1-exp(-t/\tau_1)] \quad (0\le t\le t_1) \tag{2-UP}$$

$$I(t)=I_1+\frac{KC-I_1}{1-\alpha_1}[1-exp[-(t-t_1)/\tau_2]] \tag{3-UP}$$

$$(t_1\le t\le t_P) \quad \tau_2=C(R_2+R_X)/(1-\alpha_1)$$

$$=-\frac{\alpha_1 I_1}{1-\alpha_1}+\frac{1}{1-\alpha_1}[I_1+KC[exp[(t_P-t_1)/\tau_2]-1]]exp[-(t-t_1)/\tau_2] \tag{3-DN}$$

$$(t_P\le t\le t_a)$$

$$I(t)=I_1 exp[-(t-t_a)/\tau_1] \quad (t_a\le t) \tag{4-DN}$$

By taking limit as $t\to\infty$ in Eq. (3-UP), condition $I(t)>I_2$ is reached if

$$KC>I_3 \ [=\alpha_1 I_1+(1-\alpha_1)I_2]$$

at time $t=t_2$. Time t_2 is given by

$$t_2=t_1+\tau_2 \log[(KC-I_1)/(KC-I_3)].$$

If $KC<I_3$ NPN1 turns off subsequently at time $t=t_a$. Time t_a is given from Eq. (3-DN) and $I(t_a)=I_1$ as

$$t_a=t_1+\tau_2 \log[1+(KC/I_1)[exp[(t_P-t_1)/\tau_2]-1]]$$

[C] If $KC>I_3$, PNP1 turns on at time $t=t_2$. We then have

$$I(t)=KC[1-exp(-t/\tau_1)] \quad (0\le t\le t_1) \tag{2-UP}$$

$$I(t)=I_1+\frac{KC-I_1}{1-\alpha_1}[1-exp[-(t-t_1)/\tau_2]] \quad (t_1\le t\le t_2) \tag{5-UP}$$

$$=I_2+\frac{KC-I_3}{1-(\alpha_1+\alpha_2)}[1-exp[-(t-t_2)/\tau_3]] \tag{6-UP}$$

$$(t_2\le t\le t_P) \quad \tau_3=R_XC/(1-\alpha_1-\alpha_2).$$

At $t=t_P$, power supply voltage reaches V_{DD}. After that time

$$I(t)=\frac{1}{1-(\alpha_1+\alpha_2)}[[(exp[(t_P-t_2)/\tau_3]-1)KC+I_3]exp[-(t-t_2)/\tau_3]-I_k] \quad \text{(6-DN)}$$

where t is in the range $t_P \leq t \leq t_b$ and where $I_k=\alpha_1 I_1+\alpha_2 I_2$. At time $t=t_b$ PNP1 turns off. Time t_b is given by

$$t_b=t_2+\tau_3 \log[1+(KC/I_3)[exp[(t_P-t_2)/\tau_3]-1]].$$

After PNP1 turns off

$$I(t)=(I_2+\frac{\alpha_1 I_1}{1-\alpha_1})exp[-(t-t_b)/\tau_2]-\frac{\alpha_1 I_1}{1-\alpha_1} \quad (t_b \leq t \leq t_c) \quad \text{(7-DN)}$$

and at time $t=t_c$ NPN1 turns off. Time t_c is given by

$$t_c=t_b+\tau_2 \log[1+(1-\alpha_1)[(R_1/R_2)-1]].$$

After $t>t_c$ we have

$$I(t)=I_1 exp[-(t-t_c)/\tau_1] \quad (t_c<t) \quad \text{(8-DN)}$$

[D] Finally we study how the latchup state is established. Time constant τ_3 is given by

$$\tau_3=R_X C/[1-(\alpha_1+\alpha_2)]$$

$\tau_3>0$ if $\alpha_1+\alpha_2<1$, and $\tau_3<0$ if $\alpha_1+\alpha_2>1$. The formula for $I(t)$ in region 3 [Eq. (29b)] holds irrespective of the sign of τ_3. If $\tau_3<0$ $I(t)$ increases rapidly with time. Current increase leads to latchup. Using the formula $\beta=\alpha/(1-\alpha)$ where β is the grounded-emitter current gain ($I_{collector}/I_{base}$), $\alpha_1+\alpha_2>1$ is equivalent to $\beta_1\beta_2>1$, as was derived in Eq. (25).

In region 3 we are able to derive the following general results. The circuit presents an S-shaped current-controlled negative resistance. Since $V(t)$ is given by $V(t)=V_0(t)-R_X I(t)$, the requirement

$$\frac{dV(t)}{dt}=\frac{d}{dt}[V_0(t)-R_X I(t)]=0 \quad (30)$$

gives the *turning* point in the S-shaped current-voltage characteristics of the circuit that goes into the latchup state. At that point, the voltage developed across the circuit reaches a maximum, and beyond that point voltage *decreases* as the current *increases*. Let the time when the *turning point* is reached be t_k. By substituting the solution [from Eq. (29b)]

$$I(t)=-\frac{1}{1-(\alpha_1+\alpha_2)}[I_k-I_3 exp[-(t-t_2)/\tau_3]]+\frac{1}{R_X}\int_{t_2}^{t} V_0'(\theta)exp[-(t-\theta)/\tau_3]d\theta$$

into Eq. (32) and by differentiating we obtain

$$\int_{t_2}^{t_k} V_0'(\theta)exp[-(t_2-\theta)/\tau_3]d\theta=\frac{R_X I_3}{(\alpha_1+\alpha_2)-1} \quad (31)$$

and by substituting this equation back into $I(t)$ we have current at time t_k given by

$$I(t_k)=\frac{\alpha_1 I_1+\alpha_2 I_2}{(\alpha_1+\alpha_2)-1}. \quad (32)$$

$I(t_k)$ is independent of the external source. This current is the latchup *switching* current.

We consider the case t_P is large so that the switching current is reached before the power supply voltage ramp terminates. In this case we find that latchup starts at time t_4. Time t_4 is determined by requiring that

$$I(t_4)=I_2+\frac{KC-I_3}{1-(\alpha_1+\alpha_2)}[1-exp[-(t_4-t_2)/\tau_3]]=I(t_k) \tag{33}$$

as

$$t_4=t_k=t_2+|\tau_3|\log[KC/(KC-I_3)]\,.$$

In the other case of small t_P, where the voltage ramp terminates before $I(t)$ reaches the switching current $I(t_k)$, current $I(t)$ during the period $t_P \le t \le t_b$ [Eq. (6-DN)]

$$I(t)=\frac{1}{1-(\alpha_1+\alpha_2)}[[(exp[(t_P-t_2)/\tau_3]-1)KC+I_3]exp[-(t-t_2)/\tau_3]-I_k]$$

must increase with time. The coefficient of the exponential must be positive. Then we must have

$$I_3-KC[1-exp[(t_P-t_2)/\tau_3]]<0 \text{ or } KC>\frac{I_3}{1-exp[(t_P-t_2)/\tau_3]}\,. \tag{34}$$

The theory of this section leads to the following conclusions.

[A] The circuit is unconditionally stable if $\alpha_1+\alpha_2<1$, for any values of K and C.

[B] The circuit is stable even if $\alpha_1+\alpha_2>1$ if $KC<I_1$, since the circuit is unable to enter region 2.

[C] The circuit is stable even if $\alpha_1+\alpha_2>1$, if $KC<I_3$ since the circuit is unable to enter region 3.

[D] There are cases for which the circuit is stable even if the condition [C] is not satisfied. In this case the circuit *dynamically* recovers from region 3.

Figures 31(a)–(c) show current $I(t)$ and voltage developed across the circuit given by

$$V(t)=V_0(t)-R_X I(t) \tag{35}$$

versus time t. Parameter values in Fig. 31 are the same as those used by Troutman and Zappe. The rate of increase of input voltage K ranges from $0.25\times10^9\ V/sec$ to $1.0\times10^9 V/sec$. If $K=1.0\times10^9 V/sec$, current $I(t)$ increases with time even after the power supply voltage reaches V_{DD} at time t_P. The current and voltage waveforms are shown in Fig. 31(a). Before the input voltage ramp is terminated, irreversible increase in current has already set in. Latchup occurs. Device voltage begins to decrease starting shortly before the end of the voltage ramp. The current versus voltage characteristics of the device are plotted in Fig. 32. At about $V(t)=1.25\,V$, the circuit goes into the current-controlled negative resistance region. A decrease in voltage is associated with increase in current. Figures 31(a)–(c) have indications of the boundaries between different *microstates* of the latchup circuit, that are times t_1, t_2, t_a, and t_b. Before time t_1, no BJT is turned on, and therefore current is quite small [in Fig. 31(a) and (b) currents are practically zero because of the small magnification factors]. The current is the chargeup current of capacitor C through series resistor $R_1+R_2+R_X$. Current begins to

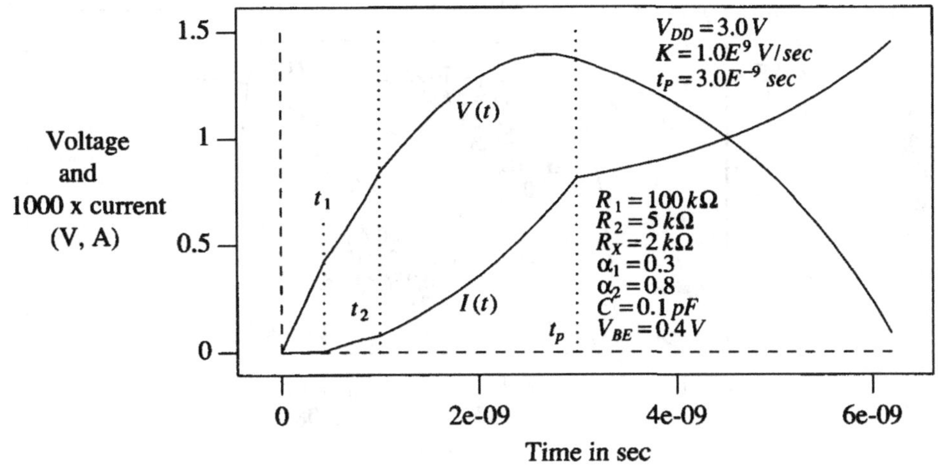

Figure 31(a) Latchup voltage and current waveforms

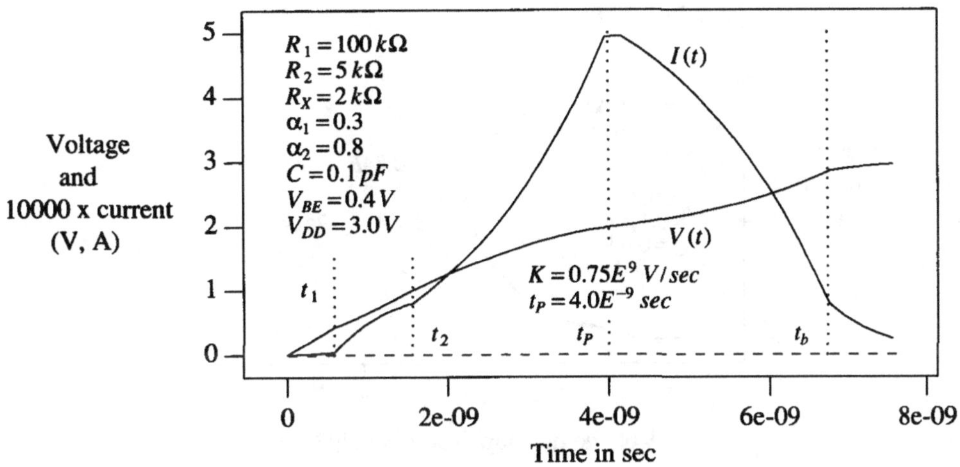

Figure 31(b) Latchup voltage and current waveforms

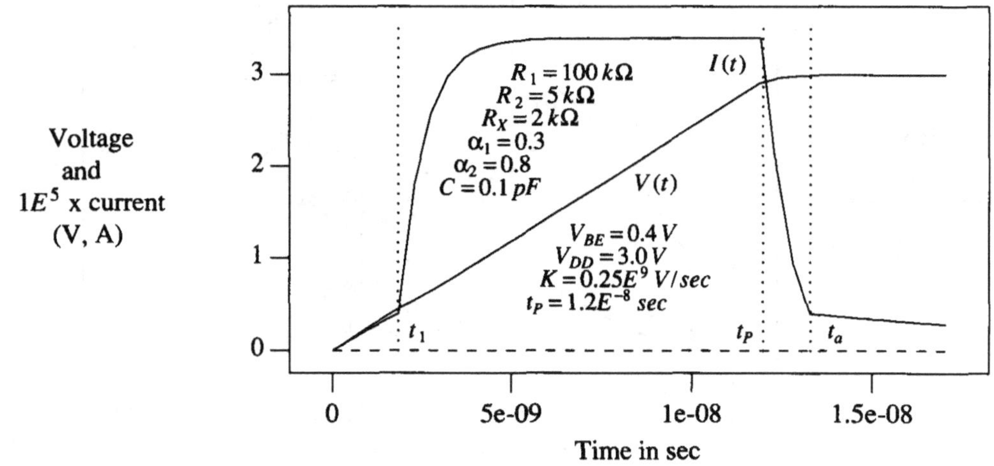

Figure 31(c) **Latchup voltage and current waveforms**

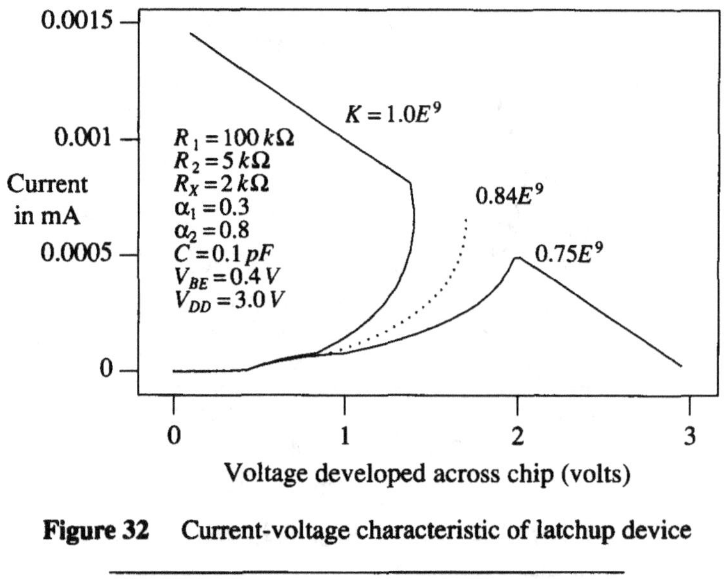

Figure 32 **Current-voltage characteristic of latchup device**

increase as NPN1 turns on. The current waveform in the period $t_1 < t < t_2$ is convex upwards, since the current is essentially the chargeup current of capacitor C through series resistance $R_2 + R_X$. Current increase is accelerated when PNP1 turns on. The current waveform in this phase is concave upwards. If K is reduced to $0.75 \times 10^9 V/sec$, device current reaches the maximum at the time when the input voltage ramp terminates, as shown in Fig. 31(b). Soon after, the device current begins to decrease, and the device returns to the nonconducting state after some time. Current through the latch circuit did not reach the switching current, and the latchup state has not been established. Since both NPN1 and PNP1 were conducting at time $t = t_P$, the latch circuit *recovered* from the *near* latchup state. Troutman and Zappe called this phenomenon a *dynamic recovery*. If K is still further reduced to $0.25 \times 10^9\ V/sec$, only the NPN1 turns on, as shown in Fig. 31(c). Then current flows only when input voltage is increasing: the current turns off immediately after termination of the input voltage ramp. Figure 32 shows current-voltage characteristic of the circuit determined by the analysis. If $K = 1.0 \times 10^9\ V/sec$, the device presents negative resistance to the power supply. Since the internal resistance of the power supply R_X is small, the device settles at the high current state, which is the latchup state. If $K = 0.25 \times 10^9\ V/sec$, the device presents voltage-controlled negative resistance at high voltage. Since the internal resistance of the power supply is small, the device can be biased stably, and therefore the device settles finally at the nonconducting state. At $K = 0.84 \times 10^9\ V/sec$ the two regimes change over.

6.9 Thermal Problems of High Frequency CMOS VLSI Chips

Circuit failures due to thermal effects could be classified into generalized noise phenomena, but we classify them into the circuit's parametric uncertainty, because what is influenced by thermal interaction or temperature variation are parameter values like resistance and transconductance, and not voltage and current (resistance fluctuation is not usually considered *noise*). First we investigate how heat is generated within a CMOS chip. Figure 33 shows where electrical energy is converted to heat in CMOS circuits. We consider a CMOS inverter that has

---

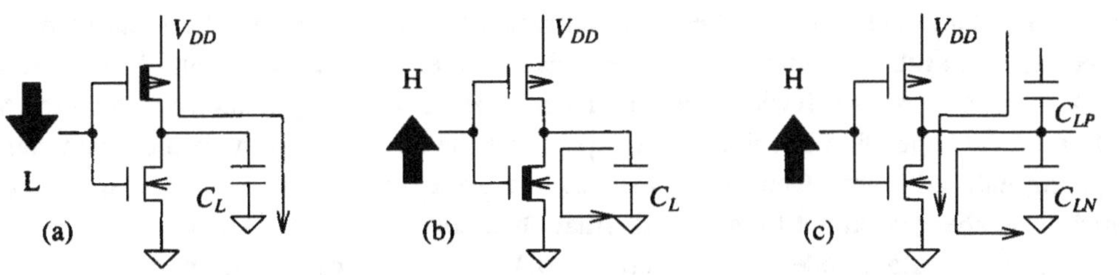

Figure 33 Energy dissipation in CMOS circuits

capacitive load C_L connected between the output node and ground. Input voltage was originally at V_{DD}, and the output node was at ground. Capacitance C_L was completely discharged. Starting from the steady state, the input voltage makes instantaneous high-to-low transition [Fig. 33(a)]. After the output node settles to the final logic voltage level, V_{DD}, the input voltage makes an instantaneous low-to-high transition [Fig. 33(b)]. During the first transient, capacitor C_L is charged to V_{DD}. Current flows from V_{DD} to ground, as shown by the arrow in Fig. 33(a). When the chargeup process is over, capacitor C_L has acquired charge Q, given by $Q = C_L V_{DD}$, and electrostatic energy

$$\text{Energy in capacitor} = \frac{C_L {V_{DD}}^2}{2} \tag{36}$$

During the second transient, the energy stored in the capacitor is dissipated by the NFET as shown in Fig. 33(b). Heat is generated within the channel resistance of the NFET (the channel shown by thick line). We assumed instantaneous input transition for simplicity. If input transition is gradual, overlap current flows, and an excess energy is dissipated by the FETs (Section 5.12).

When the two transients are completed, charge Q has been moved from the V_{DD} bus (potential V_{DD}) to V_{SS} bus (potential 0). Total electrical energy lost is given by

$$\text{Lost energy} = QV_{DD} = C_L {V_{DD}}^2 . \tag{37}$$

This is the energy dissipated by a digital signal that makes an up-and-down transition every cycle, like a clock signal. Since the total energy lost in Eq. (37) is twice the capacitor energy of Eq. (36) (that is lost in the second transient), energy $C_L {V_{DD}}^2/2$ must have been lost during the first transient, while charging up capacitor C_L. This energy loss might not be obvious: The energy is lost within the channel of the PFET [Fig. 33(a)]. The same amount of heat is generated during output up-going transition and output down-going transition, in the channel of PFET and NFET, respectively.

This conclusion may sound strange if the FET channel resistance is small (or in an idealized case of zero). In this case, energy is dissipated as electromagnetic wave: the circuit that includes load capacitances and FET oscillates, and the energy is radiated to space. As the switching speeds of integrated circuits increase, the fraction of energy radiated directly from the circuit increases. The radiation energy loss adds a resistance loading to the circuit. This *radiation resistance* becomes significant at the high frequencies where the wavelength and the conductor size are comparable. Since thin IC conductors have significant resistance and FET channel resistance is still higher, the radiation resistance is insignificant at ordinary clock frequencies, but in future, high-speed ICs the radiation effect can be significant. It is interesting to ask, If the energy used for logic operation could be dissipated by radiation, would the thermal limit of IC operation be relieved? I expect that the circuit would have too much *electromagnetic* induced noise. Conversion to heat is a noise-free way of removing used energy, which is another reason not to include thermal phenomena in noise. When the load capacitor is between the output node of the inverter and V_{DD}, energy loss is the same. Figure 33(c) shows a more general case, where both type load capacitances exist, and the input makes an up-going transition. Energy dissipated during a the up-going transition is given by

$$\text{Energy per transition} = \frac{1}{2} C_{LP} V_{DD}^2 + \frac{1}{2} C_{LN} V_{DD}^2 \tag{38}$$

where C_{LP} and C_{LN} are load capacitances that exist between the output node and V_{DD}, and between the output node and ground, respectively. Lost energy per clock cycle is twice this amount.

If the interconnect is resistive, how much energy is converted to heat in the interconnect and how much in the FET channel depends on many details. In Fig. 34(a), an instantaneous low-to-high transition of the input voltage takes place. First we assume capacitance C' is negligible. If the current of MN1 in the saturation region, $I_N = B_N(V_{DD} - V_{THN})$, is larger than the maximum current the resistor carries, $I_R = V_{DD}/R$, node A voltage drops immediately to ground. In the microstate sequence of Fig. 34(b), time spent between event MN1 (N → S) and event MN1 (S → T) is negligibly short. Then all the energy in capacitor C is dissipated by resistor R, and therefore heat is generated entirely within the interconnect. If C' is small but not zero, and if resistor R is large, the energy stored in capacitor C' is dissipated by MN1 first, along with some energy in C, and then after node A fuses to ground, the rest of the energy in capacitor C is dissipated in the interconnect resistance R. If the resistor is small, however, practically all the energy stored in both capacitors C' and C is dissipated by MN1's channel. Intermediate cases between the two extrema are also possible. As is shown by this example, determining where heat is generated requires involved study.

We make an estimate of the absolute thermal limit of a CMOS IC. Silicon has a thermal conductivity of about 1.5 $Watts/cm \cdot deg$. Suppose that heat generated in the active circuit on the top surface of the CMOS chip is removed from the bottom surface that is attached to an ideal constant temperature heat sink. If 40 degrees temperature difference is tolerated between the top and the bottom surfaces, a 500 μm thick silicon chip is able to conduct $(1.5 Watts/cm \cdot deg) \times (40 deg/0.05 cm) = 1{,}200 Watts/cm^2$ of heat energy. The effective capacitance of CMOS circuits on the chip surface is of the order of $2 \times 10^{-8} F/cm^2$. Then the power dissipated at $f\,(MHz)$ clock is estimated at $2 \times 10^{-8} \times 5^2 \times f \times 10^6/2 = 0.25f$ $Watts/cm^2$, assuming a 5-volt power supply. If $f = 300\ MHz$, the power per unit area will be $75\ Watts/cm^2$. This will be a significant overestimate because all C will not switch every clock cycle. It is quite fortunate that the high thermal conductivity of silicon allows a significant margin to the absolute thermal limit in high-speed CMOS ICs. The estimate shows

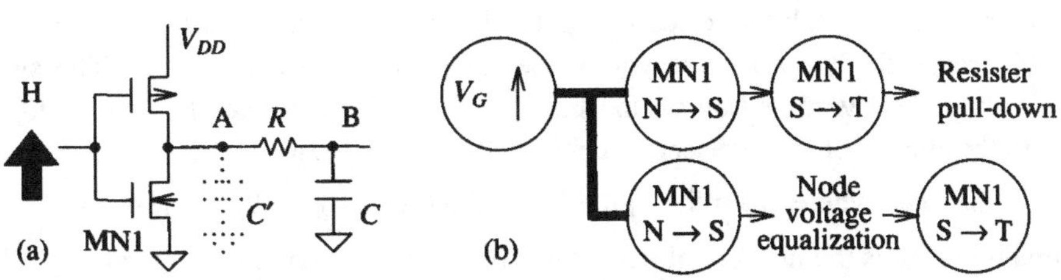

Figure 34 Energy loss in resistive interconnect

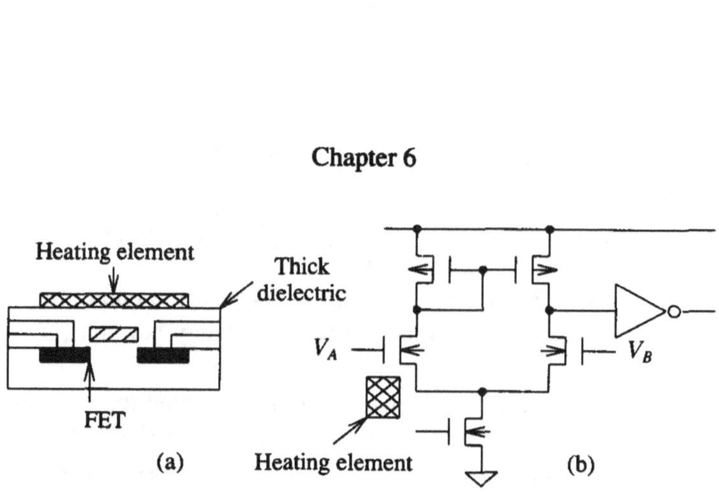

Figure 35 Thermal input data coupling to CMOS chip

clearly that an effective use of high thermal conductivity of silicon, or development of good packaging method, is a key issue in future, high-speed CMOS IC development.

Thermal effects can be applied for some unusual circuit designs. CMOS chips are not resistant to electrostatic shock or high voltage surges on input signal lines. To make chips resistant to extreme and harsh conditions is difficult. If the data rate of a signal is low, the signal can be introduced to the chip by thermal coupling. In Fig. 35(a) thick dielectric is deposited above the FET, and a heating element is deposited on top. If current is passed, the heating element warms the FET up, and the FET's characteristics change. The FET is used in the differential amplifier circuit shown in Fig. 35(b). If the heating element is turned on, one of the FETs is heated, and the balance of the differential amplifier tips to the other extremum. In this way a low rate control signal can be introduced to a CMOS chip.

6.10 Uncertainty in Switching Threshold Voltage

CMOS gates, and in general any conventional logic gates, are multi-input amplifier circuits. Since there are more than one input, signals from different inputs interact, and they create effects that do not exist in single-input amplifier circuits. This problem was briefly introduced in Sections 2.6 and 2.7. Figure 36(a) shows CMOS 3-input NAND gate. We derive the formula for the switching threshold voltage of this gate as an example. In order to make the analysis tractable, we assume all NFETs have the same size, and all PFETs have the same size. FETs are modeled using the collapsible current generator model with zero threshold voltage (we neglect the back-bias effect). Input voltage to the three inputs A, B, and C are V_A, V_B, and V_C, respectively, which are within the range between 0 and V_{DD}. The switching threshold voltage is determined by setting the current of the three parallel connected PFETs, I_P, equal to the current of the three series-connected NFETs, I_N, as

$$I_P = I_N \quad \text{where} \quad I_P = B_P(V_{DD} - V_A) + B_P(V_{DD} - V_B) + B_P(V_{DD} - V_C) \tag{39}$$

Determination of I_N is the non-trivial part of this problem. Depending on the voltages V_A, V_B, and V_C, the three FETs are either in saturation or in the triode region. There are three major cases shown in Table 2.

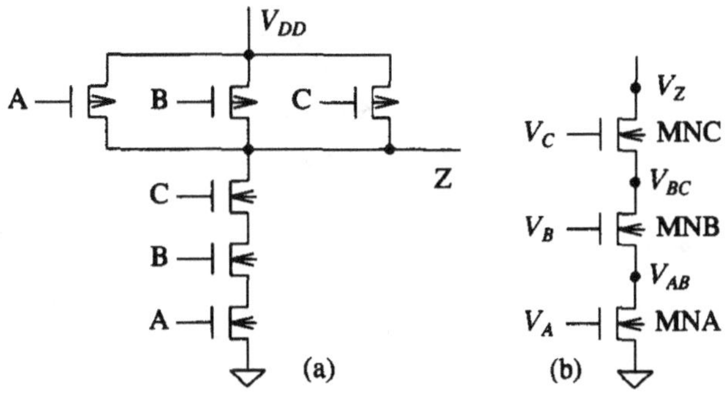

Figure 36 CMOS 3-input NAND gate switching threshold analysis

Table 2: States of 3 NFET chain				
Case	Subcase	MNA	MNB	MNC
Case 1	1	T	T	S
Case 2	21	T	S	S
	22	T	S	T
Case 3	31	S	T	T
	32	S	S	T
	33	S	T	S
	34	S	S	S

In Table 2, T means triode region and S means saturation region. In case 1, the I_N is given by

$$I_N = B_N V_C$$

and the maximum current of NFET MNA ($B_N V_A$) and of NFET MNB ($B_N V_B$) must be larger than I_N. Then we have

$$V_C < V_A \quad \text{and} \quad V_C < V_B . \tag{40}$$

The combination of the three input voltages, or the *state* of the gate's input, can be represented by a point in the three-dimensional coordinate system shown in Fig. 37. Since the range of the input voltages is between 0 and V_{DD}, the points are contained within a cube enclosed by the dotted lines of Fig. 37. The volume within the cube that satisfy the inequalities of Eq. (40) is a pyramid (exactly speaking, 1/4 of a *regular* square-base pyramid) bounded by the solid lines of Fig. 37(a).

In case 21, I_N is given by

$$I_N = B_N V_B$$

and the saturation current of MNA ($B_N V_A$) must be larger than I_N. Then we get

$$V_A > V_B . \tag{41a}$$

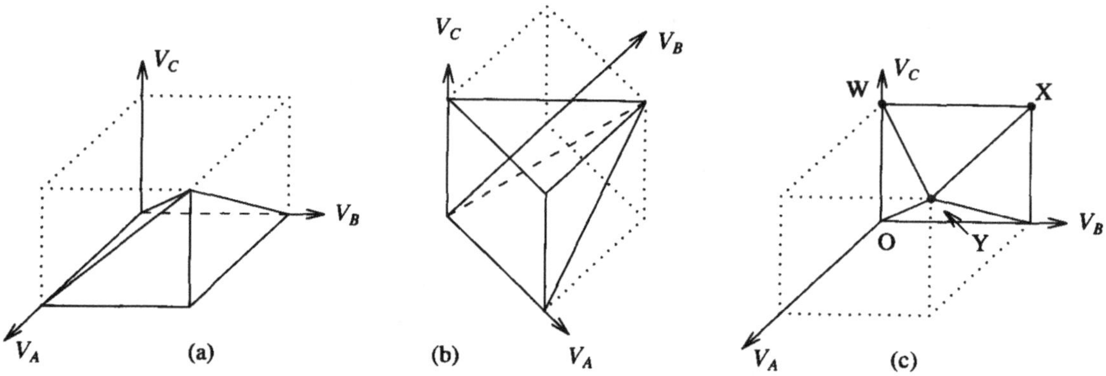

Figure 37 Guide for switching threshold analysis of NAND3 gate

Node voltage V_{BC} [Fig. 36(b)] is determined by requiring that the current of NFET MNC equals I_N. Then

$$B_N(V_C - V_{BC}) = B_N V_B \text{ or } V_{BC} = V_C - V_B$$

and since $V_{BC} > 0$, we have

$$V_C > V_B. \tag{41b}$$

In case 22, $I_N = B_N V_B$, and the current of MNC is larger than I_N, so that

$$B_N(V_C - V_{BC}) > B_N V_B \text{ or } V_C - V_B > V_{BC} > 0 \text{ and } V_A > V_B \tag{41c}$$

and we get the same result. The region in $V_A - V_B - V_C$ coordinate system, where the points satisfy inequalities Eqs. (41a)–(41c), is a pyramid whose base is on the $V_A - V_C$ coordinate plane bounded by solid lines in Fig. 37(b) (pyramid pointing the direction of V_B-axis).

The results of case 31-34 are summarized as follows:

Case 31. $I_N = B_N V_A$, $V_B > V_A$ and $V_C > V_A$.
Case 32. $I_N = B_N V_A$, $V_C > V_B > V_A$ and $V_{AB} = V_B - V_A$.
Case 33. Same I_N, $V_B > V_A$, $V_C > V_A$ and $V_{AB} = V_C - V_A$.
Case 34. Same I_N, $V_C > V_B > V_A$, $V_{AB} = V_B - V_A$ and $V_{BC} = V_C - V_A$.

The inequalities of cases 31–34 can be reduced to $V_B > V_A$ and $V_C > V_A$ for current $I_N = B_N V_A$. The region of points satisfying the inequalities is shown in Fig. 37(c). This is a pyramid based on the $V_B - V_C$ coordinate plane and pointing in the direction of the V_A axis.

Let us consider some typical cases. If $V_B = V_C = V_{DD}$ and V_A makes a low-to-high transition, the input voltage follows straight line $\overline{XY}$ of Fig. 37(c). Then

$$I_P = B_P(V_{DD} - V_A) \text{ and } I_N = B_N V_A$$

and therefore the switching threshold voltage is given by

$$V_A = V_{GSW} = \frac{B_P}{B_P + B_N} V_{DD}$$

If $V_C = V_{DD}$ and V_A and V_B make a simultaneous low-to-high transition, the input voltage follows straight line $\overline{\mathrm{WY}}$ of Fig. 37(c). Then

$$I_P = 2B_P(V_{DD} - V_A) \text{ and } I_N = B_N V_A .$$

Then the threshold voltage is given by

$$V_A = V_{GSW} = \frac{2B_P}{B_N + 2B_P} V_{DD}$$

If V_A, V_B, and V_C make a simultaneous low-to-high transition, the input voltage follows the diagonal of the cube OY of Fig. 37(c). Then

$$I_P = 3B_P(V_{DD} - V_A) \text{ and } I_N = B_N V_A .$$

Then

$$V_A = V_{GSW} = \frac{3B_P}{B_N + 3B_P} V_{DD} .$$

The analysis of 3-input NAND gate can be generalized to an N-input NAND gate subject to the same assumptions on the FET characteristics. Figure 38 shows a chain of N NFETs that form the pull-down chain of an N-input NAND gate. All the NFETs have the same transconductance (B_N) and zero threshold voltage. I_N is determined as follows. Out of the N-input voltages V_1, V_2, ..., V_N, the lowest is V_k, the second lowest is V_l, the third lowest is V_m, and so on. If the voltage applied to the entire chain, V_Z, is positive, at least one NFET of the chain must be in the saturation region. This is the k-th NFET. FETs to the left side, 1, 2, ..., $(k-1)$, are all in the triode region. If chain voltage V_Z is low, all the NFETs to the right side are in the triode region as well. The current of the chain is given by

$$I_N = B_N V_k$$

Let the second lowest input voltage to the right side of k-th NFET be V_l. If V_Z is higher than $(V_l - V_k)$, NFETs $(k+1)$, $(k+2)$, ..., $(l-1)$ are in the triode region, and the *fused* nodes k, $(k+1)$, ..., $(l-1)$ have common node voltage $(V_l - V_k)$. NFET l is in the saturation region, and the currents of NFET k and NFET l are equal. This procedure is repeated for still higher chain voltages, V_Z, and in this way the current and the node voltage profile of the NFET chain are determined.

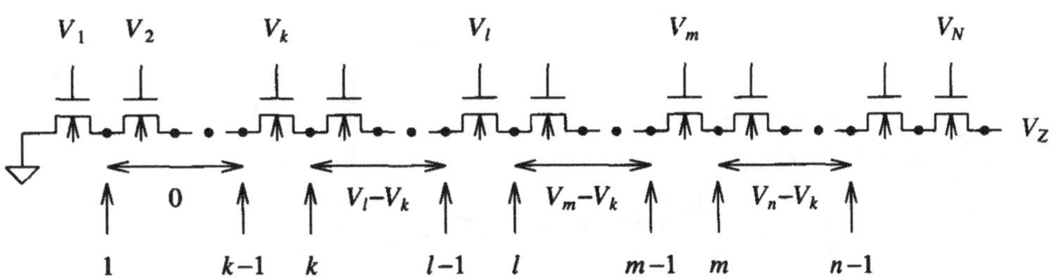

Figure 38 Static characteristics of NFET chain

If M out of N input voltages change simultaneously, and the rest $(N - M)$ stay at V_{DD}, the switching threshold voltage V_{GSW} of an N-input NAND gate is determined from the equation of current balance

$$I_N = I_P \quad \text{where} \quad I_N = B_N V_{GSW} \quad I_P = MB_P(V_{DD} - V_{GSW})$$

as

$$V_{GSW} = \frac{MB_P}{B_N + MB_P} V_{DD}. \tag{42}$$

This equation gives the essential uncertainty of the switching threshold voltage for N-input NAND gates.

Input voltage dependence of the switching threshold voltage exists in practically any logic gate circuit. Figure 39(a) shows an NMOS N-input NOR gate. If all the NFETs are conducting, the output node voltage V_0 depends on input voltages V_A, V_B, ..., by

$$V_0 = V_{DD} - R_L[B_N(V_A - V_{THN}) + B_N(V_B - V_{THN}) + \cdots]$$

where B_N is transconductance and V_{THN} is the threshold voltage of the NFETs. If the Boolean value at the output node is determined at the logic threshold voltage V_{LOGIC},

$$\frac{V_A + V_B + \cdots}{N} = V_{THN} + \frac{V_{DD} - V_{LOGIC}}{NG} \quad (V_A > V_{THN} \; V_B > V_{THN} \cdots) \tag{43a}$$

where $G = R_L B_N$ is the gain of the amplifier stage. For the NOR gate to be at the critical output node voltage V_{LOGIC}, only the average of the input voltages must be at the required level. If that is satisfied, each input voltage can be at any level. Therefore each input has a range of possible switching threshold voltages.

Figure 39(b) shows an N-input source-coupled circuit. The differential multi-input amplifier balances if

$$B_N(V_A - V_0 - V_{THN}) + B_N(V_B - V_0 - V_{THN}) + \cdots = B_N(V_{REF} - V_0 - V_{THN}) \tag{43b}$$

and by using $I_0/2 = B_N(V_{REF} - V_0 - V_{THN})$ we obtain

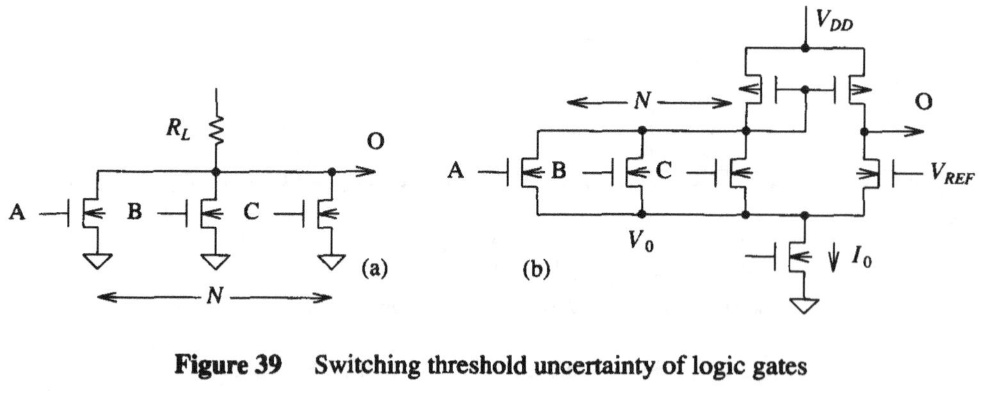

Figure 39 Switching threshold uncertainty of logic gates

$$\frac{V_A + V_B + \cdots}{N} = V_{REF} - \frac{N-1}{N}\frac{I_0}{2B_N}$$

and this gate has similar switching threshold voltage uncertainty.

If the switching threshold voltage of a gate input depends on voltages of the other input signals, the delay of the gate cannot be determined accurately beyond a certain tolerance. If the uncertainty in the gate's threshold voltage exists, the delay time of the gate is uncertain by at least *(threshold voltage uncertainty)/(input signal slew-rate)*, because the time when the gate begins switching becomes that uncertain. If the gate is a link in a critical path, the maximum operating frequency of the circuit becomes essentially uncertain. The frequency can be determined between an upper and lower bounds, but not as a single value.

6.11 Waveform Distortion

In Section 2.2 we showed that the *quality* of digital signals is determined from the signal amplitude (the difference between the signal voltages for Boolean values 1 and 0), and how early the signal is generated. We assumed signals that make step transitions. We compared digital signals A and B of Fig. 40(a), and concluded that signal A is of higher quality than signal B. Two signals have same quality if one can be generated from the other with no additional *cost*. In this section we attempt to compare quality of signals including rise and fall times, as shown by waveforms C and D of Fig. 40(b). We consider how the difference between waveforms C and D matters in digital signal processing. Suppose that a CMOS gate is driven by the input voltage waveform α (*dashed* waveform, with short rise/fall times) shown at the upper part of Fig. 40(c). We assume that the CMOS gate responds instantly to the input voltage change. The down-going output waveform of the gate is shown by dashed curve α (below). In this case the gate may have a switching threshold voltage V_{GSW} anywhere between V_{SS} and V_{DD}. If the gate is driven by voltage waveform β of Fig. 40(c), which has long rise/fall times, the response times of the gate depend on the gate's switching threshold voltage. If the threshold voltage is low (1), a wide down-going pulse (β_1) is generated, and if the threshold voltage is high (2), a narrow down-going pulse (β_2) is generated. If the switch-

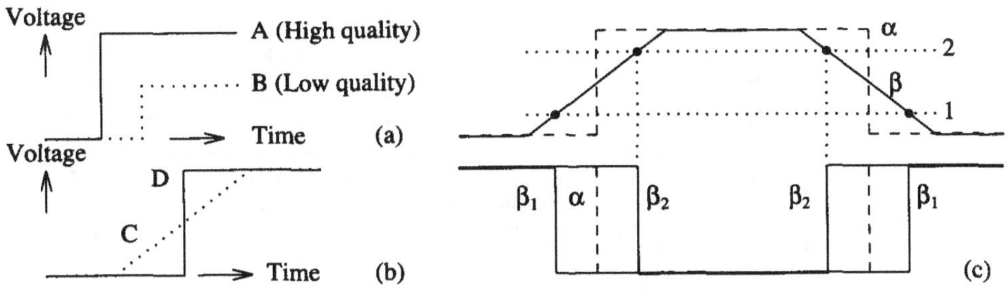

Figure 40 Node waveforms

ing threshold voltage is low (1), the output pull-down transition responds quickly, but the pull-up transition responds slowly. Therefore we may conclude as follows: for a low threshold voltage, quality of the output signal generated from the up-going edge of input signal β is higher than the quality of the output signal generated from input signal α. Quality of the output signal generated from the down-going edge of input signal β is lower than that generated from the same edge of input signal α. If the switching threshold voltage is set high (2), the delay and quality relationships reverse. Signals that make a gradual transition provide better quality signal for one transition polarity and poorer quality signal for the other. Long rise/fall times of signal make delay time of logic gates dependent on V_{GSW} of the gate, and therefore estimate of delay time becomes uncertain. Waveform distortion is defined as the modification of a digital signal waveform that makes the quality of the signal generated from the signal dependent on the parameters of the processing circuits like switching threshold voltage. Long rise/fall times of digital signal is an example of waveform distortion. In a CMOS circuit that must maintain a high degree of waveform accuracy, gate rise/fall times must be reduced to the minimum.

An interesting question arises: Is there any logic gate that gives a better quality output signal for both transition polarities? Such a gate is conceptually possible, but it has not been used because of complexity. Figure 41 shows one implementation of an inverter that sets different threshold voltages for pull-up and pull-down transitions. Structurally the circuit is a CMOS dynamic inverter. The D-latch detects the *present* output node level of the inverter and prepares for the next transition by steering the input signal to the *active* FET and by turning off the other. Since the inverter switches at the threshold voltage of the active FET, the inverter is fast. We note that this circuit is the opposite of the Schmitt trigger circuit described in Section 4.17.

The concept of *quality* of digital signals resolves a confusing puzzle of a circuit speedup technique. Sometimes a positive feedback circuit such as that shown in Fig. 42(a) is proposed for speedup. This circuit works as follows: When the driving source of node N_1 (represented by current generator I_P) pulls the node up to the switching threshold voltage of the first inverter V_{GSW}, the second inverter begins to switch, and node N_1 is pulled up strongly by MP2. This circuit can be used, for instance, to speedup signal propagation through a polysilicon con-

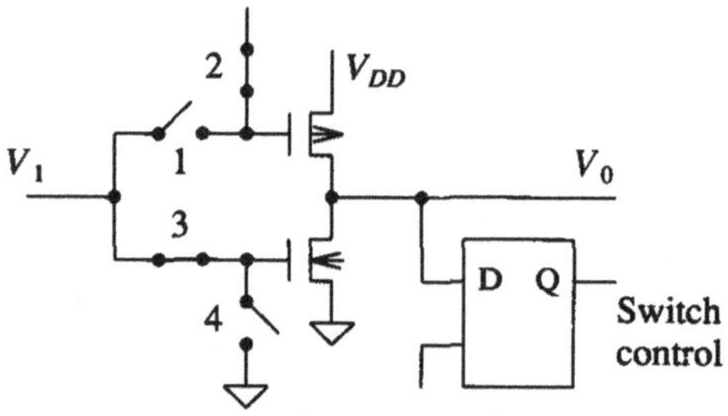

Switch setting

Operation	1	2	3	4
Pull-down	OFF	ON	ON	OFF
Pull-up	ON	OFF	OFF	ON

Figure 41 Threshold shift for pull-up and pull-down

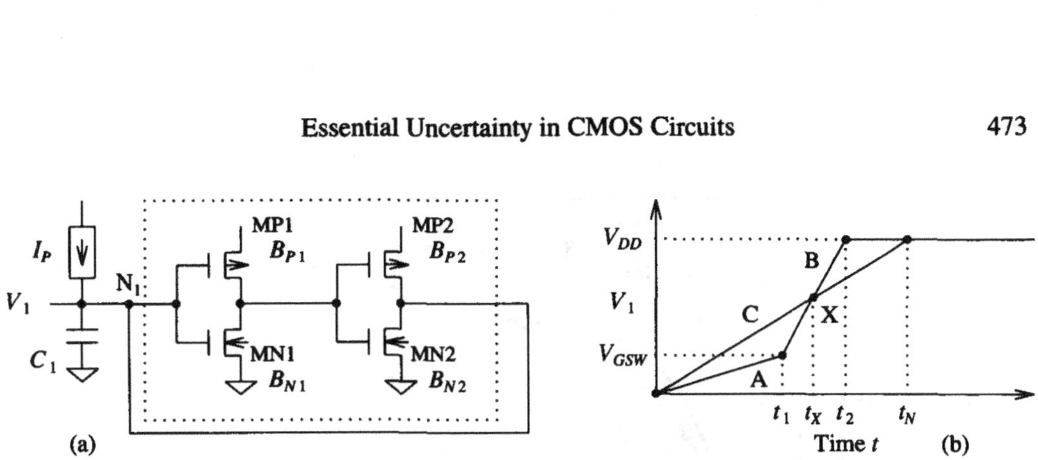

Figure 42 Effects of positive feedback on digital signal quality

trol line of a microprocessor datapath. Designer who examine this concept are often puzzled when they observe that the delay improvement is not so great as would be expected from this neat mechanism. We analyze the response of this circuit using the collapsible current generator FET model with zero threshold voltage and assume that the capacitance load of the first inverter is negligible. At time $t=0$, $V_1(t)$ is zero. If the active circuit within the dotted box of Fig. 42(a) is removed, node N_1 voltage increases with time as $V_1(t)=I_P t/C_1$. $V_1(t)$ reaches V_{DD} at $t=t_N$, where $t_N=C_1V_{DD}/I_P$. This node waveform is shown by linear ramp C of Fig. 42(b). If the active circuit is reconnected we have

$$V_1(t)=\frac{I_P-B_{N2}V_{DD}}{C_1}t \quad \text{until} \quad t_1=\frac{C_1B_{P1}V_{DD}}{(B_{P1}+B_{N1})(I_P-B_{N2}V_{DD})} \tag{44}$$

is reached. At $t=t_1$, V_1 equals the switching threshold voltage of the first inverter, $V_{GSW}=B_{P1}V_{DD}/(B_{P1}+B_{N1})$. In Eq. (44), if the current MN2 is able to sink, $B_{N2}V_{DD}$, is higher than I_P, pull-up of node N_1 never takes place. Therefore $I_P>B_{N2}V_{DD}$ is assumed. At $t=t_1$ the first stage inverter switches, MN2 turns off and MP2 turns on rapidly. Thereafter

$$V_1(t)=\frac{I_P+B_{P2}V_{DD}}{C_1}(t-t_1)+V_{GSW}\ (t_1<t<t_2)$$

$$t_2=\frac{C_1V_{DD}}{B_{P1}+B_{N1}}\ \frac{(B_{P1}+B_{N1})I_P+(B_{P1}B_{P2}-B_{N1}B_{N2})V_{DD}}{(I_P-B_{N2}V_{DD})(I_P+B_{P2}V_{DD})}$$

at time $t=t_2$, V_1 reaches V_{DD}. The node waveform $V_1(t)$ is shown by two line segments, A and B of Fig. 42(b). As is observed from this waveform, reducing delay time by applying positive feedback deforms the signal waveform. Lines A, B, and C intersect at point X at time t_X, where t_X is given by

$$t_X=t_N(B_{P1}/B_{P2})[(B_{P2}+B_{N2})/(B_{P1}+B_{N1})][1/[1-(B_{N2}V_{DD}/I_P)]].$$

If t_1, t_2 and t_X are normalized to $t_N=C_1V_{DD}/I_P$ we have

$$\theta_1=\frac{t_1}{t_N}=\frac{v_{GSW}}{1-\xi_N} \quad \theta_X=\frac{v_{GSW}(\xi_P+\xi_N)}{\xi_P(1-\xi_N)} \quad \theta_2=\frac{1+v_{GSW}\xi_P-(1-v_{GSW})\xi_N}{(1-\xi_N)(1+\xi_P)} \tag{45}$$

where $v_{GSW}=V_{GSW}/V_{DD}$, $\xi_N=B_{N2}V_{DD}/I_P$ and $\xi_P=B_{P2}V_{DD}/I_P$. Figure 43 shows θ_1, θ_X and θ_2 versus ξ_N, for given values of v_{GSW} and ξ_P.

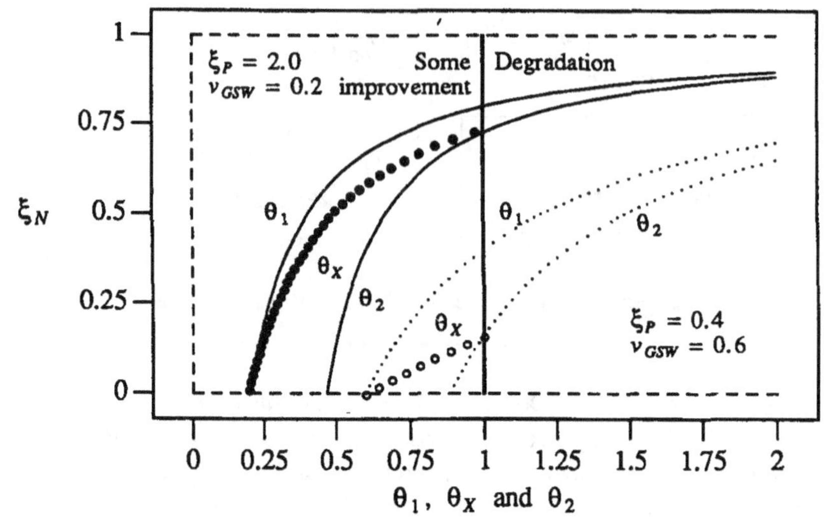

Figure 43 Effects of active pull-up for *quality* of signal

If $\theta_X < 1$, node N_1 delay decreases in the later part of the transient, $t_X < t < t_N$. This speedup is gained at the cost of the extra delay at the earlier part of the response ($0 < t < t_X$). The feedback circuit improves digital signal quality at the later part of the transient by deteriorating the quality of the signal at the beginning. This is the reason why there is a hard limit to the improvement. Delay would improve significantly if the feedback circuit had reduced delay in the *earlier* part of the transient. Delay improves by feedback technique if the node is loaded unusually heavily, and the node is driven by a small driver. If the node is already fast, there is no room for improvement. If the delay of the first inverter is included, t_X increases, and the improvement becomes even less. To optimize, response time ξ_P must be large, and v_{GSW} must be small. The delay is optimum if MP1 and MN2 are quite small or deleted. Since then the circuit will not recover by itself, the circuit must be clocked.

Nonzero rise/fall times of input digital signals and uneven and uncertain switching threshold voltages of CMOS gates [that are data-dependent (Section 6.11)] create uncertainty in delay times. Practically there is one more important mechanism that must be considered. This is Miller effect. We studied various Miller effect phenomena in Sections 3.5–3.7. Miller effect produces digital node waveforms that have a plateau as shown in Fig. 44 (dashed curve). Suppose that an inverter was driven by the dashed input waveform shown in Fig. 44. The input waveform has a plateau that lasts time T_X at voltage level V_X ($0 < V_X < V_{DD}$). The inverter driven by the distorted node waveform is modeled using the collapsible current generator model. The NFET and PFET have transconductances B_N and B_P, respectively, and the FET threshold voltages are assumed zero. The switching threshold voltage of the inverter V_{GSW} is given by

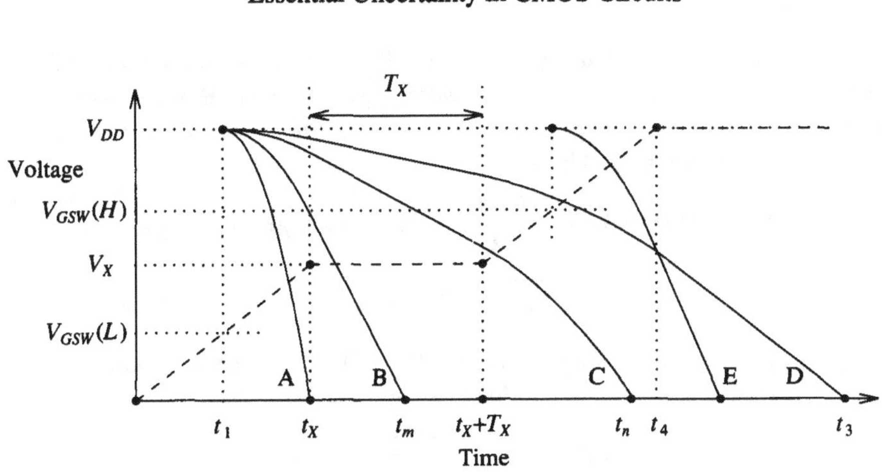

Figure 44 Miller effect waveform distortion

$$V_{GSW} = \frac{B_P}{B_P + B_N} V_{DD} .$$

If V_{GSW} is higher than V_X [as in the case $V_{GSW}(H)$ in Fig. 44], the analysis carried out in Section 2.8 applies directly. We use the notation of Section 2.8, and we refer to the circuit diagram (Fig. 21 of chapter 2). The output node waveform E of this case shows that the inverter switching takes place entirely after the Miller effect plateau.

If $V_{GSW}(L) < V_X$, we determine the node voltage waveforms as follows. At time t_1 the input voltage reaches the switching threshold voltage V_{GSW}. We define

$$t_1 = V_{GSW}(L)/\alpha \quad t_1 < t_X \quad t_X = V_X/\alpha$$

where α is the rate of increase of the input voltage (Section 2.8). The inverter is loaded by capacitance C_L. During the period of time $t_1 < t < t_X$, $V_0(t)$ changes with time as

$$V_0(t) = V_{DD} - \frac{(B_N + B_P)\alpha}{2C_L}(t - t_1)^2 . \tag{46a}$$

If the time when $V_0(t)$ becomes zero, $t_3 = t_1 + [2C_L V_{DD}/(B_N + B_P)\alpha]^{1/2}$, is less than t_X (curve A of Fig. 44), the switching transient ends before the Miller effect plateau begins. After t_3, $V_0(t) = 0$, and the switching is completed.

If this is not the case, $V_0(t_X) > 0$. $V_0(t)$ in the period $t_X < t < t_X + T_X$ is given by

$$V_0(t) = V_0(t_X) - \frac{(B_N + B_P)(V_X - V_{GSW})}{C_L}(t - t_X) \tag{46b}$$

where $V_{GSW} = V_{GSW}(L)$,

$$V_0(t_X) = V_{DD} - \frac{(B_N + B_P)\alpha}{2C_L}(t_X - t_1)^2 \quad \text{and} \quad t_m = t_X + \left[\frac{V_0(t_X)}{V_X - V_{GSW}}\right]\frac{C_L}{B_N + B_P}$$

and where $V_0(t_X)$ was computed using Eq. (46a). If $t_m < t_X + T_X$, the output node voltage reaches ground potential at time $t = t_m$, and thereafter $V_0(t) = 0$ (curve B of Fig. 44).

If $t_m > t_X + T_X$, $V_0(t)$ in the period $t_X + T_X < t < t_4$ (input voltage increases after the plateau is over, until it reaches V_{DD}) is given by

$$V_0(t) = V_0(t_X + T_X) - \frac{(B_N + B_P)}{C_L}(V_X - V_{GSW})(t - t_X - T_X) - \frac{(B_N + B_P)\alpha}{2C_L}(t - t_X - T_X)^2 \quad (46c)$$

where $V_0(t_X + T_X)$ is computed using Eq. (46b). If this $V_0(t)$ becomes zero at $t = t_n$ ($t_n < t_4$), $V_0(t) = 0$ for $t \geq t_n$ (Curve C of Fig. 44). Otherwise for $t > t_4$

$$V_0(t) = V_0(t_4) - \frac{B_N V_{DD}}{C_L}(t - t_4) \quad (46d)$$

where $V_0(t_4)$ is computed using Eq. (46c). $V_0(t)$ reaches ground potential at $t = t_3$, which is given by

$$t_3 = t_4 + \frac{V_0(t_4)}{V_{DD}} \frac{C_L}{B_N}$$

as shown by curve D of Fig. 44. We note that the output node voltage falls linearly with time while the input node voltage lingers in the Miller effect plateau. Because the Miller effect plateau stretches the transition period of the signal and increases rise/fall times, the waveform distortion is quite significant. As Miller effect theory shows, V_X is quite uncertain: V_X depends on various conditions of the gates that load the node. The output waveform delay becomes quite uncertain, by as much as the duration of the Miller effect, T_X.

The pulse width in digital circuits is set by the system clock period, T_S. Therefore an *ideal* logic waveform is an up-going or down-going pulse that has width nT_S (where n is an integer). This is because the logic variable does not change every clock cycle. Waveform distortion is measured by the difference of the processed signal waveform from the ideal *square* waveform having width nT_S. In reality the square waveform is never realized in any logic circuits. Even a lightly loaded CMOS inverter has significant rise/fall time at the operative clock period of the digital circuit, since digital circuits are operated close to the maximum operative frequency. If we consider the square waveform as the *ideal* waveform limit, there is a practical waveform limit. The practical waveform limit is the signal waveform that propagates through a long chain of lightly loaded inverters. In the practical digital signal waveform limit, pulse width determined at the typical switching threshold voltage ($V_{DD}/2$) becomes the measure of pulse waveform distortion.

The simplest design technique to maintain the pulse width against process variations is to cascade two *equivalent* stages of inverters and compensate for the effects of variations of the FET characteristics, as shown in Fig. 45. In this circuit, inverters A, B, C, and D have all the same PFET/NFET size ratio, and the size ratios of any two consecutive inverters are the same. Wiring parasitics are small. Figure 45(b) shows node waveforms. Suppose that process variation occurs such that NFET current decreases and PFET current increases. Then up-going square pulse A emerges as a narrower down-going pulse at node B, but the pulse

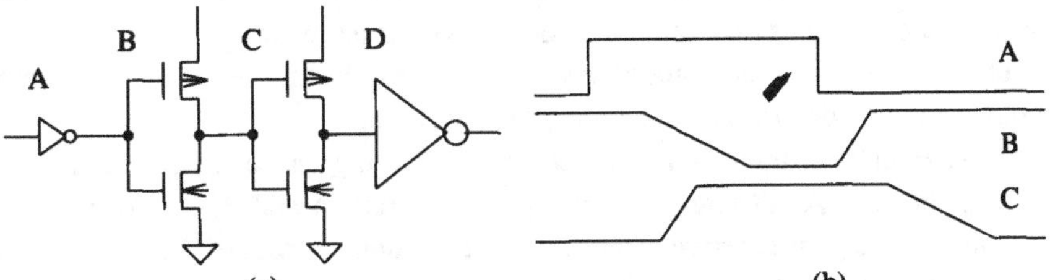

Figure 45 Compensation of skew

width is restored at node C because of slow pull-down of node C. Such stage-to-stage compensation is quite effective in maintaining pulse width in CMOS VLSI circuits.

We note that pulse width instability is an essential and fundamental characteristic of CMOS static logic circuits. This observation has the following consequence. A long chain of logic gates has a curious characteristic: Step signal propagates, but isolated pulse may not, if the chain is longer than a limit. Since pulse width may collapse, there is a certain statistically determined upper limit in the number of gates that can be cascaded. Suppose the input signal to a long chain (N-stage) of gates is a square pulse having duration T_0. There are many types of gates, whose operating conditions are all different. Then the increase or decrease in pulse width per gate has the statistical population, whose standard deviation is ΔT_0. This means exactly the following: If the gate is enabled and if an input is driven by a crisp square pulse having width T_0, the output pulse has width between $T_0 - \Delta T_0$ and $T_0 + \Delta T_0$, depending on gate types, relative timing, V_{GSW}, wafer processing variation, environmental variation, etc. The pulse width increases and decreases as it propagates through the gate chain. Suppose that the pulse width variation is random, and uncorrelated from stage to stage. Then statistically the width of the output pulse of N-stage cascaded gate chain will be between $T_0 - \sqrt{N}\Delta T_0$ and $T_0 + \sqrt{N}\Delta T_0$. If N is significantly larger than the limit N_{max} given by

$$N_{max} = (T_0/\Delta T_0)^2 \tag{47}$$

there is a probability that the signal pulse collapses before emerging from the output of the logic chain. N_{max} is the maximum number of gates that can be cascaded. In order to prevent such unpredictable pulse width, pulse shape must be restored by capturing the data signal occasionally in a latch, using precisely timed external clock. Sequential data processing has the merit of maintaining digital signal waveform *integrity*.

6.12 Essential Uncertainty in Signal Delay

It is generally (and mistakenly) believed that a circuit level simulator gives accurate signal delay times. In practice it seldom does. Simulation accuracy depends on many of details of simulator setup, but the most fundamental problem is that circuit delay depends on conditions to which the circuit is subjected, and not all these conditions can be specified when the circuit is simulated. An example is that gate delay depends on input data, and on the details of timing of the input data (how the input data changes with time). In this section, we summarize why circuit delay cannot be determined accurately by simulation.

Most fundamentally, there is a problem of identifying cause and effect in determining delay. In Fig. 46(a), an AND gate is driven by input signals A and B, both of which make up-going transitions. If signal B arrives significantly later than signal A [Fig. 46(b)], and the two transitions do not overlap, the signal B transition can be identified as the cause of the output node low-to-high transition that occurs some time later. If two input signal transitions occur almost simultaneously as shown in Fig. 46(c), however, which one causes transition can never be determined. Although signal B begins to rise earlier, signal A reaches the high logic level earlier. Therefore features of the input signal waveforms never allow clean identification of the cause. Since the cause of the output node transition cannot be identified, the delay time cannot be determined. If multiple signals are combined by a gate, cause-effect identification is generally lost, except for certain special circumstances. Delay time can be defined only from *one* input to outputs, not from many inputs to outputs. Signals at the other input terminals must stay at steady level (not necessarily at logic level). This is the problem, interestingly, shared by elementary particle physics, where identification of *particles* is lost upon their collision. A digital signal transition in a CMOS circuit is an *excitation*, like particles in elementary particle physics. If two signal transitions are separated either by time or location, their identity can be maintained. If they come close together in time and location, their identity is lost. A gate is equivalent to the spatial location where elementary particles collide, and thereafter they lose identification. I think that this analogy will produce many interesting new results.

This point has the following practical significance. When input signals to a multi-input gate change simultaneously, the definition of gate delay time becomes ambiguous, because which signal *caused* the output node transition cannot be determined. However, delay time starting from the primary clock transition that provided all the input data can be defined clearly. We considered sequential data processing as the process of splitting the single primary clock signal, processing the *split clocks* separately, and then later recombining them (Section 2.3). This viewpoint guarantees that delay times from the primary clock edge to any digital node transition can be determined. We note, however, that circuit simulation has practically never been carried out by reflecting this fundamental viewpoint. We recognize that in the level of elementary objects (elementary particles or elementary *excitation* of digital circuits)

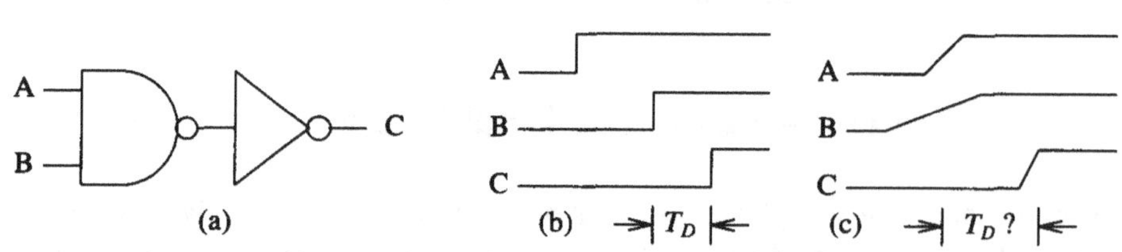

Figure 46 Cause and effect in circuit delay

cause-result relationship can not be established if a phenomenon is arbitrarily divided into sub-phenomena.

Less fundamentally, we find that there are many mechanisms that make circuit delay uncertain. Even in the case where cause is well identified, gate delay time may not be defined uniquely unless quite detailed specifications are given. Gate level specification is never sufficient. The delay of the NAND2 gate shown in Fig. 46(a) depends on which terminal, A or B, is used for switching the gate. Logic symbols do not provide clear indication (they have never been used as such). With reference to Fig. 47(a), pull-down delay by pulling input A up is longer than pull-down delay by pulling input B up. This is because the gate's conditions that existed before the input signal transition are different. If initially $A = high$ and $B = low$, node X was discharged before input signal B makes its transition. If $A = low$ and $B = high$ initially, node X was charged to a positive voltage. Which of the two conditions holds can not be determined by gate level specification. This electrical asymmetry in logically symmetric gates is a problem in precision clock circuit design. One way to avoid this problem is to use a *symmetrized* gate (the term *symmetrized* refers to gate structure and not to the pull-up/pull-down delay times). The symmetrized gate shown in Fig. 47(b) is absolutely symmetrical with respect to inputs A and B. This gate can be laid out perfectly symmetrically. The same concept works for a NOR2 gate. This scheme, however, requires $N!$ independent pull-down chain to symmetrize an N-input NAND or NOR gate completely. $N = 3$ shown in Fig. 47(c) is already too complex to be practical. $N = 2$ is the only case that is practical. A certain, highly symmetrical gate can be symmetrized as shown by the example of AOI22 gate of Fig. 47(d). A symmetrical gate must be laid out symmetrically, since otherwise the true symmetry is lost.

A *symmetrized* gate can be designed using a symmetrical resistance matrix signal combination, followed by a voltage limiter and amplifier as in DTL or RTL bipolar circuits. How precisely the voltage limiter holds the input voltage determines the degree of symmetry. This

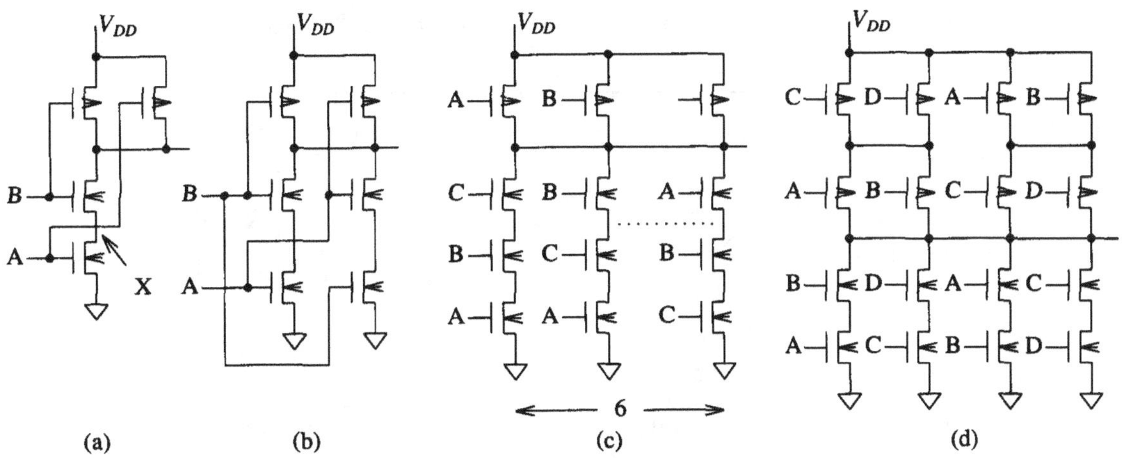

Figure 47 Symmetrized gates

is the structure of many BJT logic circuits. Compared to BJT logic circuits, conventional CMOS static logic circuits are *highly* asymmetrical.

Some CMOS circuits have data-dependent delay. Data dependence of delay originates from either of two causes: data-dependent load capacitance and data-dependent drive current of the gate. We consider data-dependent capacitive load first. The circuit shown in Fig. 48(a) has multiple transmission gates that connect the driver output to multiple destinations. When a transmission gate is turned off, the transmission gate presents small capacitive load to the driver. When the transmission gate is turned on, however, a heavy capacitive load on the output end of the transmission gate becomes a part of the driver capacitive load. The delay of the inverter depends on how many transmission gates are turned on, or the inverter delay depends on the state of a set of transmission gate control signals. This is a very undesirable circuit design. The circuit shown in Fig. 48(b) does not have this problem. In many practical cases, passive transmission gates are better replaced by active tristatable inverters.

Node capacitance uncertainty due to Miller effect has been discussed in Sections 3.5–3.7, and the subject is not repeated here. A similar problem of wiring capacitance uncertainty due to simultaneous switching was discussed in Section 1.9. The capacitive load of a node on insulating substrate ICs such as SOI CMOS ICs and GaAs MESFET ICs cannot be determined beyond a broad limit of uncertainty. In multilevel metal CMOS ICs, the interconnect that has small capacitance of its own, like an upper metal level wire having narrow width, has a similar problem. This problem has significant consequences on the accuracy of gate-level timing simulation of automatically designed ICs, which depend on wiring capacitance values extracted from layout data.

We consider the case of data-dependent drive capability. The circuit shown in Fig. 49 is a cascaded two-stage exclusive OR gate. The exclusive-OR gate is built from the parallel combination of a transmission gate and a tristatable inverter, using the following ideas [12]. The exclusive-OR of two variables, A and B is

$$EXOR(A, B) = A\ (transmission\ gate)\ \text{if}\ B = Low$$

$$EXOR(A, B) = \bar{A}\ (tristatable\ inverter)\ \text{if}\ B = High$$

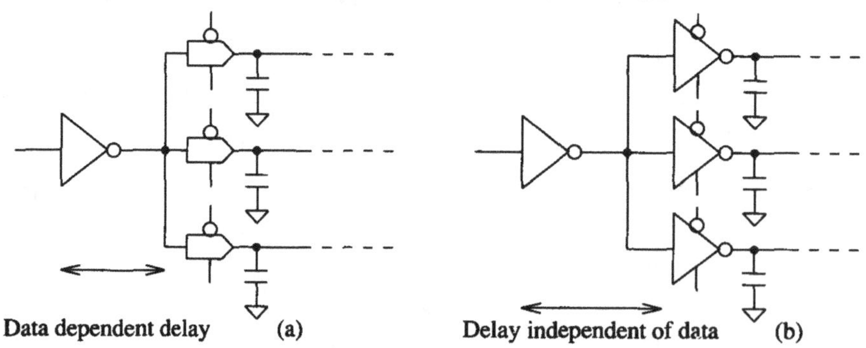

Figure 48 Data-dependent delay

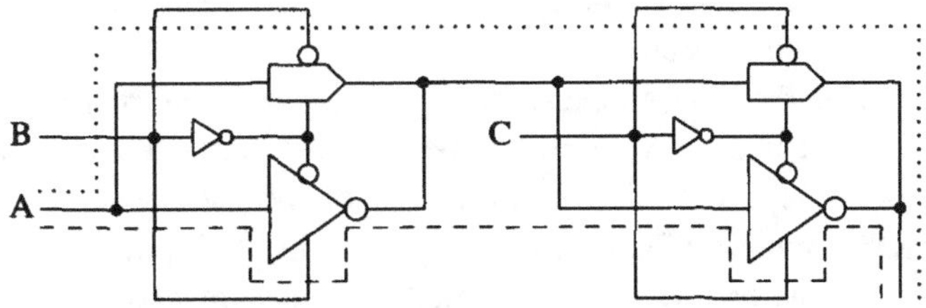

Figure 49 Data-dependent signal delay

If the exclusive OR gates are cascaded as shown in Fig. 49, the signal path depends on the input signals B and C. If $B=C=Low$, signal A goes through the *passive* path consisting of two series-connected transmission gates. If $B=C=High$, signal A goes through the active path consisting of two stages of cascaded tristatable inverters. The drive capability of the circuit depends on the control data. Directly cascading transmission gates over many stages should be avoided. Passive transmission gates must be used carefully, to avoid this kind of timing problem.

6.13 Gate Delay and Initial Conditions

Delay time of CMOS gates depends on the conditions of the gate immediately before switching. We study this problem using the equivalent circuit of an inverter shown in Fig. 50(a). In this simple equivalent circuit, the input signal makes a rapid low-to-high transition, and the PFET turns off instantly. Therefore the PFET is not shown. NFET MN1 pulls the output node down. The node voltage as a function of time is given by

$$V_0(t)=V_0(0)-\frac{B_N(V_{DD}-V_{THN})}{C_L}t. \tag{48}$$

Delay time, t_D, is determined by requiring that $V_0(t_D)=V_{DD}/2$ to obtain

$$t_D=t_{D0}(1-2\frac{\Delta V_0}{V_{DD}})$$

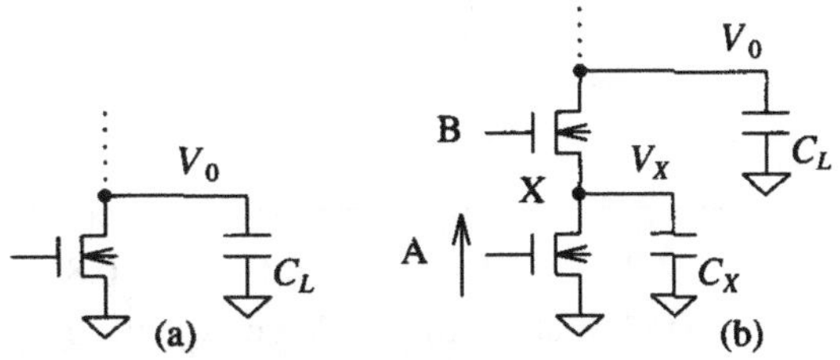

Figure 50 Analysis of initial condition dependence of gate delay

where $t_{D0} = (C_L/B_N)/[2(1 - v_{THN})]$ and $\Delta V_0 = V_{DD} - V_0(0)$. Delay time decreases linearly as the initial output node voltage decreases.

CMOS gates other than the simplest static inverter have *internal* circuit nodes that are neither input nor output nodes. Voltages of the *internal* nodes before switching cannot be specified: the voltages depend on the recent *history* of the gate switching. We consider the example of a CMOS static NAND2 gate shown in Fig. 50(b). Input node B was pulled up to V_{DD} before $t=0$. At time $t=0$ input node A makes an instantaneous low-to-high transition. At that moment the output node voltage was V_{DD}, and voltage of the internal node X was $V_X(0)$. If the NFET has no leakage current, $V_X(0)$ is in the range from 0 to $V_{DD} - V_{THN}$. If input node B was at V_{DD} for a long time, the maximum $V_{DD} - V_{THN}$ is reached, but this is not always the case. We assume that threshold voltage V_{THN} is independent of source voltage (no back-bias effect), and the two NFETs have same beta, B_N. The circuit equations are

$$C_L \frac{dV_0(t)}{dt} = -B_N[V_{DD} - V_X(t) - V_{THN}] \tag{49a}$$

$$C_X \frac{dV_X(t)}{dt} = B_N[V_{DD} - V_X(t) - V_{THN}] - B_N(V_{DD} - V_{THN}) = -B_N V_X(t). \tag{49b}$$

Equations (49a) and (49b) can be solved immediately to obtain

$$V_X(t) = V_X(0)exp[-(B_N/C_X)t] \tag{50a}$$

$$V_0(t) = V_{DD} - \frac{B_N}{C_L}(V_{DD} - V_{THN})t + \frac{C_X}{C_L}V_X(0)[1 - exp[-(B_N/C_X)t]] \tag{50b}$$

Delay time, t_D, is determined by requiring that $V_0(t_D) = V_{DD}/2$ to obtain

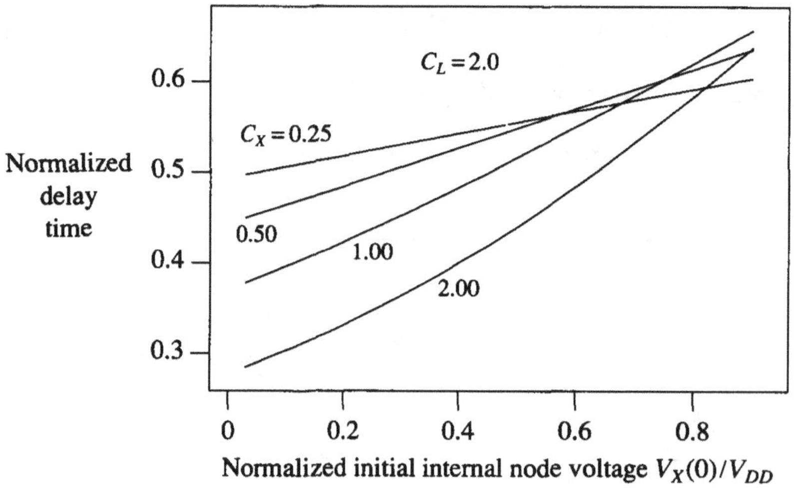

Figure 51 Dependence of delay on initial condition

$$t_D = \left(\frac{C_L + C_X}{B_N}\right) \frac{(C_X/C_L)}{1 + (C_X/C_L)} \theta \left[\frac{1 - v_{THN}}{v_X(0)}, 1 + \frac{C_L}{C_X} \frac{1}{2v_x(0)}\right] \tag{51}$$

where $\theta(..)$ is the theta function introduced in chapter 2. We use $(C_L + C_X)/B_N$ as the unit of time. Figure 51 shows normalized delay time $t_D/[(C_L + C_X)/B_N]$ versus normalized initial internal node voltage, $v_X(0) = V_X(0)/V_{DD}$, for several values of internal node capacitance C_X. If capacitance C_X is large ($C_X \approx C_L$), normalized delay time decreases quite significantly with decreasing initial internal node voltage. NAND2 gate pulls down initially by sharing charge between capacitors C_L and C_X. If C_X is small ($C_X \approx 0.25C_L$), delay time is only weakly dependent on the initial internal node voltage. We conclude that all the FETs that belong to a single gate should be placed close together: otherwise the charge stored initially in the wiring capacitance creates significant delay uncertainty.

6.14 Elementary Excitation of a Logic Chain

Uncertainty in a gate's condition prior to switching creates uncertainty in delay time. A simple theory was developed in the last section. The same uncertainty creates an essentially new effect, which will be discussed in this section. A signal propagating along a long uniform gate chain has many interesting and surprizing details that originate from uncertainty in the initial condition of gates prior to switching. This effect shows up dramatically at the maximum operating frequency of the circuit, and therefore the somewhat artificial example used to show the effect in this section is not unrealistic. The example we study is cascaded identical CMOS inverters as shown in Fig. 52(a). We assume CMOS inverters that respond identically to pull-up and to pull-down transitions.

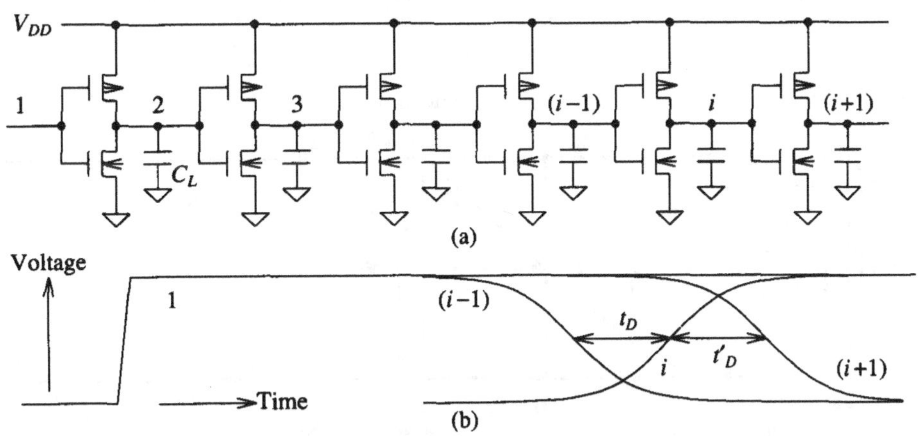

Figure 52 Chain of identical inverters and stepwave front propagating along the chain

The FET current charges and discharges node capacitance C_L, consisting of gate capacitances of the load and the drain capacitances of the driver itself. Wiring capacitance is included in C_L. Input node 1 is driven by a step voltage source shown by waveform 1 of Fig. 52(b). After some time the wave front arrives at node $(i-1)$ (even number), and node i (odd number) and the respective node voltages vary as shown by waveforms $(i-1)$ and i of Fig. 52(b). Node i waveform is delayed relative to node $(i-1)$ waveform by delay time t_D defined in Fig. 52(b). Similarly the node $(i+1)$ waveform is a down-going wave front like that of node $(i-1)$, further delayed by t'_D. If i is a large number, waveforms at nodes $(i-1)$ and $(i+1)$ are the same, except that node $(i+1)$ waveform arrives time $t_D + t'_D$ later. If we assume that each inverter is symmetrical with respect to up-going and down-going transitions, $V_i(t)$ and $V_{i+1}(t)$ satisfy

$$V_{i+1}(t) = V_{DD} - V_i(t - t_D) . \tag{52}$$

Since this is the simplest case, we investigate the propagating signal using the symmetrical inverter chain. The mathematical problem of solving Eq. (52) is difficult if a conventional FET model is used. In order to obtain a closed form solution, we assume the following. In Fig. 53(a) PFET and NFET currents, I_P and I_N, respectively, are given by the collapsible current generator model with zero threshold voltage,

$$I_P = B[V_{DD} - V_{i-1}(t)] \qquad I_N = BV_{i-1}(t)$$

if $0 < V_i(t) < V_{DD}$. If $V_i(t) = 0$, however, the NFET is able to conduct any current less than $BV_{i-1}(t)$ without developing voltage across the channel. The same is true for PFET if $V_i = V_{DD}$. Figure 53(b) shows plots of $V_{i-1}(t)$ and $V_i(t)$. At time $t = 0$ (this time origin is chosen for convenience), $V_{i-1}(t)$ reaches V_{DD}, and no further change in $V_{i-1}(t)$ after that time. Node i is pulled down by the NFET alone, which acts as a constant current generator BV_{DD}. We have,

$$V_i(t) = V_0 - V_{DD}(B/C_L)t \qquad (t > 0) \tag{53}$$

where $V_i(0) = V_0$ is the initial condition. If $t < 0$, $V_{i-1}(t)$ increases with time as

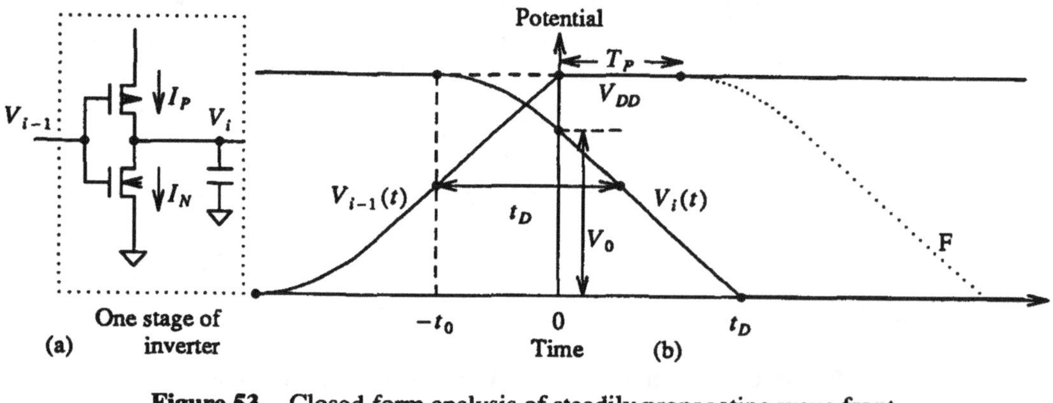

Figure 53 Closed-form analysis of steadily propagating wave front

$$V_{i-1}(t) = V_{DD} + V_{DD}(B/C_L)t\,.$$

We arrive at this expression by combining Eqs. (52) and (53). $V_i(t)$ satisfies equation

$$C_L \frac{dV_i(t)}{dt} = -2B[V_{i-1}(t) - \frac{1}{2}V_{DD}] \tag{54}$$

which is solved subject to the condition $V_i(0) = V_0$ to obtain

$$V_i(t) = -V_{DD}(B/C_L)t - V_{DD}(B/C_L)^2 t^2 + V_0$$

$V_i(t)$ must satisfy the following two conditions at time $t = -t_0$ (we choose $t_0 > 0$).

$$V_i(t) = V_{DD} \quad \text{and} \quad \frac{dV_i(t)}{dt} = 0 \quad \text{at} \quad t = -t_0$$

where the derivative must be zero since $V_{i-1}(-t_0) = V_{DD}/2$. From this requirement we have

$$t_0 = (1/2)(C_L/B) \quad \text{and} \quad V_0 = (3/4)V_{DD}\,.$$

We obtain

$$V_i(t) = \frac{3}{4}V_{DD} - V_{DD}\left(\frac{B}{C_L}\right)t \quad (t>0) \quad = V_{DD} - V_{DD}\left(\frac{B}{C_L}\right)^2 (t+t_0)^2 \quad (t<0) \tag{55}$$

and waveform delay per stage, t_D, is given by

$$t_D = \frac{3}{4}\left(\frac{C_L}{B}\right) \tag{56}$$

If more realistic FET model than the collapsible current generator model is used, this problem cannot be solved in a closed form. It is possible, however, to obtain numerical solutions. We use the standard, gradual channel, low-field FET model with negligible back-bias effect. NFET current is given by this model as

$$I_N = B_G(V_{i-1} - V_{THN} - \frac{1}{2}V_i)V_i \quad \text{if} \quad V_i < V_{i-1} - V_{THN}$$

$$= \frac{B_G}{2}(V_{i-1} - V_{THN})^2 \quad \text{if} \quad V_i > V_{i-1} - V_{THN}\,.$$

We normalized voltages to V_{DD}, and time to t_{UNIT} defined by $t_{UNIT} = C_L/B_G V_{DD}$. The time integration increment was chosen as $0.01t_{UNIT}$. Figure 54 shows travelling waveform in a symmetrical inverter chain.

Closed-form theory suggests the propagation of an *isolated* pulse made by combination of waveforms $V_{i-1}(t)$ and time-shifted $V_i(t)$ (dotted curve F) as shown in Fig. 53(b). The isolated pulse should propagate along the chain of the cascaded symmetrical inverters without changing its shape. In Fig. 53(b), the leading edge and the trailing edge are given by Eq. (55), and its mirror image, to the power supply voltages, respectively, and there is a nonzero time between the two waveform features (T_P). The closed-form theory predicts that the leading and the trailing edges do not interact, and the pulse is able to propagate along the inverter chain indefinitely. We show that this conclusion is restricted to the collapsible current generator FET model used, and if the gradual channel FET model is used, there can be no travelling isolated pulse. This is the same as that of Section 6.12, but it is a stronger conclusion than

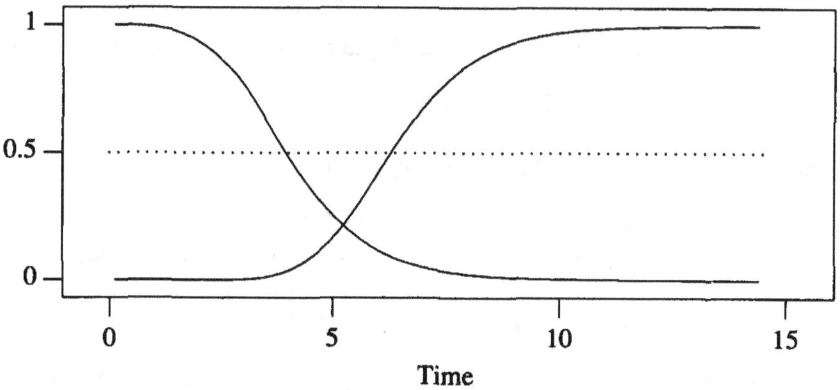

Figure 54 Stepwave front determined using gradual channel FET model

that. Even in a chain of symmetrical and identical gates an isolated pulse is *unable* to propagate indefinitely. If the inverter chain shown in Fig. 52(a) is driven by an isolated pulse, the pulse width decreases as pulse is inverted many times by the inverters. As the effect accumulates, finally the isolated pulse collapses.

This effect can be understood from Fig. 55. When the down-going edge arrives at the input of the i-th inverter $[V_{i-1}(t)]$, the output node of the inverter has been discharged completely: We have $V_i(t_A)=0$. As node voltage $V_{i-1}(t)$ goes down, node voltage $V_i(t)$ goes up, as shown by curve $\overline{ABC}$. At time t_C when $V_{i-1}(t)$ goes up again, $V_i(t)$ has not yet reached V_{DD}. Therefore the inverter pulls down from voltage $V_S(i)$, which is lower than V_{DD}. Since the inverter is symmetrical with respect to pull-up and pull-down, pull-up of node i to $(V_{DD}/2)$ takes more time than pull-down to $(V_{DD}/2)$, or the leading edge delay is more than the trailing edge delay. Therefore waveform $V_i(t)$ is narrower than waveform $V_{i-1}(t)$. Pulse width decreases every time it is inverted. If the collapsible current generator FET model is used, the output node voltage reaches ground, or V_{DD} within a finite time, and therefore, if the leading and the trailing edges of the isolated pulse do not overlap, the pulse width does not decrease. This conclusion is not realistic. Pulse width decrease is, however, very slow if the input pulse is wide. Curves in Fig. 56 show simulation results. Figure 57 shows pulse width (measured at

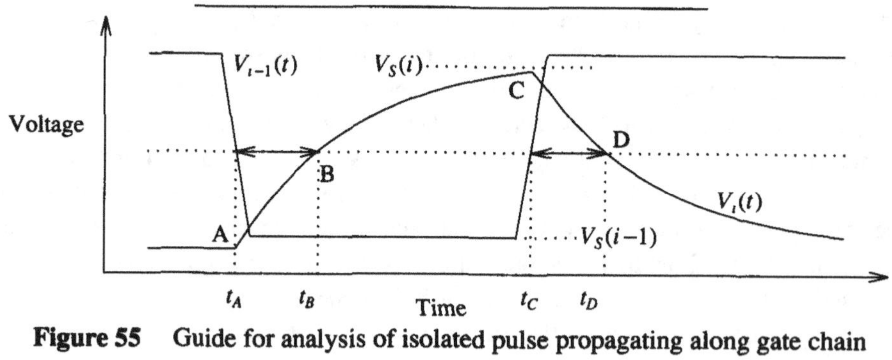

Figure 55 Guide for analysis of isolated pulse propagating along gate chain

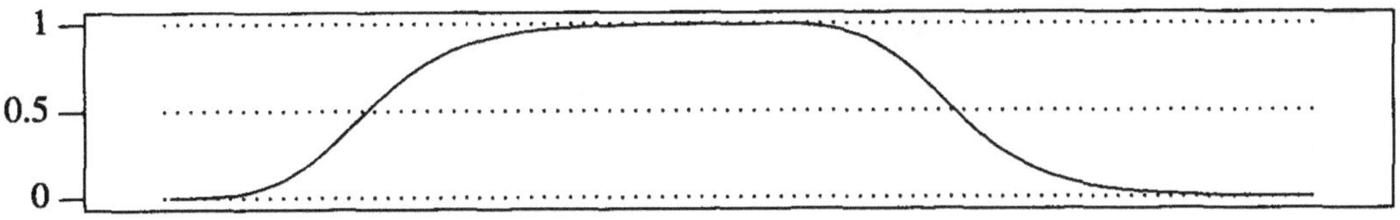

50-stages

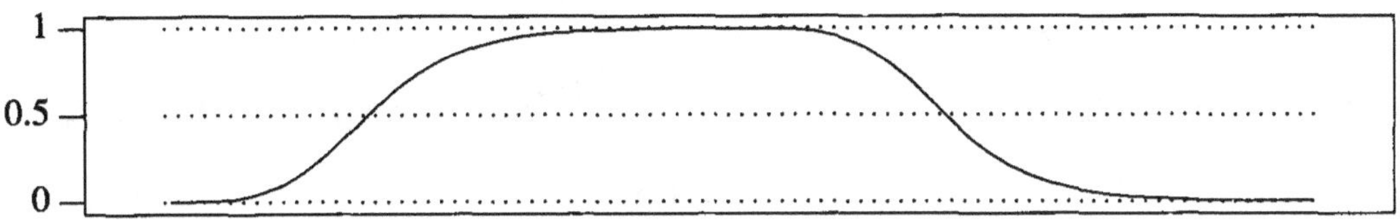

300-stages

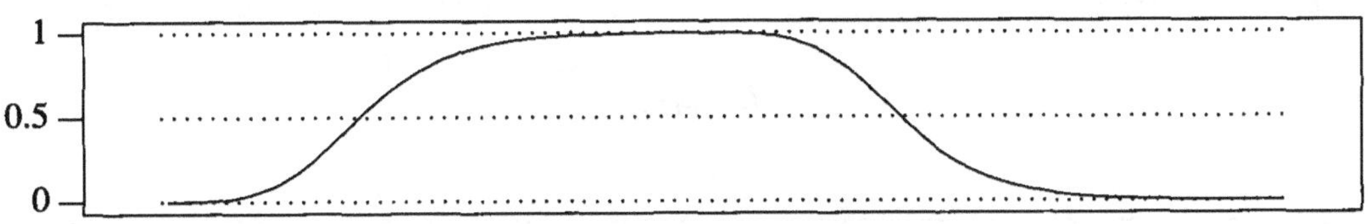

700-stages

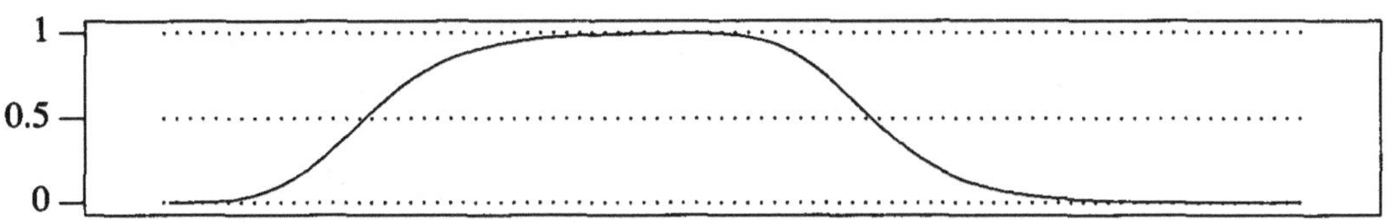

1100-stages

Figure 56 (first part)

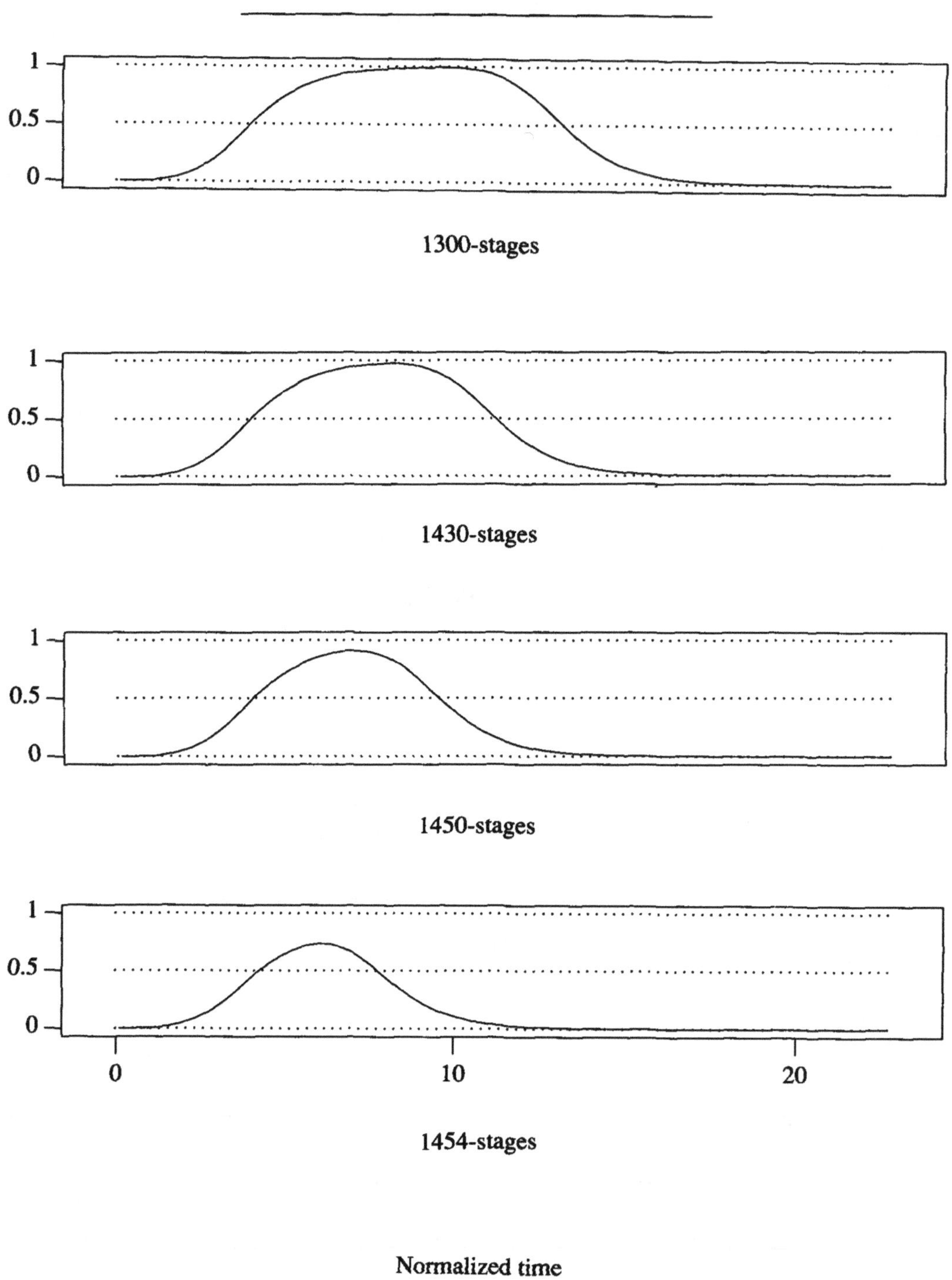

Figure 56 Collapse of pulse (original width = 11.6 times output time constant)

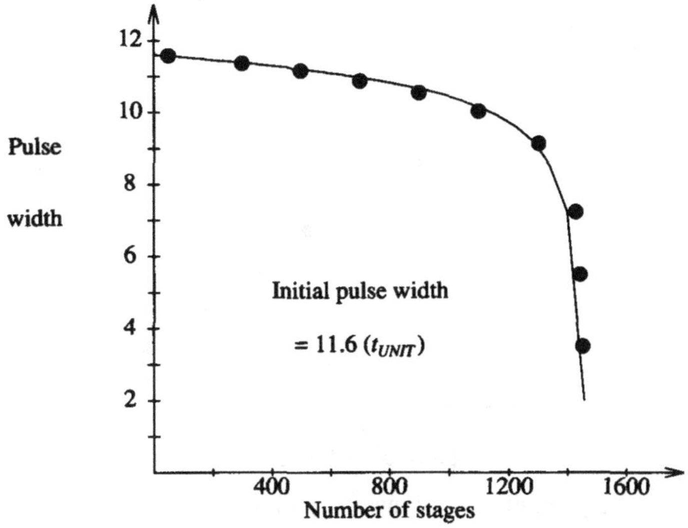

Figure 57 Pulse width versus number of stages of inverters

$V_{DD}/2$) versus number of stages from the input terminal. The initial pulse, whose width is $11.6t_{UNIT}$, propagates through 1455 stages of inverters before collapse. We note that pulse width decays very rapidly once the width is reduced to less than about 5.

The number of inverter stages, N, required to collapse a pulse that has initial width T_0, can be estimated from the following theory. With reference to Fig. 55, the pulse has width $T(i)$ at the input of i-th inverter, where

$$T(i) = t_C - t_A . \tag{57}$$

The process of switching at the leading edge (curve $\overline{ABC}$) can be separated into two phases. During the initial phase ($t_A < t < t_B$), the PFET that pulls node i up is in the saturation region. At the beginning of pull-up, the PFET carries current $I_P = (B_G/2)(V_{DD} - V_{TH})^2$ and the drain to source voltage is V_{DD}. The PFET acts approximately like an equivalent resistor having value

$$R_{SAT} = \frac{2V_{DD}}{B_G(V_{DD} - V_{TH})^2} \approx \frac{2}{B_G V_{DD}} \tag{58a}$$

if $V_{DD} \gg V_{TH}$ (saturation region). During the final phase of switching ($t_B < t < t_C$) the PFET looks approximately like a linear resistor having value (triode region)

$$R_{TRI} = \frac{1}{B_G(V_{DD} - V_{TH})} \approx \frac{1}{B_G V_{DD}} . \tag{58b}$$

Therefore $R_{SAT} = 2R_{TRI}$. Figure 58 is a logarithmic plot of the waveform of a symmetrical inverter. The horizontal axis is normalized time and the vertical axis is $[V_{DD} - V_i(t)]/V_{DD}$. The inverter input is driven by a step voltage. Figure 58 shows that the time constant changes indeed, by a factor of two from the beginning to the end of the transition. From Eqs. (58a),

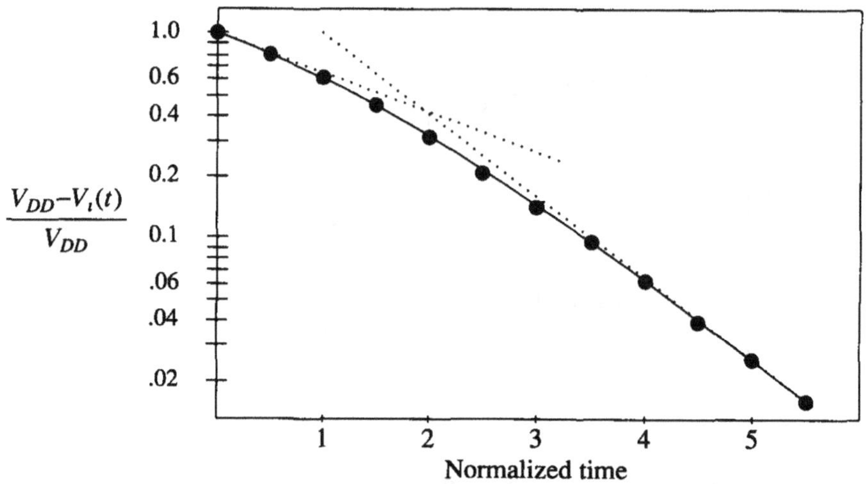

Figure 58 Logarithmic plot of inverter pull-up output node waveform

(58b) and Fig. 58 we use the following model for a switching inverter using time constant t_0 defined by

$$t_0 = \frac{C_L}{B_G(V_{DD} - V_{TH})} = \frac{t_{UNIT}}{1 - (V_{TH}/V_{DD})} \tag{59}$$

Leading edge delay [$(t_B - t_A)$ of Fig. 55] is given by

$$t_B - t_A = \log 2 \cdot 2t_0 \cdot \left\{ \frac{V_S(i)}{V_{DD}} \right\}^2 \tag{60}$$

where log 2 reflects delay of voltage change from 0 to $V_{DD}/2$, factor 2 of $(2t_0)$ for the ratio of (R_{SAT}/R_{TRI}) from Eqs. (58a) and (58b), and the extra factor $[V_S(i)/V_{DD}]^2$ for reduction in input voltage affecting t_0. $V_S(i)$ in the factor should be $V_{DD} - V_S(i-1)$, but we substituted $V_S(i)$ assuming gradual change of waveform from one stage to the next. The trailing edge delay [$(t_D - t_C)$ of Fig. 55] is given by

$$t_D - t_C = \left[\log 2 - \left\{ \frac{V_{DD} - V_S(i)}{V_{DD}} \right\} \right] \cdot 2t_0 \cdot \left\{ \frac{V_S(i)}{V_{DD}} \right\}^2 \tag{61}$$

where the negative term in the curly bracket originates from the fact that $V_i(t)$ changes from $V_S(< V_{DD})$ to $V_{DD}/2$, not from V_{DD} to $V_{DD}/2$. From Eqs. (60) and (61) pulse width shrinks in one stage of inversion by

$$(t_B - t_A) - (t_D - t_C) = 2t_0 \frac{V_{DD} - V_S(i)}{V_{DD}} \left\{ \frac{V_S(i)}{V_{DD}} \right\}^2 . \tag{62}$$

The last factor, $[V_S(i)/V_{DD}]^2$ may be set at 1.0 except at the very last phase of collapse of the pulse. From Fig. 58, we approximate the final approach to the steady state by

$$\frac{V_{DD}-V_S(i)}{V_{DD}} \approx \exp\left\{-\frac{T(i)-T_\Delta}{t_0}\right\} \tag{63}$$

where $T_\Delta \approx t_0$ is as observed from Fig. 56. It is convenient, however, to include the following in T_Δ. Figure 56 shows that the pulse has significant *fall* time T_F (at the trailing edge). Since we define pulse width using the times when the node voltage crosses $V_{DD}/2$, T_F must be subtracted before T in Fig. 56 is used in Eq. (60). We observe that $T_F \approx 2.5t_0$ from Fig. 56. We then have $T_\Delta = t_0 + 2.5t_0 = 3.5t_0$. The pulse width at the i-th inverter $T(i)$ satisfies the differential equation

$$\frac{dT(i)}{di} = -2t_0 \exp\left(\frac{T_\Delta}{t_0}\right)\exp\left\{-\frac{T(i)}{t_0}\right\} \tag{64}$$

which can be solved subject to the initial condition

$$T(0)=T_0$$

as

$$\exp\left(\frac{T_0}{t_0}\right)-\exp\left\{\frac{T(i)}{t_0}\right\}=2\exp\left(\frac{T_\Delta}{t_0}\right)i \tag{65}$$

In order to estimate the number of stages for pulse decay, the second term on the left-hand side can be neglected. We then have

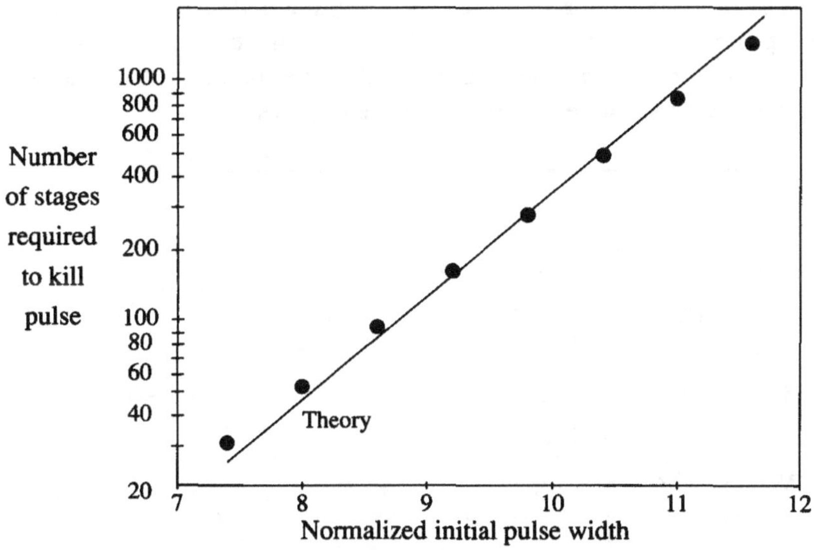

Figure 59 Number of inverter stages required to collapse pulse versus initial pulse width

$$i_{max} = \frac{1}{2}\exp\left\{-\frac{T_\Delta}{t_0}\right\}\exp\left(\frac{T_0}{t_0}\right) \approx 1.51 \times 10^{-2}\exp\left(\frac{T_0}{t_0}\right) \tag{66}$$

Figure 59 shows a comparison between numerical simulation and closed-form theory of Eq. (66). Agreement between the two is reasonable.

The pulse in an inverter chain decays because output pulse O of Fig. 60(b) is narrower than input pulse I. Let us consider two consecutive stages of inverters shown in Fig. 60(a). If the FET sizes are adjusted such that I and O have the same width, pulse decay can be prevented. Let us write the size (width) of NFET MN1 as $S(MN1)$. We scale the FETs as follows:

$$S(MN1) = S_0 \cdot \alpha \quad S(MP1) = S_0 \cdot \beta$$

$$S(MN2) = S_0 \quad S(MP2) = S_0 \cdot \beta \cdot \alpha$$

where β is the NFET/PFET beta ratio determined by device physics (approximately 2.5 in 1.75 μm CMOS), and α is a parameter larger than 1, which is chosen to maintain pulse width. Since the scaled chain responds faster to the leading edge than to the trailing edge of the isolated pulse of Fig. 60(b), pulse width does not decrease. Figure 61 shows normalized pulse width versus number of stages of inverters for different values of α. If $\alpha = 1.0$, the pulse decays, and if $\alpha = 1.01$, the pulse grows. At about $\alpha = 1.004$, the pulse stays unchanged for many stages. To maintain pulse width by scaling the chain is theoretically interesting but unpractical. A practical method is either to increase pulse width or decrease gate delay.

If more than one pulse propagates along the inverter chain, they interact and decay in a peculiar way. Figure 62 shows waveforms of two pulses that propagate along the inverter chain. The second pulse remains unaffected until the first pulse collapses. This result is consistent with the theory of the preceding sections: The first pulse trailing edge does not decay completely before the leading edge of the second pulse arrives, and the *residual* node voltage of the first pulse expedites the leading edge transition of the second pulse. This mechanism maintains the width of the second pulse until the first pulse vanishes.

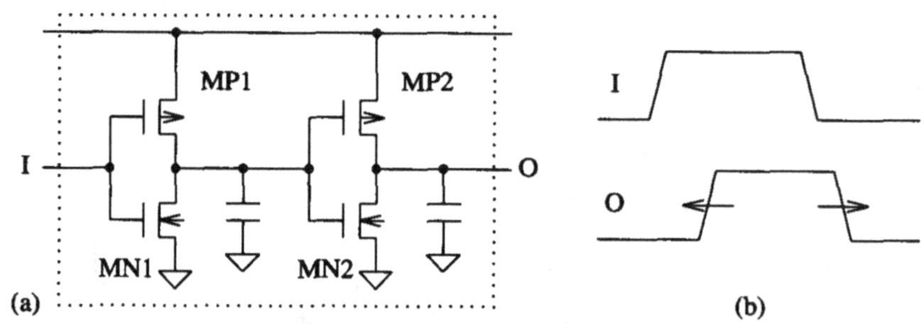

Figure 60 Isolatd pulse propagation in scaled inverter chain

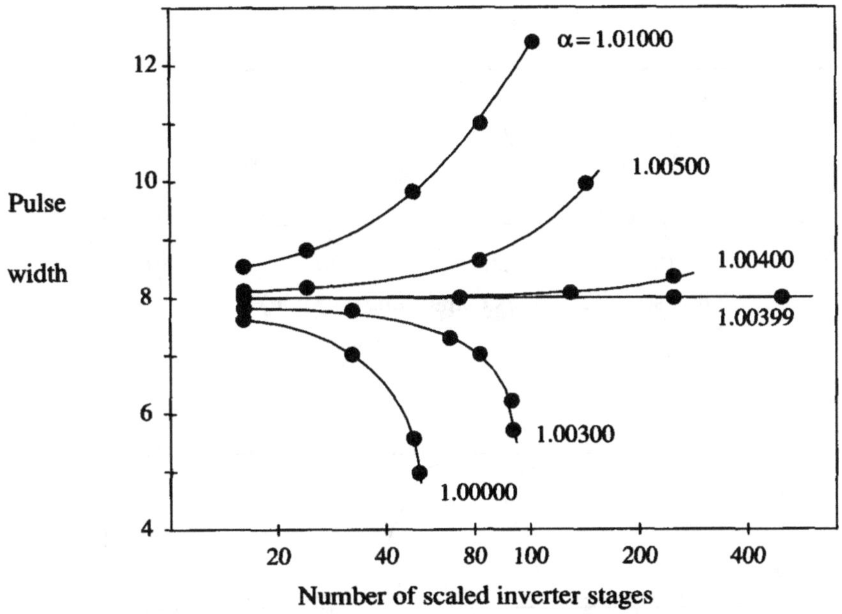

Figure 61 Normalized pulse width versus number of inverter for various scaling factor of FETs

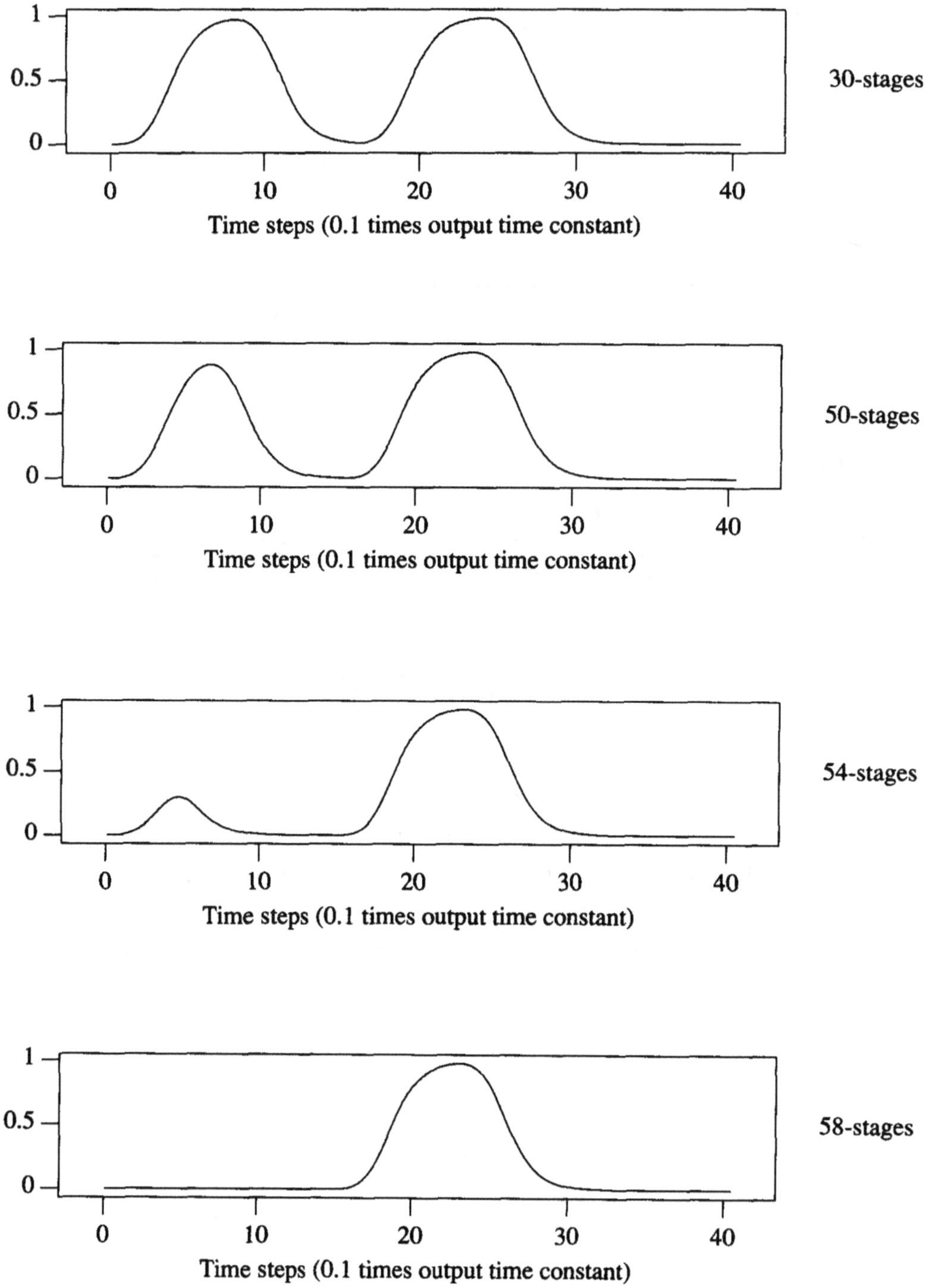

Figure 62 Double pulse propagation in a symmetrical inverter chain

6.15 Accuracy of Circuit Simulation

Circuit simulation is useful since real circuits are complex. If back-of-envelope calculation is not adequate because of circuit complexity, simulation is practically the only method of design verification. There is a fundamental limit in simulation, however. Not all the circuit parameters can be determined. These parameters cannot be determined by experiments, either.

Let us consider estimating the delay of cascaded inverter chain A shown in Fig. 63. Delay of signal from input I to output O_A, T_A, is sum of the delays of the stages, T_{A1}, T_{A2}, ..., as

$$T_A = \sum_{i=1}^{N} T_{Ai} \tag{67}$$

where each components T_{Ai} is positive. Chain delay, T_A, increases monotonically with increasing number of stages, N ($N = 4$ in Fig. 63).

In circuit simulation there are inevitable errors, originating from inaccuracies in the device model, nonuniformities in the silicon wafer, variations in processing parameters and the like. Error in T_A, ΔT_A, and errors in component delays T_{Ai}, ΔT_{Ai}, are related by

$$\Delta T_A = \sum_{i=1}^{N} \Delta T_{Ai}\,. \tag{68}$$

Let us consider the signs of ΔT_{Ai}. If the signs are all positive (or all negative), there must be *correctable* error(s) in simulation parameters, such as underestimates of FET transconductances and overestimates of the FET threshold voltages. Since simulation conditions are closely correlated to wafer process line conditions, global over- or under-estimates should not take place. Therefore ΔT_{Ai} take positive and negative signs with equal probability. The average of the absolute values of ΔT_A, $|\Delta T_A|$ is $\sqrt{N}$ times $< |\Delta T_{Ai}| >$ (average over i) because of the positive and the negative signs. Therefore the relative error

$$\text{Relative error} = |\Delta T_A|/T_A \tag{69}$$

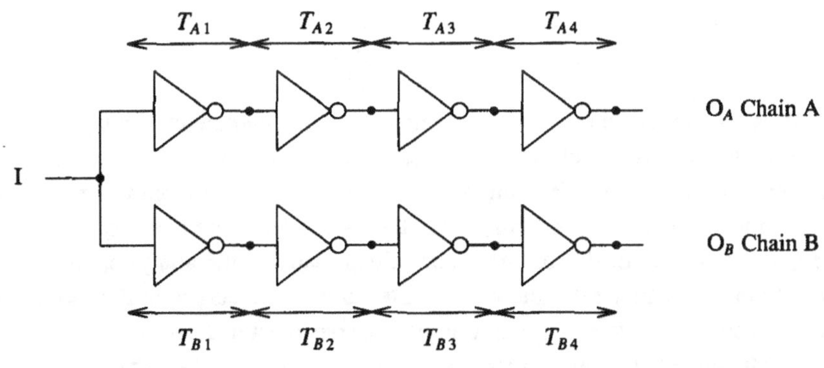

Figure 63 Chain of inverters that generate clocks

decreases gradually, as the central limit theorem of statistics dictates. Since process parameters are determined by taking the over-all consistency of all the types of circuits and operating conditions, simulation accuracy increases gradually as the circuit becomes complex. This is *not* a surprising conclusion, since more complex circuits match statistically better to the entire statistical variations. Simulation results are useful if relative error is less than a certain practical limit, typically about 10%. Simulation accuracy of about 10% is now routinely attained.

There are cases, however, in which relative error increases without limit. In Fig. 63, I is the primary clock source, chain A generates the master latch clock, and chain B generates the slave latch clock. The master latch clock must precede the slave latch clock by a certain minimum delay time, $T_{\min}$, to prevent data flowthrough, or race conditions. Difference in the delays of the two independent inverter chains A and B, T_{AB} is given by

$$T_{AB} = \sum_{i=1}^{N} T_{ABi} \quad T_{ABi} = T_{Bi} - T_{Ai} \tag{70}$$

and we require

$$T_{AB} > T_{\min}. \tag{71}$$

In order to verify the condition of Eq. (71), T_{AB} must be determined better than the standard accuracy (say, 10%). Let us consider whether or not this attempt is feasible. T_{ABi} may have a positive or negative sign. T_{AB} does not necessarily increase monotonically with increasing N. Especially, T_{ABi}s have the possibility of cancelling out such that

$$T_{AB} \approx 0. \tag{72}$$

Since the error in T_{AB}

$$\Delta T_{AB} = \sum_{i=1}^{N} (\Delta T_{Bi} - \Delta T_{Ai}) \tag{73}$$

is never zero, the relative error of T_{AB}, $|\Delta T_{AB}|/|T_{AB}|$ diverges. T_{AB} is a parameter that can be determined only under certain restrictions. If T_{AB} is computed blindly, it is not possible to specify the *a priori* limit of relative error.

This is an essential problem in circuit analysis. When a parameter P is expressed in the form

$$P = \sum_{i=1}^{N} P_i \tag{74}$$

and if P_i is able to take a positive or negative sign, the parameter can be determined by simulation only when P is not close to zero, especially when N is large. In general, any circuit parameter that can be zero, depending on the condition of the circuit, can be determined with specified relative accuracy limit only when the parameter is not close to zero. When P is close to zero, P depends critically on the assumptions made in the analysis, such as the parameter values, the circuit, and the device models. Demands for accuracy in the models and parameter values in such a case exceed realistic limits. Experimental determination of such parameters fails as well, since the measured values depend critically on the experimental conditions, not all of which can be controlled.

What was described here might seem to be special cases. In reality there are many cases where circuit parameters that are not *positive definite* do create difficulties, and that has been discussed already in this book. Figure 64(a) shows cascaded inverter stages: a pulse having width T_W drives the input. The pulse delivered by the inverter chain has width T'_W, where we have $T_W \neq T'_W$. It was discussed in Sections 5.14 and 5.15 that involved design techniques are required to maintain the pulse width. Pulse width T'_W equals initial pulse width T_W plus the sum of the differences in delays of the individual inverters for pull-up (T_U) and pull-down (T_D) transitions,

$$T'_W = T_W + \sum_i (-1)^i [T_D(i) - T_U(i)]$$

and the second term is not positive definite. This is the fundamental source of difficulty in maintaining pulse width. The technique discussed in Sections 5.14 and 5.15 gives first-order compensation to process variation, by utilizing device matching. Figure 64(b) shows a design for an *M out of N* gate. Such a gate is a key component in *majority* logic (CMOS has no good majority logic circuit; this is a rather poor design). If at least M of the total N inputs are high, the output voltage V_0 is high. Current of i-th pull-down NFET is $I_D(i)$ and that of the pull-up NFET is $I_U(i)$. Their average, I_D and I_U, respectively, should satisfy $NI_U = (M - 0.5)I_D$. Pull-down currents add and subtract as

$$\text{Pull-down current} = \sum_i \Delta I_i \quad \text{where} \quad \Delta I_i = I_D(i) - I_U(i)$$

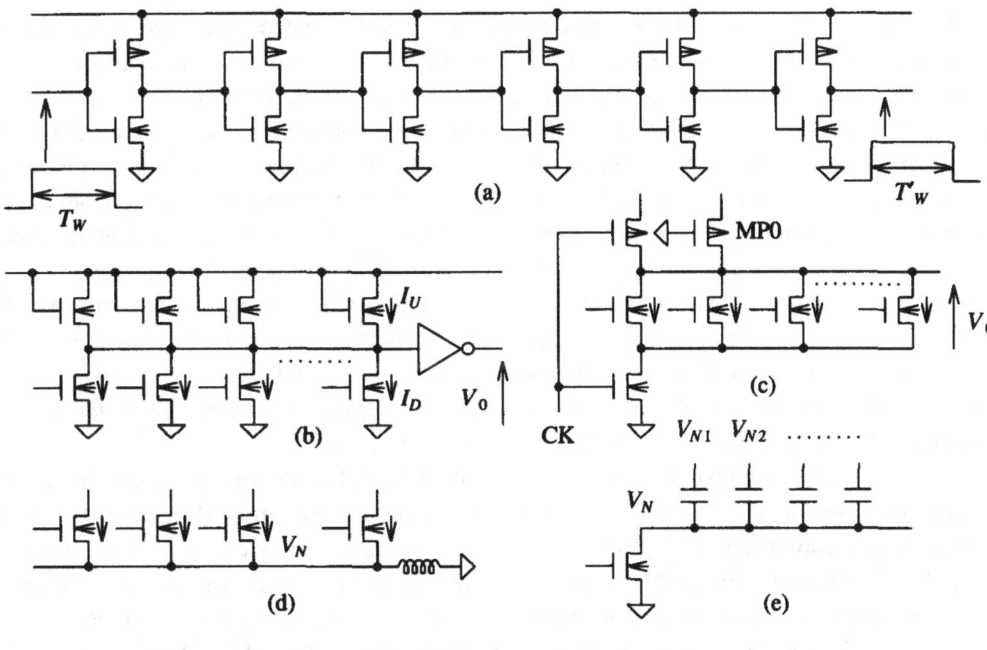

Figure 64 Examples of the cases where simulation parameters are not positive definite

is i-th FET pair current, and if the pull-down current is positive, the output is a high logic level. The total pull-down current is not positive definite. The most critical operating condition of a majority logic gate is that $[N/2]$ (where $[x]$ is the maximum integer less than real number x) inputs are low, and $[N/2]+1$ inputs are high, or vice versa (N is an odd number). Then the net current that drives the sum node down is $I_0 = < I_D(i) > /2$ (average over i), and the error of the net current, δI, is

$$\delta I = \sum_{i=1}^{N} \delta(\Delta I_i)$$

where $\delta(\Delta I_i)$ is the error of the i-th FET pair current. Error δI is statistically $\sqrt{N}\sqrt{< \delta(\Delta I_i)^2 >}$ (average over i). Then relative error of the worst-case net current is

$$\text{Relative error} = \frac{\sqrt{N}\sqrt{< \delta(\Delta I_i)^2 >}}{I_0}$$

This means that simulation error of this majority logic circuit increases with increasing number N. In this case net current does not vanish, but relative error increases as circuit complexity increases.

Figure 64(c) is one stage of a CMOS dynamic PLA. Since there are several tens to a few hundred of NFETs connected to the single output node, the NFET leakage current is uncertain. How large the leakage compensation PFET MP0 should be can never be determined accurately. As was discussed in Section 6.7, leakage current can be compensated better by an active feedback circuit. Figure 64(d) shows a power-bus noise equivalent circuit. In this case, power-bus noise, V_N, is proportional to the time derivative of the discharge current of the FETs. Since the derivative can be either positive or negative, power-bus noise is not positive definite, and it is difficult to determine noise amplitude accurately. Since the number of gates served by a single power-bus connection is huge, power-bus noise simulation has a severe accuracy problem. The problem of non-positive definite parameters occurs frequently if the parameter is a derivative of the other parameter. Figure 64(e) shows an equivalent circuit for capacitively induced noise. Node voltage V_N is the sum of the effects originating from a large number of noise sources V_{N1}, V_{N2}, ..., that have unknown occurrence time and random transition polarity. Node voltage V_N is practically a random variable: the mechanism that increases randomness of this phenomena was discussed in Section 4.3. These examples suggest that non-positive definite parameters have certain characteristics of random noise. Our observation in this section shows that difficulty in determining non-positive definite circuit parameters and the randomness mechanism of digital circuit noise are common. Noise and signal are only the two extrema, and there are intermediate states of variables.

Since uncertainty in parametric values is a fundamental and unavoidable problem, some circuits are unverifiable. This does not necessarily mean that the circuit does not work: if the parameter in question is generally confined to a harmless range, the yield of working chips can be high. Nevertheless this problem may create difficulty in development project. Selecting the proper circuit structure eases simulation and verification of circuits quite significantly. The circuit shown in Fig. 65 is intended for the same application as that of Fig. 63, but verification is trivial: cumulative delays of inverters 5 and 6 should be more than T_{min}. The simulation

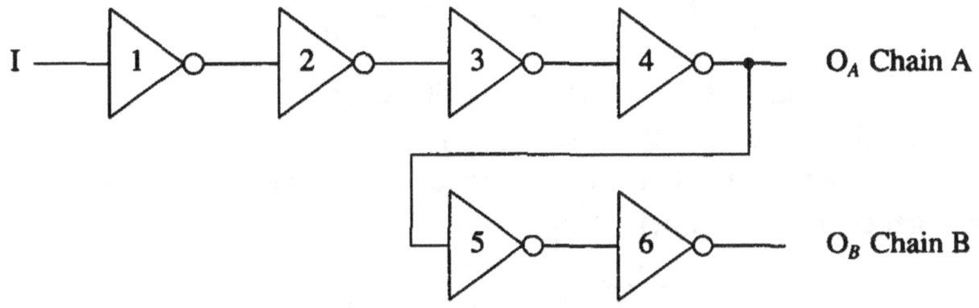

Figure 65 Different implementation of clock generator

has no accuracy problem. This scheme is called *interlock*. An important objective of circuit design is therefore to select circuit structures that are easy to verify.

Cancellation of parameter values or accumulation of component errors and resultant divergence of error occurs not only in DC parameters but also in the Fourier components of a signal. In the next section we show that higher harmonic amplitudes suffer from the same problem.

6.16 Harmonic Distortion

Higher harmonic amplitudes are the measures of circuit nonlinearity [13]. The accuracy of computing higher harmonic amplitudes depends on the structure of the circuit (the particular way the devices are connected), which may affect cancellation of device nonlinearities [14]. We study this problem using a simple closed-form circuit model. Vanishing parameter values occur even if a more sophisticated model is used: Harmonic cancellation is primarily determined by how the devices are connected.

Figure 66(a) shows a CMOS linear amplifier. Resistor R biases the amplifier to the high-gain operating point. MN1 has the drain current/drain voltage characteristic given by the simple analytical formula

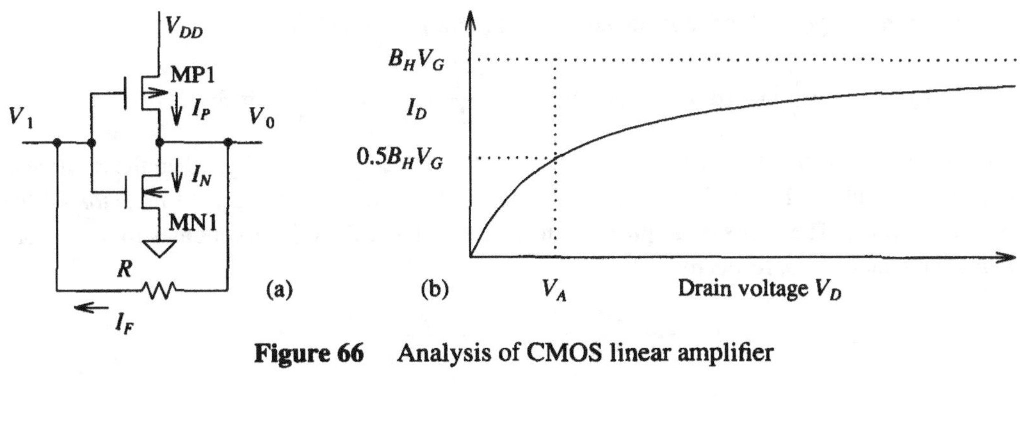

Figure 66 Analysis of CMOS linear amplifier

$$I_{DN} = B_H(V_G - V_{TH})\frac{V_D}{V_A + V_D} \tag{75a}$$

where V_G is the gate voltage and V_D is the drain voltage, both measured relative to the source, and V_{TH} is the threshold voltage. If $V_G < V_{TH}$, I_D vanishes. V_A is the parameter that determines sharpness of current saturation as drain voltage increases. If $V_A \to 0$ we have the collapsible current generator FET model. If $V_D \to \infty$ we have

$$\frac{\partial I_{DN}}{\partial V_G} \to B_H \quad (V_D \to \infty).$$

MP1 of Fig. 66(a) has a characteristic analogous to Eq. (75a), given by

$$I_{DP} = B_H(V_{DD} - V_G - V_{TH})\frac{V_{DD} - V_D}{V_A + (V_{DD} - V_D)}. \tag{75b}$$

In the following analysis we set $V_{TH} = 0$.

We assume that amplifier input voltage is

$$V_1 = V_S + v_1$$

where V_S is DC bias voltage and v_1 is a small AC variation. Using Eqs. (75a) and (75b) we find the output voltage V_0 as

$$V_0 = V_S + \alpha v_1 + \beta {v_1}^2 + \gamma {v_1}^3 + \cdots \tag{76}$$

We assume that variation of the voltages with time is slow, so that static circuit analysis is valid. In Eq. (76), α, β, γ, ... are coefficients that depend on the circuit parameters and biasing conditions of the amplifier. Expressions for α, β, ... can be found using Eqs. (75a) and (75b), by expanding the current conservation equation

$$I_F = I_{DP} - I_{DN}$$

where I_F is defined in Fig. 66(a), into a power series of v_1. We write Eq. (76) up to the cubic term in v_1 explicitly, as

$$V_0 = V_S + \alpha v_1 + \beta {v_1}^2 + \gamma {v_1}^3 + o({v_1}^4)$$

and assume

$$v_1 = A \sin \omega t$$

where A is the amplitude and ω is the angular frequency. We then have

$$V_0 = \left(V_S + \frac{\beta A^2}{2}\right) + \left(\alpha A + \frac{3\gamma A^3}{4}\right)\sin \omega t - \frac{\beta A^2}{2}\cos 2\omega t - \frac{\gamma A^3}{4}\sin 3\omega t + o(A^4).$$

The second term within the first bracket is a correction to the DC level, V_S, when the *nonlinear* amplifier is driven. This term means that the amplifier operates as a square *detector* of the input sine wave. The ratios of amplitudes of the second and the third harmonics to the fundamental component are, respectively,

$$-\frac{2\beta A}{4\alpha + 3\gamma A^2} \qquad -\frac{\gamma A^2}{4\alpha + 3\gamma A^2}$$

We note that amplitudes of the second and the third harmonics are approximately proportional to the expansion coefficients β and γ, respectively. If β or γ should become zero, the harmonic vanishes, and that makes accurate determination of the amplitude impossible. In many amplifiers, α and γ have opposite sign. This is because the output voltage range is limited by power supply voltage (saturation of amplifier). When the signal amplitude increases saturation of the amplifier dominates ultimately over all the other mechanisms of nonlinearity. If this characteristic is approximated by a cubic equation, it has an inflection point, or β becomes zero at that point. Cancellation of the second harmonic is generally inevitable. Then the harmonic amplitude cannot be determined unconditionally by numerical analysis.

The harmonic analysis is tedious. Because of the complexity, closed-form analysis is limited to the second harmonic. The circuit shown in Fig. 66(a) is analyzed as follows. The circuit equation is

$$I_P = I_F + I_N$$

where

$$I_N = B_{HN} V_1 \frac{V_0}{V_A + V_0} \quad I_F = \frac{1}{R}(V_0 - V_1) \text{ and } I_P = B_{HP}(V_{DD} - V_1)\frac{V_{DD} - V_0}{V_A + (V_{DD} - V_0)}$$

where B_{HP} and B_{HN} are transconductances of MP1 and MN1, respectively. We have

$$V_1 = V_S + v_1 \quad V_0 = V_S + \alpha v_1 + \beta {v_1}^2 \tag{77}$$

where V_S is DC bias point and v_1 is a small variation of input voltage. V_S satisfies

$$B_{HP}(V_{DD} - V_S)\frac{V_{DD} - V_S}{V_A + (V_{DD} - V_S)} = B_{HN} V_S \frac{V_S}{V_A + V_S}.$$

We have

$$\begin{aligned}(V_0 - V_1)(V_A + V_{DD} - V_0)(V_A + V_0) = (RB_{HP})(V_{DD} - V_1)(V_{DD} - V_0)(V_A + V_0) \\ -(RB_{HN})V_1 V_0 (V_A + V_{DD} - V_0).\end{aligned} \tag{78}$$

Equation (77) is substituted into Eq. (78) and terms having the same power of v_1 on both sides of the equal sign, are set equal.

The zeroth order term

$$(RB_{HP})(V_{DD} - V_S)^2(V_A + V_S) - (RB_{HN}){V_S}^2(V_A + V_{DD} - V_S) = 0 \tag{79}$$

The expansion coefficients are as follows.

$$\alpha = \frac{\phi_1\phi_2 - (RB_{HP})\phi_2\phi_3 - (RB_{HN})\phi_1\phi_4}{\phi_1\phi_2 - (RB_{HP})\phi_3(\phi_3 - \phi_2) + (RB_{HN})\phi_4(\phi_1 - \phi_4)} \tag{80}$$

$$\beta = \frac{-[(RB_{HP})(\phi_3 - \phi_2) + (RB_{HN})(\phi_1 - \phi_4) + \phi_4 - \phi_3]\alpha + [-(RB_{HP})\phi_3 + (RB_{HN})\phi_4 + \phi_4 - \phi_3]\alpha^2}{\phi_1\phi_2 - (RB_{HP})\phi_3(\phi_3 - \phi_2) + (RB_{HN})\phi_4(\phi_1 - \phi_4)}$$

where

$$\phi_1 = V_A + V_{DD} - V_S \quad \phi_2 = V_A + V_S \quad \phi_3 = V_{DD} - V_S \quad \phi_4 = V_S.$$

From Eqs. (77)–(80) we learn the following. If $B_{HP} = B_{HN}$, $V_S = V_{DD}/2$ is the solution of Eq.

(79). If $B_{HP} = B_{HN}$ and $V_S = V_{DD}/2$ are substituted into Eq. (80), we find $\beta = 0$. The second harmonic vanishes. This means that the transfer characteristic of the amplifier is point-symmetrical with respect to the operating point, and the second harmonic vanishes if the amplifier is operated at the bias point.

References

[01] Kazimierczuk, M., and Sokol, M. O. "Cause of instability of power amplifier with parallel-connected power transistors." *IEEE J. of Solid-State Circuits* vol. SC-19 (August, 1984): 541–42.

[02] Edson, W. A. *Vacuum-tube oscillators*. New York: John Wiley & Sons, 1953.

[03] Uda, S. *Short wave projector -historical records of my studies in early days*. Japan: Tohoku University, 1974.

[04] Svensson, C., and Tjarnstrom, R. "Switch-level simulation and the pass transistor Exor gate." *IEEE Trans. on Computer-aided Design* 7 (September, 1988): 994–97.
Professor F. Rosenberger and Professor C. Zhukowski of Columbia University suggested that this subject be added. I am grateful for their comments.

[05] Chirlian, P. M. *Analysis and design of electronic circuits*. New YorK: McGraw Hill, 1965.

[06] Clayton, G. B. *Operational amplifiers*. London: Butterworth, 1971.

[07] Ziman, J. M. *Electrons and phonons*. Oxford: Clarendon Press, 1960.

[08] Shockley, W. *Electrons and holes in semiconductors*. D. van Nostrand Company, 1950.

[09] Troutman, R. R. *Latchup in CMOS technology*. Boston: Kluwer Academic Publishers, 1986.

[10] Troutman, R. R., and Zappe, H. P. "A transient analysis of latchup in bulk CMOS." *IEEE Trans. on Electron Devices* vol. ED-30 (February, 1983): 170–79.

[11] Sparkes, J. J. *Junction transistors*. Oxford: Pergamon Press, 1966.

[12] RCA solid-state "COS/MOS integrated circuits," 1980 (Slight modification of the circuit of CD4070B)

[13] Miller, G. M. *Linear circuits for electronics technology*. Englewood Cliffs, NJ: Prentice-Hall Inc., 1974.

[14] Khorramabadi, H. "A highly-efficient CMOS line driver with 80dB linearity for ISDN U-interface applications." *ISSCC92 digest* (February, 1992): 192–93.
Difficulties in determining higher harmonics amplitude by simulation was first recognized by Ms. Khorramabadi. The work of this section was intended to explain her observations.

7 Design Failures of CMOS Systems

7.1 Introduction

In this chapter CMOS circuit failures that cannot be traced back directly to circuit problems are discussed. The failures are, on one hand, related to the fundamental nature of CMOS technology. The failures are, on the other hand, due to the enormous complexity of large systems such as main-frame computers and electronic switching machines, which never existed in semiconductor devices before the advent of VLSI technology. The new problem was *imported* into semiconductor technology from system technology, because single IC chips became a significant fraction of large systems. Since CMOS is the front-runner of VLSI technology, the complexity problem is more serious there than in any other IC technology.

We consider problems of technology selection first. One obvious and frequent scenario of such a problem is as follows: Chips or subsystems that should not have been built using CMOS were built using CMOS, and the opposite problem where subsystems that should have been built using CMOS were not. If this happened, the problem traces back to a wrong decision made at the very high level of project management, at a very early phase of the project, and the cost is quite high. To make the crucial decision correctly, pros and cons of CMOS technology in assembling large data-processing systems must be critically evaluated. This is the first subject discussed in this chapter.

Success of large system development depends on the quality of many decisions made in the course of the project. Project members and management must make decisions that have technical, human, and business aspects, and a wide range of decisions must be made correctly. This is obviously a very demanding task. In this chapter we review what sorts of decisions must be made, and we discuss general guidelines for making such decisions on a solid technical background, based on the author's own experience. The author's intention is to limit these nontechnical (human, business) aspects of discussion to the minimum, but total elimination is impractical.

7.2 Integrated Circuits That Should Not Be Designed Using CMOS

Although CMOS is a very versatile integrated circuit technology, CMOS is not suitable for certain applications. In this section we begin with discussion of the type of ICs that should and should not be built using CMOS.

1) CMOS is a slow technology

CMOS is significantly slower than other technologies that stand in competition. Since CMOS static gates must drive both an NFET and a PFET (typically twice larger as an NFET) of the gates that accept the signal, a CMOS gate is as much as 3 times slower than the equivalent NMOS gate in pull-down transition. Since the pull-up transition of NMOS gates cannot be as fast as the pull-down transition, the cumulative delay ratio of CMOS to NMOS is from 1.5 to 2, as will be shown in Section 7.3. This estimate shows that, within the compatible MOS technology family, NMOS has a significant speed advantage.

The problem of NMOS is asymmetry of pull-up and pull-down transitions. Because of the waveform asymmetry, it is difficult to attain high precision in timing of clock or other critical signals. Often timing imprecision defeats the speed advantage of NMOS. It is necessary to determine for what objective high switching speed is required, whether it is for precision or for raw processing power. If precision in timing is not of primarily importance, and if only very high processing throughput is desired, the chip should be designed using NMOS. Such NMOS circuits effectively work as linear or nonlinear amplifiers. In circuits that operate at the ultimate highest frequency limit of technology, voltages of nodes do not swing completely from low level voltage to high level, as we saw in Section 6.14. Such *rounded* or *filtered* waveforms deteriorate rapidly with increasing gate delay (Section 6.14), and therefore the faster gate always wins. Such a critical circuit should be built using NMOS.

CMOS and NMOS have logic swings that are directly determined by the power supply voltage. The logic swings are typically more than 80–90% of the power supply voltage or more (CMOS, NMOS). Then CMOS or NMOS gates must drive their nodes between ground and V_{DD}. CMOS and NMOS are *large logic swing* technologies. Logic swings of the competitive technologies, GaAs MESFET technology and bipolar ECL technology are determined from the energy gap of the semiconductor material used to fabricate the active devices. Logic swings are significantly smaller than the power supply voltage (in GaAs MESFET technology logic swing is about 30–50% of supply voltage, and in bipolar ECL technology the logic swing is about 10% of the supply voltage). They are *small logic swing* technologies. Although the development of a new generation of small logic swing technologies is behind CMOS at present, their gate delays are significantly shorter than that of CMOS. Once fully developed (the time of maturity is expected to be the early 1990s), gate delays of GaAs MESFET ICs and bipolar ECL ICs will range from 5 to 20 picoseconds, 3–5 times shorter than that of CMOS expected at about the same time. Therefore if raw switching speed of *static* logic gates is required, GaAs MESFET technology or bipolar ECL technology should be used.

Using CMOS dynamic logic circuits, the speed of CMOS circuits can be enhanced to within a factor of about three of the small logic swing, high-speed technologies. Dynamic design is, however, not always allowed.

The delay of a CMOS gate can be reduced quite significantly by lowering the operating temperature from normal 300–400 degrees Kelvin to 77 degrees Kelvin (liquid nitrogen temperature). There are a number of details to be worked out to make the best use of the low temperatures. Acceptance of cryogenic CMOS by the system's customers is still quite limited, but some encouraging changes are expected to take place.

2) Scaled-down CMOS circuits have only limited methods of speedup

Channel current of scaled-down CMOS FETs is proportional to gate voltage. The discharge delay time constant of FETs is then independent of logic amplitude. Delay of a gate is determined from the single, amplitude-independent, FET time constant. This leads to an inevitable conclusion that scaled-down CMOS circuits have no method of delay minimization, except for logic type (static or dynamic) selection, logic design optimization, and fanout optimization. If a CMOS circuit is optimized using these three basic methods, the delay cannot be reduced any more, even if higher power and larger silicon area for larger devices are allowed. This is a very significant disadvantage of scaled-down CMOS.

3) CMOS chips do not communicate efficiently

If a large system is assembled from many ICs, how to arrange interchip communication is a matter of fundamental concern. CMOS logic-level, interchip communication at high data rates is, unfortunately, very inefficient in power, in delay time, and in silicon area. Power is high because CMOS is a large logic swing technology. Output driver delay is long since MOSFET transconductance is small. If transconductance of the output driver is increased, the output driver becomes large. Extensive CMOS logic-level, interchip communication on large circuit boards in high-speed systems, including backplane interconnects, should be avoided. In data-bus communication there is a limit to delay improvement that is determined by the ratio of transconductance to output capacitance of the bus driver device. In CMOS, the ratio is lower than any small logic swing technology. If a large system is built from high-speed ICs whose level of integration is not high enough to integrate a significant fraction of the entire system in a single chip, either CMOS should not be used or special interface circuits that convert internal CMOS logic levels to ECL logic levels for interchip communication should be used. For high-speed interchip communication, ECL logic levels are the best choice. Circuits that interface CMOS signals with ECL signals, designed using MOSFET only, do not have optimum performance. Another difficulty associated with CMOS interchip communication is that high speed CMOS output drivers generate high current spikes. It is impossible to strobe input data while driving data out from the chip. Circuits that generate smaller noise (like bipolar ECL) are definitely more desirable than CMOS.

4) Other peculiarities of CMOS

DC power consumed by CMOS chips decreases drastically if clock frequency and power supply voltage are reduced. Systems that run from limited energy sources like batteries or solar cells, or a system that operates very infrequently, should be designed using CMOS.

CMOS ICs have latchup susceptibility. Hot insertion (plugging or unplugging ICs with power supply on) may induce latchup. Although there are several circuit techniques to prevent latchup, the problem must be considered carefully. Similar caution must be exercised if CMOS devices are exposed to ionizing radiation.

CMOS and generally MOS technologies of small feature size are susceptible to ESD (electrostatic discharge) failures. Circuits that are exposed to harsh environments (such as *smart* cards) must be designed using sufficient protection mechanisms.

Certain logic circuit types, like majority logic circuits, cannot be implemented efficiently using CMOS. If extensive majority logic is required, bipolar ECL should be used. We begin this chapter with discussions of the problems of CMOS technology that are identified in this section. The first subject is how CMOS circuit delays compare with NMOS circuit delays.

7.3 CMOS and NMOS

CMOS and NMOS are compatible MOS technologies, both in processing and in circuits, but there are significant differences in their circuit structures [01]. In this section we study the differences in two respects: first by comparing switching characteristics of inverters, and second, by examining preferred usage of gate types.

7.3.1 *Switching Characteristics of CMOS and NMOS*

CMOS and NMOS inverter circuits are shown in Figs. 1(a) and (b), respectively. The load resistance, R_L, in the NMOS inverter is in reality a depletion mode NFET (or grounded-gate

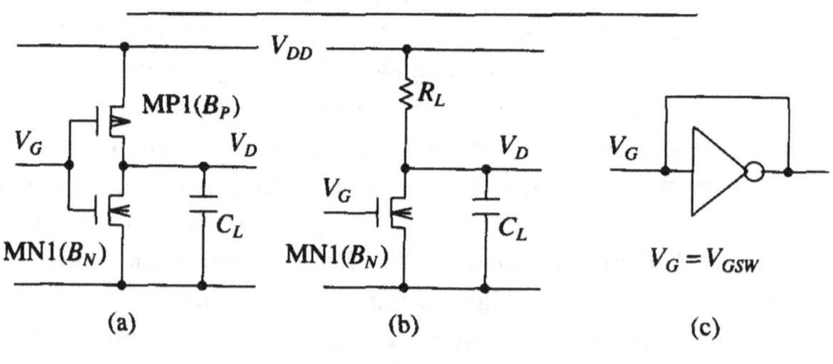

Figure 1 CMOS and NMOS inverters

PFET in pseudo NMOS), that is approximated by a resistor. We use the collapsible current generator FET model. Simple theory based on these two approximations reveals clearly the essential differences between CMOS and NMOS circuits.

In the analysis of CMOS circuits presented in the preceding chapters, we used the gate delay time defined as the interval between the time when input voltage is equal to $V_{DD}/2$ and the time when the output voltage equals $V_{DD}/2$. $V_{DD}/2$ is the *conveniently defined* threshold voltage of CMOS gates. Justification of this *defined* threshold voltage is given in the author's previous book, and the discussion is not repeated here [02]. To compare CMOS and NMOS gates, however, we must use a *consistent* definition of the threshold voltages. Figure 1(c) shows how a gate's switching threshold voltage that is consistent for both NMOS and CMOS gates is defined. If the input and output of an inverter are connected together, the *common* node settles at the high gain operating point (Section 2.2). This *static* bias point is stable, as we learned in Section 6.3. This voltage, V_{GSW}, is the switching threshold voltage we use. This definition is not convenient for CMOS circuits in general, but it is useful for the present theoretical study.

We assume enhancement mode NFETs and PFETs have the following characteristics

$$I_N = I_{N0} = B_N(V_G - V_{THN}) \quad (V_G > V_{THN}, V_D > 0) \quad V_D = 0 \quad (I_N < I_{N0}) \tag{1a}$$

$$I_P = I_{P0} = B_P(V_{DD} - V_G - V_{THP}) \tag{1b}$$

$$(V_G < V_{DD} - V_{THP}, V_D < V_{DD}) \quad V_D = V_{DD} \quad (I_P < I_{P0}).$$

The switching threshold voltages V_{GSW} of CMOS and NMOS inverters are given, respectively, by

$$V_{GSW}(CMOS) = \frac{B_P V_{DD} - B_P V_{THP} + B_N V_{THN}}{B_N + B_P} \tag{2}$$

$$V_{GSW}(NMOS) = \frac{V_{DD} + R_L B_N V_{THN}}{1 + R_L B_N}$$

We note that $V_{GSW}(NMOS) \to V_{THN}$ if $R_L B_N \to \infty$. $R_L B_N$ is the gain of the inverter. High gain is required to obtain crisp switching characteristics (Section 2.4). NMOS inverters must have gain significantly higher than unity. In the high-gain limit, the NMOS V_{GSW} decreases to a small value. $V_{GSW}(CMOS)$ is about $[B_P/(B_P + B_N)]V_{DD}$. Since CMOS inverters are usually designed so that $B_P \approx B_N$, the switching threshold voltage of CMOS inverters is usually close to $V_{DD}/2$. Since we consider only enhancement mode MOS logic (Section 2.2), the high level of NMOS inverters (V_H) is V_{DD}. The low level (V_L) of NMOS is the output voltage of the inverter of Fig. 1(b) when the input voltage is V_{DD}. V_L is given by

$$V_L = 0 \quad \text{if} \quad (R_L B_N) > V_{DD}/(V_{DD} - V_{THN}) > 1. \tag{3}$$

Suppose that CMOS inverter input of Fig. 1(a) makes an instantaneous transition from low (ground) to high (V_{DD}). The output node voltage $V_D(t)$ is given by

$$V_D(t) = V_{DD} - \frac{B_N(V_{DD} - V_{THN})}{C_L} t \quad (0 < t < t_C) \tag{4}$$

$$V_D(t) = 0 \quad (t > t_C)$$

where

$$t_C = \frac{C_L V_{DD}}{B_N(V_{DD} - V_{THN})}.$$

By requiring that $V_D(t_D) = V_{GSW}(CMOS)$, we find the pull-down delay time t_D as

$$t_D(CMOS) = \frac{C_L}{B_N(V_{DD} - V_{THN})} \times \frac{B_N V_{DD} + B_P V_{THP} - B_N V_{THN}}{B_N + B_P} \tag{5}$$

$$\rightarrow \frac{C_L}{B_N + B_P} \quad (V_{THP} \rightarrow 0 \text{ and } V_{THN} \rightarrow 0)$$

If the CMOS inverter input makes an instantaneous high-to-low transition at $t = 0$, V_D is given by

$$V_D(t) = \frac{B_P(V_{DD} - V_{THP})}{C_L} t \quad (t < t_C) \qquad V_D(t) = V_{DD} \quad (t > t_C)$$

where

$$t_C = \frac{C_L V_{DD}}{B_P(V_{DD} - V_{THP})}.$$

By requiring that $V_D(t_U) = V_{GSW}(CMOS)$, we find the pull-up delay time t_U as

$$t_U(CMOS) = \frac{C_L}{B_P(V_{DD} - V_{THP})} \times \frac{B_P V_{DD} - B_P V_{THP} + B_N V_{THN}}{B_N + B_P} \tag{6}$$

$$\rightarrow \frac{C_L}{B_N + B_P} \quad (V_{THP} \rightarrow 0 \text{ and } V_{THN} \rightarrow 0).$$

Pull-up and pull-down delay times are equal if the FET threshold voltages are small. This conclusion was derived using the *consistent* gate switching threshold voltage. We note that pull-down and pull-up delay times defined using $V_{DD}/2$ as the *defined* logic threshold voltage are not equal even in the limit of small FET threshold voltages. $B_N = B_P$ is also required to make the inverter symmetrical. $V_{GSW} \rightarrow B_P/(B_N + B_P)$ in the limit as $V_{THN}, V_{THP} \rightarrow 0$. B_N and B_P are usually about the same. Then the switching threshold voltage stays about halfway between V_{DD} and V_{SS}. A CMOS gate is *essentially* symmetrical with respect to pull-up and pull-down transitions.

We repeat the same analysis for the NMOS inverter shown in Fig. 1(b). For the pull-down transition,

$$V_D(t) = [V_{DD} - (R_L B_N)(V_{DD} - V_{THN})] + (R_L B_N)(V_{DD} - V_{THN}) exp(-t/R_L C_L)$$

and the pull-down delay time is given by

$$t_D(NMOS) = (R_L C_L) \log(1 + \frac{1}{R_L B_N}). \tag{7}$$

For a pull-up transition

$$V_D(t) = V_{DD}[1 - exp(-t/R_L C_L)]$$

and the pull-up delay time is given by

$$t_U(NMOS) = (R_L C_L)\log\left[[1 + \frac{1}{R_L B_N}]\frac{V_{DD}}{V_{DD} - V_{THN}}\right] \tag{8}$$

$$\rightarrow (R_L C_L)\log(1 + \frac{1}{R_L B_N}) \quad (V_{THN} \rightarrow 0)$$

From Eqs. (7) and (8)

$$t_U(NMOS)/t_D(NMOS) \rightarrow 1 \qquad V_{THN} \rightarrow 0 \tag{9}$$

Since high gain is required, this limit is reached with $V_{GSW} \rightarrow V_{THN} \rightarrow 0$. Delay symmetry in NMOS is achieved by creating significant switching threshold voltage asymmetry relative to the high and low logic levels. We conclude the following.

1) CMOS inverters are essentially *symmetrical*. A symmetrical inverter pulls up and pulls down with the same delay times, the switching threshold voltage is halfway between the logic level voltages, and the switching waveforms are symmetrical with respect to the power supply voltage levels.

2) An NMOS inverter can never be *symmetrical* in this sense. In order to secure noise immunity by crisp switching, the gain $R_L B_N$ must be chosen significantly higher than unity. Then, the switching threshold voltage of the inverter approaches the threshold voltage of the NFET. At this low switching threshold voltage, the pull-up and pull-down delay times can be made equal by further reducing the FET threshold voltage, but if that is done, the switching waveform becomes very asymmetrical with respect to the boundary of logic 1 and 0. A precise, skew-free clock signal can be generated easily using CMOS symmetrical inverters. In NMOS, skew-free clock circuit design is much more difficult than CMOS.

CMOS and NMOS delays can be compared as follows. We define the average of pull-up and pull-down delay times by

$$t_{DAV} = \frac{t_D + t_U}{2}.$$

In a CMOS inverter we have

$$t_{DAV}(CMOS) = C_L(CMOS)/2B \tag{10}$$

where $B = 0.5(B_N + B_P)$ is the averaged transconductance and the threshold voltages were assumed small ($V_{DD} \gg V_{THN}, V_{THP}$). The factor of 2 in the denominator of Eq. (10) is because both PFET and NFET are active. In NMOS,

$$t_{DAV}(NMOS) = R_L C_L(NMOS) f(R_L B_N) \quad \text{where } f(x) = \log[1 + (1/x)] \tag{11}$$

assuming $V_{DD} \gg V_{THN}$. If gain of the NMOS inverter is high ($R_L B_N \gg 1$),

$$t_{DAV}(NMOS) \approx R_L C_L(NMOS) \cdot \frac{1}{R_L B_N} = \frac{C_L(NMOS)}{B_N} \tag{12}$$

Equation (12) is interpreted as follows. If wiring capacitance and the capacitance of depletion mode load devices in NMOS (which is much smaller than the enhancement mode NFET) are neglected,

$$C_L(CMOS) \approx (1 + \beta)C_L(NMOS). \tag{13}$$

Only the input capacitances of the fanouts are included in Eq. (13), and β is the ratio of current of NFET and PFET that have the same size (beta ratio). If $\beta = 2.5$, $C_L(CMOS)$ is 3.5 times higher than $C_L(NMOS)$, $C_L(CMOS)/C_L(NMOS) \approx 3.5$. Then from Eqs. (12) and (13),

$$\frac{t_{DAV}(CMOS)}{t_{DAV}(NMOS)} \approx 1.8 . \tag{14}$$

NMOS can be approximately twice as fast as CMOS. We note again that *consistent* definitions of switching threshold voltage were used for CMOS and NMOS, to derive this conclusion. The ratio of Eq. (14) depends on β.

7.3.2 CMOS and NMOS Gates

CMOS and NMOS have significantly different preferred gate types in logic design. In CMOS static logic circuits, NAND gates having a small number of inputs are preferred. Since NFET/PFET current ratio, β, ranges from 2 to 3 in 1–2 μm CMOS, 2–3 input NAND gates are approximately symmetrical with respect to pull-up and pull-down, if PFETs and NFETs have the same size. A symmetrical NOR gate requires large PFETs, and this is inconvenient for layout. Use of many-input NOR gates is discouraged in CMOS design. NAND gates assembled from equal size PFET and NFET are convenient in gate-array and standard-cell ICs. Because of equal FET sizes, CMOS gate array layout has quite a simple structure. In NMOS, NOR gates having a small number of inputs are preferred. Although NMOS gate pull-up delay is longer than pull-down delay, pull-up delay does not increase rapidly with an increasing number of inputs as in CMOS NOR gates [02]. NMOS NAND gates require careful NFET sizing to maintain low level within the desired range. Because of this problem, NAND gates are not used in high-performance NMOS design. Many NMOS circuits are fast partially because of preferential use of fast NOR gates.

CMOS transmission gates are able to drive the output node to any voltages between V_{SS} and V_{DD} (Section 1.10). This means that digitally controlled analog on-off switches are available in CMOS. Using the on-off switches, the simple D-latch shown in Fig. 2(a) can be designed. This simple CMOS D-latch is inserted anywhere in a logic circuit to capture data and to store it temporarily. Because of this convenience, CMOS system timing design is more

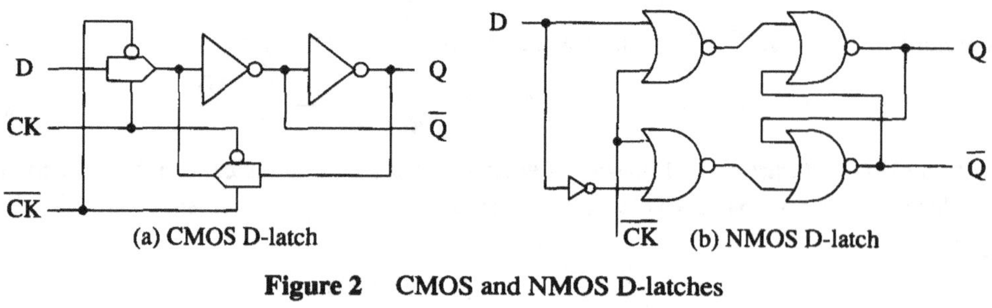

Figure 2 CMOS and NMOS D-latches

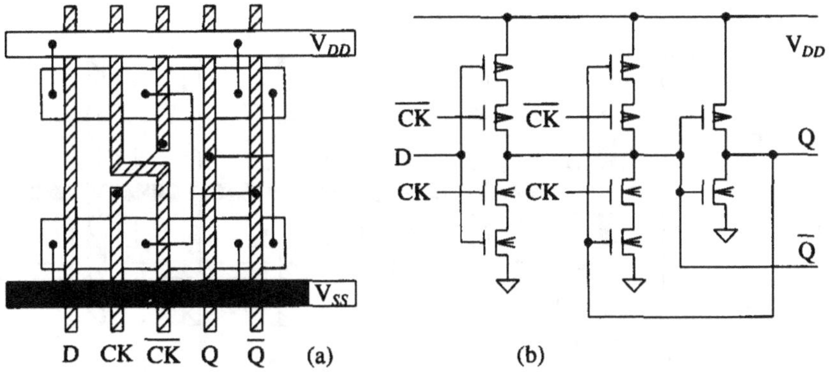

Figure 3 Compact layout of CMOS D-latch

flexible and easier than NMOS. The CMOS D-latch shown in Fig. 3(b) is laid out compactly, as shown in Fig. 3(a) [03]. Width of the latch circuit is only about 10 to 12 times the feature size.

An NMOS transmission gate consisting of single NFET (often called *pass* transistor) is unable to drive the output to V_{DD}. A *pass* transistor is not an on-off switch (Section 1.10). The only way to get around this problem is to drive the gate above V_{DD} potential using charge-pump circuit. Therefore simple NMOS D-latch circuit must tolerate NFET threshold drop: otherwise an NMOS D-latch is a conventional gate design, as shown in Fig. 2(b). The circuit requires 14 NFETs [the CMOS circuit of Fig. 2(a) requires only 8–10], and therefore it is not convenient to use D-latches lavishly in NMOS circuits. Simplicity of NMOS logic gates compared with CMOS is partially offset by complexity of the NMOS D-latch.

Availability of nearly ideal on-off switches makes CMOS an ideal technology for linear circuits. Since CMOS circuit has complementary-symmetry, high-performance operational amplifiers, which are the key components in analog circuits, can be designed. Operational amplifiers designed using NMOS are quite cumbersome. If analog circuits are required on chip, CMOS is definitely a better choice than NMOS. Since analog circuit controlled digital CMOS VLSI circuits have many advantages, CMOS is definitely more flexible than NMOS.

In CMOS gates, input signals must drive both PFET and NFET. Because of that requirement, CMOS gates are laid out almost exclusively in the dual in-line structure shown in Fig. 4(a). A certain minimum distance must be maintained between PFETs and NFETs to allow room for tubs, but the tub boundaries have usually simple shape. Simplicity in layout is a significant advantage of CMOS. In NMOS, input signals need to drive NFETs only, and the substrate is P-type everywhere. NMOS does not have the layout constraints, but pull-down FETs and pull-up FET must be sized. Therefore NMOS gate cells have very complex, manually laid out internal structure.

To create the most compact and regular layout is a kind of engineering art. For example, high-performance memory that has static column switching operation requires a static column address decoder. Static address decoders can be designed using only NMOS gates, or *pseudo*

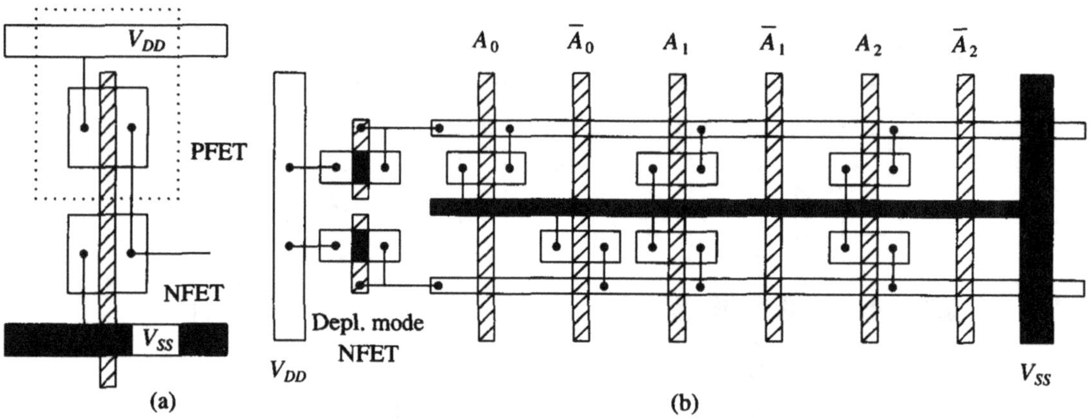

Figure 4 CMOS and NMOS layout structures

NMOS gates compatible with CMOS (load device is a PFET instead of a depletion mode NFET). The structure of the address decoder is shown in Fig. 4(b).

7.4 Low-temperature CMOS Circuits

Within the last five years the first serious efforts have been directed to capitalizing on the advantages of liquid nitrogen temperature electronics, and interestingly enough, in CMOS technology. This new trend stands against the traditional opinion that cooling an electronic system below the temperature of tap water is the absolute last measure. This opinion, supported by the arguments from energy and maintenance cost, has been invincible until now, to the point that it has been considered as *common sense* for electronic designers. During the last thirty years, many new low-temperature semiconductor devices were proposed but later scrapped because of the prevalent opinion. I do not disagree with this established view, but simultaneously I believe it necessary to investigate what sort of assumptions the historically long-lasting opinion stands. To do so, CMOS circuits are the most realistic vehicle. A fair and technically sound evaluation must be given to this new trend, and this section is intended to be the first step to this objective.

First we review the effects of reduced temperature on CMOS circuits. Figure 5 shows the delay of CMOS inverters versus temperature, measured by Laramee, Aubin, and Cheeke using several commercial CMOS ICs. This data appears to reflect the fundamental physical mechanisms most clearly. Temperature dependence consists of three distinct temperature regions, A, B, and C. From room temperature down to about 100 degrees K (Kelvin) (region A) delay of inverters decreases significantly with decreasing temperature. The decrease is caused by increase in carrier mobility, which is the result of reduced lattice scattering of the carriers with decreasing temperature. In the temperature range from 100 degrees K down to about 30

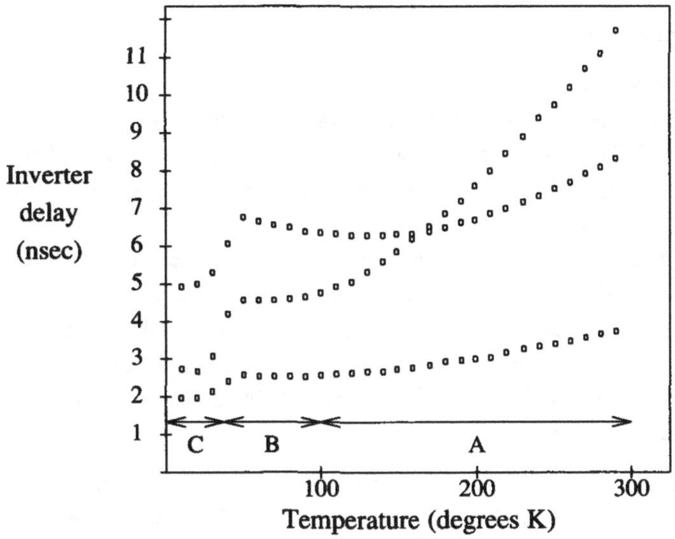

Figure 5 Temperature dependence of delay of CMOS inverters

degrees K (region B) delay is approximately independent of temperature, or delay may even increase slightly, with decreasing temperature. Increase in carrier mobility by decreasing temperature in range A is limited ultimately by the ionized impurity scattering that dominates in temperature range B. Since ionized impurity scattering is collisions between fixed and moving charged particles, the effect of collisions is less if energies of the moving particles are higher. Then reducing the number of high energy electrons by reducing temperature increases the effects of scattering, thereby reducing device current, and increasing delay.

In the temperature range from 30 degrees K down to 4.2 degrees K (region C) delay decreases again with decreasing temperature. There are two effects operative in this temperature range that cause delay to decrease. Laramee et al. proposed decrease in ionized impurity scattering [04], and Kamger proposed reduced drain/source island capacitance [05], both due to carrier freezeout. Capacitance of CMOS is practically temperature independent by mechanisms other than carrier freezeout. This is the only known effect that gives significant capacitance temperature dependence. This effect suggests that at liquid Helium temperature, bulk CMOS ICs resemble SOI CMOS ICs closely. Either way, carrier freezeout appears to play a significant role on CMOS at or near liquid Helium temperatures. Since the substrate has high resistance at liquid Helium temperature, substrate potential is unable to follow gate or drain potential changes. As a result "kinks" similar to that observed in SOI FETs appear in the current-voltage characteristics of FETs [06]. Because of many additional new effects at liquid Helium temperature, and because of the cooling costs, the present trend of low-temperature CMOS is to work out technical details at liquid nitrogen temperature (77 degrees Kelvin).

Improvement in circuit delay by cooling CMOS devices from room temperature to liquid nitrogen temperatures has been reported by many investigators, but the data is quite scattered. The sources of uncertainties are:

1) FET structure: Whether or not the FET has lightly doped drain (LDD) is crucial, since an LDD region that is intended to prevent hot-carrier effects adds significant series resistance to the FET channel at low temperatures. Whether or not a FET has a buried-channel structure matters, because carrier freezeout creates temperature dependence in the FET threshold voltage.
2) Degree of velocity saturation of carriers in the channel: Electron mobility (at low fields) increases 5 to 6 times by cooling NFETs from 300 degrees K to 77 degrees K. Saturation electron drift velocity, however, increases by only 10% or less. Then long-channel NFETs (having low channel field) improve significantly by cooling, but short-channel, scaled-down FETs that have high channel fields do not improve very much.
3) Type of the circuit used for test.

Drift velocity saturation of channel carriers is a crucial factor that determines delay characteristics of low-temperature CMOS circuits. The velocity-field characteristic of channel carriers is determined by the saturation drift velocity at high fields (v_∞) and mobility at low fields (μ). The two parameters have different temperature dependence. It is useful to determine how gate delay depends on temperature indirectly through the two parameters. We consider the circuit of Fig. 6 and derive a formula that gives the discharge delay time in terms of v_∞ and μ. Suppose that load capacitance C_L is charged to V_{DD}, and at time $t=0$, the FET gate voltage makes an instantaneous low-to-high transition. MN1 current with zero threshold voltage is given by Eq. (22a) of chapter 1 as

$$I_D = \frac{\varepsilon_{OX}\mu W}{T_{OX}L}\,\frac{V_{DD}-(V_D/2)}{1+(\mu/Lv_\infty)V_D}\,V_D \quad (V_D < V_{Dmax})$$

$$= I_{Dmax} = \frac{\varepsilon_{OX}\mu W}{2T_{OX}L}\,V_{Dmax}{}^2 \quad (V_{Dmax} < V_D)$$

Zero threshold voltage simplifies the analysis but retains the essential details of the circuit (threshold control is the issue of technology optimization). We defined $V_{Dmax} = V_0[\sqrt{1+(2V_{DD}/V_0)}-1]$ and $V_0 = Lv_\infty/\mu$ in Chapter 1. The circuit equation for Fig. 6 is

$$C_L\frac{dV_D(t)}{dt} + I_D[V_D(t)] = 0 \tag{15}$$

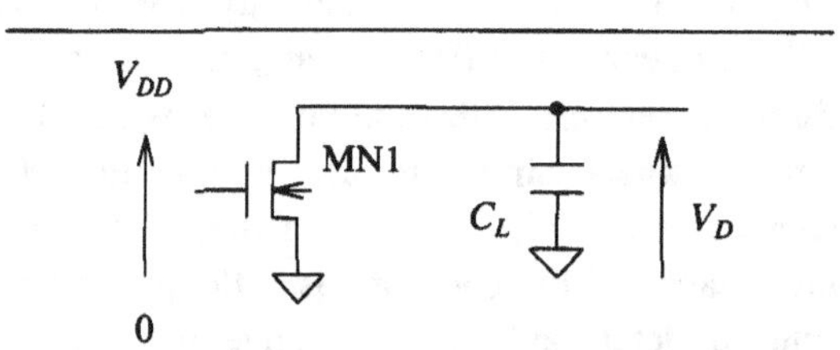

Figure 6 Low-temperature CMOS delay analysis

which is solved subject to the initial condition $V_D(0)=V_{DD}$. Since the current-voltage characteristic consists of two sections [for $V_D > V_{Dmax}$ (saturation region) and for $V_D < V_{Dmax}$ (triode region)] solutions of the two regions are connected at time t_1 and voltage $V_D=V_{Dmax}$. Discharge delay time, T_D, is found from the solution of Eq. (15) by solving for $V_D(T_D)=V_{DD}/2$. The details of the algebra are not given here. A closed-form solution is available, which is in the form solved for time t. Delay time, T_D, is determined by setting $V_D/V_{DD}=1/2$ in the closed-form equation. We define $v_0=V_0/V_{DD}$. Then we have

$$T_D=\left(\frac{T_{OX}\cdot C_L}{\varepsilon_{OX}v_\infty W}\right)\frac{1}{(\sqrt{2+v_0}-\sqrt{v_0})^2} \quad \text{if} \quad v_0<\frac{1}{4} \tag{16}$$

$$T_D=\left(\frac{T_{OX}\cdot C_L}{\varepsilon_{OX}v_\infty W}\right)\cdot(1+A)+\left(\frac{T_{OX}L\cdot C_L}{\varepsilon_{OX}\mu W V_{DD}}\right)\cdot B \quad \text{if} \quad v_0\geq\frac{1}{4} \quad \text{where} \tag{17}$$

$$A=-2\log\frac{4-2v_0[\sqrt{1+(2/v_0)}-1]}{3} \qquad B=\log\frac{3v_0[\sqrt{1+(2/v_0)}-1]}{2-v_0[\sqrt{1+(2/v_0)}-1]}. \tag{18}$$

T_D is independent of low-field mobility, μ, if the effect of velocity saturation is large. If the velocity saturation effect is moderate, however, delay time T_D is the sum of terms including mobility, μ, and saturation drift velocity, v_∞, both characterizing the velocity-field characteristic. The two parameters have different temperature dependence. Mobility increases quite significantly, by a factor of 5–7, by decreasing temperature from 300 degrees Kelvin to 77 degrees Kelvin. This large temperature dependence is not effective, however, in the high-field limit. Saturation drift velocity that is the characteristic of *hot* carriers is essentially temperature independent, and it increases by only 10%. Equations (16)–(18) show that saturation drift velocity is more crucial in determining the delay time. High-field operation of FETs is characterized by small V_0. Then cooling FETs to reduce delay is not as effective as at low fields.

CMOS FETs that have optimized performance at room temperature are not optimum at liquid nitrogen temperatures. Performance comparisons should be made between two different technologies, one optimized at room temperature and the other at 77 degrees K. Sun et al. reported that many process details must be adjusted to obtain a factor of 2 improvement in delay time defined in this way. The details include dual polysilicon gate (P+ polysilicon for PFETs and N+ polysilicon for NFETs) to produce surface-channel PFETs and NFETs (buried-channel FETs have carrier freezeout problems), recessed doping of tubs to minimize the effects of ionized impurity scattering within the surface channel, reduced V_{DD} (to the range of 2.5 volts) to eliminate the need for LDD, and to minimize the effects of carrier velocity saturation.

Reducing power supply voltage is likely to open new possibilities for low temperature CMOS, but reducing V_{DD} increases delay. In Eq. (16), T_D increases with decreasing V_{DD}. Reduced field in the channel, however, makes performance improvement by lowering temperature more significant than keeping the power supply voltage at $V_{DD}=5$ *volts*. This is because the second term of Eq. (17) becomes more significant than the first term. By lowering the power supply voltage, hot-carrier trapping effects by gate oxide can be eliminated. In addition to the beneficial effects, interchip communication between reduced voltage CMOS chips consumes drastically less power (power is proportional to V_{DD}^2 — see Section 7.7), and the reduced voltage CMOS chips even become compatible with bipolar ECL chips. As will be discussed later, interchip communication has been the most fundamental problem of conventional CMOS at high clock frequencies. Low-temperature CMOS at low power supply voltage is able to offer a free solution to many system integration problems.

There are other significant advantages in reducing the temperature of CMOS circuits to 77 degrees Kelvin, which are summarized as follows.

1) Current gain of parasitic bipolar transistors decreases, and as a consequence, CMOS devices become practically latchup free at liquid nitrogen temperatures. By the same mechanism, BiMOS does not have good performance at 77 degrees Kelvin (although it is possible to *optimize* bipolar transistor at 77 degrees Kelvin).

2) Leakage current of MOSFETs decreases drastically. At liquid nitrogen temperature, CMOS DRAMs maintain data practically indefinitely. Since subthreshold current decreases drastically, dynamic CMOS circuits become noise-resistant.

3) Series resistance of interconnects (aluminum and polysilicon) decreases significantly (by a factor of 5–7 times). The technology is potentially compatible with the new, high-temperature superconductor technology.

4) Noise of linear amplifiers is proportional to the temperature.

7.5 IC Technology Comparison

In this section we compare CMOS device technology with non-MOS device technologies. It is generally believed that bipolar transistors and field-effect transistors are completely different semiconductor devices. Field-effect transistors were once called *unipolar* transistors. The name sounds like the two transistors are really different. If this were the case, however, a natural question would be whether or not another three-terminal semiconductor device has yet to be invented. Then there would be a lot of uncertainties in technology assessment. I show that there is only one fundamental structure of *electron triode*, and bipolar transistors and field-effect transistors are the two special cases of the fundamental device structure [07]. This is the basic philosophy of this book: electronic circuit theory can be built by abstracting out the detailed device characteristics, and the microstate theory is an example. From this view, it will become clear what is the limit of a variety of semiconductor triode devices.

Bipolar transistors and MOS field-effect transistors have the following in common. There is a terminal common to the input and to the output of the device (we assume that the device is used as voltage amplifier). They are emitter of a bipolar transistor and source of a field-effect transistor. There is an input terminal that can be driven by the signal source. This is the base and gate of the respective transistors. The last is the output terminal, the collector, or drain, respectively. Current from the output electrode to the common electrode is controlled by the input voltage applied between the input electrode and the common electrode. That the control voltage is referenced to the *common* electrode, and not to any other *global* reference

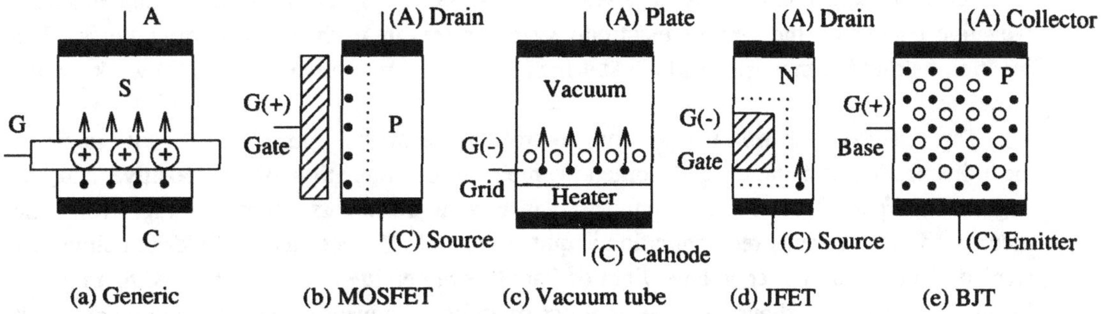

Figure 7 Structure of electronic triodes

voltage like ground, creates significant complexity (Section 1.10). The general structure of a semiconductor triode is shown in Fig. 7(a). Let us use generic names in the following. The common electrode C is called the cathode, which is grounded. Output A is the electrode called anode, which is biased positively with respect to the grounded cathode C. G is the control electrode (gate, or grid), whose potential is controlled by the input voltage. S is the space between cathode C and anode A, where electrons that are emitted from cathode C travel to the anode. Whether or not electrons are emitted from cathode C to the inter-electrode space S is determined by the voltage of the control electrode G. In Fig. 7(a), if positive voltage is applied to the control electrode G, the positive charge on the control electrode (indicated by + sign inside circle) pulls electrons (shown by closed circles) from cathode C by electrostatic attraction between charges having opposite sign. Since anode A is biased positively, no electrons are pulled out from anode A. In the space, S, electrons feel the force from anode A, as well as from electrode G. The electron device is designed such that the influence on electrons from anode A dominates over the influence from control electrode G, once electrons are pulled out from cathode C. This is necessary for electrons to move to and be absorbed by anode A, thereby creating current to anode A and not to control electrode G. Once electrons are pulled out to space S, electrons move to anode A and not to control electrode G. This is the crucial point of the mechanism, and we do need ingenuity to assure it. Electron devices can be classified by what sort of ingenuity is exercised to attain this rather tricky objective.

The best way to attain this objective would be to insulate control electrode G entirely from space S using an insulator. The structure shown in Fig. 7(b) shows a MOS field-effect transistor. In this device structure, no DC current flows through the gate electrode: all the electrons injected into space S arrive at anode A. The second way is somehow to separate control (positive) charge from electron flow, so that they do not mix together. This can be achieved by biasing the control electrode negatively with respect to the cathode. Then negative control charge on the control electrode repels negatively charged electrons from the cathode, and they do not mix. The problem with this scheme is that electrons cannot be pulled out from cathode C if the control electrode is negatively biased. This difficulty can be resolved by heating the cathode to a high temperature, thereby evaporating electrons from cathode. Vacuum triodes use this method, as shown in Fig. 7(c). The second way to get electrons to S is to give nonzero conductivity to space S by filling the space with conducting material. Negative charge on the control electrode *squeezes* the area where the current flows. The junction field-effect transistor and GaAs MESFET transistors shown in Fig. 7(d) work by this principle.

The last way is to mix the positive control charge of the input electrode with negative electrons from the cathode, and somehow prevent recombination of the positive and the negative charges. This is the scheme used in bipolar transistors, shown in Fig. 7(e). The mechanism that prevents recombination is quite involved. If electrons and holes in column IV semiconductor material recombine, light (infrared) is *not* emitted. It is established that recombination of charges without emitting light occurs quite infrequently, because the process must be catalized by minute impurity or defect centers in the semiconductor. Semiconductor material like germanium or silicon is purified to the point that concentration of such a catalizer is quite small.

The four devices have different current dependence (between anode A and cathode C) on the voltage of control electrode G (voltage applied between control electrode G and cathode C). The difference originates from the material that fills space S, and from the mechanism of electron emission from the cathode. In NMOS field-effect transistors [Fig. 7(b)] and bipolar transistors [Fig. 7(e)], the material that fills the space is P-type semiconductor. In the vacuum tube [Fig. 7(c)], the space is vacuum, and in the junction field-effect transistor [Fig. 7(d)] the material is N-type semiconductor. In the vacuum tube, the cathode is heated by passing current through its filament (tungsten wire). The cathode itself is a thorium oxide layer coated on the surface of a metal tube that is heated by the filament. Electrons evaporate from thorium oxide into the vacuum surrounding it, and they form a cloud of electrons, or space-charge, around the cathode. The control grid is a cylindrical metal cage that surrounds the cathode. Through the gap of the metal cage, electrons flow out, and fly to the anode. How many electrons are extracted from the pool of space charge depends on how much negative voltage is applied to the grid. A more negative voltage makes the gap of the grid cage through which electrons escape narrower. Grid voltage versus current is approximately a linear relationship. For a higher grid voltage (means *less* negative), more current flows. The grid cannot be biased positively: then electrons flow into the grid and are wasted.

Junction field-effect transistors work somewhat similarly to vacuum tubes. The space between anode and cathode, S, is filled with conductive N-type semiconductor material. If the entire space, S, is available for conduction, the maximum device current flows. Control of a junction field-effect transistor is exercised by reducing the portion of S available for conduction. Electrons are squeezed out from S in the immediate vicinity of the control electrode (called *gate*) by applying a negative voltage to the gate. This is analogous to the vacuum tube that opens and closes the path of electrons to the anode. Current and control voltages are related approximately linearly, since the gate voltage controls the size of the *escape hatch* for electrons. There is one difference between junction field-effect transistors and vacuum tubes. Junction field-effect transistors work even if their cathode (source) and the anode (drain) are exchanged.

Space S consists of P-type semiconductor material in NMOS field-effect transistors and in NPN bipolar transistors. The way the P-type semiconductor space works is, however, significantly different in the two devices. In MOS field-effect transistors the control gate electrode is biased positively, and electrons are attracted to the surface of the P-type semiconductor. The electrons originate from the source, which is strongly N-type region and contains many free electrons. If the drain is grounded, the electrons flow in from the drain region also. As we see, identification of source and drain in MOS field-effect transistors depends on the biasing conditions of the circuit. The device is symmetrical with respect to the source and the drain. Physically it may be said that surface of the P-type semiconductor volume is converted to N-type, and together with the strongly N-type semiconductor areas of the source and the drain, the two contacts are bridged by single and continuous N-type semiconductor. The current path is like a linear resistor. Obviously the number of electrons attracted to the surface of the P-type semiconductor is proportional to the voltage of the control gate. Therefore current increases proportionally to the gate input voltage. In this case, except for the thin surface layer, the P-type semiconductor is not affected at all.

Bipolar transistors, shown in Fig. 7(e), are significantly different from the other devices. Positive mobile charge must coexist with negative mobile electrons injected from the (opposite type) emitter within space S, and therefore the space should be P-type semiconductor. What is different from MOS field-effect transistor is that electrons from the cathode are physically *injected* into space S. How can we inject electrons into P-type semiconductor? We may understand this problem as follows. The fact that mobile electrons do not exist in P-type semiconductor means that electrons have higher energy in P-type semiconductor than in N-type semiconductor. If N-type and P-type semiconductors are connected, only those electrons in N-type material that have high energy are able to move into the P-type semiconductor. The energy of electrons and the temperature of electrons are a measure of the same. Therefore, injection of electrons from an N-type cathode (called emitter) to P-type base region is effected by increasing *temperature* of electrons inside the emitter, thereby evaporating them. This is carried out by applying a negative voltage to the emitter, thereby providing more energy to the electrons inside the emitter. This technique is somewhat analogous to heating electrons and evaporating them. This mechanism has the following details. Not all the electrons in the emitter are able to *evaporate* simultaneously to the base region. Only those electrons that have higher than a certain threshold energy are able to evaporate into the base region. According to the law of Maxwell, the number of such electrons is proportional to the inverse of the exponent of the energy. Then by applying voltage from base to emitter, device current increases exponentially: bipolar transistors are practically nonconducting up to a certain base voltage, and then they turn on very rapidly. This rapid increase in device current is the most distinguishing feature of bipolar transistors. It might be thought that the number of electrons injected into the P-type base by this mechanism is rather small. In a semiconductor, there are indeed significantly fewer electrons than in ordinary metal. The electrons in a semiconductor move, however, much easier than electrons in metal by applied electric field. Therefore even by modest injection of electrons into the base, quite a significant current flows through bipolar transistor.

Someone may wonder how far the similarities extend between a vacuum tube that heats a cathode to evaporate electrons and a bipolar transistor. If a vacuum tube is modified such that the cathode temperature is controlled by the input electrode, and if the vacuum tube is operated at extreme high anode voltage where the space-charge surrounding the cathode is totally stripped off, operation of a vacuum tube resembles that of a bipolar transistor. We do not operate vacuum tubes under such conditions, however. The only case where a vacuum tube is used in this way is to generate noise current of known magnitude (called noise diode). As this application suggests, space-charge-less operation of a vacuum tube is undesirable for general applications. By a similar mechanism, field-effect transistors are much less noisy than bipolar transistors [08].

7.5.1 Gate Delay Characteristics

Technology advance is primarily measured by the photolithography feature size. Since the minimum feature size is dictated by photolithography and by the processing equipment available, different technologies at the same time have approximately the same feature size, with a slight lead for silicon MOS technology. How the delay depends on feature size is crucial in assessing the degree of development of each technology. Basic physics of devices determines the circuit performance through parameters that characterize logic voltage swing, conductance, and parasitic capacitance of the device. We study how delay time of a gate is related to the device parameters. Delay time of a gate is a fundamental parameter, but it requires clarification. Figure 8 shows the most fundamental inverter structure, made using any electron triode device. The device can be a vacuum tube, bipolar transistor (BJT), field-effect transistor (FET), or any other three-terminal electron triode. The input signal at the device terminal propagates through the input circuit, and it arrives at the *internal* control area of the device, where the real control function is exercised. The input circuit consists of the device parasitics:

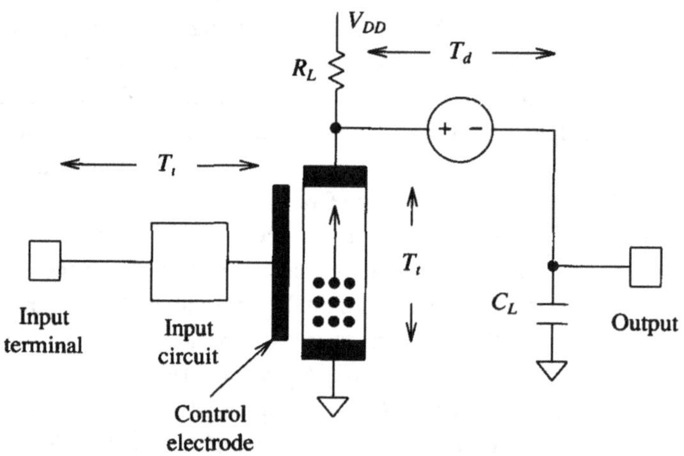

Figure 8 Breakdown of gate delay time

In a vacuum tube it is metal wire that connects the pin to the control grid; in a BJT it is an equivalent RC low-pass circuit that consists of base series resistance (resistance between the base terminal and the *internal* base) and various device capacitances; and in a FET it is the similar RC low-pass circuit that consists of gate capacitance and resistance of the gate material. The input circuit delays signal by T_i, which is called the input delay.

A signal that arrives at the internal control area pulls charged carriers (electrons) out from the cathode. As soon as the charged carriers are pulled out to the space between the cathode and the anode, current begins to flow. Current, however, does not build up instantly to the steady-state current level determined by the control voltage. In the transient period, the current is proportional to the number of carriers that are pulled out to the cathode-anode space. Suppose that the control electrode voltage makes a step increase at time $t=0$. The current is zero at that moment. If the charged carriers travel with a constant velocity v_∞ from the cathode to the anode, and if cathode-anode distance is L, time

$$T_t = \frac{L}{v_\infty} \tag{19}$$

is required to fill up the entire inter-electrode space by moving carriers and to complete current buildup. T_t is called the transit time.

The current drains the charge stored in output capacitor C_L. This capacitor is not just the parasitic capacitance of the device itself, but it includes capacitance of the load and the interconnects. Time required for discharge, T_d, is determined by current of the device, I_D, and the required voltage change at the capacitor terminal, ΔV_L to complete a logic operation. This time is given by

$$T_d = \frac{C_L \Delta V_L}{I_D}. \tag{20}$$

Voltage $\Delta V_L = V_1 - V_0$ is the logic voltage swing, which was defined previously in Section 2.2. This is the difference between the voltages that represent digital 1 (or high) and 0 (or low). Equation (20) shows that T_d is proportional to ΔV_L. Since I_D depends on ΔV_L as well, the dependence is complicated.

We define I_{D1} and I_{D0}—the currents of the device when the input voltage is V_1 (logic high level) and V_0 (logic low level), respectively. Then the *excess* current that discharges output capacitor C_L when the input voltage is high [I_D of Eq. (20)] is

$$I_D = I_{D1} - I_{D0}$$

and therefore

$$T_d = C_L\left(\frac{V_1 - V_0}{I_{D1} - I_{D0}}\right). \tag{21}$$

How T_d will change by changing the logic amplitude can be understood from Fig. 9, which shows device current, I_D, versus input voltage, V_{IN}.

When the characteristic is linear as shown by curve B of Fig. 9, the ratio of Eq. (21) does not change if logic amplitude changes. Scaled-down MOS devices are an example of this case. In scaled-down CMOS, device delay depends weakly on power supply voltage (which is

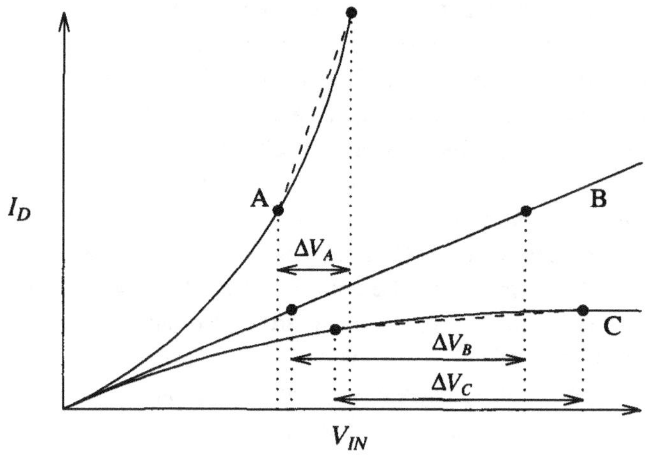

Figure 9 I-V characteristics of device

the logic amplitude) if the ratio of the FET threshold voltages to the power supply voltage is small (Section 5.16). Let us consider the sublinear characteristic shown by curve C. Suppose that V_1 is held constant, and V_0 is changed. The ratio $(V_1 - V_0)/(I_{D1} - I_{D0})$ decreases by increasing the logic swing. The delay of the gate using this device decreases by increasing the logic amplitude. This sublinear characteristic does not occur in conventional circuits except when the bias point is set wrong. In superlinear characteristics shown by curve A, the ratio $(V_1 - V_0)/(I_{D1} - I_{D0})$ decreases as logic amplitude $(V_1 - V_0)$ decreases. The limit of delay reduction is reached when

$$\frac{V_1 - V_0}{I_{D1} - I_{D0}} \approx \left(\frac{dV_{IN}}{dI_D}\right)_{V_{IN}=V_1}$$

is satisfied. In a bipolar transistor,

$$I_D = I_0 \exp\left(\frac{qV_{IN}}{kT}\right) \tag{22}$$

and therefore $dI_D/dV_{IN} = I_D/V_{THERM}$, where $V_{THERM} = kT/q$ is the thermal voltage ($25mV$ at room temperature). Then reduction in discharge delay time is expected by reducing the logic amplitude down to this voltage level. Superior delay characteristic of bipolar digital circuits originates from this mechanism.

Delay time of a gate is the sum of three delay components as

$$T_D = T_i + T_t + T_d\,. \tag{23}$$

We study the three delay components, input delay T_i, transit time delay T_t, and discharge delay T_d for MOSFET, GaAs MESFET, and for bipolar transistor circuits.

7.5.2 Gate Delays of Various Technologies

We study gate delays of several common IC technologies. The three delay components introduced in the last section are estimated using the simplest physical models of the devices.

[1] Si MOSFETs

A silicon MOSFET has the structure shown in Figs. 10(a) and (b). The FET channel has *electrical* length L, *designed* length L' (where polysilicon width $L' > L$) and device width W. If the carriers in the channel move at the saturation drift velocity, v_∞ *(cm/sec)*, the time required to fill the entire channel by carriers emitted from the source (the transit time T_t) is given by

$$T_t = \frac{L}{v_\infty} . \tag{24}$$

In silicon, the saturation velocity is reached when the channel field is from 10^4 to $10^5 (volt/cm)$. The saturation velocity v_∞ ranges from $3 \times 10^6 (cm/sec)$ to $1 \times 10^7 (cm/sec)$, depending on the reports. Assuming $v_\infty = 5 \times 10^6\ (cm/sec)$ and $L = 1\ \mu$m, transit time is

$$T_t = 20\ picosec .$$

In the near future, channel length will scale down to 0.5 μm. Then transit time will be 10 picoseconds.

The gate of a MOSFET is usually connected from only one side, as shown in Fig. 10(b). The length of the polysilicon gate approximately equals the width (often called size) of the FET, W. The FET size is chosen by the designer. The minimum size FET has a width from 2 to 2.5 times the feature size. In typical CMOS circuits, up to 4 to 5 times minimum size FETs are most frequently used. Effective gate series resistance of a FET is then given by

$$R_{IN} = \frac{\rho_{sq} W}{2L'} \tag{25}$$

where ρ_{sq} is resistance per square of gate polysilicon or silicide material and L' is the *designed* gate polysilicon width. In silicided technology $\rho_{sq} = 1$ (ohm/square). Assuming $W = 10\ \mu$m, and $L' = 1.5 \mu$m, R_{IN} is 13 ohms. If the technology has a doped polysilicon gate,

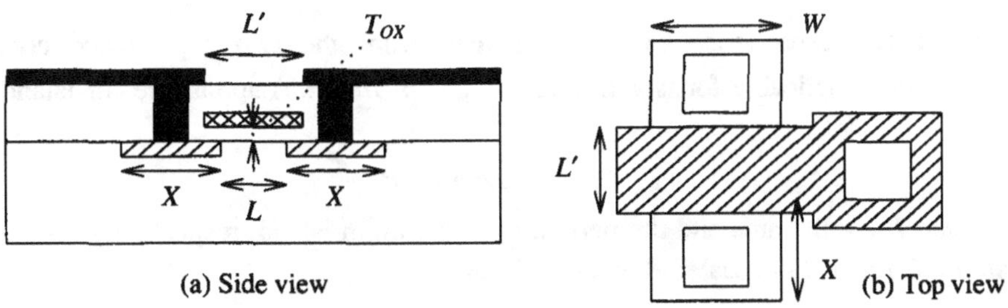

Figure 10 MOS field-effect transistor

$\rho_{sq} = 25$ *ohms per square*. Then R_{IN} is 83 ohms. Contact resistance from metal to polysilicon is 5 to 10 ohms. Then the typical input resistance of MOSFET ranges from 20 ohms in silicided technology to 90 ohms in doped polysilicon technology.

The gate capacitance is estimated from

$$C_G = \frac{\varepsilon_{OX} L'W}{T_{OX}} = 0.345\,(pF/cm)\,\frac{L'W}{T_{OX}}\,. \tag{26}$$

Assuming $L' = 1.5 \times 10^{-4} cm$, $W = 10 \times 10^{-4} cm$ and $T_{OX} = 2.5 \times 10^{-6} cm$, $C_G = 0.021\ pF$. The time required for an input signal to reach the central gate area, T_i, is the product of R_{IN} and C_G as follows.

$$T_i = R_{IN} C_G = 20 \times 0.021 = 0.4\ picosec\ \text{(silicide)}$$

$$= 90 \times 0.021 = 1.8\ picosec\ \text{(doped polysilicon)}$$

In MOSFETs, the input delay time is an order of magnitude smaller than the transit time. Input delay time of MOSFETs is practically negligible, even if the feature size is reduced. Even in a large FET, input delay can be made small by using layout techniques.

The current of a scaled-down MOSFET, I_D, is estimated by the product of the average induced charge on the channel, Q, and the saturation drift velocity, v_∞, divided by channel length L. Q is given by (factor 2 in the denominator is to average over the length of channel)

$$Q = \frac{\varepsilon_{OX} V_G}{2T_{OX}}\, LW$$

and therefore

$$I_D = \frac{Q v_\infty}{L} = \frac{\varepsilon_{OX} V_G}{2T_{OX}}\, W v_\infty \tag{27}$$

and therefore transconductance G_m is estimated as

$$G_m = \frac{\partial I_D}{\partial V_G} = \frac{\varepsilon_{OX} v_\infty W}{2T_{OX}}\,. \tag{28}$$

Assuming the same parameter values as before, and $V_G = V_{DD}$ (power supply voltage = 5 V),

$$I_D = \frac{0.345 \times 10^{-12}(F/cm) \times 5(volts) \times 10 \times 10^{-4}(cm) \times 5 \times 10^{6}(cm/sec)}{2 \times 2.5 \times 10^{-6}(cm)} = 1.7 \times 10^{-3}(amp)$$

$$G_m = 0.35 \times 10^{-3}\ (mho)$$

Drain island capacitance can be estimated from the two capacitance coefficients, $C_A(pF/micron^2)$ applicable for island area and $C_P(pF/micron)$ applicable for island periphery, as

$$C_D = C_A A + C_P P \tag{29}$$

where A and P are the area and the periphery of the drain island, respectively. A reasonable estimate for C_A and C_P in scaled-down CMOS are

$$C_A = 3 \times 10^{-4}(pF/micron^2) \quad C_P = 3 \times 10^{-4}(pF/micron)\,.$$

Assuming $W = 10$ *microns* and $X = 4$ *microns* gives

$$C_D = 2.0 \times 10^{-2} (pF)$$

In a MOSFET, gate and drain parasitic capacitances are about the same. Therefore, the time required for FET to discharge its own capacitance equals

$$T_d = (C_G + C_D) \frac{V_{DD}}{I_D} = \frac{0.04(pF) \times 5(volts)}{1.7 \times 10^{-3}(amp)} = 110 \, picosec . \quad (30)$$

This discharge delay time is an order of magnitude longer than the transit time. Discharge delay time increases if the device is integrated into a circuit, since capacitance of wiring and fanout is added. From the estimates, we conclude that

$$T_d > T_t > T_i \quad (31)$$

in silicon MOSFET circuits. Discharge delay time of FETs is 1 to 2 orders of magnitude longer than the *intrinsic* delay times such as T_t and T_i. Scaledown of technology improves performance of MOSFETs as follows: If W and T_{OX} are reduced while keeping ratio W/T_{OX} unchanged, channel current stays the same. Since W, X, and L' decrease, C_G and C_D decrease. Then T_d decreases. Around 1 μm feature size, CMOS gate delay time decreases a little faster than proportionally to the feature size. Figure 11 shows CMOS gate delay subject to typical process and operating conditions versus feature size. This trend is likely to continue down to 1/4 μm feature size. In the range of sub 1/2 micron feature size, power supply voltage will decrease to 3 volts. Below 0.1 μm feature size (if such CMOS technology is viable), a new effect is expected to set in. The carriers injected from the source to channel arrive at the drain without being scattered. The carriers reach significantly higher velocity than that

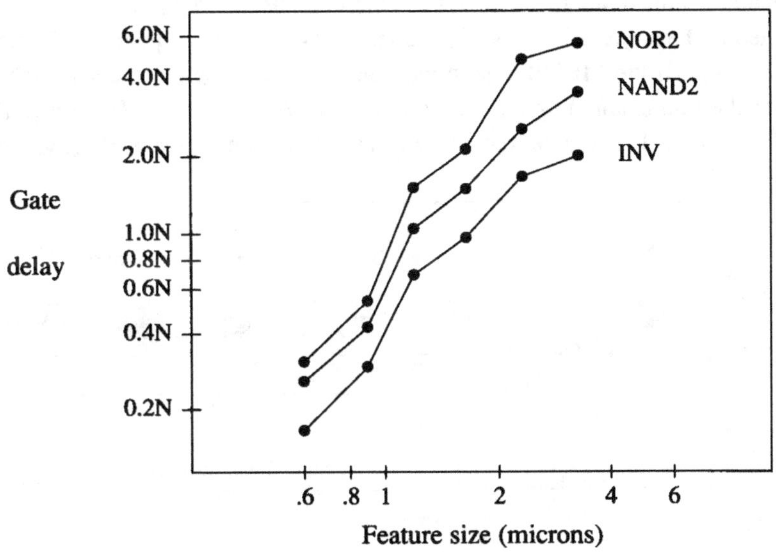

Figure 11 CMOS technology feature size versus gate delay time

achievable in larger devices. The velocity overshoot effect will maintain improvement in FET characteristic with further scaledown. In this *ballistic* transport regime, FETs will resemble vacuum tubes.

Scaledown below 1 micron feature size, however, brings many problems: subthreshold leakage current increases, capture of hot carriers by gate oxide causes reliability problems, punchthrough (between source and drain) and breakdown (of drain) must be controlled, and improvement in FET characteristic is not as significant as scaledown from 10 micron feature size to 1 micron feature size, because of high-field carrier velocity saturation effects. Although the scaling law holds down to 1/2 μm feature size, the intrinsic limit of device capability may not be reachable because of many practical problems.

[2] GaAs MESFETs

A GaAs MESFET is a junction field-effect transistor fabricated using a GaAs epitaxial layer grown on semi-insulating (SI) GaAs substrate [09], [10]. With reference to the cross-sectional drawing of Fig. 12(a), we may show the following. If channel length is L, width is W, channel (N-layer) thickness is b_0, carrier mobility is μ, and epitaxial layer doping is N_D, channel current I_D at low drain voltage, V_D, is given by

$$I_D = G_0 V_D [1 - \sqrt{(V_T - V_G)/V_P}] \tag{32}$$

$$\text{where} \quad G_0 = \frac{qN_D \mu W b_0}{L} \quad \text{and} \quad V_P = \frac{qN_D {b_0}^2}{2\varepsilon_S}$$

where V_D is the drain voltage and V_G is the gate voltage, respectively, both of which are measured relative to the grounded source. V_T is the work function difference between the gate material and the channel semiconductor. V_P is the potential sustained by the space-charge layer created by depleting the entire thickness of the conducting channel. Equation (32) means that $I_D = 0$ if $V_G \le V_T - V_P$. If the MESFET is in the nonconducting region, the space-charge layer spreads all over the thin channel layer, and no N-type GaAs bulk is left for conduction. Suppose that $V_G \ge V_T - V_P$. If V_D is increased, current saturation takes place at V_{Dsat} given by

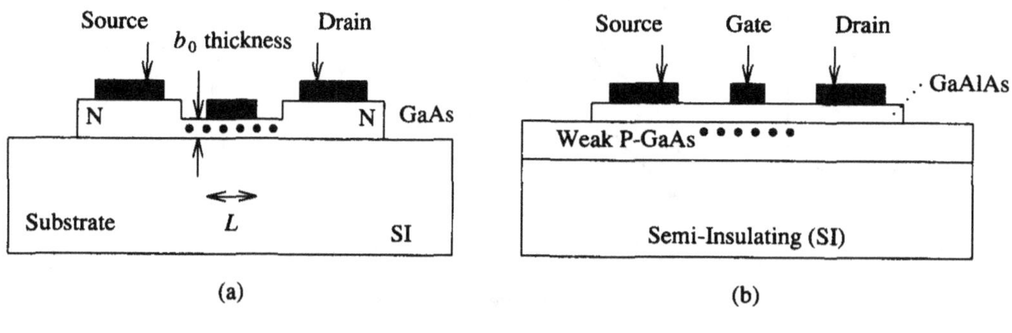

Figure 12 GaAs MESFET (a) and HEMT (b)

$$V_D = V_{Dsat} = V_G - V_T + V_P \tag{33}$$

and the saturation current I_{sat} is given by

$$I_{sat} = G_0 \left[V_G - V_T + V\frac{P}{3} + \frac{2}{3}\frac{(V_T - V_G)^{3/2}}{\sqrt{V_P}} \right]. \tag{34}$$

Equation (34), however, overestimates GaAs MESFET channel current. Velocity-field characteristics of N-GaAs have a drift velocity maximum at about $3.5\ kV/cm$, and at higher fields electron drift velocity decreases and tends to the saturation velocity v_∞ at very high fields, which is about half the maximum drift velocity. Bulk GaAs shows negative differential conductivity above the drift velocity maximum field. This drift velocity anomaly is due to the transferred-electron effect, often referred to as Gunn effect, which had been studied extensively in the 1960s. The negative differential conductivity is not effective in improving MESFET characteristics in any way, since response time of the intervalley electron transfer mechanism that creates the negative differential conductivity is comparable or even longer than the switching delay time of present MESFETs. The negative differential conductivity, however, helps MESFET characteristics in one interesting way: a stationary high-field layer is formed at the drain's end of the channel [11], and the capacitance of the space-charge layer is in series with the geometrically *inevitable* gate to drain coupling capacitance. This effect reduces the Miller capacitance of GaAs MESFETs quite significantly.

The saturation drift velocity of electrons in N-type GaAs, v_∞, has been measured by several different methods. High-field domain (Gunn domain) theories predict that the saturation velocity equals drift velocity of high-field domains. The domain velocity can be measured precisely by probe measurements. By the author's measurements, $v_\infty = 9 \times 10^6 cm/sec$ [11]. Since my material was oxygen-doped bulk material of the 1960s, the data is considered as a lower limit. $v_\infty = 10^7 cm/sec$ appears to be a reasonable estimate. This velocity is not significantly higher than the saturation drift velocity in silicon, which ranges 7–8 $\times 10^6 cm/sec$ in the recent measurements [12].

The transit time of electrons is given by

$$T_t = \frac{L}{v_\infty} = \frac{0.8 \times 10^{-4}}{10^7} = 8\ picosec$$

assuming 0.8 micron channel length, and uniform high field in the channel. If we assume that the gate capacitance equals the capacitance of a parallel plate capacitor whose gap is the depletion layer thickness b_0,

$$C_G = \frac{\varepsilon_S}{b_0} WL .$$

Since this is the minimum capacitance during the switching opertation of the MESFET, the real capacitance is larger than that, by as much as several times. Gate series resistance is $R_{IN} = \rho_{sq} W/2L$ where ρ_{sq} is the sheet resistance of the gate material. For aluminum, $\rho_{sq} = 0.05\ ohms/square$. Then

$$T_i = R_{IN} C_G = \frac{\rho_{sq} \varepsilon_S W^2}{2b_0}.$$

Assuming $b_0 = 10^{-5}\ cm$, $\varepsilon_S = 10^{-12} F/cm$, and $W = 10 \times 10^{-4}\ cm$, we have input delay time, $T_i = 0.03\ psec$. Input delay time of ordinary GaAs MESFETs is entirely negligible.

Control of MESFET channel current is achieved by depleting part of the conducting channel. The voltage required to deplete the entire channel of thickness b_0 and doping N_D is given by V_P of Eq. (32). By *pinching* the channel, current drops to zero: the rate of change of channel current by gate voltage change (averaged over range V_P) is,

$$\frac{qN_D v_\infty b_0 W}{2V_P} = \frac{\varepsilon_S v_\infty W}{b_0} \approx G_m \tag{35}$$

where factor 2 in the denominator is average over the channel. Transconductance G_m increases by decreasing the thickness of channel b_0 and simultaneously increasing doping N_D such that V_P or $N_D {b_0}^2$ remains unchanged. A thinner and more heavily doped channel is able to control channel current more effectively. The ratio of GaAs MESFET transconductance to that of a silicon MOSFET is

$$\frac{G_m(MESFET)}{G_m(MOSFET)} = 2 \frac{\varepsilon_S(GaAs)}{\varepsilon_{OX}(SiO_2)} \frac{v_\infty(GaAs)}{v_\infty(Si)} \frac{T_{OX}}{b_0} \tag{36}$$

where $\varepsilon_S(GaAs) \approx 3\varepsilon_{OX}(SiO_2)$. If we use $v_\infty(GaAs) = 1 \times 10^7 cm/sec$, $v_\infty(Si) = 5 \times 10^6 cm/sec$, T_{OX} = 250 *angstroms*, and b_0 = 1000 *angstroms*, a GaAs MESFET has about 3 times more transconductance than a Si MOSFET. Depending on the parameter values used (especially parameter v_∞), the ratio can be as low as 2. This comparison is valid at the high-field limit.

At low fields, the ratio of the drain conductances is given by

$$\frac{G_D(MESFET)}{G_D(MOSFET)} = 2 \frac{\mu(GaAs)}{\mu(Si)} \frac{\varepsilon_S(GaAs)}{\varepsilon_{OX}(SiO_2)} \frac{V_P}{V_{DD}} \frac{T_{OX}}{b_0} \tag{37}$$

and since $\mu(GaAs)/\mu(Si) \approx 10$ and $V_P/V_{DD} \approx 1/6$, the ratio is approximately the same as that given by Eq. (36). In a field-effect transistor at low fields, the maximum drain conductance (at $V_D = 0$) equals the transconductance in saturation region (at $V_D > V_{Dsat}$). Then it is not surprising that the two estimates give similar results. Conductance advantage of GaAs MESFET over Si MOSFET is at most 2 to 3, which is rather small.

The ratio of MESFET gate capacitance to that of a MOSFET can be estimated from

$$\frac{C_G(MESFET)}{C_G(MOSFET)} = \sqrt{2} \frac{\varepsilon_S(GaAs)}{\varepsilon_{OX}(SiO_2)} \frac{T_{OX}}{b_0} \tag{38}$$

where the voltage sustained by the MESFET channel space-charge layer was assumed to be $V_P/2$. We have $\varepsilon_S(GaAs) \approx 3\varepsilon_{OX}(SiO_2)$. If we assume $4T_{OX} \approx b_0$, a MESFET and MOSFET have about the same gate capacitance.

Equations (36) and (38) show that if the gate oxide thickness of a silicon MOSFET T_{OX} and the channel thickness of a GaAs MESFET b_0 are used as parameters characterizing the essential *vertical* dimension of the respective devices, transconductances of the two devices are of the *same order of magnitude*. The few times higher transconductance of a GaAs MESFET originates from the slightly higher saturation velocity of electrons in GaAs and from various minor factors that make MESFETs different from MOSFETs. The ratio of dielectric constants of GaAs and of silicon dioxide has neutral effect on advantage or disadvantage of the two devices, since higher transconductance is associated with higher gate capacitance.

GaAs MESFET devices are faster than silicon bulk MOS devices primarily by another mechanism: a GaAs device does not have a conducting substrate. Source, drain, and wiring capacitances of GaAs MESFET devices are significantly smaller than the corresponding capacitances of bulk silicon MOS devices. We consider the source or drain island. Drain or source island capacitance of a MESFET can be estimated by approximating the island by an ellipsoid having half axis a, b, and c, where $a > b \gg c$. Figures 13(a) and (b) show the source or the drain islands of a bulk CMOS MOSFET and a GaAs MESFET, respectively. Capacitance of an ellipsoidal conductor imbedded in a uniform dielectric material having dielectric constant ε_S is given by [13]

$$C_D = 8\pi\varepsilon_S \left[\int_0^\infty \frac{d\xi}{\sqrt{(a^2+\xi)(b^2+\xi)(c^2+\xi)}} \right]^{-1} \tag{39}$$

Parameter $2c$ is the vertical dimension (thickness perpendicular to the substrate surface), and therefore $c \ll a, b$. If we assume that geometrical average and arithmetic average of a^2 and b^2 are approximately equal $[(a^2+b^2)/2 \approx ab]$

$$\int_0^\infty \frac{d\xi}{\sqrt{(a^2+\xi)(b^2+\xi)(c^2+\xi)}} \to \int_0^\infty \frac{d\xi}{(\xi+ab)\sqrt{\xi}} = \frac{\pi}{\sqrt{ab}} \quad (c \to 0). \tag{40}$$

With reference to Fig. 13(a), the capacitance of a bulk CMOS FET drain island is approximated by the capacitance of parallel plates having area $2a \times 2b$. The capacitance is $4\varepsilon_S ab/d$, where d is the gap between the two electrodes, which is the depletion layer width. The ratio of GaAs MESFET drain capacitance to that of a MOSFET can be estimated from

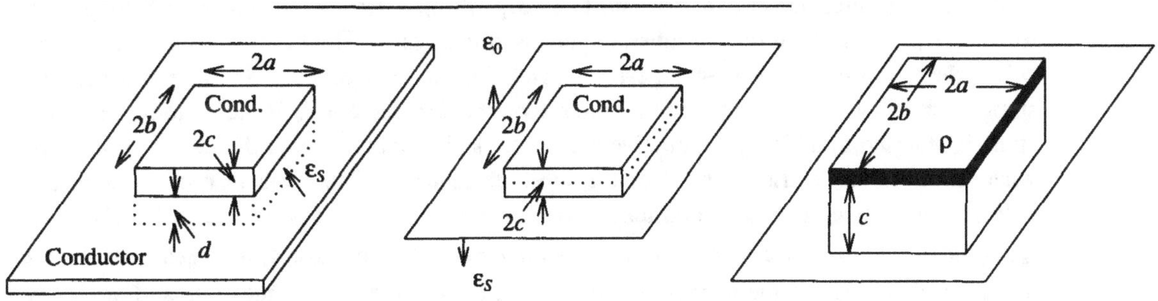

(a) Capacitance of Si MOSFET ICs

(b) Capacitance of GaAs MESFET ICs

(c) Quality of substrate

Figure 13 Capacitance of MOSFET ICs and MESFET ICs

$$\frac{C_D(MESFET)}{C_D(MOSFET)} = \frac{2d}{\sqrt{ab}} \tag{41}$$

where the dielectric constants of the two semiconductors are about the same. Since a and b are horizontal dimensions, and d is a vertical dimension, drain island capacitance of a MESFET can be quite significantly less than that of a bulk MOSFET. The source-drain island is not completely surrounded by material having dielectric constant ε_S. Then the capacitance ratio can be smaller by another factor of 2.

Reduced drain capacitance of GaAs MESFETs, and reduced wiring capacitance of MESFET ICs, both originating from the insulating substrate, are the most important advantages of GaAs MESFET technology. The good performance in relatively small-scale integrated GaAs MESFET ICs compared with silicon MOSFET ICs originates jointly from the reduced capacitance and a factor of 2 to 3 higher transconductance of GaAs MESFETs. Absence of a conducting substrate, however, makes GaAs MESFET device more sensitive to induced noise, and therefore the speed advantage comes with the higher cost of more involved design.

HEMTs or MODFETs use GaAlAs material deposited on P-type GaAs layer, as shown in Fig. 12(b). Because of the larger band gap of GaAlAs, electrons spill over to the P-type GaAs interface, thereby creating surface conducting channel. The spilled electrons have higher mobility than electrons in doped GaAs, since they do not suffer from ionized impurity scattering. This mechanism, however, does not change the estimate of device capability very significantly. What matters in circuit performance is the drift velocity of the carriers at high electric fields, and not the low-field mobility (Section 7.4). The device structure of Fig. 12(b), however, has smaller source/drain series resistance than the structure of Fig. 12(a), and this effect is important in short-channel MESFETs.

In conclusion, the relationship between the three delay components of GaAs MESFET logic circuits is similar to that of silicon MOSFET circuits

$$T_d > T_t > T_i \tag{42}$$

except that the ratio of T_d to T_t is smaller than that of MOSFET ICs.

It is interesting to note that at high device speeds, quality of the substrate, and not quality of the active semiconductor, matters most. In Fig. 13(c), a semiconductor layer having sheet resistance ρ is deposited on the substrate having thickness c. Thickness c cannot be reduced below the width of the space-charge layer insulating the component. Suppose that the substrate thickness was reduced to the minimum required for insulation. If the bottom of the substrate is attached to a highly conductive metal heat sink, capacitance of the device is $4ab\varepsilon_S/c$, where ε_S is the dielectric constant of the substrate material. The resistance of the device is $(b/a)\rho$, and therefore the minimum time constant of the device, $t_{\min}$, is given by $t_{\min} = 4b^2\varepsilon_S\rho/c$. If power supply voltage V_{DD} is applied to the resistor, heat generated by the device is $aV_{DD}^2/b\rho$. If the temperature difference between the metal substrate and the active layer should be less than a given maximum $\Delta T_{\max}$, we have $\Delta T_{\max} = cV_{DD}^2/4b^2\rho\kappa_S$, where κ_S is the thermal conductivity of the substrate. Then we have

$$t_{\min} = (\varepsilon_S/\kappa_S)(V_{DD}^2/\Delta T_{\max})$$

and therefore the merit of the substrate in creating small RC time constant is measured by the ratio of the dielectric constant and the thermal conductivity. Since silicon has ε_S/κ_S about 1/4 of III–V compounds, silicon is indeed a very good high-speed substrate. It is suspected that the superiority of III–V devices over silicon devices exists only for small scale integration, where the thermal constraint can be relieved by package design. This point was first recognized in the late 1960s, when Read diodes (silicon) and Gunn diodes (GaAs) competed as commercial microwave generators. If high DC power is pumped into the devices, Gunn diodes burn out easily, but Read diodes do not burn out that easily.

[3] Bipolar Transistors

Estimation of BJT device performance is more difficult than for MOSFETs [14]. Assumptions like uniform-base create many uncertainties. Here we use the simplest device model to derive the approximate relationships. Transit time of electrons that diffuse through the base of a bipolar transistor is estimated from the standard formula for a uniform-base BJT as $T_t = L^2/2D_N$, where L is the base width and D_N is the diffusion coefficient of the carriers. Assuming $D_N = 5\,(\mathrm{cm}^2/\mathrm{sec})$ and base thickness of 0.1 micron, $T_t = (1 \times 10^{-5}\,cm)^2/[2 \times 5(cm^2/sec)] =$ 10 picosec. This estimate is comparable to that of the present-day MOSFETs. The transit time is, however, an overestimate, since the effects of the accelerating field created by the base doping gradient (doping higher in emitter side and lower in collector side) is not included [14]. Therefore the estimate is an upper limit. Transit time is really less than 10 picoseconds, which is shorter than that of MOSFETs.

BJT base capacitance consists of collector-base junction capacitance, base-emitter junction capacitance, and the capacitance contributed by the excess majority carriers in the base region, which are required to neutralize the minority carriers injected from the emitter during the switching transient. We note that collector junction capacitance and emitter junction capacitance are analogous to the gate-to-drain and gate-to-source *overlap* capacitances of MOSFETs (Section 1.5), but they are higher than those of MOSFETs. Although the capacitances are the integral parts of any BJT structure and minimization is hard, the capacitances have no *essential* role to play in the conduction mechanisms of BJTs. The third capacitance component is associated with the flow of collector current, and that is essential to the working mechanisms. This capacitance is called diffusion capacitance.

To gain insight into diffusion capacitance, we compare a BJT with a MOSFET. Figures 14(a) and (b) show schematics of charge distribution within an N-channel MOSFET and an NPN BJT, respectively. In the MOSFET [Fig. 14(a)], positive charge on the gate electrode induces negative charge on the surface channel, that carries current. Positive and negative charges are separated by the insulating gate oxide, but if we disregard the distance between them (thickness of gate oxide), charge neutrality is maintained in the device: within the dotted box of Fig. 14(a), an equal amount of positive and negative charge exists. In the BJT, current in the base is carried by diffusion of electrons that are injected into the P-type base region from the strongly N-type emitter. The density of the minority electrons is higher on the emitter's end of the base [E of Fig. 14(b)] than on the collector's end (C). If the collector-base PN

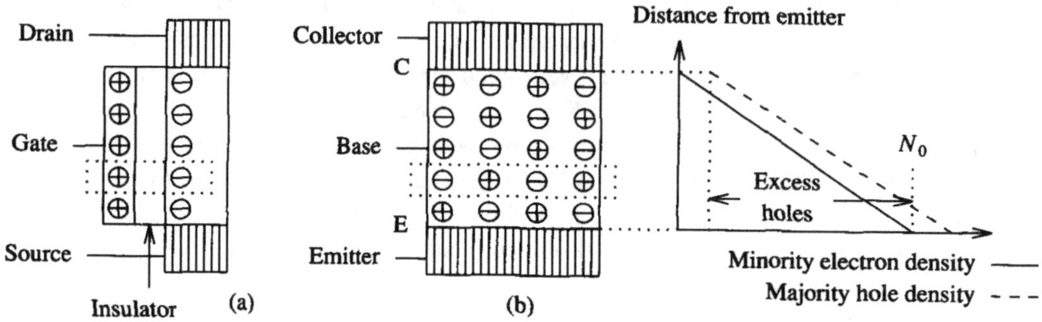

Figure 14 FET and BJT

junction is reverse-biased, electron density at the collector's end of the base is practically zero. Electron density versus distance within the base measured from the emitter's edge is shown by the solid line of Fig. 14(b). Charge neutrality must be maintained everywhere within the base. In the small section of base within the dotted box of Fig. 14(b), equal positive and negative charges exist. Therefore hole density is as shown by the dashed line of Fig. 14(b). The excess hole density (above the base doping level) must be established by injection of *majority* holes into the base region from the base terminal. We now see that the excess hole density in the base is similar to the positive gate charge of the MOSFET. The base charge is, however, in contact with the electrons that are injected from the emitter. Then some holes recombine with the electrons and are lost. This loss shows up as the base current, in the range 0.1 to 1 percent of the collector current. In MOSFETs, the gate charge is isolated from mobile channel charge, and therefore no charge is lost; the DC gate current is practically zero. This is a significant difference between BJT and MOSFET.

Figure 14(b) shows the minority electron distribution within the base. Collector current, I_C, of the uniform-base BJT is given by the simple diffusion mechanism

$$I_C = qD_N\left(\frac{N_0}{L}\right) \tag{43}$$

where N_0 is the electron density at the emitter side edge of the base. Electrical neutrality within the base requires that equal hole charge, Q_P, must be injected into the base. We have

$$Q_P = \frac{qLN_0}{2}.$$

N_0 increases exponentially with increasing base-emitter voltage V_B as $N_0 \approx exp(V_B/V_{THERM})$ where $V_{THERM} = kT/q$. The capacitance looking into the base terminal is the rate of increase of Q_P with increasing V_B as

$$C_{DIFF} = \frac{\partial Q_P}{\partial V_B} = \frac{qN_0L}{2V_{THERM}} = \frac{T_tI_C}{V_{THERM}} \tag{44}$$

where L and N_0 were eliminated using BJT current I_C and the transit time T_t. The physical meaning of this formula is quite obvious: If current I_C flows during the interval T_t, the base charge is established, and because of the exponential dependence of current on V_B, a base voltage change of about V_{THERM} creates current I_C.

This argument may be strange to the readers who consider that a capacitor is made of a pair of flat electrodes and an insulating gap. Especially the following question may be asked: What is the *gap* between the charges of the diffusion capacitance? Electrons and holes in the base of BJT have thermal energy, that maintains nonzero distance between them, even though attractive force works between dissimilar charges. The distance maintained by thermal energy is the Debye length (Section 1.8). This mechanism is easy to see from an analogy: If all the material that presently makes the solar system had no initial velocity at the Genesis, only the sun, but no planet, would exist now.

There are uncertainties in the base doping level estimate. Here we consider that the emitter is doped close to the solubility limit of boron (estimated at 3×10^{20} *impurities*/*cm*^{3}), and the base doping is 1/100 of the emitter doping for high emitter efficiency. Then base doping is 3×10^{18} *acceptors*/*cm*^{3} and the resistivity is 0.03 ohm-cm.

The resistance between the base contact and the region where the minority electrons flow is $r_{bb'}$. If base width L reduces, $r_{bb'}$ inevitably increases. If base doping is $3 \times 10^{18}/cm^3$, P-type base sheet resistance is 3 kohms/square, if $L = 0.1\ \mu$m. To attain $r_{bb'} \approx 100$ ohms or less, the number of squares must be quite small. The emitter is designed long and narrow to reduce the number of squares, and the base contacts are provided on both sides of the emitter. In spite of such device design innovations, reduction of $r_{bb'}$ is not easy.

We have the estimate of the input delay

$$T_i \geq r_{bb'} C_{DIFF} = T_t (I_C / I_{BT}) \quad \text{where} \quad I_{BT} = V_{THREM} / r_{bb'} \tag{45}$$

where the right-hand side of inequality is an underestimate, since only the diffusion capacitance is included. If we use a crude estimate that $I_C = 1$ mA, $V_{THERM} = 25$ mV and $r_{bb'} = 100$ ohms, $I_C/I_{BT} = 4$. We may conclude that T_i is the same or longer than T_t. The MOSFET's T_i can be made as short as the designer wishes, since MOSFET gate is exposed. This is a fundamental difference between BJT and MOSFET.

The transconductance of a bipolar transistor is estimated as

$$g_m = \partial I_C / \partial V_B = I_C / V_{THERM}\ .$$

Transconductance of BJT is significantly higher than that of MOSFET. This is another fundamental difference between BJT and MOSFET. If we consider that a BJT of one gate drives a BJT of the other gate, and the base capacitance is given by Eq. (44), the discharge delay time of the driver stage is estimated by

$$T_d \geq C_{DIFF} / g_m = (T_t I_C / V_{THERM})(V_{THERM} / I_C) = T_t \tag{46}$$

where T_d is again an underestimate, since the other capacitances are neglected. T_d is again the same or longer than T_t, but the three delay components are essentially the same order of magnitude. We note that there are significant uncertainties, depending on the BJT structure. Even allowing wide uncertainties, all three delay times fall approximately within an order of magnitude of each other,

$$T_i \approx T_t \approx T_d\ . \tag{47}$$

In bipolar transistor technology, the *natural* technological limit attained by device scaledown has been reached already, and therefore straightforward scaledown of device dimensions does not result in further improvement of the performance. This situation is clear from the estimate that shows that input delay can be the major delay of bipolar transistor circuits. Improvements of bipolar devices must be made to minimize T_i and T_t simultaneously, since T_d is determined primarily by current. A device structure that minimizes the series resistance of the thin base must be developed. Modern BJTs have a very thin base, typically 1000 angstrom or less, and the emitter is also very narrow, typically 0.5 to 0.3 microns wide. The problem is how to fabricate such a narrow emitter. In the structure of Japanese SST (Super Self-aligned Transistor) the only critically small feature of photolithography is the emitter strip. Insulation between emitter and base contacts is secured by the oxide layer grown on the base contacting polysilicon.

Figures 14(a) and (b) suggest that MOSFETs and BJTs have similar principles of operation. The similarity helps to understand a confusing issue in the simple MOSFET theory, namely, the mechanism of current saturation at drain voltages higher than the pinch-off voltage. Figure 15(a) shows a schematic of a MOSFET biased in the saturation region. The region of the channel close to the source works effectively as a resistor that is controlled by the gate voltage. If drain voltage is low, the resistor region extends over the entire channel, and the FET is a nonlinear resistor. The gradual-channel, low-field FET model is valid in this region of FET operation. If the drain voltage is higher than the saturation voltage of the FET (this is the voltage at which the induced charge density at the drain's end of the channel vanishes), a space-charge layer is formed in the immediate vicinity of the drain. The channel vanishes some distance before the drain. The area of the channel where the inversion layer vanishes is essentially P-type semiconductor to which electrons are injected from the conducting channel. The vanishing channel and the gate positive charge work jointly as a *simulated* base, and the base-emitter voltage works as the threshold voltage of the surface channel. The end region of the channel, with its steep gradient of mobile electron density, works as the *pseudo* base of a BJT. Therefore, an N-channel MOSFET in saturation is equivalent to a series connection of an NPN BJT and a FET as shown in Fig. 15(b). The FET in this equivalent circuit is biased slightly under saturation: the FET works essentially as a nonlinear emitter-degeneration resistor in the bipolar equivalent circuit. The current of the BJT in the equivalent circuit is determined by the emitter degeneration resistor of the circuit. The BJT is in the

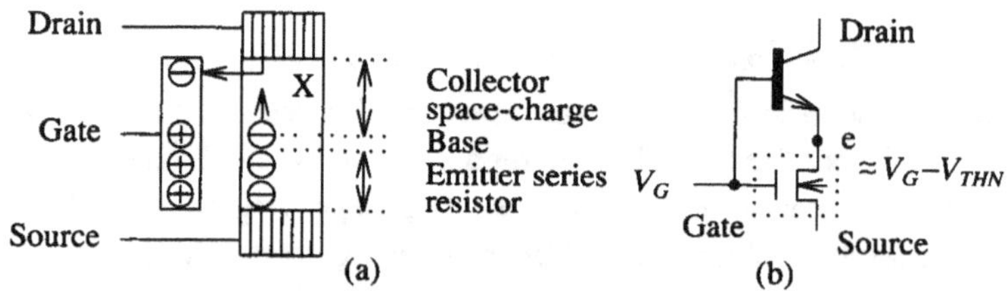

Figure 15 FET and BJT equivalence

emitter-follower configuration, and therefore the BJT is unable to determine its own current: the BJT's role is to absorb the extra drain voltage of FET as BJT's collector voltage. This model explains clearly how a FET works beyond saturation. Figure 15 may have an alternative interpretation: the BJT may be considered as an NFET, whose gate oxide thickness is Debye length (Fig. 14). Then it is easy to understand why the collector voltage of a BJT in saturation is lower than the base voltage (drain voltage of MOSFET lower than gate voltage is not surprising at all!). In this way the BJT and MOSFET models help each other to clarify the working mechanisms of semiconductor triodes.

As we saw from the discussions on the three device technologies, the three component delays T_i, T_t and T_d have different dependence on scaledown of technology. Input delay depends more on details of the technology than on feature size of the technology. Device scaledown often tends to increase input delay in bipolar transistors. Transit time, T_t, of MOSFETs is determined by feature size, and of BJTs by process details. Discharge delay, T_d, of MOSFETs is determined by the feature size, and that of BJTs by current and the technology details. When the natural, or *scaledown* limit of technology is reached, the three component delays T_i, T_t, and T_d become comparable. This limit has been reached in bipolar technology. Then various improvements in the *details* of the technology are more effective in further improving BJT performance. Technologies whose gate delays depend strongly on the feature size like CMOS have not yet reached the *scaledown* limit. GaAs technology stands somewhere in between the two extrema. The effects of scaling is observable, but it is not so conspicuous as in MOS technology.

Although the limit of MOS technology has not yet been reached, we cannot conclude that the future of MOS technology is the most promising. The limit of MOS technology *may not be reachable*, due to *many* secondary reasons such as subthreshold leakage, hot carrier trapping by oxide, drain PN junction breakdown, source-drain punch-through, and dominance of sidewall capacitance for both wiring and drain diffusion islands. The scaling law of MOSFETs dictates that gate delay of submicron CMOS is proportional to feature size. This relationship holds well, at least down to 0.6 μm CMOS. That the scaledown limit of bipolar technology has already been reached is not entirely a negative indication, since the mechanisms that limits the performance are now studied intensively, and many new techniques to increase this performance limit are emerging.

7.5.3 Merits of Logic Circuits

We summarize briefly the merits of each technology. CMOS technology has the merit of very large scale integration. As scaledown of feature size continues, CMOS will set every milestone in the scale of integration. Although the capability of driving heavy capacitive loads is improving with scaledown, CMOS circuits need very large drivers for an interface. Assembling a large system is difficult for that reason. This problem will be discussed in detail in the next section. Emerging BiMOS technology may play a significant role by providing the design option of using the more desirable ECL logic level interface. CMOS is a relatively slow technology. The present customer base shows that many chips for telecommunication/computer customers cannot be designed using CMOS because of its low speed. To meet the need, a technology that is faster than bulk CMOS and that is directly compatible with CMOS is required. The technology can be either NMOS, SOS CMOS, or low-temperature CMOS. The delay time of the three technologies are about the same. NMOS is the practical choice.

As for bipolar and GaAs MESFET technologies, the following points must be noted. Bipolar ECL/EFL circuits have a number of merits. Figure 16 shows bipolar ECL and EFL logic circuits [15]. The circuits use a differential amplifier as the key element, which has outstanding circuit flexibility—flexibility in achieving the highest speed by trading logic swing for delay, the capability of combining signals in a very flexible way, absence of switching noise, and the ease of chip-to-chip communication. At present, the highest logic speed appears to be achieved by ECL devices, at least at the system level. The recent revival of bipolar device technology has direct impact on the future of bipolar integrated circuits. It is expected that the technology will have applications in the area of high-speed optical fiber communication systems as well. This technology will hold the market in high-end computers, as far as we can see now.

The power of the bipolar ECL gate has been reduced by a factor of 10 (from a few mW to a few hundred μW) in recent years, but the power consumption is still the major problem of bipolar ECL ICs. The power, however, is justifiable to attain high speed. When power is not justifiable, however, bipolar ECL technology must have a compatible alternative, and the role will be played by some form of GaAs MESFET technologies. Figure 17 shows the two dominant versions of GaAs MESFET logic circuits [09], [10]. The logic-level compatibility between ECL and GaAs MESFET technologies is quite fortunate, and this is an advantage that should be exploited to the full extent. Less flexible, relatively small-power MESFET circuits can be complementary to highly flexible but power consuming bipolar ECL circuits. This combination is very powerful. A misleading trend in the development of GaAs technology is to attempt impractically small power to compete in the market that belongs to CMOS technol-

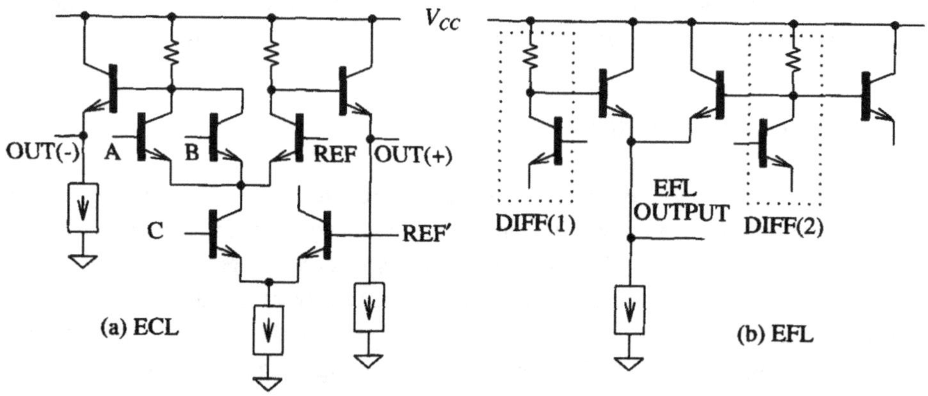

Figure 16 Bipolar ECL and EFL logic circuits

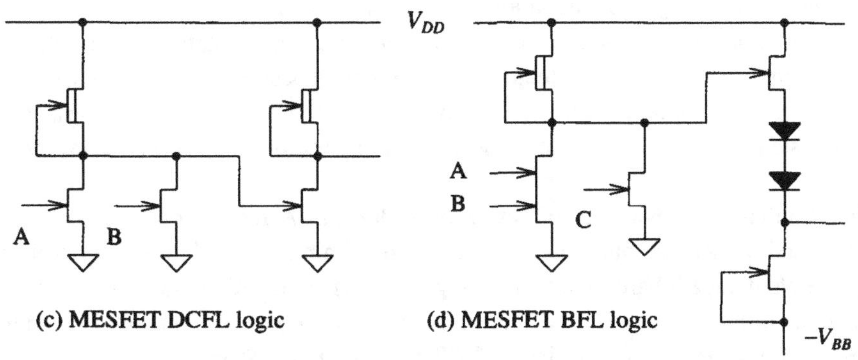

Figure 17 GaAs MESFET logic circuits

ogy. If power-delay tradeoff is to be made, the base of reference should be bipolar ECL, and not CMOS. The development strategy of GaAs and bipolar ECL technologies should be to avoid competition with CMOS technology. The high-speed technologies should be developed with an explicit objective of attaining significantly less than 100 picosecond gate delay in mass-produced chips within five years. No futile competition should be attempted with CMOS in gate delays above 100 picoseconds; this application area belongs to CMOS or CMOS/BiMOS. An interesting new possibility is GaAs bipolar technology using heterojunction emitters. The technology allows integration of MESFET technology and ECL technologies. A chip whose interfaces are ECL, but whose power-consuming internal logic is MESFET circuits is very attractive. This is the GaAs version of BiMOS technology.

7.6 CMOS Circuits and System-Level Performance

The fundamental but subtle decision in a system development project is whether or not CMOS should be used for over-all performance reasons. If this decision is incorrectly made, there will be far-reaching consequences. The crucial issue is the following: CMOS has the highest scale of integration, low power, and logic reliability (noise immunity). CMOS is, however, a relatively slow technology. Slow CMOS speed does not matter in short logic paths, however, and most of the logic paths in a VLSI chip are short. The small number of logic paths that consist of complex logic gates or of many stages of cascaded gates have long delays, and these delays determine system performance. Such logic paths are called critical paths. Whether or not critical paths can be successfully dealt with is crucial in high performance system design.

CMOS circuits have only limited techniques available to speedup critical paths. Logic optimization and fanout ratio optimization are the only general and practical techniques. Fundamental to this limitation is the linear dependence of FET current on gate voltage in scaled-down CMOS. Gate delay is determined from FET discharge time constants only, and it is independent of any other parameter like logic swing. There is almost no tradeoff between delay and power, or delay and silicon area: spending 10 times more power or silicon area provides no significant improvement in delay. This fundamental limitation of CMOS technology has a direct impact on system architecture. CMOS works best if many operations are executed in parallel, using a large number of gates, and at relatively low clock frequencies. A high performance system must be architectured so that one operation is split into many parallel processes.

This problem is relevant to microprocessor design. A microprocessor accepts a binary instruction and generates microinstructions in the chip's control area. Microinstructions directly operate datapath hardware such as registers and arithmetic logic unit, thereby processing data given to the processor. Width N (in number of bits) of the datapath is used to specify a processor (for example, $N = 8, 16$, and 32 bit microprocessors). By increasing the data width, data processing power of a microprocessor increases. The binary instruction that is given to the processor is coded in formats consisting of the same N-bit-wide words. Depending on how the instruction is coded in N-bit word(s), the efficiency of converting a binary instruction into microinstructions is significantly different. Microprocessor binary instructions consist of three components. The operation code (opcode) specifies the type of operation executed (such as ADD, SUBTRACT, AND, OR, JUMP, etc.). Any instruction has an opcode, and some instructions consist of opcode alone (operation that needs no other specification, such as a NOP-no operation). To have enough variety of operations, at least 1 byte of width is required. The second component, addressing mode, specifies the algorithm to find location of the source data, or the location of the destination where the processed data is stored. The address is a binary number. In early 8-bit microprocessors the range was usually from 0 to $(2^{16} - 1)$. In the state-of-art 32-bit microprocessors, the range is from 0 to $(2^{32} - 1)$. The binary numbers specifying the address are stored in an internal register of the processor, or in the memory. An address may be generated through one or more steps of referencing the memory or the register, and by adding an *offset* address to the content of the referenced register. The addressing requires specification of modes (address computation algorithm) and identification of the register to be referenced. If there are 16 addressing modes and 16 registers, one byte is necessary to store one set of addressing information. The offset address, if any, is given with the instruction. The offset is a binary number 8, 16, or 32 bits wide. As we saw, an instruction consists of the part that requires decoding (opcode and addressing mode) and numbers (data directly given, immediate address data and address offset).

Since a complete instruction is at least several bytes long, how to squeeze a binary instruction into a regular format of N-bit words matters essentially to the efficiency of converting an instruction into microinstructions. Figures 18 and 19 show two examples. Figure 18(a) shows *byte-oriented* instructions. Opcode (1) is 1 byte, addressing mode of the source data (2) is 1 byte, and the extention (or offset, 3) is one, two or four bytes (We assume a 32-bit processor). Extention (3) may or may not exist. The second addressing mode (4) is either to specify the second data source, or to specify the destination. An extension may follow the second addressing mode as well. This byte-oriented instruction type is used in DEC's VAX 780 processor. Figure 18(b) shows the block diagram of the decoder of this type binary instruction [16]. The data-bus consists of four 8-bit buses, D1, D2, D3, and D4. A multiplexer selects data on D1 first. Immediately following the reset, D1 must be the opcode of the first instruc-

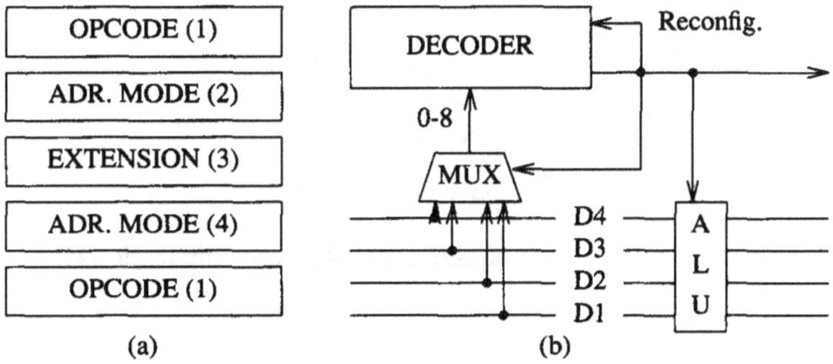

Figure 18 Sequential decoding of opcode

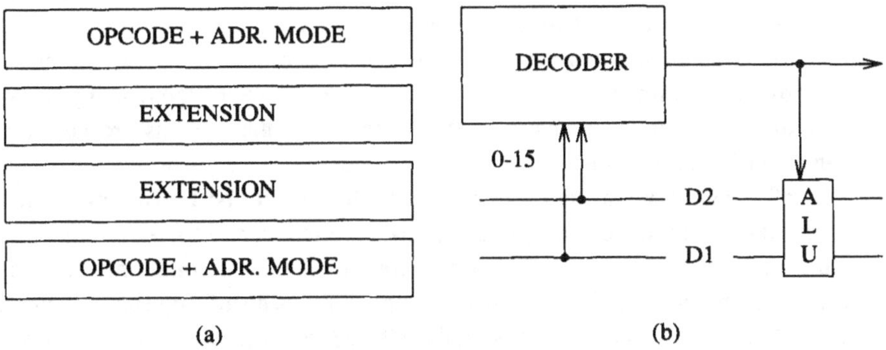

Figure 19 Parallel decoding of opcode

tion. The 8-bit data are gated to the decoder, and the type of operation, and how many inputs are required, are determined. The information determines whether or not the second byte [2 of Fig. 18(a)] is an addressing mode. If so, the decoder reconfigures itself to decode the addressing mode information. Simultaneously the multiplexer selects D2 data. After decoding the D2 data, whether or not the third byte [3 of Fig. 18(a)] is an extension, is determined. As is shown by this example, the instruction is decoded 1 byte at a time. Four cycles of the instruction bus are approximately equivalent to 1 cycle of the main data-bus (D1–D4), which is the instruction cycle. This type of decoding is hardware efficient if the multiplexer and the decoder are built using gates 4 times faster than the rest of the processor. If the processor is built from many integrated circuits on a PC board, it is possible to use fast bipolar gates for the decoder, and slow MOS gates to the rest of the processor. If the entire processor must be built on a single CMOS chip, however, the gates have the same delay everywhere on the chip, and the throughput of the processor is limited by the capability of decoding instructions, not by processing the data. Therefore the instruction format of Fig. 18(a) is inefficient for a CMOS implementation [16].

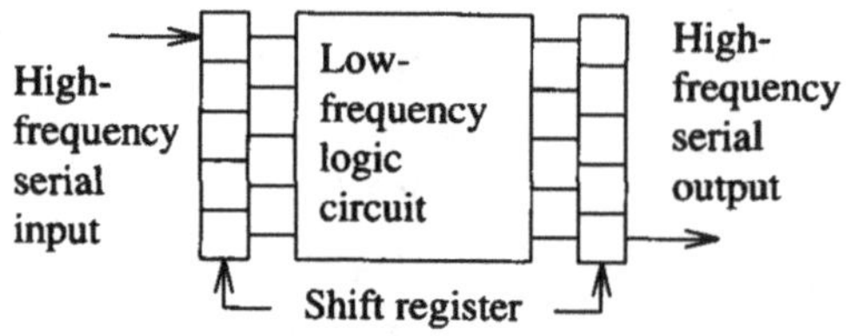

Figure 20 Structure of high-speed data communication ICs

Figure 19(a) shows the instruction format of a 16-bit microprocessor, the MC68000 [17]. Opcode and addressing mode are squeezed in a single 16-bit word, and extensions follow. A single instruction consists of one, three, or five 16-bit words, depending on the number of extensions. In Fig. 19(b), as soon as the first word is delivered to the decoder and decoded, what is to be done about the extension words is determined, and the microinstruction is already available. This simple instruction format allows execution of many simple instructions with an absolute minimum number of machine cycles, thereby contributing to excellent throughput of the processor. A problem of this approach is that 16 bits are not enough to squeeze in an extensive instruction set and addressing modes.

One way of using slow but highly integrable CMOS to the best advantage is to down-convert the data rate to the range manageable by CMOS circuits, and to use multiple data processing circuits at this lower data rate. The architecture of Fig. 20 is commonly used in digital data transmission systems. A high data rate serial signal drives a shift register, and the register sends out a parallel set of lower data rate signals. CMOS data processing is carried out at the lower data rate. At the chip's output, the multiple low data rate signals are converted back to a high data rate serial signal using a similar shift register. In this architecture, the shift register is the crucial circuit that must be fast. Various techniques of designing fast shift register were discussed in Section 5.2.

7.7 Design Problems of CMOS-based Systems

In Section 7.2 we saw that logic circuits in electronic data-processing systems can be classified into two distinct groups using logic amplitude. Logic amplitude is the difference between the logic 1 voltage and logic 0 voltage. In the first group of circuits, logic amplitude is determined essentially by power supply voltage, and the amplitude is equal to, or a large fraction (40% or more) of, the power supply voltage. MOS logic circuits (CMOS, NMOS), bipolar TTL logic circuits, and GaAs MESFET BFL (Buffered FET Logic) are examples of the *large amplitude* logic circuits. The maximum attainable logic amplitude in a large amplitude logic circuit is the power supply voltage, and CMOS is unique because its logic amplitude is the maximum attainable. In the second group of logic circuits, logic amplitude is determined by physics of the semiconductor material used to build the active device. Logic amplitude is approximately equal to the energy gap of the semiconductor material (measured in electron volts) that creates the *nonlinearity*. The *band gap* voltage of commonly used semiconductors (Si, GaAs and Ge) is equal or less than 1.5 volts. If devices and circuits are designed properly, the band-gap voltage is sufficient to create significant nonlinearity in circuit characteristics. Nonlinearity is required to identify the two distinct regions of logic device characteristic that can be assigned to Boolean values 0 and 1. In this type of logic circuit, regions of voltage below and above the large nonlinearity are assigned to Boolean logic values 0 and 1. Bipolar ECL/EFL logic circuits, bipolar IIL logic circuits and GaAs MESFET DCFL (Directly Coupled FET Logic) circuits are the examples of the *small amplitude* logic circuits. Bipolar TTL logic circuits use twice the bandgap voltage, and therefore the circuits stand somewhere between the two groups.

Logic amplitude has significant impact on technology selection. Reliability of CMOS logic circuits (noise immunity) and absence of standby power, both originating from the maximum logic amplitude, allow the largest number of FETs integrated into a single chip. If the entire system can be integrated into a single chip, CMOS is the technology of choice. If the system is much larger than the limit of single-chip integration, and if a large number of chips must be integrated into the system using a *higher level* interconnecting media (like printed circuit boards), the problems of interchip communication becomes the crucial issue. CMOS circuits have a fundamental difficulty in designing efficient interchip communication circuits. The most fundamental reason is the mismatch of CMOS device impedance and the impedance of the transmission media.

A rough estimate of impedance of the active device is as follows. We note that a *typical* sized single gate in any IC chip consumes approximately the same maximum current, I_0, which is approximately independent of technology. This is because of the following reason: Current I_0 charges and discharges capacitance of circuit nodes. Node capacitance depends on the dimension of the node, and that is determined by the minimum feature size of photolithography. For different IC technologies to be competitive in delay, charge stored in the same node capacitance must be discharged at approximately the same rate. The current required to charge-discharge the capacitance, I_0, determines the rate. In this scenario the only difference between the technologies is logic amplitude. Small-amplitude technology needs less time to charge-discharge the capacitance than large-amplitude technology. Then ICs fabricated using the same photolithography must have approximately the same I_0. At the end of the 1980s, I_0 was around 1 mA. The current of CMOS FETs changes by I_0 when the output voltage changes by V_{DD}. Similarly, the current of GaAs MESFETs changes by I_0 when the output voltage changes by V_{SB}, where V_{SB} is the conduction threshold voltage of the Schottky barrier junction. Current of bipolar ECL/EFL emitter-follower drivers changes by I_0 when the output voltage changes by $f \cdot V_{THERM}$, where $V_{THERM} = kT/q$ is the thermal voltage (25 mV at room temperature), and f is a factor of about 2. Output impedance of a gate can be characterized by using the characteristic voltage V_C as

$$V_C = V_{DD} = 5 \ (CMOS) \quad = V_{SB} = 0.75 \ (GaAs\ MESFET)$$

$$= fV_{THERM} = 0.050 \ (BJT\ ECL/EFL)$$

$$Z_{OUT} = \frac{V_C}{I_0} = 5\,kohm\ (CMOS) \quad = 750\,ohm\ (GaAs\ MESFET)$$

$$= 50\,ohm\ (BJT\ \ ECL/EFL)$$

Typical MOSFETs used in a CMOS VLSI chip have two orders of magnitude higher internal impedance than that of conventional signal transmission media like twisted pair wires ($Z_0 = 90\,ohms$) and coaxial cables ($Z_0 = 50\,ohms$). The internal resistance of the output driver must be significantly less than Z_0 to present a constant voltage source to the transmission line. If characteristic impedance Z_0 were several kohms, a typical CMOS driver would be able to drive terminated transmission lines, and therefore the interchip communication problem would not be the concern. Since Z_0 is much lower than that, there is a problem of matching the two impedances for efficient signal transmission. About 100 times the typical size CMOS driver must be used to drive the transmission line. Typical devices in small amplitude technologies (BJTs and GaAs MESFETs) have internal impedance about the same, or only by one order of magnitude higher than that of transmission media, and therefore impedance matching is easier.

The practical problem of CMOS output drivers can be understood from Fig. 21(a). The output driver of chip A sends a CMOS-level signal by driving interchip wire that has capacitance C_L. C_L includes input capacitances of the chips that receive the signal, capacitances of the interconnecting wires, and all the parasitic capacitances of the output driver and the packages. For simplicity, suppose that the signal is a train of square pulses that change state every T_P seconds [clock frequency $f_P = 1/(2T_P)$]. We assume that the length of the wire is much less than $c \cdot T_P$, where c is the velocity of the electromagnetic wave. This is to assume that the voltage of the entire transmission line reaches equilibrium every time the driver switches. The wire may be considered as a capacitive load. Since energy stored in load capacitor C_L charged to V_{DD} volts is $(C_L/2)V_{DD}^2$, and an equal amount of energy is lost in the driver when the capacitor is charged (Section 6.9), power dissipated by output driver, P_O, is given by (Section 6.10)

$$P_O = \frac{C_L V_{DD}^2}{2T_P} = C_L V_{DD}^2 f_P \quad \text{where} \quad f_P = 1/(2T_P)\,. \tag{48}$$

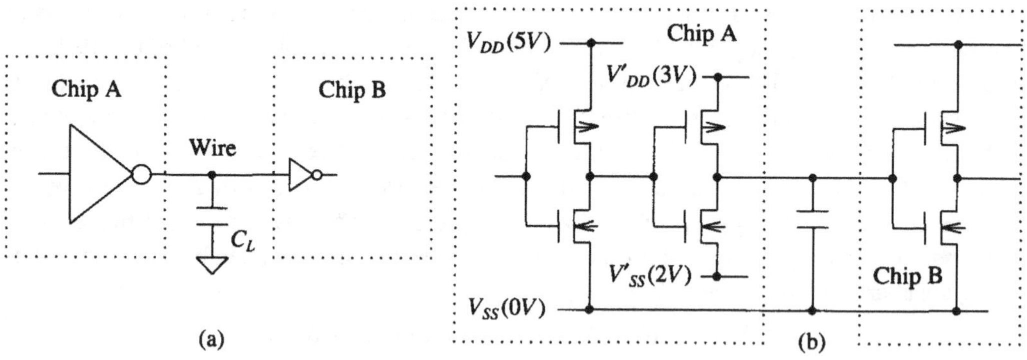

Figure 21 CMOS level interchip communication

In a microprocessor system on a PC board, if standard dual in-line packages are used, and if the interconnect has a realistic number of fanout destinations, $C_L = 150pF$ must be tolerated. If $f_P = 20MHz$, $P_O = 7.5 \times 10^{-2}W = 75mW$. If there are 32 such drivers sending out a set of 32-bit data (like microprocessor's address bus) and if 1/2 of the drivers change state every time, the consumed power is given by $75(mW) \times 32 \times (1/2) = 1.2W$. The output drivers consume quite significant power. Gate delay is decreasing rapidly by scaledown of the feature size of CMOS technology. As far as a chip's internal logic delay is concerned, it is feasible to design 32-bit microprocessors operating with 4-phase 60MHz fundamental clock. If this processor is packaged in a conventional package and integrated with other chips on a PC board, power spent for interchip communication is quite high. CMOS chips used in digital communication systems have clock frequencies ranging up to several hundred megahertz. Power consumed by 5-volt CMOS logic level output drivers is then impossibly high.

CMOS level interchip communication consumes high power because P_O in Eq. (48) is proportional to V_{DD}^2. In order to reduce power, interchip logic amplitude must be reduced. If the power supply voltage of the output driver is reduced by using a pair of lower voltage power supplies as shown in Fig. 21(b), power needed for interchip communication is reduced quite significantly. This interchip communication scheme creates, however, the following problem. Since the voltage swing of the interchip wire is only 1 volt, the input buffer of chip B must amplify the input signal to recover the CMOS logic amplitude, which is 5 volts. This amplification costs time. Furthermore the switching threshold voltage of the DC-coupled input buffer of the receiving chip B must be maintained within the narrow range of the interchip signal voltage swing, and this is a complicated design objective to achieve, due to the process, temperature, and voltage variations. The single wire interchip signal with reduced logic voltage swing is susceptible to external noise (Section 4.15), and the generation of additional DC voltages is costly.

A significant source of noise in CMOS systems is charge-discharge current spikes dumped to the power- and ground-buses by the switching output drivers. Noise current spikes are generated every time when output drivers charge (or discharge) the interchip communication wires. An effective way of reducing sensitivity to the power-bus noise is to use a pair of drivers that send out the signal, and its complement, simultaneously. Then only the difference between the voltages of the two wires, and not the voltage referenced to ground, carries logic information. Interconnect wires of the circuits of Figs. 21(a) and (b) are not impedance matched or *terminated*. If the wire is long (propagation time at the light velocity across the length of wires is longer than the time required for the output driver to change the voltage of the input end of the communication wire), the waveform on the wire is distorted by *ringing*, or multiple voltage changes. Therefore to reduce noise sensitivity and ringing, interchip driver circuits must be a pair of drivers that send out a signal and its complement simultaneously, and the interconnecting pair of wires must be impedance matched and terminated. We need to review the properties of the signal transmission media.

The media of interchip communication can be traces on printed circuit board (which is the microstrip lines), coaxial cable, or twisted pair. They are, in general, *lossy* transmission lines. Figure 22 shows two chips that are connected by such a *transmission line*. A transmis-

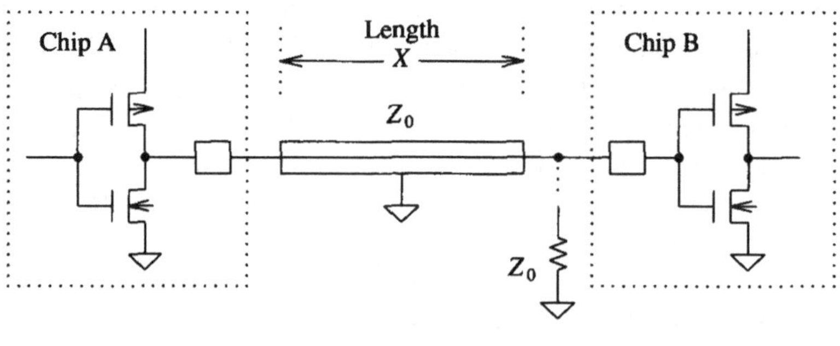

Figure 22 CMOS interchip communication

sion line can be modeled by many sections of capacitance C, inductance L, and resistance R, all measured per unit length of the transmission line [18]. Capacitance and inductance are the essential components of a transmission line, and the resistance means loss, which is a parasitic. When a lossless transmission line having length X is driven from its end by a constant voltage source (V_{DD}), the length of section that is charged up, x, increases with time t as

$$x = ct \quad \text{where} \quad c = \frac{1}{\sqrt{LC}}$$

where c is the velocity of propagation of electromagnetic waves along the transmission line. Since charge into the transmission line, Q, is given by $Q = xCV_{DD}$, current at the driving end of the transmission line is $j = Q/t = V_{DD}/\sqrt{L/C}$. Therefore the transmission line is equivalent to resistance Z_0 given by

$$Z_0 = \sqrt{L/C}\,. \tag{49}$$

Equation (49) was derived assuming that the electrical energy that is fed into the transmission line does not return the driving point. If the transmission line is infinitely long, the signal never returns. Since we assumed that the line has finite length X, the signal returns after round trip delay time given by $2X/c$. After that time, impedance looking into the transmission line is not pure resistance Z_0 any more. If the far end of the line is *terminated* by a resistor having resistance Z_0, however, the line looks as if it were infinitely long at the source.

Suppose we have a terminated transmission line. If a conventional (high internal resistance) CMOS output buffer is to drive the transmission line, amplitude of the voltage wave that is launched on the transmission line is reduced because the internal resistance of the CMOS driver (R_I) and the characteristic impedance of the transmission line (Z_0) work effectively as a voltage divider. The voltage amplitude on the transmission line, V_{TRANS}, is reduced to

$$V_{TRANS} = \frac{Z_0}{Z_0 + R_I} V_{DD}\,.$$

If $V_{TRANS} \approx V_{DD}$ is required to maintain CMOS logic amplitude, R_I must be reduced to quite a low level: Practically, the driver must have internal impedance 10% or less than Z_0. A CMOS driver would then be 100 to 1000 times the minimum size inverter in 1 μm CMOS.

If the left end of an infinitely long or terminated transmission line is driven up to V_{DD} and then driven down to ground by a large CMOS driver every T_P seconds, the power the transmission line consumes is given by

$$P_O = \frac{V_{DD}^2}{2Z_0} \tag{50}$$

which is independent of period T_P or frequency of switching f_P (the factor 2 in the denominator is to include the assumed duty cycle of 50%). In contrast to Eq. (48), the power does not increase with increasing frequency. This is because capacitance of the part of the transmission line more than distance $c \cdot T_P$ away from the driver is ineffective: power given by Eq. (48) has an upper bound given by Eq. (50). This is the power consumed by a voltage source like a low-impedance CMOS buffer driving an infinitely long transmission line. The maximum power is finite but is very high, since V_{DD} is high and Z_0 is low.

To design an ideal interchip communication circuit using only CMOS alone is impossible. One compromise design uses the circuit shown in Fig. 23. In this circuit, CMOS drivers D and $\bar{\text{D}}$ are loaded by resistors R_S and R_D, so that effective output impedances of the drivers are reduced. Since resistances R_S and R_D are in parallel with the output drivers, small (high internal resistance) CMOS drivers can be matched to the impedance of the transmission line. Since both ends of the transmission line are matched, the waveform is clean. Amplitude of the signal is reduced, however, from V_{DD} to a small amplitude, typically 0.5–1.0 volt. The two CMOS drivers send out a pair of complementary signals to a twisted pair transmission line. The signal is received by the chip through a differential amplifier that recovers CMOS logic levels. Amplification, however, costs time.

We have many reasons to use a pair of *reduced amplitude* interchip drivers and terminated transmission lines. The nearly ideal scheme is used in bipolar ECL logic circuits, which

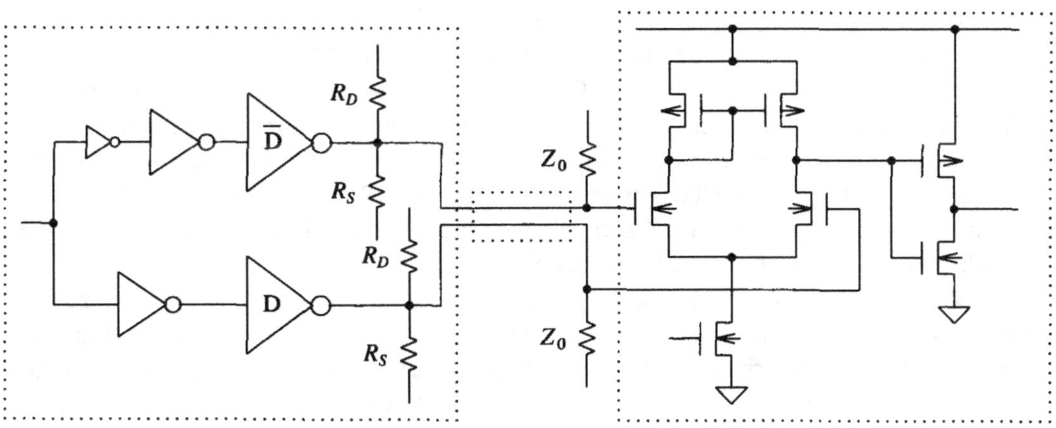

Figure 23 CMOS reduced swing interchip communication scheme

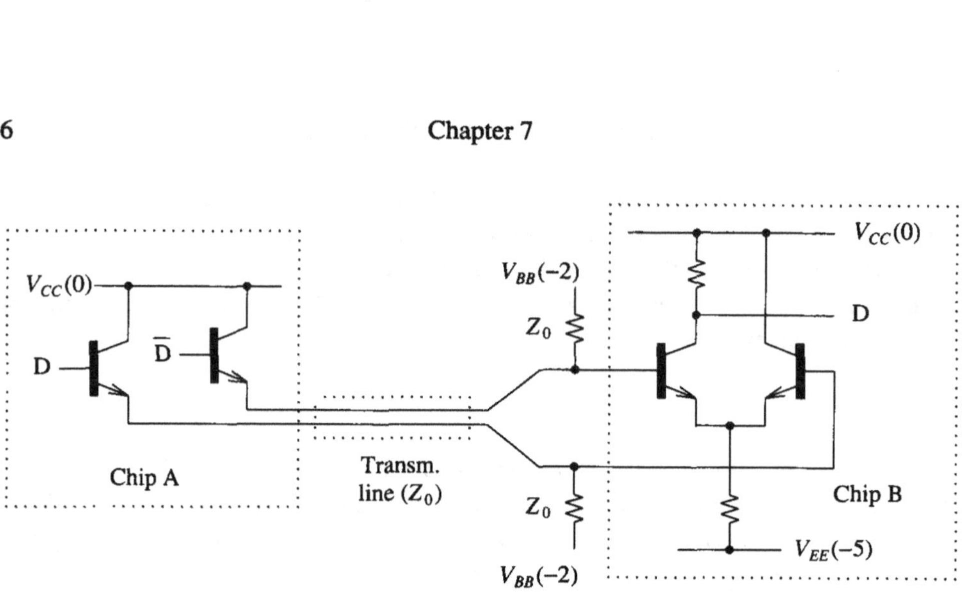

Figure 24 Bipolar ECL logic interchip communication

is shown in Fig. 24. A pair of balanced (differential) transmission lines that connect the two chips are terminated at the receiving end by termination resistors (Z_0). An auxiliary power supply, V_{BB}, (–2 volts), is used to bias the termination resistors, which are the loads of the emitter-follower drivers.

ECL interchip circuits benefit from the advantages of bipolar transistors, which are summarized as follows.

1) BJT transconductance is high. A small base voltage change (only a few times the thermal voltage $V_{THERM} = kT/q = 25mV$ at room temperature) is sufficient to switch a bipolar transistor between the conducting and nonconducting states.

2) Output impedance of an emitter follower, R_O, is given by

$$R_O(ohms) = \frac{V_{THERM}}{I} = \frac{25(mV)}{I(mA)} \tag{51}$$

and if $I = 2\,mA$, $R_O = 12.5\,ohms$. Transmission lines of ordinary impedance level ($Z_0 = 50 - 100\,ohms$) can be driven by an emitter follower driver.

3) BJTs in integrated circuits are better matched than MOSFETs (V_{BE} of BJTs are matched to $1\,mV$, while V_{TH} of MOSFET is matched only to about $10\,mV$). Furthermore MOSFETs suffer from the back-bias effect. Better input interface circuits (differential amplifier) can be designed using BJTs than using MOSFETs. It is a recent trend to upgrade CMOS technology by adding compatible bipolar NPN transistors [19]. Then the I/O interface circuit of Fig. 23 can be replaced by the ECL circuit, shown in Fig. 25. The circuit of Fig. 25 is compatible with other ECL chips, and interface delay is short.

Equation (48) shows that the power required in CMOS logic-level interchip communication is reduced if interchip capacitance C_L is reduced, for instance, by removing the chip package and extensive interconnects. Interchip capacitance can be reduced if the chip is attached directly to a high-level interconnect media that is significantly smaller than conventional PC boards [20]. Recently, advanced packaging technology that provides dense high-level interconnecting media as shown in Fig. 26(a) has been developed. Silicon chips have *area* contact

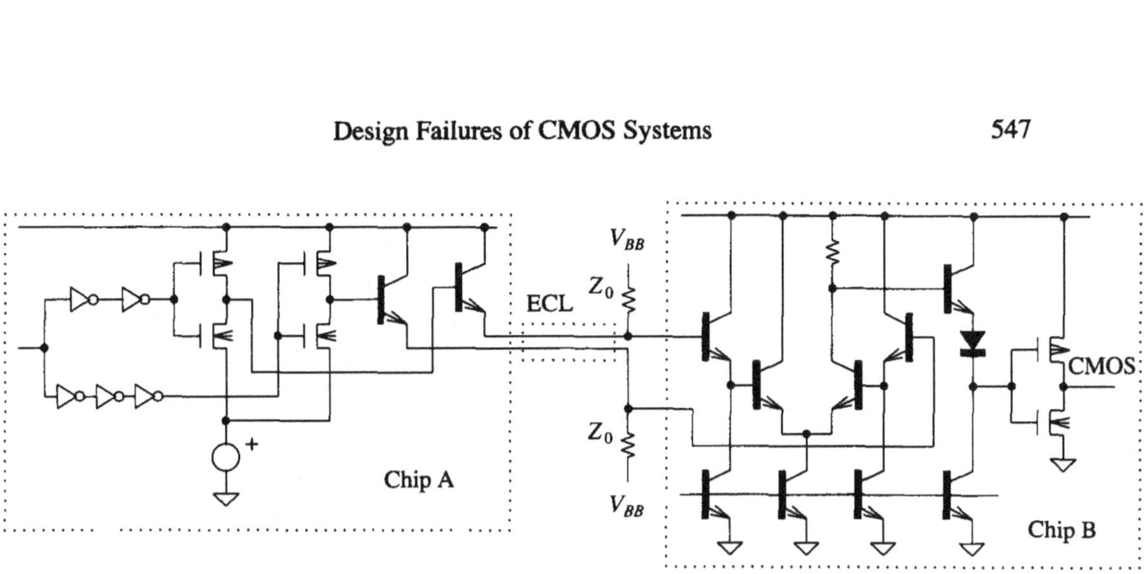

Figure 25 BiMOS interchip communication circuit

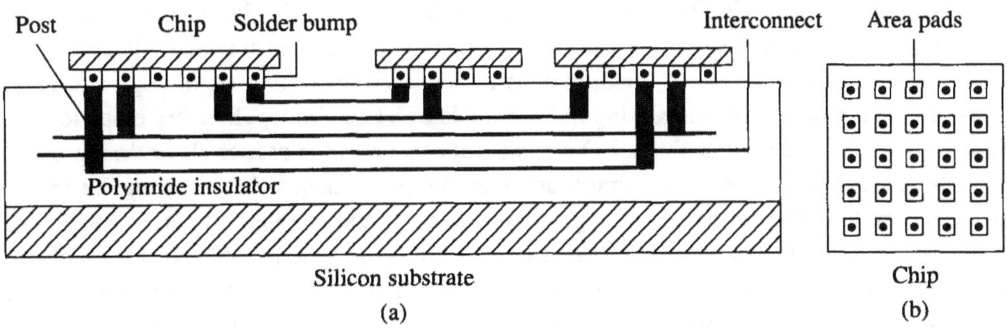

Figure 26 Advanced packaging interconnect media

pads shown in Fig. 26(b). Silicon chips are soldered face down on the interconnecting media. This silicon interconnecting media is fabricated on a wafer-size silicon substrate using the same microfabrication technology as used in silicon chips. This technology allows very high wiring density. The chips can be placed close together, thereby reducing system size and interchip capacitance quite significantly. Multilevel wires in the interconnecting media may be insulated by organic materials like polyimide, which has a small dielectric constant and has capability of planarization of multilevel wiring. Interchip capacitance can be reduced to several pF, which is an order of magnitude less than that of PC board circuits. In this structure, CMOS logic level interchip communication is practical.

Narrow interconnecting wires on the advanced packaging substrate have significant series resistance, which can be used beneficially. Waveform distortion of interchip signals by reflection and ringing can be reduced to tolerable levels if the total parasitic series resistance of *unterminated* wires is about equal to the characteristic impedance of the wire. This will be discussed in the next section.

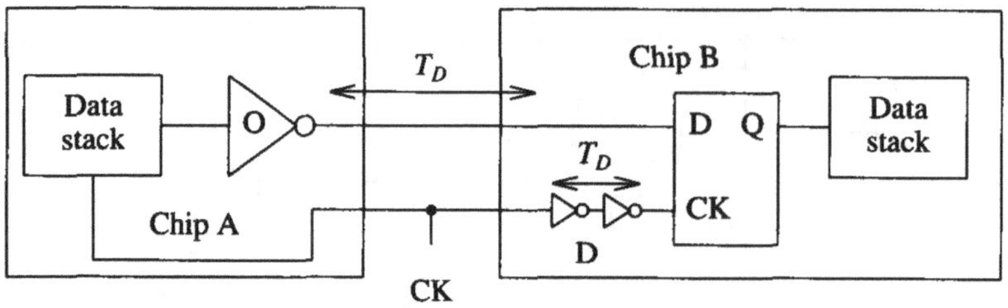

Figure 27 Serial interchip communication

A MOS system assembled on advanced interconnecting media looks like a large monolithic CMOS chip if internal clocks of all the chips on the media are well synchronized. This is an interesting new trend in CMOS technology that makes the best use of the merits of CMOS. A preliminary estimate shows that integration of a main-frame computer system is feasible using 0.5 μm CMOS technology and the advanced packaging technology. Such a gigantic *chip* is, however, very design intensive, and this narrows the area of application.

Two-dimensional arrays of bonding pads in advanced packaging technology are used to resolve the problem of ever-increasing I/O terminals of VLSI chips. There are other solutions to this problem. If a CMOS chip's I/O timing is the same as the master-slave timing used in the internal logic of the chip, I/O circuits are able to input or output data only one bit per port per clock cycle. State-of-the-art microprocessor clock period is about 50 nanoseconds. This scheme is then inefficient, because scaled-down CMOS drivers are able to transmit pulses whose width is 5 nanoseconds or less if the port's capacitive load is small. If many bits of data must be transmitted from one chip to another, and if only a small number of I/O ports are available, the scheme shown in Fig. 27 can be used. Multibit data are sent out from chip A to chip B. The data are originally stored in the data stack of chip A, which is a circuit similar to a shift register. A high-speed external clock, CK, controls the data stack. Data arrives at chip B at time T_D after the clock edge. The input latch clock of chip B is delayed by T_D through clock delay buffer D. In this arrangement, CK may have many times higher frequency than the fundamental clock of the chip, and the number of I/O pins can be reduced significantly. Techniques of maintaining such a high frequency clock at low skew and in good synchronization with the system's clock are now well established. Only one clock, CK, is needed for many pairs of such *serial* I/O ports.

7.8 Interchip Communication in CMOS Systems

The characteristic impedance of ordinary transmission lines is in the range of 50 to 100 ohms. If the end of the transmission line is terminated by a matched resistor, the characteristic impedance and the driver's internal impedance form a voltage divider, and the voltage swing of the driver is attenuated. Since reduced signal amplitude is not desirable, we can remove the termination resistor. If the resistor is removed, however, reflections from the far end of the transmission line deform the signal waveform. The entire line may go into multiple *ringing*. In order to control ringing, resistance in series with inductance of the transmission line works effectively. Such resistance exists naturally, since any transmission line has parasitic resistance. Ohmic loss of the transmission line can be taylored accurately to suppress ringing, thereby optimizing the response of the transmission line. The transmission line shown in Fig. 28(a) can be approximated by the one-stage section shown in Fig. 28(b). Series-connected r, l, and c are related to R, L, and C (per unit length) of Fig. 28(a) by

$$r = r_S + r_D \quad r_S = XR \quad l = XL \quad \text{and} \quad c = XC + C_L$$

where C_L is the input capacitance of the receiver, and where X is the length of the transmission line. Current through the series resonant circuit, I, satisfies the equation

$$l\frac{d^2I}{dt^2} + r\frac{dI}{dt} + \frac{I}{c} = 0. \tag{52}$$

If we assume a solution of the form $exp(St)$, S is given by

$$S = \alpha + \beta i \quad \text{where} \quad \alpha = -r/2l \quad \text{and} \quad \beta = \pm(1/2l)\sqrt{4(l/c) - r^2}.$$

Critical damping is the condition beyond which ringing (or multiple voltage swing) takes place. Critical damping is reached when S has a double root. We have

$$r = r_S + r_D = 2\left(\frac{l}{c}\right)^{1/2} = 2\left(\frac{XL}{XC + C_L}\right)^{1/2} \rightarrow 2\left(\frac{L}{C}\right)^{1/2} \quad (X \rightarrow \infty). \tag{53}$$

If r_S is adjusted to the critical value, the transmission line is in the state of critical damping that transmits good digital waveform with minimum delay [21]. Equation (53) shows that at the critical damping, *total* resistance of the transmission line is twice the characteristic impedance. The current that flows through the series-connected l, r, c circuit is given by

$$I(t) = \frac{V_{DD}}{\beta l} exp(\alpha t) \sin \beta t \rightarrow \frac{V_{DD}}{l} t \quad (t \rightarrow 0).$$

The total energy dissipated by the circuit is given by

$$P_{TOTAL} = \int_0^\infty P(t)dt \quad \text{where} \quad P(t) = rI(t)^2$$

is the power dissipated at time t. By carrying out integration we find that $P_{TOTAL} = cV_{DD}{}^2/2$,

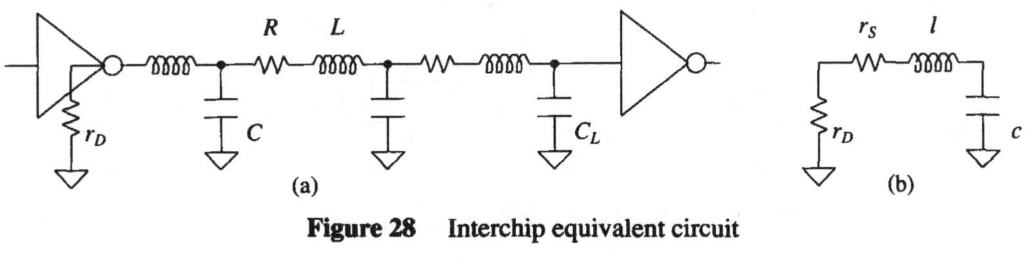

Figure 28 Interchip equivalent circuit

which is the energy stored in the capacitor. Most of the energy is dissipated as heat in the resistance, but some is radiated into space (radiation resistance, Section 6.10).

Numerical analysis of a lossy transmission line in an advanced packaging substrate can be carried out as follows. Figure 29 shows the equivalent circuit of a lossy transmission line that is divided into N equal sections. Each section consists of series inductance L, series (loss) resistance R, and parallel capacitance C. Capacitance C_L at the right end of the transmission line represents input capacitance of the CMOS receiver chip. The transmission line is *not* terminated by a matched resistor. The transmission line was originally charged to V_{DD}, and the driver at the left end of the line (represented by *internal* resistance R_D) discharges the line. The design objective is as follows: By properly choosing values of R and R_D, the voltage at the right end of the transmission line should have no overshoot or excessive resistance-capacitance delay.

The following algorithm is convenient for non-specialists of numerical analysis. Voltage of i-th node, $V_i(t)$, and current from node $i-1$ to node i, $I_i(t)$, are defined as shown in Fig. 29. $V_i(t)$ and $I_i(t)$ satisfy

$$L\frac{dI_i(t)}{dt} + RI_i(t) = V_{i-1}(t) - V_i(t) \qquad C\frac{dV_i(t)}{dt} = I_i(t) - I_{i+1}(t). \tag{54}$$

It is convenient to normalize Eq. (54) by using

$$t = \sqrt{LC}\theta \qquad V_i(t) = V_{DD}\phi_i(\theta) \qquad I_i(t) = \frac{V_{DD}}{Z_0}J_i(\theta)$$

where $Z_0 = \sqrt{L/C}$ is the characteristic impedance of the transmission line. We have

$$\frac{dJ_i}{d\theta} = \phi_{i-1} - \phi_i - \frac{R}{Z_0}J_i \qquad \frac{d\phi_i}{d\theta} = J_i - J_{i+1}. \tag{55}$$

We write current, J_i, and voltage, ϕ_i, at time $\theta + \Delta\theta$ in terms of the current and the voltage at time θ as

$$J_i(\theta+\Delta\theta) = J_i(\theta) + J_i'(\theta)\Delta\theta + \frac{J_i''(\theta)}{2!}(\Delta\theta)^2 + \frac{J_i'''(\theta)}{3!}(\Delta\theta)^3 + o(\Delta\theta)^4$$

$$\phi_i(\theta+\Delta\theta) = \phi_i(\theta) + \phi_i'(\theta)\Delta\theta + \frac{\phi_i''(\theta)}{2!}(\Delta\theta)^2 + \frac{\phi_i'''(\theta)}{3!}(\Delta\theta)^3 + o(\Delta\theta)^4$$

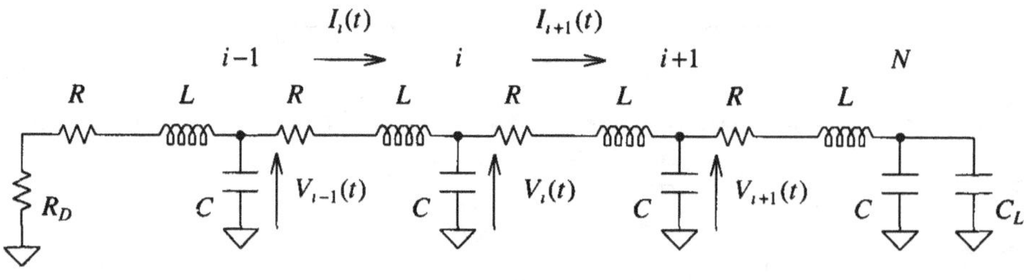

Figure 29 Equivalent circuit of advanced packaging interconnect

and determine the derivatives from $J_i'(\theta)$ to $\phi_i'''(\theta)$ in terms of voltages and currents at time θ. From Eq. (55)

$$J_i''(\theta)=\phi_{i-1}'-\phi_i'-\frac{R}{Z_0}J_i'.$$

ϕ_{i-1}', ϕ_i' and J_i' are substituted from Eq. (55) as follows:

$$J_i''(\theta)=J_{i+1}+J_{i-1}-2J_i+\left(\frac{R}{Z_0}\right)^2 J_i-\frac{R}{Z_0}(\phi_{i-1}-\phi_i). \tag{56a}$$

The other derivatives can be found by similar manipulations. For nodes i $(2 \le i \le N-2)$ we have

$$\phi_i''(\theta)=\phi_{i+1}+\phi_{i-1}-2\phi_i-\frac{R}{Z_0}(J_i-J_{i+1}) \tag{56b}$$

$$J_i'''(\theta)=-\phi_{i+1}+3\phi_i-3\phi_{i-1}+\phi_{i-2} \tag{56c}$$

$$-2\frac{R}{Z_0}(J_{i+1}+J_{i-1}-2J_i)+\left(\frac{R}{Z_0}\right)^2[\phi_{i-1}-\phi_i-(R/Z_0)J_i]$$

$$\phi_i'''(\theta)=-J_{i+2}+3J_{i+1}-3J_i+J_{i-1} \tag{56d}$$

$$-\frac{R}{Z_0}(\phi_{i+1}+\phi_{i-1}-2\phi_i)+\left(\frac{R}{Z_0}\right)^2(J_i-J_{i+1})$$

For nodes $i=1$, 2, $(N-1)$ and N, a similar formula can be derived. It is possible to include still higher derivatives of time, but up to the third order is usually enough to give reasonable simulation accuracy. If only the first order derivative is included, the simulation results may oscillate in the vicinity of a steep transition. Using Eqs. (56a)–(56d) and a similar formula derived for the nodes close to the two ends, and assuming the following initial conditions

$$J_i(0)=0 \quad \phi_i(0)=1.0 \quad (1 \le i \le N)$$

evolution of the node voltages and the currents can be determined by repeating the numerical integration.

Figure 30 shows voltage waveforms at the right end of the transmission line (the receiving end). The parameter value in the figure is $r=R/Z_0$ (normalized series resistance), and the other simulation parameters are given in the figure. When $r=0.03$, total resistance of the transmission line is $rN=1.5$. The transmission line is approximately critically damped. If r is smaller (r = 0.0025), the transmission line goes into damped oscillation. If r is larger (r = 0.1), the transmission line is overdamped, and delay time increases with increasing r. T_D of Fig. 30 is the delay time of a signal propagating from the left end of the transmission line to the right end by the velocity of light. This time is given by

$$T_D=\sqrt{LC}\cdot N$$

Evolution of potential profile within the transmission line for underdamped, critically damped and overdamped cases are shown in Figs. 31(a), (b), and (c), respectively. If the transmission line is underdamped, a step voltage wave propagates to the right, is reflected by the receiving

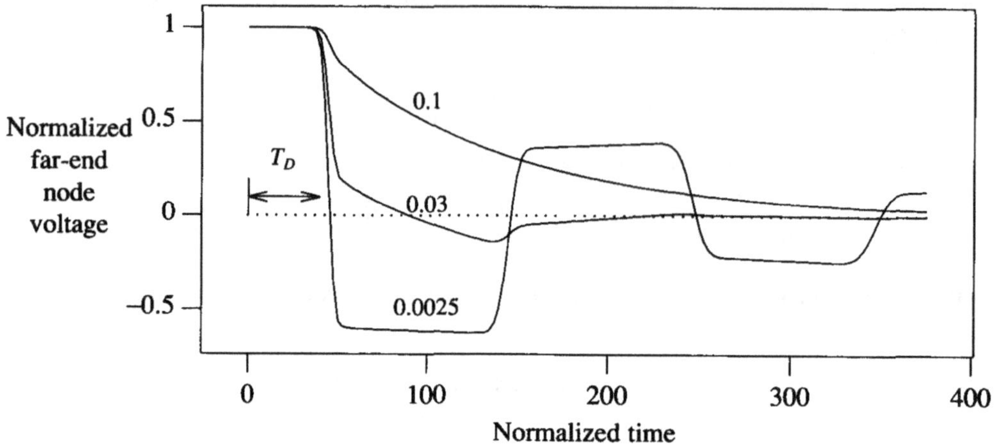

Figure 30 Evolution of node voltage versus time
$r_d = (R_D/Z_0 = 0.3, N = 50$ stages, and $\Delta\theta = 0.25$

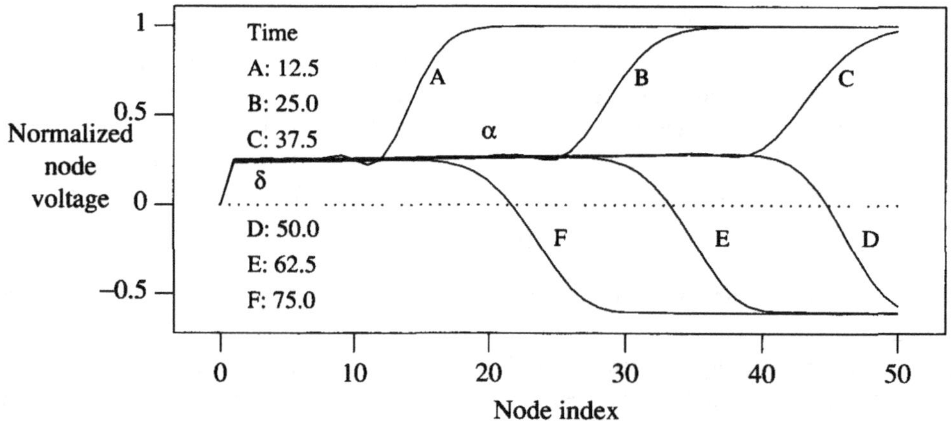

Figure 31(a) Evolution of node voltage
$r = 0.0025$, $r_d = 0.3$, $N = 50$ stages, and $\Delta\theta = 0.25$

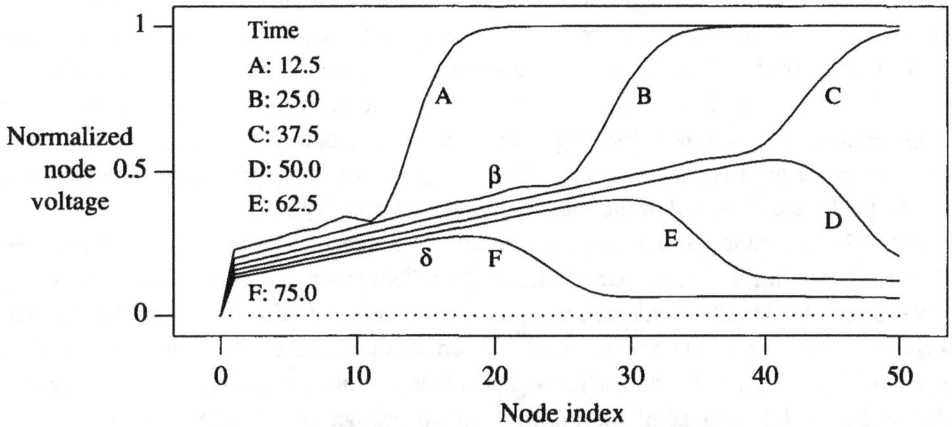

Figure 31(b) Evolution of node voltage
$r = 0.030$, $r_d = 0.3$, $N = 50$ stages, and $\Delta\theta = 0.25$

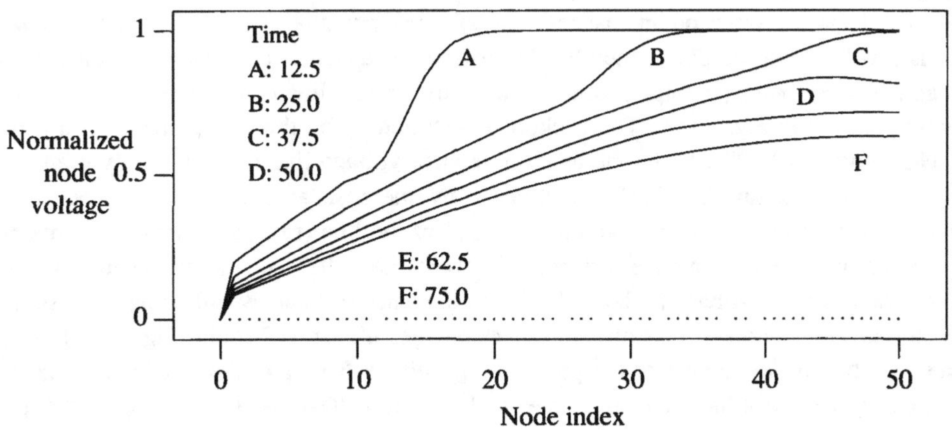

Figure 31(c) Evolution of node voltage
$r = 0.1$, $r_d = 0.3$, $N = 50$ stages, and $\Delta\theta = 0.25$

end with inverted polarity, and then returns to the driver's end as shown in Fig. 31(a). A small voltage step at the driving end (δ) is the voltage developed across the CMOS driver, whose internal resistance is R_D. The potential profile has a conspicuous *backbone*, the straight line labeled α, that has small slope. The backbone shows a slowly varying discharge mechanism based on resistance-capacitance discharge, and the small slope in Fig. 31(a) reflects small series resistance of the transmission line. The voltage wavefront *slides* along the backbone, and is reflected by the right end of the line, with reversed polarity.

If the series resistance of the transmission line is increased, the potential profile evolves with time as shown in Fig. 31(b). Amplitude of the reflected wave is reduced, and the voltage developed across the length of the transmission line becomes significant. Therefore the voltage of the receiving end of the transmission line remains positive all the time. The backbone of the potential profile, the bundle of lines β, is clearly visible, and they have steeper slope than that in Fig. 31(a) because of the higher series resistance of the transmission line. Discharge by the diffusion mechanism is now clearly operating, and therefore the single *backbone* of Fig. 31(a) breaks into a bundles of curves. When the series resistance is increased still further [Fig. 31(c)] large voltage is developed within the transmission line by resistance, and the potential profile evolves in a similar manner to an RC chain, rather than an LC chain. The backbone itself is now broken down entirely, as shown in Fig. 31(c).

Figures 31(a)–(c) show the following mechaniasms operative in a lossy transmission line. A wave in a lossy transmission line has internal *reference* potential profile we called *backbone*, which is a voltage profile determined by the resistance-capacitance discharge mechanism of the chain. The *reference* potential profile is shown by straight line α in Fig. 31(a), or a bundle of straight lines β in Fig. 30(b). The backbone is shifted up by the voltage developed across the CMOS driver (δ). The backbone or the reference voltage line decreases only gradually: this is the transmission line discharge process by the resistance-capacitance mechanism. Amplitude of the wave reflected from the right end of the line (from an open-end in inverted polarity) approximately equals the difference between the voltage of the propagating voltage front and the reference voltage at the end. If the difference becomes small, there is no significant reflected wave, and the transmission line discharges like an RC chain [Fig. 31(c)]. If the difference between the voltage of the propagating voltage front and the reference voltage is large, the reflected wave has large amplitude, and the line voltage oscillates. Critical damping occurs when reflection is not large enough.

Before concluding this section, the following related problem is considered. If a CMOS VLSI chip is packaged in a conventional pin-grid array (PGA) package having a large number of pins (usually over 100), not all the bonding wires have minimum length. To reduce power-bus noise, ground and V_{DD} pins are assigned to the package terminals that are closest to the chip. Since many power/ground pins exist on high-frequency chips, power/ground inductance can be made small. Ordinary signal interconnects have longer bonding wires and even longer package traces to the pin, and they have large inductance. The difference can be quite sifgnificant: power/ground pins can have 2–3 nH per bonding wire, but signal wires may have 10–15nH. Inductance of the output driver circuit cannot be reduced below a certain limit. The equivalent circuit of the output driver is shown in Fig. 32. Then if

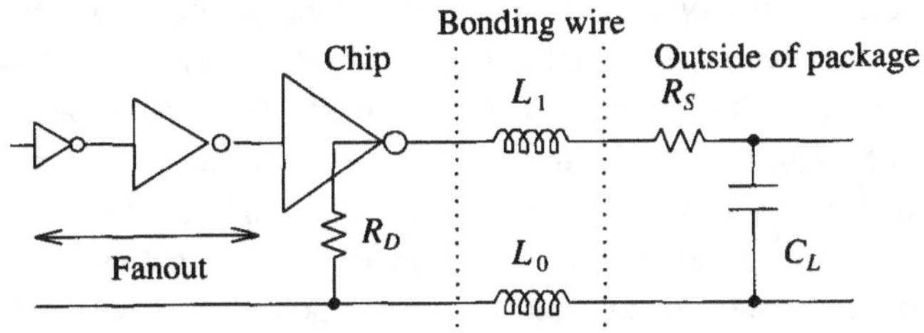

Figure 32 Ringing of bonding wire

$$R_S + R_D < 2\sqrt{(L_0 + L_1)/C_L}$$

the circuit is underdamped, and ringing occurs. The problem is aggravated if process, temperature, and power supply voltage variations are included.

The best way to control this problem is to use as small an output driver as possible that meets design specification. The logic circuit that feeds the signal to the driver is designed carefully to minimize delay, and the delay time saved there is given to the driver stage. It often helps to insert a small, process-independent resistor in series with the output terminal. Two bonding wires side by side are often used in exceptionally large drivers (like clock buffers) to reduce inductance. If the problem is really serious, the last stage of the output buffer is split into two parts, and one part is tristated after the external load is driven close to the final voltage level.

7.9 Design Methodology — Large and Small System Cultures

In this section we change our subject and discuss problems related to design methodology and design project control. The issue is how to carry huge design projects to successful completion, and with minimum expense of resources. The author wishes to point out that there is a significant difference in design methodology used by small-system customers (who design and use monolithic microprocessors, for instance) and large-system customers (who assemble large mainframe CPUs or large electronic telephone switching machines, primarily from non-VLSI integrated circuits). Small-system culture is a recent development, and it is characterized with optimism for an ever increasing scale of integration, adoption of high-level software design aids, bold adoption of new architectures, and strong interdisciplinary cooperation.

Large-system culture is characterized with conservatism: increasing scale of integration is not an unconditional positive attribute of technology if not accompanied with design flexibility allowed for the system architects. Simulation of large systems is more difficult than simulation of VLSI chips, and the designers rely more on hardware experiments than on simulation. Use of ICs having smaller scale of integration than VLSI microprocessors makes hardware experiments easier. Architectural flexibility is restricted because many large systems have to be retrofitted into the existing system environments, often maintaining software compatibility. Systems are assembled from a huge number of individual custom chips. Cooperation between system architects and circuit engineers is less frequent. Interaction between architects and semiconductor process engineers, a common event in microprocessor development projects, is practically non-existent. System architects start, execute, and complete development projects practically unaided: the IC designer's role is secondary, often invisible.

Small-system culture developed rapidly with the advent of CMOS VLSI technology in the 1980s. In addition to process, logic design, circuit design, and testing research, the method of organizing chip design project — the so called design methodology — was investigated intensely during the last decade. At present, every VLSI development organization has its own established and fine-tuned design strategy. It is unfortunate that design methodology of *large* systems, which should be able to take full advantage of the progressing VLSI technology, has not been developed as vigorously as VLSI chip design methodology. It is time that IC engineers consider the best way to help architects of large systems in order to exploit the best performance of VLSI technology. Here we wish to discuss a few key issues of this problem.

Any large system customers, be they mainframe computer customers or telephone switching machine customers, demand a high degree of flexibility in controlling their chip design, most notably chip delivery schedule. This demand was considered so important that practically all large systems are assembled from gate-array or standard-cell like devices that are automatically designed from logic specifications without intensive optimization of the circuits. The problem is, however, that those ICs are not the top of performance in any (CMOS, ECL, or GaAs MESFET) technologies. In scaled-down CMOS, the difference in performance of optimized and unoptimized chips is quite large. In the case of bipolar ECL or GaAs MESFET technologies, the performance lost by not optimizing chip performance is not so great as in CMOS. Two problems in CMOS — difficulty in interchip communication and that high-performance CMOS chips always use circuit design intensive dynamic logic — make CMOS not conform with the present large-system customer's needs. Bipolar ECL ICs and GaAs MESFET ICs, however, can never reach the scale of integration that CMOS VLSI is able to reach. Therefore we came to a contradiction, namely, that high speed and a high degree of system integration are incompatible in this particular sense. This problem can be resolved only if CMOS circuit engineers are able to offer computer architects new design methodology that allows them to assemble a large system using high-performance circuits. This is a new direction that must be actively investigated. A few fundamental requirements are as follows:

1) Assuming logic change is a fact of life, it must be considered how to adapt CMOS circuit design to constantly changing logic design specifications.
2) It must be considered how to determine the best circuits to implement the system, and how to interface components built using different technologies.
3) When logic design *intention* is given, it must be considered how to implement the function using the best and the fastest circuit with the absolute minimum design time.

These problems have no solution at present. If this objective were somehow achieved, we would have truly amazing data processing systems.

7.10 Project Organization Problems

Circuit failures are not the only cause of failure in VLSI projects. If a custom VLSI chip is designed by a group of designers, there is a problem of managing the design group. VLSI chip development is a large-scale development project involving staffs from diverse disciplines. The modes of project organizational failures depend on the culture of the development organization. Organizational failure originates from the huge size of projects. Since VLSI chip development is dedicated to attain the same final objective as designing and producing a significant fraction of a large system such as a computer, there are many aspects that are common to any large system project. The most fundamental similarities between the two are that both projects must seek *over-all* success of the entire project, rather than spectacular success of some of the subprojects while allowing failure in others. This viewpoint is applicable to every level of the project. Therefore VLSI chip development should be organized in a flexible structure.

It is always difficult to estimate the workforce required for a large development project. Since VLSI chip development demands highly skilled workers, the number of project workers is never an accurate measure of the workforce. Talent, and especially accumulated design experiences of the project members must be taken into consideration. One experienced designer is equivalent to *at least* three to five inexperienced designers. Some experienced designers prefer to work alone, with only minimum interaction with the rest of the project members. Some others prefer to unload "dog-works" to junior members of the team, to make time for themselves for thinking at a higher level. Such individual preferences must be considered as the most important matter in project management. Generally, a small number of experienced designers produce much more and much better designs than a large number of inexperienced designers. The reason for this difference is that experienced designers know how to solve design problems efficiently by using ingenious shortcuts, and furthermore, communication between experienced designers is efficient. Project management must understand that experienced designers can be easily demoralized by bad working environment. In the hard, long working hour environment, humane concern to the working members is the basic requirement of project management. As far as I know, only those managers who are obviously concerned with this issue seem to have success in the long run.

Underestimating workload results in project delays. There is a certain time limit beyond which maintenance of design team morale becomes difficult (designers have rumor machines, and they are very sensitive to the reputation of their project). A development project must be finished, at least up to the point of the first working model, within two years. If longer time is mandatory, the initial project phase (that defines design objectives) should be carried out by a loosely organized, unofficial group.

When a large number of individuals handle a large data base, the following problems inevitably creep in. Not all the design information is filed in the central site, like a computer database. This is especially true for designer's intents: why certain circuits were chosen, why certain layout was used, etc., are often kept within the designer's memory and are never documented. Therefore if an experienced designer is lost, his part of the design information is essentially lost. It is vital to keep experienced designers working in the project, at least until the first working model is produced. This viewpoint is consistent with the idea that a development project cannot be carried out for longer than a certain maximum in the present social climate of U.S. industry.

Because of huge design database in a computer, and because of the interactive design with customers who often change requirements and specifications, the design database can be lost and destroyed by updating. Some mechanism of maintaining a historical record of the development project and of verifying the current database in *historical perspective*, even under most complicated circumstances, is required.

It is often necessary to educate customers, so that the customer-designer interface does not deteriorate into a master-slave relationship. Although some specification changes inevitably occur in the course of VLSI design project, the difficulties and problems that chip designers must cope with due to specification changes must be explained to the customers, and such incidence should be minimized. Furthermore, a judgment on the side of the design manager to evaluate whether or not the customer's initial design concepts and business plans are solid, stable, and mature is a prerequisite to a successful project. The designer-customer relation is bidirectional. For example, I experienced the following: How strict the customer's specifications are depends on how confident the customer is of the chip design organization. When an unestablished design organization works for a skeptical customer, it is impossible to design practically anything (skeptical customers multiply *guardband* specifications). Ironically, the author, as a chip designer, experienced that a desperate customer is always the most cooperative, and over-all success is greatest because of the mutual sympathy.

A design team must have a conservative design practice. Too aggressive designs, risky designs, and unproven circuits must be avoided. Stress must be placed on creation of a VLSI chip using as many proven techniques as possible and by executing meticulous design work in detail. Although design errors can be weeded out by verification software, the design team must maintain high morale to avoid sloppy design. It is necessary to set up design quality standards to avoid obviously sloppy design, and in some instances too meticulous a design does not pay for the effort. Practically, design quality depends on the nature of compromise among technical, economical, and psychological factors. Although some compromise is necessary, too much compromise in design must be avoided by design review mechanisms.

The major workhorse in VLSI design is CAD. There are many problems that originate from CAD. The most serious problem is compatibility between programs and design data, and, in addition, insufficient computing power. As for incompatibility, it is best not to use logic or gate circuits that software cannot handle. A blind trust in the capability of CAD usually results in a disaster. *Dry runs* of software before use is absolutely necessary in design. If a CMOS VLSI chip is assembled only from CMOS static logic, Domino CMOS logic, and dynamic PLAs, many existing software products understand the circuits. Multi-output gates, like carry generation trees, are often too difficult for software to deal with. Gate-level timing simulators do not have the capability of modeling extensive delays of resistive polysilicon wiring. The best solution is to use two-level metal CMOS technology to make simulation accurate. The number of possible design iterations depends on the run-time of the design verification software (usually all-chip gate-level timing simulator). The cost can be prohibitively high. Therefore, it is important that the design that is simulated by verification software be as close to the final design as possible.

We examined a chip development project highlighting problems that required the manager's attention. Technically a VLSI chip development project goes out of control when any of the following *symptoms* emerges: (1) chip size becomes too large, (2) chip timing goes out of control, (3) the chip's viability is questioned because of too many freaky designs or too many aggressive designs, and (4) because of too many changes and too sloppy design data management, designers do not know what they are designing or have designed. To avoid design failure, these symptoms must be watched closely by tightly under the project managers, and therefore a strategy to administrate a development project becomes important. The so-called top-down design method is a systematization of this strategy [22].

7.11 Top-down VLSI Chip Design Methods

Figure 33 shows a flow chart of the *top-down* design methodology for VLSI microprocessor. Starting design data are architectural specifications of the chip that are enclosed in dotted box A. These specifications come from the customer, who is not familiar with the problems of CMOS circuits on silicon chips. Therefore original design concepts may not be suitable for

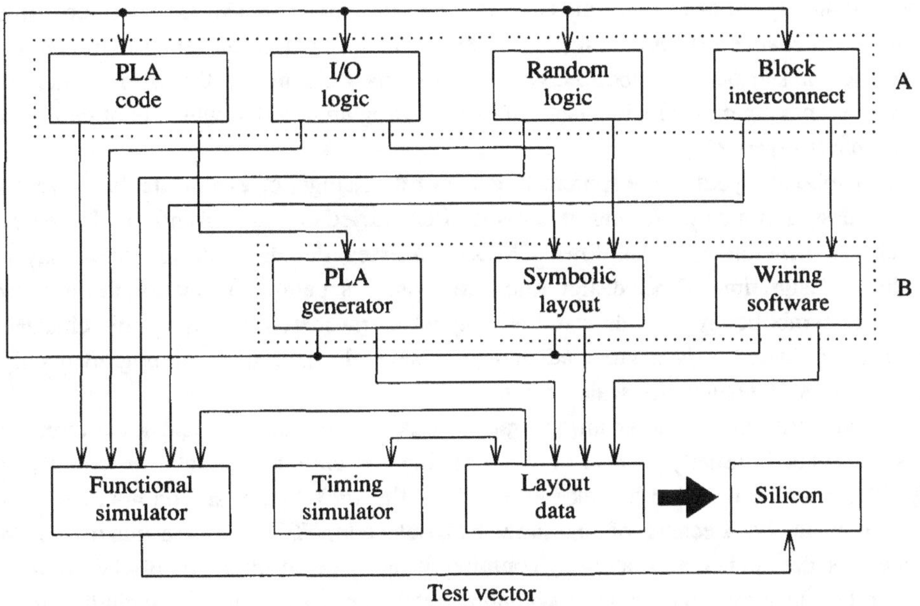

Figure 33 Top-down design method

silicon implementation. For instance, the logic operation and the required clock frequency of the chip may not be compatible. Although large-scale modification of architectural concepts by circuit designers will never be accepted by the customer, many small changes can be made. The other common problem is that the required hardware may not fit in a silicon chip of a size having practical yield. The design manager's task is to maintain the two problems under tight control, through the entire process of chip design.

The major design *labor* is chip layout. The basic drafting work is shown in dotted box B. Layouts of structurally regular circuit components like PLA, RAM, and ROM are carried out using dedicated software aids. Most frequently the procedure is as follows: area-critical *unit cell(s)* are manually laid out, and software is used to assemble many cells into functional subsystems. This procedure is straightforward and leaves little room for layout error to creep in. What is important in this design procedure is that the size of active components, like PFETs and NFETs, and parasitics of interconnect, like series resistance and parasitic capacitance, be consistent with the timing of the logic block. Active involvement of a circuit designer is essential, so that an automatically assembled logic block does not require additional timing improvements. The area and shape of the block assembled by this method is immediately available, and the information is directly fed back to determine the chip structure.

Layout of ICs is now carried out routinely using symbolic layout techniques, or software-controlled layout techniques. Traditional manual layout, in which each feature of the photomask is specified by many (X,Y) coordinates is too inflexible. There are several reasons that symbolic layout is preferred.

1) Updating layout for logic design change. Logic design that is delivered from the customer is seldom final. It is common that logic flaws are discovered in a silicon chip that met every customer specification. In this case, the chip designer is the weaker partner of the project: he has to change the logic design until the customer's real need, rather than his mistaken specification, is satisfied. Symbolic layout, whose data are stored in the form of stick figures, is flexible in making minor logic changes (like inserting an inverter, which is often nicknamed the *last minute inverter* !).

2) Symbolic layout is transparent to design rule change, or even to feature size change. Since the design of high performance chips is often carried out concurrently with the development of new generation, scaled-down CMOS processes, details of design rules may not be available at design time. Since design rules are given as separate information to the compiling software of symbolic layout code, symbolic layout is transparent to design rule change. This *technology updatability* is a great advantage if several generations of upgraded chips are required as CMOS feature size scales down.

3) Symbolic layout allows an accurate estimate of silicon area and logic circuit delay. Symbolic layout is usually carried out by drawing the circuit schematic on equally spaced grids. The drawing is called a stick figure. From the stick figure, silicon area can be determined immediately. Because of regularized layout style, FET sizes are quantized. Wiring length can be determined accurately by counting grids, since wiring is usually laid out in *Manhattan* style. Once the stick figure is available, detailed circuit simulation including real parasitic values can be run. This design procedure may go one step further: it is feasible to create a layout that satisfies design objectives by repeating a few iterative steps of symbolic layout and

circuit performance evaluation. This advantage is not insignificant: I consider that circuit optimization using this *short-loop* design feedback is essential in CMOS devices whose clock frequency is several hundred megahertz.

4) Symbolic layout has a clean interface between designer and draftsmen. Since circuit designers are able to determine circuit performance directly from stick figures, they need not sit at computer terminals all day. I consider this aspect of symbolic layout to require special attention for best use. It is a fact of life that a design group consists of experienced and inexperienced personnel, and it is very difficult to divide the workload. An easy interface between designer and draftsmen solves many project organizational problems.

5) Symbolic layout code can be examined directly for logic integrity by using connectivity extraction software, thereby adding one extra level of design verification.

The block interconnect information in Fig. 33 assumes that the original logic design specification comes in many logic blocks that are interconnected on the chip. Block interconnect is the highest level of chip assembly. This type of specification is conventional in microprocessors that contain more than several thousand gates. Signals represented by interconnect wires have well-defined meanings in the system function. Therefore design data that are separated into logic blocks that have well-defined functions are convenient for both design and verification. Interconnection of logic blocks is carried out using routing software. At least two levels of interconnects are necessary to interconnect freely the logic blocks. Routing software takes logic blocks that are specified in the shape of *Manhattan* geometry polygons, and on the sides of which locations of signals are specified by names and coordinates. Software places the polygonal functional blocks and connects terminals of different blocks that have the same name. Ingenuity in computational geometry is exhaustively applied to create routing software to achieve dense and short routing. Some software provides capability of block placement optimization, so that interconnect is automatically minimized. Some preferential signals, like system clock, can be routed without extensive crossovers and bending (nicknamed *dog-legging*). In spite of all these features in routing software, many problems are left for designers. If signals connecting distant blocks are routed automatically, routing length depends critically on how a wire goes around logic blocks between the source and the destination (they are *road blocks*). In order to force certain desired routing, it is often necessary to provide a conduit that induces the software to take the preferred route.

An important subfunction of routing software is to compute routing parasitics. This information is returned to the designer of individual logic blocks to check whether or not signal drivers are adequate. If a resistive interconnect like polysilicon, polysilicide, or diffusion is used, series resistance delay of the interconnect must be evaluated. Conventional routing software uses standard (usually minimum size) wires. For some critical signals, minimum size wire has too much resistive delay. This is the case even in metal wires. A minimum width unloaded metal wire 1cm long has almost 1 ns delay. Delay increases if wire is loaded by gate capacitance. It is necessary to use wide wires for certain critical signals like clocks. About 10 μm wide wire is conventionally used to distribute clock signals. These specifications can be given to the routing software.

It is desirable that layout data that are dumped in the box *Layout data* of Fig. 33 need minimum modification at the time of final chip assembly. To achieve this objective, circuit designers must oversee layout operation and must weed out timing problems before layout data are deposited there. This operation must cover the following details.

1) Screening of logic. It is possible to combine only a certain number of signals within the given timing budget. If logic design does not satisfy this criterion the logic design should be sent back for reconsideration.

2) Scaling of FETs. At least rough scaling of FETs must be carried out during the first layout. If FETs are approximately scaled, most of the signal path delays are close to the optimum. It is very cumbersome to readjust FET size later, using global timing simulation. Rough FET scaling can be carried out using the number of fanouts, and approximate length of wiring (in 1.75 μm CMOS, 200 μm of wire is approximately equivalent to one extra minimum size inverter load).

3) Structuring chip. Although a rough idea about how the chip is laid out exists at the beginning of the project, detailed chip structure emerges concurrently with layout. Chip structure specification includes placement of individual logic block and distribution of utilities (V_{DD}, V_{SS} and clocks). In order to arrive at the optimum chip layout, it is necessary that chip structure satisfies noise-free chip operation. Structuring I/O frame and the power-bus of noisy components like memories and output drivers is the key issue. Chip structure is tested by simulating noise voltage generated on I/O frame and internal power-bus. One important design consideration is the following: If static CMOS and NFET-based dynamic logic circuits are used, ground (V_{SS}) bus must have small noise, but noise of V_{DD} bus can be more than that of V_{SS} bus. If PFET-based dynamic logic circuits are used in addition, V_{DD} bus noise must be controlled as well. This additional design requirement may be difficult to meet, unless multilevel metal is available.

Layout data of logic subblocks are assembled to the entire chip layout data, and those are submitted for photomask generation, with additional features like registration marks and process test patterns. Before mask submission, a verification procedure is carried out. *Layout extraction* software is used to extract circuit connectivity information from the photomask data. Connectivity information includes sizes of PFETs, NFETs, and capacitive loading of the nodes. Extracted circuits are exercised with digital data given to the chip's terminals, very much in the same way as real chip testing. Whether or not desired digital outputs come out from the terminal verifies design, or it points out the direction of debugging work. A timing simulator computes signal delays. Since series resistances of interconnects are neglected, and since various simplifications are used to compute gate delay, delay accuracy is not as high as circuit-level simulation, but neverthless the simulation points out potential timing problems of the chip. Once such problems are discovered, circuits are reexamined using the circuit-level simulator. The problem of this scheme is that if the first design is not proper, a huge number of critical paths are detected by gate-level timing simulator, and design optimization becomes practically impossible. To generate a nearly optimum first design is therefore crucial in VLSI microprocessor development projects, since it is not easy to make large-scale design changes at the verification phase. Serious design problems, if any, must be weeded out before the first design is examined using the global timing/functional simulator. If timing design is verified, the layout design geometry rule is checked before photomask submission.

The final verification procedure is an examination of the entire chip layout drawing by designers. This step is necessary to gain an idea about sanity of the verification software, and to detect problems that are not covered by the software. A functional simulator generates test vectors that are used by automatic test equipment to screen defective chips. Testing of VLSI chips at high frequency (above 80–100 MHz clock) is difficult. Commercial test machines are unable to test chips in the realistic environment of the customer's circuit, since pin card electronics are used to force/sense the chip's terminals. The present trend is to test the chip's functionality at low frequencies. The final test at the required clock frequency is carried out using packaged devices, and in an environment that simulates the customer's system.

CAD is the crucial issue in top-down design methodology. A circuit's structural information is extracted from the photomask data, and the design verification software *exercises* the circuit model to check whether or not the logic function is correct. If the exercise is able to detect *potential* circuit failures as well as real logic failures caused by layout error, the verification is effective. Attaining this objective is, however, quite difficult. In the top-down design method, software that carries out design verification is the gate-level timing simulator with logic exercising capability. The simulator must be able to deal with the huge amount of data extracted from VLSI chip layout, and therefore the processing is already very extensive. Adding functionality for circuit failure study to already complex software is not desirable. Therefore circuit failure detection must be carried out with proper strategy.

Gate-level timing simulators compute gate delay efficiently using a *macromodel* of gate circuits. A macromodel, which is not a detailed circuit model, uses various simplifying assumptions and, above all, unidirectional signal flow. To analyze the effects of anomalous small-signal flow that is bidirectional, the model is not adequate. The possibility of spontaneous oscillation can be evaluated by first determining the DC operating point and then by carrying out small-signal stability analysis. These two capabilities are the functions of the circuit-level simulator.

As for induced noise, there is a fundamental difficulty to detect this problem by software. It is practically impossible to identify which circuit node is affected by which circuit node, and under what circuit excitation conditions. Coupling capacitances between nodes must be determined. Carrying out the capacitance computation (multi-dimensional field analysis) and then analyzing induced noise for all possible pairs of circuit nodes for all possible excitation is quite impossible. It is necessary that chip structure must provide resistance to induced noise. The induced noise can be minimized only if the technology provides secure device structure (bulk CMOS, and not SOS CMOS) and if the circuit designer uses noise-immune circuits (properly designed Domino CMOS, and not NORA).

Detection of RACE conditions using gate-level timing simulators is conventionally carried out. Improvement in a gate's macromodels will make hazard detection possible. For this purpose, however, gate macromodels based on Elmore's formula are not adequate, since node voltage change in hazards is not unidirectional. In RACE analysis, skewed process conditions (when NFET/PFET current ratio is too high or too low) should be included.

Uncertainty in circuit parameters is relevant to accuracy of circuit models used in the gate-level timing simulation. The macromodel should be able to reflect gate circuit details, excitation sequence, and gate condition prior to switching. Although rapid progress in macromodel development is unlikely, many improvements in the macromodel can be made that will improve gate-level timing simulation accuracy. This is the area where research would be the most rewarding. The details that must be included are as follows:

1) If more than one input switches, the effects of relative delay of input signals on the gate delay
2) Effects of input signal-dependent switching threshold voltage of gate
3) Inclusion of Miller effect, especially its data dependence
4) Inclusion of the feedforward effects associated with Miller effect
5) Effects of input signal slew rate, and generally of input waveform
6) Internal state of the gate (initial condition of gate prior to switching) that depends on the time elapsed since the last switching events.

Circuit configurational uncertainty can be detected only if circuit structural information, including all the parasitics, is extracted from the layout data, and the circuits including parasitics are examined for their likely mode of malfunction. It is not enough to recognize the existence of all the parasitic components (active or passive). The parameter values characterizing the components must be determined for evaluation of the extent of the effect. Even simple determination of parasitic capacitance requires multi-dimensional field analysis. Computer analysis of this type of circuit failure is not practical.

7.12 Future Directions of CMOS Circuit Failure Research

Before concluding this book, it will be useful to point out the directions of research into CMOS circuit failure, and the possible directions of future CMOS IC development that provide the area of application. The most fundamental unsolved problem in CMOS digital circuit theory is that not *all* the CMOS digital circuits have been invented, studied, and documented. Invention of new circuit structures has been carried out largely by chance. It is interesting to note that still new and useful circuits that contain only a small number of FETs are occasionally invented and reported. How do we deal with the issue of new circuit invention? From the mathematical point of view, CMOS circuits are connected *graphs* of *atoms* that have, at the most, four terminals, two of which are often interchangeable (resistance, capacitance, and FET). It is tedious, but mathematically a straightforward matter, to make a list of all the possible circuit configurations of N (small integer) components. If such configurations are examined exhaustively for possible applications, and if all the problems are cataloged, the problems of circuit failures can be solved once and for all. This is an extensive but attractive new area of research, since there is a possibility of discovering new and useful circuits. This is an interesting area of application of the computer-based *elephant* (versus traditional *elegant*) mathematics by which the cerebrated four-color mapping problem was resolved.

An important conclusion that the present work on CMOS circuit failure research created is classification of the CMOS circuit into digital and analog circuits. Although the classification is practically clear in the CMOS IC business, there has been no clear theoretical distinction. I provided a clear theoretical distinction between the two by using the concept of microstates (Section 1.15). We need to recognize, however, that this is just one way of doing so. It is possible to develop a different classification between analog and digital circuits. One direction of development is to make the boundary gradual. A hint is available in the theory of NMOS circuits, where microstate sequences are not a very clean way of describing the circuit operation. Instead of grouping CMOS circuits into digital and analog circuits, I would proceed as follows. The fundamental parameter of device technology that determines the circuit performance is the time constant of the device (time required for a device to charge or discharge itself). The ratio of the clock period of the circuit and the device time constant is the measure of the *margin* of the device capability in the circuit. Digital circuits generally have small or no margin at all, and analog circuits have large margin. The margin of device capability in circuits is necessary to clear the effects of the initial condition of the circuit, and therefore an essential prerequisite to achieve high precision. In this definition, one circuit works as analog and as digital circuits at different operating conditions (static versus dynamic switching, Section 2.8). A CMOS circuit theory could be built on this viewpoint.

An idea of using many *local* times instead of a single *global* time in CMOS circuit theory was introduced in Section 2.8. This concept appears to open many new possibilities in circuit representation. I intend to develop a theory of digital CMOS circuits at the highest frequency limit based on this idea. This kind of new prospect indicates that the science of electronic circuits has not reached the terminal point: the science will grow well into the 21st century.

CMOS technology reached maturity in the middle 1980s, but it has not reached its limit. CMOS acquired a permanent place in electronics technology. In certain important application areas, including small- to medium-sized computers, CMOS's position is practically unchallenged. This domination will continue well into the 21st century. We learned, however, that certain areas of application belong to technology other than CMOS. This is the area of large systems, like mainframe computers, electronic switching machines, and data communication systems. In these areas bipolar ECL technology will dominate for at least some time.

We recognize that integrated circuit technology entered into a quiet phase of growth some time ago. Drastic new inventions are now heard of less frequently than in the stormy period of growth between 1970 and 1985. We old-timers feel that the situation is somewhat similar to vacuum-tube electronics in the mid-1950s. There is difference, however. This time there is no strong alternative to MOS technology. Clearly there is a risk of falling into happy stagnation. We need to have a vision to the 21st century, to break out from the prevailing prosperous mannerism. Before concluding this book, I wish to suggest three areas of new developments that will add significant vitality to CMOS technology.

1) Integrated circuit technologies of 1960–1990 provided only two-dimensional device structures. Now the time is ripe to consider three-dimensional device structures seriously. The past trend in this direction has been the extention of planar technology to fabricate several layers of active devices. This approach is by no means the only one. I wish to point out that three-dimensional integrated circuits can be fabricated using the technology of advanced pack-

aging. This approach will perhaps materialize practical devices earlier than the approach using planar technology.

2) Integration of more than five million random logic FETs in a single CMOS chip is now feasible. If several tens of such chips are integrated on an advanced packaging substrate, it is possible to integrate 100 million random logic FETs in a pseudo monolithic chip. All the chips on the substrate work from a synchronous clock. Even a mainframe computer or supercomputer can be integrated in such a *megachip*. CMOS circuit technology to achieve this objective exists already. It is necessary to resolve the cultural difference between large system designers and CMOS chip designers to initiate such an ambitious project.

3) Low-temperature CMOS is a promising area of development, but it is difficult to persuade customers to pay for the development and operational costs of the system. It is necessary to upgrade low-temperature CMOS so that the scheme becomes more attractive. It is possible to get more advantage from low-temperature CMOS by reducing power supply voltage to the level of 1 volt or less. Then the essential difficulty in interchip communication is removed, and power required to cool CMOS systems will be reduced by one to two orders of magnitude. Low-temperature, low-voltage CMOS has practically all the desirable attributes of integrated circuit technology. It is attractive in the following respects.

i) Since power is proportional to power supply voltage squared, DC power consumption will be made very small, typically 16–40 times less than present CMOS chip.
ii) Since channel carriers have less energy available, there is no risk of hot-carrier trapping by gate oxides.
iii) If FETs are operated at low fields, decrease of delay by reducing temperature is quite significant.
iv) Technology can be scaled down below 0.1 μm without serious difficulty, where beneficial effects of electron ballistic transport emerges.
v) Wiring can be scaled down proportionally to feature size, since resistance of aluminum decreases drastically with reducing temperature.
vi) There will be no latchup.
vii) Logic swing is directly compatible with bipolar ECL devices.
viii) Leakage current decreases so much by cooling that dynamic memory works effectively as static memory.
ix) Because of leakage current reduction, a dynamic logic circuit can be used freely.
x) Thermal conductivity of silicon increases, and a stable temperature environment can be maintained.
xi) Thermal noise of MOSFETs decreases.

The low-temperature system must be assembled on an advanced packaging substrate to control power-bus noise. The three examples show that future CMOS technology grows with concurrent development of advanced packaging technology.

This prospect shows that CMOS technology will grow strong into the 21st century, will eventually dominate in *all* the areas of electronics, and there will be a wide area of application of CMOS circuit failure research. Although the subject of this book is a practical one, some readers might be displeased that my approach is very much like "circuit theory for circuit theory alone." I believe that CMOS technology is large enough that this heresy can be accommodated.

References

[01] Elmasry, M. I. *Digital MOS integrated circuits: a tutorial.* Digital MOS integrated circuits. New York: IEEE Press, 1981.

[02] Shoji, M. *CMOS digital circuit technology.* Englewood Cliffs, NJ: Prentice Hall, 1988.

[03] Lopez, A. D., and Law, H. F. "A dense gate-matrix layout style for MOS LSI." *ISSCC80 digest* (February, 1980): 212–13.

[04] Laramee, J., Aubin, M. J., and Cheeke, J. D. N. "Behavior of CMOS inverters at cryogenic temperatures." *Solid-State Electronics* 28 (1985): 453–56.

[05] Kamgar, A., and Johnston, R. L. "Delay times in Si MOSFETs in the 4.2-400 K temperature range." *Solid-State Electronics* 26 (1983): 291–94.

[06] Ipri, A. C. "The properties of silicon-on-sapphire substrates, devices, and integrated circuits" in *Silicon integrated circuits.* ed. D. Kahng. New York: Academic Press, 1981.

[07] The following view originated from a discussion with late W. Shockley in 1965.

[08] Ziel, A. van der. "Thermal noise in field-effect transistors." *Proc.IRE* vol. 50 (August, 1962): 1808–12.

[09] Wing, O. *Gallium arsenide digital circuits.* Boston: Kluver Academic Publishers, 1990.

[10] Soares, R., ed. *Application of GaAs MESFETs.* Dedham, MA: Artech House Inc., 1983.

[11] Shoji, M. "Two-dimensional Gunn-domain dynamics." *IEEE Trans. on Electron Devices* vol. ED-16 (September, 1969): 748–58.

[12] Cooper, J. A., Jr., and Nelson, D. F. "Measurement of the high-field drift velocity of electrons in inversion layers of silicon." *IEEE Electron Device Letters* vol. EDL-2 (July, 1981): 171–73.

[13] Stratton, J. A. *Electromagnetic theory.* New York: McGraw-Hill, Inc., 1941.

[14] Sparkes, J. J. *Junction transistors.* Oxford: Pergamon Press, 1966.

[15] Elmasry, M. I. *Digital bipolar integrated circuits.* New York: Wiley-Interscience, 1983.

[16] Murphy, B. T., Edwards, R., Thomas, L. C., and Molinelli, J. "A CMOS 32b single chip microprocessor." *ISSCC81 digest* (February, 1981): 230–31.

[17] Stritter, E., and Gunter, A. "A microprocessor architecture for a changing world: the Motorola 68000" in *Digital MOS integrated circuits.* ed. M. I. Elmasry. IEEE Press, 1981, pp. 364–72.

[18] Metzger, G., and Vabre, J.-P. *Transmission lines with pulse excitation.* New York: Academic Press, 1969.

[19] Alvarez, A. R. *BiCMOS technology and applications.* Boston: Kluwer Academic Publishers, 1989.

[20] Tewksbury, S. K. *Wafer level integrated systems: implementation issues.* Boston: Kluver Academic Publishers, 1988.

[21] Frye, R. C., and Chen, H. Z. "High speed interconnections using self-damped lossy transmission lines." IEEE Symposium on high density integration in communication and computer systems, Waltham, MA, 17–18 October 1991.

[22] Krambeck, R. H., Blahut, D. E., Law, S. F., Colbry, B. W., So, H. C., and Harrison, M. L. "Top-down design of one-chip 32 bit CPU." VLSI81, Edinbrough, Scotland, September, 1982.

Index

GPSR Authorized Representative: Easy Access System Europe - Mustamäe tee 50, 10621 Tallinn, Estonia, gpsr.requests@easproject.com

www.ingramcontent.com/pod-product-compliance
Lightning Source LLC
LaVergne TN
LVHW080113011026
849160LV00013B/1083
* 9 7 8 0 6 9 1 6 0 3 0 1 8 *